THIRD EDITION

COMMUNICATION SYSTEMS

THIRD
EDITION

COMMUNICATION SYSTEMS

SIMON HAYKIN
McMaster University

JOHN WILEY & SONS, INC.
New York • Chichester • Brisbane • Toronto • Singapore

ACQUISITIONS EDITOR	Steven Elliot
MARKETING MANAGER	Susan Elbe
SENIOR PRODUCTION EDITOR	Richard Blander
DESIGNER	David Levy
MANUFACTURING MANAGER	Andrea Price
ILLUSTRATION COORDINATOR	Jaime Perea

This book was set in New Baskerville by CRWaldman Graphic Communications and printed and bound by Hamilton Printing Company. The cover was printed by Lehigh.

Recognizing the importance of preserving what has been written, it is a policy of John Wiley & Sons, Inc. to have books of enduring value published in the United States printed on acid-free paper, and we exert our best efforts to that end.

Library of Congress Cataloging in Publication Data:
Haykin, Simon, 1931-
　　Communication systems / Simon Haykin, — 3rd ed.
　　　　p.　　cm.

Includes bibliographical references and index.
ISBN 0-471-57176-8
　　1. Telecommunication. 2. Signal theory (Telecommunication)
I. Title.

TK5101.H37　1994　　93-47663
621.382—dc20　　CIP
ISBN 0-471-57178-8
ISBN 0–471-xxxxx-x (pbk)

Printed in the United States of America

10 9 8

In loving memory of my wife Vera,
and to Michael and Juanné

Preface

The study of communication systems is basic to an undergraduate program in electrical engineering. In this third edition of the book *Communication Systems*, I have preserved the most relevant parts of the previous edition of the book and made ample space for an expanded treatment of digital communications. The objective of the book is to present a study of *classical communication theory* in a logical and interesting manner. The material is illustrated with examples and computer-oriented experiments (using Matlab) intended to help the reader develop an intuitive grasp of the theory under discussion. Except for the introductory and final chapters, each chapter ends with numerous problems designed not only to help readers test their understanding of the material covered in the chapter, but to also challenge them to extend this material. Every chapter includes notes and references that provide suggestions for further reading.

The book is organized into 12 chapters and 11 appendices, as outlined in what follows.

Chapters 2 through 5 constitute the first part of the book, devoted to *introductory material* and *analog communications*. Chapter 1, the *Introduction*, presents an overview of communication systems and finishes with some historical notes on the subject. Chapter 2 on *Representation of Signals and Systems*, emphasizes the interplay between time and frequency using the Fourier transform. The Hilbert transform is introduced in this chapter to develop the complex low-pass representation of band-pass signals and systems. The chapter concludes with a discussion of the discrete Fourier transform and its numerical computation using the fast Fourier transform algorithm. With this background at hand, the stage is set for a detailed discussion of *Continuous-Wave Modulation* in Chapter 3. Specifically, amplitude modulation (and its variants) and frequency modulation are studied. Chapter 4 on *Random Processes* reviews probabilistic concepts and develops the partial characterization (i.e., second-order statistics) of random processes; particular emphasis is given to narrow-band noise and its representations. Chapter 5 on *Noise in Continuous-Wave Modulation Systems* resumes the study of amplitude and frequency modulation techniques by evaluating the effects of noise on their performance.

Chapters 6 through 9 constitute the second part of the book, which is devoted to *digital communications*. Chapter 6 on *Pulse Modulation* discusses the processes of sampling, quantization, and coding that are fundamental to the digital representation of analog signals; this chapter may be viewed as the transition from analog to digital communications. Chapter 7 on *Baseband Pulse Transmission* develops the matched filter for the detection of known signals in noise, and presents techniques for combatting the effects of intersymbol interference in dispersive channels. Chapter 8 on *Digital Passband Transmission* introduces a concept of profound theoretical importance, namely, the geometric interpretation of signals, and uses it to study various digital modulation techniques. Chapter 9 on *Spread-Spectrum Modulation* discusses the use of pseudonoise sequences for secure digital communications.

Chapters 10 through 12 constitute the third and final part of the book, which is devoted to *advanced topics*. Chapter 10 on *Fundamental Limits in Information Theory* develops Shannon's classic theorems for data compression, data compaction, and data transmission. These theorems provide upper bounds on the performance of information sources and communication channels. Chapter 11 on *Error-Control Coding* presents various techniques for the encoding and decoding of digital data streams for their reliable transmission over noisy channels. The final chapter on *Advanced Communication Systems* discusses satellite communications, mobile radio, and optical communications, and it concludes with an emerging viewpoint of telecommunications.

Appendices 1 through 10 present ancillary material on the production of *speech and television signals,* a review of *Fourier series,* an introductory treatment of *time-frequency analysis,* including *wavelets,* a review of *Bessel functions,* a derivation of *Schwarz's inequality,* a discussion of *noise figure* for assessing how noisy electronic devices are, definitions of the *error function,* a statistical characterization of *complex random processes,* a review of *binary arithmetic,* and an introductory treatment of *cryptography* for secret communications. To the best of my knowledge, this is the first undergraduate book on communication systems to include introductory treatments of time-frequency analysis and cryptography. Appendix 11, the last one in the set, presents a compilation of useful *mathematical* tables.

The book is essentially self-contained, and suited for a one-academic year or one-semester undergraduate course in communication theory. It is expected that the reader has a knowledge of electronics, circuit theory, and probability theory. The make-up of the material for the course is naturally determined by the background of the students and the interests of the teacher involved. However, the material covered in the book is both broad and deep enough to satisfy a variety of backgrounds and interests, thereby allowing considerable flexibility in making up the course material. As an aid to the teacher of the course, a detailed *Solutions Manual* for all the problems in the book is available from the publisher.

November 1993   SIMON HAYKIN

Acknowledgments

I would like to express my deep gratitude to Dr. Greg Pottie, University of California at Los Angeles; Dr. Dilip Sarwate, University of Illinois at Urbana-Champaign; Dr. Peter Willett, University of Connecticut, Storrs; and Dr. Mike Sablatash and Dr. Norman Secord, both of the Communications Research Centre, Ottawa for their critical reviews of the entire book and the many constructive suggestions for improving the book. I am deeply indebted to them all for their kind help. I am also grateful to Dr. Ian Blake, University of Waterloo, Ontario; Mr. Michael Moher, Communications Research Centre, Ottawa; Dr. David Parsons, University of Liverpool, United Kingdom; and Dr. Andrew Viterbi, Qualcomm, for their helpful inputs and suggestions on selected chapters of the book.

I am grateful to my graduate students Robert Dony, Andrew Ukrainec, and Paul Yee for the various computer-oriented experiments included in the book.

I wish to thank Dr. David Parsons, The Institute of Electrical and Electronic Engineers (IEEE), AT&T, Macmillan Publishing, and McGraw-Hill for their permission to reproduce certain figures acknowledged in the book.

I wish to thank my Editor, Steven Elliot, for his constant encouragement and strong support in writing the book. The help provided by Richard Blander and other members of the production staff at Wiley is most appreciated.

I am indebted to my librarian, Elaine Tooke, for checking the bibliography. Last, but by no means least, I am grateful to my Secretary, Lola Brooks, for her tireless effort in typing so many different versions of the manuscript.

Needless to say, without all the above help and support, the writing and production of this book would not have been possible.

SH

Contents

CHAPTER 1
INTRODUCTION 1

1.1 The Communication Process 1
1.2 Sources of Information 3
1.3 Communication Channels 6
1.4 Baseband and Passband Signals 9
1.5 Representation of Signals and Systems 10
1.6 Probabilistic Considerations 11
1.7 The Modulation Process 12
1.8 Primary Communication Resources 14
1.9 Information Theory and Coding 14
1.10 Analog Versus Digital Communications 17
1.11 Networks 19
1.12 Some Historical Notes 23
Notes and References 26

CHAPTER 2
REPRESENTATION OF SIGNALS AND SYSTEMS 27

2.1 Introduction 27
2.2 The Fourier Transform 27

2.3   Properties of the Fourier Transform   33
2.4   Rayleigh's Energy Theorem   46
2.5   The Inverse Relationship Between Time and Frequency   48
2.6   Dirac Delta Function   51
2.7   Fourier Transforms of Periodic Signals   59
2.8   Transmission of Signals Through Linear Systems   62
2.9   Filters   69
2.10   Hilbert Transform   79
2.11   Pre-Envelope   83
2.12   Canonical Representations of Band-Pass Signals   85
2.13   Band-Pass Systems   91
2.14   Phase and Group Delay   98
2.15   Numerical Computation of the Fourier Transform   100
2.16   Summary   109
       Notes and References   109
       Problems   111

CHAPTER 3
CONTINUOUS-WAVE MODULATION   121

3.1   Introduction   121
3.2   Amplitude Modulation   122
3.3   Virtues, Limitations, and Modifications of Amplitude Modulation   131
3.4   Double Sideband–Suppressed Carrier Modulation   132
3.5   Filtering of Sidebands   140
3.6   Vestigial Sideband Modulation   144
3.7   Single Sideband Modulation   147
3.8   Frequency Translation   151
3.9   Frequency-Division Multiplexing   152
3.10   Angle Modulation   154
3.11   Frequency Modulation   158
3.12   Phase-Locked Loop   181
3.13   Nonlinear Effects in FM Systems   191
3.14   The Superheterodyne Receiver   195
3.15   Summary and Discussion   197
       Notes and References   200
       Problems   201

CHAPTER 4
RANDOM PROCESSES   218

4.1   Introduction   218
4.2   Probability Theory   219
4.3   Random Variables   226
4.4   Statistical Averages   230
4.5   Transformations of Random Variables   235
4.6   Random Processes   239
4.7   Stationarity   241
4.8   Mean, Correlation, and Covariance Functions   242
4.9   Ergodicity   249

4.10 Transmission of a Random Process Through a Linear Filter 250
4.11 Power Spectral Density 252
4.12 Gaussian Process 264
4.13 Noise 269
4.14 Narrow-Band Noise 282
4.15 Sine Wave Plus Narrow-Band Noise 295
4.16 Summary 299
      Notes and References 299
      Problems 301

CHAPTER 5
NOISE IN CW MODULATION SYSTEMS   313

5.1 Introduction 313
5.2 Receiver Model 314
5.3 Noise in DSB-SC Receivers 317
5.4 Noise in SSB Receivers 319
5.5 Noise in AM Receivers 322
5.6 Noise in FM Receivers 326
5.7 Pre-emphasis and De-emphasis in FM 340
5.8 Summary and Discussion 344
     Notes and References 346
     Problems 346

CHAPTER 6
PULSE MODULATION   351

6.1 Introduction 351
6.2 The Sampling Process 352
6.3 Pulse-Amplitude Modulation 357
6.4 Time-Division Multiplexing 362
6.5 Pulse-Position Modulation 364
6.6 Bandwidth-Noise Trade-off 373
6.7 The Quantization Process 374
6.8 Pulse-Code Modulation 378
6.9 Noise Considerations in PCM Systems 387
6.10 Virtues, Limitations, and Modifications of PCM 390
6.11 Delta Modulation 391
6.12 Differential Pulse-Code Modulation 396
6.13 Coding Speech at Low Bit Rates 399
6.14 Summary and Discussion 404
      Notes and References 405
      Problems 406

CHAPTER 7
BASEBAND PULSE TRANSMISSION   412

7.1 Introduction 412
7.2 Matched Filter 413
7.3 Error Rate Due to Noise 418

7.4   Intersymbol Interference   424
7.5   Nyquist's Criterion for Distortionless Baseband Binary Transmission   427
7.6   Correlative-Level Coding   434
7.7   Baseband *M*-ary PAM Transmission   446
7.8   Tapped-Delay-Line Equalization   448
7.9   Adaptive Equalization   452
7.10   Eye Pattern   461
7.11   Summary and Discussion   464
        Notes and References   465
        Problems   466

CHAPTER 8
DIGITAL PASSBAND TRANSMISSION   473

8.1   Introduction   473
8.2   Passband Transmission Model   474
8.3   Gram-Schmidt Orthogonalization Procedure   478
8.4   Geometric Interpretation of Signals   483
8.5   Response of Bank of Correlators to Noisy Input   486
8.6   Coherent Detection of Signals in Noise   491
8.7   Probability of Error   495
8.8   Correlation Receiver   501
8.9   Detection of Signals with Unknown Phase   503
8.10   Hierarchy of Digital Modulation Techniques   507
8.11   Coherent Binary PSK   508
8.12   Coherent Binary FSK   511
8.13   Coherent Quadriphase-Shift Keying   516
8.14   Coherent Minimum Shift Keying   523
8.15   Noncoherent Orthogonal Modulation   532
8.16   Noncoherent Binary Frequency-Shift Keying   539
8.17   Differential Phase-Shift Keying   540
8.18   Comparison of Binary and Quaternary Modulation Schemes   544
8.19   *M*-ary Modulation Techniques   546
8.20   Power Spectra   555
8.21   Bandwidth Efficiency   561
8.22   Synchronization   564
8.23   Summary and Discussion   568
        Notes and References   569
        Problems   570

CHAPTER 9
SPREAD-SPECTRUM MODULATION   578

9.1   Introduction   578
9.2   Pseudo-Noise Sequences   579
9.3   A Notion of Spread Spectrum   586
9.4   Direct-Sequence Spread-Spectrum with Coherent Binary
        Phase-Shift Keying   589
9.5   Signal-Space Dimensionality and Processing Gain   592
9.6   Probability of Error   597
9.7   Frequency-Hop Spread Spectrum   599

9.8   Code-Division Multiplexing   605
9.9   Summary and Discussion   609
      Notes and References   610
      Problems   611

CHAPTER 10
FUNDAMENTAL LIMITS IN INFORMATION THEORY   614

10.1   Introduction   614
10.2   Uncertainty, Information, and Entropy   615
10.3   Source Coding Theorem   621
10.4   Data Compaction   623
10.5   Discrete Memoryless Channels   631
10.6   Mutual Information   634
10.7   Channel Capacity   637
10.8   Channel Coding Theorem   640
10.9   Differential Entropy and Mutual Information for
       Continuous Ensembles   644
10.10 Information Capacity Theorem   648
10.11 Implications of the Information Capacity Theorem   652
10.12 Rate Distortion Theory   657
10.13 Compression of Information   659
10.14 Summary and Discussion   661
       Notes and References   662
       Problems   664

CHAPTER 11
ERROR CONTROL CODING   670

11.1   Introduction   670
11.2   Discrete Memoryless Channels   672
11.3   Linear Block Codes   675
11.4   Cyclic Codes   686
11.5   Convolutional Codes   700
11.6   Maximum Likelihood Decoding of Convolutional Codes   707
11.7   Trellis-Coded Modulation   715
11.8   Coding for Compound-Error Channels   721
11.9   Summary and Discussion   723
       Notes and References   723
       Problems   724

CHAPTER 12
ADVANCED COMMUNICATION SYSTEMS   730

12.1   Introduction   730
12.2   Satellite Communications   731
12.3   Mobile Radio   735
12.4   Optical Communications   753
12.5   Broadband Integrated Services Digital Network   759
12.6   An Emerging View of Telecommunications   762
       Notes and References   763

Appendix 1
Speech and Television as Sources of Information   765

Appendix 2
Fourier Series   773

Appendix 3
Time-Frequency Analysis   781

Appendix 4
Bessel Functions   793

Appendix 5
Schwarz's Inequality   799

Appendix 6
Noise Figure   801

Appendix 7
Error Function   806

Appendix 8
Statistical Characterization of Complex Random Processes 810

Appendix 9
Binary Arithmetic   813

Appendix 10
Cryptography   815

Appendix 11
Mathematical Tables   837

Glossary   845

Abbreviations   848

Bibliography   851

Index   867

CHAPTER 1

Introduction

1.1 THE COMMUNICATION PROCESS[1]

Today, *communication* enters our daily lives in so many different ways that it is very easy to overlook the multitude of its facets. The telephones at our hands, the radios and televisions in our living rooms, the computer terminals in our offices and homes, and our newspapers are all capable of providing rapid communications from every corner of the globe. Communication provides the senses for ships on the high seas, aircraft in flight, and rockets and satellites in space. Communication through a cordless telephone keeps a car driver in touch with the office or home miles away. Communication keeps a weather forecaster informed of conditions measured by a multitude of sensors. Indeed, the list of applications involving the use of communication in one way or another is almost endless.

In the most fundamental sense, communication involves implicitly the transmission of *information* from one point to another through a succession of processes, as described here:

1. The generation of a thought pattern or image in the mind of an originator.
2. The description of that image, with a certain measure of precision, by a set of aural or visual symbols.

3. The encoding of these symbols in a form that is suitable for transmission over a physical medium of interest.

4. The transmission of the encoded symbols to the desired destination.

5. The decoding and reproduction of the original symbols.

6. The recreation of the original thought pattern or image, with a definable degradation in quality, in the mind of a recipient; the degradation is caused by imperfections in the system.

There are, of course, many other forms of communication that do not directly involve the human mind in real time. For example, in *computer communications* involving communication between two or more computers, human decisions may enter only in setting up the programs or commands for the computer, or in monitoring the results.

Irrespective of the form of communication process being considered, there are three basic elements to every communication system, namely, *transmitter*, *channel*, and *receiver*, as depicted in Fig. 1.1. The transmitter is located at one point in space, the receiver is located at some other point separate from the transmitter, and the channel is the physical medium that connects them together. The purpose of the transmitter is to transform the *message signal* produced by the *source of information* into a form suitable for transmission over the channel. However, as the transmitted signal propagates along the channel, it is distorted due to channel imperfections. Moreover, noise and interfering signals (originating from other sources) are added to the channel output, with the result that the *received signal* is a corrupted version of the *transmitted signal*. The receiver has the task of operating on the received signal so as to reconstruct a recognizable form of the original message signal and to deliver it to the *user destination*. The signal processing role of the receiver is thus the reverse of that of the transmitter.

In this book we present a study of communication systems. Our approach will be from a *systems viewpoint*, emphasizing mathematical descriptions, representations, and processing of electrical signals that characterize such systems. In this introductory chapter, we present an overview of the various issues involved in the communication process, ending with some historical notes on the subject.

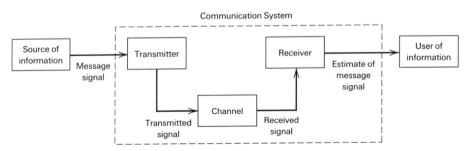

Figure 1.1 Elements of a communication system.

1.2 SOURCES OF INFORMATION

The telecommunications environment is dominated by four important sources of information: *speech, television, facsimile,* and *personal computers.* A source of information may be characterized in terms of the signal that carries the information. A *signal* is defined as a single-valued function of *time* that plays the role of the dependent variable; at every instant of time, the function has a unique value.

Speech is the primary method for human communication. Specifically, the speech communication process involves the transfer of information from a speaker to a listener, which takes place in three successive stages:

- *Production.* An intended message in the speaker's mind is represented by a *speech signal* that consists of sounds (i.e., pressure waves) generated inside the speaker's mouth and whose arrangement is governed by the rules of language.
- *Propagation.* The sound waves propagate through the air, reaching the listener's ears.
- *Perception.* The incoming sounds are deciphered by the listener into a received message, thereby completing the chain of events that culminate in the transfer of information from the speaker to the listener.

The speech-production process may be viewed as a form of filtering, in which a *sound source excites a vocal tract filter.* The vocal tract consists of a tube of non-uniform cross-sectional area, beginning at the *glottis* (i.e., the opening between the vocal cords) and ending at the *lips.* As the sound propagates along the vocal tract, the spectrum (i.e., frequency content) is shaped by the frequency selectivity of the vocal tract; this effect is somewhat similar to the resonance phenomenon observed in organ pipes. The important point to note here is that the power spectrum (i.e., the distribution of long-term average power versus frequency) of speech approaches zero for zero frequency and reaches a peak in the neighborhood of a few hundred hertz. To put matters into proper perspective, however, we have to keep in mind that the hearing mechanism is highly sensitive to frequency. Moreover, the type of communication system being considered has an important bearing on the band of frequencies considered to be "essential" for the communication process. For example, a frequency band from 300 to 3100 Hz is considered adequate for telephonic communication on a commercial basis.

The second source of information, *television* (TV), refers to the transmission of pictures in motion by means of electrical signals. To accomplish this transmission, each complete picture has to be *sequentially scanned.* The scanning process is carried out in a TV *camera.* In a *black-and-white TV,* the camera contains optics designed to focus an image on a *photocathode* consisting of a large number of photosensitive elements. The charge pattern so generated on the photosensitive surface is scanned by an *electron beam,* thereby producing an output current that varies *temporally* in accordance with the way in which the brightness of the original picture varies *spatially* from one point to another. The resulting output current is called a video signal. The type of scanning used in television is a form of spatial sampling called *raster scanning,* the purpose of which is to convert a two-dimensional image intensity into a one-dimensional waveform; it is somewhat analogous to the manner in which we read a printed paper in that the scanning is performed from left to right on a line-by-line basis. In television, a picture is divided into 525 lines, which constitute a *frame.* Each frame is decomposed into

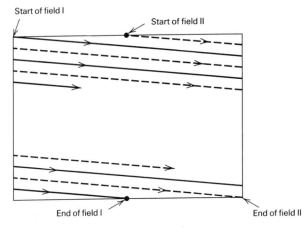

Start of field I

Start of field II

End of field I

End of field II

Figure 1.2 Interlaced raster scan.

two *interlaced fields*, each one of which consists of 262.5 lines. For convenience of presentation, we will refer to the two fields as I and II. The scanning procedure is illustrated in Fig. 1.2. The lines of field I are depicted as solid lines, and those of field II are depicted as dashed lines. The *start* and *end* of each field are also included in the figure. Field I is scanned first. The scanning spot of the TV camera moves with constant velocity across each line of the field from left to right, and the image intensity at the center of the spot is measured; the scanning spot itself is partly responsible for local spatial averaging of the image. When the end of a particular line is reached, the scanning spot quickly flies back (in a horizontal direction) to the start of the next line down in the field. This flyback is called the *horizontal retrace*. The scanning process described here is continued until the whole field has been accounted for. When this condition is reached, the scanning spot moves quickly (in a vertical direction) from the end of field I to the start of field II. This second flyback is called the *vertical retrace*. Field II is treated in the same fashion as field I. The time taken for each field to be scanned is 1/60 second. Correspondingly, the time taken for a frame or a complete picture to be scanned is 1/30 second. With 525 lines in a frame, the *line-scanning frequency* equals 15.75 kHz. Thus, by flashing 30 still pictures per second on the display tube of the TV receiver, the human eye perceives them to be moving pictures. This effect is due to a phenomenon known as the *persistence of vision*. During the horizontal- and vertical-retrace intervals, the picture tube is made inoperative by means of *blanking pulses* that are generated at the transmitter. Moreover, synchronization between the various scanning operations at both transmitter and receiver is accomplished by means of special pulses that are transmitted during the blanking periods; thus, the synchronizing pulses do not show on the reproduced picture. The reproduction quality of a TV picture is limited by two basic factors:

1. The number of lines available in a raster scan, which limits resolution of the picture in the vertical direction.
2. The channel bandwidth available for transmitting the video signal, which limits resolution of the picture in the horizontal direction.

For each direction, *resolution* is expressed in terms of the maximum number of lines alternating between black and white that can be resolved in the TV image along the pertinent direction by a human observer. In the NTSC (National Television System Committee) system, which is the North American Standard, the parameter values used result in a *video bandwidth* of 4.2 MHz, which extends down to zero frequency. This bandwidth is orders of magnitude bigger than that of a speech signal.

The purpose of the third source of information, a *facsimile (fax) machine*, is to transmit still pictures over a communication channel (most notably, a telephone channel). Such a machine provides a highly popular facility for the transmission of handwritten or printed text from one point to another; transmitting text by facsimile is treated simply like transmitting a picture. The basic principle employed for signal generation in a facsimile machine is to scan an original document (picture) and use an image sensor to convert the light to an electrical signal.

Finally, *personal computers* (PCs) are becoming increasingly an important part of our daily lives. We use them for electronic mail, exchange of software, and sharing of resources. It is estimated that over 30 percent of the personal computers in use today are already networked, and the number is increasing rapidly. The text transmitted by a PC is usually encoded using the *American Standard Code for Information Interchange* (ASCII), which is the first code developed specifically for computer communications. Each character in ASCII is represented by seven *data bits* constituting a unique binary pattern made up of 0s and 1s. Thus a total of $2^7 = 128$ different characters can be represented in ASCII. The characters are various lowercase and uppercase letters, numbers, special control symbols, and punctuation symbols commonly used such as @, $, and %. Some of the special "control" symbols, such as BS (backspace) and CR (carriage return), are used to control the printing of characters on a page. Other symbols, such as ENQ (enquiry) and ETB (end of transmission block), are used for communication purposes. (A complete listing of ASCII characters is given in Table 9 of Appendix 11 at the end of this book.) The seven data bits are ordered starting with the most significant bit b_7 down to the least significant bit b_1, as illustrated in Fig. 1.3. At the end of the data bits, an extra bit b_8 is appended for the purpose of *error detection*. This error-detection bit is called a *parity bit*. A sequence of eight bits is referred to as a *byte* or an *octet*. The parity bit is set in such a way that the total number of 1s in each byte is odd for odd parity, or even for even parity. Suppose, for example, the communicators agree to use even parity; then the parity bit will be a 0 when the number of 1s in the data bits is even and a 1 when it is odd. Hence, if a single bit in a byte is received in error and thereby violates the even parity rule, it can be detected and then corrected. Personal computers are often connected via their RS (recommended standard)-232 ports. When ASCII data

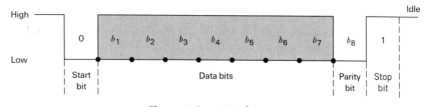

Figure 1.3 ASCII format.

(in fact, all character data) are transmitted through these ports a start bit, set to 0, and one or more stop bits, set to 1, as shown in Fig. 1.3, are added to provide character framing. When the transmission is idle, a long series of 1s is sent so as to keep the circuit connection alive. In Fig. 1.3, symbols 0 and 1 are designated as "low" and "high," respectively. They are also sometimes referred to as "space" and "mark," respectively; the latter terminology comes from the days of telegraphy. The text prepared on a PC is usually stored and then transmitted over a communication channel (commonly, a telephone channel) with a single character being sent at a time. This form of data transmission is called *asynchronous transmission*, as opposed to *synchronous transmission* in which a whole sequence of encoded characters is sent over the channel in one long transmission. Encoded characters produced by a mixture of asynchronous and synchronous terminals are combined by means of *data multiplexers*. The multiplexed stream of data so formed is then applied to a device called a *modem* (modulator–demodulator) for the purpose of transmission over the channel.

In summary, computer-generated data and television signals are both *wideband signals*, in that their power content occupies a wide range of frequencies. Another important characteristic of data communication between personal computers is *burstiness*, which means that information is usually transmitted from one terminal to another in bursts with silent periods between bursts. Indeed, data traffic involving computers in one form or another tends to be of a bursty nature. This is to be contrasted with traffic in a digital transmission network due to voice or interactive video, which, relatively speaking, is of a *continuous* nature.

1.3   COMMUNICATION CHANNELS

In the previous section we discussed various sources of information responsible for the generation of message signals that need to be delivered to their destination. We next briefly discuss various *communication channels* (i.e., physical media) that do the transmission. Specifically, we consider *telephone channels*, *optical fibers*, *mobile radio channels*, and *satellite channels*. We have chosen these four different channels for special attention because of their important roles in today's telecommunications environment, which can only grow with time.

A *telephone network* uses a switching mechanism called *circuit switching* to establish an end-to-end communication link on a temporary basis. The primary purpose of the network is to ensure that the telephone transmission between a speaker at one end of the link and a listener at the other end is an acceptable substitute for face-to-face conversation. In this form of communication, the message source is the sound produced by the speaker's voice, and the ultimate destination is the listener's ear. The telephone channel, however, supports only the transmission of electrical signals. Accordingly, appropriate transducers are used at the transmitting and receiving ends of the system. Specifically, a *microphone* is placed near the speaker's mouth to convert sound waves into an electrical signal, and the electrical signal is converted back into acoustic form by means of a *moving-coil receiver* placed near the listener's ear. Present-day designs of these two transducers have been perfected so as to respond well to frequencies from 20 to 8000 Hz; moreover, a pair of them are compactly packaged inside a single telephone set that is easy to speak into or listen from. The telephone channel is essentially a *linear, bandwidth-limited channel*. The restriction on bandwidth arises from the requirement of sharing the channel among a multitude of users at any one time. A practical solution to the telephonic communication problem must therefore minimize the channel bandwidth requirement, subject to a satisfactory transmission of human voice. To meet this requirement, the transducers and channel specifications must conform to standards based on subjective tests that are performed on the intelligibility, or articulation, of telephone signals by representative male and female speakers. A speech signal (male or female) is essentially limited to a band from 300 to 3100 Hz in the sense that frequencies outside this band do not contribute much to articulation efficiency. This frequency band may therefore be viewed as a ''rough guideline'' for the passband of a telephone channel that provides a satisfactory service.

The second communication channel, an *optical fiber*, is a dielectric waveguide that transports light signals from one place to another just as a metallic wire pair or a coaxial cable transports electrical signals. It consists of a central *core* within which the propagating electromagnetic field is confined and which is surrounded by a *cladding* layer, which is itself surrounded by a thin protective *jacket*. The core and cladding are both made of pure silica glass, whereas the jacket is made of plastic. Optical fibers have unique characteristics that make them highly attractive as a transmission medium. In particular, they offer the following unique advantages:

- *Enormous potential bandwidth*, resulting from the use of optical carrier frequencies around 2×10^{14} Hz; with such a high carrier frequency and a bandwidth roughly equal to 10 percent of the carrier frequency, the theoretical bandwidth of a light-wave system is around 2×10^{13} Hz, which is very large indeed.
- *Low transmission losses*, as low as 0.2 dB/km.
- *Immunity to electromagnetic interference*, which is an inherent characteristic of an optical fiber viewed as a dielectric waveguide.
- *Small size and weight*, characterized by a diameter no greater than that of a human hair.
- *Ruggedness and flexibility*, exemplified by very high tensile strengths and the possibility of being bent or twisted without damage.

Last, but by no means least, optical fibers offer the potential for low-cost line communications since they are fabricated from sand, which, unlike copper used in metallic conductors, is not a scarce resource. The unique properties of optical fibers have fueled phenomenal advances in light-wave systems technology during the past two decades, which have, in turn, revolutionized long-distance communications.

The third communication channel, *mobile radio channel*, extends the capability of the public telecommunications network by introducing *mobility* into the network by virtue of its ability to *broadcast*. The term ''mobile radio'' is usually meant to encompass terrestrial situations where a radio transmitter or receiver is capable of being moved, regardless of whether it actually moves or not. The major propagation effects encountered in the use of a mobile radio in built-up areas are due to the fact that the antenna of the mobile unit may lie well below the surrounding buildings. Simply put, there is no ''line-of-sight'' path for communication; rather, radio propagation takes place mainly by way of scattering from the surfaces of the surrounding buildings and by diffraction over and/or around them. The end result is that energy reaches the receiving antenna via more than one path. In a mobile radio environment, we thus speak of a *multipath phenomenon* in that the various incoming radio waves reach their destination from different directions and with different time delays. Indeed, there may be a multitude of propagation paths with different electrical lengths, and their contributions to the received signal could combine in a variety of ways. Consequently, the received signal strength varies with location in a very complicated fashion, and so a mobile radio channel may be viewed as a *linear time-varying channel* that is statistical in nature.

Finally, a *satellite channel* adds another invaluable dimension to the public telecommunications network by providing *broad-area coverage* in a continental as well as intercontinental sense. Moreover, access to remote areas not covered by conventional cable or fiber communications is also a distinct feature of satellites. In almost all satellite communication systems, the satellites are placed in *geostationary orbit*. For the orbit to be geostationary, it has to satisfy two requirements. First, the orbit is *geosynchronous*, which requires the satellite to be at an altitude of 22,300 miles; a geosynchronous satellite orbits the Earth in exactly 24 hours (i.e., the satellite is synchronous with the Earth's rotation). Second, the satellite is placed in orbit directly above the equator on an eastward heading (i.e., it has zero inclination). Viewed from Earth, a satellite in geostationary orbit appears to be stationary in the sky. Consequently, an Earth station does *not* have to track the satellite; rather, it merely has to point its antenna along a fixed direction, pointing toward the satellite. By so doing, the system design is simplified considerably. Communications satellites in geostationary orbit offer the following unique system capabilities:

• Broad-area coverage.
• Reliable transmission links.
• Wide transmission bandwidths.

In terms of services, satellites can provide fixed point-to-point links extending over long distances and into remote areas, communication to mobile platforms (e.g., aircraft, ships), or broadcast capabilities. Indeed, communications satellites play a key role in the notion of the whole world being viewed as a "global village." In a typical satellite communication system, a message signal is transmitted from an Earth station via an *uplink* to a satellite, amplified in a *transponder* (i.e., electronic circuitry) on board the satellite, and then retransmitted from the satellite via a *downlink* to another Earth station. With the satellite positioned in geostationary orbit, it is always visible to all the Earth stations located inside the satellite antenna's coverage zones on the Earth's surface. In effect, the satellite acts as a powerful *repeater* in the sky. The most popular frequency band for satellite communications is 6 GHz for the uplink and 4 GHz for the downlink. The use of this frequency band offers the following advantages:

• Relatively inexpensive microwave equipment.
• Low attenuation due to rainfall; rainfall is a primary atmospheric cause of signal loss.
• Insignificant sky background noise; the sky background noise (due to random noise emissions from galactic, solar, and terrestrial sources) reaches its lowest level between 1 and 10 GHz.

In the 6/4-GHz band, a typical satellite is assigned a 500 MHz bandwidth that is divided among 12 transponders on board the satellite. Each transponder, using approximately 36 MHz of the satellite bandwidth, corresponds to a specific radio channel. A single transponder can carry at least one color television signal, 1200 voice circuits, or digital data at a rate of 50 Mb/s (megabits per second).

To summarize, a communication channel is central to the operation of a communication system. Its properties determine both the information-carrying capacity of the system and the quality of service offered by the system. We may classify communication channels in different ways:

- A channel may be *linear* or *nonlinear*; a telephone channel is linear, whereas a satellite channel is usually (but not always) nonlinear.
- A channel may be *time invariant* or *time varying*; an optical fiber is time invariant, whereas a mobile radio channel is time varying.
- A channel may be *bandwidth limited* or *power limited* (i.e., limited in the available transmitted power); a telephone channel is bandwidth limited, whereas an optical fiber link and a satellite channel are both power limited.

Now that we have some understanding of sources of information and communication channels, we may return to the block diagram of a communication system shown in Fig. 1.1.

1.4 BASEBAND AND PASSBAND SIGNALS

In the context of a communication system, a signal of primary interest is the *message signal* delivered by a source of information. This signal is also referred to as a *baseband signal*, with the term "baseband" being used to designate the band of frequencies representing the message signal. Baseband signals can be of an analog or digital type. In an *analog signal*, time takes on values in a continuum, and so does the amplitude of the signal. Analog baseband (message) signals arise when a physical waveform such as an acoustic or light wave is converted into an electrical signal. In a digital signal, on the other hand, *both* time and the signal's amplitude take on discrete values only. The output of a digital computer is an example of a baseband signal of the digital type. In this particular example, there are only two values, commonly represented by binary symbols 0 and 1, and so the signal is referred to as a *binary signal*.

Another signal of primary interest is the transmitted signal, the characterization of which is determined by the type of channel used in the communication system. In this context, we speak of *baseband* or *passband transmission*. In baseband transmission, as the name implies, the band of transmission frequencies supported by the channel closely matches the band of frequencies occupied by the message signal. In passband transmission, on the other hand, the transmission band of the channel is centered at a frequency much higher than the highest frequency component of the message signal. In the latter case, the transmitted signal is said to be a *passband signal*, the generation of which is accomplished in the transmitter using a process known as modulation; we will have more to say on modulation in Section 1.7.

1.5 REPRESENTATIONS OF SIGNALS AND SYSTEMS

In order to develop insight into the operation of a communication system, we need mathematical tools for the representations of signals and systems. The most useful method of signal representation depends on the particular type of signal being considered. Depending on the feature of interest, we may distinguish three different classes of signals that arise in the study of communication systems:

1. *Periodic and nonperiodic signals.* A periodic signal $g(t)$ is a function of time that satisfies the condition

$$g(t) = g(t + T_0) \qquad \text{for all } t \qquad (1.1)$$

   where t denotes time and T_0 is a constant. The smallest value of T_0 for which this condition is satisfied is called the *period* of $g(t)$. Accordingly, the period T_0 defines the duration of one complete cycle of $g(t)$. Any signal for which Eq. (1.1) does not hold for any T_0 is called a *nonperiodic* or *aperiodic signal.*

2. *Deterministic and random signals.* A *deterministic signal* is a signal about which there is no uncertainty with respect to its value at any time, past, present, or future. Accordingly, we find that deterministic signals may be modeled as completely specified functions of time. On the other hand, a *random signal* is a signal about which there is some degree of uncertainty before it actually occurs.

3. *Energy and power signals.* In electrical systems, a signal may represent a voltage or a current. Consider a voltage $v(t)$ developed across a resistor R, producing a current $i(t)$. The *instantaneous power* dissipated in this resistor is defined by

$$p(t) = \frac{|v(t)|^2}{R} \qquad (1.2)$$

   or, equivalently,

$$p(t) = R|i(t)|^2 \qquad (1.3)$$

   In both cases, the instantaneous power $p(t)$ is proportional to the squared amplitude of the signal. Furthermore, for a resistor R equal to 1 ohm, we see that Eqs. (1.2) and (1.3) take on the same mathematical form. Accordingly, in *signal analysis* it is customary to *normalize* the calculations by working with a 1-ohm resistor, so that, regardless of whether a given signal $g(t)$ represents a voltage or a current, we may express the instantaneous power associated with the signal simply as

$$p(t) = |g(t)|^2 \qquad (1.4)$$

Based on this convention, the *total energy* of a signal $g(t)$ is defined by

$$E = \lim_{T \to \infty} \int_{-T}^{T} |g(t)|^2 \, dt$$
$$= \int_{-\infty}^{\infty} |g(t)|^2 \, dt \qquad (1.5)$$

Correspondingly, the *average power* of a signal $g(t)$ is defined by

$$P = \lim_{T \to \infty} \frac{1}{2T} \int_{-T}^{T} |g(t)|^2 \, dt \tag{1.6}$$

We say that the signal $g(t)$ is an *energy signal* if and only if the total energy of the signal satisfies the condition

$$0 < E < \infty$$

We say that the signal $g(t)$ is a *power signal* if and only if the average power of the signal satisfies the condition

$$0 < P < \infty$$

The energy and power classifications of signals are mutually exclusive. In particular, an energy signal has zero average power, whereas a power signal has infinite energy. Also, it is of interest to note that, usually, periodic signals and random signals are power signals, and signals that are both deterministic and nonperiodic are energy signals.

In theory, there are many possible methods for the representation of signals. In practice, however, we find that *Fourier analysis*, involving the resolution of signals into sinusoidal components, overshadows all other methods in usefulness. Basically, this is a consequence of the well-known fact that the response of a system to a sine-wave input is another sine wave of the same frequency (but with a different phase and amplitude) under two conditions:

1. The system is *linear* in that it obeys the *principle of superposition*. That is, if $y_1(t)$ and $y_2(t)$ denote the responses of a system to the inputs $x_1(t)$ and $x_2(t)$, respectively, the system is linear if the response to the composite input $a_1 x_1(t) + a_2 x_2(t)$ is equal to $a_1 y_1(t) + a_2 y_2(t)$, where a_1 and a_2 are arbitrary constants.

2. The system is *time invariant*. That is, if $y(t)$ is the response of a system to the input $x(t)$, the system is time invariant if the response to the time-shifted input $x(t - t_0)$ is equal to $y(t - t_0)$ where t_0 is constant.

There are several methods of Fourier analysis available for the representation of signals. The particular version that is used in practice depends on the type of signal being considered. For example, if the signal is periodic, then the logical choice is to use the *Fourier series* to represent the signal in terms of a set of harmonically related sine waves. On the other hand, if the signal is an energy signal, then it is customary to use the *Fourier transform* to represent the signal. By using the Fourier series or the Fourier transform, we obtain the *frequency-domain description* or *spectrum* of the signal, by means of which we are often able to discern important characteristics of the signal in a way that may otherwise be difficult.

1.6   PROBABILISTIC CONSIDERATIONS

Intrinsically, the communication process is of a *probabilistic* nature. To appreciate this fundamental property of the communication process, we merely have to recognize that if the receiver of a communication system were to know the composition of a message exactly, there would be no point in having the system transmit that message.

A major source of "uncertainty" in the operation of a communication system is *noise* that originates naturally at the front end of the receiver. The two most common forms of noise are *thermal noise*, produced by the random motion of electrons in conducting media, and *shot noise*, produced by random fluctuations of current flow in electronic devices. The received signal may be further corrupted by *interference* due to undesirable sources or external effects. The net result is that the received signal is randomlike in appearance. To be more precise, we cannot predict the exact value of the received signal. Nevertheless, the received signal can be described in terms of its statistical properties such as the average power or the spectral distribution of the average power. The mathematical discipline that deals with the statistical characterization of random signals and noise is *probability theory*.

Another factor contributing to uncertainty in the communication process may be traced to the source of information itself. Consider, for example, English text in which each of the letters is known to have a certain probability of occurrence. Here we find that although each letter of the English alphabet can be represented by a deterministic waveform that is distinctive, the composite waveform representing a sequence of letters emitted by the relevant source is random, because we do not know exactly which particular letter will be emitted in advance of its actual occurrence.

A random signal, irrespective of its origin, may be viewed as belonging to an ensemble that is collectively called a *random* or *stochastic process*; the term "stochastic" is of Greek origin. The characterization of a random process and the evaluation of the effect produced by passing it through a linear system are essential to a thorough understanding of the operation of a communication system.

1.7 THE MODULATION PROCESS

The purpose of a communication system is to deliver a message signal from an information source in recognizable form to a user destination, with the source and the user being physically separated from each other. To do this, the transmitter modifies the message signal into a form suitable for transmission over the channel. This modification is achieved by means of a process known as *modulation*, which involves varying some parameter of a *carrier wave* in accordance with the message signal. The receiver recreates the original message signal from a degraded version of the transmitted signal after propagation through the channel. This recreation is accomplished by using a process known as *demodulation*, which is the reverse of the modulation process used in the transmitter. However, owing to the unavoidable presence of noise and distortion in the received signal, we find that the receiver cannot recreate the original message signal exactly. The resulting degradation in overall system performance is influenced by the type of modulation scheme used. Specifically, we find that some modulation schemes are less sensitive to the effects of noise and distortion than others.

We may classify the modulation process into *continuous-wave modulation* and *pulse modulation*. In continuous-wave (CW) modulation, a sinusoidal wave is used as the carrier. When the amplitude of the carrier is varied in accordance with the message signal, we have *amplitude modulation* (AM), and when the angle of the carrier is varied, we have *angle modulation*. The latter form of CW modulation may be further subdivided into *frequency modulation* (FM) and *phase modulation* (PM), in which the instantaneous frequency and phase of the carrier, respectively, are varied in accordance with the message signal.

In pulse modulation, on the other hand, the carrier consists of a periodic sequence of rectangular pulses. Pulse modulation can itself be of an analog or digital type. In *analog pulse modulation*, the amplitude, duration, or position of a pulse is varied in accordance with sample values of the message signal. In such a case, we speak of *pulse-amplitude modulation* (PAM), *pulse-duration modulation* (PDM), and *pulse-position modulation* (PPM).

The standard digital form of pulse modulation is known as *pulse-code modulation* (PCM) that has no CW counterpart. PCM starts out essentially as PAM, but with an important modification: The amplitude of each modulated pulse (i.e., *sample* of the original message signal) is *quantized* or *rounded off* to the nearest value in a prescribed set of *discrete* amplitude levels and then *coded* into a corresponding sequence of binary symbols. The binary symbols 0 and 1 are themselves represented by pulse signals that are suitably shaped for transmission over the channel. In any event, as a result of the quantization process, some information is always lost and the original message signal cannot therefore be reconstructed exactly. However, provided that the number of quantizing (discrete amplitude) levels is large enough, the distortion produced by the quantization process is not discernible to the human ear in the case of a speech signal or the human eye in the case of a two-dimensional image. Among all the different modulation schemes, pulse-code modulation has emerged as the preferred method of modulation for the transmission of analog message signals for the following reasons:

- *Robustness* in noisy environment by *regenerating* the transmitted signal at regular intervals.
- *Flexible* operation.
- *Integration* of diverse sources of information into a common format.
- *Security* of information in its transmission from source to destination.

In introducing the idea of modulation, we stressed its importance as a process that ensures the transmission of a message signal over a prescribed channel. There is another important benefit, namely, *multiplexing*, that results from the use of modulation. Multiplexing is the process of combining several message signals for their simultaneous transmission over the same channel. Two commonly used methods of multiplexing are as follows:

- *Frequency-division multiplexing* (FDM) in which CW modulation is used to translate each message signal to reside in a specific frequency slot inside the passband of the channel by assigning it a distinct carrier frequency; at the receiver, a bank of filters is used to separate the different modulated signals and prepare them individually for demodulation.
- *Time-division multiplexing* (TDM), in which pulse modulation is used to position samples of the different message signals in nonoverlapping time slots.

Thus, in FDM the message signals overlap with each other in time, raising the possibility of *cross-talk* due to nonlinearity of the channel. On the other hand, in TDM the message signals exploit the full passband of the channel, but on a time-shared basis.

1.8   PRIMARY COMMUNICATION RESOURCES

In a communication system, there are two primary resources to be employed: *transmitted power* and *channel bandwidth*. The transmitted power refers to the average power of the transmitted signal. The channel bandwidth is defined as the band of frequencies allocated for the transmission of the message signal. A general system design objective is to use these two resources as efficiently as possible. In most communication channels, one resource may be considered more important than the other. We may therefore classify communication channels as *power limited* or *band limited*. For example, the telephone circuit is a typical band-limited channel, whereas a space communication link or a satellite channel is typically power limited.

For the case when the spectrum of a message signal extends down to zero or low frequencies, we define the bandwidth of the signal as that upper frequency above which the spectral content of the signal is negligible and therefore unnecessary for transmitting information. For example, the average voice spectrum extends well beyond 10 kHz, though most of the average power is concentrated in the range of 100 to 600 Hz, and a band from 300 to 3100 Hz gives good articulation. Accordingly, we find that telephone circuits that respond well to this latter range of frequencies give quite satisfactory commercial telephone service.

Another important point that we have to keep in mind is the unavoidable presence of noise in a communication system. A quantitative way of accounting for the effect of noise is to introduce *signal-to-noise ratio* (SNR) as a system parameter. For example, we may define the SNR at the receiver input as *the ratio of the average signal power to the average noise power*, both being measured at the same point. The customary practice is to express the SNR in *decibels* (dBs), defined as 10 times the logarithm (to base 10) of the power ratio.

A fundamental question that arises in the study of modulation schemes is the following: With channel bandwidth and signal-to-noise ratio being the two principal parameters that are available to the designer of a communication system, which particular modulation scheme provides for their most efficient use? Stated in another way, for a specified channel bandwidth, which modulation scheme requires the smallest signal-to-noise ratio for a prescribed level of system performance? The answer to this important question lies in information theory, which we briefly review next.

1.9   INFORMATION THEORY AND CODING[2]

Information theory applies the laws of probability, and mathematics in general, to the study of the collection and manipulation of information. In the context of communication theory, it answers two fundamental questions:

1. What is the fundamental limit on the compression and refinement of information generated by a source?
2. What is the fundamental limit on the transmission rate of information over a noisy communication channel?

These two questions are aimed at what the communication process is really all about.

In the case of a *discrete source* (i.e., a source whose amplitude is in discrete form), the answer to the first question is embodied in the *source coding theorem*. This theorem involves the use of a quantity known as *entropy*, which provides a quantitative measure of information; entropy is usually measured in units of *bits*, which is an acronym for binary digits. According to the source coding theorem, the average number of bits per symbol required to encode the output of a discrete (memoryless) source with arbitrarily small probability of decoding error is lower bounded by the entropy of the source. The coding efficiency of a source is measured in bits per symbol. The source encoding is said to be *exact* or *lossless* if the original information source output can be reconstructed without loss of entropy (i.e., information). The operation used to do this form of source encoding is called *data compaction*. To handle the case of a *continuous* source (i.e., a source whose amplitude has a continuous value), the source coding theorem is generalized by introducing the information *rate-distortion function* that represents a lower bound on the rate at which the source output can be transmitted over a noiseless channel and still be reconstructed with a prescribed level of distortion. The important point to note here is that the digital (discrete) representation of an analog (continuous) signal is necessarily accompanied by destruction (loss) of information. In information-theoretic terms, the conversion of an analog signal into a digital signal is called *data compression*; unlike data compaction, data compression is lossy. Although data compression and data compaction may be combined in practice, we usually think of them as two consecutive functions, as shown in Fig. 1.4. The *data compression encoder* reduces the intrinsic information content of the source by introducing a tolerable amount of distortion into the data stream. The *data compaction encoder* then represents the compressed data efficiently, producing a new data stream that has the smallest number of bits per second. Finally, the *data encryption encoder* disguises the data (bit) stream in such a way that it has no meaning to an unauthorized receiver. At the receiving end, a corresponding sequence of *decoding* operations, in the reverse order to that in the transmitter, is performed to reconstruct the original information source output. The operations described in Fig. 1.4 represent the dissection of *source coding*, so called because it is related to the source of information exclusively.

In the case of a *discrete channel*, the answer to the second question we raised earlier is embodied in the *channel coding theorem*, which involves the use of an-

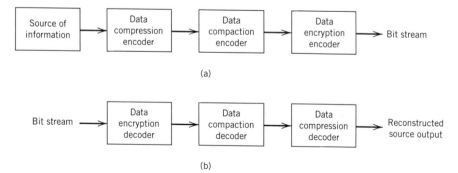

Figure 1.4 Dissection of source coding. (a) Source encoding at the transmitter. (b) Source decoding at the receiver.

Figure 1.5 Dissection of channel coding.

other quantity called *channel capacity*. The channel coding theorem states that as long as the rate of information transmission is below the channel capacity, it is possible to have error-free transmission over the channel. This is achieved by using a *channel encoder* in the transmitter and a *channel decoder* in the receiver, as shown in Fig. 1.5. The channel coding theorem is in reality a nonconstructive existence proof that error-free communication is possible over a noisy channel; however, the theorem does not tell us how to design the best channel encoder and decoder, nor does it tell us how complex they must be. Nevertheless, a considerable amount of work has gone into an attempt to solve these problems. Although the final solutions are still unknown, many good answers have already been obtained, and work inspired by information theory continues to improve the design of communication systems.

The channels used for communication in practice are of a continuous (analog) type. To accommodate the transmission of bit streams over such channels, we require the use of modulation. Thus, combining the operations of source coding, channel coding, and modulation together, we may envision the block diagram of a digital communication system as shown in Fig. 1.6. The operations performed in the transmitter are as follows. The message signal from a source

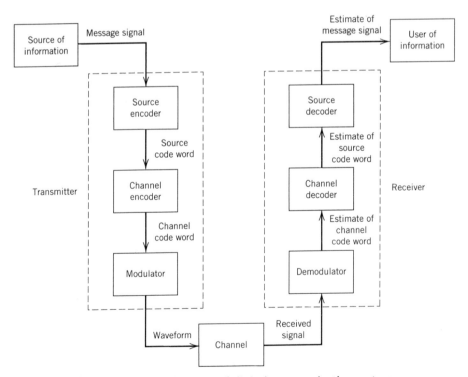

Figure 1.6 Block diagram of digital communication system.

of information is first processed by the source encoder, whose function is to provide a compact representation of the message signal. The resulting sequence of symbols is called the *source code word*. The data stream is processed next by the channel encoder, which produces a new sequence of symbols called the *channel code word*. The channel code word is longer than the source code word by virtue of the redundancy built into its construction. Finally, the modulator represents each symbol of the channel code word by a corresponding analog symbol, appropriately selected from a finite set of possible analog symbols. The sequence of analog symbols produced by the modulator is called a *waveform*, suitable for transmission over the channel. At the receiver, the channel output (received signal) is processed in the reverse order to that in the transmitter, thereby reconstructing a recognizable version of the original message signal. The reconstructed message signal is finally delivered to the user of information at the destination. Note that insofar as the discrete model of Fig. 1.5 is concerned, the cascade combination of the modulator, channel, and demodulator in the expanded model of Fig. 1.6 may be viewed as a discrete channel, the input and output of which are both in discrete form.

The source coding and channel coding theorems, both due to Shannon, deal with a discrete source and a discrete channel, respectively. The information capacity theorem, also due to Shannon, deals with a continuous channel; according to this latter theorem, channel bandwidth and signal-to-noise ratio are *exchangeable* in that we may trade off one for the other for a prescribed system performance. The choice of one modulation scheme over another for the transmission of a message signal in the communication system of Fig. 1.6 is often dictated by the nature of this trade-off. Indeed, the interplay between channel bandwidth and signal-to-noise ratio, and the limitation that they impose on communication, is highlighted most vividly by the *information capacity theorem*. Let B denote the channel bandwidth, and SNR denote the received signal-to-noise ratio. The information capacity theorem states that ideally these two parameters are related by

$$C = B \log_2(1 + \text{SNR}) \text{ b/s} \qquad (1.7)$$

where C is the information capacity. The *information capacity* is defined as the maximum rate at which information may be transmitted without error through the channel; it is measured in *bits per second* (b/s). Equation (1.7) clearly shows that for a prescribed information capacity, we may reduce the required SNR by increasing the channel bandwidth, hence the advantage of using a broad bandwidth to transmit messages. Moreover, Eq. (1.7) provides an idealized framework for comparing the noise performance of one modulation scheme against another.

1.10 ANALOG VERSUS DIGITAL COMMUNICATIONS

In the design of a communication system, be it analog or digital, the source of information, the communication channel, and the user of information are all given. The communication channel makes certain resources available for information transmission, subject to specific limitations. The challenge is to design the transmitter and the receiver with the following guidelines in mind:

- Take the message signal generated by the source of information, transmit it over the channel, and produce an "estimate" of it at the receiver output that satisfies the expectation of the user block.
- Do all of this at an affordable cost.

Consider first the case of a *digital communication system* represented by the block diagrams of Figs. 1.4 and 1.6, the rationale for both of which is rooted in information theory. The functional blocks of the transmitter and the receiver, starting from the far ends of the channel, are paired as follows:

- Source encoder–decoder:
  - Data compression encoder–decoder.
  - Data compaction encoder–decoder.
  - Data encryption encoder–decoder.
- Channel encoder–decoder.
- Modulator–demodulator.

Accordingly, we have several signal processing operations available to us, thereby making the sophisticated design of a digital communication system possible. In general, the design philosophy of the system is influenced by three rules of thumb[3], which may be described as follows:

1. The techniques for source coding (data compression and compaction) should be performed in a completely separate manner from those for channel transmission (channel coding and modulation), even though source coding removes redundancy from the data stream and channel coding inserts redundancy.
2. In designing the receiver, information contained in the received signal should be processed with care. In particular, in making a decision about which particular symbol was transmitted, information should never be discarded prematurely that may be useful to that decision.
3. In the presence of interference, intentional or otherwise, signal processing can be used in both the transmitter and receiver to ensure that degradation in the performance of the system owing to the interference will be no worse than that caused by *Gaussian noise* at the same power level as the interference; by Gaussian noise we mean a noise process whose amplitude has a Gaussian (normal) probability distribution.

In contrast, the design of an *analog communication system* is quite simple: In signal-processing terms, the transmitter consists of a modulator and the receiver consists of a demodulator, the details of which are determined by the type of CW modulation used. In comparing analog versus digital communications, the first impression created in our minds is that an analog communication system is much simpler to design than a digital communication system. This simplicity is due to the fact that analog modulation techniques, exemplified by their wide use in radio and television, make relatively superficial changes to the message signal in order to prepare it for transmission over the channel. More specifically, there is no significant effort made by the system designer to tailor the waveform of the transmitted signal to suit the channel at any deeper level. On the other hand, digital communication theory endeavors to find a finite set of waveforms that are closely matched to the characteristics of the channel and which are therefore more tolerant of channel impairments. In so doing, reliable communication is established over the channel. In the selection of good waveforms for digital communication over a noisy channel, the design is influenced solely by the channel characteristics. However, once the appropriate set of waveforms for transmission over the channel has been selected, the source information can be encoded into the channel waveforms, and the efficient transmission of information from the source to the user is thereby ensured. In summary, the use of digital communications provides the capability for information transmission that is both *efficient* and *reliable*.

From this discussion, it is apparent that the use of digital communications may require a considerable amount of electronic circuitry. But we have to recognize that nowadays electronics is relatively cheap, due to the ever-increasing availability of very-large-scale integrated (VLSI) circuits in the form of silicon chips. Indeed, with continuing improvements in the semiconductor industry, the technology favors digital communications over analog communications. Thus, although cost considerations used to be a factor in selecting analog communications over digital communications in the past, that is no longer the case today.

1.11   NETWORKS[4]

A *communication network* or simply *network*, illustrated in Fig. 1.7, consists of an interconnection of a number of *nodes* made up of intelligent processors (e.g., microcomputers). The primary purpose of these nodes is to route data through the network. Each node has one or more *stations* attached to it; stations refer to devices wishing to communicate. The network is designed to serve as a shared resource for moving data exchanged between stations in an efficient manner and also to provide a framework to support new applications and services. The telephone network is an example of a communication network in which *circuit switching* is used to provide a dedicated communication path or *circuit* between two stations. The circuit consists of a connected sequence of links from source to destination. The links may consist of time slots in a time-division multiplexed

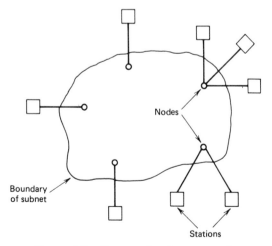

Figure 1.7   Communication network.

(TDM) system or frequency slots in a frequency-division multiplexed (FDM) system. The circuit, once in place, remains uninterrupted for the entire duration of transmission. Circuit switching is usually controlled by a centralized hierarchical control mechanism with knowledge of the network's organization. To establish a circuit-switched connection, an available path through the network is seized and then dedicated to the exclusive use of the two stations wishing to communicate. In particular, a call-request signal must propagate all the way to the destination, and be acknowledged, before transmission can begin. Then, the network is effectively transparent to the users. This means that during the connection time, the bandwidth and resources allocated to the circuit are essentially "owned" by the two stations, until the circuit is disconnected. The circuit thus represents an efficient use of resources only to the extent that the allocated bandwidth is properly utilized. Although the telephone network is used to transmit data, voice constitutes the bulk of the network's traffic. Indeed, circuit switching is well suited to the transmission of voice signals, since voice conversations tend to be of long duration (about 2 minutes on the average) compared to the time required for setting up the circuit (about 0.1–0.5 seconds). Moreover, in most voice conversations, there is information flow for a relatively large percentage of the connection time, which makes circuit switching all the more suitable for voice conversations.

In circuit switching, a communication link is shared between the different sessions using that link on a *fixed* allocation basis. In *packet switching*, on the other hand, the sharing is done on a *demand* basis, and therefore it has an advantage over circuit switching in that when a link has traffic to send, the link may be more fully utilized.

The network principle of packet switching is "store and forward." Specifically, in a *packet-switched network*, any message larger than a specified size is subdivided prior to transmission into segments not exceeding the specified size. The segments are commonly referred to as *packets*. The original message is reassembled at the destination on a packet-by-packet basis. The network may be viewed as a distributed pool of *network resources* (i.e., channel bandwidth, buffers, and switching processors) whose capacity is *shared dynamically* by a community of competing users (stations) wishing to communicate. In contrast, in a circuit-switched network, resources are dedicated to a pair of stations for the entire period they are in session. Accordingly, packet switching is far better suited to a computer-communication environment in which "bursts" of data are exchanged between stations on an occasional basis. The use of packet switching, however, requires that careful *control* be exercised on user demands; otherwise, the network may be seriously abused.

The design of a *data network* (i.e., a network in which the stations are all made up of computers and terminals) may proceed in an orderly way by looking at the network in terms of a *layered architecture*, regarded as a hierarchy of nested layers. A *layer* refers to a process or device inside a computer system, designed to perform a specific function. Naturally, the designers of a layer will be intimately familiar with its internal details and operation. At the system level, however, a user views the layer merely as a "black box" that is described in terms of the inputs, the outputs, and the functional relation between outputs and inputs. In a layered architecture, each layer regards the next lower layer as one or more black boxes with some given functional specification to be used by the given higher layer. Thus, the highly complex communication problem in data networks is resolved as a manageable set of well-defined interlocking functions. It is this line of reasoning that has led to the development of the *open systems interconnection* (OSI)[5] *reference model* by a subcommittee of the International Organization for Standardization. The term "open" refers to the ability of any two systems conforming to the reference model and its associated standards to interconnect.

In the OSI reference model, the communications and related-connection functions are organized as a series of *layers* or *levels* with well-defined *interfaces*, and with each layer built on its predecessor. In particular, each layer performs a related subset of primitive functions, and it relies on the next lower layer to perform additional primitive functions. Moreover, each layer offers certain services to the next higher layer and shields the latter from the implementation details of those services. Between each pair of layers, there is an *interface*. It is the interface that defines the services offered by the lower layer to the upper layer.

The OSI model is composed of seven layers, as illustrated in Fig. 1.8; this figure also includes a description of the functions of the individual layers of the model. Layer k on system A, say, communicates with layer k on some other system B in accordance with a set of rules and conventions, collectively constituting the layer k *protocol*, where $k = 1, 2, \ldots, 7$. (The term "protocol" has been borrowed from common usage, describing conventional social behavior between human beings.) The entities that comprise the corresponding layers on different systems are referred to as *peer processes*. In other words, communication is achieved by having the peer processes in two different systems communicate via a protocol, with the protocol itself being defined by a set of rules of procedure. Physical communication between peer processes exists only at layer 1. On the other hand, layers 2 through 7 are in *virtual communication* with their distant peers. However, each of these six layers can exchange data and control information with its neighboring layers (below and above) through layer-to-layer interfaces. In Fig. 1.8 physical communication is shown by solid lines and virtual communication by dashed lines. The major principles involved in arriving at seven layers of the OSI reference model are as follows:

1. Each layer performs well-defined functions.
2. A boundary is created at a point where the description of services offered is small and the number of interactions across the boundary is the minimum possible.
3. A layer is created from easily localized functions, so that the architecture of the model may permit modifications to the layer protocol to reflect changes in technology without affecting the other layers.
4. A boundary is created at some point with an eye toward standardization of the associated interface.
5. A layer is created only when a different level of abstraction is needed to handle the data.
6. The number of layers employed should be large enough to assign distinct functions to different layers, and yet small enough to maintain a manageable architecture for the model.

Note that the OSI reference model is not a network architecture; rather, it is an international standard for computer communications, which just tells what each layer should do.

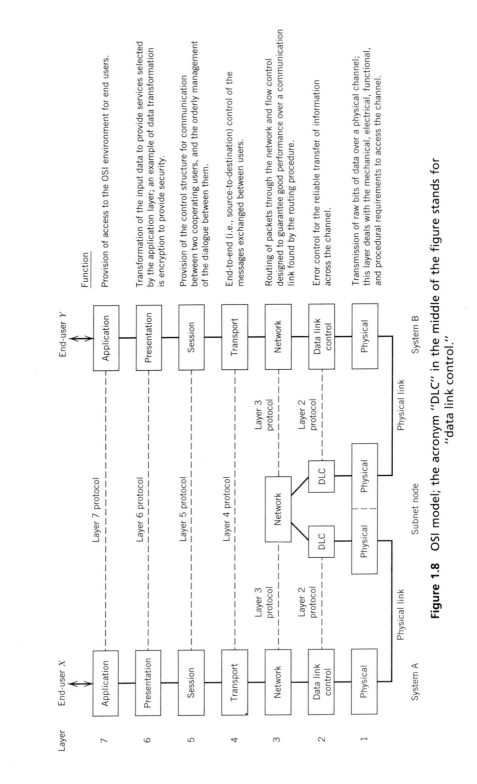

Figure 1.8 OSI model; the acronym "DLC" in the middle of the figure stands for "data link control."

1.12 SOME HISTORICAL NOTES[6]

An overview of communications would be incomplete without a history of the subject. In this final section of this introductory chapter we present some historical notes on communications; each set of notes focuses on some important and related events. It is hoped that this material will provide a sense of inspiration and motivation for the reader.

In 1837, the telegraph was perfected by Samuel Morse, a painter. With the words "What hath God wrought," transmitted by Morse's electric telegraph between Washington, D.C., and Baltimore, Maryland, in 1844, a completely revolutionary means of real-time, long-distance communications was triggered. The telegraph is the forerunner of digital communications in that the *Morse code* is a *variable-length* ternary code using an alphabet of four symbols: a dot, a dash, a letter space, and a word space; short sequences represent frequent letters, whereas long sequences represent infrequent letters. This type of signaling is ideal for manual keying. Subsequently, Emile Baudot developed a *fixed-length* binary code for telegraphy in 1875. In Baudot's telegraphic code, well-suited for use with teletypewriters, each code word consists of five-equal-length code elements, and each element is assigned one of two possible states: a *mark* or a *space* (i.e., symbol 1 or 0 in today's terminology).

In 1864, James Clerk Maxwell formulated the electromagnetic theory of light and predicted the existence of radio waves. The existence of radio waves was established experimentally by Heinrich Hertz in 1887. In 1894, Oliver Lodge demonstrated wireless communication over a relatively short distance (150 yards). Then, on December 12, 1901, Guglielmo Marconi received a radio signal at Signal Hill in Newfoundland; the radio signal had originated in Cornwall, England, 1700 miles away across the Atlantic. The way was thereby opened toward a tremendous broadening of the scope of communications. In 1905, Reginald Fessenden demonstrated wireless telephony by transmitting speech and music over a radio channel.

In 1875, the *telephone* was invented by Alexander Graham Bell, a teacher of the deaf. The telephone made real-time transmission of speech by electrical encoding and replication of sound a practical reality. The first version of the telephone was crude and weak, enabling people to talk over short distances only. When telephone service was only a few years old, interest developed in automating it. Notably, in 1897, A. B. Strowger, an undertaker from Kansas City, Missouri, devised the automatic *step-by-step switch* that bears his name; of all the electromechanical switches devised over the years, the Strowger switch was the most popular and widely used.

In 1904, John Ambrose Fleming invented the *vacuum-tube diode*, which paved the way for the invention of the *vacuum-tube triode* by Lee de Forest in 1906. The discovery of the triode was instrumental in the development of transcontinental telephony in 1913 and signaled the dawn of wireless voice communications. Indeed, until the invention and perfection of the transistor, the triode was the supreme device for the design of electronic amplifiers.

In 1918, Edwin H. Armstrong invented the *superheterodyne radio* receiver; even to this day, almost all radio receivers are of this type. Then, in 1933, Armstrong demonstrated another revolutionary concept, namely, a modulation scheme that he called *frequency modulation* (FM); Armstrong's paper making the case for FM radio was published in 1936.

The first all-electronic *television* system was demonstrated by Philo T. Farnsworth in 1928, and then by Vladimir K. Zworykin in 1929. By 1939, the British Broadcasting Corporation (BBC) was broadcasting television on a commercial basis.

In 1928, Harry Nyquist published a classic paper on the theory of signal transmission in telegraphy. In particular, Nyquist developed criteria for the correct reception of telegraph signals transmitted over dispersive channels in the absence of noise. Much of Nyquist's early work was applied later to the transmission of digital data over dispersive channels.

In 1937, Alec Reeves invented *pulse-code modulation* (PCM) for the digital encoding of speech signals. The technique was developed during World War II to enable the encryption of speech signals; indeed, a full-scale, 24-channel system was used in the field by the United States military at the end of the war. However, PCM had to await the discovery of the transistor and the subsequent development of large-scale integration of circuits for its commercial exploitation.

In 1943, D. O. North devised the *matched filter* for the optimum detection of a known signal in additive white noise. A similar result was obtained in 1946 independently by J. H. Van Vleck and D. Middleton, who coined the term "matched filter."

In 1947, the geometric representation of signals was developed by V. A. Kotel'nikov in a doctoral dissertation presented before the Academic Council of the Molotov Energy Institute in Moscow. This method was subsequently brought to full fruition by John M. Wozencraft and Irwin M. Jacobs in a landmark textbook published in 1965.

In 1948, the theoretical foundations of digital communications were laid by Claude Shannon in a paper entitled "A Mathematical Theory of Communication." Shannon's paper was received with immediate and enthusiastic acclaim. It was perhaps this response that emboldened Shannon to amend the title of his paper to "The Mathematical Theory of Communication" when it was reprinted a year later in a book co-authored with Warren Weaver. It is noteworthy that prior to the publication of Shannon's 1948 classic paper, it was believed that increasing the rate of information transmission over a channel would increase the probability of error; the communication theory community was taken by surprise when Shannon proved that this was not true, provided that the transmission rate was below the channel capacity.

The *transistor* was invented in 1948 by Walter H. Brattain, John Bardeen, and William Shockley at Bell Laboratories. The first silicon integrated circuit (IC) was produced by Robert Noyce in 1958. These landmark innovations in solid-state devices and integrated circuits led to the development of *very-large-scale integrated* (VLSI) circuits and single-chip *microprocessors*, and with them the nature of the telecommunications industry changed forever.

The invention of the transistor in 1948 spurred the application of electronics to switching and digital communications. The motivation was to improve reliability, increase capacity, and reduce cost. The first call through a stored-program system was placed in March 1958 at Bell Laboratories; and the first commercial telephone service with digital switching began in Morris, Illinois, in June 1960. The first *T-1 carrier system* transmission was installed in 1962 by Bell Laboratories in the United States.

During the period 1943 to 1946, the first electronic digital computer, called the ENIAC, was built at the Moore School of Electrical Engineering of the University of Pennsylvania under the technical direction of J. Presper Eckert, Jr., and John W. Mauchly. However, John von Neumann's contributions were among the earliest and most fundamental to the theory, design, and application of digital computers, which go back to the first draft of a report written in 1945. Computers and terminals started communication with each other over long distances in the early 1950s. The links used were initially voice-grade telephone channels operating at low speeds (300 to 1200 b/s). Various factors have contributed to a dramatic increase in data transmission rates; notable among them are the idea of *adaptive equalization*, pioneered by Robert Lucky in 1965, and efficient modulation techniques, pioneered by G. Ungerboeck in 1982. Another idea widely employed in computer communications is that of *automatic repeat-request* (ARQ). The ARQ method was originally devised by H. C. A. van Duuren during World War II and published in 1946. It was used to improve radio-telephony for telex transmission over long distances. During 1950–1970, various studies were made on *computer networks*. However, the most significant of them in terms of impact on computer communications was the *Advanced Research Project Agency Network* (ARPANET), first put into service in 1971. The development of ARPANET was sponsored by the Advanced Research Projects Agency of the U. S. Department of Defense. The pioneering work in *packet switching* was done on ARPANET.

In 1955, John R. Pierce proposed the use of satellites for communications. This proposal was preceded, however, by an earlier paper by Arthur C. Clark that was published in 1945, also proposing the idea of using an *Earth-orbiting* satellite as a relay point for communication between two Earth stations. In 1957, the Soviet Union launched Sputnik I, which transmitted telemetry signals for 21 days. This was followed shortly by the launching of Explorer I by the United States in 1958, which transmitted telemetry signals for about five months. A major experimental step in communications satellite technology was taken with the launching of Telstar I from Cape Canaveral on July 10, 1962. The Telstar satellite was built by the Bell Laboratories, which had acquired considerable knowledge from pioneering work by Pierce. The satellite was capable of relaying TV programs across the Atlantic; this was made possible only through the use of maser receivers and large antennas.

The use of optical means (e.g., smoke and fire signals) for the transmission of information dates back to prehistoric times. However, no major breakthrough in optical communications was made until 1966, when K. C. Kao and G. A. Hockham of Standard Telephone Laboratories, U. K., proposed the use of a clad glass fiber as a dielectric waveguide, The *laser* (acronym for *l*ight *a*mplification by *s*timulated *e*mission of *r*adiation) had been invented and developed in 1959 and 1960. Kao and Hockham pointed out that (1) the attenuation in an optical fiber was due to impurities in the glass, and (2) the intrinsic loss, determined by Rayleigh scattering, is very low. Indeed, they predicted that a loss of 20 dB/km should be attainable. This remarkable prediction, made at a time when the power loss in a glass fiber was about 1000 dB/km, was to be demonstrated later. Nowadays, transmission losses as low as 0.2 dB/km are achievable.

The spectacular advances in microelectronics, digital computers, and lightwave systems that we have witnessed to date, and that will continue into the future, are all responsible for dramatic changes in the telecommunications environment; many of these changes are already in place, and more changes will evolve as time goes on.

NOTES AND REFERENCES

1. For essays on communications and other related disciplines (e.g., electronics, computers, radar, radio astronomy, satellites), see Overhage (1962); in particular, see the chapter on "Communications" by L. V. Berkner, pp. 35–50.

2. The material presented in Section 1.9 on Information Theory and Coding and Section 1.10 on Analog Versus Digital Communications is influenced by the introductory chapters of the books by Blahut (1987, 1990).

3. For an insightful discussion of the rules of thumb learned from the applications of information theory to communication systems, see the paper by Viterbi (1991).

4. For a detailed discussion of communication networks, see Stallings (1985, 1987), and Tanenbaum (1988).

5. The OSI reference model was developed by a subcommittee of the International Organization for Standardization (ISO) in 1977. For a discussion of the principles involved in arriving at the seven layers of the OSI model, and a description of the layers themselves, see Tanenbaum (1988).

6. Historical accounts of telecommunications are presented in Stark, Tuteur, and Anderson (1988), Carlson (1986), Couch (1990), Blahut (1990), Pierce and Noll (1990); see also *A History of Engineering and Science in the Bell System* (Bell Laboratories, 1975).

CHAPTER 2

Representation of Signals and Systems

2.1 INTRODUCTION

In the previous chapter we identified *deterministic signals* as a class of signals whose waveforms are defined exactly as functions of time. In this chapter we study the mathematical description of such signals using the *Fourier transform* that provides the link between the time-domain and frequency-domain descriptions of a signal. The waveform of a signal and its spectrum (i.e., frequency content) are two natural vehicles to understand the signal.

Another related issue that we study in this chapter is the representation of linear time-invariant systems. Here also we find that the Fourier transform plays a key role. Filters of different kinds and certain communication channels are important examples of this class of systems.

We begin the study by presenting a formal definition of the Fourier transform, followed by a discussion of its important properties.

2.2 THE FOURIER TRANSFORM[1]

Let $g(t)$ denote a *nonperiodic deterministic signal*, expressed as some function of time t. By definition, the *Fourier transform* of the signal $g(t)$ is given by the integral

$$G(f) = \int_{-\infty}^{\infty} g(t)\exp(-j2\pi ft)\ dt \qquad (2.1)$$

where $j = \sqrt{-1}$, and the variable f denotes *frequency*. Given the Fourier transform $G(f)$, the original signal $g(t)$ is recovered exactly using the formula for the *inverse Fourier transform*:

$$g(t) = \int_{-\infty}^{\infty} G(f)\exp(j2\pi ft)\ df \qquad (2.2)$$

Note that in Eqs. (2.1) and (2.2) we have used a lowercase letter to denote the time function and an uppercase letter to denote the corresponding frequency function. The functions $g(t)$ and $G(f)$ are said to constitute a Fourier-transform pair.[2]

For the Fourier transform of a signal $g(t)$ to exist, it is sufficient, but not necessary, that $g(t)$ satisfies three conditions known collectively as *Dirichlet's conditions*:

1.   The function $g(t)$ is single-valued, with a finite number of maxima and minima in any finite time interval.
2.   The function $g(t)$ has a finite number of discontinuities in any finite time interval.
3.   The function $g(t)$ is absolutely integrable, that is,

$$\int_{-\infty}^{\infty} |g(t)|\ dt < \infty$$

We may safely ignore the question of the existence of the Fourier transform of a time function $g(t)$ when it is an accurately specified description of a physically realizable signal. In other words, physical realizability is a sufficient condition for the existence of a Fourier transform. Indeed, we may go one step further and state that all energy signals, that is, signals $g(t)$ for which

$$\int_{-\infty}^{\infty} |g(t)|^2\ dt < \infty$$

are Fourier transformable.[3]

Notations

The formulas for the Fourier transform and the inverse Fourier transform presented in Eqs. (2.1) and (2.2) are written in terms of two variables: *time t* measured in *seconds* (s), and frequency f measured in *Hertz* (Hz). The frequency f is related to the *angular frequency* ω as

$$\omega = 2\pi f$$

which is measured in *radians per second* (rad/s). We may simplify the expressions for the exponents in the integrands of Eqs. (2.1) and (2.2) by using ω instead of f. However, the use of f is preferred over ω for two reasons. First, the use of frequency results in mathematical *symmetry* of Eqs. (2.1) and (2.2) with respect to each other in a natural way. Second, the frequency contents of communication signals (i.e., speech and video signals) are usually expressed in Hertz.

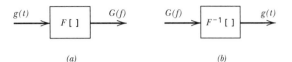

Figure 2.1 (a) Fourier transformation and (b) inverse Fourier transformation depicted as linear operators.

A convenient *shorthand* notation for the transform relations of Eqs. (2.1) and (2.2) is

$$G(f) = F[g(t)] \tag{2.3}$$

and

$$g(t) = F^{-1}[G(f)] \tag{2.4}$$

where $F[\ ]$ and $F^{-1}[\ ]$ play the roles of *linear operators*, as depicted in Fig. 2.1.

Another convenient shorthand notation for the *Fourier-transform pair*, represented by $g(t)$ and $G(f)$, is

$$g(t) \rightleftharpoons G(f) \tag{2.5}$$

The shorthand notations described in (2.3) to (2.5) are used in the text where appropriate.

Continuous Spectrum

By using the Fourier transform operation, a pulse signal $g(t)$ of finite energy is expressed as a continuous sum of exponential functions with frequencies in the interval $-\infty$ to ∞. The amplitude of a component of frequency f is proportional to $G(f)$, where $G(f)$ is the Fourier transform of $g(t)$. Specifically, at any frequency f, the exponential function $\exp(j2\pi ft)$ is weighted by the factor $G(f)\ df$, which is the contribution of $G(f)$ in an infinitesimal interval df centered at the frequency f. Thus we may express the function $g(t)$ in terms of the continuous sum of such infinitesimal components, as shown by the integral

$$g(t) = \int_{-\infty}^{\infty} G(f)\exp(j2\pi ft)\ df$$

The Fourier transformation provides us with a tool to resolve a given signal $g(t)$ into its complex exponential components occupying the entire frequency interval from $-\infty$ to ∞. In particular, the Fourier transform $G(f)$ of the signal defines the frequency-domain representation of the signal in that it specifies relative amplitudes of the various frequency components of the signal. We may equivalently define the signal in terms of its time-domain representation by specifying the function $g(t)$ at each instant of time t. The signal is uniquely defined by either representation.

In general, the Fourier transform $G(f)$ is a complex function of frequency f, so that we may express it in the form

$$G(f) = |G(f)|\exp[j\theta(f)] \qquad (2.6)$$

where $|G(f)|$ is called the *continuous amplitude spectrum* of $g(t)$, and $\theta(f)$ is called the *continuous phase spectrum* of $g(t)$. Here, the spectrum is referred to as a *continuous spectrum* because both the amplitude and phase of $G(f)$ are defined for all frequencies.

For the special case of a real-valued function $g(t)$, we have

$$G(-f) = G^*(f)$$

where the asterisk denotes complex conjugation. Therefore, it follows that if $g(t)$ is a real-valued function of time t, then

$$|G(-f)| = |G(f)|$$

and

$$\theta(-f) = -\theta(f)$$

Accordingly, we may make the following statements on the spectrum of a *real-valued signal*:

1. The amplitude spectrum of the signal is an even function of the frequency; that is, the amplitude spectrum is *symmetric* about the vertical axis.
2. The phase spectrum of the signal is an odd function of the frequency; that is, the phase spectrum is *antisymmetric* about the vertical axis.

These two statements are summed up by saying that the spectrum of a real-valued signal exhibits *conjugate symmetry*.

EXAMPLE 1   Rectangular Pulse

Consider a *rectangular pulse* of duration T and amplitude A, as shown in Fig. 2.2a. To define this pulse mathematically in a convenient form, we use the following

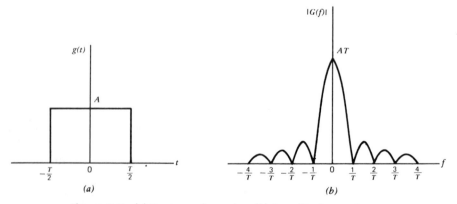

Figure 2.2 (a) Rectangular pulse. (b) Amplitude spectrum.

notation[4]

$$\text{rect}(t) = \begin{cases} 1, & -\dfrac{1}{2} < t < \dfrac{1}{2} \\[2ex] 0, & t > \dfrac{1}{2} \end{cases} \qquad (2.7)$$

which stands for a *rectangular function* of unit amplitude and unit duration centered at $t = 0$. Then, in terms of this "standard" function, we may express the rectangular pulse of Fig. 2.2a simply as follows:

$$g(t) = A \, \text{rect}\left(\frac{t}{T}\right)$$

The Fourier transform of the rectangular pulse $g(t)$ is given by

$$G(f) = \int_{-T/2}^{T/2} A \exp(-j2\pi f t) \, dt$$

$$= AT\left(\frac{\sin(\pi f T)}{\pi f T}\right) \qquad (2.8)$$

To simplify the notation in the preceding and subsequent results, we introduce another standard function, namely, the *sinc function* defined by

$$\text{sinc}(\lambda) = \frac{\sin(\pi\lambda)}{\pi\lambda} \qquad (2.9)$$

where λ is the independent variable. The sinc function plays an important role in communication theory. As shown in Fig. 2.3, it has its maximum value of unity

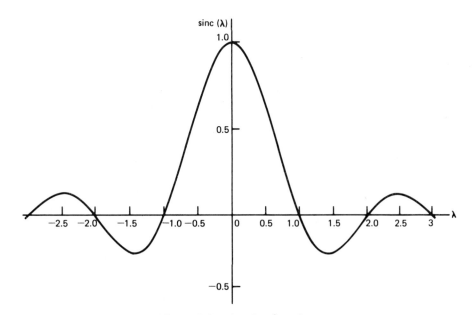

Figure 2.3 The sinc function.

at $\lambda = 0$, and approaches zero as λ approaches infinity, oscillating through positive and negative values. It goes through zero at $\lambda = \pm 1, \pm 2, \ldots$, and so on.

Thus, in terms of the sinc function, we may rewrite Eq. (2.8) as

$$G(f) = AT \, \text{sinc}(fT)$$

We thus have the Fourier-transform pair:

$$A \, \text{rect}\left(\frac{t}{T}\right) \rightleftharpoons AT \, \text{sinc}(fT) \qquad (2.10)$$

The amplitude spectrum $|G(f)|$ is shown plotted in Fig. 2.2b. The first zero-crossing of the spectrum occurs at $f = \pm 1/T$. As the pulse duration T is decreased, this first zero-crossing moves up in frequency. Conversely, as the pulse duration T is increased, the first zero-crossing moves toward the origin.

This example shows that the relationship between the time-domain and frequency-domain descriptions of a signal is an *inverse* one. That is, a pulse, narrow in time, has a significant frequency description over a wide range of frequencies, and vice versa. We shall have more to say on the inverse relationship between time and frequency in Section 2.5.

Note also that in this example the Fourier transform $G(f)$ is a real-valued and symmetric function of frequency f. This is a direct consequence of the fact that the rectangular pulse $g(t)$ shown in Fig. 2.2a is a symmetric function of time t.

EXAMPLE 2   Exponential Pulse

A truncated form of a decaying *exponential pulse* is shown in Fig. 2.4a. We may define this pulse mathematically in a convenient form using the *unit step function*:

$$u(t) = \begin{cases} 1, & t > 0 \\[2mm] \dfrac{1}{2}, & t = 0 \\[2mm] 0, & t < 0 \end{cases} \qquad (2.11)$$

We may then express the decaying exponential pulse of Fig. 2.4a as

$$g(t) = \exp(-at)\,u(t)$$

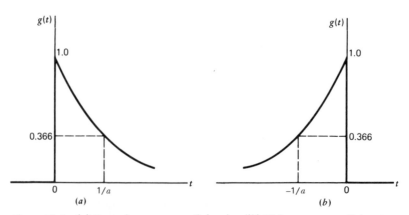

Figure 2.4   (a) Decaying exponential pulse. (b) Rising exponential pulse.

The Fourier transform of this pulse is

$$G(f) = \int_0^\infty \exp(-at)\exp(-j2\pi ft) \, dt$$

$$= \int_0^\infty \exp[-t(a + j2\pi f)] \, dt$$

$$= \frac{1}{a + j2\pi f}$$

The Fourier-transform pair for the decaying exponential pulse of Fig. 2.4a is therefore

$$\exp(-at)u(t) \rightleftharpoons \frac{1}{a + j2\pi f} \qquad (2.12)$$

A truncated rising exponential pulse is shown in Fig. 2.4b, which is defined by

$$g(t) = \exp(at)u(-t)$$

Note that $u(-t)$ is equal to unity for $t < 0$, one-half at $t = 0$, and zero for $t > 0$. The Fourier transform of this pulse is

$$G(f) = \int_{-\infty}^0 \exp(at)\exp(-j2\pi ft) \, dt$$

$$= \int_{-\infty}^0 \exp[t(a - j2\pi f)] \, dt$$

$$= \frac{1}{a - j2\pi f}$$

The Fourier-transform pair for the rising exponential pulse of Fig. 2.4b is therefore

$$\exp(at)u(-t) \rightleftharpoons \frac{1}{a - j2\pi f} \qquad (2.13)$$

The decaying and rising exponential pulses of Fig. 2.4 are both asymmetric functions of time t. Their Fourier transforms are therefore complex valued, as shown in Eqs. (2.12) and (2.13). Moreover, from these Fourier-transform pairs, we readily see that truncated decaying and rising exponential pulses have the same amplitude spectrum, but the phase spectrum of the one is the negative of that of the other.

2.3   PROPERTIES OF THE FOURIER TRANSFORM

It is useful to have insight into the relationship between a time function $g(t)$ and its Fourier transform $G(f)$, and also into the effects that various operations on the function $g(t)$ have on the transform $G(f)$. This may be achieved by examining certain properties of the Fourier transform. In this section we describe 12 of these properties, which we will prove, one by one. These properties are summarized in Table 1 of Appendix 11 at the end of the book.

PROPERTY 1 Linearity (Superposition)

Let $g_1(t) \rightleftharpoons G_1(f)$ and $g_2(t) \rightleftharpoons G_2(f)$. Then for all constants c_1 and c_2, we have

$$c_1 g_1(t) + c_2 g_2(t) \rightleftharpoons c_1 G_1(f) + c_2 G_2(f) \tag{2.14}$$

The proof of this property follows simply from the linearity of the integrals defining $G(f)$ and $g(t)$.

This property permits us to find the Fourier transform $G(f)$ of a function $g(t)$ that is a linear combination of two other functions $g_1(t)$ and $g_2(t)$ whose Fourier transforms $G_1(f)$ and $G_2(f)$ are known, as illustrated in the following example.

EXAMPLE 3 Combinations of Exponential Pulses

Consider a *double exponential pulse* defined by (see Fig. 2.5a)

$$g(t) = \begin{cases} \exp(-at), & t > 0 \\ 1, & t = 0 \\ \exp(at), & t < 0 \end{cases}$$

$$= \exp(-a|t|) \tag{2.15}$$

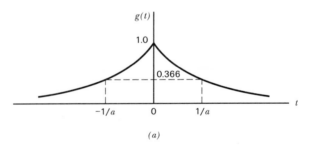

(a)

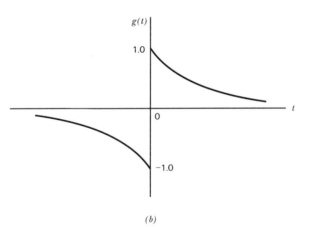

(b)

Figure 2.5 (a) Double-exponential pulse (symmetric). (b) Another double-exponential pulse (antisymmetric).

This pulse may be viewed as the sum of a truncated decaying exponential pulse and a truncated rising exponential pulse. Therefore, using the linearity property and the Fourier-transform pairs of Eqs. (2.12) and (2.13), we find that the Fourier transform of the double exponential pulse of Fig. 2.5a is

$$G(f) = \frac{1}{a + j2\pi f} + \frac{1}{a - j2\pi f}$$

$$= \frac{2a}{a^2 + (2\pi f)^2}$$

We thus have the following Fourier-transform pair for the double exponential pulse of Fig. 2.5a:

$$\exp(-a|t|) \rightleftharpoons \frac{2a}{a^2 + (2\pi f)^2} \tag{2.16}$$

Note that because of the symmetry in the time domain, as in Fig. 2.5a, the spectrum is real and symmetric; this is a general property of such Fourier-transform pairs.

Another interesting combination is the difference between a truncated decaying exponential pulse and a truncated rising exponential pulse, as shown in Fig. 2.5b. Here we have

$$g(t) = \begin{cases} \exp(-at), & t > 0 \\ 0, & t = 0 \\ -\exp(at), & t < 0 \end{cases} \tag{2.17}$$

We may formulate a compact notation for this composite signal by using the *signum function* that equals $+1$ for positive time and -1 for negative time, as shown by

$$\text{sgn}(t) = \begin{cases} +1, & t > 0 \\ 0, & t = 0 \\ -1, & t < 0 \end{cases} \tag{2.18}$$

The signum function is shown in Fig. 2.6. Accordingly, we may reformulate the composite signal $g(t)$ defined in Eq. (2.17) simply as

$$g(t) = \exp(-a|t|)\ \text{sgn}(t)$$

Hence, applying the linearity property of the Fourier transform, we readily find that in light of Eqs. (2.12) and (2.13), the Fourier transform of the signal $g(t)$ is given by

$$F[\exp(-a|t|)\ \text{sgn}(t)] = \frac{1}{a + j2\pi f} - \frac{1}{a - j2\pi f}$$

$$= -\frac{-j4\pi f}{a^2 + (2\pi f)^2}$$

We thus have the Fourier-transform pair:

$$\exp(-a|t|)\ \text{sgn}(t) \rightleftharpoons \frac{-j4\pi f}{a^2 + (2\pi f)^2} \tag{2.19}$$

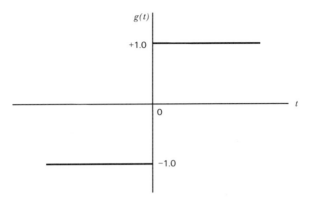

Figure 2.6 The signum function.

In contrast to the Fourier-transform pair of Eq. (2.16), the Fourier transform in Eq. (2.19) is odd and purely imaginary. It is a general property of Fourier-transform pairs that a real antisymmetric time function, as in Fig. 2.5*b*, has an odd and purely imaginary function as its Fourier transform.

PROPERTY 2   **Time Scaling**

Let $g(t) \rightleftharpoons G(f)$. Then,

$$g(at) \rightleftharpoons \frac{1}{|a|} G\left(\frac{f}{a}\right) \tag{2.20}$$

To prove this property, we note that

$$F[g(at)] = \int_{-\infty}^{\infty} g(at)\exp(-j2\pi ft) \; dt$$

Set $\tau = at$. There are two cases that can arise, depending on whether the scaling factor a is positive or negative. If $a > 0$, we get

$$F[g(at)] = \frac{1}{a} \int_{-\infty}^{\infty} g(\tau)\exp\left[-j2\pi\left(\frac{f}{a}\right)\tau\right] \; d\tau$$

$$= \frac{1}{a} G\left(\frac{f}{a}\right)$$

On the other hand, if $a < 0$, the limits of integration are interchanged so that we have the multiplying factor $-(1/a)$ or, equivalently, $1/|a|$. This completes the proof of Eq. (2.20).

Note that the function $g(at)$ represents $g(t)$ compressed in time by a factor a, whereas the function $G(f/a)$ represents $G(f)$ expanded in frequency by the same factor a. Thus, the scaling property states that the compression of a function $g(t)$ in the time domain is equivalent to the expansion of its Fourier transform $G(f)$ in the frequency domain by the same factor, or vice versa.

For the special case when $a = -1$, we readily find from Eq. (2.20) that

$$g(-t) \rightleftharpoons G(-f) \qquad (2.21)$$

In words, if a function $g(t)$ has a Fourier transform given by $G(f)$, then the Fourier transform of $g(-t)$ is $G(-f)$.

PROPERTY 3   Duality

If $g(t) \rightleftharpoons G(f)$, then

$$G(t) \rightleftharpoons g(-f) \qquad (2.22)$$

This property follows from the relation defining the inverse Fourier transform by writing it in the form:

$$g(-t) = \int_{-\infty}^{\infty} G(f)\exp(-j2\pi ft)\ df$$

and then interchanging t and f.

EXAMPLE 4   Sinc Pulse

Consider a signal $g(t)$ in the form of a sinc function, as shown by

$$g(t) = A\ \text{sinc}(2Wt)$$

To evaluate the Fourier transform of this function, we apply the duality and time-scaling properties to the Fourier-transform pair of Eq. (2.10). Then, recognizing that the rectangular function is an even function of time, we obtain the following result:

$$A\ \text{sinc}(2Wt) \rightleftharpoons \frac{A}{2W}\ \text{rect}\left(\frac{f}{2W}\right) \qquad (2.23)$$

which is illustrated in Fig. 2.7. We thus see that the Fourier transform of a sinc pulse is zero for $|f| > W$. Note also that the sinc pulse itself is only asymptotically limited in time in the sense that it approaches zero as time t approaches infinity.

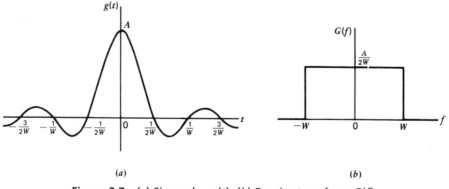

Figure 2.7   (a) Sinc pulse $g(t)$. (b) Fourier transform $G(f)$.

PROPERTY 4   Time Shifting

If $g(t) \rightleftharpoons G(f)$, then

$$g(t - t_0) \rightleftharpoons G(f)\exp(-j2\pi f t_0) \qquad (2.24)$$

To prove this property, we take the Fourier transform of $g(t - t_0)$ and then set $\tau = (t - t_0)$ to obtain

$$F[g(t - t_0)] = \exp(-j2\pi f t_0) \int_{-\infty}^{\infty} g(\tau)\exp(-j2\pi f \tau) \, d\tau$$

$$= \exp(-j2\pi f t_0) G(f)$$

The time-shifting property states that if a function $g(t)$ is shifted in the positive direction by an amount t_0, the effect is equivalent to multiplying its Fourier transform $G(f)$ by the factor $\exp(-j2\pi f t_0)$. This means that the amplitude of $g(f)$ is unaffected by the time shift, but its phase is changed by the linear factor $-2\pi f t_0$.

PROPERTY 5   Frequency Shifting

If $g(t) \rightleftharpoons G(f)$, then

$$\exp(j2\pi f_c t)g(t) \rightleftharpoons G(f - f_c) \qquad (2.25)$$

where f_c is a real constant.

This property follows from the fact that

$$F[\exp(j2\pi f_c t)g(t)] = \int_{-\infty}^{\infty} g(t)\exp[-j2\pi t(f - f_c)] \, dt$$

$$= G(f - f_c)$$

That is, multiplication of a function $g(t)$ by the factor $\exp(j2\pi f_c t)$ is equivalent to shifting its Fourier transform $G(f)$ in the positive direction by the amount f_c. This property is called the *modulation theorem*, because a shift of the range of frequencies in a signal is accomplished by using modulation. Note the duality between the time-shifting and frequency-shifting operations described in Eqs. (2.24) and (2.25).

EXAMPLE 5   Radio Frequency (RF) Pulse

Consider the pulse signal $g(t)$ shown in Fig. 2.8a which consists of a sinusoidal wave of amplitude A and frequency f_c, extending in duration from $t = -T/2$ to $t = T/2$. This signal is sometimes referred to as an *RF pulse* when the frequency f_c falls in the radio-frequency band. The signal $g(t)$ of Fig. 2.8a may be expressed mathematically as follows:

$$g(t) = A \, \text{rect}\left(\frac{t}{T}\right) \cos(2\pi f_c t) \qquad (2.26)$$

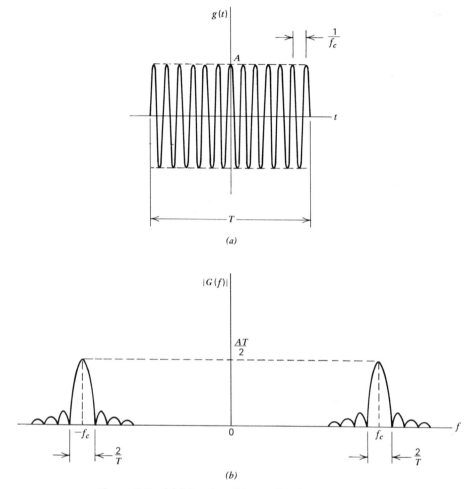

Figure 2.8 (a) RF pulse. (b) Amplitude spectrum.

To find the Fourier transform of this signal, we note that

$$\cos(2\pi f_c t) = \frac{1}{2}\left[\exp(j2\pi f_c t) + \exp(-j2\pi f_c t)\right]$$

Therefore, applying the frequency-shifting property to the Fourier-transform pair of Eq. (2.10), we get the desired result

$$G(f) = \frac{AT}{2}\{\text{sinc}[T(f - f_c)] + \text{sinc}[T(f + f_c)]\} \qquad (2.27)$$

In the special case of $f_c T \gg 1$, we may use the approximate result

$$G(f) \simeq \begin{cases} \dfrac{AT}{2}\,\text{sinc}[T(f - f_c), & f > 0 \\[2mm] 0, & f = 0 \\[2mm] \dfrac{AT}{2}\,\text{sinc}[T(f + f_c)], & f < 0 \end{cases} \qquad (2.28)$$

The amplitude spectrum of the RF pulse is shown in Fig. 2.8b. This diagram, in relation to Fig. 2.2b, clearly illustrates the frequency-shifting property of the Fourier transform.

PROPERTY 6 Area Under $g(t)$

If $g(t) \rightleftharpoons G(f)$, then

$$\int_{-\infty}^{\infty} g(t) \ dt = G(0) \tag{2.29}$$

That is, the area under a function $g(t)$ is equal to the value of its Fourier transform $G(f)$ at $f = 0$.

This result is obtained simply by putting $f = 0$ in the formula defining the Fourier transform of the function $g(t)$.

PROPERTY 7 Area Under $G(f)$

If $g(t) \rightleftharpoons G(f)$, then

$$g(0) = \int_{-\infty}^{\infty} G(f) \ df \tag{2.30}$$

That is, the value of a function $g(t)$ at $t = 0$ is equal to the area under its Fourier transform $G(f)$.

The result is obtained simply by putting $t = 0$ in the formula defining the inverse Fourier transform of $G(f)$.

PROPERTY 8 Differentiation in the Time Domain

Let $g(t) \rightleftharpoons G(f)$, and assume that the first derivative of $g(t)$ is Fourier transformable. Then

$$\frac{d}{dt} g(t) \rightleftharpoons j2\pi f G(f) \tag{2.31}$$

That is, differentiation of a time function $g(t)$ has the effect of multiplying its Fourier transform $G(f)$ by the factor $j2\pi f$.

This result is obtained simply by taking the first derivative of both sides of the integral defining the inverse Fourier transform of $G(f)$, and then interchanging the operations of integration and differentiation.

We may generalize Eq. (2.31) as follows:

$$\frac{d^n}{dt^n} g(t) \rightleftharpoons (j2\pi f)^n G(f) \tag{2.32}$$

Equation (2.32) assumes that the Fourier transform of the higher-order derivative exists.

EXAMPLE 6   Gaussian Pulse

In this example we use the differentiation property of the Fourier transform to derive the particular form of a pulse signal that has the same mathematical form as its own Fourier transform.

Let $g(t)$ denote the pulse expressed as a function of time, and $G(f)$ its Fourier transform. We note that by differentiating the formula for the Fourier transform $G(f)$ with respect to f, we have

$$-j2\pi tg(t) \rightleftharpoons \frac{d}{df} G(f) \tag{2.33}$$

which expresses the effect of differentiation in the frequency domain. From Eqs. (2.32) and (2.33) we thus deduce that if

$$\frac{d}{dt} g(t) = -2\pi tg(t) \tag{2.34}$$

then

$$\frac{d}{df} G(f) = -2\pi fG(f) \tag{2.35}$$

which means that the pulse signal and its transform are the same function. In other words, provided that the pulse signal $g(t)$ satisfies the differential equation (2.34), then $G(f) = g(f)$, where $g(f)$ is obtained from $g(t)$ by substituting f for t. Solving Eq. (2.34) for $g(t)$, we obtain

$$g(t) = \exp(-\pi t^2) \tag{2.36}$$

The pulse defined by Eq. (2.36) is called a *Gaussian pulse*, the name being derived from the similarity of the function to the Gaussian probability density function of probability theory (see Chapter 4). It is shown plotted in Fig. 2.9. By applying Eq. (2.29), we find that the area under this Gaussian pulse is unity, as shown by

$$\int_{-\infty}^{\infty} \exp(-\pi t^2)\, dt = 1 \tag{2.37}$$

When the central ordinate and the area under the curve of a pulse are both unity, as in the case of the Gaussian pulse of Eq. (2.36), we say that the pulse is

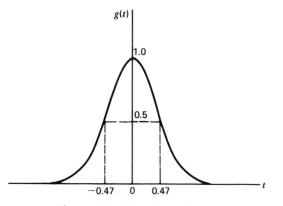

Figure 2.9 Gaussian pulse.

normalized. We conclude therefore that the normalized Gaussian pulse is its own Fourier transform, as shown by

$$\exp(-\pi t^2) \rightleftharpoons \exp(-\pi f^2) \tag{2.38}$$

PROPERTY 9   Integration in the Time Domain

Let $g(t) \rightleftharpoons G(f)$. Then provided that $G(0) = 0$, we have

$$\int_{-\infty}^{t} g(\tau)\ d\tau \rightleftharpoons \frac{1}{j2\pi f}\ G(f) \tag{2.39}$$

That is, integraion of a time function $g(t)$ has the effect of dividing its Fourier transform $G(f)$ by the factor $j2\pi f$, assuming that $G(0)$ is zero.

The result is obtained by expressing $g(t)$ as

$$g(t) = \frac{d}{dt}\left[\int_{-\infty}^{t} g(\tau)\ d\tau\right]$$

and then applying the time-differentiation property of the Fourier transform to obtain

$$G(f) = j2\pi f\left\{F\left[\int_{-\infty}^{t} g(\tau)\ d\tau\right]\right\}$$

from which Eq. (2.39) follows readily.

It is a straightforward matter to generalize Eq. (2.39) to multiple integration; however, the notation becomes rather cumbersome.

Equation (2.39) assumes that $G(0)$, that is, the area under $g(t)$, is zero. The more general case pertaining to $G(0) \neq 0$ is considered in Section 2.6.

EXAMPLE 7   Triangular Pulse

Consider the *doublet pulse* $g_1(t)$ shown in Fig. 2.10a. By integrating this pulse with respect to time, we obtain the *triangular pulse* $g_2(t)$ shown in Fig. 2.10b. We note that the doublet pulse $g_1(t)$ consists of two rectangular pulses: one of amplitude A, defined for the interval $-T \leq t \leq 0$; and the other of amplitude $-A$, defined for the interval $0 \leq t \leq T$. Applying the time-shifting property of the Fourier transform to Eq. (2.10), we find that the Fourier transforms of these two rectangular pulses are equal to $AT\operatorname{sinc}(fT)\exp(j\pi fT)$ and $-AT\operatorname{sinc}(fT)\exp(-j\pi fT)$, respectively. Hence, invoking the linearity property of the Fourier transform, we find that the Fourier transform $G_1(f)$ of the doublet pulse $g_1(t)$ of Fig. 2.10a is given by

$$\begin{aligned}
G_1(f) &= AT\operatorname{sinc}(fT)[\exp(j\pi fT) - \exp(-j\pi fT)] \\
&= 2jAT\operatorname{sinc}(fT)\sin(\pi fT)
\end{aligned} \tag{2.40}$$

We further note that $G_1(0)$ is zero. Hence, using Eqs. (2.39) and (2.40), we find that the Fourier transform $G_2(f)$ of the triangular pulse $g_2(t)$ of Fig. 2.10b is

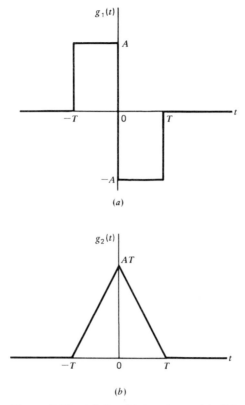

Figure 2.10 (a) Doublet pulse $g_1(t)$. (b) Triangular pulse $g_2(t)$ obtained by integrating $g_1(t)$.

given by

$$G_2(f) = \frac{1}{j2\pi f}\, G_1(f)$$

$$= AT\frac{\sin(\pi f T)}{\pi f}\, \mathrm{sinc}(f T) \tag{2.41}$$

$$= AT^2\, \mathrm{sinc}^2(f T)$$

Note that the doublet pulse of Fig. 2.10a is real and antisymmetric and its Fourier transform is therefore odd and purely imaginary, whereas the triangular pulse of Fig. 2.10b is real and symmetric and its Fourier transform is therefore symmetric and purely real.

PROPERTY 10   Conjugate Functions

If $g(t) \rightleftharpoons G(f)$, then for a complex-valued time function $g(t)$ we have

$$g^*(t) \rightleftharpoons G^*(-f), \tag{2.42}$$

where the asterisk denotes the complex conjugate operation.

To prove this property, we know from the inverse Fourier transform that

$$g(t) = \int_{-\infty}^{\infty} G(f)\exp(j2\pi ft)\ df$$

Taking the complex conjugates of both sides yields

$$g^*(t) = \int_{-\infty}^{\infty} G^*(f)\exp(-j2\pi ft)\ df$$

Next, replacing f with $-f$ gives

$$g^*(t) = -\int_{\infty}^{-\infty} G^*(-f)\exp(j2\pi ft)\ df$$

$$= \int_{-\infty}^{\infty} G^*(-f)\exp(j2\pi ft)\ df$$

That is, $g^*(t)$ is the inverse Fourier transform of $G^*(-f)$, which is the desired result.

As a corollary to Property 10, we may state that

$$g^*(-t) \rightleftharpoons G^*(f) \tag{2.43}$$

This result follows directly from Eq. (2.42) by applying the special form of the scaling property described in Eq. (2.21).

EXAMPLE 8 Real and Imaginary Parts of a Time Function

Expressing a complex-valued function $g(t)$ in terms of its real and imaginary parts, we may write

$$g(t) = \text{Re}[g(t)] + j\,\text{Im}[g(t)] \tag{2.44}$$

where Re denotes "the real part of" and Im denotes the "imaginary part of." The complex conjugate of $g(t)$ is

$$g^*(t) = \text{Re}[g(t)] - j\,\text{Im}[g(t)] \tag{2.45}$$

Adding Eqs. (2.44) and (2.45) gives

$$\text{Re}[g(t)] = \frac{1}{2}[g(t) + g^*(t)] \tag{2.46}$$

and subtracting them yields

$$\text{Im}[g(t)] = \frac{1}{2j}[g(t) - g^*(t)] \tag{2.47}$$

Therefore, applying Property 10, we obtain the following two Fourier-transform pairs:

$$\text{Re}[g(t)] \rightleftharpoons \frac{1}{2}[G(f) + G^*(-f)]$$

$$\text{Im}[g(t)] \rightleftharpoons \frac{1}{2j}[G(f) - G^*(-f)] \tag{2.48}$$

From Eq. (2.48), it is apparent that in the case of a real-valued time function $g(t)$, we have $G(f) = G^*(-f)$, that is, $G(f)$ exhibits *conjugate symmetry*, confirming a result that we stated previously in Section 2.2.

PROPERTY 11   Multiplication in the Time Domain

Let $g_1(t) \rightleftharpoons G_1(f)$ and $g_2(t) \rightleftharpoons G_2(f)$. Then

$$g_1(t)g_2(t) \rightleftharpoons \int_{-\infty}^{\infty} G_1(\lambda) G_2(f - \lambda) \ d\lambda \tag{2.49}$$

To prove this property, we first denote the Fourier transform of the product $g_1(t)g_2(t)$ by $G_{12}(f)$, so that we may write

$$g_1(t)g_2(t) \rightleftharpoons G_{12}(f)$$

where

$$G_{12}(f) = \int_{-\infty}^{\infty} g_1(t)g_2(t)\exp(-j2\pi ft) \ dt$$

For $g_2(t)$, we next substitute the inverse Fourier transform

$$g_2(t) = \int_{-\infty}^{\infty} G_2(f')\exp(j2\pi f't) \ df'$$

in the integral defining $G_{12}(f)$ to obtain

$$G_{12}(f) = \int_{-\infty}^{\infty} \int_{-\infty}^{\infty} g_1(t) G_2(f')\exp[-j2\pi(f - f')t] \ df' \ dt$$

Define $\lambda = f - f'$. Then, interchanging the order of integration, we obtain

$$G_{12}(f) = \int_{-\infty}^{\infty} d\lambda\, G_2(f - \lambda) \int_{-\infty}^{\infty} g_1(t)\exp(-j2\pi\lambda t) \ dt$$

The inner integral is recognized simply as $G_1(\lambda)$, and so we may write

$$G_{12}(f) = \int_{-\infty}^{\infty} G_1(\lambda) G_2(f - \lambda) \ d\lambda$$

which is the desired result. This integral is known as the *convolution integral* expressed in the frequency domain, and the function $G_{12}(f)$ is referred to as the *convolution* of $G_1(f)$ and $G_2(f)$. We conclude that *the multiplication of two signals in the time domain is transformed into the convolution of their individual Fourier transforms in the frequency domain*. This property is known as the *multiplication theorem*.

In a discussion of convolution, the following shorthand notation is frequently used:

$$G_{12}(f) = G_1(f) \star G_2(f)$$

Accordingly, we may reformulate Eq. (2.49) in the following symbolic form:

$$g_1(t)g_2(t) \rightleftharpoons G_1(f) \bigstar G_2(f) \tag{2.50}$$

Note that convolution is *commutative*, that is,

$$G_1(f) \bigstar G_2(f) = G_2(f) \bigstar G_1(f)$$

which follows directly from Eq. (2.50).

PROPERTY 12   Convolution in the Time Domain

Let $g_1(t) \rightleftharpoons G_1(f)$ and $g_2(t) \rightleftharpoons G_2(f)$. Then

$$\int_{-\infty}^{\infty} g_1(\tau)g_2(t - \tau) \, d\tau \rightleftharpoons G_1(f)G_2(f) \tag{2.51}$$

This result follows directly by combining Property 3 (duality) and Property 11 (time-domain multiplication). We may thus state that the *convolution of two signals in the time domain is transformed into the multiplication of their individual Fourier transforms in the frequency domain.* This property is known as the *convolution theorem.* Its use permits us to exchange a convolution operation for a transform multiplication, an operation that is ordinarily easier to manipulate.

Using the shorthand notation for convolution, we may rewrite Eq. (2.51) in the form

$$g_1(t) \bigstar g_2(t) \rightleftharpoons G_1(f)G_2(f) \tag{2.52}$$

where the symbol $\bigstar$ denotes convolution.

Note that Properties 11 and 12, described by Eqs. (2.49) and (2.51), respectively, are the dual of each other.

2.4   RAYLEIGH'S ENERGY THEOREM

Traditionally, the energy of a signal is defined using the time-domain description of the signal. In this section we develop another procedure for calculating the energy of a signal using the Fourier transform of the signal. In so doing, we introduce the notion of energy spectral density.

Consider a signal $g(t)$ defined over the entire interval $-\infty < t < \infty$, and let it be assumed that its Fourier transform, denoted by $G(f)$, exists. The signal $g(t)$ may be complex valued. The total energy of the signal is defined by the usual formula

$$E = \int_{-\infty}^{\infty} |g(t)|^2 \, dt \tag{2.53}$$

According to this definition, the integrand $|g(t)|^2$ may be viewed as an *energy intensity* that varies with time. The total energy under the curve of energy intensity, plotted as a function of time, equals the total energy of the signal.

The energy intensity $|g(t)|^2$ may be expressed as the product of two time functions, namely, $g(t)$ and its complex conjugate $g^*(t)$. The Fourier transform of $g^*(t)$ is equal to $G^*(-f)$, by virtue of Property 10 (complex conjugation). Then, applying Property 11 (the multiplication theorem) or, more specifically, applying Eq. (2.49) to the product $g(t)g^*(t)$ and evaluating the result for $f = 0$, we obtain the relation

$$\int_{-\infty}^{\infty} g(t)g^*(t) \ dt = \int_{-\infty}^{\infty} G(\lambda) \, G^*(\lambda) \ d\lambda$$

Replacing λ with f in the right-hand side of this relation, and noting that $|G(f)|^2 = G(f)G^*(f)$, we may redefine the total energy of the signal $g(t)$ as

$$E = \int_{-\infty}^{\infty} |g(t)|^2 \ dt = \int_{-\infty}^{\infty} |G(f)|^2 \ df \tag{2.54}$$

This result is known as *Rayleigh's energy theorem*. To apply the theorem, we need only know the amplitude spectrum $|G(f)|$ of the signal.

Let $\mathscr{E}_g(f)$ denote the squared amplitude spectrum of the signal $g(t)$, as shown by

$$\mathscr{E}_g(f) = |G(f)|^2 \tag{2.55}$$

Accordingly, we may express the total energy of the signal $g(t)$ in terms of $\mathscr{E}_g(f)$ as follows:

$$E = \int_{-\infty}^{\infty} \mathscr{E}_g(f) \ df \tag{2.56}$$

The quantity $\mathscr{E}_g(f)$ is referred to as the *energy spectral density*[5] of the signal $g(t)$. To explain the meaning of this definition, suppose that $g(t)$ denotes the voltage of a source connected across a 1-ohm load resistor. Then the quantity

$$\int_{-\infty}^{\infty} |g(t)|^2 \ dt$$

equals the total energy E delivered by the source. According to Rayleigh's theorem, this energy equals the total area under the $\mathscr{E}_g(f)$ curve. It follows therefore that the function $\mathscr{E}_g(f)$ is a measure of the density of the energy contained in $g(t)$ in joules per Hertz. Note that since in the special case of a real-valued signal the amplitude spectrum is an even function of f, the energy spectral density of such a signal is symmetrical about the vertical axis passing through the origin.

EXAMPLE 9   Sinc Pulse (continued)

Consider again the sinc pulse $A \operatorname{sinc}(2Wt)$. The energy of this pulse equals

$$E = A^2 \int_{-\infty}^{\infty} \operatorname{sinc}^2(2Wt) \ dt$$

The integral in the right-hand side of this equation is rather difficult to evaluate. However, we note from Example 4 that the Fourier transform of the sinc pulse $A \operatorname{sinc}(2Wt)$ is equal to $(A/2W)\operatorname{rect}(f/2W)$; hence, applying Rayleigh's energy theorem to the probelm at hand, we readily obtain the desired result:

$$
\begin{aligned}
E &= \left(\frac{A}{2W}\right)^2 \int_{-\infty}^{\infty} \operatorname{rect}^2\left(\frac{f}{2W}\right) df \\
&= \left(\frac{A}{2W}\right)^2 \int_{-W}^{W} df \\
&= \frac{A^2}{2W}
\end{aligned}
\tag{2.57}
$$

This example clearly illustrates the usefulness of Rayleigh's energy theorem.

2.5   THE INVERSE RELATIONSHIP BETWEEN TIME AND FREQUENCY

The properties of the Fourier transform discussed in Section 2.3 clearly show that the time-domain and frequency-domain descriptions of a signal are *inversely* related. In particular, we may make the following important statements:

1. If the time-domain description of a signal is changed, the frequency-domain description of the signal is changed in an *inverse* manner, and vice versa. This inverse relationship prevents arbitrary specifications of a signal in both domains. In other words, *we may specify an arbitrary function of time or an arbitrary spectrum, but we cannot specify both of them together.*

2. If a signal is strictly limited in frequency, the time-domain description of the signal will trail on indefinitely, even though its amplitude may assume a progressively smaller value. We say a signal is *strictly limited in frequency* or *strictly band limited* if its Fourier transform is exactly zero outside a finite band of frequencies. The sinc pulse is an example of a strictly band-limited signal, as illustrated in Fig. 2.7. This figure also shows that the sinc pulse is only *asymptotically limited in time*, which confirms the opening statement we made for a strictly band-limited signal. In an inverse manner, if a signal is *strictly limited in time* (i.e., the signal is exactly zero outside a finite time interval), then the spectrum of the signal is infinite in extent, even though the amplitude spectrum may assume a progressively smaller value. This behavior is exemplified by both the rectangular pulse (described in Fig. 2.2) and the triangular pulse (described in Fig. 2.10*b*). Accordingly, we may state that *a signal cannot be strictly limited in both time and frequency.*

Bandwidth

The *bandwidth* of a signal provides a measure of the *extent of significant spectral content of the signal for positive frequencies.* When the signal is strictly band limited, the bandwidth is well defined. For example, the sinc pulse described in Fig. 2.7 has a bandwidth equal to W. When, however, the signal is not strictly band limited, which is generally the case, we encounter difficulty in defining the bandwidth of the signal. The difficulty arises because the meaning of "significant" attached to the spectral content of the signal is mathematically imprecise. Consequently, there is no universally accepted definition of bandwidth.

Nevertheless, there are some commonly used definitions for bandwidth. In this section, we consider three such definitions; the formulation of each definition depends on whether the signal is low-pass or band-pass. A signal is said to be *low-pass* if its significant spectral content is centered around the origin. A signal is said to be *band-pass* if its significant spectral content is centered around $\pm f_c$, where f_c is a nonzero frequency.

When the spectrum of a signal is symmetric with a *main lobe* bounded by well-defined *nulls* (i.e., frequencies at which the spectrum is zero), we may use the main lobe as the basis for defining the bandwidth of the signal. Specifically, if the signal is low-pass, the bandwidth is defined as one half the total width of the main spectral lobe, since only one half of this lobe lies inside the positive frequency region. For example, a rectangular pulse of duration T seconds has a main spectral lobe of total width $2/T$ Hertz centered at the origin, as depicted in Fig. 2.2. Accordingly, we may define the bandwidth of this rectangular pulse as $1/T$ Hertz. If, on the other hand, the signal is band-pass with main spectral lobes centered around $\pm f_c$, where f_c is large, the bandwidth is defined as the width of the main lobe for positive frequencies. This definition of bandwidth is called the *null-to-null bandwidth*. For example, an RF pulse of duration T seconds and frequency f_c has main spectral lobes of width $2/T$ Hertz centered around $\pm f_c$, as depicted in Fig. 2.8. Hence, we may define the null-to-null bandwidth of this RF pulse as $2/T$ Hertz. On the basis of the definitions presented here, we may state that shifting the spectral content of a low-pass signal by a sufficiently large frequency, has the effect of doubling the bandwidth of the signal; such a frequency translation is attained by using modulation.

Another popular definition of bandwidth is the *3-dB bandwidth*. Specifically, if the signal is low-pass, the 3-dB bandwidth is defined as the separation between zero frequency, where the amplitude spectrum attains its peak value, and the *positive frequency* at which the amplitude spectrum drops to $1/\sqrt{2}$ of its peak value. For example, the decaying exponential and rising exponential pulses defined in Fig. 2.4 have a 3-dB bandwidth of $a/2\pi$ Hertz. If, on the other hand, the signal is band-pass, centered at $\pm f_c$, the 3-dB bandwidth is defined as the separation (along the positive frequency axis) between the two frequencies at which the amplitude spectrum of the signal drops to $1/\sqrt{2}$ of the peak value at f_c. The 3-dB bandwidth has the advantage in that it can be read directly from a plot of the amplitude spectrum. However, it has the disadvantage in that it may be misleading if the amplitude spectrum has slowly decreasing tails.

Yet another measure for the bandwidth of a signal is the *root mean square (rms) bandwidth*, defined as the square root of the second moment of a properly normalized form of the squared amplitude spectrum of the signal about a suitably chosen point. We assume that the signal is low-pass, so that the second moment may be taken about the origin. As for the normalized form of the squared amplitude spectrum, we use the nonnegative function $|G(f)|^2/\int_{-\infty}^{\infty}|G(f)|^2\,df$, in which the denominator applies the correct normalization in the sense that the integrated value of this ratio over the entire frequency axis is unity. We may thus formally define the rms bandwidth of a low-pass signal $g(t)$ with Fourier transform $G(f)$ as follows:

$$W_{\text{rms}} = \left(\frac{\int_{-\infty}^{\infty} f^2 |G(f)|^2 \, df}{\int_{-\infty}^{\infty} |G(f)|^2 \, df} \right)^{1/2} \tag{2.58}$$

An attractive feature of the rms bandwidth W_{rms} is that it lends itself more readily to mathematical evaluation than the other two definitions of bandwidth, but it is not as easily measurable in the laboratory.

Time–Bandwidth Product

For any family of pulse signals that differ by a time-scaling factor, the product of the signal's duration and its bandwidth is always a constant, as shown by

$$(\text{duration}) \cdot (\text{bandwidth}) = \text{constant}$$

The product is called the *time–bandwidth product* or *bandwidth–duration product*. The constancy of the time–bandwidth product is another manifestation of the inverse relationship that exists between the time-domain and frequency-domain descriptions of a signal. In particular, if the duration of a pulse signal is decreased by reducing the time scale by a factor a, the frequency scale of the signal's spectrum, and therefore the bandwidth of the signal, is increased by the same factor a, by virtue of Property 2 (time scaling), and the time–bandwidth product of the signal is thereby maintained constant. For example, a rectangular pulse of duration T seconds has a bandwidth (defined on the basis of the positive-frequency part of the main lobe) equal to $1/T$ Hertz, making the time–bandwidth product of the pulse equal unity. Whatever definition we use for the bandwidth of a signal, the time–bandwidth product remains constant over certain classes of pulse signals. The choice of a particular definition for bandwidth merely changes the value of the constant.

To be more specific, consider the rms bandwidth defined in Eq. (2.58). The corresponding definition for the *rms duration* of the signal $g(t)$ is

$$T_{\text{rms}} = \left(\frac{\int_{-\infty}^{\infty} t^2 |g(t)|^2 \, dt}{\int_{-\infty}^{\infty} |g(t)|^2 \, dt} \right)^{1/2} \tag{2.59}$$

where it is assumed that the signal $g(t)$ is centered around the origin. It may be shown that using the rms definitions of Eqs. (2.58) and (2.59), the time–bandwidth product has the following from:

$$T_{\text{rms}} \, W_{\text{rms}} \geq \frac{1}{4\pi} \tag{2.60}$$

where the constant is $1/4\pi$. It may also be shown that the Gaussian pulse satisfies this condition with the equality sign. For the details of these calculations, the reader is referred to Problem 2.15.

2.6   DIRAC DELTA FUNCTION

Strictly speaking, the theory of the Fourier transform, as described in Sections 2.2 to 2.4, is applicable only to time functions that satisfy the Dirichlet conditions. Such functions include energy signals. However, it would be highly desirable to extend this theory in two ways:

1. To combine the Fourier series and Fourier transform into a unified theory, so that the Fourier series may be treated as a special case of the Fourier transform. (A review of the Fourier series is presented in Appendix 2.)
2. To include power signals in the list of signals to which we may apply the Fourier transform.

It turns out that both of these objectives can be met through the "proper use" of the *Dirac delta function* or *unit impulse.*

The Dirac delta function,[6] denoted by $\delta(t)$, is defined as having zero amplitude everywhere except at $t = 0$, where it is infinitely large in such a way that it contains unit area under its curve; this is,

$$\delta(t) = 0, \qquad t \neq 0 \tag{2.61}$$

and

$$\int_{-\infty}^{\infty} \delta(t) \ dt = 1 \tag{2.62}$$

An implication of this pair of relations is that the delta function $\delta(t)$ must be an even function of time t.

For the delta function to have meaning, however, it has to appear as a factor in the integrand of an integral with respect to time and then, strictly speaking, only when the other factor in the integrand is a continuous function of time. Let $g(t)$ be such a function, and consider the product of $g(t)$ and the time-shifted delta function $\delta(t - t_0)$. In light of the two defining equations (2.61) and (2.62), we may express the integral of this product as follows

$$\int_{-\infty}^{\infty} g(t)\delta(t - t_0) \ dt = g(t_0) \tag{2.63}$$

The operation indicated on the left-hand side of this equation sifts out the value $g(t_0)$ of the function $g(t)$ at time $t = t_0$, where $-\infty < t < \infty$. Accordingly, Eq. (2.63) is referred to as the *sifting property* of the delta function. This property is sometimes used as the defining equation of a delta function; in effect, it incorporates Eqs. (2.61) and (2.62) into a single relation.

Noting that the delta function $\delta(t)$ is an even function of t, we may rewrite Eq. (2.63) in a way emphasizing its resemblance to the convolution integral, as shown by

$$\int_{-\infty}^{\infty} g(\tau)\delta(t - \tau) \ d\tau = g(t) \tag{2.64}$$

or, using the notation for convolution:

$$g(t) \star \delta(t) = g(t)$$

In words, the convolution of any function with the delta function leaves that function unchanged. We refer to this statement as the *replication property* of the delta function.

By definition, the Fourier transform of the delta function is given by

$$F[\delta(t)] = \int_{-\infty}^{\infty} \delta(t)\exp(-j2\pi ft) \ dt$$

Hence, using the sifting property of the delta function and noting that $\exp(-j2\pi ft)$ is equal to unity at $t = 0$, we obtain

$$F[\delta(t)] = 1$$

We thus have the Fourier-transform pair for the Dirac delta function:

$$\delta(t) \rightleftharpoons 1 \qquad\qquad (2.65)$$

This relation states that the spectrum of the delta function $\delta(t)$ extends uniformly over the entire frequency interval, as shown in Fig. 2.11.

It is important to realize that the Fourier-transform pair of Eq. (2.65) exists only in a limiting sense. The point is that no function in the ordinary sense has the two properties of Eqs. (2.61) and (2.62) or the equivalent sifting property of Eq. (2.63). However, we can imagine a sequence of functions that have progressively taller and thinner peaks at $t = 0$, with the area under the curve remaining equal to unity, whereas the value of the function tends to zero at every point except $t = 0$, where it tends to infinity. That is, we may view the delta function as *the limiting form of a pulse of unit area as the duration of the pulse approaches zero.* It is immaterial what sort of pulse shape is used.

In a rigorous sense, the Dirac delta function belongs to a special class of functions known as *generalized functions* or *distributions*. Indeed, in some situations its use requires that we exercise considerable care. Nevertheless, one beautiful aspect of the Dirac delta function lies precisely in the fact that a rather intuitive treatment of the function along the lines described herein often gives the correct answer.

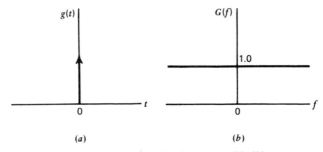

Figure 2.11 (a) The Dirac delta function $\delta(t)$. (b) Spectrum of $\delta(t)$.

EXAMPLE 10 The Delta Function as a Limiting Form of the Gaussian Pulse

Consider a Gaussian pulse of unit area, defined by

$$g(t) = \frac{1}{\tau}\exp\left(-\frac{\pi t^2}{\tau^2}\right) \tag{2.66}$$

where τ is a variable parameter. The Gaussian function $g(t)$ has two useful properties: (1) its derivatives are all continuous, and (2) it dies away more rapidly than any power of t. The delta function $\delta(t)$ is obtained by taking the limit $\tau \to 0$. The Gaussian pulse then becomes infinitely narrow in duration and infinitely large in amplitude, and yet its area remains finite and fixed at unity. Figure 2.12a illustrates the sequence of such pulses as the parameter τ varies.

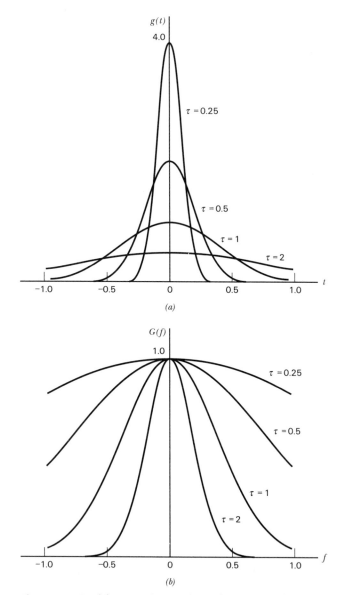

(a)

(b)

Figure 2.12 (a) Gaussian pulse of varying duration. (b) Corresponding spectrum.

The Gaussian pulse $g(t)$, defined here, is the same as the normalized Gaussian pulse $\exp(-\pi t^2)$ derived in Example 6, except for the fact that it is now expanded in time by the factor τ and compressed in amplitude by the same factor. Therefore, applying the linearity and time-scaling properties of the Fourier transform to the transform pair of Eq. (2.38), we find that the Fourier transform of the Gaussian pulse $g(t)$ defined in Eq. (2.66) is also Gaussian, as shown by

$$G(f) = \exp(-\pi\tau^2 f^2)$$

Figure 2.12b illustrates the effect of varying the parameter τ on the spectrum of the Gaussian pulse $g(t)$. Thus, putting $\tau = 0$, we find, as expected, that the Fourier transform of the delta function is unity.

Applications of the Delta Function

1. dc Signal By applying the duality property to the Fourier-transform pair of Eq. (2.65) and noting that the delta function is an even function, we obtain

$$1 \rightleftharpoons \delta(f) \tag{2.67}$$

Equation (2.67) states that a *dc signal* is transformed in the frequency domain into a delta function $\delta(f)$ occurring at zero frequency, as shown in Fig. 2.13. Of course, this result is intuitively satisfying.

Invoking the definition of Fourier transform, we readily deduce from Eq. (2.67) the useful relation

$$\int_{-\infty}^{\infty} \exp(-j2\pi ft) \, dt = \delta(f)$$

Recognizing that the delta function $\delta(f)$ is real valued, we may simplify this relation as follows:

$$\int_{-\infty}^{\infty} \cos(2\pi ft) \, dt = \delta(f) \tag{2.68}$$

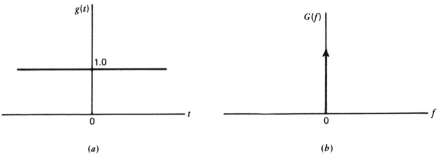

Figure 2.13 (a) dc signal. (b) Spectrum.

which provides yet another definition for the delta function, albeit in the frequency domain.

2. Complex Exponential Function Next, by applying the frequency-shifting property to Eq. (2.67), we obtain the Fourier-transform pair

$$\exp(j2\pi f_c t) \rightleftharpoons \delta(f - f_c) \tag{2.69}$$

for a complex exponential function of frequency f_c. Equation (2.69) states that the complex exponential function $\exp(j2\pi f_c t)$ is transformed in the frequency domain into a delta function $\delta(f - f_c)$ occurring at $f = f_c$.

3. Sinusoidal Functions Consider next the problem of evaluating the Fourier transform of the cosine function $\cos(2\pi f_c t)$. We first use Euler's formula to write

$$\cos(2\pi f_c t) = \frac{1}{2}\left[\exp(j2\pi f_c t) + \exp(-j2\pi f_c t)\right] \tag{2.70}$$

Therefore, using Eq. (2.69), we find that the cosine function $\cos(2\pi f_c t)$ is represented by the Fourier-transform pair

$$\cos(2\pi f_c t) \rightleftharpoons \frac{1}{2}\left[\delta(f - f_c) + \delta(f + f_c)\right] \tag{2.71}$$

In other words, the spectrum of the cosine function $\cos(2\pi f_c t)$ consists of a pair of delta functions occurring at $f = \pm f_c$, each of which is weighted by the factor $1/2$, as shown in Fig. 2.14.

Similarly, we may show that the sine function $\sin(2\pi f_c t)$ is represented by the Fourier-transform pair

$$\sin(2\pi f_c t) \rightleftharpoons \frac{1}{2j}\left[\delta(f - f_c) - \delta(f + f_c)\right] \tag{2.72}$$

which is illustrated in Fig. 2.15.

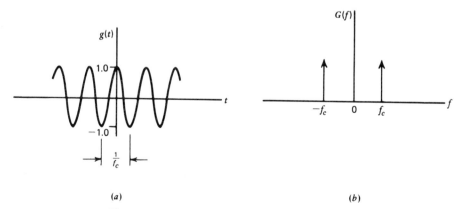

(a)

(b)

Figure 2.14 (a) Cosine function. (b) Spectrum.

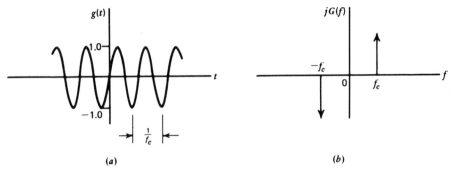

Figure 2.15 (a) Sine function. (b) Spectrum.

4. Signum Function The *signum function* sgn(t) equals $+1$ for positive time and -1 for negative time, as shown by the solid curve in Fig. 2.16a. The signum function was defined previously in Eq. (2.18); this definition is reproduced here for convenience of presentation:

$$\operatorname{sgn}(t) = \begin{cases} +1, & t > 0 \\ 0, & t = 0 \\ -1, & t < 0 \end{cases}$$

The signum function does not satisfy the Dirichlet conditions and therefore, strictly speaking, it does not have a Fourier transform. However, we may define a Fourier transform for the signum function by viewing it as the limiting form of the antisymmetric double-exponential pulse

$$g(t) = \begin{cases} \exp(-at), & t > 0 \\ 0, & t = 0 \\ -\exp(at), & t < 0 \end{cases} \tag{2.73}$$

as the parameter a approaches zero. The signal $g(t)$, shown as the dashed curve in Fig. 2.16a, does satisfy the Dirichlet conditions. Its Fourier transform was derived in Example 3; the result is given by [see Eq. (2.19)]:

$$G(f) = \frac{-j4\pi f}{a^2 + (2\pi f)^2}$$

The amplitude spectrum $|G(f)|$ is shown as the dashed curve in Fig. 2.16b. In the limit as a approaches zero we have

$$F[\operatorname{sgn}(t)] = \lim_{a \to 0} \frac{-4j\pi f}{a^2 + (2\pi f)^2}$$

$$= \frac{1}{j\pi f}$$

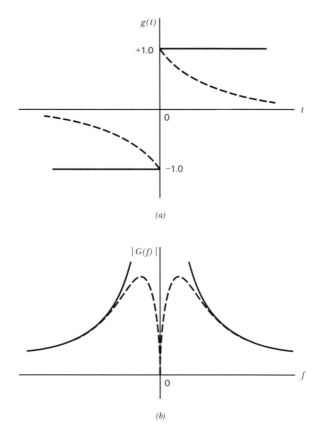

Figure 2.16 (*a*) Signum function (continuous curve), and double-exponential pulse (dashed curve). (*b*) Amplitude spectrum of signum function (continuous curve), and that of double-exponential pulse (dashed curve).

That is,

$$\text{sgn}(t) \rightleftharpoons \frac{1}{j\pi f} \tag{2.74}$$

The amplitude spectrum of the signum function is shown as the continuous curve in Fig. 2.16*b*. Here we see that for small *a*, the approximation is very good except near the origin on the frequency axis. At the origin, the spectrum of the approximating function $g(t)$ is zero for $a > 0$, whereas the spectrum of the signum function goes to infinity. It should also be noted that although the Fourier-transform pair of Eq. (2.74) does not involve a delta function, the Fourier transform $1/j\pi f$ can be obtained from the signum function $\text{sgn}(t)$ only if it is given a special meaning that implies the use of the delta function.

5. Unit Step Function The *unit step function* $u(t)$ equals $+1$ for positive time and zero for negative time. Previously defined in Eq. (2.11), it is reproduced here

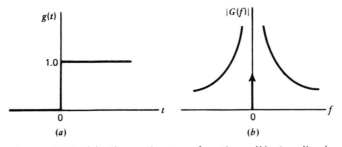

Figure 2.17 (a) The unit step function. (b) Amplitude spectrum.

for convenience:

$$u(t) = \begin{cases} 1, & t > 0 \\ \dfrac{1}{2}, & t = 0 \\ 0, & t < 0 \end{cases}$$

The waveform of the unit step function is shown in Fig. 2.17a. From this defining equation and that of the signum function, or from the waveforms of Figs. 2.16a and 2.17a, we see that the unit step function and signum function are related by

$$u(t) = \frac{1}{2}\,[\mathrm{sgn}(t) + 1] \tag{2.75}$$

Hence, using the linearity property of the Fourier transform and the Fourier-transform pairs of Eqs. (2.67) and (2.75), we find that the unit step function is represented by the Fourier-transform pair

$$u(t) \rightleftharpoons \frac{1}{j2\pi f} + \frac{1}{2}\,\delta(f) \tag{2.76}$$

This means that the spectrum of the unit step function contains a delta function weighted by a factor of $1/2$ and occurring at zero frequency, as shown in Fig. 2.17b.

6. Integration in the Time Domain (Revisited) The relation of Eq. (2.39) describes the effect of integration on the Fourier transform of a signal $g(t)$, assuming that $G(0)$ is zero. We now consider the more general case, with no such assumption made.

Let

$$y(t) = \int_{-\infty}^{t} g(\tau)\,d\tau \tag{2.77}$$

The integrated signal $y(t)$ can be viewed as the convolution of the original signal $g(t)$ and the unit step function $u(t)$, as shown by

$$y(t) = \int_{-\infty}^{\infty} g(\tau) u(t - \tau) \, d\tau$$

where the time-shifted unit step function $u(t - \tau)$ is defined by

$$u(t - \tau) = \begin{cases} 1, & \tau < t \\ \dfrac{1}{2}, & \tau = t \\ 0, & \tau > t \end{cases}$$

Recognizing that convolution in the time domain is transformed into multiplication in the frequency domain in accordance with Property 12, and using the Fourier-transform pair of Eq. (2.76) for the unit step function $u(t)$, we find that the Fourier transform of $y(t)$ is

$$Y(f) = G(f) \left[\frac{1}{j2\pi f} + \frac{1}{2} \delta(f) \right] \qquad (2.78)$$

where $G(f)$ is the Fourier transform of $g(t)$. Since

$$G(f) \delta(f) = G(0) \delta(f)$$

we may rewrite Eq. (2.77) in the equivalent form:

$$Y(f) = \frac{1}{j2\pi f} G(f) + \frac{1}{2} G(0) \delta(f)$$

In general, the effect of integrating the signal $g(t)$ is therefore described by the Fourier-transform pair

$$\int_{-\infty}^{t} g(\tau) \, d\tau \rightleftharpoons \frac{1}{j2\pi f} G(f) + \frac{1}{2} G(0) \delta(f) \qquad (2.79)$$

This is the desired result, which includes Eq. (2.39) as a special case (i.e., $G(0) = 0$).

2.7 FOURIER TRANSFORMS OF PERIODIC SIGNALS

It is well known that by using the Fourier series, a periodic signal can be represented as a sum of complex exponentials; for a review of the Fourier series, see Appendix 2. Also, in a limiting sense, Fourier transforms can be defined for complex exponentials. Therefore, it seems reasonable that a periodic signal can be represented in terms of a Fourier transform, provided that this transform is permitted to include delta functions.

Consider then a periodic signal $g_{T_0}(t)$ of *period* T_0. We can represent $g_{T_0}(t)$ in terms of the *complex exponential Fourier series*:

$$g_{T_0}(t) = \sum_{n=-\infty}^{\infty} c_n \exp(j2\pi n f_0 t) \qquad (2.80)$$

where c_n is the *complex Fourier coefficient* defined by

$$c_n = \frac{1}{T_0} \int_{-T_0/2}^{T_0/2} g_{T_0}(t) \exp(-j2\pi n f_0 t) \ dt \tag{2.81}$$

and f_0 is the *fundamental frequency* defined as the reciprocal of the period T_0; that is,

$$f_0 = \frac{1}{T_0} \tag{2.82}$$

Let $g(t)$ be a pulselike function, which equals $g_{T_0}(t)$ over one period and is zero elsewhere; that is,

$$g(t) = \begin{cases} g_{T_0}(t), & -\dfrac{T_0}{2} \le t \le \dfrac{T_0}{2} \\ \\ 0, & \text{elsewhere} \end{cases} \tag{2.83}$$

The periodic signal $g_{T_0}(t)$ may now be expressed in terms of the function $g(t)$ as an infinite summation, as shown by

$$g_{T_0}(t) = \sum_{m=-\infty}^{\infty} g(t - mT_0) \tag{2.84}$$

Based on this representation, we may view $g(t)$ as a *generating function*, which generates the periodic signal $g_{T_0}(t)$.

The function $g(t)$ is Fourier transformable. Accordingly, we may rewrite the formula for the complex Fourier coefficient as follows:

$$\begin{aligned} c_n &= f_0 \int_{-\infty}^{\infty} g(t) \exp(-j2\pi n f_0 t) \ dt \\ &= f_0 G(nf_0) \end{aligned} \tag{2.85}$$

where $G(nf_0)$ is the Fourier transform of $g(t)$, evaluated at the frequency nf_0. We may thus rewrite the formula for the reconstruction of the periodic signal $g_{T_0}(t)$ as

$$g_{T_0}(t) = f_0 \sum_{n=-\infty}^{\infty} G(nf_0) \exp(j2\pi n f_0 t) \tag{2.86}$$

or, equivalently, in light of Eq. (2.84)

$$\sum_{m=-\infty}^{\infty} g(t - mT_0) = f_0 \sum_{n=-\infty}^{\infty} G(nf_0) \exp(j2\pi n f_0 t) \tag{2.87}$$

Equation (2.87) is one form of *Poisson's sum formula*.

Finally, using Eq. (2.69), which defines the Fourier transform of a complex exponential function, and Eq. (2.87), we deduce the following Fourier-transform pair for a periodic signal $g_{T_0}(t)$ with a generating function $g(t)$ and period T_0:

$$\sum_{m=-\infty}^{\infty} g(t - mT_0) \rightleftharpoons f_0 \sum_{n=-\infty}^{\infty} G(nf_0)\delta(f - nf_0) \qquad (2.88)$$

where f_0 is the fundamental frequency. This relation simply states that the Fourier transform of a periodic signal consists of delta functions occurring at integer multiples of the fundamental frequency $f_0 = 1/T_0$, including the origin, and that each delta function is weighted by a factor equal to the corresponding value of $G(nf_0)$. Indeed, this relation merely provides a method to display the frequency content of a periodic signal $g_{T_0}(t)$.

It is of interest to observe that the function $g(t)$, constituting one period of the periodic signal $g_{T_0}(t)$, has a continuous spectrum defined by $G(f)$. On the other hand, the periodic signal $g_{T_0}(t)$ itself has a discrete spectrum. We conclude, therefore, that *periodicity in the time domain has the effect of changing the frequency-domain description or spectrum of the signal into a discrete form defined at integer multiples of the fundamental frequency.*

EXAMPLE 11 Ideal Sampling Function

An *ideal sampling function,* or *Dirac comb,* consists of an infinite sequence of uniformly spaced delta functions, as shown in Fig. 2.18a. We denote this waveform by

$$\delta_{T_0}(t) = \sum_{m=-\infty}^{\infty} \delta(t - mT_0) \qquad (2.89)$$

We observe that the generating function $g(t)$ for the ideal sampling function $\delta_{T_0}(t)$ consists simply of the delta function $\delta(t)$. We therefore have $G(f) = 1$, and

$$G(nf_0) = 1 \qquad \text{for all } n$$

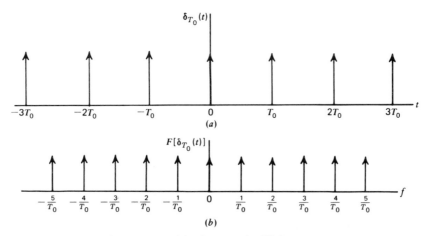

Figure 2.18 (a) Dirac comb. (b) Spectrum.

Thus the use of Eq. (2.88) yields the result

$$\sum_{m=-\infty}^{\infty} \delta(t - mT_0) \rightleftharpoons f_0 \sum_{n=-\infty}^{\infty} \delta(f - nf_0) \qquad (2.90)$$

Equation (2.90) states that the Fourier transform of a periodic train of delta functions, spaced T_0 seconds apart, consists of another set of delta functions weighted by the factor $f_0 = 1/T_0$ and regularly spaced f_0 Hz apart along the frequency axis as in Fig. 2.18b. In the special case of $T_0 = 1$, a periodic train of delta functions is, like a Gaussian pulse, its own transform.

We also deduce from Poisson's sum formula, Eq. (2.87), the following useful relation:

$$\sum_{m=-\infty}^{\infty} \delta(t - mT_0) = f_0 \sum_{n=-\infty}^{\infty} \exp(j2\pi nf_0 t) \qquad (2.91)$$

The dual of this relation is as follows

$$\sum_{m=-\infty}^{\infty} \exp(j2\pi mfT_0) = f_0 \sum_{n=-\infty}^{\infty} \delta(f - nf_0) \qquad (2.92)$$

2.8   TRANSMISSION OF SIGNALS THROUGH LINEAR SYSTEMS

With the Fourier transform theory presented in the previous sections at our disposal, we are ready to turn our attention to the study of a special class of systems known to be linear. A *system* refers to any physical device that produces an output signal in response to an input signal. It is customary to refer to the input signal as the *excitation* and to the output signal as the *response*. In a *linear* system, the *principle of superposition* holds; that is, the response of a linear system to a number of excitations applied simultaneously is equal to the sum of the responses of the system when each excitation is applied individually. Important examples of linear systems include *filters* and *communication channels* operating in their linear region. A filter refers to a frequency-selective device that is used to limit the spectrum of a signal to some band of frequencies. A channel refers to a transmission medium that connects the transmitter and receiver of a communication system. We wish to evaluate the effects of transmitting signals through linear filters and communication channels. This evaluation may be carried out in two ways, depending on the description adopted for the filter or channel. That is, we may use time-domain or frequency-domain ideas, as described below.

Time Response

In the time domain, a linear system is described in terms of its *impulse response, which is defined as the response of the system (with zero initial conditions) to a unit impulse or delta function $\delta(t)$ applied to the input of the system.* If the system is *time invariant,* then the shape of the impulse response is the same no matter when the unit impulse is applied to the system. Thus, assuming that the unit impulse or delta function is applied at time $t = 0$, we may denote the impulse response of a linear time-invariant system by $h(t)$. Let this system be subjected to an arbitrary exci-

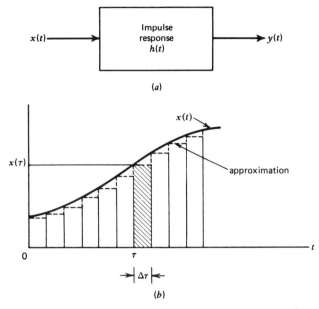

Figure 2.19 (a) Linear system. (b) Approximation of input $x(t)$.

tation $x(t)$, as in Fig. 2.19a. To determine the response $y(t)$ of the system, we begin by first approximating $x(t)$ by a staircase function composed of narrow rectangular pulses, each of duration $\Delta\tau$, as shown in Fig. 2.19b. Clearly the approximation becomes better for smaller $\Delta\tau$. As $\Delta\tau$ approaches zero, each pulse approaches, in the limit, a delta function weighted by a factor equal to the height of the pulse times $\Delta\tau$. Consider a typical pulse, shown shaded in Fig. 2.19b, which occurs at $t = \tau$. This pulse has an area equal to $x(\tau)\Delta\tau$. By definition, the response of the system to a unit impulse or delta function $\delta(t)$, occurring at $t = 0$, is $h(t)$. It follows, therefore, that the response of the system to a delta function, weighted by the factor $x(\tau)\Delta\tau$ and occurring at $t = \tau$, must be $x(\tau)h(t - \tau)\Delta\tau$. To find the total response $y(t)$ at some time t, we apply the principle of superposition. Thus, summing the various infinitesimal responses due to the various input pulses, we obtain in the limit, as $\Delta\tau$ approaches zero,

$$y(t) = \int_{-\infty}^{\infty} x(\tau) h(t - \tau) \, d\tau \qquad (2.93)$$

This relation is called the *convolution integral*.

In Eq. (2.93), three different time scales are involved: *excitation time τ, response time t*, and *system-memory time $t - \tau$*. This relation is the basis of time-domain analysis of linear time-invariant systems. It states that the present value of the response of a linear time-invariant system is a weighted integral over the past history of the input signal, weighted according to the impulse response of the system. Thus, the impulse response acts as a *memory function* for the system.

In Eq. (2.93), the excitation $x(t)$ is convolved with the impulse response $h(t)$ to produce the response $y(t)$. Since convolution is commutative, it follows that we may also write

$$y(t) = \int_{-\infty}^{\infty} h(\tau) x(t - \tau) \, d\tau \tag{2.94}$$

where $h(t)$ is convolved with $x(t)$.

EXAMPLE 12   Tapped-Delay-Line Filter

Consider a linear time-invariant filter with impulse response $h(t)$. We make the following two assumptions:

1. The impulse response $h(t) = 0$ for $t < 0$.
2. The impulse response of the filter is of some finite duration T_f, so that we may write $h(t) = 0$ for $t \geq T_f$.

Then we may express the filter output $y(t)$ produced in response to the input $x(t)$ as follows:

$$y(t) = \int_{0}^{T_f} h(\tau) x(t - \tau) \, d\tau \tag{2.95}$$

Let the input $x(t)$, impulse response $h(t)$, and output $y(t)$ be *uniformly sampled* at the rate $1/\Delta\tau$ samples per second, so that we may put

$$t = n\Delta\tau$$

and

$$\tau = k\Delta\tau$$

where k and n are integers, and $\Delta\tau$ is the *sampling period*. Assuming that $\Delta\tau$ is small enough for the product $h(\tau) x(t - \tau)$ to remain essentially constant for $k\Delta\tau \leq \tau \leq (k + 1) \Delta\tau$ for all values of k and t of interest, we may approximate Eq. (2.95) by the *convolution sum*:

$$y(n\Delta\tau) = \sum_{k=0}^{N-1} h(k\Delta\tau) x(n\Delta\tau - k\Delta\tau) \Delta\tau \tag{2.96}$$

where $N\Delta\tau = T_f$. Define

$$w_k = h(k\Delta\tau) \Delta\tau$$

We may then rewrite Eq. (2.96) as

$$y(n\Delta\tau) = \sum_{k=0}^{N-1} w_k x(n\Delta\tau - k\Delta\tau) \tag{2.97}$$

Equation (2.97) may be realized using the circuit shown in Fig. 2.20, which consists of a set of *delay elements* (each producing a delay of $\Delta\tau$ seconds), a set of *multipliers* connected to the *delay-line taps*, a corresponding set of *weights* applied to the multipliers, and a *summer* for adding the multiplier outputs. This circuit is known as a *tapped-delay-line filter* or *transversal filter*. Note that in Fig. 2.20 the tap-spacing or basic increment of delay is equal to the sampling period of the input sequence $\{x(n\Delta\tau)\}$.

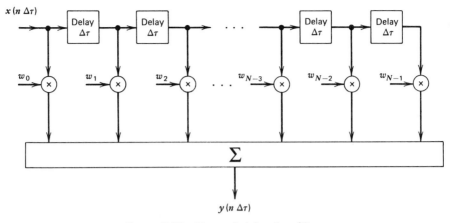

Figure 2.20   Tapped-delay-line filter.

Causality and Stability

A system is said to be *causal* if it does not respond before the excitation is applied. For a linear time-invariant system to be causal, it is clear that the impulse response $h(t)$ must vanish for negative time. That is, the necessary and sufficient condition for causality is

$$h(t) = 0, \qquad t < 0 \tag{2.98}$$

Clearly, for a system operating in *real time* to be physically realizable, it must be causal. However, there are many applications in which the signal to be processed is available in stored form; in these situations the system can be noncausal and yet physically realizable.

The system is said to be *stable* if the output signal is bounded for all bounded input signals; we refer to this as the *bounded input-bounded output* (BIBO) *stability criterion*, which is well suited for the analysis of linear time-invariant systems. Let the input signal $x(t)$ be bounded, as shown by

$$|x(t)| \le M \tag{2.99}$$

where M is a positive real finite number. Substituting Eq. (2.99) in (2.94), we get

$$|y(t)| \le M \int_{-\infty}^{\infty} |h(\tau)| \, d\tau$$

It follows therefore that for a linear time-invariant system to be stable, the impulse response $h(t)$ must be absolutely integrable. That is, the necessary and sufficient condition for BIBO stability is

$$\int_{-\infty}^{\infty} |h(t)| \, dt < \infty \tag{2.100}$$

Frequency Response

Consider a linear time-invariant system of impulse response $h(t)$ driven by a complex exponential input of unit amplitude and frequency f, that is,

$$x(t) = \exp(j2\pi f t) \tag{2.101}$$

Using Eq. (2.101) in (2.94), the response of the system is obtained as

$$y(t) = \int_{-\infty}^{\infty} h(\tau) \exp[j2\pi f(t - \tau)] \, d\tau$$
$$= \exp(j2\pi f t) \int_{-\infty}^{\infty} h(\tau) \exp(-j2\pi f \tau) \, d\tau \tag{2.102}$$

Define the *transfer function* of the system as the Fourier transform of its impulse response, as shown by

$$H(f) = \int_{-\infty}^{\infty} h(t) \exp(-j2\pi f t) \, dt \tag{2.103}$$

The integral in the last line of Eq. (2.102) is the same as that of Eq. (2.103), except that τ is used in place of t. Hence, we may rewrite Eq. (2.102) in the form

$$y(t) = H(f) \exp(j2\pi f t) \tag{2.104}$$

The response of a linear time-invariant system to a complex exponential function of frequency f is, therefore, the same complex exponential function multiplied by a constant coefficient $H(f)$.

An alternative definition of the transfer function may be deduced by dividing Eq. (2.104) by (2.101) to obtain

$$H(f) = \left. \frac{y(t)}{x(t)} \right|_{x(t) = \exp(j2\pi f t)} \tag{2.105}$$

Consider next an arbitrary signal $x(t)$ applied to the system. The signal $x(t)$ may be expessed in terms of its inverse Fourier transform as

$$x(t) = \int_{-\infty}^{\infty} X(f) \exp(j2\pi f t) \, df \tag{2.106}$$

or, equivalently, in the limiting form

$$x(t) = \lim_{\substack{\Delta f \to 0 \\ f = k\Delta f}} \sum_{k=-\infty}^{\infty} X(f) \exp(j2\pi f t) \Delta f \tag{2.107}$$

That is, the input signal $x(t)$ may be viewed as a superposition of complex exponentials of incremental amplitude. Because the system is linear, the response to this superposition of complex exponential inputs is

$$y(t) = \lim_{\substack{\Delta f \to 0 \\ f = k\Delta f}} \sum_{k=-\infty}^{\infty} H(f)X(f)\exp(j2\pi ft)\Delta f$$

$$= \int_{-\infty}^{\infty} H(f)X(f)\exp(j2\pi ft)\ df \tag{2.108}$$

The Fourier transform of the output signal $y(t)$ is therefore

$$Y(f) = H(f)X(f) \tag{2.109}$$

A linear time-invariant system may thus be described quite simply in the frequency domain by noting that the Fourier transform of the output is equal to the product of the transfer function of the system and the Fourier transform of the input.

The result of Eq. (2.109) may, of course, be deduced directly by recognizing that the response $y(t)$ of a linear time-invariant system of impulse response $h(t)$ to an arbitrary input $x(t)$ is obtained by convolving $x(t)$ with $h(t)$, or vice versa, and by the fact that the convolution of a pair of time functions is transformed into the multiplication of their Fourier transforms. The derivation above is presented primarily to develop an understanding of why the Fourier representation of a time function as a superposition of complex exponentials is so useful in analyzing the behavior of linear time-invariant systems.

The transfer function $H(f)$ is a characteristic property of a linear time-invariant system. It is, in general, a complex quantity, so that we may express it in the form

$$H(f) = |H(f)|\exp[j\beta(f)] \tag{2.110}$$

where $|H(f)|$ is called the *amplitude response*, and $\beta(f)$ the *phase* or *phase response*. In the special case of a linear system with a real-valued impulse response $h(t)$, the transfer function $H(f)$ exhibits conjugate symmetry, which means that

$$|H(f)| = |H(-f)|$$

and

$$\beta(f) = -\beta(-f)$$

That is, the amplitude response $|H(f)|$ of a linear system with real-valued impulse response is an even function of frequency, whereas the phase $\beta(f)$ is an odd function of frequency.

In some applications it is preferable to work with the logarithm of $H(f)$, expressed in polar form, rather than with $H(f)$ itself. Define

$$\ln H(f) = \alpha(f) + j\beta(f) \tag{2.111}$$

where

$$\alpha(f) = \ln|H(f)| \tag{2.112}$$

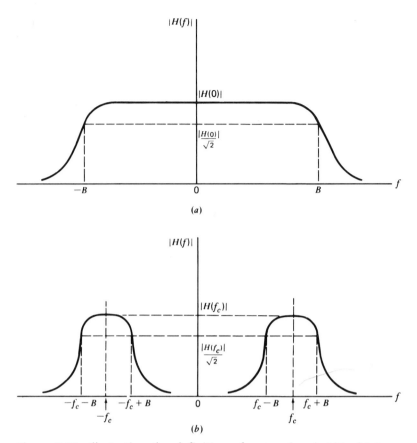

Figure 2.21 Illustrating the definition of system bandwidth. (*a*) Low-pass system. (*b*) Band-pass system.

The function $\alpha(f)$ is called the *gain* of the system. It is measured in *nepers*, whereas $\beta(f)$ is measured in *radians*. Equation (2.111) indicates that the gain $\alpha(f)$ and phase $\beta(f)$ are the real and imaginary parts of the (natural) logarithm of the transfer function $H(f)$, respectively. The gain may also be expressed in *decibels* (dB) by using the definition

$$\alpha'(f) = 20 \log_{10}|H(f)| \tag{2.113}$$

The two gain functions $\alpha(f)$ and $\alpha'(f)$ are related by

$$\alpha'(f) = 8.69\alpha(f) \tag{2.114}$$

That is, 1 neper is equal to 8.69 dB.

As a means of specifying the constancy of the amplitude response $|H(f)|$ or gain $\alpha(f)$ of a system, we use a parameter called the *bandwidth* of the system. In the case of a low-pass system, the bandwidth is customarily defined as the frequency at which the amplitude response $|H(f)|$ is $1/\sqrt{2}$ times its value at zero frequency or, equivalently, the frequency at which the gain $\alpha'(f)$ drops by 3 dB below its value at zero frequency, as illustrated in Fig. 2.21*a*. In the case of a band-pass system, the bandwidth is defined as the range of frequencies over which the amplitude response $|H(f)|$ remains within $1/\sqrt{2}$ times its value at the mid-band frequency, as illustrated in Fig. 2.21b.

Paley–Wiener Criterion

A necessary and sufficient condition for a function $\alpha(f)$ to be the gain of a causal filter is the convergence of the integral

$$\int_{-\infty}^{\infty} \frac{|\alpha(f)|}{1 + f^2} \, df < \infty \tag{2.115}$$

This condition is known as the *Paley–Wiener criterion*.[7] It states that provided the gain $\alpha(f)$ satisfies the condition of Eq. (2.115), then we may associate with this gain a suitable phase $\beta(f)$, such that the resulting filter has a causal impulse response that is zero for negative time. In other words, the Paley–Wiener criterion is the frequency-domain equivalent of the causality requirement. A system with a realizable gain characteristic may have infinite attenuation for a discrete set of frequencies, but it cannot have infinite attenuation over a band of frequencies; otherwise, the Paley–Wiener criterion is violated.

2.9 FILTERS

As previously mentioned, a *filter* is a frequency-selective device that is used to limit the spectrum of a signal to some specified band of frequencies. Its frequency response is characterized by a *passband* and a *stopband*. The frequencies inside the passband are transmitted with little or no distortion, whereas those in the stopband are rejected. The filter may be of the *low-pass, high-pass, band-pass*, or *band-stop* type, depending on whether it transmits low, high, intermediate, or all but intermediate frequencies, respectively. We have already encountered examples of low-pass and band-pass systems in Fig. 2.21.

Filters, in one form or another, represent an important functional block in building communication systems. In this book, we will be largely concerned with the use of low-pass and band-pass filters.

In this section we study the time response of the *ideal low-pass filter*, which transmits, without any distortion, all frequencies inside the passband and completely rejects all frequencies inside the stopband, as illustrated in Fig. 2.22. The

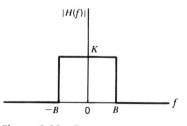

Figure 2.22 Frequency response of ideal low-pass filter.

transfer function of an ideal low-pass filter is therefore defined by

$$H(f) = \begin{cases} \exp(-j2\pi f t_0), & -B \le f \le B \\ 0, & |f| > B \end{cases} \quad (2.116)$$

The parameter B defines the bandwidth of the filter. The ideal low-pass filter is, of course, noncausal because it violates the Paley–Wiener criterion. This observation may also be confirmed by examining the impulse response $h(t)$. Thus, by evaluating the inverse Fourier transform of the transfer function of Eq. (2.116), we get

$$h(t) = \int_{-B}^{B} \exp[j2\pi f(t - t_0)] \, df \quad (2.117)$$

where the limits of integration have been reduced to the frequency band inside which $H(f)$ does not vanish. Equation (2.117) is readily integrated, yielding

$$h(t) = \frac{\sin[2\pi B(t - t_0)]}{\pi(t - t_0)}$$
$$= 2B \, \text{sinc}[2B(t - t_0)] \quad (2.118)$$

The impulse response has a peak amplitude of $2B$ centered on time t_0, as shown in Fig. 2.23 for $t_0 = 1/B$. The duration of the main lobe of the impulse response is $1/B$, and the build-up time from the zero at the beginning of the main lobe to the peak value is $1/2B$. We see from Fig. 2.23 that, for any finite value of t_0, there is some response from the filter before the time $t = 0$ at which the unit impulse is applied to the input, confirming that the ideal low-pass filter is noncausal. Note, however, that we can always make the delay t_0 large enough for the condition

$$|\text{sinc}[2B(t - t_0)]| \ll 1 \qquad \text{for } t < 0$$

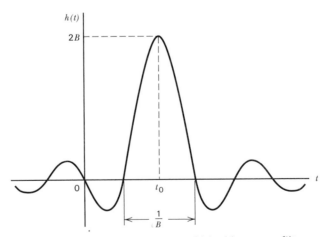

Figure 2.23 Impulse response of ideal low-pass filter.

to be satisfied. By so doing, we are able to build a causal filter that closely approximates an ideal low-pass filter.

COMPUTER EXPERIMENT I   Pulse Response of Ideal Low-pass Filter

Consider a rectangular pulse $x(t)$ of unit amplitude and duration T, which is applied to an ideal low-pass filter of bandwidth B. The problem is to determine the response $y(t)$ of the filter.

The impulse response $h(t)$ of the filter is defined by Eq. (2.118). The delay t_0 has no effect on the *shape* of the filter response $y(t)$. Without loss of generality, we may therefore simplify the exposition by setting $t_0 = 0$, in which case the impulse response of Eq. (2.118) reduces to

$$h(t) = 2B \, \text{sinc}(2Bt) \tag{2.119}$$

The resulting filter response is thus given by the convolution integral

$$
\begin{aligned}
y(t) &= \int_{-\infty}^{\infty} x(\tau) h(t - \tau) \, d\tau \\
&= 2B \int_{-T/2}^{T/2} \frac{\sin[2\pi B(t - \tau)]}{2\pi B(t - \tau)} \, d\tau
\end{aligned}
\tag{2.120}
$$

Define

$$\lambda = 2\pi B(t - \tau)$$

Then, changing the integration variable from τ to λ, we may rewrite Eq. (2.120) as

$$
\begin{aligned}
y(t) &= \frac{1}{\pi} \int_{2\pi B(t - T/2)}^{2\pi B(t + T/2)} \frac{\sin \lambda}{\lambda} \, d\lambda \\
&= \frac{1}{\pi} \left[\int_{0}^{2\pi B(t + T/2)} \frac{\sin \lambda}{\lambda} \, d\lambda - \int_{0}^{2\pi B(t - T/2)} \frac{\sin \lambda}{\lambda} \, d\lambda \right] \\
&= \frac{1}{\pi} \{ \text{Si}[2\pi B(t + T/2)] - \text{Si}[2\pi B(t - T/2)] \}
\end{aligned}
\tag{2.121}
$$

where the *sine integral* is defined by[8]

$$\text{Si}(u) = \int_{0}^{u} \frac{\sin x}{x} \, dx \tag{2.122}$$

The sine integral $\text{Si}(u)$ cannot be evaluated in closed form in terms of elementary functions, but it can be integrated in a power series. It is shown plotted in Fig. 2.24. We see that (1) the sine integral $\text{Si}(u)$ has odd symmetry about $u = 0$; (2) it has its maxima and minima at multiples of π; and (3) it approaches the limiting value $\pi/2$ for large values of u.

Since the sine integral $\text{Si}(u)$ oscillates at a frequency of $1/2\pi$, the filter response $y(t)$ will oscillate at a frequency equal to the cutoff frequency (i.e., bandwidth) B of the low-pass filter, as indicated in Fig. 2.25. The maximum value of $\text{Si}(u)$ occurs at $u_{\text{max}} = \pi$ and is equal to 1.8519 or $(1.179)(\pi/2)$. The filter response $y(t)$, then, has maxima at

$$t_{\text{max}} = \pm \left(\frac{T}{2} - \frac{1}{2B} \right)$$

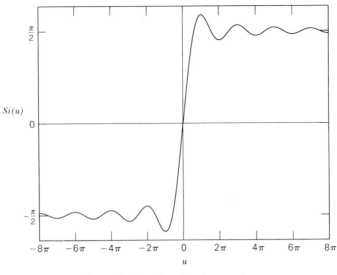

Figure 2.24   The sine integral.

with

$$y(t_{\max}) = \frac{1}{\pi} [\text{Si}(\pi) - \text{Si}(\pi - 2\pi BT)]$$

$$= \frac{1}{\pi} [\text{Si}(\pi) + \text{Si}(2\pi BT - \pi)]$$

Let

$$\text{Si}(2\pi BT - \pi) = \frac{\pi}{2} (1 \pm \Delta)$$

where Δ is the absolute value of the deviation in the value of $\text{Si}(2\pi BT - \pi)$ expressed as a fraction of the final value $+\pi/2$. Thus, recognizing that

$$\text{Si}(\pi) = (1.179)(\pi/2)$$

we may redefine $y(t_{\max})$ as

$$y(t_{\max}) = \frac{1}{2} (1.179 + 1 \pm \Delta)$$

$$= 1.09 \pm \frac{1}{2} \Delta$$

(2.123)

For a time–bandwidth product $BT \gg 1$, the fractional deviation Δ has a very small value, in which case we may make two observations from Eq. (2.123):

- The percentage *overshoot* in the filter response is approximately 9 percent.
- The overshoot is practically *independent* of the filter bandwidth B.

The basic phenomenon underlying these two observations is called the *Gibb's phenomenon*.[9] Figure 2.25 shows the oscillatory nature of the filter response and

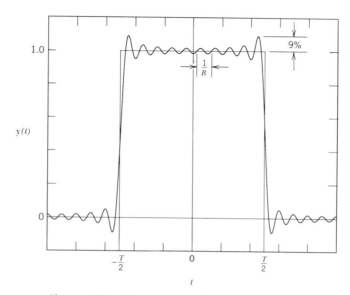

Figure 2.25 Filter response for a square pulse.

the 9 percent overshoot characterizing the response, assuming that $BT \gg 1$.

Figure 2.26 shows the filter response for four time–bandwidth products: $BT = 5, 10, 20, 100$, assuming that the pulse duration T is 15. Table 2.1 shows the corresponding frequencies of oscillations and percentage overshoots for these time–bandwidth products, confirming the observations made above.

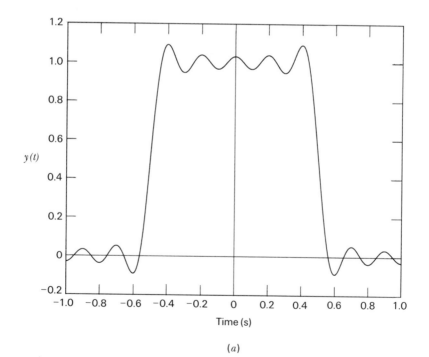

(a)

Figure 2.26 Pulse response of ideal low-pass filter for pulse duration $T = 15$ and varying time-bandwidth (BT) product: (a) BT = 5.

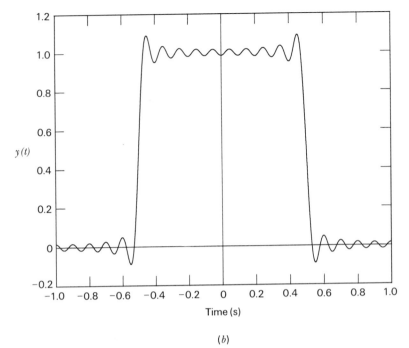

(*b*)

Figure 2.26 (*continued*)   (*b*) BT = 10.

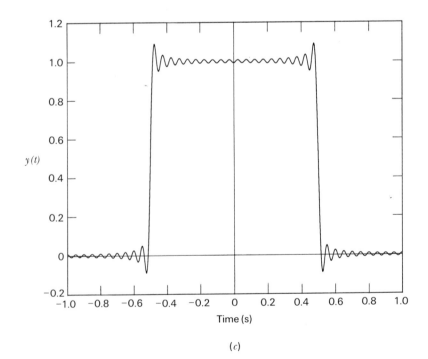

(*c*)

Figure 2.26 (*continued*)   (*c*) BT = 20.

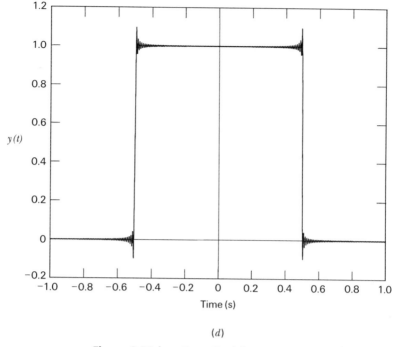

(d)

Figure 2.26 (*continued*)   (d) BT = 100.

Table 2.1 Oscillation Frequency and Percentage
Overshoot for Varying Time–Bandwidth Product

BT	Oscillation Frequency	Percentage Overshoot
5	5 Hz	9.11
10	10 Hz	8.98
20	20 Hz	8.99
100	100 Hz	9.63

Figure 2.27 shows the filter response for periodic square-wave inputs of different fundamental frequencies: $f_0 = 0.1, 0.25, 0.5$, and 1 Hz, and with the bandwidth of the low-pass filter being fixed at $B = 1$ Hz. From Fig. 2.27 we may make the following observations:

- For $f_0 = 0.1$, corresponding to a time–bandwidth product $BT = 5$, the filter somewhat distorts the input pulse, but the shape of the input square wave is still evident at the filter output. Unlike the input, the filter output has nonzero rise and fall times that are inversely proportional to the filter bandwidth. Also, the output exhibits oscillations (ringing) at both the leading and trailing edges.

- As the fundamental frequency f_0 of the input square wave increases, the low-pass filter cuts off more of the higher-frequency components of the input. Thus, when $f_0 = 0.25$ Hz, corresponding to $BT = 2$, only the fundamental

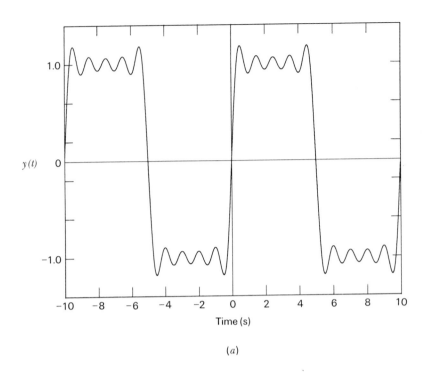

(a)

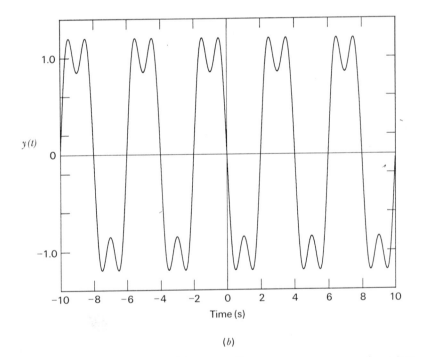

(b)

Figure 2.27 Response of ideal low-pass filter to a square wave of varying frequency f_o: (a) f_o = 0.1 Hz; (b) f_o = 0.25 Hz.

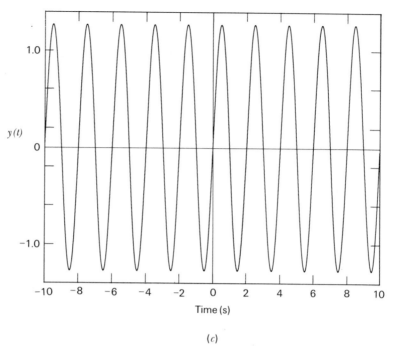

(c)

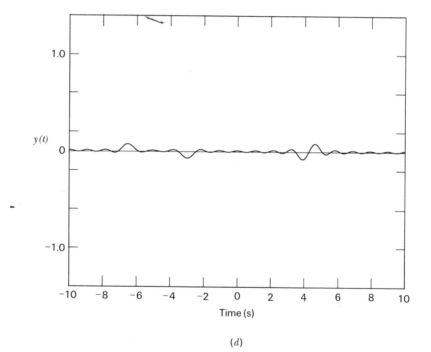

(d)

Figure 2.27 (*continued*)   (c) $f_o = 0.5$ Hz; (d) $f_o = 1$ Hz.

frequency and the first harmonic component pass through the filter; the rise and fall times of the output are now significant compared with the input pulse duration T. When $f_0 = 0.5$ Hz, corresponding to $BT = 1$, only the fundamental frequency component of the input square wave is preserved by the filter, resulting in an output that is essentially sinusoidal.

- When the fundamental frequency of the input square wave is much higher than the bandwidth of the low-pass filter, exemplified by the case of $f_0 = 1$ Hz, (i.e., time–bandwidth product $BT = 0.5$), the dc component becomes the dominant output, and the shape of the input square wave is completely destroyed by the filter.

From these results, we draw the conclusion that we must use a time–bandwidth product $BT \geq 1$ in order to ensure that the filter output is recognizable.

Design of Filters

A filter may be characterized by specifying its impulse response $h(t)$ or, equivalently, its transfer function $H(f)$. However, the application of a filter usually involves the separation of signals on the basis of their spectra (i.e., frequency contents). This, in turn, means that the design of filters is usually carried out in the frequency domain. There are two basic steps involved in the design of a filter:

1. The *approximation* of a prescribed frequency response (i.e., amplitude response, phase response, or both) by a realizable transfer function.
2. The *realization* of the approximating transfer function by a physical device.

For an approximating transfer function $H(f)$ to be physically realizable, it must represent a *stable* system. Stability is defined here on the basis of the bounded input–bounded output criterion described in Eq. (2.100) that involves the impulse response $h(t)$. To specify the corresponding condition for stability in terms of the transfer function, the traditional approach is to replace $j2\pi f$ with s and recast the transfer function in terms of s. The new variable s is permitted to have a real part as well as an imaginary part. Accordingly, we refer to s as the *complex frequency*. Let $H'(s)$ denote the transfer function of the system, defined in the manner described herein. Ordinarily, the approximating transfer function $H'(s)$ is a rational function, which may therefore be expressed in a *factored* form as

$$H'(s) = H(f)\big|_{j2\pi f = s}$$

$$= K \frac{(s - z_1)(s - z_2) \cdots (s - z_m)}{(s - p_1)(s - p_2) \cdots (s - p_n)}$$

where K is a scaling factor; $z_1, z_2, \ldots, z_m$ are the *zeros* of the transfer function; $p_1, p_2, \ldots, p_n$ are its *poles*. For low-pass and band-pass filters, the number of zeros, m, is less than the number of poles, n. If the system is causal, then the bounded input–bounded output condition for stability of the system is satisfied by restricting all the poles of the transfer function $H'(s)$ to be inside the left half of the s-plane; that is to say,

$$\text{Re}[p_i] < 0 \qquad \text{for all } i$$

Note that the condition for stability involves only the poles of the transfer function $H'(s)$; the zeros may indeed lie anywhere in the s-plane. Two types of systems may be distinguished here:

- *Minimum-phase systems*, characterized by a transfer function whose poles and zeros are all restricted to lie inside the left half of the s-plane.
- *Nonminimum-phase systems*, whose transfer functions are permitted to have zeros on the imaginary axis as well as the right half of the s-plane.

In the case of low-pass filters where the principal requirement is to approximate the ideal amplitude response shown in Fig. 2.22, we may mention two popular families of filters: *Butterworth filters* and *Chebyshev filters*, both of which have all their zeros at $s = \infty$. In a Butterworth filter, the poles of the transfer function $H'(s)$ lie on a circle with origin as the center and $2\pi B$ as the radius, where B is the 3-dB bandwidth of the filter. In a Chebyshev filter, on the other hand, the poles lie on an ellipse. In both cases, of course, the poles are confined to the left half of the s-plane.

Turning next to the issue of physical realization of the filter, there are several options available to us, depending on the technology of choice:

- *Analog filters*, built using (a) inductors and capacitors, or (b) capacitors, resistors, and operational amplifiers.
- *Discrete-time filters*, for which the signals are sampled in time but their amplitude is continuous. These filters include switched-capacitor filters, and surface-acoustic wave (SAW) filters.
- *Digital filters*, for which the signals are sampled in time and their amplitude is also quantized. These filters are built using digital hardware; hence the name. An important feature of a digital filter is that it is *programmable*, thereby offering a high degree of flexibility in design.

2.10   HILBERT TRANSFORM

The Fourier transform, which has occupied so much of our attention thus far, is particularly useful for evaluating the frequency content of an energy signal or, in a limiting sense, that of a power signal. As such, it provides the mathematical basis for analyzing and designing frequency-selective filters for the separation of signals on the basis of their frequency contents. Another method of separating signals is based on *phase selectivity*, which uses phase shifts between the pertinent signals to achieve the desired separation. The simplest phase shift is that of 180 degrees, which is merely a polarity reversal in the case of a sinusoidal signal. Shifting the phase angles of all components of a given signal by 180 degrees requires the use of an *ideal transformer*. Another phase shift of interest is that of ± 90 degrees. In particular, when the phase angles of all components of a given signal are shifted by ± 90 degrees, the resulting function of time is known as the Hilbert transform of the signal.

To be specific, consider a signal $g(t)$ with Fourier transform $G(f)$. The *Hilbert transform* of $g(t)$, which we shall denote by $\hat{g}(t)$, is defined by[10]

$$\hat{g}(t) = \frac{1}{\pi} \int_{-\infty}^{\infty} \frac{g(\tau)}{t - \tau} d\tau \qquad (2.124)$$

Clearly, the Hilbert transformation of $g(t)$ is a linear operation. The *inverse Hilbert transform*, by means of which the original signal $g(t)$ is recovered from $\hat{g}(t)$, is defined by

$$g(t) = -\frac{1}{\pi} \int_{-\infty}^{\infty} \frac{\hat{g}(\tau)}{t - \tau} \, d\tau \qquad (2.125)$$

The functions $g(t)$ and $\hat{g}(t)$ are said to constitute a *Hilbert-transform pair*. A short table of Hilbert-transform pairs is given in Table 3 in Appendix 11 at the end of the book.

We note from the definition of the Hilbert transform that $\hat{g}(t)$ may be interpreted as the convolution of $g(t)$ with the time function $1/(\pi t)$. We also know from the convolution theorem (Property 12 of the Fourier transform) that the convolution of two functions in the time domain is transformed into the multiplication of their Fourier transforms in the frequency domain. For the time function $1/\pi t$, we have

$$\frac{1}{\pi t} \rightleftharpoons -j \, \mathrm{sgn}(f) \qquad (2.126)$$

where $\mathrm{sgn}(f)$ is the signum function, defined in the frequency domain as

$$\mathrm{sgn}(f) = \begin{cases} 1, & f > 0 \\ 0, & f = 0 \\ -1, & f < 0 \end{cases} \qquad (2.127)$$

The Fourier-transform pair of Eq. (2.126) is obtained by applying the duality property of the Fourier transform to Eq. (2.74). In light of Eq. (2.126), it follows therefore that the Fourier transform $\hat{G}(f)$ of $\hat{g}(t)$ is given by

$$\hat{G}(f) = -j \, \mathrm{sgn}(f) G(f) \qquad (2.128)$$

Equation (2.128) states that given a signal $g(t)$, we may obtain its Hilbert transform $\hat{g}(t)$ by passing $g(t)$ through a linear two-port device whose transfer function is equal to $-j \, \mathrm{sgn}(f)$. This device may be considered as one that produces a phase shift of -90 degrees for all positive frequencies of the input signal and $+90$ degrees for all negative frequencies, as in Fig. 2.28. The amplitudes of all frequency components in the signal, however, are unaffected by transmission through the device. Such an ideal device is referred to as a *Hilbert transformer*.

The Hilbert transform has several important applications, which include the following:

1. It can be used to realize phase selectivity in the generation of a special kind of modulation known as *single sideband modulation*. We shall have more to say about this application in Chapter 3.
2. It provides the mathematical basis for the representation of band-pass signals. This application is discussed in Section 2.12.

The Hilbert transform, as defined above, applies to any signal that is Fourier transformable. Accordingly, it may be applied to energy signals as well as power signals.

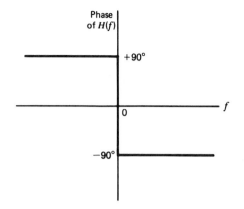

Figure 2.28 Phase characteristic of linear two-port device for obtaining the Hilbert transform of a real-valued signal.

EXAMPLE 13   Sinusoidal Functions

Consider the cosine function

$$g(t) = \cos(2\pi f_c t)$$

whose Fourier transform is

$$G(f) = \frac{1}{2}[\delta(f - f_c) + \delta(f + f_c)]$$

Using this Fourier transform in Eq. (2.128), we get

$$\hat{G}(f) = -j\,\mathrm{sgn}(f)\,G(f)$$

$$= -\frac{j}{2}[\delta(f - f_c) + \delta(f + f_c)]\mathrm{sgn}(f)$$

$$= \frac{1}{2j}[\delta(f - f_c) - \delta(f + f_c)]$$

which represents the Fourier transform of the sine function $\sin(2\pi f_c t)$. The Hilbert transform of the cosine function is therefore equal to $\sin(2\pi f_c t)$.

In a similar way, we find that the sine function $\sin(2\pi f_c t)$ has a Hilbert transform equal to $-\cos(2\pi f_c t)$.

Properties of the Hilbert Transform

The Hilbert transform differs from the Fourier transform in that it operates exclusively in the time domain. It has a number of useful properties, some of which are listed below. To derive these properties, we make use of Eq. (2.128), which defines the relationship between the Fourier transform of a signal $g(t)$ and that of its Hilbert transform $\hat{g}(t)$. The signal $g(t)$ is assumed to be *real valued*, which is the usual domain of application of the Hilbert transform.

PROPERTY 1

A signal $g(t)$ and its Hilbert transform $\hat{g}(t)$ have the same amplitude spectrum.

To prove this property, we observe that the Fourier transform of $\hat{g}(t)$ is equal to $-j\,\text{sgn}(f)$ times the Fourier transform of $g(t)$, and since the magnitude of $-j\,\text{sgn}(f)$ is equal to one for all f, then $g(t)$ and $\hat{g}(t)$ will both have the same amplitude specturm.

As a corollary to this property, we may state that if a signal $g(t)$ is band limited, then its Hilbert transform $\hat{g}(t)$ will also be band limited.

PROPERTY 2

If $\hat{g}(t)$ is the Hilbert transform of $g(t)$, then the Hilbert transform of $\hat{g}(t)$ is $-g(t)$.

To prove this property, we note that the process of Hilbert transformation is equivalent to passing $g(t)$ through a linear two-port device with a transfer function equal to $-j\,\text{sgn}(f)$. A double Hilbert transformation is therefore equivalent to passing $g(t)$ through a cascade of two such devices. The overall transfer function of such a cascade is equal to

$$[-j\,\text{sgn}(f)]^2 = -1 \qquad \text{for all } f$$

The resulting output is thus $-g(t)$; that is, the Hilbert transform of $\hat{g}(t)$ is equal to $-g(t)$. This result is subject to the requirement that $G(0) = 0$, where $G(0)$ is the Fourier transform of $g(t)$ evaluated at $f = 0$.

PROPERTY 3

A signal $g(t)$ and its Hilbert transform $\hat{g}(t)$ are orthogonal.

To prove this property, we use a special case of the multiplication theorem described by Eq. (2.49). In particular, for a signal $g(t)$ multiplied by its Hilbert transform $\hat{g}(t)$ we may write

$$\int_{-\infty}^{\infty} g(t)\hat{g}(t)\,dt = \int_{-\infty}^{\infty} G(f)\hat{G}(-f)\,df \qquad (2.129)$$

Using Eq. (2.128) in (2.129), we get

$$\int_{-\infty}^{\infty} g(t)\hat{g}(t)\,dt = j\int_{-\infty}^{\infty} \text{sgn}(f)G(f)G(-f)\,df$$

$$= j\int_{-\infty}^{\infty} \text{sgn}(f)G(f)G^*(f)\,df \qquad (2.130)$$

$$= j\int_{-\infty}^{\infty} \text{sgn}(f)\,|G(f)|^2\,df$$

where, in the second line, we have used the fact that for a real-valued signal, $G(-f) = G^*(f)$. The integrand in the right-hand side of Eq. (2.130) is an odd function of f, being the product of the odd function $\text{sgn}(f)$ and the even function $|G(f)|^2$. Hence, the integral is zero, yielding the final result

$$\int_{-\infty}^{\infty} g(t)\hat{g}(t) \ dt = 0 \qquad (2.131)$$

This shows that an energy signal $g(t)$ and its Hilbert transform $\hat{g}(t)$ are orthogonal over the entire interval $(-\infty, \infty)$. Similarly, we may show that a power signal $g(t)$ and its Hilbert transform $\hat{g}(t)$ are orthogonal over one period, as shown by

$$\lim_{T \to \infty} \frac{1}{2T} \int_{-T}^{T} g(t)\hat{g}(t) \ dt = 0 \qquad (2.132)$$

The validity of Eq. (2.132) is readily demonstrated by the results presented in Example 13 on the Hilbert transforms of sinusoidal functions.

2.11   PRE-ENVELOPE

Consider a real-valued signal $g(t)$. We define the *pre-envelope* of the signal $g(t)$ as the complex-valued function[11]

$$g_+(t) = g(t) + j\hat{g}(t) \qquad (2.133)$$

where $\hat{g}(t)$ is the Hilbert transform of $g(t)$. We note that the given signal $g(t)$ is the real part of the pre-envelope $g_+(t)$, and the Hilbert transform of the signal is the imaginary part of the pre-envelope. Just as the use of phasors simplifies manipulations of alternating currents and voltages, so we find that the pre-envelope is particularly useful in handling band-pass signals and systems. The reason for the name ''pre-envelope'' is explained later in Section 2.12.

One of the important features of the pre-envelope $g_+(t)$ is the behavior of its Fourier transform. Let $G_+(f)$ denote the Fourier transform of $g_+(t)$. Then, we may write

$$G_+(f) = G(f) + j[-j \, \text{sgn}(f)]G(f)$$

Using the definition of the signum function $\text{sgn}(f)$ given in Eq. (2.127), we readily find that

$$G_+(f) = \begin{cases} 2G(f), & f > 0 \\ G(0), & f = 0 \\ 0, & f < 0 \end{cases} \qquad (2.134)$$

where $G(0)$ is the value of $G(f)$ at frequency $f = 0$. This means that the pre-envelope of a signal has no frequency content (i.e., its Fourier transform vanishes) for all negative frequencies, as illustrated in Fig. 2.29b for the case of a low-pass signal. Note that the use of triangular spectrum for a low-pass signal in Fig. 2.29a is intended only for the purpose of illustration.

From the foregoing analysis it is apparent that for a given signal $g(t)$ we may determine its pre-envelope $g_+(t)$ in one of two equivalent ways:

1.   We determine the Hilbert transform $\hat{g}(t)$ of the signal $g(t)$, and then use Eq. (2.133) to compute the pre-envelope $g_+(t)$.

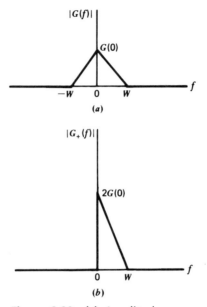

Figure 2.29 (a) Amplitude spectrum of low-pass signal $g(t)$. (b) Amplitude spectrum of pre-envelope $g_+(t)$.

2. We determine the Fourier transform $G(f)$ of the signal $g(t)$, use Eq. (2.134) to determine $G_+(f)$, and then evaluate the inverse Fourier transform of $G_+(f)$ to obtain

$$g_+(t) = 2\int_0^\infty G(f)\exp(j2\pi f t)\ df \qquad (2.135)$$

For a particular signal $g(t)$ of Fourier transform $G(f)$, one way may be better than the other.

For the purpose of illustration in Fig. 2.29, we have used a *low-pass signal* with its spectrum limited to the band $-W \leq f \leq W$ and centered at the origin. Nevertheless, it should be emphasized that the pre-envelope can be defined for any signal, be it low-pass or band-pass, so long as it possesses a spectrum.

Equation (2.133) defines the pre-envelope $g_+(t)$ for positive frequencies. Symmetrically, we may define the pre-envelope for *negative frequencies* as

$$g_-(t) = g(t) - j\hat{g}(t) \qquad (2.136)$$

The two pre-envelopes $g_+(t)$ and $g_-(t)$ are simply the complex conjugate of each other, as shown by

$$g_-(t) = g_+^*(t) \qquad (2.137)$$

The spectrum of the pre-envelope $g_+(t)$ is nonzero only for *positive* frequencies, as emphasized in Eq. (2.134); hence, the use of a plus sign as the subscript. In contrast, the spectrum of the other pre-envelope $g_-(t)$ is nonzero only for *nega-*

tive frequencies, as shown by the Fourier transform

$$G_-(f) = \begin{cases} 0, & f > 0 \\ G(0), & f = 0 \\ 2G(f), & f < 0 \end{cases} \qquad (2.138)$$

Thus, the pre-envelopes $g_+(t)$ and $g_-(t)$ constitute a complementary pair of complex-valued signals. Note also that the sum of $g_+(t)$ and $g_-(t)$ is exactly twice the original signal $g(t)$.

2.12 CANONICAL REPRESENTATIONS OF BAND-PASS SIGNALS

We say that a signal $g(t)$ is a *band-pass signal* if its Fourier transform $G(f)$ is non-negligible only in a band of frequencies of total extent $2W$, say, centered about some frequency $\pm f_c$. This is illustrated in Fig. 2.30a. We refer to f_c as the *carrier frequency*. In the majority of communication signals, we find that the bandwidth $2W$ is small compared with f_c, and so we refer to such a signal as a *narrow-band signal*. However, a precise statement about how small the bandwidth must be in order for the signal to be considered narrow-band is not necessary for our present discussion.

Let the pre-envelope of a narrow-band signal $g(t)$, with its Fourier transform $G(f)$ centered about some frequency $\pm f_c$, be expressed in the form

$$g_+(t) = \tilde{g}(t)\exp(j2\pi f_c t) \qquad (2.139)$$

We refer to $\tilde{g}(t)$ as the *complex envelope* of the signal. Equation (2.139) may be viewed as the basis of a definition for the complex envelope $\tilde{g}(t)$ in terms of the pre-envelope $g_+(t)$. We note that the spectrum of $g_+(t)$ is limited to the frequency band $f_c - W \le f \le f_c + W$, as illustrated in Fig. 2.30b. Therefore, applying the frequency-shifting property of the Fourier transform to Eq. (2.139), we find that the spectrum of the complex envelope $\tilde{g}(t)$ is limited to the band $-W \le f \le W$ and centered at the origin as illustrated in Fig. 2.30c. That is, the complex envelope $\tilde{g}(t)$ of a band-pass signal $g(t)$ is a *low-pass signal*, which is an important result.

By definition, the given signal $g(t)$ is the real part of the pre-envelope $g_+(t)$. We may thus express the original band-pass signal $g(t)$ in terms of the complex envelope $\tilde{g}(t)$ as follows:

$$g(t) = \text{Re}[\tilde{g}(t)\exp(j2\pi f_c t)] \qquad (2.140)$$

In general, $\tilde{g}(t)$ is a complex-valued quantity; to emphasize this property, we may express it in the form

$$\tilde{g}(t) = g_I(t) + jg_Q(t) \qquad (2.141)$$

where $g_I(t)$ and $g_Q(t)$ are both real-valued low-pass functions; their low-pass property is inherited from the complex envelope $\tilde{g}(t)$. We may therefore use Eqs.

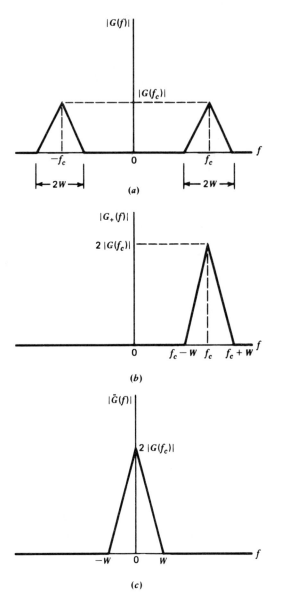

Figure 2.30 (a) Amplitude spectrum of band-pass signal $g(t)$. (b) Amplitude spectrum of pre-envelope $g_+(t)$. (c) Amplitude spectrum of complex envelope $\tilde{g}(t)$.

(2.140) and (2.141) to express the original band-pass signal $g(t)$ in the *canonical form*:

$$g(t) = g_I(t)\cos(2\pi f_c t) - g_Q(t)\sin(2\pi f_c t) \qquad (2.142)$$

We refer to $g_I(t)$ as the *in-phase component* of the band-pass signal $g(t)$ and to $g_Q(t)$ as the *quadrature component* of the signal; this nomenclature recognizes that

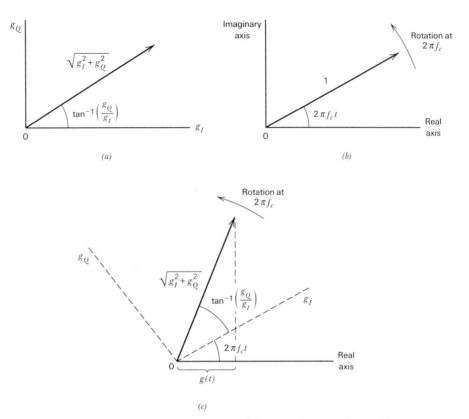

Figure 2.31 Illustrating an interpretation of the complex envelope $\tilde{g}(t)$ and its multiplication by exp $(j2\pi f_c t)$.

$\sin(2\pi f_c t)$ [i.e., the multiplying factor of $g_Q(t)$] is in phase-quadrature with respect to $\cos(2\pi f_c t)$ [i.e., the multiplying factor of $g_I(t)$].

According to Eq. (2.141), the complex envelope $\tilde{g}(t)$ may be pictured as a *time-varying phasor* positioned at the origin of the $g_I g_Q$-plane, as indicated in Fig. 2.31a. With time t varying, the end of the phasor moves about in the plane. Figure 2.31b shows the phasor representation of the complex exponential $\exp(j2\pi f_c t)$. In the definition given in Eq. (2.140), the complex envelope $\tilde{g}(t)$ is multiplied by the complex exponential $\exp(j2\pi f_c t)$. The angles of these two phasors therefore add and their lengths multiply, as shown in Fig. 2.31c. Moreover, in this latter figure, we show the $g_I g_Q$-plane rotating with an angular velocity equal to $2\pi f_c$ radians per second. Thus, in the picture portrayed here, the phasor representing the complex envelope $\tilde{g}(t)$ moves in the $g_I g_Q$-plane and at the same time the plane itself rotates about the origin. The original band-pass signal $g(t)$ is the projection of this time-varying phasor on a *fixed line* representing the real axis, as indicated in Fig. 2.31c.

Both $g_I(t)$ and $g_Q(t)$ are low-pass signals limited to the band $-W \le f \le W$. Hence, except for scaling factors, they may be derived from the band-pass signal $g(t)$ using the scheme shown in Fig. 2.32a, where both low-pass filters are identical, each having a bandwidth equal to W (see Problem 2.29). To reconstruct $g(t)$ from its in-phase and quadrature components, we may use the scheme shown in Fig. 2.32b.

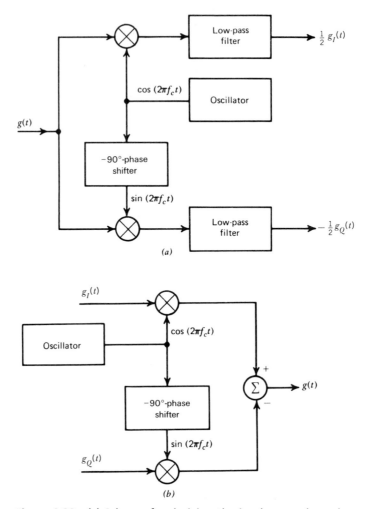

Figure 2.32 (a) Scheme for deriving the in-phase and quadrature components of a band-pass signal. (b) Scheme for reconstructing the band-pass signal from its in-phase and quadrature components.

The two schemes shown in Fig. 2.32 are basic to the study of *linear modulation systems*. The multiplication of the low-pass in-phase component $g_I(t)$ by $\cos(2\pi f_c t)$ and the multiplication of the low-pass quadrature component $g_Q(t)$ by $\sin(2\pi f_c t)$ represent linear forms of modulation. Given that the carrier frequency f_c is sufficiently large, the resulting band-pass function $g(t)$ defined in Eq. (2.142) is referred to as a *passband signaling waveform*. Correspondingly, the mapping from $g_I(t)$ and $g_Q(t)$ into $g(t)$ is known as *passband modulation*.

Equation (2.141) is the "Cartesian" form of expressing the complex envelope $\tilde{g}(t)$. Alternatively, we may express it in the "polar" form

$$\tilde{g}(t) = a(t)\exp[\,j\phi(t)\,] \tag{2.143}$$

where $a(t)$ and $\phi(t)$ are both real-valued low-pass functions. Based on this polar representation, the original band-pass signal $g(t)$ is defined by

$$g(t) = a(t)\cos[2\pi f_c t + \phi(t)] \tag{2.144}$$

We refer to $a(t)$ as the *natural envelope* or simply the *envelope* of the band-pass signal $g(t)$ and to $\phi(t)$ as the *phase* of the signal. Equation (2.144) represents a *hybrid form of amplitude modulation and angle modulation*; indeed, it includes amplitude modulation, frequency modulation, and phase modulation as special cases.

From this discussion it is apparent that whether we represent a band-pass (modulated) signal $g(t)$ in terms of its in-phase and quadrature components as in Eq. (2.142) or in terms of its envelope and phase as in Eq. (2.144), the information content of the signal $g(t)$ is completely represented by the complex envelope $\tilde{g}(t)$. The particular virtue of using the complex envelope $\tilde{g}(t)$ to represent the band-pass signal is an analytical one, and will become evident in Section 2.13.

Terminology

The distinctions among the three different envelopes that we have introduced to describe a band-pass signal $g(t)$ should be carefully noted. We may summarize their definitions as follows:

1. The pre-envelope $g_+(t)$ for positive frequencies is defined by

$$g_+(t) = g(t) + j\hat{g}(t)$$

where $\hat{g}(t)$ is the Hilbert transform of the signal $g(t)$. According to this representation, $\hat{g}(t)$ may be viewed as the quadrature function of $g(t)$. Correspondingly, in the frequency domain we have

$$G_+(f) = \begin{cases} 2G(f), & f > 0 \\ G(0), & f = 0 \\ 0, & f < 0 \end{cases}$$

Symmetrically, the pre-envelope for negative frequencies is defined by

$$g_-(t) = g(t) - j\hat{g}(t)$$

the Fourier transform of which is

$$G_-(f) = \begin{cases} 0, & f > 0 \\ G(0), & f = 0 \\ 2G(f), & f < 0 \end{cases}$$

2. The complex envelope $\tilde{g}(t)$ equals a frequency-shifted version of the pre-envelope $g_+(t)$, as shown by

$$\tilde{g}(t) = g_+(t)\exp(-j2\pi f_c t)$$

where f_c is the carrier frequency of the band-pass signal $g(t)$.

3. The envelope $a(t)$ equals the magnitude of the complex envelope $\tilde{g}(t)$ and also that of the pre-envelope $g_+(t)$, as shown by

$$a(t) = |\tilde{g}(t)| = |g_+(t)|$$

Note that for a band-pass signal $g(t)$, the pre-envelope $g_+(t)$ is a complex band-pass signal whose value depends on the carrier frequency f_c. On the other hand, the envelope $a(t)$ is always a real low-pass signal and, in general, the complex envelope $\tilde{g}(t)$ is a complex low-pass signal; the values of the latter two envelopes are independent of the choice of the carrier frequency f_c.

In Chapter 3 it is shown that the signal $a(t)$ results from envelope detection (i.e., rectification and low-pass filtering) of the band-pass signal $g(t)$. For this reason we call $a(t)$ the envelope of $g(t)$, and call the complex signals $\tilde{g}(t)$ and $g_+(t)$ the complex envelope and pre-envelope of $g(t)$, respectively.

The envelope $a(t)$ and phase $\phi(t)$ of $g(t)$ are related to its quadrature components $g_I(t)$ and $g_Q(t)$ as follows (see the time-varying phasor representation of Fig. 2.31a):

$$a(t) = \sqrt{g_I^2(t) + g_Q^2(t)}$$

$$\phi(t) = \tan^{-1}\left(\frac{g_Q(t)}{g_I(t)}\right)$$

Conversely, we may write

$$g_I(t) = a(t) \cos[\phi(t)]$$

$$g_Q(t) = a(t) \sin[\phi(t)]$$

Thus, each of the quadrature components of a band-pass signal contains both amplitude and phase information. Both components are required for a unique definition of the phase $\phi(t)$, modulo 2π.

EXAMPLE 14   RF Pulse (continued)

Suppose we wish to determine the different envelopes of the RF pulse defined by

$$g(t) = A \operatorname{rect}\left(\frac{t}{T}\right) \cos(2\pi f_c t)$$

We assume that $f_c T \gg 1$, so that the RF pulse $g(t)$ may be considered narrow-band.

From Example 5, we recall that the Fourier transform of $g(t)$ is given by

$$G(f) = \begin{cases} \dfrac{AT}{2} \operatorname{sinc}[T(f - f_c)], & f > 0 \\[2mm] 0, & f = 0 \\[2mm] \dfrac{AT}{2} \operatorname{sinc}[T(f + f_c)], & f < 0 \end{cases}$$

Hence,

$$G_+(f) = \begin{cases} AT \operatorname{sinc}[T(f - f_c)], & f > 0 \\ 0, & f \leq 0 \end{cases}$$

Taking the inverse Fourier transform of $G_+(f)$, we obtain the pre-envelope

$$g_+(t) = A \operatorname{rect}\left(\frac{t}{T}\right) \exp(j2\pi f_c t)$$

Correspondingly, the complex envelope equals

$$\tilde{g}(t) = A \operatorname{rect}\left(\frac{t}{T}\right)$$

and the envelope equals

$$a(t) = |\tilde{g}(t)| = A \operatorname{rect}\left(\frac{t}{T}\right)$$

The latter result is intuitively satisfying. Note that in this example the complex envelope is real valued and has the same value as the envelope.

2.13 BAND-PASS SYSTEMS

Now that we know how to handle the representation of band-pass signals, it is logical that we develop a corresponding procedure for handling the analysis of band-pass systems. Specifically, we wish to show that the analysis of band-pass systems can be greatly simplified by establishing an analogy (or, more precisely, an isomorphism) between low-pass and band-pass systems. This analogy is based on the use of the Hilbert transform for the representation of band-pass signals.

Consider a narrow-band signal $x(t)$, with its Fourier transform denoted by $X(f)$. We assume that the spectrum of the signal $x(t)$ is limited to frequencies within $\pm W$ Hz of the carrier frequency f_c. Also, we assume that $W < f_c$. Let this signal be represented in terms of its in-phase and quadrature components as follows:

$$x(t) = x_I(t)\cos(2\pi f_c t) - x_Q(t)\sin(2\pi f_c t) \tag{2.145}$$

where $x_I(t)$ is the in-phase component and $x_Q(t)$ is the quadrature component. Then, using $\tilde{x}(t)$ to denote the complex envelope of $x(t)$, we may write

$$\tilde{x}(t) = x_I(t) + jx_Q(t) \tag{2.146}$$

Let the signal $x(t)$ be applied to a linear time-invariant band-pass system with impulse response $h(t)$ and transfer function $H(f)$. We assume that the frequency response of the system is limited to frequencies within $\pm B$ of the carrier frequency f_c. The system bandwidth $2B$ is usually narrower than or equal to the input signal bandwidth $2W$. We wish to represent the band-pass impulse response $h(t)$ in terms of two quadrature components, designated as $h_I(t)$ and $h_Q(t)$. Thus, by analogy to the representation of band-pass signals, we may express $h(t)$ in the form

$$h(t) = h_I(t)\cos(2\pi f_c t) - h_Q(t)\sin(2\pi f_c t) \tag{2.147}$$

Define the *complex impulse response* of the band-pass system as

$$\tilde{h}(t) = h_I(t) + jh_Q(t) \tag{2.148}$$

Hence, we have the complex representation

$$h(t) = \text{Re}[\tilde{h}(t)\exp(j2\pi f_c t)] \qquad (2.149)$$

Note that $h_I(t)$, $h_Q(t)$, and $\tilde{h}(t)$ are all low-pass functions limited to the frequency band $-B \leq f \leq B$.

We may determine the complex impulse response $\tilde{h}(t)$ in terms of the quadrature components $h_I(t)$ and $h_Q(t)$ of the band-pass impulse response $h(t)$ by using Eq. (2.148). Alternatively, we may determine it from the band-pass transfer function $H(f)$ in the following way. We first note from Eq. (2.149) that

$$2h(t) = \tilde{h}(t)\exp(j2\pi f_c t) + \tilde{h}^*(t)\exp(-j2\pi f_c t) \qquad (2.150)$$

where $\tilde{h}^*(t)$ is the complex conjugate of $\tilde{h}(t)$. Therefore, applying the Fourier transform to Eq. (2.150), and using the complex-conjugation property of the Fourier transform, we get

$$2H(f) = \tilde{H}(f - f_c) + \tilde{H}^*(-f - f_c) \qquad (2.151)$$

where $H(f)$ is the Fourier transform of $h(t)$, and $\tilde{H}(f)$ is the Fourier transform of $\tilde{h}(t)$. Equation (2.151) satisfies the requirement that $H^*(f) = H(-f)$ for a real impulse response $h(t)$. Since $\tilde{H}(f)$ represents a low-pass transfer function limited to $|f| \leq B$ with $B < f_c$, we deduce from Eq. (2.151) that

$$\tilde{H}(f - f_c) = 2H(f), \qquad f > 0 \qquad (2.152)$$

Equation (2.152) indicates that for a specified band-pass transfer function $H(f)$, we may determine $\tilde{H}(f)$ by taking the part of $H(f)$ corresponding to positive frequencies shifting it to the origin and then scaling it by the factor 2. To determine the complex impulse response $\tilde{h}(t)$, we take the inverse Fourier transform of $\tilde{H}(f)$, obtaining

$$\tilde{h}(t) = \int_{-\infty}^{\infty} \tilde{H}(f)\exp(j2\pi f t) \, df \qquad (2.153)$$

The representations just described for band-pass signals and systems provide the basis of an efficient method for determining the output of a band-pass system driven by a band-pass signal. We assume that the spectrum of the input signal $x(t)$ and the transfer function $H(f)$ of the system are both centered around the same frequency f_c. In practice, there is no need to consider a situation in which the carrier frequency of the input signal is not aligned with the mid-band frequency of the band-pass system, since we have considerable freedom in choosing the carrier or mid-band frequency. Thus, changing the carrier frequency of the input signal by an amount Δf_c, say, simply corresponds to absorbing (or removing) the factor $\exp(\pm j2\pi\Delta f_c t)$ in the complex envelope of the input signal or the complex impulse response of the band-pass system. We are therefore justified in proceeding on the assumption that $X(f)$ and $H(f)$ are both centered around f_c. Suppose then we use $y(t)$ to denote the output signal of the system. It is clear that $y(t)$ is also a band-pass signal, so that we may represent it in terms of its low-pass complex envelope $\tilde{y}(t)$, as follows:

$$y(t) = \text{Re}[\tilde{y}(t)\exp(j2\pi f_c t)] \qquad (2.154)$$

The output signal $y(t)$ is related to the input signal $x(t)$ and impulse response $h(t)$ of the system in the usual way by the convolution integral

$$y(t) = \int_{-\infty}^{\infty} h(\tau) x(t - \tau) \, d\tau \qquad (2.155)$$

In terms of pre-envelopes, we have $h(t) = \text{Re}[h_+(t)]$ and $x(t) = \text{Re}[x_+(t)]$. We may therefore rewrite Eq. (2.155) in terms of the pre-envelopes $x_+(t)$ and $h_+(t)$ as follows:

$$y(t) = \int_{-\infty}^{\infty} \text{Re}[h_+(\tau)] \text{Re}[x_+(t - \tau)] \, d\tau \qquad (2.156)$$

To proceed further, we make use of a basic property of pre-envelopes that is described by the relation (see Problem 2.27)

$$\int_{-\infty}^{\infty} \text{Re}[h_+(\tau)] \text{Re}[x_+(\tau)] \, d\tau = \frac{1}{2} \text{Re} \left[\int_{-\infty}^{\infty} h_+(\tau) x_+^*(\tau) \, d\tau \right] \qquad (2.157)$$

where we have used τ as the integration variable to be consistent with that in Eq. (2.156). Next, we note that using $x(-\tau)$ in place of $x(\tau)$ has the effect of removing the complex conjugation on the right-hand side of Eq. (2.157). Hence, bearing in mind the algebraic difference between the argument of $x_+(\tau)$ in Eq. (2.157) and that of $x_+(t - \tau)$ in Eq. (2.156), and using the relationship between the pre-envelope and complex envelope of a band-pass function, we get

$$\begin{aligned} y(t) &= \frac{1}{2} \text{Re} \left[\int_{-\infty}^{\infty} h_+(\tau) x_+(t - \tau) \, d\tau \right] \\ &= \frac{1}{2} \text{Re} \left[\int_{-\infty}^{\infty} \tilde{h}(\tau) \exp(j2\pi f_c \tau) \tilde{x}(t - \tau) \exp(j2\pi f_c(t - \tau)) \, d\tau \right] \qquad (2.158) \\ &= \frac{1}{2} \text{Re} \left[\exp(j2\pi f_c t) \int_{-\infty}^{\infty} \tilde{h}(\tau) \tilde{x}(t - \tau) \, d\tau \right] \end{aligned}$$

Thus, comparing the right-hand sides of Eqs. (2.154) and (2.158), we readily deduce that for a large enough carrier frequency f_c, the complex envelope $\tilde{y}(t)$ of the output signal is related to the complex envelope $\tilde{x}(t)$ of the input signal and the complex impulse response $\tilde{h}(t)$ of the band-pass system as follows:

$$2\tilde{y}(t) = \int_{-\infty}^{\infty} \tilde{h}(\tau) \tilde{x}(t - \tau) \, d\tau \qquad (2.159)$$

or, using the shorthand notation for convolution,

$$2\tilde{y}(t) = \tilde{h}(t) \bigstar \tilde{x}(t) \qquad (2.160)$$

In other words, except for the scaling factor 2, the complex envelope $\tilde{y}(t)$ of the output signal of a band-pass system is obtained by convolving the complex impulse response $\tilde{h}(t)$ of the system with the complex envelope $\tilde{x}(t)$ of the input

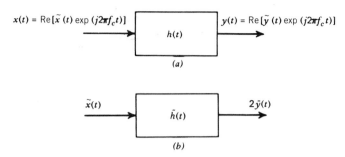

Figure 2.33 (a) Narrow-band filter of impulse response $h(t)$ with narrow-band input signal $x(t)$. (b) Equivalent low-pass filter of complex impulse response $\tilde{h}(t)$ with complex low-pass input $\tilde{x}(t)$.

band-pass signal. Equation (2.160) is the result of the isomorphism, for convolution, between band-pass functions and the corresponding low-pass functions.

The significance of this result is that in dealing with band-pass signals and systems, we need only concern ourselves with the low-pass functions $\tilde{x}(t)$, $\tilde{y}(t)$, and $\tilde{h}(t)$, representing the excitation, the response, and the system, respectively. That is, the analysis of a band-pass system, which is complicated by the presence of the multiplying factor $\exp(j2\pi f_c t)$, is replaced by an equivalent but much simpler low-pass analysis that completely retains the essence of the filtering process. This procedure is illustrated schematically in Fig. 2.33.

The complex envelope $\tilde{x}(t)$ of the input band-pass signal and the complex impulse response $\tilde{h}(t)$ of the band-pass system are defined in terms of their respective in-phase and quadrature components by Eqs. (2.146) and (2.148), respectively. Substituting these relations in Eq. (2.160), we get

$$2\tilde{y}(t) = [h_I(t) + jh_Q(t)] \bigstar [x_I(t) + jx_Q(t)] \qquad (2.161)$$

Because convolution is distributive, we may rewrite Eq. (2.161) in the form

$$
\begin{aligned}
2\tilde{y}(t) = {} & [h_I(t) \bigstar x_I(t) - h_Q(t) \bigstar x_Q(t)] \\
& + j[h_Q(t) \bigstar x_I(t) + h_I(t) \bigstar x_Q(t)]
\end{aligned}
\qquad (2.162)
$$

Left the complex envelope $\tilde{y}(t)$ of the response be defined in terms of its in-phase and quadrature components as

$$\tilde{y}(t) = y_I(t) + jy_Q(t) \qquad (2.163)$$

We therefore have for the in-phase component $y_I(t)$ the relation

$$2y_I(t) = h_I(t) \bigstar x_I(t) - h_Q(t) \bigstar x_Q(t) \qquad (2.164)$$

and for the quadrature component $y_Q(t)$ the relation

$$2y_Q(t) = h_Q(t) \bigstar x_I(t) + h_I(t) \bigstar x_Q(t) \qquad (2.165)$$

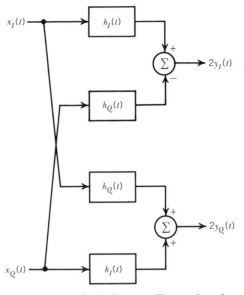

Figure 2.34 Block diagram illustrating the relationships between the in-phase and quadrature components of the response of a band-pass filter and those of the input signal.

Thus, for the purpose of evaluating the in-phase and quadrature components of the complex envelope $\tilde{y}(t)$ of the system output, we may use the *low-pass equivalent model* shown in Fig. 2.34. All the signals and impulse responses shown in this model are low-pass functions. Accordingly, this equivalent model provides a practical basis for the simulation of band-pass filters or communication channels on a digital computer.

To sum up, the procedure for evaluating the response of a band-pass system (with mid-band frequency f_c) to an input band-pass signal (of carrier frequency f_c) is as follows:

1. The input band-pass signal $x(t)$ is replaced by its complex envelope $\tilde{x}(t)$, which is related to $x(t)$ by

$$x(t) = \mathrm{Re}[\tilde{x}(t)\exp(j2\pi f_c t)]$$

2. The band-pass system, with impulse response $h(t)$, is replaced by a low-pass analog, which is characterized by a complex impulse response $\tilde{h}(t)$ related to $h(t)$ by

$$h(t) = \mathrm{Re}[\tilde{h}(t)\exp(j2\pi f_c t)]$$

3. The complex envelope $\tilde{y}(t)$ of the output band-pass signal $y(t)$ is obtained by convolving $\tilde{h}(t)$ with $\tilde{x}(t)$, as shown by

$$2\tilde{y}(t) = \tilde{h}(t) \bigstar \tilde{x}(t)$$

4. The desired output $y(t)$ is finally derived from the complex envelope $\tilde{y}(t)$ by using the relation

$$y(t) = \mathrm{Re}[\tilde{y}(t)\exp(j2\pi f_c t)]$$

COMPUTER EXPERIMENT II Response of an Ideal Band-pass
Filter to a Pulsed RF Wave

Consider an ideal band-pass filter of mid-band frequency f_c, the amplitude response of which is band limited to $f_c - B \le |f| \le f_c + B$, as in Fig. 2.35a, with $f_c > B$. To simplify the exposition, the effect of delay in the filter is ignored, as it has no effect on the shape of the filter response. We wish to compute the response of this filter to an RF pulse of duration T and carrier frequency f_c; it is defined by (see Fig. 2.36a)

$$x(t) = A \operatorname{rect}\left(\frac{t}{T}\right) \cos(2\pi f_c t)$$

where $f_c T \gg 1$.

Retaining the positive frequency part of the transform function $H(f)$, defined in Fig. 2.35a, and then shifting it to the origin, we find that the transfer function $\tilde{H}(f)$ of the low-pass equivalent filter is given by (see Fig. 2.35b)

$$\tilde{H}(f) = \begin{cases} 2, & -B < f < B \\ 0, & |f| > B \end{cases}$$

The complex impulse response in this example has only a real component, as shown by

$$\tilde{h}(t) = 4B \operatorname{sinc}(2Bt)$$

From Example 14, we recall that the complex envelope $\tilde{x}(t)$ of the input RF pulse also has only a real component, as shown by (see Fig. 2.36b):

$$\tilde{x}(t) = A \operatorname{rect}\left(\frac{t}{T}\right)$$

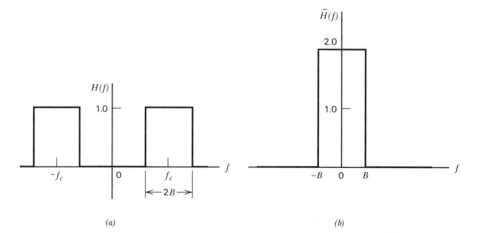

(a)   (b)

Figure 2.35 (a) Amplitude response $H(f)$ of an ideal band-pass filter. (b) Corresponding complex transfer function $\tilde{H}(f)$.

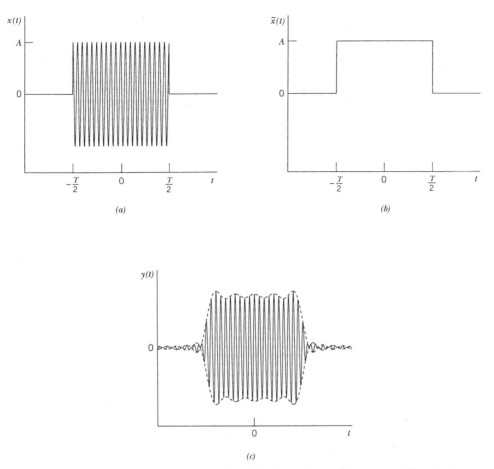

Figure 2.36 Illustrating the response of an ideal band-pass filter to an RF pulse input. (a) RF pulse input $x(t)$. (b) Complex envelope $\tilde{x}(t)$ of RF pulse. (c) Filter response $y(t)$.

The complex envelope $\tilde{y}(t)$ of the filter output is obtained by convolving $\tilde{h}(t)$ with $\tilde{x}(t)$ and then scaling it by the factor $1/2$. This convolution is exactly the same as the low-pass filtering operation that we studied in Computer Experiment I. Thus, using Eq. (2.121), we may write

$$\tilde{y}(t) = \frac{A}{\pi}\left\{ \text{Si}\left[2\pi B\left(t + \frac{T}{2}\right)\right] - \text{Si}\left[2\pi B\left(t - \frac{T}{2}\right)\right]\right\} \qquad (2.166)$$

As expected, the complex envelope $\tilde{y}(t)$ of the output has only a real component. Accordingly, from Eqs. (2.154) and (2.166), the output is obtained as

$$y(t) = \frac{A}{\pi}\left\{ \text{Si}\left[2\pi B\left(t + \frac{T}{2}\right)\right] - \text{Si}\left[2\pi B\left(t - \frac{T}{2}\right)\right]\right\}\cos\left(2\pi f_c t\right) \qquad (2.167)$$

which is the desired result. Equation (2.167) is shown plotted in Fig. 2.36c for the case of a time-bandwidth product $BT = 5$.

2.14 PHASE AND GROUP DELAY

Whenever a signal is transmitted through a dispersive (frequency-selective) device such as a filter or communication channel, some *delay* is introduced into the output signal in relation to the input signal. In an ideal low-pass filter or ideal band-pass filter, the phase response varies *linearly* with frequency inside the passband of the filter, in which case the filter introduces a constant delay equal to t_0, say; in effect, the parameter t_0 controls the slope of the linear phase response of the filter. Now, what if the phase response of the filter is nonlinear, which is frequently the case in practice? The purpose of this section is to address this important question.

To begin the discussion, suppose that a steady sinusoidal signal at frequency f_c is transmitted through a dispersive channel or filter that has a total phase-shift of $\beta(f_c)$ radians at that frequency. By using two phasors to represent the input signal and the received signal, we see that the received signal phasor lags the input signal phasor by $\beta(f_c)$ radians. The time taken by the received signal phasor to sweep out this phase lag is simply equal to $\beta(f_c)/2\pi f_c$ seconds. This time is called the *phase delay* of the channel.

It is important to realize, however, that the phase delay is not necessarily the true signal delay. This follows from the fact that a steady sinusoidal signal does not carry information, and so it would be incorrect to deduce from the above reasoning that the phase delay is the true signal delay. In actual fact, as we will see in subsequent chapters, information can be transmitted only by applying some appropriate form of modulation to the sinusoidal wave. Suppose then that a slowly varying signal is multiplied by a sinusoidal carrier wave, so that the resulting modulated signal consists of a narrow group of frequencies centered around the carrier frequency; the waveform of Fig. 2.36*c* illustrates such a modulated signal. When this modulated signal is transmitted through a communication channel, we find that there is a delay between the envelope of the input signal and that of the received signal. This delay is called the *envelope* or *group delay* of the channel; it represents the true signal delay.

Assume that the dispersive channel is described by the transfer function

$$H(f) = K \exp[j\beta(f)] \tag{2.168}$$

where the amplitude K is a constant and the phase $\beta(f)$ is a nonlinear function of frequency. The input signal $x(t)$ consists of a narrow-band signal defined by

$$x(t) = m(t)\cos(2\pi f_c t)$$

where $m(t)$ is a low-pass (information-bearing) signal with its spectrum limited to the frequency interval $|f| \le W$. We assume that $f_c \gg W$. By expanding the phase $\beta(f)$ in a Taylor series about the point $f = f_c$, and retaining only the first two terms, we may approximate $\beta(f)$ as

$$\beta(f) \simeq \beta(f_c) + (f - f_c) \left. \frac{\partial \beta(f)}{\partial f} \right|_{f=fc} \tag{2.169}$$

Define

$$\tau_p = -\frac{\beta(f_c)}{2\pi f_c} \tag{2.170}$$

and

$$\tau_g = -\frac{1}{2\pi} \frac{\partial \beta(f)}{\partial f}\bigg|_{f=f_c} \tag{2.171}$$

Then we may rewrite Eq. (2.169) in the simple form

$$\beta(f) \simeq -2\pi f_c \tau_p - 2\pi(f - f_c)\tau_g \tag{2.172}$$

Correspondingly, the transfer function of the channel takes the form

$$H(f) \simeq K \exp[-j2\pi f_c \tau_p - j2\pi(f - f_c)\tau_g]$$

Following the procedure described in Section 2.13, in particular, using Eq. (2.152), we may replace the channel described by $H(f)$ by an equivalent low-pass filter whose transfer function is approximately given by

$$\tilde{H}(f) \simeq 2K \exp(-j2\pi f_c \tau_p - j2\pi f \tau_g)$$

Similarly, we may replace the input narrow-band signal $x(t)$ by its low-pass complex envelope $\tilde{x}(t)$ (for the problem at hand), which equals

$$\tilde{x}(t) = m(t)$$

The Fourier transform of $\tilde{x}(t)$ is simply

$$\tilde{X}(f) = M(f)$$

where $M(f)$ is the Fourier transform of $m(t)$. Therefore, the Fourier transform of the complex envelope of the received signal is given by

$$\tilde{Y}(f) = \frac{1}{2}\tilde{H}(f)\tilde{X}(f)$$

$$\simeq K \exp(-j2\pi f_c \tau_p)\exp(-j2\pi f \tau_g)M(f)$$

We note that the multiplying factor $K\exp(-j2\pi f_c \tau_p)$ is a constant for fixed values of f_c and τ_p. We also note, from the time-shifting property of the Fourier transform, that the term $\exp(-j2\pi f \tau_g)M(f)$ represents the Fourier transform of the delayed signal $m(t - \tau_g)$. Accordingly, the complex envelope of the received signal is

$$\tilde{y}(t) \simeq K \exp(-j2\pi f_c \tau_p)m(t - \tau_g)$$

Finally, we find that the received signal is itself given by

$$\begin{aligned} y(t) &= \text{Re}[\tilde{y}(t)\exp(j2\pi f_c t)] \\ &= Km(t - \tau_g)\cos[2\pi f_c(t - \tau_p)] \end{aligned} \tag{2.173}$$

Equation (2.173) shows that, as a result of transmission through the channel, two delay effects occur:

1. The sinusoidal carrier wave $\cos(2\pi f_c t)$ is delayed by τ_p seconds; hence τ_p represents the *phase delay*. Sometimes, τ_p is also referred to as the *carrier delay*.

2. The envelope $m(t)$ is delayed by τ_g seconds; hence, τ_g represents the *envelope* or *group delay*.

Note that τ_g is related to the slope of the phase $\beta(f)$, measured at $f = f_c$, as in Eq. (2.171). Note also that when the phase response $\beta(f)$ varies linearly with frequency and $\beta(f_c)$ is zero, the phase delay and group delay assume a common value.

2.15 NUMERICAL COMPUTATION OF THE FOURIER TRANSFORM

The material presented in this chapter clearly testifies to the importance of the Fourier transform as a theoretical tool for the representation of deterministic signals and linear time-invariant systems. The importance of the Fourier transform is further enhanced by the fact that there exists a class of algorithms called fast Fourier transform algorithms[12] for the numerical computation of the Fourier transform in a highly efficient manner.

The fast Fourier transform algorithm is derived from the discrete Fourier transform in which, as the name implies, both time and frequency are represented in discrete form. The discrete Fourier transform provides an *approximation* to the Fourier transform. In order to properly represent the information content of the original signal, we have to take special care in performing the sampling operations involved in defining the discrete Fourier transform. A detailed treatment of the sampling process is presented in Chapter 6. For the present, it suffices to say that given a band-limited signal, the sampling rate should be greater than twice the highest frequency component of the input signal. Moreover, if the samples are uniformly spaced by T_s seconds, the spectrum of the signal becomes periodic, repeating every $f_s = (1/T_s)$ Hz. Let N denote the number of frequency samples contained in an interval f_s. Hence, the *frequency resolution* involved in the numerical computation of the Fourier transform is defined by

$$\Delta f = \frac{f_s}{N} = \frac{1}{NT_s} = \frac{1}{T} \qquad (2.174)$$

where T is the total duration of the signal.

Consider then a *finite data sequence* $\{g_0, g_1, \ldots, g_{N-1}\}$. For brevity, we refer to this sequence as g_n, in which the subscript is the *time index* $n = 0, 1, \ldots, N - 1$. Such a sequence may represent the result of sampling an *analog signal* $g(t)$ at times $t = 0, T_s, \ldots, (N - 1)T_s$, where T_s is the sampling interval. The ordering of the data sequence defines the sample time in that $g_0, g_1, \ldots, g_{N-1}$ denote samples of $g(t)$ taken at times $0, T_s, \ldots, (N - 1)T_s$, respectively. Thus we have

$$g_n = g(nT_s) \qquad (2.175)$$

We formally define the *discrete Fourier transform* (DFT) of g_n as

$$G_k = \sum_{n=0}^{N-1} g_n \exp\left(-\frac{j2\pi}{N} kn\right) \qquad k = 0, 1, \ldots, N - 1 \qquad (2.176)$$

The sequence $\{G_0, G_1, \ldots, G_{N-1}\}$ is called the *transform sequence*. For brevity, we refer to this sequence as G_k, in which the subscript is the *frequency index* $k = 0$, $1, \ldots, N - 1$. Correspondingly, we define the *inverse discrete Fourier transform* (IDFT) of G_k as

$$g_n = \frac{1}{N} \sum_{k=0}^{N-1} G_k \exp\left(\frac{j2\pi}{N} kn\right) \qquad n = 0, 1, \ldots, N - 1 \qquad (2.177)$$

The DFT and the IDFT form a transform pair. Specifically, given a data sequence g_n, we may use the DFT to compute the transform sequence G_k; and given the transform sequence G_k, we may use the IDFT to recover the original data sequence g_n.

A distinctive feature of the DFT is that for the finite summations defined in Eqs. (2.176) and (2.177), there is no question of convergence.

When discussing the DFT (and algorithms for its computation), the words "sample" and "point" are used interchangeably to refer to a sequence value. Also, it is common practice to refer to a sequence of length N as an *N-point sequence*, and refer to the DFT of a data sequence of length N as an *N-point DFT*.

Interpretation of the DFT and the IDFT

We may visualize the DFT process, described in Eq. (2.176), as a collection of N *complex heterodyning* and *averaging* operations, as shown in Fig. 2.37a. We say that the heterodyning is complex in that samples of the data sequence are multiplied by *complex exponential sequences*. There are a total of N complex exponential sequences to be considered, corresponding to the frequency index $k = 0, 1, \ldots,$ $N - 1$. Their periods have been selected in such a way that each complex exponential sequence has precisely an integer number of cycles in the total interval 0 to $N - 1$. The zero-frequency response, corresponding to $k = 0$, is the only exception.

For the interpretation of the IDFT process, described in Eq. (2.177), we may use the scheme shown in Fig. 2.37b. Here we have a collection of N *complex signal generators*, each of which produces a complex exponential sequence:

$$
\begin{aligned}
\exp\left(\frac{j2\pi}{N} kn\right) &= \cos\left(\frac{2\pi}{N} kn\right) + j \sin\left(\frac{2\pi}{N} kn\right) \\
&= \left\{ \cos\left(\frac{2\pi}{N} kn\right), \sin\left(\frac{2\pi}{N} kn\right) \right\}
\end{aligned} \qquad (2.178)
$$

where $k = 0, 1, \ldots, N - 1$.

Thus, each complex signal generator, in reality, consists of a pair of generators that outputs a cosinusoidal and a sinusoidal sequence of k cycles per observation interval. The output of each complex signal generator is weighted by the complex Fourier coefficient G_k. At each time index n, an output is formed by summing the weighted complex generator outputs.

It is noteworthy that although the DFT and the IDFT are similar in their mathematical formulations, as described in Eqs. (2.176) and (2.177), their interpretations, as depicted in Figs. 2.37a and 2.37b, are so completely different.

Also, the addition of harmonically related periodic signals, as in Figs. 2.37a and 2.37b, suggests that their outputs G_k and g_n must be both periodic. Moreover,

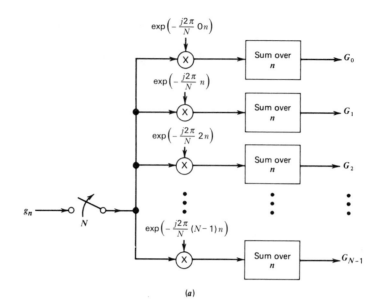

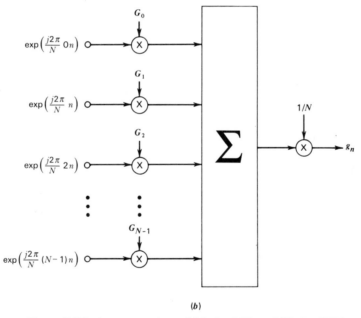

Figure 2.37   Interpretation of (a) the DFT and (b) the IDFT.

the processors shown in Figs. 2.37a and 2.37b must be linear, suggesting that the DFT and IDFT ar both linear operations. This important property is also obvious from the defining equations (2.176) and (2.177).

Fast Fourier Transform Algorithms

In the discrete Fourier transform (DFT) both the input and the output consist of sequences of numbers defined at uniformly spaced points in time and frequency, respectively. This feature makes the DFT ideally suited for direct numerical evaluation on a digital computer. Moreover, the computation can be implemented most efficiently using a class of algorithms, called *fast Fourier transform (FFT) algorithms*. An algorithm refers to a "recipe" that can be written in the form of a computer program.

FFT algorithms are efficient because they use a greatly reduced number of arithmetic operations as compared to the brute force computation of the DFT. Basically, an FFT algorithm attains its computational efficiency by following a "divide and conquer" strategy, whereby the original DFT computation is decomposed successively into smaller DFT computations. In this section, we describe one version of a popular FFT algorithm, the development of which is based on such a strategy.

To proceed with the development, we first rewrite Eq. (2.176), defining the DFT of g_n, in the convenient form

$$G_k = \sum_{n=0}^{N-1} g_n W^{nk} \qquad k = 0, 1, \ldots, N-1 \tag{2.179}$$

where

$$W = \exp\left(-\frac{j2\pi}{N}\right) \tag{2.180}$$

We readily see that

$$W^N = 1$$

$$W^{N/2} = -1$$

$$W^{(k+lN)(n+mN)} = W^{kn} \qquad m, l = 0, \pm 1, \pm 2, \ldots$$

That is, W^{kn} is periodic with period N. The periodicity of W^{kn} is a key feature in the development of FFT algorithms.

Let N, the number of points in the data sequence, be an integer power of two, as shown by

$$N = 2^L$$

where L is an integer. Since N is an even integer, $N/2$ is an integer, and so we may divide the data sequence into the first half and the last half of the points. Thus, we may rewrite Eq. (2.179) as

$$
\begin{aligned}
G_k &= \sum_{n=0}^{(N/2)-1} g_n W^{kn} + \sum_{n=N/2}^{N-1} g_n W^{kn} \\
&= \sum_{n=0}^{(N/2)-1} g_n W^{kn} + \sum_{n=0}^{(N/2)-1} g_{n+N/2} W^{k(n+N/2)} \\
&= \sum_{n=0}^{(N/2)-1} (g_n + g_{n+N/2} W^{kN/2}) W^{kn} \qquad k = 0, 1, \ldots, N-1
\end{aligned}
\tag{2.181}
$$

Since $W^{N/2} = -1$, we have

$$W^{kN/2} = (-1)^k$$

Accordingly, the factor $W^{kN/2}$ in Eq. (2.181) takes on only one of two possible values, $+1$ or -1, depending on whether the frequency index k is even or odd, respectively. These two cases are considered in what follows.

First, let k be *even*, so that $W^{kN/2} = 1$. Also let

$$k = 2l, \qquad l = 0, 1, \ldots, \frac{N}{2} - 1$$

and let

$$x_n = g_n + g_{n+N/2} \tag{2.182}$$

Then, we may put Eq. (2.181) into the form

$$
\begin{aligned}
G_{2l} &= \sum_{n=0}^{(N/2)-1} x_n W^{2ln} \\
&= \sum_{n=0}^{(N/2)-1} x_n (W^2)^{ln} \qquad l = 0, 1, \ldots, \frac{N}{2} - 1
\end{aligned}
\tag{2.183}
$$

From the definition of W given in Eq. (2.180), we readily see that

$$
\begin{aligned}
W^2 &= \exp\left(-\frac{j4\pi}{N}\right) \\
&= \exp\left(-\frac{j2\pi}{N/2}\right)
\end{aligned}
$$

Hence, we recognize the sum on the right-hand side of Eq. (2.183) as the $(N/2)$-point DFT of the sequence x_n.

Next, let k be *odd*, so that $W^{kN/2} = -1$. Also, let

$$k = 2l + 1, \qquad l = 0, 1, \ldots, \frac{N}{2} - 1$$

and let

$$y_n = g_n - g_{n+N/2} \tag{2.184}$$

Then, we may put Eq. (2.181) into the corresponding form

$$
\begin{aligned}
G^{2l+1} &= \sum_{n=0}^{(N/2)-1} y_n W^{(2l+1)n} \\
&= \sum_{n=0}^{(N/2)-1} [y_n W^n](W^2)^{ln} \qquad l = 0, 1, \ldots, \frac{N}{2} - 1
\end{aligned}
\tag{2.185}
$$

We recognize the sum on the right-hand side of Eq. (2.185) as the $(N/2)$-point DFT of the sequence $y_n W^n$. The parameter W^n associated with y_n is called a *twiddle factor*.

Equations (2.183) and (2.185) show that the even- and odd-valued samples of the transform sequence G_k can be obtained from the $(N/2)$-point DFTs of the sequences x_n and $y_n W^n$, respectively. The sequences x_n and y_n are themselves related to the original data sequence g_n by Eqs. (2.182) and (2.184), respectively. Thus, the problem of computing an N-point DFT is reduced to that of computing two $(N/2)$-point DFTs. The procedure is repeated a second time, whereby an $(N/2)$-point is decomposed into two $(N/4)$-point DFTs. The decomposition procedure is continued in this fashion until (after $L = \log_2 N$ stages) we reach the trivial case of N single-point DFTs.

Figure 2.38 illustrates the computations involved in applying the formulas of Eqs. (2.183) and (2.185) to an 8-point data sequence; that is, $N = 8$. In constructing the left-hand portions of the figure, we have used signal-flow graph notation. A *signal-flow graph* consists of an interconnection of *nodes* and *branches*. The *direction* of signal transmission along a branch is indicated by an arrow. A branch multiplies the variable at a node (to which it is connected) by the branch *transmittance*. A node sums the outputs of all incoming branches. The convention used for branch transmittances in Fig. 2.38 is as follows. When no coefficient is indicated on a branch, the transmittance of that branch is assumed to be unity. For other branches, the transmittance of a branch is indicated by -1 or an integer power of W, placed alongside the arrow on the branch.

Thus, in Fig. 2.38a the computation of an 8-point DFT is reduced to that of two 4-point DFTs. The procedure for the 8-point DFT may be mimicked to simplify the computation of the 4-point DFT. This is illustrated in Fig. 2.38b, where the computation of a 4-point DFT is reduced to that of two 2-point DFTs. Finally, the computation of a 2-point DFT is shown in Fig. 2.38c.

Combining the ideas described in Fig. 2.38, we obtain the complete signal-flow graph of Fig. 2.39 for the computation of the 8-point DFT. A repetitive structure, called a *butterfly*, can be discerned in the FFT algorithm of Fig. 2.39; a butterfly has two inputs and two outputs. Examples of butterflies (for the three stages of the algorithm) are shown by the bold-faced lines in Fig. 2.39.

For the general case of $N = 2^L$, the algorithm requires $L = \log_2 N$ stages of computation. Each stage requires $(N/2)$ butterflies. Each butterfly involves one complex multiplication and two complex additions (to be precise, one addition and one subtraction). Accordingly, the FFT structure described here requires $(N/2)\log_2 N$ complex multiplications and $N\log_2 N$ complex additions. (Actually, the number of multiplications quoted is pessimistic, because we may omit all twiddle factors $W^0 = 1$ and $W^{N/2} = -1$, $W^{N/4} = j$, $W^{3N/4} = -j$.) This computational complexity is significantly smaller than that of the N^2 complex multiplications and $N(N - 1)$ complex additions required for the *direct* computation of the DFT. The computational savings made possible by the FFT algorithm become more substantial as we increase the data length N.

We may establish two other important features of the FFT algorithm by carefully examining the signal-flow graph shown in Fig. 2.39:

1. At each stage of the computation, the new set of N complex numbers resulting from the computation can be stored in the same memory locations used to store the previous set. This kind of computation is referred to as *in-place computation*.

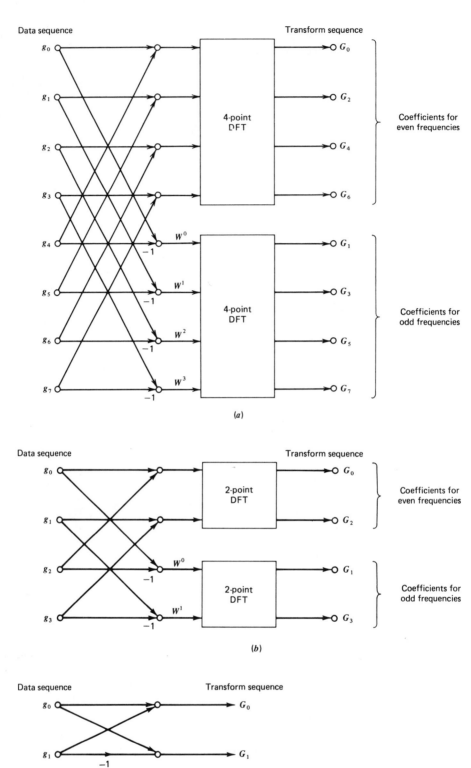

Figure 2.38 (*a*) Reduction of 8-point DFT into two 4-point DFTs. (*b*) Reduction of 4-point DFT into two 2-point DFTS. (*c*) Trivial case of 2-point DFT.

Data sequence   Transform sequence

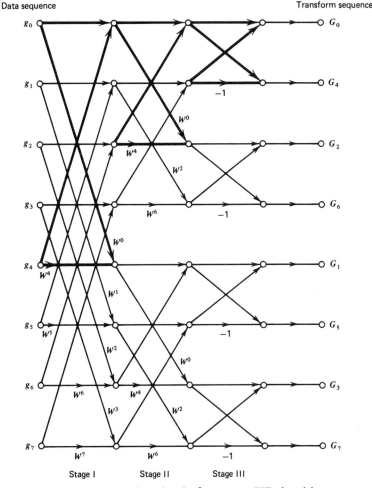

Figure 2.39 Decimation-in-frequency FFT algorithm.

2. The samples of the transform sequence G_k are stored in a *bit-reversed order*. To illustrate the meaning of this terminology, consider Table 2.2 constructed for the case of $N = 8$. At the left of the table, we show the eight possible values of the frequency index k (in their natural order) and their 3-bit binary representations. At the right of the table, we show the corresponding bit-reversed binary representations and indices. We observe that the bit-reversed indices in the right-most column of Table 2.2 appear in the same order as the indices at the output of the FFT algorithm in Fig. 2.39.

The FFT algorithm depicted in Fig. 2.39 is referred to as a *decimation-in-frequency algorithm,* because the transform (frequency) sequence G_k is divided successively into smaller subsequences. In another popular FFT algorithm, called a *decimation-in-time* algorithm, the data (time) sequence g_n is divided successively into smaller subsequences. Both algorithms have the same computational complexity. They differ from each other in two respects. First, for decimation-in-frequency, the input is in natural order, whereas the output is in bit-reversed

Table 2.2   Illustrating Bit Reversal

Frequency Index, k	Binary Representation	Bit-Reversed Binary Representation	Bit-Reversed Index
0	000	000	0
1	001	100	4
2	010	010	2
3	011	110	6
4	100	001	1
5	101	101	5
6	110	011	3
7	111	111	7

order. The reverse is true for decimation-in-time. Second, the butterfly for decimation-in-time is slightly different from that for decimation-in-frequency. The reader is invited to derive the details of the decimation-in-time algorithm using the divide-and-conquer strategy that led to the development of the algorithm described in Fig. 2.39.

Computation of the IDFT

The IDFT of the transform G_k is defined by Eq. (2.177). We may rewrite this equation in terms of the complex parameter W as

$$g_n = \frac{1}{N} \sum_{k=0}^{N-1} G_k W^{-kn} \qquad n = 0, 1, \ldots, N-1 \qquad (2.186)$$

Taking the complex conjugate of Eq. (2.186) and multiplying by N, we get

$$N g_n^* = \sum_{k=0}^{N-1} G_k^* W^{kn} \qquad n = 0, 1, \ldots, N-1 \qquad (2.187)$$

The right-hand side of Eq. (2.187) is recognized as the N-point DFT of the complex-conjugated sequence G_k^*. Accordingly, Eq. (2.187) suggests that we may compute the desired sequence g_n using the scheme shown in Fig. 2.40, based on an N-point FFT algorithm. Thus, the same FFT algorithm can be used to handle the computation of both the IDFT and the DFT.

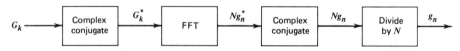

Figure 2.40   Use of the FFT algorithm for computing the IDFT.

2.16   SUMMARY

In this chapter we have described the Fourier transform as a fundamental tool for relating the time-domain and frequency-domain descriptions of a deterministic signal. The signal of interest may be an energy signal or a power signal. The Fourier transform includes the exponential Fourier series as a special case, provided that we permit the use of the Dirac delta function.

An inverse relationship exists between the time-domain and frequency-domain descriptions of a signal. Whenever an operation is performed on the waveform of a signal in the time domain, a corresponding modification is applied to the spectrum of the signal in the frequency domain. An important consequence of this inverse relationship is the fact that the time–bandwidth product of an energy signal is a constant; the definitions of signal duration and bandwidth merely affect the value of the constant.

An important signal processing operation frequently encountered in communication systems is that of linear filtering. This operation involves the convolution of the input signal with the impulse response of the filter or, equivalently, the multiplication of the Fourier transform of the input signal by the transfer function (i.e., Fourier transform of the impulse response) of the filter. Low-pass and band-pass filters represent two commonly used types of filters. Band-pass filtering is usually more complicated than low-pass filtering. However, through the combined use of a complex envelope for the representation of an input band-pass signal and the complex impulse response for the representation of a band-pass filter, we may formulate a low-pass equivalent for the band-pass filtering problem and thereby replace a difficult problem with a much simpler one. It is also important to note that there is no loss of information in establishing this equivalence. The concepts of complex envelope and complex impulse response are rooted in Hilbert transformation.

The final part of the chapter was concerned with the discrete Fourier transform and its numerical computation. Basically, the discrete Fourier transform is obtained from the standard Fourier transform by uniformly sampling both the input and the output. The fast Fourier transform algorithm provides a practical means for the efficient implementation of the discrete Fourier transform on a digital computer. This makes the fast Fourier transform algorithm a powerful computational tool for spectral analysis and linear filtering.

NOTES AND REFERENCES

1. The origin of the theory of Fourier series and Fourier transform is to be found in J. B. J. Fourier, *The Analytical Theory of Heat*, translated from French by A. Freeman (Cambridge University Press, 1878). The books by Bracewell (1986) and Champeney (1973) provide detailed treatments of the Fourier transform with emphasis on the physical aspects of the subject.

2. For an extensive list of Fourier-transform pairs, see the book by Campbell and Foster (1948).

3. If a time function $g(t)$ is such that the value of the energy $\int_{-\infty}^{\infty} |g(t)|^2 \, dt$ is defined and finite, then the Fourier transform $G(f)$ of the function $g(t)$ exists and

$$\lim_{A \to \infty} \left[\int_{-\infty}^{\infty} |g(t) - \int_{-A}^{A} G(f) \exp(j2\pi ft) \, df|^2 \right] = 0$$

This result is known as *Plancherel's theorem.* For a proof of this theorem, see Titchmarsh (1950, p. 69) and Wiener (1958, pp. 46–71).

4. The notation for rect(*t*) defined in Eq. (2.7) originated in the book by Woodward (1964, p. 29).

5. The energy intensity $|g(t)|^2$ and the energy spectral density $\mathscr{E}_g(f) = |G(f)|^2$ do *not* always tell the whole story about the energy content of a signal $g(t)$. This is particularly so when the spectral characteristics of the signal (e.g., speech signal) vary with time. Such signals are often referred to as *time-varying* or *nonstationary signals.* For an accurate spectral analysis of this important class of signals, we cannot use the standard Fourier transform. Rather, we need to use *time-frequency analysis,* a brief exposition of which is presented in Appendix 3.

6. The notation $\delta(t)$ for a delta function was first introduced into quantum mechanics by Dirac. This notation is now in general use in the signal processing literature. For detailed discussions of the delta function, see Bracewell (1986, Chapter 5), and Papoulis (1962, pp. 35–52 and pp. 269–282). For a rigorous treatment of the subject, see the book by Lighthill (1958).

7. The Paley–Wiener criterion is named in honor of the authors of the paper by Paley and Wiener (1934). For a discussion of the Paley–Wiener criterion, see also the book by Papoulis (1962, pp. 215–217).

8. For tables of values of the sine integral defined in Eq. (2.122), see the well-known handbook by Abramowitz and Stegun (1965).

9. Gibb's phenomenon is observed not only in the Fourier transform (as shown in Computer Experiments I and II) but also in the Fourier series. For a detailed discussion of this phenomenon, see Guillemin (1958, pp. 485–496).

10. The integral in Eq. (2.124), defining the Hilbert transform of a signal, is an improper integral in that the integrand has a singularity at $\tau = t$. In order to avoid this singularity, the integration must be carried out in a symmetrical manner about the point $\tau = t$. For this purpose, we use the definition

$$P \int_{-\infty}^{\infty} \frac{g(\tau)}{t - \tau} \, d\tau = \lim_{\varepsilon \to 0} \left[\int_{-\infty}^{t-\varepsilon} \frac{g(\tau)}{t - \tau} \, d\tau + \int_{t+\varepsilon}^{\infty} \frac{g(\tau)}{t - \tau} \, d\tau \right]$$

where the symbol P denotes *Cauchy's principal value of the integral.* For notational simplicity, the symbol P has been omitted from Eqs. (2.124) and (2.125). For a rigorous discussion of the Hilbert transform, see Guillemin (1958).

11. The complex representation of an arbitrary signal defined in Eq. (2.133) was first described by Gabor in 1946. Gabor used the term "analytic signal." The term "preenvelope" was used in Arens (1957) and Dungundji (1958). For a review of the different envelopes, see the paper by Rice (1982).

12. Fast Fourier transform (FFT) algorithms were brought into prominence by the publication of the paper by Cooley and Tukey (1965). For discussions of FFT algorithms, see the books by Oppenheim and Schafer (1975, pp. 290–321), Rabiner and Gold (1975, pp. 357–381), and Elliott and Rao (1982, pp. 58–177). The last reference includes treatment of the newer FFT algorithms that have a reduced number of multiplications; it also includes a detailed mathematical analysis of the relation between the discrete Fourier transform and continuous Fourier transform. For a discussion of how the FFT algorithm may be used to perform linear filtering, see Oppenheim and Shafer (1975, pp. 110–115) and Rabiner and Gold (1975, pp. 63–67). A useful treatment of the subject is also presented in the book *Numerical recipes in C* by Press et al. (1988).

PROBLEMS

Problem 2.1

(a) Find the Fourier transform of the half-cosine pulse shown in Fig. P2.1a.

(b) Apply the time-shifting property to the result obtained in part (a) to evaluate the spectrum of the half-sine pulse shown in Fig. P2.1b.

(c) What is the spectrum of a half-sine pulse having a duration equal to aT?

(d) What is the spectrum of the negative half-sine pulse shown in Fig. P2.1c?

(e) Find the spectrum of the single sine pulse shown in Fig. P2.1d.

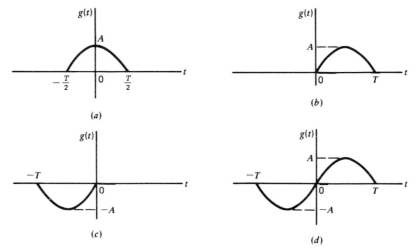

(a)

(b)

(c)

(d)

Figure P2.1

Problem 2.2 Evaluate the Fourier transform of the damped sinusoidal wave

$$g(t) = \exp(-t)\sin(2\pi f_c t)u(t)$$

where $u(t)$ is the unit step function.

Problem 2.3 Any function $g(t)$ can be split unambiguously into an *even part* and an *odd part*, as shown by

$$g(t) = g_e(t) + g_o(t)$$

The even part is defined by

$$g_e(t) = \frac{1}{2}[g(t) + g(-t)]$$

and the odd part is defined by

$$g_o(t) = \frac{1}{2}[g(t) - g(-t)]$$

(a) Evaluate the even and odd parts of a rectangular pulse defined by

$$g(t) = A \operatorname{rect}\left(\frac{t}{T} - \frac{1}{2}\right)$$

(b) What are the Fourier transforms of these two parts of the pulse?

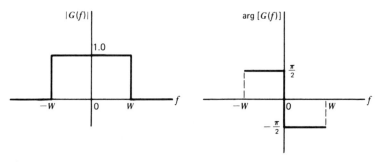

Figure P2.2

Problem 2.4   Determine the inverse Fourier transform of the frequency function $G(f)$ defined by the amplitude and phase spectra shown in Fig. P2.2.

Problem 2.5   The following expression may be viewed as an approximate representation of a pulse with finite rise time:

$$g(t) = \frac{1}{\tau} \int_{t-T}^{t+T} \exp\left(-\frac{\pi u^2}{\tau^2} \right) du$$

where it is assumed that $T \gg \tau$. Determine the Fourier transform of $g(t)$. What happens to this transform when we allow τ to become zero? *Hint:* Express $g(t)$ as the superposition of two signals, one corresponding to integration from $t - T$ to 0, and the other from 0 to $t + T$.

Problem 2.6   The Fourier transform of a signal $g(t)$ is denoted by $G(f)$. Prove the following properties of the Fourier transform:

(a) If a real signal $g(t)$ is an even function of time t, the Fourier transform $G(f)$ is purely real. If a real signal $g(t)$ is an odd function of time t, the Fourier transform $G(f)$ is purely imaginary.

(b)
$$t^n g(t) \rightleftharpoons \left(\frac{j}{2\pi} \right)^n G^{(n)}(f)$$

where $G^{(n)}(f)$ is the nth derivative of $G(f)$ with respect to f.

(c)
$$\int_{-\infty}^{\infty} t^n g(t)\; dt = \left(\frac{j}{2\pi} \right)^n G^{(n)}(0)$$

(d)
$$g_1(t) g_2^*(t) \rightleftharpoons \int_{-\infty}^{\infty} G_1(\lambda) G_2^*(\lambda - f)\; d\lambda$$

(e)
$$\int_{-\infty}^{\infty} g_1(t) g_2^*(t)\; dt = \int_{-\infty}^{\infty} G_1(f) G_2^*(f)\; df$$

Problem 2.7   The Fourier transform $G(f)$ of a signal $g(t)$ is bounded by the following three inequalities:

$$|G(f)| \le \int_{-\infty}^{\infty} |g(t)|\; dt$$

$$|j2\pi f G(f)| \le \int_{-\infty}^{\infty} \left| \frac{dg(t)}{dt} \right|\; dt$$

and

$$\left|(j2\pi f)^2 G(f)\right| \le \int_{-\infty}^{\infty} \left|\frac{d^2 g(t)}{dt^2}\right| \, dt$$

where it is assumed that the first and second derivatives of $g(t)$ exist.

Construct these three bounds for the triangular pulse shown in Fig. P2.3 and compare your results with the actual amplitude spectrum of the pulse.

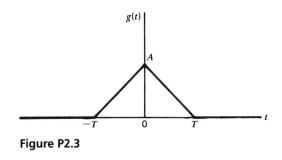

Figure P2.3

Problem 2.8 Prove the following properties of the convolution process:

(a) The commutative property:

$$g_1(t) \bigstar g_2(t) = g_2(t) \bigstar g_1(t)$$

(b) The associative property:

$$g_1(t) \bigstar [g_2(t) \bigstar g_3(t)] = [g_1(t) \bigstar g_2(t)] \bigstar g_3(t)$$

(c) The distributive property:

$$g_1(t) \bigstar [g_2(t) + g_3(t)] = g_1(t) \bigstar g_2(t) + g_1(t) \bigstar g_3(t)$$

Problem 2.9 Consider the convolution of two signals $g_1(t)$ and $g_2(t)$. Show that

(a)
$$\frac{d}{dt}[g_1(t) \bigstar g_2(t)] = \left[\frac{d}{dt} g_1(t)\right] \bigstar g_2(t)$$

(b)
$$\int_{-\infty}^{t} [g_1(\tau) \bigstar g_2(\tau)] \, d\tau = \left[\int_{-\infty}^{t} g_1(\tau) \, d\tau\right] \bigstar g_2(t)$$

Problem 2.10 A signal $x(t)$ of finite energy is applied to a square-law device whose output $y(t)$ is defined by

$$y(t) = x^2(t)$$

The spectrum of $x(t)$ is limited to the frequency interval $-W \le f \le W$. Hence, show that the spectrum of $y(t)$ is limited to $-2W \le f \le 2W$. *Hint:* Express $y(t)$ as $x(t)$ multiplied by itself.

Problem 2.11 Evaluate the Fourier transform of the delta function by considering it as the limiting form of (1) a rectangular pulse of unit area, and (2) a sinc pulse of unit area.

Problem 2.12 The Fourier transform $G(f)$ of a signal $g(t)$ is defined by

$$G(f) = \begin{cases} 1, & f > 0 \\ \dfrac{1}{2}, & f = 0 \\ 0, & f < 0 \end{cases}$$

Determine the signal $g(t)$.

Problem 2.13 Consider a pulselike function $g(t)$ that consists of a small number of straight-line segments. Suppose that this function is differentiated with respect to time t twice so as to generate a sequence of weighted delta functions, as shown by

$$\frac{d^2g(t)}{dt^2} = \sum_i k_i \delta(t - t_i)$$

where the k_i are related to the slopes of the straight-line segments.

(a) Given the values of the k_i and the t_i, show that the Fourier transform of $g(t)$ is given by

$$G(f) = -\frac{1}{4\pi^2 f^2} \sum_i k_i \exp(-j2\pi f t_i)$$

(b) Using this procedure, show that the Fourier transform of the trapezoidal pulse shown in Fig. P2.4 is given by

$$G(f) = \frac{A}{\pi^2 f^2 (t_b - t_a)} \sin[\pi f(t_b - t_a)]\sin[\pi f(t_b + t_a)]$$

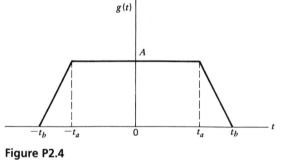

Figure P2.4

Problem 2.14 Show that the two different pulses defined in parts (a) and (b) of Fig. P2.1 have the same energy spectral density:

$$\mathscr{E}_g(f) = \frac{4A^2 T^2 \cos^2(\pi T f)}{\pi^2 (4T^2 f^2 - 1)^2}$$

Problem 2.15

(a) The *root mean-square (rms) bandwidth* of a low-pass signal $g(t)$ of finite energy is defined by

$$W_{\text{rms}} = \left[\frac{\displaystyle\int_{-\infty}^{\infty} f^2 |G(f)|^2 \, df}{\displaystyle\int_{-\infty}^{\infty} |G(f)|^2 \, df} \right]^{1/2}$$

where $|G(f)|^2$ is the energy spectral density of the signal. Correspondingly, the *root mean-square (rms) duration* of the signal is defined by

$$T_{\text{rms}} = \left[\frac{\displaystyle\int_{-\infty}^{\infty} t^2 |g(t)|^2 \, dt}{\displaystyle\int_{-\infty}^{\infty} |g(t)|^2 \, dt} \right]^{1/2}$$

Using these definitions, show that

$$T_{\text{rms}} W_{\text{rms}} \geq \frac{1}{4\pi}$$

Assume that $|g(t)| \to 0$ faster than $1/\sqrt{|t|}$ as $|t| \to \infty$.

(b) Consider a Gaussian pulse defined by

$$g(t) = \exp(-\pi t^2)$$

Show that, for this signal, the equality

$$T_{\text{rms}} W_{\text{rms}} = \frac{1}{4\pi}$$

can be reached.

Hint: Use Schwarz's inequality (see Appendix 5).

$$\left\{ \int_{-\infty}^{\infty} [g_1^*(t) g_2(t) + g_1(t) g_2^*(t)] \, dt \right\}^2 \leq 4 \int_{-\infty}^{\infty} |g_1(t)|^2 \, dt \int_{-\infty}^{\infty} |g_2(t)|^2 \, dt$$

in which we set

$$g_1(t) = tg(t)$$

and

$$g_2(t) = \frac{dg(t)}{dt}$$

Problem 2.16 Let $x(t)$ and $y(t)$ be the input and output signals of a linear time-invariant filter. Using Rayleigh's energy theorem, show that if the filter is stable and the input signal $x(t)$ has finite energy, then the output signal $y(t)$ also has finite energy. That is, given that

$$\int_{-\infty}^{\infty} |x(t)|^2 \, dt < \infty$$

then show that

$$\int_{-\infty}^{\infty} |y(t)|^2 \, dt < \infty$$

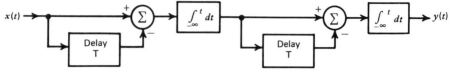

Figure P2.5

Problem 2.17 Evaluate the transfer function of a linear system represented by the block diagram shown in Fig. P2.5.

Problem 2.18

(a) Determine the overall amplitude response of the cascade connection shown in Fig. P2.6 consisting of N identical stages, each with a time constant RC equal to τ_0.

(b) Show that as N approaches infinity, the amplitude response of the cascade connection approaches the Gaussian function $\exp\left(-\dfrac{1}{2}f^2T^2\right)$, where for each value of N, the time constant τ_0 is selected so that

$$\tau_0^2 = \frac{T^2}{4\pi^2 N}$$

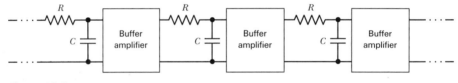

Figure P2.6

Problem 2.19 Suppose that, for a given signal $x(t)$, the integrated value of the signal over an interval T is required, as shown by

$$y(t) = \int_{t-T}^{t} x(\tau) \, d\tau$$

(a) Show that $y(t)$ can be obtained by transmitting the input signal $x(t)$ through a filter with its transfer function given by

$$H(f) = T \operatorname{sinc}(fT)\exp(-j\pi fT)$$

(b) An adequate approximation to this transfer function is obtained by using a low-pass filter with a bandwidth equal to $1/T$, passband amplitude response T, and delay $T/2$. Assuming this low-pass filter to be ideal, determine the filter output at time $t = T$ due to a unit step function applied to the filter at $t = 0$, and compare the result with the corresponding output of the ideal integrator. Note that $\operatorname{Si}(\pi) = 1.85$ and $\operatorname{Si}(\infty) = \pi/2$.

Problem 2.20 A tapped-delay-line filter consists of N weights, where N is odd. It is symmetric with respect to the center tap, that is, the weights satisfy the condition

$$w_n = w_{N-1-n} \qquad 0 \le n \le N - 1$$

(a) Find the amplitude response of the filter.
(b) Show that this filter has a linear phase response.

Problem 2.21 Let $\hat{g}(t)$ denote the Hilbert transform of $g(t)$. Derive the following set of Hilbert-transform pairs:

$g(t)$	$\hat{g}(t)$
$\dfrac{\sin t}{t}$	$\dfrac{1 - \cos t}{t}$
$\text{rect}(t)$	$-\dfrac{1}{\pi} \ln \left\| \left(t - \dfrac{1}{2}\right) \middle/ \left(t + \dfrac{1}{2}\right) \right\|$
$\delta(t)$	$\dfrac{1}{\pi t}$
$\dfrac{1}{1 + t^2}$	$\dfrac{t}{1 + t^2}$

Problem 2.22 Evaluate the inverse Fourier transform $g(t)$ of the one-sided frequency function:

$$G(f) = \begin{cases} \exp(-f), & f > 0 \\ \dfrac{1}{2}, & f = 0 \\ 0, & f < 0 \end{cases}$$

Hence, show that $g(t)$ is complex, and that its real and imaginary parts constitute a Hilbert-transform pair.

Problem 2.23 Determine the pre-envelope $g_+(t)$ corresponding to each of the following two signals:

(a) $g(t) = \text{sinc}(t)$
(b) $g(t) = [1 + k \cos(2\pi f_m t)]\cos(2\pi f_c t)$

Problem 2.24 Show that the complex envelope of the sum of two narrow-band signals (with the same carrier frequency) is equal to the sum of their individual complex envelopes.

Problem 2.25 The definition of the complex envelope $\tilde{g}(t)$ of a band-pass signal given in Eq. (2.139) is based on the pre-envelope $g_+(t)$ for positive frequencies. How is the complex envelope defined in terms of the pre-envelope $g_-(t)$ for negative frequencies? Justify your answer.

Problem 2.26 Consider the signal

$$s(t) = c(t) m(t)$$

whose $m(t)$ is a low-pass signal whose Fourier transform $M(f)$ vanishes for $|f| > W$, and $c(t)$ is a high-pass signal whose Fourier transform $C(f)$ vanishes for $|f| < W$. Show that the Hilbert transform of $s(t)$ is $\hat{s}(t) = \hat{c}(t)m(t)$, where $\hat{c}(t)$ is the Hilbert transform of $c(t)$.

Problem 2.27

(a) Consider two real-valued signals $g_1(t)$ and $g_2(t)$ whose pre-envelopes are denoted by $g_{1+}(t)$ and $g_{2+}(t)$, respectively. Show that

$$\int_{-\infty}^{\infty} \mathrm{Re}[g_{1+}(t)]\mathrm{Re}[g_{2+}(t)]\ dt = \frac{1}{2}\mathrm{Re}\left[\int_{-\infty}^{\infty} g_{1+}(t)g_{2+}^*(t)\ dt\right]$$

(b) Suppose that $g_2(t)$ is replaced by $g_2(-t)$. Show that this modification has the effect of removing the complex conjugation in the right-hand side of the relation given in part (a).

(c) Assuming that $g(t)$ is a narrow-band signal with complex envelope $\tilde{g}(t)$ and carrier frequency f_c, use the result of part (a) to show that

$$\int_{-\infty}^{\infty} g^2(t)\ dt = \frac{1}{2}\int_{-\infty}^{\infty} |\tilde{g}(t)|^2\ dt$$

Problem 2.28　Let a narrow-band signal $g(t)$ be expressed in the form

$$g(t) = g_I(t)\cos(2\pi f_c t) - g_Q(t)\sin(2\pi f_c t)$$

Using $G_+(f)$ to denote the Fourier transform of the pre-envelope of $g_+(t)$, show that the Fourier transforms of the in-phase component $g_I(t)$ and quadrature component $g_Q(t)$ are given by, respectively,

$$G_I(f) = \frac{1}{2}[G_+(f + f_c) + G_+^*(-f + f_c)]$$

$$G_Q(f) = \frac{1}{2j}[G_+(f + f_c) - G_+^*(-f + f_c)]$$

where the asterisk denotes complex conjugation.

Problem 2.29　The block diagram of Fig. 2.32a illustrates a method for extracting the in-phase component $g_I(t)$ and quadrature component $g_Q(t)$ of a narrow-band signal $g(t)$. Given that the spectrum of $g(t)$ is limited to the interval $f_c - W \le |f| \le f_c + W$, demonstrate the validity of this method. Hence, show that

$$G_I(f) = \begin{cases} G(f - f_c) + G(f + f_c), & -W \le f \le W \\ 0, & \text{elsewhere} \end{cases}$$

and

$$G_Q(f) = \begin{cases} j[G(f - f_c) - G(f + f_c)], & -W \le f \le W \\ 0, & \text{elsewhere} \end{cases}$$

where $G_I(f)$, $G_Q(f)$, and $G(f)$ are the Fourier transforms of $g_I(t)$, $g_Q(t)$, and $g(t)$, respectively.

Problem 2.30　Explain what happens to the low-pass equivalent model of Fig. 2.34 when the amplitude response of the corresponding band-pass filter has even symmetry and the phase response has odd symmetry with respect to the mid-band frequency f_c.

Problem 2.31   Consider an ideal band-pass filter with frequency mid-band f_c and bandwidth $2B$, as defined in Fig. P2.7. The carrier wave $A \cos(2\pi f_0 t)$ is suddenly applied to this filter at time $t = 0$. Assuming that $|f_c - f_0|$ is large compared to the bandwidth $2B$, determine the response of the filter.

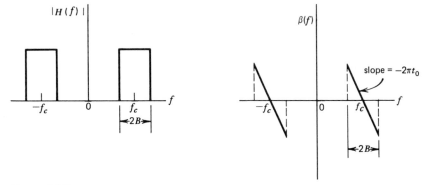

Figure P2.7

Problem 2.32   The rectangular RF pulse

$$x(t) = \begin{cases} A \cos(2\pi f_c t), & 0 \le t \le T \\ 0, & \text{elsewhere} \end{cases}$$

is applied to a linear filter with impulse response

$$h(t) = x(T - t)$$

Assume that the frequency f_c equals a large integer multiple of $1/T$. Determine the response of the filter and sketch it.

COMPUTER-ORIENTED PROBLEMS

2.1   In Computer Experiment I described in Section 2.9, we used a rectangular pulse applied to the input of an ideal low-pass filter. Repeat this experiment using an input consisting of the *raised cosine pulse*:

$$x(t) = \begin{cases} 1 + \cos\left(\dfrac{\pi t}{T}\right), & -T \le t \le T \\ 0, & \text{otherwise} \end{cases}$$

In particular, evaluate the filter response for the time–bandwidth product $BT = 5, 10, 10, 100$. Compare the results of this experiment with those of Computer Experiment I.

2.2   In Computer Experiment II described in Section 2.13, we used a rectangular RF pulse as applied to the input of an ideal band-pass filter. Repeat this experiment using an input defined by

$$x(t) = g(t)\cos(2\pi f_c t)$$

where

$$g(t) = \begin{cases} 1 + \cos\left(\dfrac{\pi t}{T}\right), & -T \leq t \leq T \\ \\ 0, & \text{elsewhere} \end{cases}$$

In particular, evaluate the filter response for the time–bandwidth product $BT = 5$, and compare the results of your experiment with those of Computer Experiment II.

CHAPTER 3

Continuous-
Wave
Modulation

3.1 INTRODUCTION

The purpose of a communication system is to transmit *information-bearing signals* or *baseband signals* through a communication channel separating the transmitter from the receiver. The term *baseband* is used to designate the band of frequencies representing the original signal as delivered by a source of information. The proper utilization of the communication channel requires a shift of the range of baseband frequencies into other frequency ranges suitable for transmission, and a corresponding shift back to the original frequency range after reception. For example, a radio system must operate with frequencies of 30 kHz and upward, whereas the baseband signal usually contains frequencies in the audio frequency range, and so some form of frequency-band shifting must be used for the system to operate satisfactorily. A shift of the range of frequencies in a signal is accomplished by using *modulation*, which is defined as *the process by which some characteristic of a carrier is varied in accordance with a modulating wave (signal)*. A common form of the carrier is a *sinusoidal wave,* in which case we speak of a *continuous-wave modulation*[1] process. The baseband signal is referred to as the *modulating wave,* and the result of the modulation process is referred to as the *modulated wave.* Modulation is performed at the transmitting end of the communication system. At the receiving end of the system, we usually require the original baseband signal to be restored. This is accomplished by using a process known as *demodulation*, which is the reverse of the modulation process.

In this chapter we study two widely used families of continuous-wave (CW) modulation systems, namely, *amplitude modulation* and *angle modulation*. In amplitude modulation, the amplitude of the sinusoidal carrier wave is varied in accordance with the baseband signal. In angle modulation, the angle of the sinusoidal carrier wave is varied in accordance with the baseband signal. Sections 3.2 through 3.7 are devoted to the standard form of amplitude modulation and its variants. In Section 3.9 we discuss the idea of *frequency-division multiplexing* for sharing a common channel among a multitude of different users. The four subsequent sections are devoted to angle modulation and related issues.

3.2 AMPLITUDE MODULATION

Consider a *sinusoidal carrier wave* $c(t)$ defined by

$$c(t) = A_c \cos(2\pi f_c t) \tag{3.1}$$

where A_c is the *carrier amplitude* and f_c is the *carrier frequency*. To simplify the exposition without affecting the results obtained and conclusions reached, we have assumed that the phase of the carrier wave is zero in Eq. (3.1). Let $m(t)$ denote the baseband signal that carries the specification of the message. The source of carrier wave $c(t)$ is physically independent of the source responsible for generating $m(t)$. *Amplitude modulation (AM) is defined as a process in which the amplitude of the carrier wave $c(t)$ is varied about a mean value, linearly with the baseband signal $m(t)$.* An amplitude-modulated (AM) wave may thus be described, in its most general form, as a function of time as follows:

$$s(t) = A_c[1 + k_a m(t)]\cos(2\pi f_c t) \tag{3.2}$$

where k_a is a constant called the *amplitude sensitivity* of the modulator responsible for the generation of the modulated signal $s(t)$. Typically, the carrier amplitude A_c and the message signal $m(t)$ are measured in volts, in which case the amplitude sensitivity k_a is measured in volt^{-1}.

Figure 3.1a shows a baseband signal $m(t)$, and Figs. 3.1b and 3.1c show the corresponding AM wave $s(t)$ for two values of amplitude sensitivity k_a and a carrier amplitude $A_c = 1$ volt. We observe that the *envelope* of $s(t)$ has essentially the same shape as the baseband signal $m(t)$ provided that two requirements are satisfied:

1. The amplitude of $k_a m(t)$ is always less than unity, that is,

$$|k_a m(t)| < 1 \qquad \text{for all } t \tag{3.3}$$

This condition is illustrated in Fig. 3.1b; it ensures that the function $1 + k_a m(t)$ is always positive, and since an envelope is a positive function, we may express the envelope of the AM wave $s(t)$ of Eq. (3.2) as $A_c[1 + k_a m(t)]$. When the amplitude sensitivity k_a of the modulator is large enough to make $|k_a m(t)| > 1$ for any t, the carrier wave becomes *overmodulated*, resulting in carrier phase reversals whenever the factor $1 + k_a m(t)$ crosses zero. The mod-

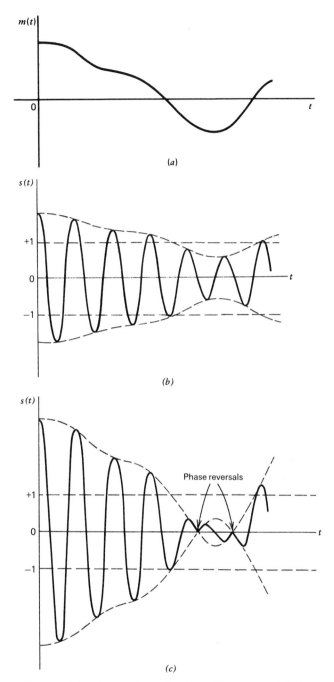

Figure 3.1 Illustrating the amplitude modulation process. (a) Baseband signal $m(t)$. (b) AM wave for $k_a m(t) < 1$ for all t. (c) AM wave for $|k_a m(t)| > 1$ for some t.

ulated wave then exhibits *envelope distortion*, as in Fig. 3.1*c*. It is therefore apparent that by avoiding overmodulation, a one-to-one relationship is maintained between the envelope of the AM wave and the modulating wave for all values of time—a useful feature, as we shall see later on. The absolute maximum value of $k_a m(t)$ multiplied by 100 is referred to as the *percentage modulation*.

2. The carrier frequency f_c is much greater than the highest frequency component W of the message signal $m(t)$, that is

$$f_c \gg W \tag{3.4}$$

We call W the *message bandwidth*. If the condition of Eq. (3.4) is not satisfied, an envelope cannot be visualized (and therefore detected) satisfactorily.

From Eq. (3.2), we find that the Fourier transform of the AM wave $s(t)$ is given by

$$S(f) = \frac{A_c}{2}[\delta(f - f_c) + \delta(f + f_c)]$$
$$+ \frac{k_a A_c}{2}[M(f - f_c) + M(f + f_c)] \tag{3.5}$$

Suppose that the baseband signal $m(t)$ is band-limited to the interval $-W \leq f \leq W$, as in Fig. 3.2*a*. The shape of the spectrum shown in this figure is intended for the purpose of illustration only. We find from Eq. (3.5) that the spectrum $S(f)$ of the AM wave is as shown in Fig. 3.2*b* for the case when $f_c > W$.

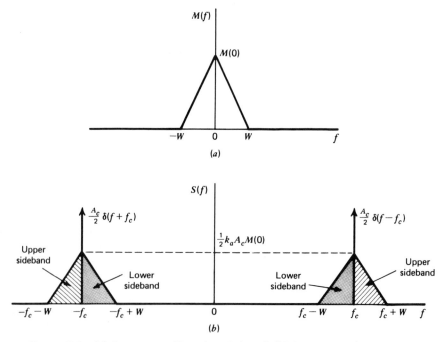

Figure 3.2 (*a*) Spectrum of baseband signal. (*b*) Spectrum of AM wave.

This spectrum consists of two delta functions weighted by the factor $A_c/2$ and occurring at $\pm f_c$, and two versions of the baseband spectrum translated in frequency by $\pm f_c$ and scaled in amplitude by $k_a A_c/2$. From the spectrum of Fig. 3.2b, we note the following:

1. As a result of the modulation process, the spectrum of the message signal $m(t)$ for negative frequencies extending from $-W$ to 0 becomes completely visible for positive (i.e., measurable) frequencies, provided that the carrier frequency satisfies the condition $f_c > W$; herein lies the importance of the idea of "negative" frequencies.

2. For positive frequencies, the portion of the spectrum of an AM wave lying above the carrier frequency f_c is referred to as the *upper sideband*, whereas the symmetric portion below f_c is referred to as the *lower sideband*. For negative frequencies, the upper sideband is represented by the portion of the spectrum below $-f_c$ and the lower sideband by the portion above $-f_c$. The condition $f_c > W$ ensures that the sidebands do not overlap.

3. For positive frequencies, the highest frequency component of the AM wave equals $f_c + W$, and the lowest frequency component equals $f_c - W$. The difference between these two frequencies defines the *transmission bandwidth* B_T for an AM wave, which is exactly twice the message bandwidth W, that is,

$$B_T = 2W \tag{3.6}$$

EXAMPLE 1   Single-Tone Modulation

Consider a modulating wave $m(t)$ that consists of a single tone or frequency component, that is,

$$m(t) = A_m \cos(2\pi f_m t)$$

where A_m is the amplitude of the sinusoidal modulating wave and f_m is its frequency (see Fig. 3.3a). The sinusoidal carrier wave has amplitude A_c and frequency f_c (see Fig. 3.3b). The corresponding AM wave is therefore given by

$$s(t) = A_c[1 + \mu \cos(2\pi f_m t)]\cos(2\pi f_c t) \tag{3.7}$$

where

$$\mu = k_a A_m$$

The dimensionless constant μ is the *modulation factor*, or the percentage modulation when it is expressed numerically as a percentage. To avoid envelope distortion due to overmodulation, the modulation factor μ must be kept below unity.

Figure 3.3c shows a sketch of $s(t)$ for μ less than unity. Let $A_{\max}$ and $A_{\min}$ denote the maximum and minimum values of the envelope of the modulated wave. Then, from Eq. (3.7) we get

$$\frac{A_{\max}}{A_{\min}} = \frac{A_c(1 + \mu)}{A_c(1 - \mu)}$$

That is,

$$\mu = \frac{A_{\max} - A_{\min}}{A_{\max} + A_{\min}}$$

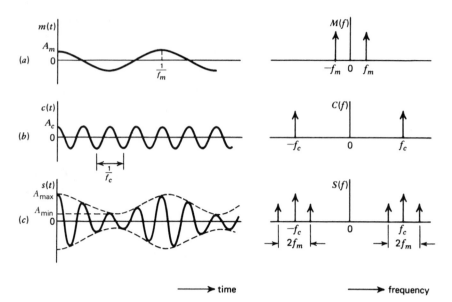

Figure 3.3 Illustrating the time-domain (on the left) and frequency-domain (on the right) characteristics of standard amplitude modulation produced by a single tone. (*a*) Modulating wave. (*b*) Carrier wave. (*c*) AM wave.

Expressing the product of the two cosines in Eq. (3.7) as the sum of two sinusoidal waves, one having frequency $f_c + f_m$ and the other having frequency $f_c - f_m$, we get

$$s(t) = A_c \cos(2\pi f_c t) + \tfrac{1}{2}\mu A_c \cos[2\pi(f_c + f_m)t] + \tfrac{1}{2}\mu A_c \cos[2\pi(f_c - f_m)t]$$

The Fourier transform of $s(t)$ is therefore

$$S(f) = \tfrac{1}{2}A_c[\delta(f - f_c) + \delta(f + f_c)]$$
$$+ \tfrac{1}{4}\mu A_c[\delta(f - f_c - f_m) + \delta(f + f_c + f_m)]$$
$$+ \tfrac{1}{4}\mu A_c[\delta(f - f_c + f_m) + \delta(f + f_c - f_m)]$$

Thus the spectrum of an AM wave, for the special case of sinusoidal modulation, consists of delta functions at $\pm f_c$, $f_c \pm f_m$, and $-f_c \pm f_m$, as in Fig. 3.3*c*.

In practice, the AM wave $s(t)$ is a voltage or current wave. In either case, the average power delivered to a 1-ohm resistor by $s(t)$ is comprised of three components:

$$\text{Carrier power} = \tfrac{1}{2}A_c^2$$
$$\text{Upper side–frequency power} = \tfrac{1}{8}\mu^2 A_c^2$$
$$\text{Lower side–frequency power} = \tfrac{1}{8}\mu^2 A_c^2$$

For a load resistor R different from 1 ohm, which is usually the case in practice, the expressions for carrier power, upper side–frequency power, and lower side–frequency power are merely scaled by the factor $1/R$ or R, depending on whether the modulated wave $s(t)$ is a voltage or a current, respectively. In any case, the ratio of the total sideband power to the total power in the modulated wave is equal to $\mu^2/(2 + \mu^2)$, which depends only on the modulation factor μ.

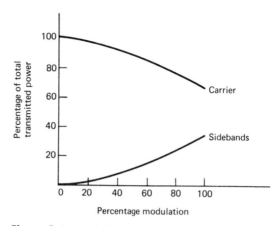

Figure 3.4 Variations of carrier power and total sideband power with percentage modulation.

If $\mu = 1$, that is, 100 percent modulation is used, the total power in the two side frequencies of the resulting AM wave is only one-third of the total power in the modulated wave.

Figure 3.4 shows the percentage of total power in both side frequencies and in the carrier plotted versus the percentage modulation. Note that when the percentage modulation is less than 20 percent, the power in one side frequency is less than 1 percent of the total power in the AM wave.

Switching Modulator

The generation of an AM wave may be accomplished using various devices; here we describe one such device called a *switching modulator.* Details of this modulator are shown in Fig. 3.5a, where it is assumed that the carrier wave $c(t)$ applied to the diode is large in amplitude, so that it swings right across the characteristic curve of the diode. We assume that the diode acts as an *ideal switch,* that is, it presents zero impedance when it is forward-biased [corresponding to $c(t) > 0$]. We may thus approximate the transfer characteristic of the diode–load resistor combination by a *piecewise-linear characteristic,* as shown in Fig. 3.5b. Accordingly, for an input voltage $v_1(t)$ consisting of the sum of the carrier and the message signal:

$$v_1(t) = A_c \cos(2\pi f_c t) + m(t) \qquad (3.8)$$

where $|m(t)| \ll A_c$, the resulting load voltage $v_2(t)$ is

$$v_2(t) \simeq \begin{cases} v_1(t), & c(t) > 0 \\ 0, & c(t) < 0 \end{cases} \qquad (3.9)$$

That is, the load voltage $v_2(t)$ varies periodically between the values $v_1(t)$ and zero at a rate equal to the carrier frequency f_c. In this way, by assuming a modulating wave that is weak compared with the carrier wave, we have effectively replaced the nonlinear behavior of the diode by an approximately equivalent piecewise-linear time-varying operation.

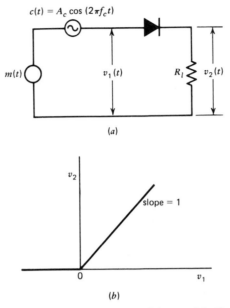

(a)

(b)

Figure 3.5 Switching modulator. (a) Circuit diagram. (b) Idealized input–output characteristic curve.

We may express Eq. (3.9) mathematically as

$$v_2(t) \simeq [A_c \cos(2\pi f_c t) + m(t)] g_{T_0}(t) \qquad (3.10)$$

where $g_{T_0}(t)$ is a periodic pulse train of duty cycle equal to one-half, and period $T_0 = 1/f_c$, as in Fig. 3.6. Representing this $g_{T_0}(t)$ by its Fourier series, we have

$$g_{T_0}(t) = \frac{1}{2} + \frac{2}{\pi} \sum_{n=1}^{\infty} \frac{(-1)^{n-1}}{2n-1} \cos[2\pi f_c t(2n-1)] \qquad (3.11)$$

Therefore, substituting Eq. (3.11) in (3.10), we find that the load voltage $v_2(t)$ consists of the sum of two components:

1. The component

$$\frac{A_c}{2}\left[1 + \frac{4}{\pi A_c} m(t)\right] \cos(2\pi f_c t)$$

which is the desired AM wave with amplitude sensitivity $k_a = 4/\pi A_c$. The switching modulator is therefore made more sensitive by reducing the carrier amplitude A_c; however, it must be maintained large enough to make the diode act like an ideal switch.

2. Unwanted components, the spectrum of which contains delta functions at $0, \pm 2f_c, \pm 4f_c$, and so on, and which occupy frequency intervals of width $2W$ centered at $0, \pm 3f_c, \pm 5f_c$, and so on, where W is the message bandwidth.

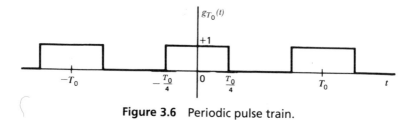

Figure 3.6 Periodic pulse train.

The unwanted terms are removed from the load voltage $v_2(t)$ by means of a band-pass filter with mid-band frequency f_c and bandwidth $2W$, provided that $f_c > 2W$. This latter condition ensures that the frequency separations between the desired AM wave and the unwanted components are large enough for the band-pass filter to suppress the unwanted components.

Envelope Detector

As mentioned in the introductory section, the process of *demodulation* is used to recover the original modulating wave from the incoming modulated wave; in effect, demodulation is the reverse of the modulation process. As with modulation, the demodulation of an AM wave can be accomplished using various devices; here, we describe a simple and yet highly effective device known as the *envelope detector*. Some version of this demodulator is used in almost all commercial AM radio receivers. For it to function properly, however, the AM wave has to be narrow-band, which requires that the carrier frequency be large compared to the message bandwidth. Moreover, the percentage modulation must be less than 100 percent.

An envelope detector of the series type is shown in Fig. 3.7a, which consists of a diode and a resistor–capacitor (RC) filter. The operation of this envelope detector is as follows. On a positive half-cycle of the input signal, the diode is forward-biased and the capacitor C charges up rapidly to the peak value of the input signal. When the input signal falls below this value, the diode becomes reverse-biased and the capacitor C discharges slowly through the load resistor R_l. The discharging process continues until the next positive half-cycle. When the input signal becomes greater than the voltage across the capacitor, the diode conducts again and the process is repeated. We assume that the diode is ideal, presenting resistance r_f to current flow in the forward-biased region and infinite resistance in the reverse-biased region. We further assume that the AM wave applied to the envelope detector is supplied by a voltage source of internal impedance R_s. The charging time constant $(r_f + R_s)C$ must be short compared with the carrier period $1/f_c$, that is,

$$(r_f + R_s)C \ll \frac{1}{f_c} \qquad (3.12)$$

so that the capacitor C charges rapidly and thereby follows the applied voltage up to the positive peak when the diode is conducting. On the other hand, the discharging time constant $R_l C$ must be long enough to ensure that the capacitor discharges slowly through the load resistor R_l between positive peaks of the car-

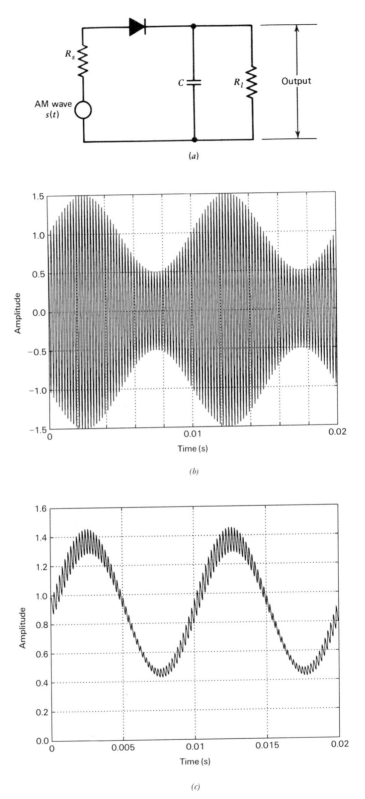

Figure 3.7 Envelope detector. (*a*) Circuit diagram.
(*b*) AM wave input. (*c*) Envelope detector output.

rier wave, but not so long that the capacitor voltage will not discharge at the maximum rate of change of the modulating wave, that is

$$\frac{1}{f_c} \ll R_l C \ll \frac{1}{W} \tag{3.13}$$

where W is the message bandwidth. The result is that the capacitor voltage or detector output is nearly the same as the envelope of the AM wave, as demonstrated next.

COMPUTER EXPERIMENT I   Sinusoidal Amplitude Modulation

Consider the sinusoidal AM wave shown in Fig. 3.7b, assuming 50 percent modulation. The envelope detector output is shown in Fig. 3.7c. This latter waveform is computed assuming that the diode is ideal, having a constant resistance r_f when forward-biased and infinite resistance when reverse-biased. The numerical values used in the computation of Fig. 3.7c are as follows:

Source resistance	$R_s = 75 \ \Omega$
Forward resistance	$r_f = 25 \ \Omega$
Load resistance	$R_l = 10 \ k\Omega$
Capacitance	$C = 0.01 \ \mu F$
Message bandwidth	$W = 1 \ kHz$
Carrier frequency	$f_c = 20 \ kHz$

Figure 3.7c shows that the envelope detector output contains a small ripple at the carrier frequency; this ripple is easily removed by low-pass filtering.

3.3   VIRTUES, LIMITATIONS, AND MODIFICATIONS OF AMPLITUDE MODULATION

Amplitude modulation is the oldest method of performing modulation. Its biggest virtue is the ease with which it is generated and reversed. Modulation is accomplished rather simply in the transmitter using a switching modulator (described earlier) or a square-law modulator (described in Problem 3.4). Demodulation is accomplished just as easily in the receiver using an envelope detector (described earlier) or a square-law detector (described in Problem 3.6). The net result is that an amplitude modulating system is relatively cheap to build, which is the reason that AM radio broadcasting has been popular for so long and is quite likely to remain so well into the future.

However, from Chapter 1 we recall that transmitted power and channel bandwidth are our two primary communication resources and they should be used efficiently. In this context, we find that the standard form of amplitude modulation defined in Eq. (3.2) suffers from two major limitations:

1. *Amplitude modulation is wasteful of power.* The carrier wave $c(t)$ is completely independent of the information-bearing signal or baseband signal $m(t)$. The transmission of the carrier wave therefore represents a waste of power, which means that in amplitude modulation only a fraction of the total transmitted power is actually affected by $m(t)$.

2. *Amplitude modulation is wasteful of bandwidth.* The upper and lower sidebands of an AM wave are uniquely related to each other by virtue of their symmetry about the carrier frequency; hence, given the amplitude and phase spectra of either sideband, we can uniquely determine the other. This means that insofar as the transmission of information is concerned, only one sideband is necessary, and the communication channel therefore needs to provide only the same bandwidth as the baseband signal. In light of this observation, amplitude modulation is wasteful of bandwidth as it requires a transmission bandwidth equal to twice the message bandwidth.

To overcome these limitations, we must make certain changes, which result in increased system complexity of the amplitude modulation process. In effect, we trade off system complexity for improved utilization of communication resources. Starting with amplitude modulation as the standard, we can distinguish three modified forms of amplitude modulation:

1. *Double sideband–suppressed carrier (DSB-SC) modulation,* in which the transmitted wave consists of only the upper and lower sidebands. Transmitted power is saved here through the suppression of the carrier wave, but the channel bandwidth requirement is the same as before (i.e., twice the message bandwidth).

2. *Vestigial sideband (VSB) modulation,* in which one sideband is passed almost completely and just a trace, or *vestige,* of the other sideband is retained. The required channel bandwidth is therefore in excess of the message bandwidth by an amount equal to the width of the vestigial sideband. This form of modulation is well suited for the transmission of wideband signals such as television signals that contain significant components at extremely low frequencies. In commercial television broadcasting, a sizable carrier is transmitted together with the modulated wave, which makes it possible to demodulate the incoming modulated signal by an envelope detector in the receiver and thereby simplify the receiver design.

3. *Single sideband (SSB) modulation,* in which the modulated wave consists only of the upper sideband or the lower sideband. The essential function of SSB modulation is therefore to translate the spectrum of the modulating signal (with or without inversion) to a new location in the frequency domain. Single sideband modulation is particularly suited for the transmission of voice signals by virtue of the *energy gap* that exists in the spectrum of voice signals between zero and a few hundred hertz. It is an optimum form of modulation in that it requires the minimum transmitted power and minimum channel bandwidth; its principal disadvantage is increased cost and complexity.

In Section 3.4 we discuss DSB-SC modulation, followed by discussions of VSB and SSB forms of modulation in later sections.

3.4   DOUBLE SIDEBAND–SUPPRESSED CARRIER MODULATION

Basically, *double sideband–suppressed carrier (DSB-SC) modulation* consists of the product of the message signal $m(t)$ and the carrier wave $c(t)$, as follows:

$$s(t) = c(t)\, m(t) \tag{3.14}$$
$$= A_c \cos(2\pi f_c t)\, m(t)$$

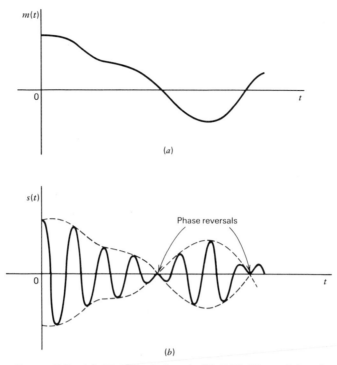

Figure 3.8 (a) Baseband signal. (b) DSB-SC modulated wave.

Consequently, the modulated signal $s(t)$ undergoes a *phase reversal* whenever the message signal $m(t)$ crosses zero, as indicated in Fig. 3.8b. The envelope of a DSB-SC modulated signal is therefore different from the message signal.

From Eq. (3.14), the Fourier transform of $s(t)$ is obtained as

$$S(f) = \frac{1}{2}A_c[M(f - f_c) + M(f + f_c)] \qquad (3.15)$$

For the case when the baseband signal $m(t)$ is limited to the interval $-W \leq f \leq W$, as in Fig. 3.9a, we thus find that the spectrum $S(f)$ of the DSB-SC wave $s(t)$ is as illustrated in Fig. 3.9b. Except for a change in scale factor, the modulation process simply *translates* the spectrum of the baseband signal by $\pm f_c$. Of course, the transmission bandwidth required by DSB-SC modulation is the same as that for amplitude modulation, namely, $2W$.

Ring Modulator[2]

One of the most useful product modulators, well suited for generating a DSB-SC wave, is the ring modulator shown in Fig. 3.10a. The four diodes form a ring in which they all point in the same way—hence the name. The diodes are controlled by a square-wave carrier $c(t)$ of frequency f_c, which is applied longitudinally by means of two center-tapped transformers. If the transformers are perfectly balanced and the diodes are identical, there is *no* leakage of the modulation (switching) frequency into the modulator output. To understand the operation of the circuit, we assume that the diodes have a constant forward

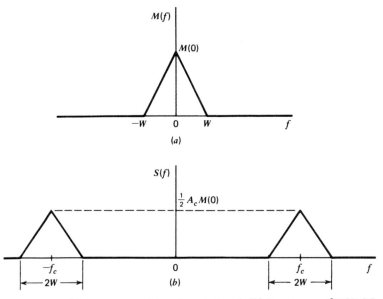

Figure 3.9 (a) Spectrum of baseband signal. (b) Spectrum of DSB-SC modulated wave.

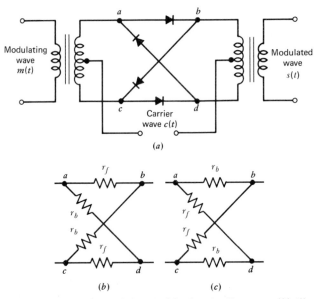

Figure 3.10 Ring modulator. (a) Circuit diagram. (b) Illustrating the condition when the outer diodes are switched on and the inner diodes are switched off. (c) Illustrating the condition when the outer diodes are switched off and the inner diodes are switched on.

resistance r_f when switched on and a constant backward resistance r_b when switched off and that they switch as the carrier wave $c(t)$ goes through zero. On one half-cycle of the carrier wave, the outer diodes are switched to their forward resistances r_f and the inner diodes are switched to their backward resistances r_b, as indicated in Fig. 3.10b. On the other half-cycle of the carrier wave, the diodes operate in the opposite condition, as shown in Fig. 3.10c. Typically, the terminating resistances at the input and output ends of the modulator are the same (assuming ideal 1 : 1 transformers). Under the conditions described herein, it is a straightforward matter to show that the output voltage in Fig. 3.10b has the same magnitude as the output voltage in Fig. 3.10c, but they have opposite polarity. In effect, the ring modulator acts as a *commutator*.

Figure 3.11c shows the idealized waveform of the modulated signal $s(t)$ produced by the ring modulator, assuming a sinusoidal modulating wave $m(t)$ as in Fig. 3.11a and a square carrier wave $c(t)$ as in Fig. 3.11b. Now, the square-wave carrier $c(t)$ can be represented by a Fourier series as follows:

$$c(t) = \frac{4}{\pi} \sum_{n=1}^{\infty} \frac{(-1)^{n-1}}{2n-1} \cos[2\pi f_c t(2n-1)] \qquad (3.16)$$

The ring modulator output is therefore

$$\begin{aligned} s(t) &= c(t)m(t) \\ &= \frac{4}{\pi} \sum_{n=1}^{\infty} \frac{(-1)^{n-1}}{2n-1} \cos[2\pi f_c t(2n-1)]m(t) \end{aligned} \qquad (3.17)$$

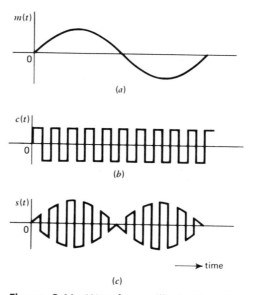

Figure 3.11 Waveforms illustrating the operation of the ring modulator for a sinusoidal modulating wave. (a) Modulating wave. (b) Square-wave carrier. (c) Modulated wave.

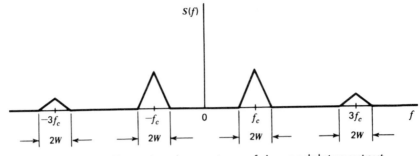

Figure 3.12   Illustrating the spectrum of ring modulator output.

We see that there is no output from the modulator at the carrier frequency; that is, the modulator output consists entirely of modulation products. The ring modulator is sometimes referred to as a *double-balanced modulator,* because it is balanced with respect to both the baseband signal and the square-wave carrier.

Assuming that $m(t)$ is limited to the frequency band $-W \leq f \leq W$, the spectrum of the modulator output consists of sidebands around each of the odd harmonics of the square-wave carrier $m(t)$, as illustrated in Fig. 3.12. Here it is assumed that $f_c > W$ so as to prevent *sideband overlap,* which arises when sidebands belonging to the adjacent harmonic frequency f_c and $3f_c$ overlap each other. Thus, provided that we have $f_c > W$, we may use a band-pass filter of mid-band frequency f_c and bandwidth $2W$ to select the desired pair of sidebands around the carrier frequency f_c. Accordingly, the circuitry needed for the generation of a DSB-SC modulated wave consists of a ring modulator followed by a band-pass filter.

Coherent Detection

The baseband signal $m(t)$ can be uniquely recovered from a DSB-SC wave $s(t)$ by first multiplying $s(t)$ with a locally generated sinusoidal wave and then low-pass filtering the product, as in Fig. 3.13. It is assumed that the local oscillator signal is exactly coherent or synchronized, in both frequency and phase, with the carrier wave $c(t)$ used in the product modulator to generate $s(t)$. This method of demodulation is known as *coherent detection* or *synchronous demodulation.*

It is instructive to derive coherent detection as a special case of the more general demodulation process using a local oscillator signal of the same fre-

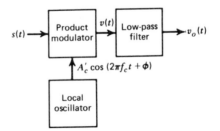

Figure 3.13   Coherent detection of DSB-SC modulated wave.

quency but arbitrary phase difference ϕ, measured with respect to the carrier wave $c(t)$. Thus, denoting the local oscillator signal by $A'_c \cos(2\pi f_c t + \phi)$, and using Eq. (3.14) for the DSB-SC wave $s(t)$, we find that the product modulator output in Fig. 3.13 is

$$
\begin{aligned}
v(t) &= A'_c \cos(2\pi f_c t + \phi) s(t) \\
&= A_c A'_c \cos(2\pi f_c t) \cos(2\pi f_c t + \phi) m(t) \\
&= \frac{1}{2} A_c A'_c \cos(4\pi f_c t + \phi) m(t) + \frac{1}{2} A_c A'_c \cos\phi \, m(t)
\end{aligned}
\tag{3.18}
$$

The first term in Eq. (3.18) represents a DSB-SC modulated signal with a carrier frequency $2f_c$, whereas the second term is proportional to the baseband signal $m(t)$. This is further illustrated by the spectrum $V(f)$ shown in Fig. 3.14, where it is assumed that the baseband signal $m(t)$ is limited to the interval $-W \leq f \leq W$. It is therefore apparent that the first term in Eq. (3.18) is removed by the low-pass filter in Fig. 3.13, provided that the cut-off frequency of this filter is greater than W but less than $2f_c - W$. This is satisfied by choosing $f_c > W$. At the filter output we then obtain a signal given by

$$
v_o(t) = \frac{1}{2} A_c A'_c \cos\phi \, m(t)
\tag{3.19}
$$

The demodulated signal $v_o(t)$ is therefore proportional to $m(t)$ when the phase error ϕ is a constant. The amplitude of this demodulated signal is maximum when $\phi = 0$, and it is minimum (zero) when $\phi = \pm \pi/2$. The zero demodulated signal, which occurs for $\phi = \pm \pi/2$, represents the *quadrature null effect* of the coherent detector. Thus the phase error ϕ in the local oscillator causes the detector output to be attenuated by a factor equal to $\cos\phi$. As long as the phase error ϕ is constant, the detector output provides an undistorted version of the original baseband signal $m(t)$. In practice, however, we usually find that the phase error ϕ varies randomly with time, due to random variations in the communication channel. The result is that at the detector output, the multiplying factor $\cos\phi$ also varies randomly with time, which is obviously undesirable. Therefore, provision must be made in the system to maintain the local oscillator in the

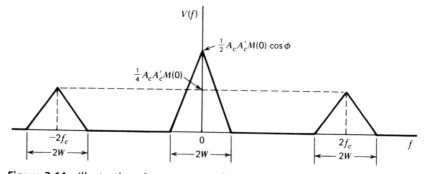

Figure 3.14 Illustrating the spectrum of a product modulator output with a DSB-SC modulated wave as input.

receiver in perfect synchronism, in both frequency and phase, with the carrier wave used to generate the DSB-SC modulated signal in the transmitter. The resulting system complexity is the price that must be paid for suppressing the carrier wave to save transmitter power.

Costas Receiver

One method of obtaining a practical synchronous receiver system, suitable for demodulating DSB-SC waves, is to use the *Costas receiver*[3] shown in Fig. 3.15. This receiver consists of two coherent detectors supplied with the same input signal, namely, the incoming DSB-SC wave $A_c \cos(2\pi f_c t) m(t)$, but with individual local oscillator signals that are in phase quadrature with respect to each other. The frequency of the local oscillator is adjusted to be the same as the carrier frequency f_c, which is assumed known *a priori*. The detector in the upper path is referred to as the *in-phase coherent detector* or *I-channel*, and that in the lower path is referred to as the *quadrature-phase coherent detector* or *Q-channel*. These two detectors are coupled together to form a *negative feedback* system designed in such a way as to maintain the local oscillator synchronous with the carrier wave.

To understand the operation of this receiver, suppose that the local oscillator signal is of the same phase as the carrier wave $A_c \cos(2\pi f_c t)$ used to generate the incoming DSB-SC wave. Under these conditions, we find that the *I*-channel output contains the desired demodulated signal $m(t)$, whereas the *Q*-channel output is zero due to the quadrature null effect of the *Q*-channel. Suppose next that the

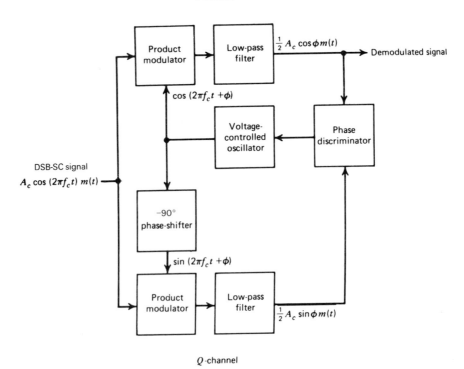

Figure 3.15 Costas receiver.

local oscillator phase drifts from its proper value by a small angle ϕ radians. The I-channel output will remain essentially unchanged, but there will now be some signal appearing at the Q-channel output, which is proportional to $\sin\phi \simeq \phi$ for small ϕ. This Q-channel output will have the same polarity as the I-channel output for one direction of local oscillator phase drift and opposite polarity for the opposite direction of local oscillator phase drift. Thus, by combining the I- and Q-channel outputs in a *phase discriminator* (which consists of a multiplier followed by a low-pass filter), as shown in Fig. 3.15, a dc control signal is obtained that automatically corrects for local phase errors in the *voltage-controlled oscillator*.

It is apparent that phase control in the Costas receiver ceases with modulation and that phase-lock has to be reestablished with the reappearance of modulation. This is not a serious problem when receiving voice transmission, because the lock-up process normally occurs so rapidly that no distortion is perceptible.

Quadrature-Carrier Multiplexing

The quadrature null effect of the coherent detector may also be put to good use in the construction of the so-called *quadrature-carrier multiplexing* or *quadrature-amplitude modulation* (QAM). This scheme enables two DSB-SC modulated waves (resulting from the application of two physically *independent* message signals) to occupy the same channel bandwidth, and yet it allows for the separation of the two message signals at the receiver output. It is therefore a *bandwidth-conservation scheme*.

A block diagram of the quadrature-carrier multiplexing system is shown in Fig. 3.16. The transmitter part of the system, shown in Fig. 3.16a, involves the use of two separate product modulators that are supplied with two carrier waves of the same frequency but differing in phase by -90 degrees. The transmitted signal $s(t)$ consists of the sum of these two product modulator outputs, as shown by

$$s(t) = A_c m_1(t)\cos(2\pi f_c t) + A_c m_2(t)\sin(2\pi f_c t) \qquad (3.20)$$

where $m_1(t)$ and $m_2(t)$ denote the two different message signals applied to the product modulators. Thus $s(t)$ occupies a channel bandwidth of $2W$ centered at the carrier frequency f_c, where W is the message bandwidth of $m_1(t)$ or $m_2(t)$. According to Eq. (3.20), we may view $A_c m_1(t)$ as the in-phase component of the multiplexed band-pass signal $s(t)$ and $-A_c m_2(t)$ as its quadrature component.

The receiver part of the system is shown in Fig. 3.16b. The multiplexed signal $s(t)$ is applied simultaneously to two separate coherent detectors that are supplied with two local carriers of the same frequency, but differing in phase by -90 degrees. The output of the top detector is $\frac{1}{2}A_c' m_1(t)$, whereas the output of the bottom detector is $\frac{1}{2}A_c' m_2(t)$. For the system to operate satisfactorily, it is important to maintain the correct phase and frequency relationships between the local oscillators used in the transmitter and receiver parts of the system.

To maintain this synchronization, we may use a Costas receiver described above. Another commonly used method is to send a *pilot signal* outside the pass-band of the modulated signal. In the latter method, the pilot signal typically consists of a low-power sinusoidal tone whose frequency and phase are related to the carrier wave $c(t)$; at the receiver, the pilot signal is extracted by means of a suitably tuned circuit and then translated to the correct frequency for use in the coherent detector.

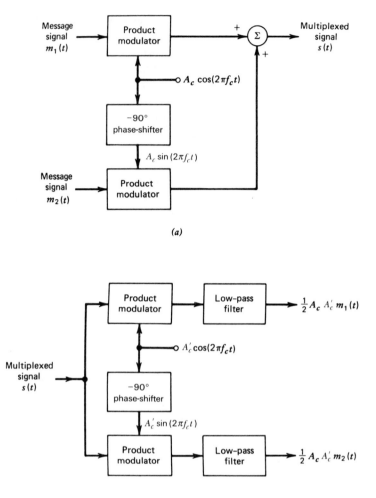

Figure 3.16 Quadrature–carrier multiplexing system. (*a*) Transmitter. (*b*) Receiver.

3.5   FILTERING OF SIDEBANDS

The next issue we wish to discuss is how to process a DSB-SC modulated signal so as to generate a vestigial sideband (VSB) or single sideband (SSB) modulated signal. An intuitively satisfying method of achieving this requirement is the *frequency-discrimination method* that involves the use of an appropriate filter following a product modulator responsible for the generation of the DSB-SC modulated signal. Naturally, the exact specification of the filter depends on the desired type of modulation.

Consider then the circuit described in Fig. 3.17*a*, where $u(t)$ denotes the product modulator output, as shown by

$$u(t) = A_c m(t) \cos(2\pi f_c t)$$

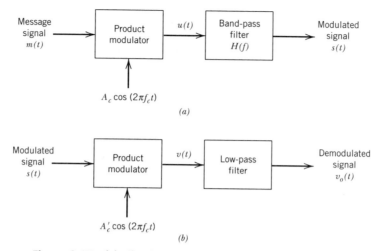

Figure 3.17 (a) Filtering scheme for processing sidebands. (b) Coherent detector for recovering the message signal.

Let $H(f)$ denote the transfer function of the filter following the product modulator. The spectrum of the modulated signal $s(t)$ produced by passing $u(t)$ through the filter is given by

$$S(f) = U(f)H(f)$$

$$= \frac{A_c}{2}\left[M(f - f_c) + M(f + f_c)\right]H(f) \tag{3.21}$$

where $M(f)$ is the Fourier transform of the message signal $m(t)$. The problem we wish to address is to determine the particular $H(f)$ required to produce a modulated signal $s(t)$ with desired spectral characteristics, such that the original message signal $m(t)$ may be recovered from $s(t)$ by coherent detection.

The first step in the coherent detection process involves multiplying the modulated signal $s(t)$ by a locally generated sinusoidal wave $A'_c\cos(2\pi f_c t)$, which is synchronous with the carrier wave $A_c\cos(2\pi f_c t)$, in both frequency and phase as in Fig. 3.17b. We may thus write

$$v(t) = A'_c\cos(2\pi f_c t)s(t)$$

Transforming this relation into the frequency domain gives the Fourier transform of $v(t)$ as

$$V(f) = \frac{A'_c}{2}[S(f - f_c) + S(f + f_c)] \tag{3.22}$$

Therefore, substitution of Eq. (3.21) in (3.22) yields

$$V(f) = \frac{A_c A'_c}{4}M(f)[H(f - f_c) + H(f + f_c)]$$

$$+ \frac{A_c A'_c}{4}[M(f - 2f_c)H(f - f_c) + M(f + 2f_c)H(f + f_c)] \tag{3.23}$$

The high-frequency components of $v(t)$ represented by the second term in Eq. (3.23) are removed by the low-pass filter in Fig. 3.17b to produce an output $v_o(t)$, the spectrum of which is given by the remaining components:

$$V_o(f) = \frac{A_c A_c'}{4} M(f)[H(f - f_c) + H(f + f_c)] \qquad (3.24)$$

For a distortionless reproduction of the original baseband signal $m(t)$ at the coherent detector output, we require $V_o(f)$ to be a scaled version of $M(f)$. This means, therefore, that the transfer function $H(f)$ must satisfy the condition

$$H(f - f_c) + H(f + f_c) = 2H(f_c) \qquad (3.25)$$

where $H(f_c)$, the value of $H(f)$ at $f = f_c$, is a constant. When the message (baseband) spectrum $M(f)$ is zero outside the frequency range $-W \leq f \leq W$, we need only satisfy Eq. (3.25) for values of f in this interval. Also, to simplify the exposition, we set $H(f_c) = 1/2$. We thus require that $H(f)$ satisfies the condition:

$$H(f - f_c) + H(f + f_c) = 1, \qquad -W \leq f \leq W \qquad (3.26)$$

There is a great deal of flexibility in the selection of $H(f)$ to satisfy this condition, as discussed later in Sections 3.6 and 3.7. In any event, under the condition described in Eq. (3.26), we find from Eq. (3.24) that the coherent detector output in Fig. 3.17b is given by

$$v_o(t) = \frac{A_c A_c'}{2} m(t) \qquad (3.27)$$

Equation (3.21) defines the spectrum of the modulated signal $s(t)$. Recognizing that $s(t)$ is a band-pass signal, we may formulate its time-domain description in terms of in-phase and quadrature components, using the passband signaling method described in Section 2.12. In particular, $s(t)$ may be expressed in the canonical form

$$s(t) = s_I(t)\cos(2\pi f_c t) - s_Q(t)\sin(2\pi f_c t) \qquad (3.28)$$

where $s_I(t)$ is the in-phase component of $s(t)$, and $s_Q(t)$ is its quadrature component. To determine $s_I(t)$, we note that its Fourier transform is related to the Fourier transform of $s(t)$ as follows (see Problem 2.29):

$$S_I(f) = \begin{cases} S(f - f_c) + S(f + f_c), & -W \leq f \leq W \\ 0, & \text{elsewhere} \end{cases} \qquad (3.29)$$

Hence, substituting Eq. (3.21) in (3.29), we find that the Fourier transform of $s_I(t)$ is given by

$$\begin{aligned} S_I(f) &= \frac{1}{2} A_c M(f)[H(f - f_c) + H(f + f_c)] \\ &= \frac{1}{2} A_c M(f), \qquad -W \leq f \leq W \end{aligned} \qquad (3.30)$$

where, in the second line, we have made use of the condition in Eq. (3.26) imposed on $H(f)$. From Eq. (3.30) we readily see that the in-phase component of the modulated signal $s(t)$ is defined by

$$s_I(t) = \frac{1}{2}A_c m(t) \qquad (3.31)$$

which, except for a scaling factor, is the same as the original message signal $m(t)$.

To determine the quadrature component $s_Q(t)$ of the modulated signal $s(t)$, we recognize that its Fourier transform is defined in terms of the Fourier transform of $s(t)$ as follows (see Problem 2.29):

$$S_Q(f) = \begin{cases} j[S(f - f_c) - S(f + f_c)], & -W \leq f \leq W \\ 0, & \text{elsewhere} \end{cases} \qquad (3.32)$$

Therefore, substituting Eq. (3.21) in (3.32), we get

$$S_Q(f) = \frac{j}{2}A_c M(f)[H(f - f_c) - H(f + f_c)] \qquad (3.33)$$

This equation suggests that we may generate $s_Q(t)$, except for a scaling factor, by passing the message signal $m(t)$ through a new filter whose transfer function is related to that of the filter in Fig. 3.17a as follows:

$$H_Q(f) = j[H(f - f_c) - H(f + f_c)], \qquad -W \leq f \leq W \qquad (3.34)$$

Let $m'(t)$ denote the output of this filter produced in response to the input $m(t)$. Hence, we may express the quadrature component of the modulated signal $s(t)$ as

$$s_Q(t) = \frac{1}{2}A_c m'(t) \qquad (3.35)$$

Accordingly, substituting Eqs. (3.31) and (3.35) in (3.28), we find that $s(t)$ may be written in the canonical form

$$m(t) = \frac{1}{2}A_c m(t)\cos(2\pi f_c t) - \frac{1}{2}A_c m'(t)\sin(2\pi f_c t) \qquad (3.36)$$

Equation (3.36), except for the scaling factor $1/2$, suggests the circuit of Fig. 3.18 as the basis of a *phase discrimination method* for the generation of the modulated wave $s(t)$. There are two important points to note here:

1. The in-phase component $s_I(t)$ is completely independent of the transfer function $H(f)$ of the band-pass filter involved in the generation of the modulated wave $s(t)$ in Fig. 3.17a, so long as it satisfies the condition of Eq. (3.26).
2. The spectral modification attributed to the transfer function $H(f)$ is confined solely to the quadrature component $s_Q(t)$.

The role of the quadrature component is merely to interfere with the in-phase component, so as to reduce or eliminate power in one of the sidebands of the modulated signal $s(t)$, depending on the application of interest.

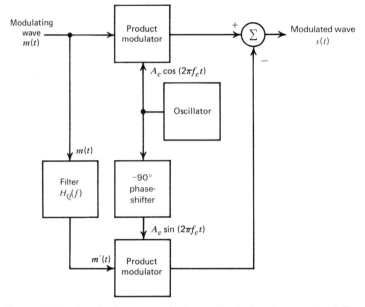

Figure 3.18 Block diagram of phase discrimination method for processing sidebands.

The discussion presented thus far has been of a fairly general nature. In the next two sections, we consider the generation of VSB and SSB modulated signals as special cases of the modulated signal $s(t)$ defined in Eq. (3.36).

3.6   VESTIGIAL SIDEBAND MODULATION[4]

Assuming that the requirement is to generate a *vestigial sideband* (VSB) modulated signal containing a vestige of the lower sideband, we find that Eq. (3.26) is satisfied by using a band-pass filter whose transfer function $H(f)$ is as shown in Fig. 3.19; to simplify matters, only the response for positive frequencies is

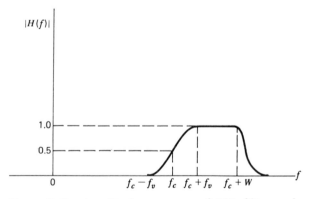

Figure 3.19 Amplitude response of VSB filter; only positive-frequency portion is shown.

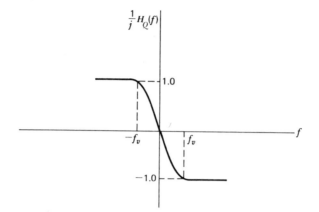

Figure 3.20 Frequency response of filter for producing the quadrature component of the VSB wave.

shown here. This frequency response is normalized, so that $|H(f)|$ is one half at the carrier frequency f_c. The important feature to note, however, is that the cutoff portion of the frequency response around the carrier frequency f_c exhibits *odd symmetry*. That is to say, inside the transition interval $f_c - f_v \leq |f| \leq f_c + f_v$ the sum of the values of $|H(f)|$ at any two frequencies equally displaced above and below f_c is unity; f_v is the width of the vestigial sideband. Note also that outside the frequency band of interest (i.e., $|f| > f_c + W$), the transfer function $H(f)$ may have an arbitrary specification.

The corresponding frequency response of the filter producing the quadrature component of the VSB modulated signal in accordance with Eq. (3.34) is as shown in Fig. 3.20.

Figures 3.19 and 3.20 apply to a VSB modulated signal containing a vestige of the lower sideband. For a VSB modulated signal containing a vestige of the upper sideband, we have similar results except for the following differences: The upper cutoff portion of $H(f)$ is controlled to exhibit odd symmetry around the carrier frequency f_c, whereas the lower cutoff portion is arbitrary. This has the effect of replacing the minus sign at the summing junction at the output of Fig. 3.18 with a plus sign.

Television Signals

A discussion of vestigial sideband modulation would be incomplete without a mention of its role in commercial television (TV) broadcasting.[5] The exact details of the modulation format used to transmit the video signal characterizing a TV system are influenced by two factors:

1. The video signal exhibits a large bandwidth and significant low-frequency content, which suggest the use of vestigial sideband modulation.
2. The circuitry used for demodulation in the receiver should be simple and therefore cheap; this suggests the use of envelope detection, which requires the addition of a carrier to the VSB modulated wave.

With regard to point 1, however, it should be stressed that although there is indeed a basic desire to conserve bandwidth, in commercial TV broadcasting the

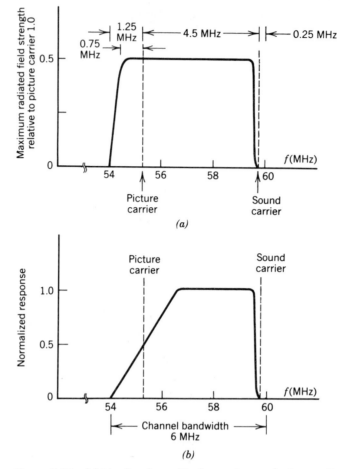

Figure 3.21 (a) Idealized amplitude spectrum of a transmitted TV signal. (b) Amplitude response of VSB shaping filter in the receiver.

transmitted signal is not quite VSB modulated. The reason is that at the transmitter the power levels are high, with the result that it would be expensive to rigidly control the filtering of sidebands. Instead, a *VSB filter* is inserted in each receiver where the power levels are low. The overall performance is the same as conventional vestigial-sideband modulation, except for some wasted power and bandwidth. These remarks are illustrated in Fig. 3.21. In particular, Fig. 3.21a shows the idealized spectrum of a transmitted TV signal. The upper sideband, 25 percent of the lower sideband, and the picture carrier are transmitted. The frequency response of the VSB filter used to do the required spectrum shaping in the receiver is shown in Fig. 3.21b.

The channel bandwidth used for TV broadcasting in North America is 6 MHz, as indicated in Fig. 3.21b. This channel bandwidth not only accommodates the bandwidth requirement of the VSB modulated video signal but also provides for the accompanying sound signal that modulates a carrier of its own. The values presented on the frequency axis in Figs. 3.21a and 3.21b pertain to a specific TV channel. According to this figure, the picture carrier frequency is at 55.25 MHz, and the sound carrier frequency is at 59.75 MHz. Note, however, that the information content of the TV signal lies in a *baseband spectrum* extending from 1.25 MHz below the picture carrier to 4.5 MHz above it.

With regard to point 2, the use of envelope detection (applied to a VSB modulated wave plus carrier) produces *waveform distortion* in the video signal recovered at the detector output. The distortion is produced by the quadrature component of the VSB modulated wave; this issue is discussed next.

Waveform Distortion

The use of the time-domain description given in Eq. (3.36) enables the determination of the waveform distortion caused by the envelope detector. Specifically, adding the carrier component $A_c \cos(2\pi f_c t)$ to Eq. (3.36), the latter being scaled by a factor k_a, modifies the modulated signal applied to the envelope detector input as

$$s(t) = A_c \left[1 + \frac{1}{2}k_a m(t) \right] \cos(2\pi f_c t) - \frac{1}{2}k_a A_c m'(t) \sin(2\pi f_c t) \quad (3.37)$$

where the constant k_a determines the percentage modulation. The envelope detector output, denoted by $a(t)$, is therefore

$$
\begin{aligned}
a(t) &= A_c \left\{ \left[1 + \frac{1}{2}k_a m(t) \right]^2 + \left[\frac{1}{2}k_a m'(t) \right]^2 \right\}^{1/2} \\
&= A_c \left[1 + \frac{1}{2}k_a m(t) \right] \left\{ 1 + \left[\frac{\frac{1}{2}k_a m'(t)}{1 + \frac{1}{2}k_a m(t)} \right]^2 \right\}^{1/2}
\end{aligned}
\quad (3.38)
$$

Equation (3.38) indicates that the distortion is contributed by $m'(t)$, which is responsible for the quadrature component of the incoming VSB modulated signal. This distortion can be reduced by using two methods:

- By reducing the percentage modulation to reduce the amplitude sensitivity k_a.
- By increasing the width of the vestigial sideband to reduce $m'(t)$.

Both methods are in fact used in practice. In commercial TV broadcasting, the width of the vestigial sideband (which is about 0.75 MHz or one-sixth of a full sideband) is determined to keep the distortion due to $m'(t)$ within tolerable limits when the percentage modulation is nearly 100.

3.7   SINGLE SIDEBAND MODULATION

Consider next the generation of a SSB modulated signal containing the upper sideband only. From a practical viewpoint, the most severe requirement of SSB generation usually arises from the unwanted sideband, the nearest frequency component of which is separated from the desired sideband by twice the lowest

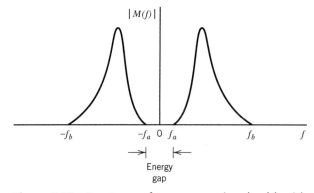

Figure 3.22 Spectrum of a message signal $m(t)$ with an energy gap centered around the origin.

frequency component of the message (modulating) signal. The implication here is that for the generation of an SSB modulated signal to be possible, the message spectrum must have an energy gap centered at the origin, as illustrated in Fig. 3.22. This requirement is naturally satisfied by voice signals, whose energy gap is about 600 Hz wide (i.e., it extends from -300 to $+300$ Hz). We may then generate an SSB signal containing the upper sideband, say, by using a band-pass filter, the frequency response of which is ideally related to the carrier frequency f_c as shown in Fig. 3.23a. Thus, given the message spectrum defined in Fig. 3.22,

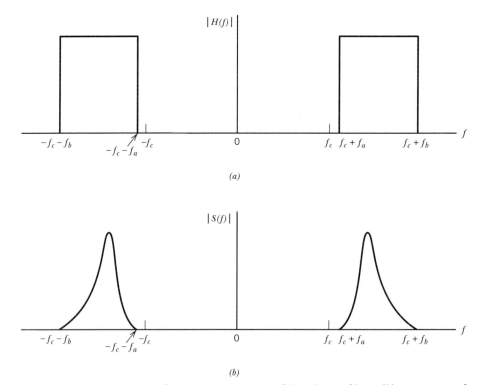

Figure 3.23 (a) Idealized frequency response of band-pass filter. (b) Spectrum of SSB signal containing the upper sideband.

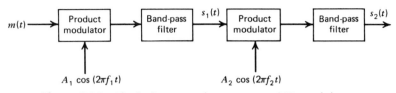

Figure 3.24 Block diagram of a two-stage SSB modulator.

we find that the corresponding spectrum of the SSB signal is as shown in Fig. 3.23*b*.

In designing the band-pass filter in Fig. 3.17*a* for the generation of an SSB modulated wave, we must satisfy three basic requirements:

- The desired sideband lies inside the passband of the filter.
- The unwanted sideband lies inside the stopband of the filter.
- The filter's transition band, separating the passband from the stopband, is twice the lowest frequency component of the message signal.

This kind of frequency discrimination usually requires the use of highly selective filters, which can only be realized in practice by means of crystal resonators.[6]

When it is necessary to generate an SSB signal occupying a frequency band that is much higher than that of the baseband signal (e.g., translating a voice signal to the high-frequency region of the radio spectrum), it becomes very difficult to design an appropriate filter that will pass the desired sideband and reject the other, using the simple arrangement of Fig. 3.17*a*. In such a situation it is necessary to resort to a multiple-modulation process so as to ease the filtering requirement. This approach, involving two stages of modulation, is illustrated in Fig. 3.24. The SSB signal $s_1(t)$ at the first band-pass filter output is used as the modulating wave for the second product modulator, which produces a DSB-SC signal with a spectrum that is symmetrically spaced about the second carrier frequency f_2. The frequency separation between the sidebands of this DSB-SC signal is effectively twice the first carrier frequency f_1, thereby making it relatively easy to remove the unwanted sideband by the second band-pass filter, and thereby generate the SSB signal $s_2(t)$.

Time-Domain Description of SSB Modulated Signal

The issue to be considered next is the time-domain description of an SSB modulated signal $s(t)$. Given the idealized frequency response of Fig. 3.23*a* as the description of $H(f)$, we find from Eq. (3.34) that the corresponding description of $H_Q(f)$ responsible for the generation of the quadrature component $s_Q(t)$ is as shown in Fig. 3.25. Using the definition of the signum function $\mathrm{sgn}(f)$, we readily see from Fig. 3.25 that

$$H_Q(f) = -j\,\mathrm{sgn}(f) \tag{3.39}$$

which is recognized as the transfer function of the Hilbert transformer (see Section 2.10). In other words, the $m'(t)$ in Eq. (3.36) is exactly the *Hilbert transform* of the original message signal $m(t)$. Thus, using $\hat{m}(t)$ to denote this Hilbert transform, we may formally describe an SSB modulated signal containing only

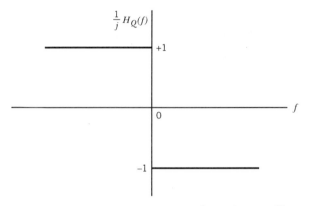

Figure 3.25 Frequency response of quadrature filter for single sideband modulation containing the upper sideband.

the upper sideband as follows:

$$s(t) = \frac{1}{2}A_c m(t) \cos(2\pi f_c t) - \frac{1}{2}A_c \hat{m}(t) \sin(2\pi f_c t) \qquad (3.40)$$

This means that the filter responsible for the generation of the quadrature component of $s(t)$ in the modulator of Fig. 3.18 consists of a Hilbert transformer. Also, if we want to generate an SSB modulated signal containing only the lower sideband, we simply replace the minus sign at the summing junction at the output of Fig. 3.18 with a plus sign. An SSB modulator based on the scheme shown in Fig. 3.18 in the manner described here is called the *Hartley modulator*.

Demodulation of SSB Signals

The demodulated signal $v_o(t)$ defined in Eq. (3.27) assumes perfect synchronism between the oscillator in the coherent detector of Fig. 3.17*b* and the oscillator in the modulator of Fig. 3.17*a*, both in phase and in frequency. This can be provided by one of two methods:

- Transmitting a low-power pilot carrier in addition to the selected sideband, or
- Using, in the receiver, a highly stable oscillator tuned to the same frequency as the carrier frequency.

In the latter method, it is inevitable that there would be some *phase error* ϕ in the local oscillator signal with respect to the carrier wave used to generate the incoming SSB modulated wave $s(t)$. Denoting the local oscillator signal by $A_c' \cos(2\pi f_c t + \phi)$, we find that (after some straightforward steps) the resulting demodulated signal is given by (for the case when the upper sideband only is transmitted)

$$v_o(t) = \frac{1}{4}A_c A_c'[m(t)\cos\phi + \hat{m}(t)\sin\phi] \qquad (3.41)$$

Unlike the idealized coherent detection process described in Eq. (3.27), the demodulated signal $v_o(t)$ of Eq. (3.41) contains an unwanted component proportional to $\hat{m}(t) \sin\phi$, which cannot be removed by filtering. This unwanted component appears as a phase distortion. To show this, we take the Fourier transform of $v_o(t)$ in Eq. (3.41) to obtain

$$V_o(f) = \frac{1}{4} A_c A_c' [M(f) \cos\phi + \hat{M}(f) \sin\phi] \tag{3.42}$$

But from the definition of the Hilbert transform $\hat{m}(t)$, we know that the Fourier transform of $\hat{m}(t)$ is related to the Fourier transform of the original signal $m(t)$ by

$$\hat{M}(f) = -j \operatorname{sgn}(f) M(f) \tag{3.43}$$

Therefore, substituting Eq. (3.43) in (3.42) and then simplifying, we get

$$V_o(f) = \begin{cases} \dfrac{1}{4} A_c A_c' M(f) \exp(-j\phi), & f > 0 \\[2mm] \dfrac{1}{4} A_c A_c' M(f) \exp(j\phi), & f < 0 \end{cases} \tag{3.44}$$

Thus the phase error ϕ in the local oscillator output results in *phase distortion*, where each frequency component of the original message signal $m(t)$ undergoes a constant phase shift ϕ in the course of demodulation at the receiver. This phase distortion may be tolerated with voice communications, because the human ear is relatively insensitive to phase distortion; the presence of phase distortion gives rise to a Donald Duck voice effect. In the transmission of music and video signals, on the other hand, the presence of phase distortion in the form of a constant phase difference is utterly unacceptable.

3.8 FREQUENCY TRANSLATION

The basic operation involved in single sideband modulation is in fact a form of *frequency translation*, which is why single sideband modulation is sometimes referred to as *frequency changing, mixing,* or *heterodyning*. This operation is clearly illustrated in the spectrum of the signal shown in Fig. 3.23b compared to that of the original message signal in Fig. 3.22. Specifically, we see that a message spectrum occupying the band from f_a to f_b for positive frequencies in Fig. 3.22 is shifted upward by an amount equal to the carrier frequency f_c in Fig. 3.23b; the message spectrum for negative frequencies is translated downward in a symmetric fashion.

The idea of frequency translation described herein may be generalized as follows. Suppose that we have a modulated wave $s_1(t)$ whose spectrum is centered on a carrier frequency f_1, and the requirement is to translate it upward in frequency such that its carrier frequency is changed from f_1 to a new value f_2. This requirement may be accomplished using the *mixer* shown in Fig. 3.26, which is similar to the scheme of Fig. 3.17a. Specifically, the *mixer* is a device that consists

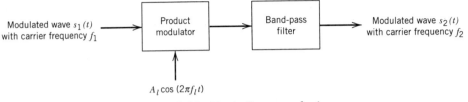

Figure 3.26   Block diagram of mixer.

of a product modulator followed by a band-pass filter. The band-pass filter is designed to have a bandwidth equal to that of the modulated signal $s_1(t)$ used as input. The key issue to be resolved is the frequency of the local oscillator connected to the product modulator. Let f_l denote this frequency. Due to the frequency translation performed by the mixer, the carrier frequency f_1 of the incoming modulated signal is changed by an amount equal to f_l; hence, we may set

$$f_2 = f_1 + f_l$$

Solving for f_l, we thus have

$$f_l = f_2 - f_1$$

This relation assumes that $f_2 > f_1$, in which case the carrier frequency is translated *upward*. If, on the other hand, we have $f_1 > f_2$, the carrier frequency is translated *downward,* for which the corresponding frequency of the local oscillator is

$$f_l = f_1 - f_2$$

It is important to note that mixing is a linear operation. Accordingly, the relation of the sidebands of the incoming modulated wave to the carrier is completely preserved at the mixer output.

3.9   FREQUENCY-DIVISION MULTIPLEXING[7]

Another important signal processing operation is *multiplexing,* whereby a number of independent signals can be combined into a composite signal suitable for transmission over a common channel. Voice frequencies transmitted over telephone systems, for example, range from 300 to 3100 Hz. To transmit a number of these signals over the same channel, the signals must be kept apart so that they do not interfere with each other, and thus they can be separated at the receiving end. This is accomplished by separating the signals either in frequency or in time. The technique of separating the signals in frequency is referred to as *frequency-division multiplexing* (FDM), whereas the technique of separating the signals in time is called *time-division multiplexing* (TDM). In this section, we discuss FDM systems, and TDM systems are discussed in Chapter 6.

A block diagram of an FDM system is shown in Fig. 3.27. The incoming

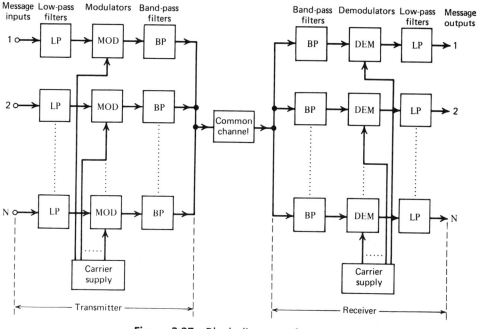

Figure 3.27 Block diagram of FDM system.

message signals are assumed to be of the low-pass type, but their spectra do not necessarily have nonzero values all the way down to zero frequency. Following each signal input, we have shown a low-pass filter, which is designed to remove high-frequency components that do not contribute significantly to signal representation but are capable of disturbing other message signals that share the common channel. These low-pass filters may be omitted only if the input signals are sufficiently band limited initially. The filtered signals are applied to modulators that shift the frequency ranges of the signals so as to occupy mutually exclusive frequency intervals. The necessary carrier frequencies needed to perform these frequency translations are obained from a carrier supply. For the modulation, we may use any one of the methods described in previous sections of this chapter. However, the most widely used method of modulation in frequency-division multiplexing is single sideband modulation which, in the case of voice signals, requires a bandwidth that is approximately equal to that of the original voice signal. In practice, each voice input is usually assigned a bandwidth of 4 kHz. The band-pass filters following the modulators are used to restrict the band of each modulated wave to its prescribed range. The resulting band-pass filter outputs are next combined in parallel to form the input to the common channel. At the receiving terminal, a bank of band-pass filters, with their inputs connected in parallel, is used to separate the message signals on a frequency-occupancy basis. Finally, the original message signals are recovered by individual demodulators. Note that the FDM system shown in Fig. 3.27 operates in only one direction. To provide for two-way transmission, as in telephony for example, we have to completely duplicate the multiplexing facilities, with the components connected in reverse order and with the signal waves proceeding from right to left.

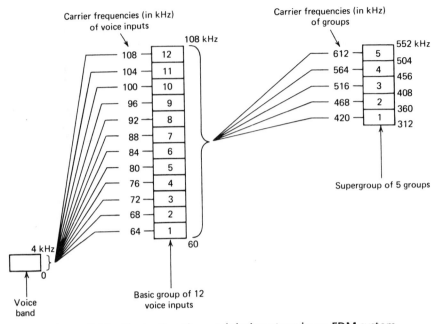

Figure 3.28   Illustrating the modulation steps in an FDM system.

EXAMPLE 2

The practical implementation of an FDM system usually involves many steps of modulation and demodulation, as illustrated in Fig. 3.28. The first multiplexing step combines 12 voice inputs into a *basic group*, which is formed by having the nth input modulate a carrier at frequency $f_c = 60 + 4n$ kHz, where $n = 1$, $2, \ldots, 12$. The lower sidebands are then selected by band-pass filtering and combined to form a group of 12 lower sidebands (one for each voice input). Thus the basic group occupies the frequency band 60–108 kHz. The next step in the FDM hierarchy involves the combination of five basic groups into a *supergroup*. This is accomplished by using the nth group to modulate a carrier of frequency $f_c = 372 + 48n$ kHz, where $n = 1, 2, \ldots, 5$. Here again the lower sidebands are selected by filtering and then combined to form a supergroup occupying the band 312–552 kHz. Thus, a supergroup is designed to accommodate 60 independent voice inputs. The reason for forming the supergroup in this manner is that economical filters of the required characteristics are available only over a limited frequency range. In a similar manner, supergroups are combined into *mastergroups,* and mastergroups are combined into *very large groups.*

3.10   ANGLE MODULATION

In the previous sections of this chapter, we investigated the effect of slowly varying the amplitude of a sinusoidal carrier wave in accordance with the baseband (information-carrying) signal. There is another way of modulating a sinusoidal carrier wave, namely, *angle modulation* in which the angle of the carrier wave is varied according to the baseband signal. In this method of modulation, the amplitude of the carrier wave is maintained constant. An important feature of angle modulation is that it can provide better discrimination against noise and interference than amplitude modulation. As will be shown in Chapter 5, however, this improvement in performance is achieved at the expense of increased transmission bandwidth; that is, angle modulation provides us with a practical means of exchanging channel bandwidth for improved noise performance. Such a trade-off is *not* possible with amplitude modulation.

Basic Definitions

Let $\theta_i(t)$ denote the *angle* of a modulated sinusoidal carrier, assumed to be a function of the message signal. We express the resulting *angle-modulated wave* as

$$s(t) = A_c \cos[\theta_i(t)] \tag{3.45}$$

where A_c is the carrier amplitude. A complete oscillation occurs whenever $\theta_i(t)$ changes by 2π radians. If $\theta_i(t)$ increases monotonically with time, the average frequency in Hertz, over an interval from t to $t + \Delta t$, is given by

$$f_{\Delta t}(t) = \frac{\theta_i(t + \Delta t) - \theta_i(t)}{2\pi\Delta t} \tag{3.46}$$

We may thus define the *instantaneous frequency* of the angle-modulated signal $s(t)$ as follows:

$$
\begin{aligned}
f_i(t) &= \lim_{\Delta t \to 0} f_{\Delta t}(t) \\
&= \lim_{\Delta t \to 0} \left[\frac{\theta_i(t + \Delta t) - \theta_i(t)}{2\pi\Delta t} \right] \\
&= \frac{1}{2\pi} \frac{d\theta_i(t)}{dt}
\end{aligned}
\tag{3.47}
$$

Thus, according to Eq. (3.45), we may interpret the angle-modulated signal $s(t)$ as a rotating phasor of length A_c and angle $\theta_i(t)$. The angular velocity of such a phasor is $d\theta_i(t)/dt$ measured in radians per second, in accordance with Eq. (3.47). In the simple case of an unmodulated carrier, the angle $\theta_i(t)$ is

$$\theta_i(t) = 2\pi f_c t + \phi_c$$

and the corresponding phasor rotates with a constant angular velocity equal to $2\pi f_c$. The constant ϕ_c is the value of $\theta_i(t)$ at $t = 0$.

There are an infinite number of ways in which the angle $\theta_i(t)$ may be varied in some manner with the message (baseband) signal. However, we shall consider only two commonly used methods, phase modulation and frequency modulation, as defined below:

1. *Phase modulation* (PM) *is that form of angle modulation in which the angle $\theta_i(t)$ is varied linearly with the message signal $m(t)$, as shown by*

$$\theta_i(t) = 2\pi f_c t + k_p m(t) \tag{3.48}$$

The term $2\pi f_c t$ represents the angle of the *unmodulated* carrier; and the constant k_p represents the *phase sensitivity* of the modulator, expressed in radians per volt on the assumption that $m(t)$ is a voltage waveform. For convenience, we have assumed in Eq. (3.48) that the angle of the unmodulated carrier is zero at $t = 0$. The phase-modulated signal $s(t)$ is thus described in the time domain by

$$s(t) = A_c \cos[2\pi f_c t + k_p m(t)] \tag{3.49}$$

2. *Frequency modulation* (FM) *is that form of angle modulation in which the instantaneous frequency $f_i(t)$ is varied linearly with the message signal $m(t)$, as shown by*

$$f_i(t) = f_c + k_f m(t) \tag{3.50}$$

The term f_c represents the frequency of the unmodulated carrier, and the constant k_f represents the *frequency sensitivity* of the modulator, expressed in Hertz per volt on the assumption that $m(t)$ is a voltage waveform. Integrating Eq. (3.50) with respect to time and multiplying the result by 2π, we get

$$\theta_i(t) = 2\pi f_c t + 2\pi k_f \int_0^t m(t)\,dt \tag{3.51}$$

where, for convenience, we have assumed that the angle of the unmodulated carrier wave is zero at $t = 0$. The frequency-modulated signal is therefore described in the time domain by

$$s(t) = A_c \cos\left[2\pi f_c t + 2\pi k_f \int_0^t m(t)\,dt\right] \tag{3.52}$$

A consequence of allowing the angle $\theta_i(t)$ to become dependent on the message signal $m(t)$ as in Eq. (3.48) or on its integral as in Eq. (3.51) is that the *zero crossings* of a PM signal or FM signal no longer have a perfect regularity in their spacing; zero crossings refer to the instants of time at which a waveform changes from a negative to a positive value or vice versa. This is one important feature that distinguishes both PM and FM signals from an AM signal. Another important difference is that the envelope of a PM or FM signal is constant (equal to the carrier amplitude), whereas the envelope of an AM signal is dependent on the message signal.

These differences between amplitude-modulated and angle-modulated signals are illustrated in Fig. 3.29 for the case of sinusoidal modulation. Figures 3.29a and 3.29b refer to the sinusoidal carrier and modulating waves, respectively. Figures 3.29c, 3.29d, and 3.29e show the corresponding AM, PM, and FM signals, respectively. These waveforms indicate that a distinction can be made between PM and FM waves only when compared with the original modulating signal, which shows that there exists a close relationship between PM and FM signals.

Furthermore, comparing Eq. (3.49) with (3.52) reveals that an FM signal may be regarded as a PM signal in which the modulating wave is $\int_0^t m(t)\,dt$ in place of $m(t)$. This means that an FM signal can be generated by first integrating $m(t)$ and then using the result as the input to a phase modulator, as in Fig. 3.30a. Conversely, a PM signal can be generated by first differentiating $m(t)$ and then using the result as the input to a frequency modulator, as in Fig. 3.30b. We may thus deduce all the properties of PM signals from those of FM signals and vice versa. Henceforth, we concentrate our attention on FM signals.

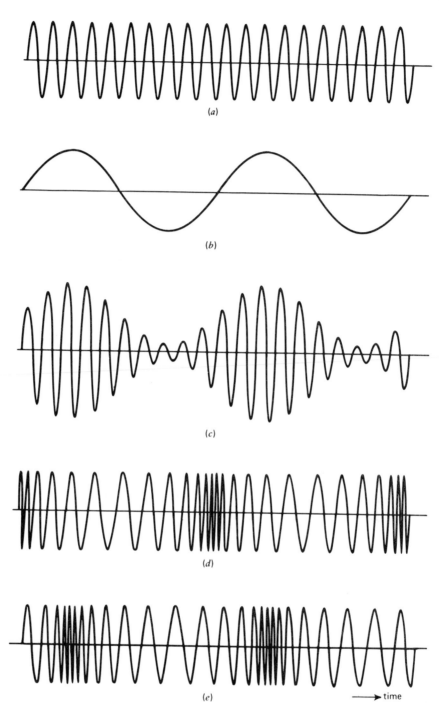

Figure 3.29   Illustrating AM, PM, and FM signals produced by a single tone. (*a*) Carrier wave. (*b*) Sinusoidal modulating signal. (*c*) Amplitude-modulated signal. (*d*) Phase-modulated signal. (*e*) Frequency-modulated signal.

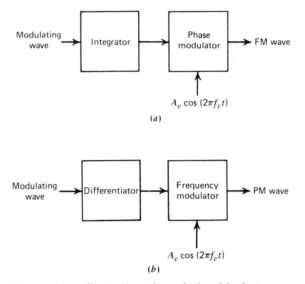

Figure 3.30 Illustrating the relationship between frequency modulation and phase modulation. (a) Scheme for generating an FM wave by using a phase modulator. (b) Scheme for generating a PM wave by using a frequency modulator.

3.11 FREQUENCY MODULATION

The FM signal $s(t)$ defined by Eq. (3.52) is a nonlinear function of the modulating signal $m(t)$, which makes frequency modulation a *nonlinear modulation process*. Consequently, unlike amplitude modulation, the spectrum of an FM signal is not related in a simple manner to that of the modulating signal; rather, its analysis is much more difficult than that of an AM signal.

How then can we tackle the spectral analysis of an FM signal? We propose to provide an empirical answer to this important question by proceeding in the following manner:

- We consider the simplest case possible, namely, that of a single-tone modulation that produces a narrow-band FM signal.
- We next consider the more general case also involving a single-tone modulation, but this time the FM signal is wide-band.

We could, of course, go on and consider the more elaborate case of a multitone FM signal.[8] However, we propose not to do so, because our immediate objective is to establish an empirical relationship between the transmission bandwidth of an FM signal and the message bandwidth. As we shall subsequently see, the two-stage spectral analysis described above provides us with enough insight to propose a solution to the problem.

Consider then a sinusoidal modulating signal defined by

$$m(t) = A_m \cos(2\pi f_m t) \tag{3.53}$$

The instantaneous frequency of the resulting FM signal equals

$$f_i(t) = f_c + k_f A_m \cos(2\pi f_m t)$$
$$= f_c + \Delta f \cos(2\pi f_m t)$$

(3.54)

where

$$\Delta f = k_f A_m$$

(3.55)

The quantity Δf is called the *frequency deviation*, representing the maximum departure of the instantaneous frequency of the FM signal from the carrier frequency f_c. A fundamental characteristic of an FM signal is that the frequency deviation Δf is proportional to the amplitude of the modulating signal and is independent of the modulating frequency.

Using Eq. (3.54), the angle $\theta_i(t)$ of the FM signal is obtained as

$$\theta_i(t) = 2\pi \int_0^t f_i(t) \, dt$$
$$= 2\pi f_c t + \frac{\Delta f}{f_m} \sin(2\pi f_m t)$$

(3.56)

The ratio of the frequency deviation Δf to the modulation frequency f_m is commonly called the *modulation index* of the FM signal. We denote it by β, and so write

$$\beta = \frac{\Delta f}{f_m}$$

(3.57)

and

$$\theta_i(t) = 2\pi f_c t + \beta \sin(2\pi f_m t)$$

(3.58)

From Eq. (3.58) we see that, in a physical sense, the parameter β represents the phase deviation of the FM signal, that is, the maximum departure of the angle $\theta_i(t)$ from the angle $2\pi f_c t$ of the unmodulated carrier; hence, β is measured in radians.

The FM signal itself is given by

$$s(t) = A_c \cos[2\pi f_c t + \beta \sin(2\pi f_m t)]$$

(3.59)

Depending on the value of the modulation index β, we may distinguish two cases of frequency modulation:

• *Narrow-band* FM, for which β is small compared to one radian.
• *Wide-band* FM, for which β is large compared to one radian.

These two cases are considered next, in that order.

Narrow-Band Frequency Modulation

Consider Eq. (3.59), which defines an FM signal resulting from the use of a sinusoidal modulating signal. Expanding this relation, we get

$$s(t) = A_c \cos(2\pi f_c t) \cos[\beta \sin(2\pi f_m t)]$$
$$- A_c \sin(2\pi f_c t) \sin[\beta \sin(2\pi f_m t)]$$

(3.60)

Assuming that the modulation index β is small compared to one radian, we may use the following approximations:

$$\cos[\beta \sin(2\pi f_m t)] \simeq 1$$

and

$$\sin[\beta \sin(2\pi f_m t)] \simeq \beta \sin(2\pi f_m t)$$

Hence, Eq. (3.60) simplifies to

$$s(t) \simeq A_c \cos(2\pi f_c t) - \beta A_c \sin(2\pi f_c t)\sin(2\pi f_m t) \qquad (3.61)$$

Equation (3.61) defines the approximate form of a narrow-band FM signal produced by a sinusoidal modulating signal $A_m \cos(2\pi f_m t)$. From this representation we deduce the modulator shown in block diagram form in Fig. 3.31. This modulator involves splitting the carrier wave $A_c \cos(2\pi f_c t)$ into two paths. One path is direct; the other path contains a -90 degree phase-shifting network and a product modulator, the combination of which generates a DSB-SC modulated signal. The difference between these two signals produces a narrow-band FM signal, but with some distortion.

Ideally, an FM signal has a constant envelope and, for the case of a sinusoidal modulating signal of frequency f_m, the angle $\theta_i(t)$ is also sinusoidal with the same frequency. But the modulated signal produced by the narrow-band modulator of Fig. 3.31 differs from this ideal condition in two fundamental respects:

1. The envelope contains a *residual* amplitude modulation and, therefore, varies with time.

2. For a sinusoidal modulating wave, the angle $\theta_i(t)$ contains *harmonic distortion* in the form of third- and higher-order harmonics of the modulation frequency f_m.

However, by restricting the modulation index to $\beta \leq 0.3$ radians, the effects of residual AM and harmonic PM are limited to negligible levels.

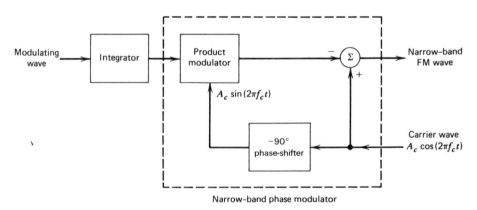

Narrow-band phase modulator

Figure 3.31 Block diagram of a method for generating a narrow-band FM signal.

Returning to Eq. (3.61), we may expand it as follows:

$$s(t) \simeq A_c \cos(2\pi f_c t) + \frac{1}{2}\beta A_c \{\cos[2\pi(f_c + f_m)t] - \cos[2\pi(f_c - f_m)t]\} \quad (3.62)$$

This expression is somewhat similar to the corresponding one defining an AM signal, which is reproduced from Example 1 as follows:

$$s_{AM}(t) = A_c \cos(2\pi f_c t) + \frac{1}{2}\mu A_c \{\cos[2\pi(f_c + f_m)t] + \cos[2\pi(f_c - f_m)t]\} \quad (3.63)$$

where μ is the modulation factor of the AM signal. Comparing Eqs. (3.62) and (3.63), we see that in the case of sinusoidal modulation, the basic difference between an AM signal and a narrow-band FM signal is that the algebraic sign of the lower side frequency in the narrow-band FM is reversed. Thus, a narrow-band FM signal requires essentially the same transmission bandwidth (i.e., $2f_m$) as the AM signal.

We may represent the narrow-band FM signal with a phasor diagram as shown in Fig. 3.32a, where we have used the carrier phasor as reference. We see that the resultant of the two side-frequency phasors is always at right angles to the carrier phasor. The effect of this is to produce a resultant phasor representing the narrow-band FM signal that is approximately of the same amplitude as the carrier phasor, but out of phase with respect to it. This phasor diagram should be contrasted with that of Fig. 3.32b, representing an AM signal. In this latter case we see that the resultant phasor representing the AM signal has an amplitude different from that of the carrier phasor, but always in phase with it.

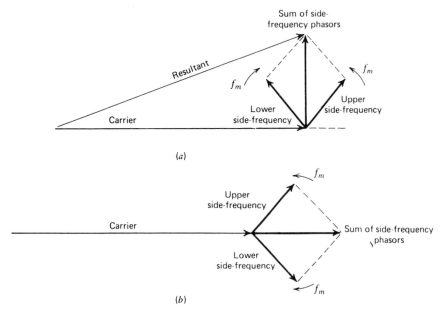

Figure 3.32 A phasor comparison of narrow-band FM and AM waves for sinusoidal modulation. (a) Narrow-band FM wave. (b) AM wave.

Wide-Band Frequency Modulation

We next wish to determine the spectrum of the single-tone FM signal of Eq.
(3.59) for an arbitrary value of the modulation index β. In general, an FM signal
produced by a sinusoidal modulating signal, as in Eq. (3.59), is in itself nonperiodic unless the carrier frequency f_c is an integral multiple of the modulation
frequency f_m. However, we may simplify matters by using the complex representation of band-pass signals described in Chapter 2. Specifically, we assume
that the carrier frequency f_c is large enough (compared to the bandwidth of the
FM signal) to justify rewriting this equation in the form

$$\begin{aligned} s(t) &= \operatorname{Re}[A_c \exp(j2\pi f_c t + j\beta \sin(2\pi f_m t))] \\ &= \operatorname{Re}[\tilde{s}(t)\exp(j2\pi f_c t)] \end{aligned} \tag{3.64}$$

where $\tilde{s}(t)$ is the complex envelope of the FM signal $s(t)$, defined by

$$\tilde{s}(t) = A_c \exp[j\beta \sin(2\pi f_m t)] \tag{3.65}$$

Thus, unlike the original FM signal $s(t)$, the complex envelope $\tilde{s}(t)$ is a periodic
function of time with a fundamental frequency equal to the modulation frequency f_m. We may therefore expand $\tilde{s}(t)$ in the form of a complex Fourier series
as follows:

$$\tilde{s}(t) = \sum_{n=-\infty}^{\infty} c_n \exp(j2\pi n f_m t) \tag{3.66}$$

where the complex Fourier coefficient c_n equals

$$\begin{aligned} c_n &= f_m \int_{-1/2f_m}^{1/2f_m} \tilde{s}(t)\exp(-j2\pi n f_m t)\ dt \\ &= f_m A_c \int_{-1/2f_m}^{1/2f_m} \exp[j\beta \sin(2\pi f_m t) - j2\pi n f_m t]\ dt \end{aligned} \tag{3.67}$$

Define a new variable:

$$x = 2\pi f_m t \tag{3.68}$$

Hence, we may rewrite Eq. (3.67) in the new form

$$c_n = \frac{A_c}{2\pi} \int_{-\pi}^{\pi} \exp[j(\beta \sin x - nx)]\ dx \tag{3.69}$$

The integral on the right-hand side of Eq. (3.69), except for a scaling factor, is
recognized as the nth *order Bessel function of the first kind*[9] and argument β. This
function is commonly denoted by the symbol $J_n(\beta)$, as shown by

$$J_n(\beta) = \frac{1}{2\pi} \int_{-\pi}^{\pi} \exp[j(\beta \sin x - nx)]\ dx \tag{3.70}$$

Accordingly, we may reduce Eq. (3.69) to

$$c_n = A_c J_n(\beta) \tag{3.71}$$

Substituting Eq. (3.71) in (3.66), we get, in terms of the Bessel function $J_n(\beta)$, the following expansion for the complex envelope of the FM signal:

$$\tilde{s}(t) = A_c \sum_{n=-\infty}^{\infty} J_n(\beta) \exp(j2\pi n f_m t) \tag{3.72}$$

Next, substituting Eq. (3.72) in (3.64), we get

$$s(t) = A_c \cdot \mathrm{Re}\left[\sum_{n=-\infty}^{\infty} J_n(\beta) \exp[j2\pi(f_c + nf_m)t] \right] \tag{3.73}$$

Interchanging the order of summation and evaluation of the real part in the right-hand side of Eq. (3.73), we get

$$s(t) = A_c \sum_{n=-\infty}^{\infty} J_n(\beta) \cos[2\pi(f_c + nf_m)t] \tag{3.74}$$

This is the desired form for the Fourier series representation of the single-tone FM signal $s(t)$ for an arbitrary value of β. The discrete spectrum of $s(t)$ is obtained by taking the Fourier transforms of both sides of Eq. (3.74); we thus have

$$S(f) = \frac{A_c}{2} \sum_{n=-\infty}^{\infty} J_n(\beta)[\delta(f - f_c - nf_m) + \delta(f + f_c + nf_m)] \tag{3.75}$$

In Fig. 3.33 we have plotted the Bessel function $J_n(\beta)$ versus the modulation index β for different positive integer values of n. We can develop further insight into the behavior of the Bessel function $J_n(\beta)$ by making use of the following properties (see Appendix 4 for more details):

1. For n even, we have $J_n(\beta) = J_{-n}(\beta)$; on the other hand, for n odd, we have $J_n(\beta) = -J_{-n}(\beta)$. That is,

$$J_n(\beta) = (-1)^n J_{-n}(\beta) \qquad \text{for all } n \tag{3.76}$$

2. For small values of the modulation index β, we have

$$\left.\begin{aligned} J_0(\beta) &\simeq 1 \\ J_1(\beta) &\simeq \frac{\beta}{2} \\ J_n(\beta) &\simeq 0, \qquad n > 2 \end{aligned}\right\} \tag{3.77}$$

3.

$$\sum_{n=-\infty}^{\infty} J_n^2(\beta) = 1 \tag{3.78}$$

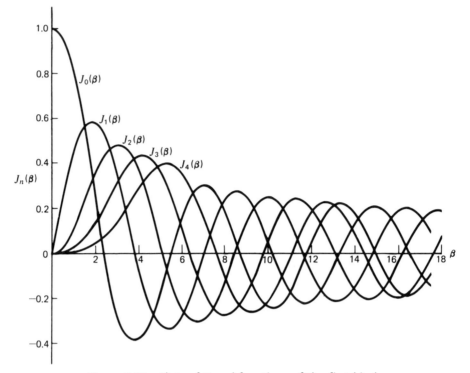

Figure 3.33 Plots of Bessel functions of the first kind.

Thus, using Eqs. (3.75) through (3.78) and the curves of Fig. 3.33, we may make the following observations:

1. The spectrum of an FM signal contains a carrier component and an infinite set of side frequencies located symmetrically on either side of the carrier at frequency separations of $f_m, 2f_m, 3f_m, \ldots$. In this respect, the result is unlike that which prevails in an AM system, since in an AM system a sinusoidal modulating signal gives rise to only one pair of side frequencies.

2. For the special case of β small compared with unity, only the Bessel coefficients $J_0(\beta)$ and $J_1(\beta)$ have significant values, so that the FM signal is effectively composed of a carrier and a single pair of side frequencies at $f_c \pm f_m$. This situation corresponds to the special case of narrow-band FM that was considered previously.

3. The amplitude of the carrier component varies with β according to $J_0(\beta)$. That is, unlike an AM signal, the amplitude of the carrier component of an FM signal is dependent on the modulation index β. The physical explanation for this property is that the envelope of an FM signal is constant, so that the average power of such a signal developed across a 1-ohm resistor is also constant, as shown by

$$P = \frac{1}{2} A_c^2 \tag{3.79}$$

When the carrier is modulated to generate the FM signal, the power in the side frequencies may appear only at the expense of the power originally in the carrier, thereby making the amplitude of the carrier component dependent on β. Note that the average power of an FM signal may also be determined from Eq. (3.74), obtaining

$$P = \frac{1}{2}A_c^2 \sum_{n=-\infty}^{\infty} J_n^2(\beta) \tag{3.80}$$

Substituting Eq. (3.78) in (3.80), the expression for the average power P reduces to Eq. (3.79), and so it should.

EXAMPLE 3

In this example, we wish to investigate the ways in which variations in the amplitude and frequency of a sinusoidal modulating signal affect the spectrum of the FM signal. Consider first the case when the frequency of the modulating signal is fixed, but its amplitude is varied, producing a corresponding variation in the frequency deviation Δf. Thus, keeping the modulation frequency f_m fixed, we find that the amplitude spectrum of the resulting FM signal is as shown plotted in Fig. 3.34 for $\beta = 1, 2,$ and 5. In this diagram we have normalized the spectrum with respect to the unmodulated carrier amplitude.

Consider next the case when the amplitude of the modulating signal is fixed; that is, the frequency deviation Δf is maintained constant, and the modulation frequency f_m is varied. In this case we find that the amplitude spectrum of the resulting FM signal is as shown plotted in Fig. 3.35 for $\beta = 1, 2,$ and 5. We see that when Δf is fixed and β is increased, we have an increasing number of spectral lines crowding into the fixed frequency interval $f_c - \Delta f < |f| < f_c + \Delta f$. That is, when β approaches infinity, the bandwidth of the FM wave approaches the limiting value of $2\Delta f$, which is an important point to keep in mind.

Transmission Bandwidth of FM Signals

In theory, an FM signal contains an infinite number of side frequencies so that the bandwidth required to transmit such a signal is similarly infinite in extent. In practice, however, we find that the FM signal is effectively limited to a finite number of significant side frequencies compatible with a specified amount of distortion. We may therefore specify an effective bandwidth required for the transmission of an FM signal. Consider first the case of an FM signal generated by a single-tone modulating wave of frequency f_m. In such an FM signal, the side frequencies that are separated from the carrier frequency f_c by an amount greater than the frequency deviation Δf decrease rapidly toward zero, so that the bandwidth always exceeds the total frequency excursion, but nevertheless is limited. Specifically, for large values of the modulation index β, the bandwidth approaches, and is only slightly greater than, the total frequency excursion $2\Delta f$. On the other hand, for small values of the modulation index β, the spectrum of the FM signal is effectively limited to the carrier frequency f_c and one pair of side frequencies at $f_c \pm f_m$, so that the bandwidth approaches $2f_m$. We may thus define an approximate rule for the transmission bandwidth of an FM signal generated by a single-tone modulating signal of frequency f_m as follows:

$$B_T \simeq 2\Delta f + 2f_m = 2\Delta f \left(1 + \frac{1}{\beta}\right) \tag{3.81}$$

This empirical relation is known as *Carson's rule*.[10]

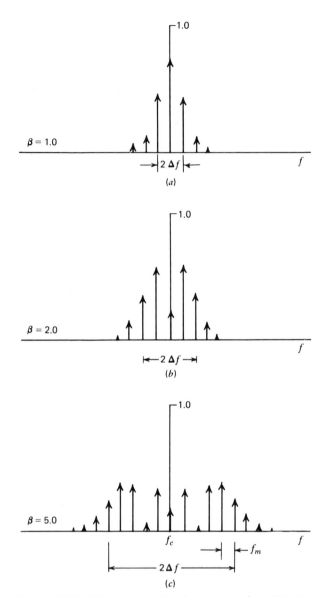

Figure 3.34 Discrete amplitude spectra of an FM signal, normalized with respect to the carrier amplitude, for the case of sinusoidal modulation of fixed frequency and varying amplitude. Only the spectra for positive frequencies are shown.

For a more accurate assessment of the bandwidth requirement of an FM signal, we may use a definition based on retaining the maximum number of significant side frequencies whose amplitudes are all greater than some selected value. A convenient choice for this value is 1 percent of the unmodulated carrier amplitude. We may thus define *the transmission bandwidth of an FM wave as the separation between the two frequencies beyond which none of the side frequencies is greater than 1 percent of the carrier amplitude obtained when the modulation is removed.* That

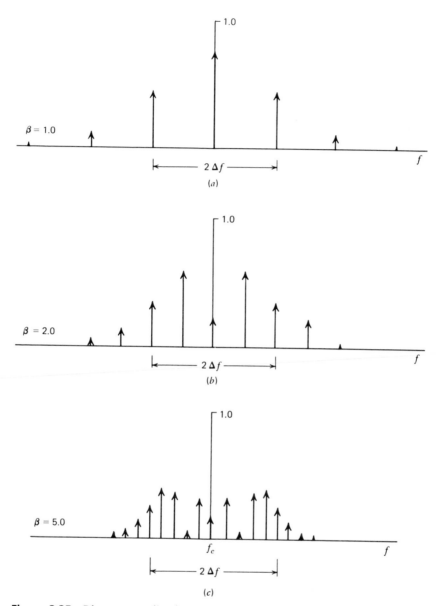

Figure 3.35 Discrete amplitude spectra of an FM signal, normalized with respect to the carrier amplitude, for the case of sinusoidal modulation of varying frequency and fixed amplitude. Only the spectra for positive frequencies are shown.

is, we define the transmission bandwidth as $2n_{max}f_m$, where f_m is the modulation frequency and n_{max} is the largest value of the integer n that satisfies the requirement $|J_n(\beta)| > 0.01$. The value of n_{max} varies with the modulation index β and can be determined readily from tabulated values of the Bessel function $J_n(\beta)$. Table 3.1 shows the total number of significant side frequencies (including both the upper and lower side frequencies) for different values of β, calculated on the 1 percent basis explained herein. The transmission bandwidth B_T calculated

Table 3.1 Number of Significant Side Frequencies
of a Wide-band FM Signal for Varying Modulation Index

Modulation Index β	Number of Significant Side Frequencies $2n_{max}$
0.1	2
0.3	4
0.5	4
1.0	6
2.0	8
5.0	16
10.0	28
20.0	50
30.0	70

using this procedure can be presented in the form of a *universal curve* by normalizing it with respect to the frequency deviation Δf and then plotting it versus β. This curve is shown in Fig. 3.36, which is drawn as a best fit through the set of points obtained by using Table 3.1. In Fig. 3.36 we note that as the modulation index β is increased, the bandwidth occupied by the significant side frequencies drops toward that over which the carrier frequency actually deviates. This means that small values of the modulation index β are relatively more extravagant in transmission bandwidth than are the larger values of β.

Consider next the more general case of an arbitrary modulating signal $m(t)$ with its highest frequency component denoted by W. The bandwidth required

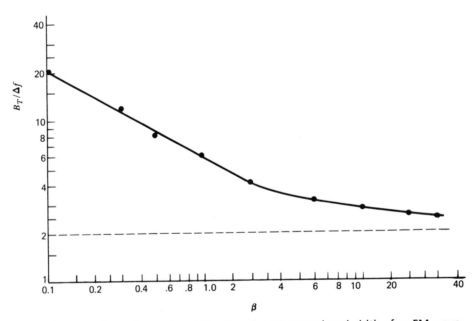

Figure 3.36 Universal curve for evaluating the 1 percent bandwidth of an FM wave.

to transmit an FM signal generated by this modulating signal is estimated by using a worst-case tone-modulation analysis. Specifically, we first determine the so-called *deviation ratio D*, defined as the ratio of the frequency deviation Δf, which corresponds to the maximum possible amplitude of the modulation signal $m(t)$, to the highest modulation frequency W; these conditions represent the extreme cases possible. *The deviation ratio D plays the same role for nonsinusoidal modulation that the modulation index β plays for the case of sinusoidal modulation.* Then, replacing β by D and replacing f_m with W, we may use Carson's rule given by Eq. (3.81) or the universal curve of Fig. 3.36 to obtain a value for the transmission bandwidth of the FM signal. From a practical viewpoint, Carson's rule somewhat underestimates the bandwidth requirement of an FM system, whereas using the universal curve of Fig. 3.36 yields a somewhat conservative result. Thus, the choice of a transmission bandwidth that lies between the bounds provided by these two rules of thumb is acceptable for most practical purposes.

EXAMPLE 4

In North America, the maximum value of frequency deviation Δf is fixed at 75 kHz for commercial FM broadcasting by radio. If we take the modulation frequency $W = 15$ kHz, which is typically the "maximum" audio frequency of interest in FM transmission, we find that the corresponding value of the deviation ratio is

$$D = \frac{75}{15} = 5$$

Using Carson's rule of Eq. (3.81), replacing β by D, and replacing f_m by W, the approximate value of the transmission bandwidth of the FM signal is obtained as

$$B_T = 2(75 + 15) = 180 \text{ kHz}$$

On the other hand, use of the curve of Fig. 3.36 gives the transmission bandwidth of the FM signal to be

$$B_T = 3.2 \ \Delta f = 3.2 \times 75 = 240 \text{ kHz}$$

Thus Carson's rule underestimates the transmission bandwidth by 25 percent compared with the result of using the universal curve of Fig. 3.36.

Generation of FM Signals

There are essentially two basic methods of generating frequency-modulated signals, namely, *indirect* FM and *direct* FM. In the direct method of producing frequency modulation, the modulating signal is first used to produce a narrow-band FM signal, and frequency multiplication is next used to increase the frequency deviation to the desired level. On the other hand, in the direct method of producing frequency modulation, the carrier frequency is directly varied in accordance with the input baseband signal. In the sequel, we describe the important features of both methods.

Indirect FM[11]   A simplified block diagram of an indirect FM system is shown in Fig. 3.37. The message (baseband) signal $m(t)$ is first integrated and then used to phase-modulate a crystal-controlled oscillator; the use of crystal control pro-

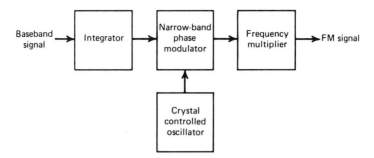

Figure 3.37 Block diagram of the indirect method of generating a wide-band FM signal.

vides *frequency stability*. In order to minimize the distortion inherent in the phase modulator, the maximum phase deviation or modulation index β is kept small, thereby resulting in a narrow-band FM signal; for the implementation of the narrow-band phase modulator, we may use the arrangement described in Fig. 3.31. The narrow-band FM signal is next multiplied in frequency by means of a frequency multiplier so as to produce the desired wide-band FM signal.

A *frequency multiplier* consists of a *memoryless* nonlinear device followed by a band-pass filter, as shown in Fig. 3.38. The implication of the nonlinear device being memoryless is that it has no energy-storage elements. The input–output relation of such a device may be expressed in the general form

$$v(t) = a_1 s(t) + a_2 s^2(t) + \ldots + a_n s^n(t) \tag{3.82}$$

where $a_1, a_2, \ldots, a_n$ are coefficients determined by the operating point of the device, and n is the *highest order of nonlinearity*. In other words, the memoryless nonlinear device is an nth power-law device. The input $s(t)$ is an FM signal defined by

$$s(t) = A_c \cos\left[2\pi f_c t + 2\pi k_f \int_0^t m(t)\, dt\right]$$

whose instantaneous frequency is

$$f_i(t) = f_c + k_f m(t) \tag{3.83}$$

The mid-band frequency of the band-pass filter in Fig. 3.38 is set equal to nf_c, where f_c is the carrier frequency of the incoming FM signal $s(t)$. Moreover, the band-pass filter is designed to have a bandwidth equal to n times the transmission bandwidth of $s(t)$. In Section 3.13 dealing with nonlinear effects in FM systems,

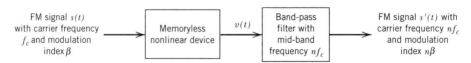

Figure 3.38 Block diagram of frequency multiplier.

we describe the spectral contributions of such nonlinear terms as the second- and third-order terms in the input–output relation of Eq. (3.82). For now it suffices to say that after band-pass filtering of the nonlinear device's output $v(t)$, we have a new FM signal defined by

$$s'(t) = A'_c \cos\left[2\pi n f_c t + 2\pi n k_f \int_0^t m(t)\ dt \right] \qquad (3.84)$$

whose instantaneous frequency is

$$f'_i(t) = n f_c + n k_f m(t) \qquad (3.85)$$

Thus, comparing Eq. (3.85) with (3.83), we see that the nonlinear processing circuit of Fig. 3.38 acts as a frequency multiplier. The frequency multiplication ratio is determined by the highest power n in the input–output relation of Eq. (3.82), characterizing the memoryless nonlinear device.

EXAMPLE 5

Figure 3.39 shows the simplified block diagram of a typical FM transmitter (based on the indirect method) used to transmit audio signals containing frequencies in the range 100 Hz to 15 kHz. The narrow-band phase modulator is supplied with a carrier signal of frequency $f_1 = 0.1$ MHz by a crystal-controlled oscillator. The desired FM signal at the transmitter output is to have a carrier frequency $f_c = 100$ MHz and a minimum frequency deviation $\Delta f = 75$ kHz.

In order to limit the harmonic distortion produced by the narrow-band phase modulator, we restrict the modulation index β_1 of this modulator to a maximum value of 0.3 radians. Consider then the value $\beta_1 = 0.2$ radians, which certainly satisfies this requirement. The lowest modulation frequencies of 100 Hz produce a frequency deviation of $\Delta f_1 = 20$ Hz at the narrow-band phase modulator output, whereas the highest modulation frequencies of 15 kHz produce a frequency deviation of $\Delta f_1 = 3$ kHz. The lowest modulation frequencies are therefore of immediate concern, as they produce a much lower frequency deviation than the highest modulation frequencies. The requirement is therefore to ensure that the frequency deviation produced by the lowest modulation frequencies of 100 Hz is raised to 75 kHz.

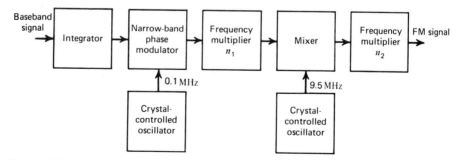

Figure 3.39 Block diagram of the wide-band frequency modulator for Example 5.

To produce a frequency deviation of $\Delta f = 75$ kHz at the FM transmitter output, the use of frequency multiplication is obviously required. Specifically, with $\Delta f_1 = 20$ Hz and $\Delta f = 75$ kHz, we require a total frequency multiplication ratio of 3750. However, using a straight frequency multiplication equal to this value would produce a much higher carrier frequency at the transmitter output than the desired value of 100 MHz. To generate an FM signal having both the desired frequency deviation and carrier frequency, we therefore need to use a *two-stage frequency multiplier* with an intermediate stage of frequency translation, as illustrated in Fig. 3.39. Let n_1 and n_2 denote the respective frequency multiplication ratios, so that

$$n_1 n_2 = \frac{\Delta f}{\Delta f_1} = \frac{75000}{20} = 3750 \tag{3.86}$$

The carrier frequency $n_1 f_1$ at the first frequency multiplier output is translated downward to $(f_2 - n_1 f_1)$ by mixing it with a sinusoidal wave of frequency $f_2 = 9.5$ MHz, which is supplied by a second crystal-controlled oscillator. However, the carrier frequency at the input of the second frequency multiplier is required to equal f_c/n_2. Equating these two frequencies, we thus get

$$f_2 - n_1 f_1 = \frac{f_c}{n_2}$$

Hence, with $f_1 = 0.1$ MHz, $f_2 = 9.5$ MHz, and $f_c = 100$ MHz, we have

$$9.5 - 0.1 n_1 = \frac{100}{n_2} \tag{3.87}$$

Solving Eqs. (3.86) and (3.87) for n_1 and n_2, we obtain

$$n_1 = 75$$

$$n_2 = 50$$

Using these frequency multiplication ratios, we get the set of values indicated in Table 3.2.

Table 3.2 Values of Carrier Frequency and Frequency Deviation at the Various Points in the Wide-band Frequency Modulator of Fig. 3.39

	At the Phase Modulator Output	At the First Frequency Multiplier Output	At the Mixer Output	At the Second Frequency Multiplier Output
Carrier frequency	0.1 MHz	7.5 MHz	2.0 MHz	100 MHz
Frequency deviation	20 Hz	1.5 kHz	1.5 kHz	75 kHz

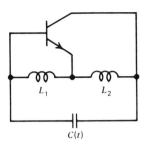

Figure 3.40 Hartley oscillator.

Direct FM In a direct FM system, the instantaneous frequency of the carrier wave is varied directly in accordance with the message signal by means of a device known as a *voltage-controlled oscillator*. One way of implementing such a device is to use a sinusoidal oscillator having a highly selective frequency-determining resonant network and to control the oscillator by symmetrical incremental variation of the reactive components of this network. An example of such a scheme is shown in Fig. 3.40, depicting a *Hartley oscillator*. We assume that the capacitive component of the frequency-determining network in the oscillator consists of a fixed capacitor shunted by a voltage-variable capacitor. The resultant capacitance is represented by $C(t)$ in Fig. 3.40. A voltage-variable capacitor, commonly called a *varactor* or *varicap*, is one whose capacitance depends on the voltage applied across its electrodes. The variable-voltage capacitance may be obtained, for example, by using a *p-n* junction diode that is biased in the reverse direction; the larger the reverse voltage applied to such a diode, the smaller the transition capacitance of the diode. The frequency of oscillation of the Hartley oscillator of Fig. 3.40 is given by

$$f_i(t) = \frac{1}{2\pi\sqrt{(L_1 + L_2)C(t)}} \tag{3.88}$$

where $C(t)$ is the total capacitance of the fixed capacitor and the variable-voltage capacitor, and L_1 and L_2 are the two inductances in the frequency-determining network of the oscillator. Assume that for a sinusoidal modulating wave of frequency f_m, the capacitance $C(t)$ is expressed as

$$C(t) = C_0 + \Delta C \cos(2\pi f_m t) \tag{3.89}$$

where C_0 is the total capacitance in the absence of modulation and ΔC is the maximum change in capacitance. Substituting Eq. (3.89) in (3.88), we get

$$f_i(t) = f_0 \left[1 + \frac{\Delta C}{C_0} \cos(2\pi f_m t) \right]^{-1/2} \tag{3.90}$$

where f_0 is the unmodulated frequency of oscillation, that is,

$$f_0 = \frac{1}{2\pi\sqrt{C_0(L_1 + L_2)}} \tag{3.91}$$

Provided that the maximum change in capacitance ΔC is small compared with the unmodulated capacitance C_0, we may approximate Eq. (3.90) as

$$f_i(t) \simeq f_0 \left[1 - \frac{\Delta C}{2C_0} \cos(2\pi f_m t) \right] \qquad (3.92)$$

Let

$$\frac{\Delta C}{2C_0} = -\frac{\Delta f}{f_0} \qquad (3.93)$$

Hence, the instantaneous frequency of the oscillator, which is being frequency-modulated by varying the capacitance of the frequency-determining network, is approximately given by

$$f_i(t) \simeq f_0 + \Delta f \cos(2\pi f_m t) \qquad (3.94)$$

Equation (3.94) is the desired relation for the instantaneous frequency of an FM wave, assuming sinusoidal modulation.

In order to generate a wide-band FM wave with the required frequency deviation, we may use the configuration shown in Fig. 3.41 consisting of a voltage-controlled oscillator, followed by a series of frequency multipliers and mixers. This configuration permits the attainment of good oscillator stability, constant proportionality between output frequency change and input voltage change, and the necessary frequency deviation to achieve wide-band FM.

An FM transmitter using the direct method as described, however, has the disadvantage that the carrier frequency is not obtained from a highly stable oscillator. It is therefore necessary, in practice, to provide some auxiliary means by which a very stable frequency generated by a crystal will be able to control the carrier frequency. One method of effecting this control is illustrated in Fig. 3.42. The output of the FM generator is applied to a mixer together with the output of a crystal-controlled oscillator, and the difference frequency term is extracted. The mixer output is next applied to a frequency discriminator and then low-pass filtered. A frequency discriminator is a device whose output voltage has an instantaneous amplitude that is proportional to the instantaneous frequency of the FM signal applied to its input; this device is described in the next subsection. When the FM transmitter has exactly the correct carrier frequency, the low-pass

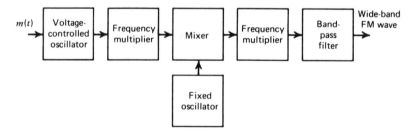

Figure 3.41 Block diagram of wide-band frequency modulator using a voltage-controlled oscillator.

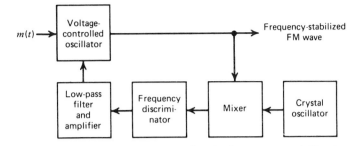

Figure 3.42 A feedback scheme for the frequency stabilization of a frequency modulator.

filter output is zero. However, deviations of the transmitter carrier frequency from its assigned value will cause the frequency discriminator–filter combination to develop a dc output voltage with a polarity determined by the sense of the transmitter frequency drift. This dc voltage, after suitable amplification, is applied to the voltage-controlled oscillator of the FM transmitter in such a way as to modify the frequency of the oscillator in a direction that tends to restore the carrier frequency to its correct value.

Demodulation of FM Signals

Frequency demodulation is the process that enables us to recover the original modulating signal from a frequency-modulated signal. The objective is to produce a transfer characteristic that is the inverse of that of the frequency modulator, which can be realized directly or indirectly. Here we describe a direct method of frequency demodulation involving the use of a popular device known as a frequency discriminator, whose instantaneous amplitude is directly proportional to the instantaneous frequency of the input FM signal. In the next section, we describe an indirect method of frequency demodulation that uses another popular device known as a phase-locked loop.

Basically, the *frequency discriminator* consists of a *slope circuit* followed by an *envelope detector*. An ideal slope circuit is characterized by a transfer function that is purely imaginary, varying linearly with frequency inside a prescribed frequency interval. Consider the transfer function depicted in Fig. 3.43a, which is defined by

$$H_1(f) = \begin{cases} j2\pi a\left(f - f_c + \dfrac{B_T}{2} \right), & f_c - \dfrac{B_T}{2} \le f \le f_c + \dfrac{B_T}{2} \\[2mm] j2\pi a\left(f + f_c - \dfrac{B_T}{2} \right), & -f_c - \dfrac{B_T}{2} \le f \le -f_c + \dfrac{B_T}{2} \\[2mm] 0, & \text{elsewhere} \end{cases} \tag{3.95}$$

where a is a constant. We wish to evaluate the response of this slope circuit, denoted by $s_1(t)$, which is produced by an FM signal $s(t)$ of carrier frequency f_c and transmission bandwidth B_T. It is assumed that the spectrum of $s(t)$ is essentially zero outside the frequency interval $f_c - B_T/2 \le |f| \le f_c + B_T/2$. For eval-

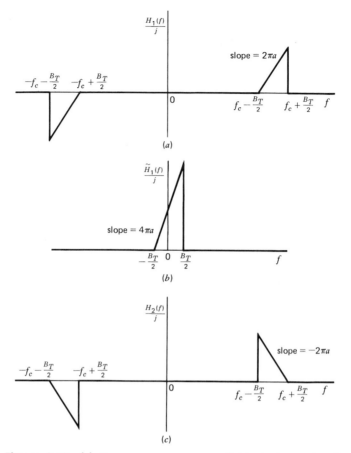

Figure 3.43 (a) Frequency response of ideal slope circuit. (b) The slope circuit's response. (c) Frequency response of the complex low-pass filter equivalent to ideal slope circuit complementary to that of part (a).

uation of the response $s_1(t)$, it is convenient to use the procedure described in Section 2.13, which involves replacing the slope circuit with an equivalent low-pass filter and driving this filter with the complex envelope of the input FM signal $s(t)$.

Let $\tilde{H}_1(f)$ denote the complex transfer function of the slope circuit defined by Fig. 3.43a. This complex transfer function is related to $H_1(f)$ by

$$\tilde{H}_1(f - f_c) = 2H_1(f), \qquad f > 0 \tag{3.96}$$

Hence, using Eqs. (3.95) and (3.96), we get

$$\tilde{H}_1(f) = \begin{cases} j4\pi a\left(f + \dfrac{B_T}{2}\right), & -\dfrac{B_T}{2} \leq f \leq \dfrac{B_T}{2} \\ 0, & \text{elsewhere} \end{cases} \tag{3.97}$$

which is shown in Fig. 3.43b.

The incoming FM signal $s(t)$ is defined by Eq. (3.52), which is reproduced here for convenience:

$$s(t) = A_c \cos\left[2\pi f_c t + 2\pi k_f \int_0^t m(t) \, dt \right]$$

Given that the carrier frequency f_c is high compared to the transmission bandwidth of the FM signal $s(t)$, the complex envelope of $s(t)$ is

$$\tilde{s}(t) = A_c \exp\left[j2\pi k_f \int_0^t m(t) \, dt \right] \tag{3.98}$$

Let $\tilde{s}_1(t)$ denote the complex envelope of the response of the slope circuit defined by Fig. 3.43b due to $\tilde{s}(t)$. Then, according to the theory described in Section 2.13, we may express the Fourier transform of $\tilde{s}_1(t)$ as follows:

$$\tilde{S}_1(f) = \frac{1}{2}\tilde{H}_1(f)\tilde{S}(f)$$

$$= \begin{cases} j2\pi a\left(f + \dfrac{B_T}{2} \right)\tilde{S}(f), & -\dfrac{B_T}{2} \leq f \leq \dfrac{B_T}{2} \\ 0, & \text{elsewhere} \end{cases} \tag{3.99}$$

where $\tilde{S}(f)$ is the Fourier transform of $\tilde{s}(t)$. Since multiplication of the Fourier transform of a signal by the factor $j2\pi f$ is equivalent to differentiating the signal in the time domain (see Section 2.3), we deduce from Eq. (3.99) that

$$\tilde{s}_1(t) = a\left[\frac{d\tilde{s}(t)}{dt} + j\pi B_T \tilde{s}(t) \right] \tag{3.100}$$

Substituting Eq.(3.98) in (3.100), we get

$$\tilde{s}_1(t) = j\pi B_T a A_c \left[1 + \frac{2k_f}{B_T}m(t) \right] \exp\left[j2\pi k_f \int_0^t m(t) \, dt \right] \tag{3.101}$$

The desired response of the slope circuit is therefore

$$s_1(t) = \text{Re}[\tilde{s}_1(t)\exp(j2\pi f_c t)]$$

$$= \pi B_T a A_c \left[1 + \frac{2k_f}{B_T}m(t) \right] \cos\left[2\pi f_c t + 2\pi k_f \int_0^t m(t) \, dt + \frac{\pi}{2} \right] \tag{3.102}$$

The signal $s_1(t)$ is a hybrid-modulated signal in which both amplitude and frequency of the carrier wave vary with the message signal $m(t)$. However, provided that we choose

$$\left| \frac{2k_f}{B_T}m(t) \right| < 1 \qquad \text{for all } t$$

then we may use an envelope detector to recover the amplitude variations and thus, except for a bias term, obtain the original message signal. The resulting envelope-detector output is therefore

$$|\tilde{s}_1(t)| = \pi B_T a A_c \left[1 + \frac{2k_f}{B_T} m(t) \right] \qquad (3.103)$$

The bias term $\pi B_T a A_c$ in the right-hand side of Eq. (3.103) is proportional to the slope a of the transfer function of the slope circuit. This suggests that the bias may be removed by subtracting from the envelope-detector output $|\tilde{s}_1(t)|$ the output of a second envelope detector preceded by the *complementary slope circuit* with a transfer function $H_2(f)$ as described in Fig. 3.43c. That is, the respective complex transfer functions of the two slope circuits are related by

$$\tilde{H}_2(f) = \tilde{H}_1(-f) \qquad (3.104)$$

Let $s_2(t)$ denote the response of the complementary slope circuit produced by the incoming FM signal $s(t)$. Then, following a procedure similar to that just described, we find that the envelope of $s_2(t)$ is

$$|\tilde{s}_2(t)| = \pi B_T a A_c \left[1 - \frac{2k_f}{B_T} m(t) \right] \qquad (3.105)$$

where $\tilde{s}_2(t)$ is the complex envelope of the signal $s_2(t)$. The difference between the two envelopes in Eqs. (3.103) and (3.105) is

$$\begin{aligned} s_o(t) &= |\tilde{s}_1(t)| - |\tilde{s}_2(t)| \\ &= 4\pi k_f a A_c m(t) \end{aligned} \qquad (3.106)$$

which is free from bias.

We may thus model the *ideal frequency discriminator* as a pair of slope circuits with their complex transfer functions related by Eq. (3.104), followed by envelope detectors and finally a summer, as in Fig. 3.44a. This scheme is called a *balanced frequency discriminator*.

The idealized scheme of Fig. 3.44a can be closely realized using the circuit shown in Fig. 3.44b. The upper and lower resonant filter sections of this circuit are tuned to frequencies above and below the unmodulated carrier frequency f_c, respectively. In Fig. 3.44c we have plotted the amplitude responses of these two tuned filters, together with their total response, assuming that both filters have a high Q-factor. The *quality factor* or *Q-factor* of a resonant circuit is a measure of goodness of the whole circuit. It is formally defined as 2π times the ratio of maximum energy stored in the circuit during one cycle to the energy dissipated per cycle. In the case of an RLC parallel (or series) resonant circuit, the Q-factor is equal to the resonant frequency divided by the 3-dB bandwidth of the circuit. In the RLC parallel resonant circuits shown in Fig. 3.44b, the resistance R is contributed largely by imperfections in the inductive elements of the circuits.

The linearity of the useful portion of the total response in Fig. 3.44c, centered at f_c, is determined by the separation of the two resonant frequencies. As illustrated in Fig. 3.44c, a frequency separation of $3B$ gives satisfactory results,

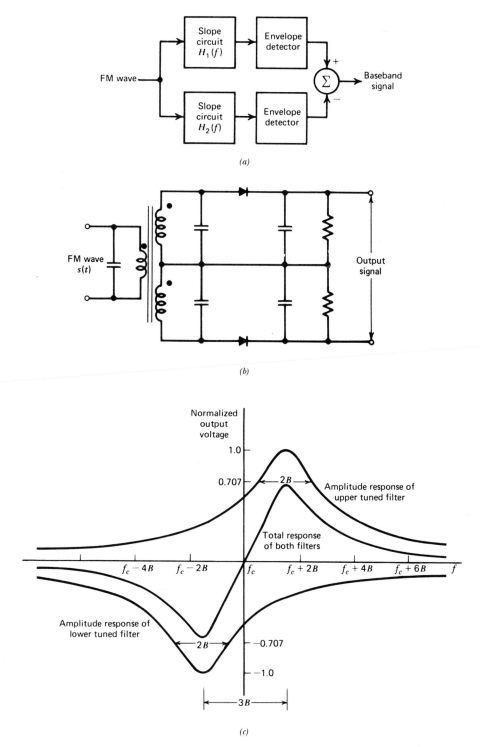

Figure 3.44 Balanced frequency discriminator. (*a*) Block diagram. (*b*) Circuit diagram. (*c*) Frequency response.

where $2B$ is the 3-dB bandwidth of either filter. However, there will be distortion in the output of this frequency discriminator due to the following factors:

1. The spectrum of the input FM signal $s(t)$ is not exactly zero for frequencies outside the range $f_c - B_T/2 \le f \le f_c + B_T/2$.

2. The tuned filter outputs are not strictly band limited, and so some distortion is introduced by the low-pass RC filters following the diodes in the envelope detectors.

3. The tuned filter characteristics are not linear over the whole frequency band of the input FM signal $s(t)$.

Nevertheless, by proper design, it is possible to maintain the FM distortion produced by these factors within tolerable limits.

FM Stereo Multiplexing[12]

Stereo multiplexing is a form of frequency-division multiplexing (FDM) designed to transmit two separate signals via the same carrier. It is widely used in FM radio broadcasting to send two different elements of a program (e.g., two different sections of an orchestra, a vocalist and an accompanist) so as to give a spatial dimension to its perception by a listener at the receiving end.

The specification of standards for FM stereo transmission is influenced by two factors:

1. The transmission has to operate within the allocated FM broadcast channels.
2. It has to be compatible with monophonic radio receivers.

The first requirement sets the permissible frequency parameters, including frequency deviation. The second requirement constrains the way in which the transmitted signal is configured.

Figure 3.45*a* shows the block diagram of the multiplexing system used in an FM stereo transmitter. Let $m_l(t)$ and $m_r(t)$ denote the signals picked up by left-hand and right-hand microphones at the transmitting end of the system. They are applied to a simple *matrixer* that generates the *sum signal*, $m_l(t) + m_r(t)$, and the *difference signal*, $m_l(t) - m_r(t)$. The sum signal is left unprocessed in its baseband form; it is available for monophonic reception. The difference signal and a 38-kHz subcarrier (derived from a 19-kHz crystal oscillator by frequency doubling) are applied to a product modulator, thereby producing a DSB-SC modulated wave. In addition to the sum signal and this DSB-SC modulated wave, the multiplexed signal $m(t)$ also includes a 19-kHz pilot to provide a reference for the coherent detection of the difference signal at the stereo receiver. Thus the multiplexed signal is described by

$$m(t) = [m_l(t) + m_r(t)] + [m_l - m_r(t)]\cos(4\pi f_c t) + K\cos(2\pi f_c t) \quad (3.107)$$

where $f_c = 19$ kHz, and K is the amplitude of the pilot tone. The multiplexed signal $m(t)$ then frequency-modulates the main carrier to produce the transmitted signal. The pilot is allotted between 8 and 10 percent of the peak frequency deviation; the amplitude K in Eq. (3.107) is chosen to satisfy this requirement.

At a stereo receiver, the multiplexed signal $m(t)$ is recovered by frequency demodulating the incoming FM wave. Then $m(t)$ is applied to the *demultiplexing system* shown in Fig. 3.45*b*. The individual components of the multiplexed signal

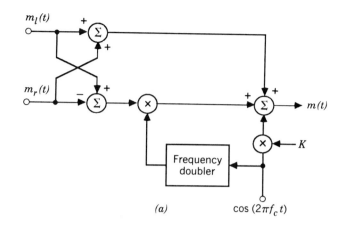

(a)

$\cos(2\pi f_c t)$

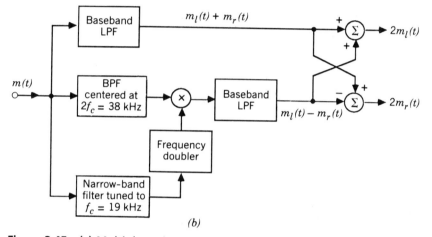

(b)

Figure 3.45 (a) Multiplexer in transmitter of FM stereo. (b) Demultiplexer in receiver of FM stereo.

$m(t)$ are separated by the use of three appropriate filters. The recovered pilot (using a narrow-band filter tuned to 19 kHz) is frequency doubled to produce the desired 38-kHz subcarrier. The availability of this subcarrier enables the coherent detection of the DSB-SC modulated wave, thereby recovering the difference signal, $m_l(t) - m_r(t)$. The baseband low-pass filter in the top path of Fig. 3.45b is designed to pass the sum signal, $m_l(t) + m_r(t)$. Finally, the simple matrixer reconstructs the left-hand signal $m_l(t)$ and right-hand signal $m_r(t)$ and applies them to their respective speakers.

3.12 PHASE-LOCKED LOOP

The *phase-locked loop* (PLL) is a negative feedback system, the operation of which is closely linked to frequency modulation. It can be used for synchronization, frequency division/multiplication, and indirect frequency demodulation. The latter application is the subject of interest here.

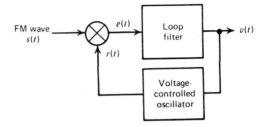

Figure 3.46 Phase-locked loop.

Basically, the phase-locked loop consists of three major components: a *multiplier*, a *loop filter*, and a *voltage-controlled oscillator* (VCO) connected together in the form of a feedback loop, as in Fig. 3.46. The VCO is a sinusoidal generator whose frequency is determined by a voltage applied to it from an external source. In effect, any frequency modulator may serve as a VCO.

We assume that initially we have adjusted the VCO so that when the control voltage is zero, two conditions are satisfied:

1. The frequency of the VCO is precisely set at the unmodulated carrier frequency f_c.
2. The VCO output has a 90 degree phase-shift with respect to the unmodulated carrier wave.

Suppose then that the input signal applied to the phase-locked loop is an FM signal defined by

$$s(t) = A_c \sin[2\pi f_c t + \phi_1(t)] \qquad (3.108)$$

where A_c is the carrier amplitude. With a modulating signal $m(t)$, the angle $\phi_1(t)$ is related to $m(t)$ by the integral

$$\phi_1(t) = 2\pi k_f \int_0^t m(t) \, dt \qquad (3.109)$$

where k_f is the frequency sensitivity of the frequency modulator. Let the VCO output in the phase-locked loop be defined by

$$r(t) = A_v \cos[2\pi f_c t + \phi_2(t)] \qquad (3.110)$$

where A_v is the amplitude. With a control voltage $v(t)$ applied to the VCO input, the angle $\phi_2(t)$ is related to $v(t)$ by the integral

$$\phi_2(t) = 2\pi k_v \int_0^t v(t) \, dt \qquad (3.111)$$

where k_v is the frequency sensitivity of the VCO, measured in Hertz per volt.

The object of the phase-locked loop is to generate a VCO output $r(t)$ that has the same phase angle (except for the fixed difference of 90 degrees) as the input FM signal $s(t)$. The time-varying phase angle $\phi_1(t)$ characterizing $s(t)$ may be due to modulation by a message signal $m(t)$ as in Eq. (3.109), in which case we wish to recover $\phi_1(t)$ in order to *estimate* $m(t)$. In other applications of the phase-locked loop, the time-varying phase angle $\phi_1(t)$ of the incoming signal $s(t)$ may be an unwanted phase shift caused by fluctuations in the communication channel; in this latter case, we wish to *track* $\phi_1(t)$ so as to produce a signal with the same phase angle for the purpose of coherent detection (synchronous demodulation).

To develop an understanding of the phase-locked loop, it is desirable to have a *model* of the loop. In what follows, we first develop a nonlinear model, which is subsequently linearized to simplify the analysis.

Nonlinear Model of the Phase-Locked Loop[13]

According to Fig. 3.46, the incoming FM signal $s(t)$ and the VCO output $r(t)$ are applied to the multiplier, producing two components:

1. A high-frequency component, represented by the *double-frequency* term

$$k_m A_c A_v \sin[4\pi f_c t + \phi_1(t) + \phi_2(t)].$$

2. A low-frequency component represented by the *difference-frequency* term

$$k_m A_c A_v \sin[\phi_1(t) - \phi_2(t)]$$

where k_m is the *multiplier gain*, measured in volt^{-1}.

The loop filter in the phase-lock loop is a low-pass filter, and its response to the high-frequency component will be negligible. The VCO also contributes to the attenuation of this component. Therefore, discarding the high-frequency component (i.e., the double-frequency term), the input to the loop filter is reduced to

$$e(t) = k_m A_c A_v \sin[\phi_e(t)] \qquad (3.112)$$

where $\phi_e(t)$ is the *phase error* defined by

$$\begin{aligned}
\phi_e(t) &= \phi_1(t) - \phi_2(t) \\
&= \phi_1(t) - 2\pi k_v \int_0^t v(t)\, dt
\end{aligned} \qquad (3.113)$$

The loop filter operates on the input $e(t)$ to produce an output $v(t)$ defined by the convolution integral:

$$v(t) = \int_{-\infty}^{\infty} e(\tau) h(t - \tau)\, d\tau \qquad (3.114)$$

where $h(t)$ is the impulse response of the loop filter.

Using Eqs. (3.111) to (3.113) to relate $\phi_e(t)$ and $\phi_1(t)$, we obtain the following nonlinear integro-differential equation as the descriptor of the dynamic behavior of the phase-locked loop:

$$\frac{d\phi_e(t)}{dt} = \frac{d\phi_1(t)}{dt} - 2\pi K_0 \int_{-\infty}^{\infty} \sin[\phi_e(\tau)] h(t - \tau)\, d\tau \qquad (3.115)$$

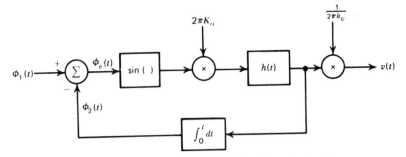

Figure 3.47 Nonlinear model of the phase-locked loop.

where K_0 is a *loop-gain parameter* defined by

$$K_0 = k_m k_v A_c A_v \qquad (3.116)$$

The amplitudes A_c and A_v are both measured in volts, the multiplier gain k_m in volt^{-1} and the frequency sensitivity k_v in Hertz per volt. Hence, it follows from Eq. (3.116) that K_0 has the dimensions of frequency. Equation (3.115) suggests the model shown in Fig. 3.47 for a phase-locked loop. In this model we have also included the relationship between $v(t)$ and $e(t)$ as represented by Eqs. (3.112) and (3.114). We see that the model resembles the block diagram of Fig. 3.46. The multiplier at the input of the phase-locked loop is replaced by a subtracter and a sinusoidal nonlinearity, and the VCO by an integrator.

The sinusoidal nonlinearity in the model of Fig. 3.47 greatly increases the difficulty of analyzing the behavior of the phase-locked loop. It would be helpful to *linearize* this model to simplify the analysis and yet give a good approximate description of the loop's behavior in certain modes of operation. This we do next.

Linear Model of the Phase-Locked Loop

When the phase error $\phi_e(t)$ is zero, the phase-locked loop is said to be in *phase-lock*. When $\phi_e(t)$ is at all times small compared with one radian, we may use the approximation

$$\sin[\phi_e(t)] \simeq \phi_e(t) \qquad (3.117)$$

which is accurate to within 4 percent for $\phi_e(t)$ less than 0.5 radians. In this case, the loop is said to be *near phase-lock*, and the sinusoidal nonlinearity of Fig. 3.47 may be disregarded. Thus, we may represent the phase-locked loop by the linearized model shown in Fig. 3.48a. According to this model, the phase error $\phi_e(t)$ is related to the input phase $\phi_1(t)$ by the linear integro-differential equation:

$$\frac{d\phi_e(t)}{dt} + 2\pi K_0 \int_{-\infty}^{\infty} \phi_e(\tau) h(t - \tau) \, d\tau = \frac{d\phi_1(t)}{dt} \qquad (3.118)$$

Transforming Eq. (3.118) into the frequency domain and solving for $\Phi_e(f)$, the

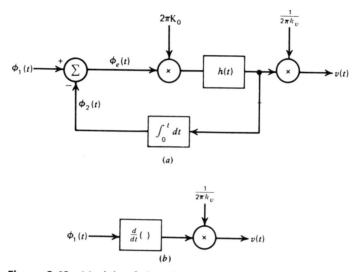

Figure 3.48   Models of the phase-locked loop. (a) Linearized model. (b) Simplified model when the loop gain is very large compared to unity.

Fourier transform of $\phi_e(t)$, in terms of $\Phi_1(f)$, the Fourier transform of $\phi_1(t)$, we get

$$\Phi_e(f) = \frac{1}{1 + L(f)} \, \Phi_1(f) \tag{3.119}$$

The function $L(f)$ in Eq. (3.119) is defined by

$$L(f) = K_0 \frac{H(f)}{jf} \tag{3.120}$$

where $H(f)$ is the transfer function of the loop filter. The quantity $L(f)$ is called the *open-loop transfer function* of the phase-locked loop. Suppose that for all values of f inside the baseband, we make the magnitude of $L(f)$ very large compared with unity. Then from Eq. (3.119) we find that $\Phi_e(f)$ approaches zero. That is, the phase of the VCO becomes asymptotically equal to the phase of the incoming signal. Under this condition, phase-lock is established, and the objective of the phase-locked loop is thereby satisfied.

From Fig. 3.48a we see that $V(f)$, the Fourier transform of the phase-locked loop output $v(t)$, is related to $\Phi_e(f)$ by

$$V(f) = \frac{K_0}{k_v} \, H(f)\Phi_e(f) \tag{3.121}$$

Equivalently, in light of Eq. (3.120), we may write

$$V(f) = \frac{jf}{k_v} \, L(f)\Phi_e(f) \tag{3.122}$$

Therefore, substituting Eq. (3.119) in (3.122), we get

$$V(f) = \frac{(jf/k_v)L(f)}{1 + L(f)} \, \Phi_1(f) \qquad (3.123)$$

Again, when we make $|L(f)| \gg 1$, we may approximate Eq. (3.123) as follows:

$$V(f) \simeq \frac{jf}{k_v} \, \Phi_1(f) \qquad (3.124)$$

The corresponding time-domain relation is

$$v(t) \simeq \frac{1}{2\pi k_v} \frac{d\phi_1(t)}{dt} \qquad (3.125)$$

Thus, provided that the magnitude of the open-loop transfer function $L(f)$ is very large for all frequencies of interest, the phase-locked loop may be modeled as a differentiator with its output scaled by the factor $1/2\pi k_v$, as in Fig. 3.48b.

The simplified model of Fig. 3.48b provides an indirect method of using the phase-locked loop as a frequency demodulator. When the input is an FM signal as in Eq. (3.108), the angle $\phi_1(t)$ is related to the message signal $m(t)$ as in Eq. (3.109). Therefore, substituting Eq. (3.109) in (3.125), we find that the resulting output signal of the phase-locked loop is approximately

$$v(t) \simeq \frac{k_f}{k_v} \, m(t) \qquad (3.126)$$

Equation (3.126) states that when the loop operates in its phase-locked mode, the output $v(t)$ of the phase-locked loop is approximately the same, except for the scale factor k_f/k_v, as the original message signal $m(t)$; frequency demodulation of the incoming FM signal $s(t)$ is thereby accomplished.

A significant feature of the phase-locked loop acting as a demodulator is that the bandwidth of the incoming FM signal can be much wider than that of the loop filter characterized by $H(f)$. The transfer function $H(f)$ can and should be restricted to the baseband. Then the control signal of the VCO has the bandwidth of the baseband (message) signal $m(t)$, whereas the VCO output is a wideband frequency-modulated signal whose instantaneous frequency tracks that of the incoming FM signal. Here we are merely restating the fact that the bandwidth of a wide-band FM signal is much larger than the bandwidth of the message signal responsible for its generation.

The complexity of the phase-locked loop is determined by the transfer function $H(f)$ of the loop filter. The simplest form of a phase-locked loop is obtained when $H(f) = 1$; that is, there is no loop filter, and the resulting phase-locked loop is referred to as a *first-order phase-locked loop*. For higher-order loops, the transfer function $H(f)$ assumes a more complex form. The order of the phase-locked loop is determined by the order of the denominator polynomial of the *closed-loop transfer function*, which defines the output transform $V(f)$ in terms of the input transform $\Phi_1(f)$, as shown in Eq. (3.123).

A major limitation of a first-order phase-locked loop is that the loop gain parameter K_0 controls both the loop bandwidth as well as the hold-in frequency range of the loop; the *hold-in frequency range* refers to the range of frequencies for which the loop remains phase-locked to the input signal. It is for this reason that a first-order phase-locked loop is seldom used in practice. Accordingly, in the remainder of this section we deal only with a second-order phase-locked loop.

Second-Order Phase-Locked Loop

To be specific, consider a *second-order phase-locked loop* using a loop filter with the transfer function

$$H(f) = 1 + \frac{a}{jf} \tag{3.127}$$

where a is a constant. The filter consists of an integrator (using an operational amplifier) and a direct connection, as shown in Fig. 3.49. For this phase-locked loop, the use of Eqs. (3.119) and (3.127) yields

$$\Phi_e(f) = \frac{(jf)^2/aK_0}{1 + [(jf)/a] + [(jf)^2/aK_0]} \, \Phi_1(f) \tag{3.128}$$

Define the *natural frequency* of the loop:

$$f_n = \sqrt{aK_0} \tag{3.129}$$

and the *damping factor*:

$$\zeta = \sqrt{\frac{K_0}{4a}} \tag{3.130}$$

Then we may recast Eq. (3.128) in terms of the parameters f_n and ζ as follows:

$$\Phi_e(f) = \left(\frac{(jf/f_n)^2}{1 + 2\zeta(jf/f_n) + (jf/f_n)^2} \right) \Phi_1(f) \tag{3.131}$$

Assume that the incoming FM signal is produced by a single-tone modulating wave, for which the phase input is

$$\phi_1(t) = \beta \sin(2\pi f_m t) \tag{3.132}$$

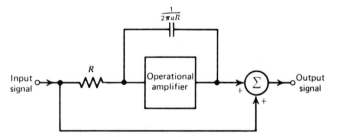

Figure 3.49 Loop filter for second-order phase-locked loop.

Hence, from Eq. (3.131) we find that the corresponding phase error is

$$\phi_e(t) = \phi_{e0} \cos(2\pi f_m t + \psi) \qquad (3.133)$$

where the amplitude ϕ_{e0} and phase ψ are, respectively, defined by

$$\phi_{e0} = \frac{(\Delta f/f_n)(f_m/f_n)}{\{[1 - (f_m/f_n)^2]^2 + 4\zeta^2(f_m/f_n)^2\}^{1/2}} \qquad (3.134)$$

and

$$\psi = \frac{\pi}{2} - \tan^{-1}\left[\frac{2\zeta(f_m/f_n)}{1 - (f_m/f_n)^2}\right] \qquad (3.135)$$

In Fig. 3.50 we have plotted the phase error amplitude ϕ_{e0}, normalized with respect to $\Delta f/f_n$, versus f_m/f_n for different values of ζ. It is apparent that for all values of the damping factor ζ, and assuming a fixed frequency deviation Δf, the phase error is small at low modulation frequencies, rises to a maximum at $f_m = f_n$, and then falls off at higher modulation frequencies. Note also that the maximum value of phase error amplitude decreases with increasing ζ.

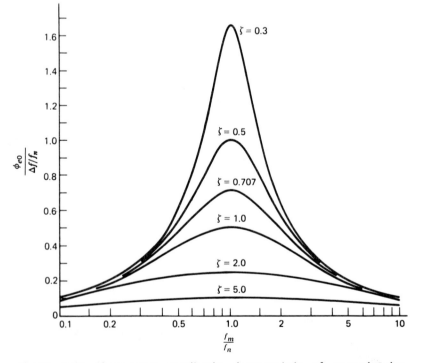

Figure 3.50 Phase-error amplitude characteristic of second-order phase-locked loop.

The Fourier transform of the loop output is related to $\Phi_e(f)$ by Eq. (3.121); hence, with $H(f)$ as defined in Eq. (3.127), we get

$$V(f) = \frac{K_0}{k_v}\left(1 + \frac{a}{jf}\right)\Phi_e(f) \tag{3.136}$$

In light of the definitions given in Eqs. (3.129) and (3.130), we have

$$V(f) = \left(\frac{f_n^2}{jf k_v}\right)\left[1 + 2\zeta\left(\frac{jf}{f_n}\right)\right]\Phi_e(f) \tag{3.137}$$

Substituting Eq. (3.131) in (3.137), we get

$$V(f) = \left(\frac{(jf/k_v)[1 + 2\zeta(jf/f_n)]}{1 + 2\zeta(jf/f_n) + (jf/f_n)^2}\right)\Phi_1(f) \tag{3.138}$$

Therefore, for the phase input $\phi_1(t)$ of Eq. (3.132), we find that the corresponding loop output is

$$v(t) = A_0 \cos(2\pi f_m t + \alpha) \tag{3.139}$$

where the amplitude A_0 and phase α are, respectively, defined by

$$A_0 = \frac{(\Delta f/k_v)[1 + 4\zeta^2(f_m/f_n)^2]^{1/2}}{\{[1 - (f_m/f_n)^2]^2 + 4\zeta^2(f_m/f_n)^2\}^{1/2}} \tag{3.140}$$

and

$$\alpha = \tan^{-1}\left[2\zeta\left(\frac{f_m}{f_n}\right)\right] - \tan^{-1}\left[\frac{2\zeta(f_m/f_n)}{1 - (f_m/f_n)^2}\right] \tag{3.141}$$

From Eq. (3.140), we see that the amplitude A_0 attains its maximum value of $\Delta f/k_v$ at $(f_m/f_n) = 0$; it decreases with increasing f_m/f_n, dropping to zero at $(f_m/f_n) = \infty$.

The important feature of the second-order phase-locked loop is that with an incoming FM signal produced by a modulating sinusoidal wave of fixed amplitude (corresponding to a fixed frequency deviation) and varying frequency, the frequency response that defines the phase error $\phi_e(t)$ is representative of a bandpass filter [see Eq. (3.134)], but the frequency response that defines the loop output $v(t)$ is representative of a low-pass filter [see Eq. (3.140)]. Therefore, by appropriately choosing the parameters ζ and f_n, which determine the frequency response of the loop, it is possible to restrain the phase error to always remain small and thereby lie within the linear range of the loop, whereas at the same time the modulating (message) signal is reproduced at the loop output with minimum distortion. This restraint is, however, conservative with respect to the hold-in capabilities of the loop. As a reasonable rule of thumb, the loop should remain locked if the maximum value of the phase error ϕ_{e0} (which occurs when the modulation frequency f_m is equal to the loop's natural frequency f_n) is always less than 90 degrees.

COMPUTER EXPERIMENT II   Acquisition Mode[14]

When a phase-locked loop is used for coherent detection (synchronous demodulation), the loop must first lock onto the input signal and then follow the variations of its phase angle with time. The process of bringing a loop into phase-lock is called *acquisition*, and the ensuing process of following angular variations in the input signal is called *tracking*. In the acquisition mode and quite possibly the tracking mode, the phase error $\phi_e(t)$ between the input signal $s(t)$ and the VCO output $r(t)$ will certainly be large, thereby invalidating the applicability of the linear model of Fig. 3.48a. In such a situation, we have to use the nonlinear model of Fig. 3.47. However, a nonlinear analysis of the acquisition process based on this latter model is beyond the scope of this book. In this experiment, we use computer simulations to study the acquisition process and thereby develop insight into some of its features.

Consider a second-order phase-locked loop having the following parameters:

$$\text{Loop-gain parameter}\quad K_0 \;=\; \frac{50}{2\pi}\,\text{Hz}$$

$$\text{Natural frequency}\quad f_n \;=\; \frac{1}{2\pi}\,\text{Hz}$$

$$\text{Damping factor}\quad \zeta \;=\; 0.3,\, 0.707,\, 1.0$$

Figure 3.51 presents the variation in the phase error $\phi_e(t)$ with time for each of these three damping factors ζ, assuming a frequency step of 0.125 Hz. These results show that the damping factor $\zeta = 0.707$ gives the best compromise between a fast response time and an underdamped oscillatory behavior.

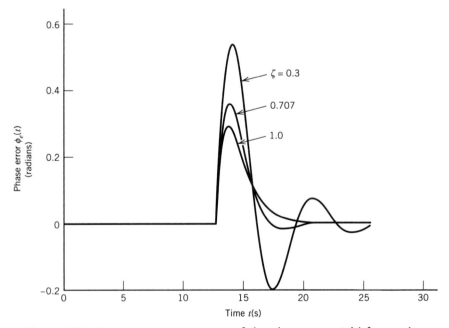

Figure 3.51 Frequency-step response of the phase error $\phi_e(t)$ for varying damping factor ζ.

Figure 3.52 presents the variations in the instantaneous frequency of the VCO for the following loop parameters for different step changes in the input frequency:

$$\text{Loop-gain parameter} \quad K_0 = \frac{50}{2\pi} \text{ Hz}$$

$$\text{Natural frequency} \quad f_n = \frac{1}{2\pi} \text{ Hz}$$

$$\text{Damping factor} \quad \zeta = 0.707$$

Figure 3.52a corresponds to a frequency step of $\Delta f = 0.125$ Hz, for which the frequency error and phase error are both zero when the acquisition process is completed (i.e., steady-state operation of the loop is restored). Figure 3.52b corresponds to a frequency step $\Delta f = 0.5$ Hz; the dynamic behavior of the loop during acquisition is now more complicated than before. Although the steady-state value of the frequency error is indeed zero, the loop has experienced a phenomenon called *cycle slipping*. Specifically, there is a phase error of 2π radians, representing a slip by one cycle. Figures 3.52c and 3.52d correspond to frequency steps of 7/12 Hz and 2/3 Hz, respectively. Now we find that the loop has slipped by two and three cycles, respectively.

Cycle slipping is an undesirable phenomenon as it causes the phase error to be a growing quantity and ultimately unbounded if it is left unchecked.

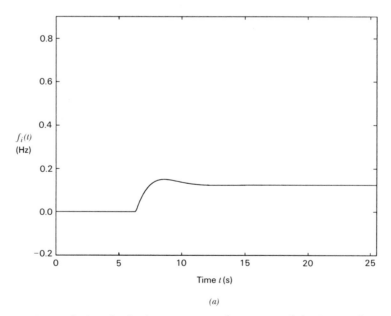

(a)

Figure 3.52 Variations in the instantaneous frequency of the PLL's voltage-controlled oscillator for varying frequency step Δf. (a) $\Delta f = 0.125$ Hz.

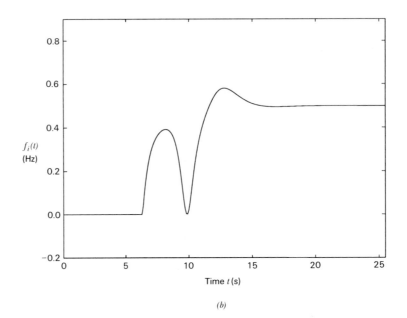

(b)

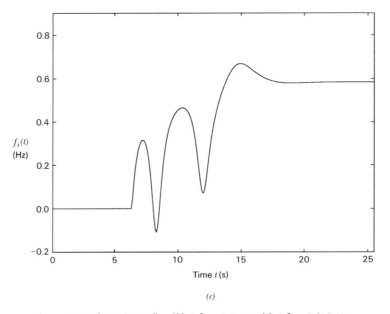

(c)

Figure 3.52 (*continued*)   (b) $\Delta f = 0.5$ Hz. (c) $\Delta f = 7/12$ Hz.

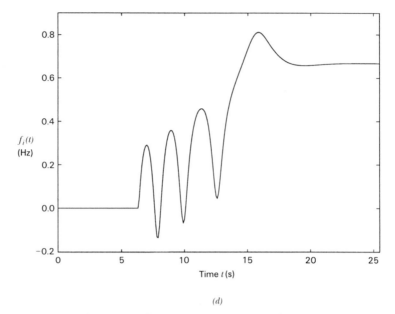

(d)

Figure 3.52 (*continued*) (*d*) $\Delta f = 2/3$ Hz.

3.13 NONLINEAR EFFECTS IN FM SYSTEMS

In the preceding three sections, we studied frequency modulation theory and methods for its generation and demodulation. We complete the discussion of frequency modulation by considering nonlinear effects in FM systems.

Nonlinearities, in one form or another, are present in all electrical networks. There are two basic forms of nonlinearity to consider;

1. The nonlinearity is said to be *strong* when it is introduced intentionally and in a controlled manner for some specific application. Examples of strong nonlinearity include square-law modulators, limiters, and frequency multipliers.

2. The nonlinearity is said to be *weak* when a linear performance is desired, but nonlinearities of a parasitic nature arise due to imperfections. The effect of such weak nonlinearities is to limit the useful signal levels in a system and thereby become an important design consideration.

In this section we examine the effects of weak nonlinearities on frequency modulation.[15]

Consider a communications channel, the transfer characteristic of which is defined by the nonlinear input–output relation

$$v_o(t) = a_1 v_i(t) + a_2 v_i^2(t) + a_3 v_i^3(t) \tag{3.142}$$

where $v_i(t)$ and $v_o(t)$ are the input and output signals, respectively, and a_1, a_2, and a_3 are constants; Eq. (3.142) is a truncated version of Eq. (3.82) used in the context of frequency multiplication. The channel described in Eq. (3.142) is said to be *memoryless* in that the output signal $v_o(t)$ is an instantaneous function of the input signal $v_i(t)$ (i.e., there is no energy storage involved in the description). We wish to determine the effect of transmitting a frequency-modulated wave through such a channel. The FM signal is defined by

$$v_i(t) = A_c \cos[2\pi f_c t + \phi(t)]$$

where

$$\phi(t) = 2\pi k_f \int_0^t m(t) \, dt$$

For this input signal, the use of Eq. (3.142) yields

$$v_o(t) = a_1 A_c \cos[2\pi f_c t + \phi(t)] + a_2 A_c^2 \cos^2[2\pi f_c t + \phi(t)]$$
$$+ a_3 A_c^3 \cos^3[2\pi f_c t + \phi(t)] \tag{3.143}$$

Expanding the squared and cubed cosine terms in Eq. (3.143) and then collecting common terms, we get

$$v_o(t) = \frac{1}{2} a_2 A_c^2 + \left(a_1 A_c + \frac{3}{4} a_3 A_c^3 \right) \cos[2\pi f_c t + \phi(t)]$$

$$+ \frac{1}{2} a_2 A_c^2 \cos[4\pi f_c t + 2\phi(t)] \tag{3.144}$$

$$+ \frac{1}{4} a_3 A_c^3 \cos[6\pi f_c t + 3\phi(t)]$$

Thus the channel output consists of a dc component and three frequency-modulated signals with carrier frequencies f_c, $2f_c$, and $3f_c$; the latter components are contributed by the linear, second-order, and third-order terms of Eq. (3.142), respectively.

To extract the desired FM signal from the channel output $v_o(t)$, that is, the particular component with carrier frequency f_c, it is necessary to separate the FM signal with this carrier frequency from the one with the closest carrier frequency, $2f_c$. Let Δf denote the frequency deviation of the incoming FM signal $v_i(t)$, and W denote the highest frequency component of the message signal $m(t)$. Then, applying Carson's rule and noting that the frequency deviation about the second harmonic of the carrier frequency is doubled, we find that the necessary condition for separating the desired FM signal with the carrier frequency f_c from that with the carrier frequency $2f_c$ is

$$2f_c - (2\Delta f + W) > f_c + \Delta f + W$$

or

$$f_c > 3\Delta f + 2W \tag{3.145}$$

Thus, by using a band-pass filter of mid-band frequency f_c and bandwidth $2\Delta f + 2W$, the channel output is reduced to

$$v'_o(t) = \left(a_1 A_c + \frac{3}{4} a_3 A_c^3\right) \cos[2\pi f_c t + \phi(t)] \qquad (3.146)$$

We see therefore that the only effect of passing an FM signal through a channel with amplitude nonlinearities, followed by appropriate filtering, is simply to modify its amplitude. That is, unlike amplitude modulation, frequency modulation is not affected by distortion produced by transmission through a channel with amplitude nonlinearities. It is for this reason that we find frequency modulation widely used in microwave radio and satellite communication systems: It permits the use of highly nonlinear amplifiers and power transmitters, which are particularly important to producing a maximum power output at radio frequencies.

An FM system is extremely sensitive to *phase nonlinearities*, however, as we would intuitively expect. A common type of phase nonlinearity that is encountered in microwave radio systems is known as *AM-to-PM conversion*. This is the result of the phase characteristic of repeaters or amplifiers used in the system being dependent on the instantaneous amplitude of the input signal. In practice, AM-to-PM conversion is characterized by a constant K, which is measured in degrees per dB and may be interpreted as the peak phase change at the output for a 1-dB change in envelope at the input. When an FM wave is transmitted through a microwave radio link, it picks up spurious amplitude variations due to noise and interference during the course of transmission, and when such an FM wave is passed through a repeater with AM-to-PM conversion, the output will contain unwanted phase modulation and resultant distortion. It is therefore important to keep the AM-to-PM conversion at a low level. For example, for a good microwave repeater, the AM-to-PM conversion constant K is less than 2 degrees per dB.

3.14 THE SUPERHETERODYNE RECEIVER[16]

In a *broadcasting* system, irrespective of whether it is based on amplitude modulation or frequency modulation, the receiver not only has the task of demodulating the incoming modulated signal, but it is also required to perform some other system functions:

- *Carrier-frequency tuning*, the purpose of which is to select the desired signal (i.e., desired radio or TV station).
- *Filtering*, which is required to separate the desired signal from other modulated signals that may be picked up along the way.
- *Amplification*, which is intended to compensate for the loss of signal power incurred in the course of transmission.

The *superheterodyne receiver*, or *superhet* as it is often referred to, is a special type of receiver that fulfils all three functions, particularly the first two, in an elegant and practical fashion. Specifically, it overcomes the difficulty of having to build a tunable high- (and variable-) Q filter. Indeed, practically all radio and TV receivers now being made are of the superheterodyne type.

Table 3.3 Typical Frequency Parameters
of AM and FM Radio Receivers

	AM Radio	FM Radio
RF carrier range	0.535–1.605 MHz	88–108 MHz
Mid-band frequency of IF section	0.455 MHz	10.7 MHz
IF bandwidth	10 kHz	200 kHz

Basically, the receiver consists of a radio-frequency (RF) section, a mixer and local oscillator, an intermediate frequency (IF) section, demodulator, and power amplifier. Typical frequency parameters of commercial AM and FM radio receivers are listed in Table 3.3. Figure 3.53 shows the block diagram of a superheterodyne receiver for amplitude modulation using an envelope detector for demodulation.

The incoming amplitude-modulated wave is picked up by the receiving antenna and amplified in the RF section that is tuned to the carrier frequency of the incoming wave. The combination of mixer and local oscillator (of adjustable frequency) provides a *heterodyning* function, whereby the incoming signal is converted to a predetermined fixed *intermediate frequency*, usually lower than the incoming carrier frequency. This frequency translation is achieved without disturbing the relation of the sidebands to the carrier. The result of the heterodyning is to produce an intermediate-frequency carrier defined by

$$f_{IF} = f_{RF} - f_{LO} \tag{3.147}$$

where f_{LO} is the frequency of the local oscillator and f_{RF} is the carrier frequency of the incoming RF signal. We refer to f_{IF} as the intermediate frequency (IF), because the signal is neither at the original input frequency nor at the final baseband frequency. The mixer–local oscillator combination is sometimes referred to as the *first detector*, in which case the demodulator is called the *second detector*.

The IF section consists of one or more stages of tuned amplification, with a bandwidth corresponding to that required for the particular type of signal that the receiver is intended to handle. This section provides most of the amplification and selectivity in the receiver. The output of the IF section is applied to a demodulator, the purpose of which is to recover the baseband signal. If coherent

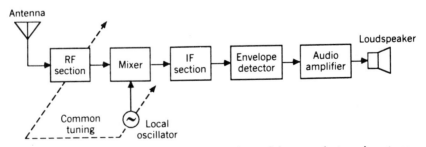

Figure 3.53 Basic elements of an AM receiver of the superheterodyne type.

detection is used, then a coherent signal source must be provided in the receiver. The final operation in the receiver is the power amplification of the recovered message signal.

In a superheterodyne receiver the mixer will develop an intermediate frequency output when the input signal frequency is greater or less than the local oscillator frequency by an amount equal to the intermediate frequency. That is, there are two input frequencies, namely, $|f_{LO} \pm f_{IF}|$, which will result in f_{IF} at the mixer output. This introduces the possibility of simultaneous reception of two signals differing in frequency by twice the intermediate frequency. For example, a receiver tuned to 1 MHz and having an IF of 0.455 MHz is subject to an *image interference* at 1.910 MHz; indeed, any receiver with this value of IF, when tuned to any station, is subject to image interference at a frequency of 0.910 MHz higher than the desired station. Since the function of the mixer is to produce the difference between two applied frequencies, it is incapable of distinguishing between the desired signal and its image in that it produces an IF output from either one of them. The only practical cure for image interference is to employ highly selective stages in the RF section (i.e., between the antenna and the mixer) in order to favor the desired signal and discriminate against the undesired or *image signal*. The effectiveness of suppressing unwanted image signals increases as the number of selective stages in the radio-frequency section increases, and as the ratio of intermediate to signal frequency increases.

The basic difference between AM and FM superheterodyne receivers lies in the use of an FM demodulator such as limiter-frequency discriminator. In an FM system, the message information is transmitted by variations of the instantaneous frequency of a sinusoidal carrier wave, and its amplitude is maintained constant. Therefore, any variations of the carrier amplitude at the receiver input must result from noise or interference. An *amplitude limiter*, following the IF section, is used to remove amplitude variations by clipping the modulated wave at the IF section output almost to the zero axis. The resulting rectangular wave is rounded off by a band-pass filter that suppresses harmonics of the carrier frequency. Thus the filter output is again sinusoidal, with an amplitude that is practically independent of the carrier amplitude at the receiver input (see Problem 3.51).

3.15 SUMMARY AND DISCUSSION

In this chapter we studied the principles of continuous-wave (CW) modulation and some important methods of generating and demodulating CW modulated signals. This analog form of modulation uses a sinusoidal carrier whose amplitude or angle is varied in accordance with a message signal. We may thus distinguish two families of CW modulation: amplitude modulation and angle modulation.

Amplitude modulation may itself be classified into four types, depending on the spectral content of the modulated signal. The four types of amplitude modulation and their practical merits are as follows:

1. Standard amplitude modulation (AM), in which the upper and lower sidebands are transmitted in full, accompanied by the carrier wave. Accordingly, demodulation of an AM signal is accomplished rather simply in the receiver by using an envelope detector, for example. It is for this reason we find that

full AM is commonly used in commercial AM radio *broadcasting*, which involves a single powerful transmitter and numerous receivers that are relatively cheap to build.

2. Double sideband-suppressed carrier (DSB-SC) modulation, in which only the upper and lower sidebands are transmitted. The suppression of the carrier wave means that DSB-SC modulation requires much less power than standard AM to transmit the same message signal; this advantage of DSB-SC modulation over full AM is, however, attained at the expense of increased receiver complexity. DSB-SC modulation is therefore well suited for *point-to-point communication* involving one transmitter and one receiver; in this form of communication, transmitted power is at a premium and the use of a complex receiver is therefore justifiable.

3. Single sideband (SSB) modulation, in which only the upper sideband or lower sideband is transmitted. It is optimum in the sense that it requires the minimum transmitted power and the minimum channel bandwidth for conveying a message signal from one point to another. We therefore find that among CW modulation techniques, SSB modulation is the preferred method of modulation for long-distance transmission of voice signals over metallic circuits, because it permits long spacing between the *repeaters*, which is a more important consideration here than simple terminal equipment. A "repeater" is a wide-band amplifier used at an intermediate point along the transmission path to make up for the attenuation incurring during the course of signal transmission. SSB is also the preferred method of CW modulation when operating over a *frequency-selective channel.* Fading arises in a radio environment due to a phenomenon known as *multipath*; in such an environment the received signal is made up of two or more components that propagate from the transmitter to the receiver along different paths, with the result that there may be cancellation of one component by another and therefore wide variations in the received signal power. Fading is said to be frequency selective when it is highly sensitive to the frequency of radio propagation. With only a single sideband transmitted in SSB modulation, it is the least affected by frequency-selective fading.

4. Vestigial sideband modulation, in which "almost" the whole of one sideband and a "vestige" of the other sideband are transmitted in a prescribed complementary fashion. VSB modulation requires a channel bandwidth that is intermediate between that required for SSB and DSB-SC systems, and the saving in bandwidth can be significant if modulating signals with large bandwidths are being handled, as in the case of television signals and high-speed data.

DSB-SC, SSB, and VSB are examples of *linear modulation*, in which the modulated signal is described by the canonical relation

$$s(t) = s_I(t)\cos(2\pi f_c t) - s_Q(t)\sin(2\pi f_c t)$$

The in-phase component $s_I(t)$ is a scaled version of the message signal $m(t)$. The quadrature component $s_Q(t)$ is derived from $m(t)$ by some form of linear filtering. Accordingly, the principle of superposition can be used to calculate the modulator output $s(t)$ as the sum of responses of the modulator to individual components of $m(t)$. In Table 3.4 we have summarized the definitions for $s_I(t)$

Table 3.4 Different Forms of Linear Modulation

Type of Modulation	In-Phase Component $s_I(t)$	Quadrature Component $s_Q(t)$	Comments
DSB-SC	$m(t)$	0	$m(t)$ = message signal
SSB:			
(a) Upper sideband transmitted	$\frac{1}{2}m(t)$	$\frac{1}{2}\hat{m}(t)$	$\hat{m}(t)$ = Hilbert transform of $m(t)$
(b) Lower sideband transmitted	$\frac{1}{2}m(t)$	$-\frac{1}{2}\hat{m}(t)$	
VSB:			
(a) Vestige of lower sideband transmitted	$\frac{1}{2}m(t)$	$\frac{1}{2}m'(t)$	$m'(t)$ = output of filter of transfer function $H_Q(f)$, produced by $m(t)$.
(b) Vestige of upper sideband transmitted	$\frac{1}{2}m(t)$	$-\frac{1}{2}m'(t)$	For the definition of $H_Q(f)$, see Eq. (3.34)

and $s_Q(t)$ in terms of $m(t)$ for DSB-SC, SSB, and VSB modulated signals, assuming a carrier of unit amplitude. In a strict sense, ordinary amplitude modulation fails to meet the definition of a linear modulator with respect to the message signal. If $s_1(t)$ is the AM signal produced by a message signal $m_1(t)$ and $s_2(t)$ is the AM signal produced by a second message signal $m_2(t)$, then the AM wave produced by $m_1(t)$ plus $m_2(t)$ is certainly not equal to $s_1(t)$ plus $s_2(t)$. Nevertheless, the departure from linearity in AM is of a rather mild sort, such that many of the mathematical procedures applicable to linear modulation may be retained. For example, the band-pass representation is still applicable to an AM wave, with the in-phase component $s_I(t) = 1 + k_a m(t)$, where k_a is the amplitude sensitivity of the modulator, and the quadrature component $s_Q(t) = 0$.

Angle modulation may be classified into frequency modulation (FM) and phase modulation (PM). In FM, the instantaneous frequency of a sinusoidal carrier is varied in proportion to the message signal. In PM, on the other hand, it is the phase of the carrier that is varied in proportion to the message signal. The instantaneous frequency is defined as the derivative of the phase with respect to time, except for the scaling factor $1/(2\pi)$. Accordingly, FM and PM are closely related to each other; if we know the properties of the one, we can determine those of the other. For this reason, and because FM is commonly used in broadcasting, much of the material on angle modulation in the chapter was devoted to FM.

Unlike amplitude modulation, FM is a nonlinear modulation process. Accordingly, spectral analysis of FM is more difficult than for AM. Nevertheless, by studying single-tone FM, we were able to develop a great deal of insight into the spectral properties of FM. In particular, we derived an empirical rule known as Carson's rule for an approximate evaluation of the transmission bandwidth B_T of FM. According to this rule, B_T is controlled by a single parameter: the modulation index β for sinusoidal FM, or the deviation ratio D for nonsinusoidal FM.

In FM, the carrier amplitude, and therefore the transmitted average power, is maintained constant. Herein lies the important advantage of FM over AM in combatting the effects of noise or interference at reception, an issue that we study in Chapter 5, after familiarizing ourselves with probability theory and random processes in the next chapter. This advantage becomes increasingly more pronounced as the modulation index (deviation ratio) is increased, which has the effect of increasing the transmission bandwidth in a corresponding way. Thus, frequency modulation provides a practical method for the trade-off of channel bandwidth for improved noise performance, which is not feasible with amplitude modulation.

NOTES AND REFERENCES

1. It appears that the terms "continous wave" and "heterodyning" were first used by Reginald Fessenden in the early 1900s.

2. For a detailed description of the ring modulator used to generate DSB-SC modulated signals, see Tucker (1953).

3. The Costas receiver is named in honor of its inventor; see Costas (1956).

4. In Section 3.6 we describe two methods for the generation of a VSB signal, one based on the scheme of Fig. 3.17a and the other based on Fig. 3.18. In an insightful article, Hill (1974) describes another time-domain method for the representation of VSB signals. Specifically, the VSB signal is expressed as the product of a narrow-band "envelope" function and an SSB signal.

5. For a collection of papers on television technology, see the book edited by Rzeszewski (1984).

6. For a discussion of the filtering requirements in generating SSB modulated signals, see the paper by Kurth (1976).

7. For a discussion of the performance of multiplex transmission, see Bennett (1970, pp. 213–218). For additional information on FDM systems, see "Transmission Systems for Communications," Bell Telephone Laboratories, pp. 128–137 (Western Electric, 1971); and "Reference Data for Radio Engineers," International Telephone and Telegraph Corporation, pp. 30-23 to 30-27 (H. W. Sams, 1968).

8. In a multitone FM signal, the modulating signal consists of a group of frequencies that are completely unrelated or harmonically related to each other. For a spectral analysis of multitone FM signals, see the books by Black (1953) and Panter (1965).

9. Bessel functions play an important role in the solution of a certain differential equation and also in the mathematical formulation of many physical problems. For a detailed treatment of the subject, see Wylie and Barrett (1982, pp. 572–625). A table of Bessel functions is given in Appendix 4 at the end of the book.

10. Carson's rule for the bandwidth of FM signals is named in honor of its originator; Carson and Fry (1937) wrote one of the early classic papers on frequency modulation theory.

11. The indirect method of generating a wide-band FM wave was first proposed by Armstrong (1936). Armstrong was also the first to recognize the noise-robustness properties of frequency modulation.

12. Stereo multiplexing usually involves the use of frequency modulation for radio transmission. However, it may also be transmitted using amplitude modulation; for details, see the paper by Mennie (1978).

13. When a phase-locked loop is used to demodulate an FM wave, the loop must first lock onto the incoming FM wave and then follow the variations in its phase. During the lock-up operation, the phase error $\phi_e(t)$ between the incoming FM wave and the VCO output will be large, thus requiring the use of the nonlinear model of Fig. 3.46. For a full treatment of the nonlinear analysis of a phase-locked loop, see Gardner (1979), Lindsey (1972), and Viterbi (1966). An introductory treatment of the phase-

locked loop is presented in Sakrison (1968, Chapter 10). For a collection of papers on phase-locked loops and their applications, see the book edited by Lindsey and Simon (1977).

14. For the computer simulation of a phase-locked loop, see the book by Jeruchim, Balaban, and Shanmugan (1992, pp. 578–597).

15. For a detailed discussion of the characterization and system effects of weak nonlinearities, see "Transmission Systems for Communication," Bell Telephone Laboratories, pp. 237–278 (Western Electric, 1971).

16. For detailed description of the superheterodyne receiver, see the *Radio Engineering handbook* edited by Henney (1959, pp. 19-34 to 19-41).

PROBLEMS

Problem 3.1 A carrier wave of frequency 1 MHz is modulated 50 percent by a sinusoidal frequency 5 kHz. The resulting AM signal is transmitted through the resonant circuit of Fig. P3.1, which is tuned to the carrier frequency and has a Q-factor of 175. Determine the modulated signal after transmission through this circuit. What is the percentage modulation of this modulated signal?

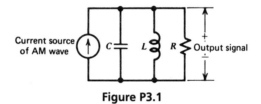

Figure P3.1

Problem 3.2 For a *p-n* junction diode, the current i through the diode and the voltage v across it are related by

$$i = I_0 \left[\exp\left(-\frac{v}{V_T} \right) - 1 \right]$$

where I_0 is the reverse saturation current, and V_T is the volt-equivalent of temperature defined by

$$V_T = \frac{kT}{e}$$

where k is Boltzmann's constant in joules per degree Kelvin, T is the absolute temperature in degrees Kelvin, and e is the charge of an electron. At room temperature, $V_T = 0.026$ volts.

(a) Expand i as a power series in v, retaining terms up to v^3.

(b) Let

$$v = 0.01 \cos(2\pi f_m t) + 0.01 \cos(2\pi f_c t) \text{ volts}$$

where $f_m = 1$ kHz and $f_c = 100$ kHz. Determine the spectrum of the resulting diode current i.

(c) Specify the band-pass filter required to extract from the diode current an AM signal with carrier frequency f_c.

(d) What is the percentage modulation of this AM signal?

Problem 3.3   Suppose that nonlinear devices are available for which the output current i_o and input voltage v_i are related by

$$i_o = a_1 v_i + a_3 v_i^3$$

where a_1 and a_3 are constants. Explain how these devices may be used to provide: (a) a product modulator and (b) an amplitude modulator.

Problem 3.4   Figure P3.2 shows the circuit diagram of a *square-law modulator*. The signal applied to the nonlinear device is relatively weak, such that it can be represented by a square law:

$$v_2(t) = a_1 v_1(t) + a_2 v_1^2(t)$$

where a_1 and a_2 are constants, $v_1(t)$ is the input voltage, and $v_2(t)$ is the output voltage. The input voltage is defined by

$$v_1(t) = A_c \cos(2\pi f_c t) + m(t)$$

where $m(t)$ is a message signal and $A_c \cos(2\pi f_c t)$ is the carrier wave.

(a)   Evaluate the output voltage $v_2(t)$.

(b)   Specify the frequency response that the tuned circuit in Fig. P3.2 must satisfy in order to generate an AM signal with f_c as the carrier frequency.

(c)   What is the amplitude sensitivity of this AM signal?

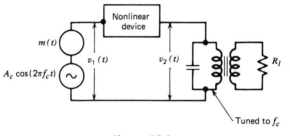

Figure P3.2

Problem 3.5   Consider the AM signal

$$s(t) = A_c[1 + \mu \cos(2\pi f_m t)]\cos(2\pi f_c t)$$

produced by a sinusoidal modulating signal of frequency f_m. Assume that the modulation factor is $\mu = 2$, and the carrier frequency f_c is much greater than f_m. The AM signal $s(t)$ is applied to an ideal envelope detector, producing the output $v(t)$.

(a)   Determine the Fourier series representation of $v(t)$.

(b)   What is the ratio of second-harmonic amplitude to fundamental amplitude in $v(t)$?

Problem 3.6   Consider a *square-law detector*, using a nonlinear device whose transfer characteristic is defined by

$$v_2(t) = a_1 v_1(t) + a_2 v_1^2(t)$$

where a_1 and a_2 are constants, $v_1(t)$ is the input, and $v_2(t)$ is the output. The input consists of the AM wave

$$v_1(t) = A_c[1 + k_a m(t)]\cos(2\pi f_c t)$$

(a) Evaluate the output $v_2(t)$.

(b) Find the conditions for which the message signal $m(t)$ may be recovered from $v_2(t)$.

Problem 3.7 The AM signal

$$s(t) = A_c[1 + k_a m(t)]\cos(2\pi f_c t)$$

is applied to the system shown in Fig. P3.3. Assuming that $|k_a m(t)| < 1$ for all t and the message signal $m(t)$ is limited to the interval $-W \le f \le W$, and that the carrier frequency $f_c > 2W$, show that $m(t)$ can be obtained from the square-rooter output $v_3(t)$.

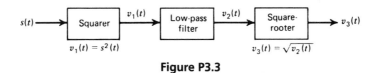

Figure P3.3

Problem 3.8 Consider a message signal $m(t)$ with the spectrum shown in Fig. P3.4. The message bandwidth $W = 1$ kHz. This signal is applied to a product modulator, together with a carrier wave $A_c \cos(2\pi f_c t)$, producing the DSB-SC modulated signal $s(t)$. The modulated signal is next applied to a coherent detector. Assuming perfect synchronism between the carrier waves in the modulator and detector, determine the spectrum of the detector output when: (a) the carrier frequency $f_c = 1.25$ kHz and (b) the carrier frequency $f_c = 0.75$ kHz. What is the lowest carrier frequency for which each component of the modulated signal $s(t)$ is uniquely determined by $m(t)$?

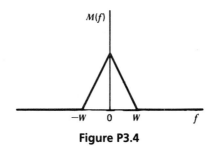

Figure P3.4

Problem 3.9 Figure P3.5 shows the circuit diagram of a *balanced modulator*. The input applied to the top AM modulator is $m(t)$, whereas that applied to the lower AM modulator is $-m(t)$; these two modulators have the same amplitude sensitivity. Show that the output $s(t)$ of the balanced modulator consists of a DSB-SC modulated signal.

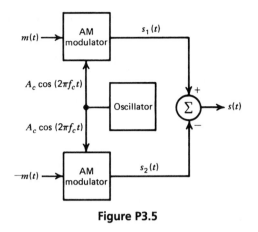

Figure P3.5

Problem 3.10   Figure 3.10 shows the circuit details of the ring modulator. Assume that the diodes are identical and the transformers are perfectly balanced. Let R denote the terminating resistance at the input end and output end of the modulator (assuming ideal 1 : 1 transformers). Determine the output voltage of the modulator for each of the two conditions described in Figs. 3.10*b* and 3.10*c*. Hence, show that these two output voltages are equal in magnitude and opposite in polarity.

Problem 3.11   A DSB-SC modulated signal is demodulated by applying it to a coherent detector.

(a) Evaluate the effect of a frequency error Δf in the local carrier frequency of the detector, measured with respect to the carrier frequency of the incoming DSB-SC signal.

(b) For the case of a sinusoidal modulating wave, show that because of this frequency error, the demodulated signal exhibits *beats* at the error frequency. Illustrate your answer with a sketch of this demodulated signal.

Problem 3.12   Consider the DSB-SC signal

$$s(t) = A_c \cos(2\pi f_c t)\, m(t)$$

where $A_c \cos(2\pi f_c t)$ is the carrier wave and $m(t)$ is the message signal. This modulated signal is applied to a square-law device characterized by

$$y(t) = s^2(t)$$

The output $y(t)$ is next applied to a narrow-band filter with a passband amplitude response of one, mid-band frequency $2f_c$, and bandwidth Δf. Assume that Δf is small enough to treat the spectrum of $y(t)$ as essentially constant inside the passband of the filter.

(a) Determine the spectrum of the square-law device output $y(t)$.

(b) Show that the filter output $v(t)$ is approximately sinusoidal, given by

$$v(t) \simeq \frac{A_c^2}{2}\, E\, \Delta f\, \cos(4\pi f_c t)$$

where E is the energy of the message signal $m(t)$.

Problem 3.13 Consider the quadrature–carrier multiplex system of Fig. 3.16. The multiplexed signal $s(t)$ produced at the transmitter output in Fig. 3.16a is applied to a communication channel of transfer function $H(f)$. The output of this channel is in turn applied to the receiver input in Fig. 3.16b. Prove that the condition

$$H(f_c + f) = H^*(f_c - f), \qquad 0 \le f \le W$$

is necessary for recovery of the message signals $m_1(t)$ and $m_2(t)$ at the receiver outputs; f_c is the carrier frequency, and W is the message bandwidth. *Hint:* Evaluate the spectra of the two receiver outputs.

Problem 3.14 Suppose that in the receiver of the quadrature–carrier multiplex system of Fig. 3.16 the local carrier available for demodulation has a phase error ϕ with respect to the carrier source used in the transmitter. Assuming a distortionless communication channel between transmitter and receiver, show that this phase error will cause *cross-talk* to arise between the two demodulated signals at the receiver outputs. By cross-talk we mean that a portion of one message signal appears at the receiver output belonging to the other message signal, and vice versa.

Problem 3.15 A particular version of *AM stereo* uses quadrature multiplexing. Specifically, the carrier $A_c\cos(2\pi f_c t)$ is used to modulate the sum signal

$$m_1(t) = V_0 + m_\ell(t) + m_r(t)$$

where V_0 is a dc offset included for the purpose of transmitting the carrier component, m_ℓ is the left-hand audio signal, and $m_r(t)$ is the right-hand audio signal. The quadrature carrier $A_c\sin(2\pi f_c t)$ is used to modulate the difference signal

$$m_2(t) = m_\ell(t) - m_r(t)$$

(a) Show that an envelope detector may be used to recover the sum $m_r(t) + m_\ell(t)$ from the quadrature-multiplexed signal. How would you minimize the signal distortion produced by the envelope detector?

(b) Show that a coherent detector can recover the difference $m_\ell(t) - m_r(t)$.

(c) How are the desired $m_\ell(t)$ and $m_r(t)$ finally obtained?

Problem 3.16 The single tone modulating signal $m(t) = A_m\cos(2\pi f_m t)$ is used to generate the VSB signal

$$s(t) = \frac{1}{2}aA_m A_c \cos[2\pi(f_c + f_m)t] + \frac{1}{2}A_m A_c(1 - a)\cos[2\pi(f_c - f_m)t]$$

where a is a constant, less than unity, representing the attenuation of the upper side frequency.

(a) Find the quadrature component of the VSB signal $s(t)$.

(b) The VSB signal, plus the carrier $A_c\cos(2\pi f_c t)$, is passed through an envelope detector. Determine the distortion produced by the quadrature component.

(c) What is the value of constant a for which this distortion reaches its worst possible condition?

Problem 3.17 Show that the quadrature component of a VSB signal may be derived by passing the Hilbert transform of the message signal through a high-pass filter. Sketch the frequency response of this filter.

Problem 3.18   Using the message signal

$$m(t) = \frac{1}{1 + t^2}$$

determine and sketch the modulated waves for the following methods of modulation:

(a) Amplitude modulation with 50 percent modulation.
(b) Double sideband-suppressed carrier modulation.
(c) Single sideband modulation with only the upper sideband transmitted.
(d) Single sideband modulation with only the lower sideband transmitted.

Problem 3.19   Figure 3.24b shows the block diagram of a two-stage SSB modulator. The input signal $m(t)$ consists of a voice signal occupying the frequency band 0.3 to 3.4 kHz. The oscillator frequencies are $f_1 = 100$ kHz and $f_2 = 10$ MHz.

(a) Specify the sidebands of all the modulated signals appearing in this system.
(b) Specify the filtering requirements of the system so that the output $s_2(t)$ consists of an SSB signal containing an upper sideband only. What is the frequency band occupied by $s_2(t)$?

Problem 3.20   The local oscillator used for the demodulation of an SSB signal $s(t)$ has a frequency error Δf measured with respect to the carrier frequency f_c used to generate $s(t)$. Otherwise, there is perfect synchronism between this oscillator in the receiver and the oscillator supplying the carrier wave in the transmitter. Evaluate the demodulated signal for the following two situations:

(a) The SSB signal $s(t)$ consists of the upper sideband only.
(b) The SSB signal $s(t)$ consists of the lower sideband only.

Problem 3.21   Figure P3.6 shows the block diagram of *Weaver's method* for generating SSB modulated waves. The message (modulating) signal $m(t)$ is limited to the frequency band $f_a \leq |f| \leq f_b$. The auxiliary carrier applied to the first pair of product modulators has a frequency f_0, which lies at the center of this band, as shown by

$$f_0 = \frac{f_a + f_b}{2}$$

The low-pass filters in the in-phase and quadrature channels are identical, each with a cutoff frequency equal to $(f_b - f_a)/2$. The carrier applied to the second pair of product modulators has a frequency f_c that is greater than $(f_b - f_a)/2$. Sketch the spectra at the various points in the modulator of Fig. P3.6, and hence show that:

(a) For the lower sideband, the contributions of the in-phase and quadrature channels are of opposite polarity, and by adding them at the modulator output, the lower sideband is suppressed.
(b) For the upper sideband, the contributions of the in-phase and quadrature channels are of the same polarity, and by adding them, the upper sideband is transmitted.
(c) How would you modify the modulator of Fig. P3.6, so that only the lower sideband is transmitted?

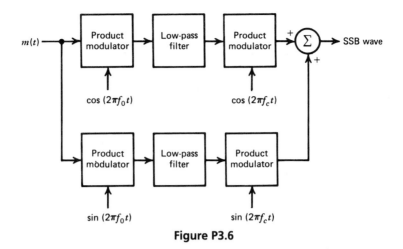

Figure P3.6

Problem 3.22

(a) Let $s_u(t)$ denote the SSB signal obtained by transmitting only the upper sideband, and let $\hat{s}_u(t)$ denote its Hilbert transform. Show that

$$m(t) = \frac{2}{A_c}[s_u(t)\cos(2\pi f_c t) + \hat{s}_u(t)\sin(2\pi f_c t)]$$

and

$$\hat{m}(t) = \frac{2}{A_c}[\hat{s}_u(t)\cos(2\pi f_c t) - s_u(t)\sin(2\pi f_c t)]$$

where $m(t)$ is the message signal, $\hat{m}(t)$ its Hilbert transform, f_c the carrier frequency, and A_c the carrier amplitude.

(b) Show that the corresponding equations in terms of the SSB signal $s_l(t)$ obtained by transmitting only the lower sideband are

$$m(t) = \frac{2}{A_c}[s_l(t)\cos(2\pi f_c t) + \hat{s}_l(t)\sin(2\pi f_c t)]$$

and

$$\hat{m}(t) = \frac{2}{A_c}[s_l(t)\sin(2\pi f_c t) - \hat{s}_l(t)\cos(2\pi f_c t)]$$

(c) Using the results of (a) and (b) above, set up the block diagram of a receiver for demodulating an SSB signal.

Problem 3.23   Consider the modulated wave

$$s(t) = A_c \cos(2\pi f_c t) + m(t)\cos(2\pi f_c t) - \hat{m}(t)\sin(2\pi f_c t)$$

which represents a carrier plus an SSB signal, with $m(t)$ denoting the message signal and $\hat{m}(t)$ its Hilbert transform. Determine the conditions for which an ideal envelope detector, with $s(t)$ as input, would produce a good approximation to the message signal $m(t)$.

Problem 3.24

(a) Consider a message signal $m(t)$ containing frequency components at 100, 200, and 400 Hz. This signal is applied to an SSB modulator together with a carrier at 100 kHz, with only the upper sideband retained. In the coherent detector used to recover $m(t)$, the local oscillator supplies a sine wave of frequency 100.02 kHz. Determine the frequency components of the detector output.

(b) Repeat your analysis, assuming that only the lower sideband is transmitted.

Problem 3.25   The spectrum of a voice signal $m(t)$ is zero outside the interval $f_a \leq |f| \leq f_b$. In order to ensure communication privacy, this signal is applied to a *scrambler* that consists of the following cascade of components: a product modulator, a high-pass filter, a second product modulator, and a low-pass filter. The carrier wave applied to the first product modulator has a frequency equal to f_c, whereas that applied to the second product modulator has a frequency equal to $f_b + f_c$; both of them have unit amplitude. The high-pass and low-pass filters have the same cutoff frequency at f_c. Assume that $f_c > f_b$.

(a) Derive an expression for the scrambler output $s(t)$, and sketch its spectrum.

(b) Show that the original voice signal $m(t)$ may be recovered from $s(t)$ by using an *unscrambler* that is identical to the unit described above.

Problem 3.26   Consider the SSB wave

$$s(t) = m(t)\cos(2\pi f_c t) - \hat{m}(t)\sin(2\pi f_c t)$$

where f_c is the carrier frequency, $m(t)$ is the message signal, and $\hat{m}(t)$ is its Hilbert transform. This modulated wave is applied to a square-law device characterized by

$$y(t) = s^2(t)$$

Show that the output $y(t)$ contains a frequency component at twice the carrier frequency but that it has a time-varying phase, which makes it impractical to recover the carrier by squaring.

Problem 3.27   A method that is used for carrier recovery in SSB modulation systems involves transmitting two pilot frequencies that are appropriately positioned with respect to the transmitted sideband. This is illustrated in Fig. P3.7a for the case when only the lower sideband is transmitted. In this case, the two pilot frequencies f_1 and f_2 are defined by

$$f_1 = f_c - W - \Delta f$$

and

$$f_2 = f_c + \Delta f$$

where f_c is the carrier frequency and W is the message bandwidth. The Δf is chosen so as to satisfy the relation

$$n = \frac{W}{\Delta f}$$

where n is an integer. Carrier recovery is accomplished by using the scheme shown in Fig. P3.7b. The outputs of the two narrow-band filters centered at f_1 and f_2 are defined by, respectively,

$$v_1(t) = A_1 \cos(2\pi f_1 t + \phi_1)$$

and

$$v_2(t) = A_2 \cos(2\pi f_2 t + \phi_2)$$

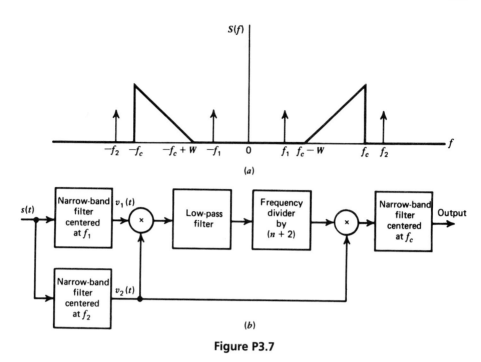

Figure P3.7

The low-pass filter is designed to select the difference frequency component of the first multiplier output due to $v_1(t)$ and $v_2(t)$.

(a) Show that the output signal of the circuit in Fig. P3.7b is proportional to the carrier wave $A_c \cos(2\pi f_c t)$ if the phase angles ϕ_1 and ϕ_2 satisfy the relation

$$\phi_2 = -\frac{\phi_1}{1 + n}$$

(b) For the case when only the upper sideband is transmitted, the two pilot frequencies are defined by

$$f_1 = f_c - \Delta f$$

and

$$f_2 = f_c + W + \Delta f$$

How would you modify the carrier recovery circuit of Fig. P3.7b in order to deal with this case? What is the corresponding relation between ϕ_1 and ϕ_2 for the circuit output to be proportional to the carrier wave?

Problem 3.28 Figure P3.8 shows the block diagram of a *frequency synthesizer*, which enables the generation of many frequencies, each with the same high accuracy as the *master oscillator*. The master oscillator of frequency 1 MHz feeds two *spectrum generators*, one directly and the other through a *frequency divider*. Spectrum generator 1 produces a signal rich in the following harmonics: 1, 2, 3, 4, 5, 6, 7, 8, and 9 MHz. The frequency divider provides a 100-kHZ output, in response to which spectrum generator 2 produces a second signal rich in the following harmonics: 100, 200, 300, 400, 500, 600, 700, 800, and 900 kHz. The harmonic selectors are designed to feed two signals into the mixer, one from

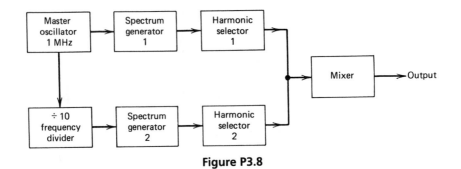

Figure P3.8

spectrum generator 1 and the other from spectrum generator 2. Find the range
of possible frequency outputs of this synthesizer and its resolution.

Problem 3.29 Consider a multiplex system in which four input signals $m_1(t)$,
$m_2(t)$, $m_3(t)$, and $m_4(t)$ are respectively multiplied by the carrier waves

$$[\cos(2\pi f_a t) + \cos(2\pi f_b t)]$$

$$[\cos(2\pi f_a t + \alpha_1) + \cos(2\pi f_b t + \beta_1)]$$

$$[\cos(2\pi f_a t + \alpha_2) + \cos(2\pi f_b t + \beta_2)]$$

$$[\cos(2\pi f_a t + \beta_3) + \cos(2\pi f_b t + \beta_3)]$$

and the resulting DSB-SC signals are summed and then transmitted over a common channel. In the receiver, demodulation is achieved by multiplying the sum
of the DSB-SC signals by the four carrier waves separately and then using filtering
to remove the unwanted components.

(a) Determine the conditions that the phase angles α_1, α_2, α_3 and β_1, β_2, β_3
    must satisfy in order that the output of the kth demodulator is $m_k(t)$, where
    $k = 1, 2, 3, 4$.
(b) Determine the minimum separation of carrier frequencies f_a and f_b in re-
    lation to the bandwidth of the input signals so as to ensure a satisfactory
    operation of the system.

Problem 3.30 In this problem we study the idea of mixing in a superheterodyne
receiver. To be specific, consider the block diagram of the *mixer* shown in Fig.
P3.9 that consists of a product modulator with a local oscillator of *variable frequency* f_l, followed by a band-pass filter. The input signal is an AM wave of bandwidth 10 kHz and carrier frequency that may lie anywhere in the range
0.535–1.605 MHz; these parameters are typical of AM radio broadcasting. It is
required to translate this signal to a frequency band centered at a fixed *intermediate frequency* (IF) of 0.455 MHz. Find the range of tuning that must be provided
in the local oscillator in order to achieve this requirement.

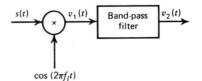

Figure P3.9

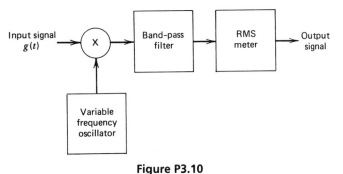

Figure P3.10

Problem 3.31 Figure P3.10 shows the block diagram of a *heterodyne spectrum analyzer*. It consists of a variable-frequency oscillator, multiplier, band-pass filter, and root mean square (rms) meter. The oscillator has an amplitude A and operates over the range f_0 to $f_0 + W$, where f_0 is the mid-band frequency of the filter and W is the signal bandwidth. Assume that $f_0 = 2W$, the filter bandwidth Δf is small compared with f_0 and that the passband amplitude response of the filter is one. Determine the value of the rms meter output for a low-pass input signal $g(t)$.

Problem 3.32 Sketch the PM and FM waves produced by the sawtooth wave shown in Fig. P3.11.

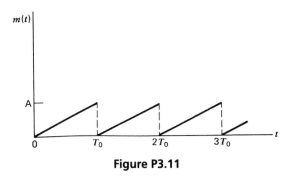

Figure P3.11

Problem 3.33 In a *frequency-modulated radar*, the instantaneous frequency of the transmitted carrier is varied as in Fig. P3.12, which is obtained by using a triangular modulating signal. The instantaneous frequency of the received echo

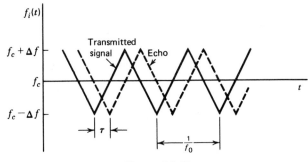

Figure P3.12

signal is shown dashed in Fig. P3.12, where τ is the round-trip delay time. The transmitted and received echo signals are applied to a mixer, and the difference frequency component is retained. Assuming that $f_0\tau \ll 1$, determine the number of beat cycles at the mixer output, averaged over one second, in terms of the peak deviation Δf of the carrier frequency, the delay τ, and the repetition frequency f_0 of the transmitted signal.

Problem 3.34 The instantaneous frequency of a sine wave is equal to $f_c - \Delta f$ for $|t| \leq T/2$, and f_c for $|t| > T/2$. Determine the spectrum of this frequency-modulated wave. *Hint:* Divide up the time interval of interest into three regions: $-\infty < t < -T/2$, $-T/2 \leq t \leq T/2$, and $T/2 < t < \infty$.

Problem 3.35 Single-sideband modulation may be viewed as a hybrid form of amplitude modulation and frequency modulation. Evaluate the envelope and instantaneous frequency of an SSB wave for the following two cases:

(a) When only the upper sideband is transmitted.
(b) When only the lower sideband is transmitted.

Problem 3.36 Consider a narrow-band FM signal approximately defined by

$$s(t) \simeq A_c \cos(2\pi f_c t) - \beta A_c \sin(2\pi f_c t)\sin(2\pi f_m t)$$

(a) Determine the envelope of this modulated signal. What is the ratio of the maximum to the minimum value of this envelope? Plot this ratio versus β, assuming that β is restricted to the interval $0 \leq \beta \leq 0.3$.
(b) Determine the average power of the narrow-band FM signal, expressed as a percentage of the average power of the unmodulated carrier wave. Plot this result versus β, assuming that β is restricted to the interval $0 \leq \beta \leq 0.3$.
(c) By expanding the angle $\theta_i(t)$ of the narrow-band FM signal $s(t)$ in the form of a power series, and restricting the modulation index β to a maximum value of 0.3 radians, show that

$$\theta_i(t) \simeq 2\pi f_c t + \beta \sin(2\pi f_m t) - \frac{\beta^3}{3}\sin^3(2\pi f_m t)$$

What is the value of the harmonic distortion for $\beta = 0.3$?

Problem 3.37 The sinusoidal modulating wave

$$m(t) = A_m \cos(2\pi f_m t)$$

is applied to a phase modulator with phase sensitivity k_p. The unmodulated carrier wave has frequency f_c and amplitude A_c.

(a) Determine the spectrum of the resulting phase-modulated signal, assuming that the maximum phase deviation $\beta_p = k_p A_m$ does not exceed 0.3 radians.
(b) Construct a phasor diagram for this modulated signal, and compare it with that of the corresponding narrow-band FM signal.

Problem 3.38 Suppose that the phase-modulated signal of Problem 3.37 has an arbitrary value for the maximum phase deviation β_p. This modulated signal is applied to an ideal band-pass filter with mid-band frequency f_c and a passband extending from $f_c - 1.5f_m$ to $f_c + 1.5f_m$. Determine the envelope, phase, and instantaneous frequency of the modulated signal at the filter output as functions of time.

Problem 3.39 A carrier wave is frequency-modulated using a sinusoidal signal of frequency f_m and amplitude A_m.

(a) Determine the values of the modulation index β for which the carrier component of the FM signal is reduced to zero. For this calculation you may use the values of $J_0(\beta)$ given in Table 1 of Appendix 4.

(b) In a certain experiment conducted with $f_m = 1$ kHz and increasing A_m (starting from 0 volts), it is found that the carrier component of the FM signal is reduced to zero for the first time when $A_m = 2$ volts. What is the frequency sensitivity of the modulator? What is the value of A_m for which the carrier component is reduced to zero for the second time?

Problem 3.40 An FM signal with modulation index $\beta = 1$ is transmitted through an ideal band-pass filter with mid-band frequency f_c and bandwidth $5f_m$, where f_c is the carrier frequency and f_m is the frequency of the sinusoidal modulating wave. Determine the amplitude spectrum of the filter output.

Problem 3.41 A carrier wave of frequency 100 MHz is frequency-modulated by a sinusoidal wave of amplitude 20 volts and frequency 100 kHz. The frequency sensitivity of the modulator is 25 kHz per volt.

(a) Determine the approximate bandwidth of the FM signal, using Carson's rule.

(b) Determine the bandwidth by transmitting only those side frequencies whose amplitudes exceed 1 percent of the unmodulated carrier amplitude. Use the universal curve of Fig. 3.36 for this calculation.

(c) Repeat your calculations, assuming that the amplitude of the modulating signal is doubled.

(d) Repeat your calculations, assuming that the modulation frequency is doubled.

Problem 3.42 Consider a wide-band PM signal produced by a sinusoidal modulating wave $A_m \cos(2\pi f_m t)$, using a modulator with a phase sensitivity equal to k_p radians per volt.

(a) Show that if the maximum phase deviation of the PM signal is large compared with one radian, the bandwidth of the PM signal varies linearly with the modulation frequency f_m.

(b) Compare this characteristic of a wide-band PM signal with that of a wide-band FM signal.

Problem 3.43 Figure P3.13 shows the block diagram of a real-time *spectrum analyzer* working on the principle of frequency modulation. The given signal $g(t)$ and a frequency-modulated signal $s(t)$ are applied to a multiplier and the output $g(t)s(t)$ is fed into a filter of impulse response $h(t)$. The $s(t)$ and $h(t)$ are *linear*

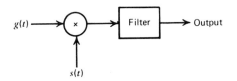

Figure P3.13

FM signals whose instantaneous frequencies vary at opposite rates, as shown by

$$s(t) = \cos(2\pi f_c t - \pi k t^2),$$

$$h(t) = \cos(2\pi f_c t + \pi k t^2)$$

where k is a constant. Show that the envelope of the filter output is proportional to the amplitude spectrum of the input signal $g(t)$ with kt playing the role of frequency f. *Hint:* Use the complex notations described in Chapter 2 for the analysis of band-pass signals and band-pass filters.

Problem 3.44 An FM signal with a frequency deviation of 10 kHz at a modulation frequency of 5 kHz is applied to two frequency multipliers connected in cascade. The first multiplier doubles the frequency and the second multiplier triples the frequency. Determine the frequency deviation and the modulation index of the FM signal obtained at the second multiplier output. What is the frequency separation of the adjacent side frequencies of this FM signal?

Problem 3.45 An FM signal is applied to a square-law device with output voltage v_2 related to input voltage v_1 by

$$v_2 = a v_1^2$$

where a is a constant. Explain how such a device can be used to obtain an FM signal with a greater frequency deviation than that available at the input.

Problem 3.46 Figure P3.14 shows the frequency-determining network of a voltage-controlled oscillator. Frequency modulation is produced by applying the modulating signal $A_m \sin(2\pi f_m t)$ plus a bias V_b to a pair of varactor diodes connected across the parallel combination of a 200-μH inductor and 100-pF capacitor. The capacitor of each varactor diode is related to the voltage V (in volts) applied across its electrodes by

$$C = 100 V^{-1/2} \text{ pF}$$

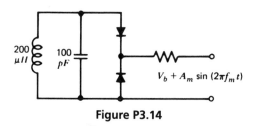

Figure P3.14

The unmodulated frequency of oscillation is 1 MHz. The VCO output is applied to a frequency multiplier to produce an FM signal with a carrier frequency of 64 MHz and a modulation index of 5. Determine (a) the magnitude of the bias voltage V_b and (b) the amplitude A_m of the modulating wave, given that $f_m = 10$ kHz.

Problem 3.47 The FM signal

$$s(t) = A_c \cos\left[2\pi f_c t + 2\pi k_f \int_0^t m(t) \, dt\right]$$

is applied to the system shown in Fig. P3.15 consisting of a high-pass RC filter

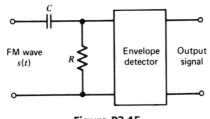

Figure P3.15

and an envelope detector. Assume that (a) the resistance R is small compared with the reactance of the capacitor C for all significant frequency components of $s(t)$ and (b) the envelope detector does not load the filter. Determine the resulting signal at the envelope detector output, assuming that $k_f|m(t)| < f_c$ for all t.

Problem 3.48 In the frequency discriminator of Fig. 3.44b, let the frequency separation between the resonant frequencies of the two parallel tuned LC filters be denoted by $2kB$, where $2B$ is the 3-dB bandwidth of either filter and k is a scaling factor. Assume that both filters have a high Q-factor.

(a) Show that the total response of both filters has a slope equal to $2k/B(1 + k^2)^{3/2}$ at the center frequency f_c.
(b) Let D denote the deviation of the total response measured with respect to a straight line passing through $f = f_c$ with this slope. Plot D versus δ for $k = 1.5$ and $-kB \le \delta \le kB$, where $\delta = f - f_c$.

Problem 3.49 Consider the frequency demodulation scheme shown in Fig. P3.16 in which the incoming FM signal $s(t)$ is passed through a delay line that produces a phase-shift of $\pi/2$ radians at the carrier frequency f_c. The delay-line output is subtracted from the incoming FM signal, and the resulting composite signal is then envelope-detected. This demodulator finds application in demodulating microwave FM signals. Assuming that

$$s(t) = A_c \cos[2\pi f_c t + \beta \sin(2\pi f_m t)]$$

analyze the operation of this demodulator when the modulation index β is less than unity and the delay T produced by the delay line is sufficiently small to justify making the approximations

$$\cos(2\pi f_m T) \simeq 1$$

and

$$\sin(2\pi f_m T) \simeq 2\pi f_m T$$

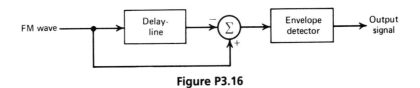

Figure P3.16

Problem 3.50 Figure P3.17 shows the block diagram of a *zero-crossing detector* for demodulating an FM signal. It consists of a limiter, a pulse generator for pro-

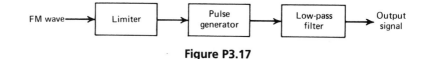

ducing a short pulse at each zero-crossing of the input, and a low-pass filter for extracting the modulating wave.

(a) Show that the instantaneous frequency of the input FM signal is proportional to the number of zero crossings in the time interval $t - (T_1/2)$ to $t + (T_1/2)$, divided by T_1. Assume that the modulating signal is essentially constant during this time interval.

(b) Illustrate the operation of this demodulator, using the sawtooth wave of Fig. P3.11 as the modulating wave.

Problem 3.51 Suppose that the received signal in an FM system contains some residual amplitude modulation of positive amplitude $a(t)$, as shown by

$$s(t) = a(t)\cos[2\pi f_c t + \phi(t)]$$

where f_c is the carrier frequency. The phase $\phi(t)$ is related to the modulating signal $m(t)$ by

$$\phi(t) = 2\pi k_f \int_0^t m(t) \, dt$$

where k_f is a constant. Assume that the signal $s(t)$ is restricted to a frequency band of width B_T, centered at f_c, where B_T is the transmission bandwidth of the FM signal in the absence of amplitude modulation, and that the amplitude modulation is slowly varying compared with $\phi(t)$. Show that the output of an ideal frequency discriminator produced by $s(t)$ is proportional to $a(t)\,m(t)$. *Hint:* Use the complex notation described in Chapter 2 to represent the modulated wave $s(t)$.

Problem 3.52

(a) Let the modulated wave $s(t)$ in Problem 3.51 be applied to an ideal amplitude *limiter*, whose output $z(t)$ is defined by

$$z(t) = \text{sgn}[s(t)]$$

$$= \begin{cases} +1, & s(t) > 0 \\ -1, & s(t) < 0 \end{cases}$$

Show that the limiter output may be expressed in the form of a Fourier series as follows:

$$z(t) = \frac{4}{\pi} \sum_{n=0}^{\infty} \frac{(-1)^n}{2n+1} \cos[2\pi f_c t(2n+1) + (2n+1)\phi(t)]$$

(b) Suppose that the limiter output is applied to a band-pass filter with a passband amplitude response of one and bandwidth B_T centered about the carrier frequency f_c, where B_T is the transmission bandwidth of the FM signal in the absence of amplitude modulation. Assuming that f_c is much greater than B_T, show that the resulting filter output equals

$$y(t) = \frac{4}{\pi} \cos[2\pi f_c t + \phi(t)]$$

By comparing this output with the original modulated signal $s(t)$ defined in Problem 3.51, comment on the practical usefulness of the result.

Problem 3.53

(a) Consider an FM signal of carrier frequency f_c, which is produced by a modulating signal $m(t)$. Assume that f_c is large enough to justify treating this FM signal as a narrow-band signal. Find an approximate expression for its Hilbert transform.

(b) For the special case of a sinusoidal modulating wave $m(t) = A_m \cos(2\pi f_m t)$, find the exact expression for the Hilbert transform of the resulting FM signal. For this case, what is the error in the approximation used in part (a)?

Problem 3.54   The *single sideband version of angle modulation* is defined by

$$s(t) = \exp[-\hat{\phi}(t)]\cos[2\pi f_c t + \phi(t)]$$

where $\hat{\phi}(t)$ is the Hilbert transform of the phase function $\phi(t)$, and f_c is the carrier frequency.

(a) Show that the spectrum of the modulated signal $s(t)$ contains no frequency components in the interval $-f_c < f < f_c$, and is of infinite extent.

(b) Given that the phase function

$$\phi(t) = \beta \sin(2\pi f_m t)$$

where β is the modulation index and f_m is the modulation frequency, derive the corresponding expression for the modulated wave $s(t)$.

Random
Processes

4.1   INTRODUCTION

In Chapter 2, we dealt with the Fourier transform as the mathematical tool for the representation of *deterministic signals* and the transmission of such signals through linear time-invariant filters; by deterministic signals we mean the class of signals that may be modeled as completely specified functions of time. In this chapter, we resume the development of background material necessary for a more detailed understanding of communication systems. Specifically, we deal with the statistical characterization of *random signals,* which may be viewed as the second pillar of communication theory.

Examples of *random signals* are encountered in every practical communication system. We say a signal is "random" if it is not possible to predict its precise value in advance. Consider, for example, a radio communication system. The received signal in such a system usually consists of an *information-bearing signal* component, a random *interference* component, and *receiver noise.* The information-bearing signal component may represent, for example, a voice signal that, typically, consists of randomly spaced bursts of energy of random duration. The interference component may represent spurious electromagnetic waves produced by other communication systems operating in the vicinity of the radio receiver. A major source of receiver noise is *thermal noise,* which is caused by the random motion of the electrons in conductors and devices at the front end of the receiver. We thus find that the received signal is completely random in nature.

Although it is not possible to predict the precise value of a random signal in advance, it may be described in terms of its *statistical* properties such as the average power in the random signal, or the spectral distribution of this power on the average. The mathematical discipline that deals with the statistical characterization of random signals is *probability theory.*

We begin this chapter on random processes by reviewing some basic definitions in probability theory, followed by a review of the notions of a random variable and random process. A random process consists of an ensemble (family) of sample functions, each of which varies randomly with time. A random variable is obtained by observing a random process at a fixed instant of time.

4.2 PROBABILITY THEORY

Probability theory[1] is rooted in phenomena that, explicitly or implicitly, can be modeled by an experiment with an outcome that is subject to *chance.* Moreover, if the experiment is repeated, the outcome can differ because of the influence of an underlying random phenomenon or chance mechanism. Such an experiment is referred to as a *random experiment.* For example, the experiment may be the observation of the result of tossing a fair coin. In this experiment, the possible outcomes of a trial are "heads" or "tails."

To be more precise in the description of a random experiment, we ask for three features:

1. The experiment is repeatable under identical conditions.
2. On any trial of the experiment, the outcome is unpredictable.
3. For a large number of trials of the experiment, the outcomes exhibit *statistical regularity;* that is, a definite *average* pattern of outcomes is observed if the experiment is repeated a large number of times.

Relative-Frequency Approach

Let *event A* denote one of the possible outcomes of a random experiment. For example, in the coin-tossing experiment, event A may represent "heads." Suppose that in n trials of the experiment, event A occurs $N_n(A)$ times. We may then assign the ratio $N_n(A)/n$ to the event A. This ratio is called the *relative frequency* of the event A. Clearly, the relative frequency is a *nonnegative real number less than or equal to one.* That is to say,

$$0 \leq \frac{N_n(A)}{n} \leq 1 \tag{4.1}$$

If event A occurs in none of the trials, $N_n(A)/n = 0$. If, on the other hand, event A occurs in all the n trials, $N_n(A)/n = 1$.

We say that the experiment exhibits *statistical regularity* if for *any* sequence of n trials the relative frequency $N_n(A)/n$ converges to the same limit as n becomes large. It thus seems natural for us to define the *probability of event A* as

$$P(A) = \lim_{n \to \infty} \left(\frac{N_n(A)}{n} \right) \tag{4.2}$$

The limit shown in Eq. (4.2) should not be viewed in a mathematical sense. Rather, we think of Eq. (4.2) as a statement that the probability of an event is the long-term proportion of times that a particular event A occurs in a long sequence of trials. For example, in the coin-tossing experiment, we may expect that out of a million tosses of a fair coin, about one half of them will show up heads.

The probability of an event is intended to represent the *likelihood* that a trial of the experiment will result in the occurrence of that event. For many engineering applications and games of chance, the use of Eq. (4.2) to define the probability of an event is acceptable. However, for many other applications, this definition is inadequate. Consider, for example, the statistical analysis of the stock market: How are we to achieve repeatability of such an experiment? A more satisfying approach is to state the properties that any measure of probability is expected to have, postulate them as *axioms,* and then use relative-frequency interpretations to justify them.

Axioms of Probability

When we perform a random experiment, it is natural for us to be aware of the various outcomes that are likely to arise. In this context, it is convenient to think of an experiment and its possible outcomes as defining a space and its points. With the kth outcome of the experiment, say, we associate a point called the *sample point,* which we denote by s_k. The totality of sample points corresponding to the aggregate of all possible outcomes of the experiment is called the *sample space,* which we denote by S. An event corresponds to either a single sample point or a set of sample points. In particular, the entire sample space S is called the *sure event;* the null set $\varnothing$ is called the *null* or *impossible event;* and a single sample point is called an *elementary event.*

Consider, for example, an experiment that involves the throw of a die. In this experiment there are six possible outcomes: the showing of one, two, three, four, five, and six dots on the upper face of the die. By assigning a sample point to each of these possible outcomes, we have a one-dimensional sample space that consists of six sample points, as shown in Fig. 4.1. The elementary event describing the statement "a six shows" corresponds to the sample point {6}. On the other hand, the event describing the statement "an even number of dots shows"

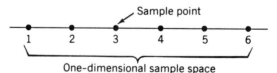

Figure 4.1 Sample space for the experiment of throwing a die.

corresponds to the subset {2,4,6} of the sample space. Note that the term "event" is used interchangeably to describe the subset or the statement.

We are now ready to make a formal definition of probability. A *probability system* consists of the triple:

1. *A sample space* S *of elementary events (outcomes).*
2. *A class* $\mathscr{E}$ *of events* that are subsets of S.
3. *A probability measure* $P(\cdot)$ assigned to each event A in the class $\mathscr{E}$, which has the following properties:

   (i)
   $$P(S) = 1 \tag{4.3}$$

   (ii)
   $$0 \le P(A) \le 1 \tag{4.4}$$

   (iii) If $A + B$ is the *union of two mutually exclusive events* in the class $\mathscr{E}$, then

   $$P(A + B) = P(A) + P(B) \tag{4.5}$$

Properties (i), (ii), and (iii) are known as the *axioms of probability*. Axiom (i) states that the probability of the sure event is unity. Axiom (ii) states that the probability of an event is a nonnegative real number that is less than or equal to unity. Axiom (iii) states that the probability of the union of two mutually exclusive events is the sum of the probabilities of the individual events. These three axioms are sufficient to deal with experiments with finite sample spaces.

Although the axiomatic approach to probability theory is abstract in nature, all three axioms have relative-frequency interpretations of their own. Axiom (ii) corresponds to Eq. (4.1). Axiom (i) corresponds to the limiting case of Eq. (4.1) when the event A occurs in all the n trials. To interpret axiom (iii), we note that if event A occurs $N_n(A)$ times in n trials and event B occurs $N_n(B)$ times, then the union event "A or B" occurs in $N_n(A) + N_n(B)$ trials (since A and B can never occur on the same trial). Hence, $N_n(A + B) = N_n(A) + N_n(B)$, and so we have

$$\frac{N_n(A + B)}{n} = \frac{N_n(A)}{n} + \frac{N_n(B)}{n}$$

which has a mathematical form similar to that of axiom (iii).

Axioms (i), (ii), and (iii) constitute an implicit definition of probability. We may use these axioms to develop some other basic properties of probability, as described next.

PROPERTY 1

$$P(\overline{A}) = 1 - P(A) \tag{4.6}$$

where $\overline{A}$ *(denoting "not A") is the complement of event A.*

The use of this property helps us investigate the *nonoccurrence of an event*. To prove it, we express the sample space S as the union of two mutually exclusive events A and $\bar{A}$:

$$S = A + \bar{A}$$

Then, the use of axioms (i) and (iii) yields

$$1 = P(A) + P(\bar{A})$$

from which Eq. (4.6) follows directly.

PROPERTY 2

If M mutually exclusive events $A_1, A_2, \ldots, A_M$ have the exhaustive property

$$A_1 + A_2 \cdots + A_M = S \qquad (4.7)$$

then

$$P(A_1) + P(A_2) + \cdots + P(A_M) = 1 \qquad (4.8)$$

To prove this property, we first use axiom (i) in Eq. (4.7) and so write

$$P(A_1 + A_2 + \cdots + A_M) = 1$$

Next, we generalize axiom (iii) by writing

$$P(A_1 + A_2 + \cdots + A_M) = P(A_1) + P(A_2) + \cdots + P(A_M)$$

Hence, the result of Eq. (4.8) follows. When the M events are *equally likely* (i.e., they have equal probabilities of occurrence), then Eq. (4.8) simplifies as

$$P(A_i) = \frac{1}{M}, \qquad i = 1, 2, \ldots, M$$

PROPERTY 3

When events A and B are not mutually exclusive, then the probability of the union event "A or B" equals

$$P(A + B) = P(A) + P(B) - P(AB) \qquad (4.9)$$

where $P(AB)$ is the probability of the joint event "A and B."

The probability $P(AB)$ is called a *joint probability*. It has the following relative-frequency interpretation:

$$P(AB) = \lim_{n \to \infty} \left(\frac{N_n(AB)}{n} \right) \qquad (4.10)$$

where $N_n(AB)$ denotes the number of times the events A and B occur simultaneously in n trials of the experiment. Axiom (iii) is a special case of Eq. (4.9); when A and B are mutually exclusive, $P(AB)$ is zero, and Eq. (4.9) reduces to the same form as Eq. (4.5).

Conditional Probability

Suppose we perform an experiment that involves a pair of events A and B. Let $P(B|A)$ denote the probability of event B, given that event A has occurred. The probability $P(B|A)$ is called the *conditional probability of B given A*. Assuming that A has nonzero probability, the conditional probability $P(B|A)$ is defined by

$$P(B|A) = \frac{P(AB)}{P(A)} \tag{4.11}$$

where $P(AB)$ is the joint probability of A and B.

We justify the definition of conditional probability given in Eq. (4.11) by presenting a relative-frequency interpretation of it. Suppose that we perform an experiment and examine the occurrence of a pair of events A and B. Let $N_n(AB)$ denote the number of times the joint event AB occurs in n trials. Suppose that in the same n trials, the event A occurs $N_n(A)$ times. Since the joint event AB corresponds to both A and B occurring, it follows that $N_n(A)$ must include $N_n(AB)$. In other words, we have

$$\frac{N_n(AB)}{N_n(A)} \leq 1$$

The ratio $N_n(AB)/N_n(A)$ represents the relative frequency of B given that A has occurred. For large n, this ratio equals the conditional probability $P(B|A)$; that is,

$$P(B|A) = \lim_{n \to \infty} \left(\frac{N_n(AB)}{N_n(A)} \right)$$

or equivalently,

$$P(B|A) = \lim_{n \to \infty} \left(\frac{N_n(AB)/n}{N_n(A)/n} \right)$$

Recognizing that

$$P(AB) = \lim_{n \to \infty} \left(\frac{N_n(AB)}{n} \right)$$

and

$$P(A) = \lim_{n \to \infty} \left(\frac{N_n(A)}{n} \right)$$

the result of Eq. (4.11) follows.

We may rewrite Eq. (4.11) as

$$P(AB) = P(B|A)P(A) \tag{4.12}$$

It is apparent that we may also write

$$P(AB) = P(A|B)P(B) \tag{4.13}$$

Accordingly, we may state that *the joint probability of two events may be expressed as the product of the conditional probability of one event given the other, and the elementary probability of the other.* Note that the conditional probabilities $P(B|A)$ and $P(A|B)$ have essentially the same properties as the various probabilities previously defined.

Situations may exist where the conditional probability $P(A|B)$ and the probabilities $P(A)$ and $P(B)$ are easily determined directly, but the conditional probability $P(B|A)$ is desired. From Eqs. (4.12) and (4.13), it follows that, provided $P(A) \neq 0$, we may determine $P(B|A)$ by using the relation

$$P(B|A) = \frac{P(A|B)P(B)}{P(A)} \tag{4.14}$$

This relation is a special form of *Bayes' rule.*

Suppose that the conditional probability $P(B|A)$ is simply equal to the elementary probability of occurrence of event B, that is,

$$P(B|A) = P(B)$$

Under this condition, the probability of occurrence of the joint event AB is equal to the product of the elementary probabilities of the events A and B:

$$P(AB) = P(A)P(B)$$

so that

$$P(A|B) = P(A)$$

That is, the conditional probability of event A, assuming the occurrence of event B, is simply equal to the elementary probability of event A. We thus see that in this case a knowledge of the occurrence of one event tells us no more about the probability of occurrence of the other event than we knew without that knowledge. Events A and B that satisfy this condition are said to be *statistically independent.*

EXAMPLE 1   Binary Symmetric Channel

Consider a *discrete memoryless channel* used to transmit binary data. The channel is said to be *discrete* in that it is designed to handle discrete messages. It is *memoryless* in the sense that the channel output at any time depends only on the channel input at that time. Owing to the unavoidable presence of *noise* in the channel, *errors* are made in the received binary data stream. Specifically, when symbol 1 is sent, *occasionally* an error is made and symbol 0 is received and vice versa. The channel is assumed to be symmetric, which means that the probability of receiving symbol 1 when symbol 0 is sent is the same as the probability of receiving symbol 0 when symbol 1 is sent.

To describe the probabilistic nature of this channel fully, we need two sets of probabilities:

1.   The *a priori probabilities* of sending binary symbols 0 and 1: They are

$$P(A_0) = p_0$$

and

$$P(A_1) = p_1$$

where A_0 and A_1 denote the events of transmitting symbols 0 and 1, respectively. Note that $p_0 + p_1 = 1$.

2. The *conditional probabilities of error*: They are

$$P(B_1|A_0) = P(B_0|A_1) = p$$

where B_0 and B_1 denote the events of receiving symbols 0 and 1, respectively. The conditional probability $P(B_1|A_0)$ is the probability of receiving symbol 1, given that symbol 0 is sent. The second conditional probability $P(B_0|A_1)$ is the probability of receiving symbol 0, given that symbol 1 is sent.

The requirement is to determine the *a posteriori probabilities* $P(A_0|B_0)$ and $P(A_1|B_1)$. The conditional probability $P(A_0|B_0)$ is the probability that symbol 0 was sent, given that symbol 0 is received. The second conditional probability $P(A_1|B_1)$ is the probability that symbol 1 was sent, given that symbol 1 is received. Both these conditional probabilities refer to events that are observed "after the fact"; hence, the name "a posteriori" probabilities.

Since the events B_0 and B_1 are mutually exclusive, we have from axiom (iii)

$$P(B_0|A_0) + P(B_1|A_0) = 1$$

That is to say,

$$P(B_0|A_0) = 1 - p$$

Similarly, we may write

$$P(B_1|A_1) = 1 - p$$

Accordingly, we may use the *transition probability diagram* shown in Fig. 4.2 to represent the binary communication channel specified in this example; the term "transition probability" refers to the conditional probability of error. Figure 4.2 clearly depicts the (assumed) symmetric nature of the channel; hence, the name "binary symmetric channel."

From Fig. 4.2, we deduce the following results:

1. The probability of receiving symbol 0 is given by

$$P(B_0) = P(B_0|A_0)P(A_0) + P(B_0|A_1)P(A_1)$$
$$= (1 - p)p_0 + pp_1$$

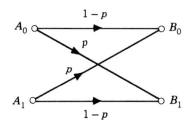

Figure 4.2 Transition probability diagram of binary symmetric channel.

2. The probability of receiving symbol 1 is given by

$$P(B_1) = P(B_1|A_0)P(A_0) + P(B_1|A_1)P(A_1)$$

$$= pp_0 + (1 - p)p_1$$

Therefore, applying Bayes' rule, we obtain

$$P(A_0|B_0) = \frac{P(B_0|A_0)P(A_0)}{P(B_0)}$$

$$= \frac{(1 - p)p_0}{(1 - p)p_0 + pp_1}$$

$$P(A_1|B_1) = \frac{P(B_1|A_1)P(A_1)}{P(B_1)}$$

$$= \frac{(1 - p)p_1}{pp_0 + (1 - p)p_1}$$

These two *a posteriori* probabilities are the desired results.

4.3 RANDOM VARIABLES

It is customary, particularly when using the language of sample space, to think of the outcome of an experiment as a variable that can wander over the set of sample points and whose value is determined by the experiment. A *function whose domain is a sample space and whose range is some set of real numbers is called a random variable of the experiment.* However, the term "random variable" is somewhat confusing. First, the word "random" is not used in the sense of equal probability of occurrence, for which it should be reserved. Second, the word "variable" does not imply dependence (on the experimental outcome), which is an essential part of the meaning. Nevertheless, the term is so deeply imbedded in the literature of probability that its usage has persisted.

When the outcome of an experiment is s, the random variable is denoted as $X(s)$ or simply X. For example, the sample space representing the outcomes of the throw of a die is a set of six sample points that may be taken to be the integers $1, 2, \ldots, 6$. Then if we identify the sample point k with the event that k dots show when the die is thrown, the function $X(k) = k$ is a random variable such that $X(k)$ equals the number of dots that show when the die is thrown. In this example, the random variable takes on only a discrete set of values. In such a case, we say that we are dealing with a *discrete random variable.* More precisely, the random variable X can take on only a finite number of values in any finite observation interval. If, however, the random variable X can take on any value in a whole observation interval, X is called a *continuous random variable.* For example, the random variable that represents the amplitude of a noise voltage at a particular instant of time is a continuous random variable because it may take on any value between plus and minus infinity.

To proced further, we need a probabilistic description of random variables that works equally well for discrete as well as continuous random variables. Let us consider the random variable X and the probability of the event $X \leq x$. We denote this probability by $P(X \leq x)$. It is apparent that this probability is a function of the *dummy variable x*. To simplify our notation, we write

$$F_X(x) = P(X \leq x) \qquad (4.15)$$

The function $F_X(x)$ is called the *cumulative distribution function* (cdf) or simply the *distribution function* of the random variable X. Note that $F_X(x)$ is a function of x, not of the random variable X. However, it depends on the assignment of the random variable X, which accounts for the use of X as subscript. For any point x, the distribution function $F_X(x)$ expresses a probability.

The distribution function $F_X(x)$ has the following properties, which follow directly from Eq. (4.15):

1. The distribution function $F_X(x)$ is bounded between zero and one.
2. The distribution function $F_X(x)$ is a monotone-nondecreasing function of x; that is,

$$F_X(x_1) \leq F_X(x_2) \qquad \text{if } x_1 < x_2 \qquad (4.16)$$

An alternative description of the probability of the random variable X is often useful. This is the derivative of the distribution function, as shown by

$$f_X(x) = \frac{d}{dx}F_X(x) \qquad (4.17)$$

which is called the *probability density function* (pdf) of the random variable X. Note that the differentiation in Eq. (4.17) is with respect to the dummy variable x. The name, density function, arises from the fact that the probability of the event $x_1 < X \leq x_2$ equals

$$
\begin{aligned}
P(x_1 < X \leq x_2) &= P(X \leq x_2) - P(X \leq x_1) \\
&= F_X(x_2) - F_X(x_1) \qquad (4.18) \\
&= \int_{x_1}^{x_2} f_X(x)\, dx
\end{aligned}
$$

The probability of an interval is therefore the area under the probability density function in that interval. Putting $x_1 = -\infty$ in Eq. (4.18), and changing the notation somewhat, we readily see that the distribution function is defined in terms of the probability density function as follows:

$$F_X(x) = \int_{-\infty}^{x} f_X(\xi)\, d\xi \qquad (4.19)$$

Since $F_X(\infty) = 1$, corresponding to the probability of a certain event, and $F_X(-\infty) = 0$, corresponding to the probability of an impossible event, we also find from Eq. (4.18) that

$$\int_{-\infty}^{\infty} f_X(x)\, dx = 1 \qquad (4.20)$$

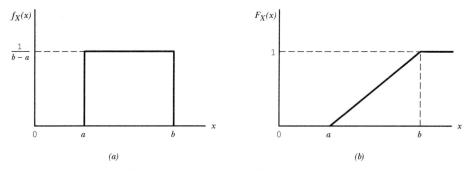

Figure 4.3 Uniform distribution. (*a*) Probability density function. (*b*) Cumulative distribution function.

Earlier we mentioned that a distribution function must always be monotone nondecreasing. This means that its derivative or the probability density function must always be nonnegative. Accordingly, we may state that a *probability density function must always be a nonnegative function, and with a total area of one.*

EXAMPLE 2   Uniform Distribution

A random variable X is said to be *uniformly distributed* over the interval (a,b) if its probability density function is

$$f_X(x) = \begin{cases} 0, & x \leq a \\[2mm] \dfrac{1}{b-a}, & a < x \leq b \\[2mm] 0, & x > b \end{cases} \tag{4.21}$$

The cumulative distribution function of X is therefore

$$F_X(x) = \begin{cases} 0, & x \leq a \\[2mm] \dfrac{x-a}{b-a}, & a < x \leq b \\[2mm] 0, & x > b \end{cases} \tag{4.22}$$

Figure 4.3 shows plots of the probability density function and the cumulative distribution function of the uniformly distributed random variable X.

Several Random Variables

Thus far we have focused attention on situations involving a single random variable. However, we find frequently that the outcome of an experiment requires several random variables for its description. We now consider situations involving two random variables. The probabilistic description developed in this way may be readily extended to any number of random variables.

Consider two random variables X and Y. We define *the joint distribution function* $F_{X,Y}(x,y)$ *as the probability that the random variable X is less than or equal to a specified value x and that the random variable Y is less than or equal to a specified value*

y. The variables X and Y may be two separate one-dimensional random variables or the components of a single two-dimensional random variable. In either case, the joint sample space is the xy-plane. The joint distribution function $F_{X,Y}(x,y)$ is the probability that the outcome of an experiment will result in a sample point lying inside the quadrant $(-\infty < X \leq x, -\infty < Y \leq y)$ of the joint sample space. That is,

$$F_{X,Y}(x, y) = P(X \leq x, Y \leq y) \tag{4.23}$$

Suppose that the joint distribution function $F_{X,Y}(x, y)$ is continuous everywhere, and that the partial derivative

$$f_{X,Y}(x, y) = \frac{\partial^2 F_{X,Y}(x,y)}{\partial x \, \partial y} \tag{4.24}$$

exists and is continuous everywhere. We call the function $f_{X,Y}(x,y)$ the *joint probability density function* of the random variables X and Y. The joint distribution function $F_{X,Y}(x,y)$ is a monotone-nondecreasing function of both x and y. Therefore, from Eq. (4.24) it follows that the joint probability density function $f_{X,Y}(x,y)$ is always nonnegative. Also the total volume under the graph of a joint probability density function must be unity, as shown by

$$\int_{-\infty}^{\infty} \int_{-\infty}^{\infty} f_{X,Y}(\xi, \eta) \, d\xi \, d\eta = 1 \tag{4.25}$$

The probability density function for a single random variable (X, say) can be obtained from its joint probability density function with a second random variable (Y, say) in the following way. We first note that

$$F_X(x) = \int_{-\infty}^{\infty} \int_{-\infty}^{x} f_{X,Y}(\xi, \eta) \, d\xi \, d\eta \tag{4.26}$$

Therefore, differentiating both sides of Eq. (4.26) with respect to x, we get the desired relation:

$$f_X(x) = \int_{-\infty}^{\infty} f_{X,Y}(x, \eta) \, d\eta \tag{4.27}$$

Thus the probability density function $f_X(x)$ is obtained from the joint probability density function $f_{X,Y}(x,y)$ by simply integrating it over all possible values of the undesired random variable, Y. The use of similar arguments in the other dimension yields $f_Y(y)$. The probability density functions $f_X(x)$ and $f_Y(y)$ are called *marginal densities*. Hence, the joint probability density function $f_{X,Y}(x,y)$ contains all the possible information about the joint random variables X and Y.

Suppose that X and Y are two continuous random variables with joint probability density function $f_{X,Y}(x,y)$. The *conditional probability density function* of Y given that $X = x$ is defined by

$$f_Y(y|x) = \frac{f_{X,Y}(x,y)}{f_X(x)} \tag{4.28}$$

provided that $f_X(x) > 0$, where $f_X(x)$ is the marginal density of X. The function $f_Y(y|x)$ may be thought of as a function of the variable y, with the variable x arbitrary, but *fixed*. Accordingly, it satisfies all the requirements of an ordinary probability density function, as shown by

$$f_Y(y|x) \geq 0$$

and

$$\int_{-\infty}^{\infty} f_Y(y|x) \, dy = 1$$

If the random variables X and Y are *statistically independent*, then knowledge of the outcome of X can in no way affect the distribution of Y. The result is that the conditional probability density function $f_Y(y|x)$ reduces to the marginal density $f_Y(y)$, as shown by

$$f_Y(y|x) = f_Y(y)$$

In such a case, we may express the joint probability density function of the random variables X and Y as the product of their respective marginal densities, as shown by

$$f_{X,Y}(x,y) = f_X(x) f_Y(y)$$

Equivalently, we may state that if the joint probability density function of the random variables X and Y equals the product of their marginal densities, then X and Y are statistically independent.

4.4   STATISTICAL AVERAGES

Having discussed probability and some of its ramifications, we now seek ways for determining the *average* behavior of the outcomes arising in random experiments.

The *expected value* or *mean* of a random variable X is defined by

$$\mu_X = E[X] = \int_{-\infty}^{\infty} x f_X(x) \, dx \qquad (4.29)$$

where E denotes the *statistical expectation operator*. That is, the mean μ_X locates the center of gravity of the area under the probability density curve of the random variable X. To interpret the expected value μ_X, we write the integral in the defining equation (4.29) as the limit of an approximating sum formulated as follows. Let $\{x_k | k = 0, \pm 1, \pm 2, \ldots,\}$ denote a set of uniformly spaced points on the real line:

$$x_k = \left(k + \frac{1}{2} \right) \Delta, \qquad k = 0, \pm 1, \pm 2, \ldots \qquad (4.30)$$

where Δ is the spacing between adjacent points. We may then rewrite Eq. (4.29) as

$$
\begin{aligned}
E[X] &= \lim_{\Delta \to 0} \sum_{k=-\infty}^{\infty} \int_{k\Delta}^{(k+1)\Delta} x_k f_X(x)\, dx \\
&= \lim_{\Delta \to 0} \sum_{k=-\infty}^{\infty} x_k P\left(x_k - \frac{\Delta}{2} < X \le x_k + \frac{\Delta}{2} \right)
\end{aligned}
\tag{4.31}
$$

For a physical interpretation of the sum on the right-hand side of Eq. (4.31), suppose that we make n independent observations of the random variable X. Let $N_n(k)$ denote the number of times that the random variable X falls inside the kth bin:

$$
x_k - \frac{\Delta}{2} < X \le x_k + \frac{\Delta}{2}, \qquad k = 0, \pm 1, \pm 2, \ldots
$$

Then, as the number of observations, n, is made large, the ratio $N_n(k)/n$ approaches the probability $P(x_k - \Delta/2 < X \le x_k + \Delta/2)$. Accordingly, we may approximate the expected value of the random variable X as

$$
\begin{aligned}
E[X] &\simeq \sum_{k=-\infty}^{\infty} x_k \left(\frac{N_n(k)}{n} \right) \\
&= \frac{1}{n} \sum_{k=-\infty}^{\infty} x_k N_n(k), \qquad n \text{ large}
\end{aligned}
\tag{4.32}
$$

We now recognize the quantity on the right-hand side of Eq. (4.32) simply as the "sample average." The sum is taken over all the values x_k, each of which is weighted by the number of times it occurs; the sum is then divided by the total number of observations to give the sample average. Indeed, Eq. (4.32) provides the basis for computing the expected value $E[X]$.

Function of a Random Variable

Let X denote a random variable, and let $g(X)$ denote a real-valued function defined on the real line. The quantity obtained by letting the argument of the function $g(X)$ be a random variable is also a random variable, which we denote as

$$
Y = g(X)
\tag{4.33}
$$

To find the expected value of the random variable Y, we could of course find the probability density function $f_Y(y)$ and then apply the standard formula

$$
E[Y] = \int_{-\infty}^{\infty} y f_Y(y)\, dy
$$

A simpler procedure, however, is to write

$$
E[g(X)] = \int_{-\infty}^{\infty} g(x) f_X(x)\, dx
\tag{4.34}
$$

Indeed, Eq. (4.34) may be viewed as generalizing the concept of expected value to an arbitrary function $g(X)$ of a random variable X.

EXAMPLE 3

Let

$$Y = g(X) = \cos(X)$$

where X is a random variable uniformly distributed in the interval $(-\pi, \pi)$; that is,

$$f_X(x) = \begin{cases} \dfrac{1}{2\pi}, & -\pi < x < \pi \\ 0, & \text{otherwise} \end{cases}$$

According to Eq. (4.34), the expected value of Y is

$$E[Y] = \int_{-\pi}^{\pi} (\cos x) \left(\frac{1}{2\pi}\right) dx$$

$$= -\frac{1}{2\pi} \sin x \Big|_{x=-\pi}^{\pi}$$

$$= 0$$

Moments

For the special case of $g(X) = X^n$, using Eq. (4.34) we obtain the nth *moment* of the probability distribution of the random variable X; that is,

$$E[X^n] = \int_{-\infty}^{\infty} x^n f_X(x) \, dx \tag{4.35}$$

By far the most important moments of X are the first two moments. Thus putting $n = 1$ in Eq. (4.35) gives the mean of the random variable as discussed above, whereas putting $n = 2$ gives the *mean-square value* of X:

$$E[X^2] = \int_{-\infty}^{\infty} x^2 f_X(x) \, dx \tag{4.36}$$

We may also define *central moments*, which are simply the moments of the difference between a random variable X and its mean μ_X. Thus, the nth central moment is

$$E[(X - \mu_X)^n] = \int_{-\infty}^{\infty} (x - \mu_X)^n f_X(x) \, dx \tag{4.37}$$

For $n = 1$, the central moment is, of course, zero, whereas for $n = 2$ the second central moment is referred to as the *variance* of the random variable X, written as

$$\text{var}[X] = E[(X - \mu_X)^2] = \int_{-\infty}^{\infty} (x - \mu_X)^2 f_X(x) \, dx \tag{4.38}$$

The variance of a random variable X is commonly denoted as σ_X^2. The square root of the variance, namely, σ_X, is called the *standard deviation* of the random variable X.

The variance σ_X^2 of a random variable X in some sense is a measure of the variable's "randomness." By specifying the variance σ_X^2, we essentially constrain the effective width of the probability density function $f_X(x)$ of the random variable X about the mean μ_X. A precise statement of this constraint is due to Chebyshev. The *Chebyshev inequality* states that for any positive number ϵ, we have

$$P(|X - \mu_X| \geq \epsilon) \leq \frac{\sigma_X^2}{\epsilon^2} \tag{4.39}$$

From this inequality we see that the mean and variance of a random variable give a *partial description* of its probability distribution.

We note from Eqs. (4.36) and (4.38) that the variance σ_X^2 and mean-square value $E[X^2]$ are related by

$$\begin{aligned} \sigma_X^2 &= E[X^2 - 2\mu_X X + \mu_X^2] \\ &= E[X^2] - 2\mu_X E[X] + \mu_X^2 \\ &= E[X^2] - \mu_X^2 \end{aligned} \tag{4.40}$$

where, in the second line, we have used the *linearity* property of the statistical expectation operator E. Equation (4.40) shows that if the mean μ_X is zero, then the variance σ_X^2 and the mean-square value $E[X^2]$ of the random variable X are equal.

Characteristic Function

Another important statistical average is the *characteristic function* $\phi_X(v)$ of the probability distribution of the random variable X, which is defined as the expectation of the complex exponential function $\exp(jvX)$, as shown by

$$\begin{aligned} \phi_X(v) &= E[\exp(jvX)] \\ &= \int_{-\infty}^{\infty} f_X(x)\exp(jvx)\, dx \end{aligned} \tag{4.41}$$

where v is real and $j = \sqrt{-1}$. In other words, the characteristic function $\phi_X(v)$ is (except for a sign change in the exponent) the Fourier transform of the probability density function $f_X(x)$. In this relation we have used $\exp(jvx)$ rather than $\exp(-jvx)$, so as to conform with the convention adopted in probability theory. Recognizing that v and x play analogous roles to the variables $2\pi f$ and t of Fourier transforms, respectively, we deduce the following inverse relation from analogy with the inverse Fourier transform:

$$f_X(x) = \frac{1}{2\pi} \int_{-\infty}^{\infty} \phi_X(v)\exp(-jvx)\, dv \tag{4.42}$$

This relation may be used to evaluate the probability density function $f_X(x)$ of the random variable X from its characteristic function $\phi_X(v)$.

EXAMPLE 4   Gaussian Random Variable

The *Gaussian random variable* is commonly encountered in the statistical analysis of a large variety of physical systems, including communication systems. Let X denote a Gaussian-distributed random variable of mean μ_X and variance σ_X^2. The probability density function of such a random variable is defined by

$$f_X(x) = \frac{1}{\sqrt{2\pi}\,\sigma_X} \exp\left(-\frac{(x - \mu_X)^2}{2\sigma_X^2}\right), \quad -\infty < x < \infty \quad (4.43)$$

Given this probability density function, we may readily show that the mean of the random variable X so defined is indeed μ_X and its variance is σ_X^2; these evaluations are left as an exercise for the reader. In this example, we wish to use the characteristic function to evaluate the higher-order moments of the Gaussian random variable X.

Differentiating both sides of Eq. (4.41) with respect to v a total of n times, and then setting $v = 0$, we get the result

$$\left.\frac{d^n}{dv^n}\,\phi_X(v)\right|_{v=0} = (j)^n \int_{-\infty}^{\infty} x^n f_X(x)\,dx$$

The integral on the right-hand side of this relation is recognized as the nth moment of the random variable X. Accordingly, we may write

$$E[X^n] = (-j)^n \left.\frac{d^n}{dv^n}\,\phi_X(v)\right|_{v=0} \quad (4.44)$$

Now, the characteristic function of a Gaussian random variable X of mean μ_X and variance σ_X^2 is given by (see Problem 4.1)

$$\phi_X(v) = \exp\left(jv\mu_X - \frac{1}{2}v^2\sigma_X^2\right) \quad (4.45)$$

Equations (4.44) and (4.45) show clearly that the higher-order moments of the Gaussian random variable are uniquely determined by the mean μ_X and variance σ_X^2. Indeed, a straightforward manipulation of this pair of equations shows that the central moments of X are as follows:

$$E[(X - \mu_X)^n] = \begin{cases} 1 \times 3 \times 5 \ldots (n-1)\,\sigma_X^n & \text{for } n \text{ even} \\ 0 & \text{for } n \text{ odd} \end{cases} \quad (4.46)$$

Joint Moments

Consider next a pair of random variables X and Y. A set of statistical averages of importance in this case are the *joint moments*, namely, the expected value of $X^i Y^k$, where i and k may assume any positive integer values. We may thus write

$$E[X^i Y^k] = \int_{-\infty}^{\infty} \int_{-\infty}^{\infty} x^i y^k f_{X,Y}(x, y)\,dx\,dy \quad (4.47)$$

A joint moment of particular importance is the *correlation* defined by $E[XY]$, which corresponds to $i = k = 1$ in Eq. (4.47).

The correlation of the centered random variables $X - E[X]$ and $Y - E[Y]$, that is, the joint moment

$$\text{cov}[XY] = E[(X - E[X])(Y - E[Y])] \qquad (4.48)$$

is called the *covariance* of X and Y. Letting $\mu_X = E[X]$ and $\mu_Y = E[Y]$, we may expand Eq. (4.48) to obtain the result

$$\text{cov}[XY] = E[XY] - \mu_X \mu_Y \qquad (4.49)$$

Let σ_X^2 and σ_Y^2 denote the variances of X and Y, respectively. Then the covariance of X and Y, normalized with respect to $\sigma_X \sigma_Y$, is called the *correlation coefficient* of X and Y:

$$\rho = \frac{\text{cov}[XY]}{\sigma_X \sigma_Y} \qquad (4.50)$$

We say that the two random variables X and Y are *uncorrelated* if and only if their covariance is zero, that is, if and only if

$$\text{cov}[XY] = 0$$

We say that they are *orthogonal* if and only if their correlation is zero, that is, if and only if

$$E[XY] = 0$$

From Eq. (4.49) we observe that if one of the random variables X and Y or both have zero means, and if they are orthogonal, then they are uncorrelated, and vice versa. Note also that if X and Y are statistically independent, then they are uncorrelated; however, the converse of this statement is not necessarily true.

4.5 TRANSFORMATIONS OF RANDOM VARIABLES

A problem that often arises in the statistical characterization of communication systems is that of determining the probability density function of a random variable Y related to another random variable X by the transformation

$$Y = g(X)$$

There are two distinct cases that need to be considered here: one-to-one (monotone) transformations and many-to-one (nonmonotone) transformations. These two cases are considered in turn.

Monotone Transformations

Let X be a continuous random variable with probability density function $f_X(x)$. Let $Y = g(X)$ be a monotone differentiable function of X, as depicted in Fig. 4.4. Suppose we make infinitesimal changes dx and dy in x and y, respectively.

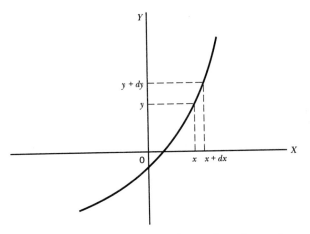

Figure 4.4 A one-to-one transformation of a random variable X.

Then, given that

$$y = g(x) \tag{4.51}$$

we may write

$$y + dy = g(x + dx) \tag{4.52}$$

$$\simeq g(x) + \frac{dg}{dx}\, dx$$

where dg/dx is the derivative of the function $g(x)$ with respect to x.

Consider now the event $(y < Y \leq y + dy)$. From Fig. 4.4 we note that this event contains the same outcomes as the event $(x < X \leq x + dx)$. Accordingly, the probabilities of these two events must be equal, as shown by

$$P(y < Y \leq y + dy) = P(x < X \leq x + dx)$$

In terms of the probability density functions $f_X(x)$ and $f_Y(y)$, we may reformulate the equality of probabilities of the events $(x < X \leq x + dx)$ and $(y < Y \leq y + dy)$ as follows:

$$f_Y(y)\, dy = f_X(x)\, dx$$

where it is assumed that the function $g(x)$ is a monotone-increasing function as indicated in Fig. 4.4. On the other hand, if $g(x)$ is a monotone-decreasing function, we have

$$f_Y(y)\, dy = -f_X(x)\, dx$$

We may combine both of these results by writing

$$f_Y(y)|dy| = f_X(x)|dx| \tag{4.53}$$

The important relation described by Eq. (4.53) may be viewed as an expression of the *conservation of probability* in the one-to-one transformation from a random variable X to another random variable Y defined as a function of the random variable X.

To obtain the probability density function $f_Y(y)$, we do two things:

- Divide both sides of Eq. (4.53) by $|dy|$.
- Substitute the inverse relation $x = g^{-1}(y)$ in the resulting expression.

We may thus write

$$f_Y(y) = \frac{f_X(x)}{|dy/dx|}$$
$$= \frac{f_X(x)}{|dg/dx|}\bigg|_{x=g^{-1}(y)} \tag{4.54}$$

which is the desired formula for the one-to-one transformation of a random variable.

Many-to-One Transformations

Consider next the more general case of a many-to-one transformation $y = g(x)$, in which several values of x can be transformed into one value of y. In this case, the formula for determining the probability density function of the random variable Y in terms of the probability density function of the random variable X is as follows:

$$f_Y(y) = \sum_k \frac{f_X(x_k)}{|dg/dx_k|}\bigg|_{x_k=g^{-1}(y)} \tag{4.55}$$

where the x_k are solutions of the equation $g(x) = y$, and absolute value signs are used because a probability must be positive. Clearly, if for a particular value of y this equation has n solutions, then the expression for the probability density function $f_Y(y)$ at that point contains n terms.

To illustrate the validity of Eq. (4.55), consider the situation described in Fig. 4.5, where we see that the equation $g(x) = y$ has three roots x_1, x_2, and x_3. Similarly, the equation

$$g(x + dx) = y + dy$$

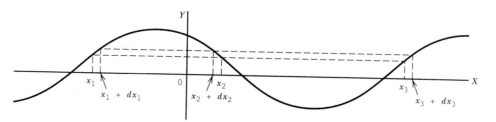

Figure 4.5 One-to-many transformation.

has three roots $x_1 + dx_1$, $x_2 + dx_2$, and $x_3 + dx_3$. The event $(y < Y \le y + dy)$ occurs when any of the three events $(x_1 < X \le x_1 + dx_1)$, $(x_2 < X \le x_2 + dx_2)$, or $(x_3 < X \le x_3 + dx_3)$ occurs. Provided that dy is infinitesimally small, the three events involving the random variable X are mutually exclusive, and so we may write

$$P(y < Y \le y + dy) = P(x_1 < X \le x_1 + dx_1)$$
$$+ P(x_2 < X \le x_2 + dx_2)$$
$$+ P(x_3 < X \le x_3 + dx_3)$$

Correspondingly, in terms of the probability density functions $f_X(x)$ and $f_Y(y)$, we may write

$$f_Y(y) \, dy = f_X(x_1) \, dx_1 + f_X(x_2) \, |dx_2| + f_X(x_3) \, dx_3$$

where, in the second term, we have used the absolute value of dx_2 because in Fig. 4.5 the event represented by this component is "backwards" since the derivative dy/dx at the point x_2 is negative. Dividing both sides of this equation by dy, we readily see that the resulting expression is a special case of the formula described in Eq. (4.55).

EXAMPLE 5

Consider again the transformation $Y = \cos X$, where the random variable X is uniformly distributed in the interval $(-\pi, \pi)$. The problem is to find the probability density function of Y.

From Fig. 4.6 showing a plot of $Y = \cos X$ over the interval of interest, we see that for $-1 < Y \le 1$ the equation $\cos x = y$ has two solutions, namely, $x_1 = -\cos^{-1}(y)$, and $x_2 = \cos^{-1}(y)$. With $y = \cos x$, we have

$$\frac{dy}{dx} = -\sin x$$

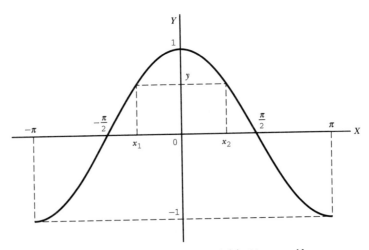

Figure 4.6 The random variable $Y = \cos X$.

The probability density function of X is given as

$$f_X(x) = \begin{cases} \dfrac{1}{2\pi}, & -\pi < x \leq \pi \\ 0, & \text{otherwise} \end{cases}$$

Therefore, the use of Eq. (4.55) for the problem at hand yields

$$f_Y(y) = \frac{1}{2\pi\sqrt{1 - y^2}} + \frac{1}{2\pi\sqrt{1 - y^2}}$$

$$= \frac{1}{\pi\sqrt{1 - y^2}} \quad \text{for } -1 < y \leq 1$$

It is of interest to use this expression to evaluate the expected value of Y. Specifically, we have

$$E[Y] = \int_{-\infty}^{\infty} y f_Y(y) \, dy$$

$$= \int_{-1}^{1} \frac{y}{\pi\sqrt{1 - y^2}} \, dy$$

$$= -\frac{1}{\pi} \sqrt{1 - y^2} \, \Big|_{y=-1}^{1}$$

$$= 0$$

which agrees with the result we obtained in Example 3, and so it should.

4.6 RANDOM PROCESSES

A basic concern in the statistical analysis of communication systems is the characterization of random signals such as voice signals, television signals, digital computer data, and electrical noise. These random signals have two properties. First, the signals are functions of time, defined on some observation interval. Second, the signals are random in the sense that before conducting an experiment, it is not possible to describe exactly the waveforms that will be observed. Accordingly, in describing random signals, we find that each sample point in our sample space is a function of time. The sample space or ensemble comprised of functions of time is called a *random* or *stochastic process*. As an integral part of this notion, we assume the existence of a probability distribution defined over an appropriate class of sets in the sample space, so that we may speak with confidence of the probability of various events.

Consider then a random experiment specified by the outcomes s from some *sample space* S, by the events defined on the sample space S, and by the probabilities of these events. Suppose that we assign to each sample point s a function of time in accordance with the rule:

$$X(t,s), \quad -T \leq t \leq T \tag{4.56}$$

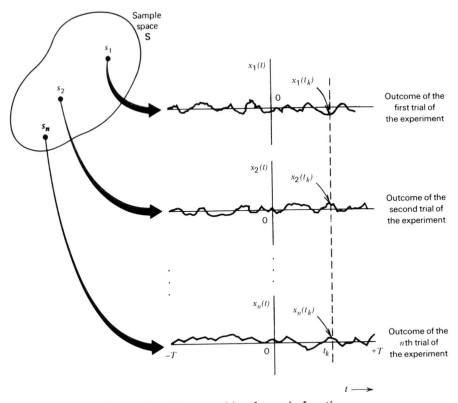

Figure 4.7 An ensemble of sample functions.

where $2T$ is the *total observation interval.* For a fixed sample point s_j, the graph of the function $X(t,s_j)$ versus time t is called a *realization* or *sample function* of the random process. To simplify the notation, we denote this sample function as

$$x_j(t) = X(t,s_j) \tag{4.57}$$

Figure 4.7 illustrates a set of sample functions $\{x_j(t)|j = 1, 2, \ldots, n\}$. From this figure, we note that for a fixed time t_k inside the observation interval, the set of numbers

$$\{x_1(t_k), x_2(t_k), \ldots, x_n(t_k)\} = \{X(t_k,s_1), X(t_k,s_2), \ldots, X(t_k,s_n)\}$$

constitutes a *random variable.* Thus we have an indexed ensemble (family) of random variables $\{X(t,s)\}$, which is called a *random process.* To simplify the notation, the customary practice is to suppress the s and simply use $X(t)$ to denote a random process. We may then formally define a random process $X(t)$ as *an ensemble of time functions together with a probability rule that assigns a probability to any meaningful event associated with an observation of one of the sample functions of the random process.* Moreover, we may distinguish between a random variable and a random process as follows:

- For a random variable, the outcome of a random experiment is mapped into a number.
- For a random process, the outcome of a random experiment is mapped into a waveform that is a function of time.

4.7   STATIONARITY

Many random processes have an important property: *the statistical characterization of the process is time invariant.* To be more precise, consider a random process $X(t)$ that is initiated at $t = -\infty$. Let $X(t_1)$, $X(t_2)$, . . . , $X(t_k)$ denote the random variables obtained by observing the random process $X(t)$ at times $t_1, t_2, \ldots, t_k$, respectively. The joint distribution function of this set of random variables is $F_{X(t_1),\ldots,\,X(t_k)}(x_1, \ldots, x_k)$.

Suppose next we shift all the observation times by a fixed amount τ, thereby obtaining a new set of random variables $X(t_1 + \tau)$, $X(t_2 + \tau)$, . . . , $X(t_k + \tau)$. The joint distribution function of this latter set of random variables is $F_{X(t_1+\tau),\ldots,\,X(t_k+\tau)}(x_1, \ldots, x_k)$. The random process $X(t)$ is said to be *stationary in the strict sense* if the following condition holds:

$$F_{X(t_1+\tau),\ldots,X(t_k+\tau)}(x_1, \ldots, x_k) = F_{X(t_1),\ldots,X(t_k)}(x_1, \ldots, x_k) \qquad (4.58)$$

for all time shifts τ, all k, and all possible choices of observation times $t_1, \ldots, t_k$. In other words, *a random process $X(t)$, initiated at time $t = -\infty$, is strictly stationary if the joint distribution of any set of random variables obtained by observing the random process $X(t)$ is invariant with respect to the location of the origin $t = 0$.*

Similarly, two random processes $X(t)$ and $Y(t)$, both initiated at $t = -\infty$, are said to be *jointly stationary* if the joint distribution functions of the random variables $X(t_1), \ldots, X(t_k)$ and $Y(t_1'), \ldots, Y(t_j')$ are invariant with respect to the location of the origin $t = 0$ for all k and j, and all choices of observation times $t_1, \ldots, t_k$ and $t_1', \ldots, t_j'$.

Returning to Eq. (4.58), we may distinguish two situations of special interest:

1. For $k = 1$, we have

$$F_{X(t)}(x) = F_{X(t+\tau)}(x) = F_X(x) \qquad \text{for all } t \text{ and } \tau \qquad (4.59)$$

That is, *the first-order distribution function of a stationary random process is independent of time.*

2. For $k = 2$ and $\tau = -t_1$, we have

$$F_{X(t_1),X(t_2)}(x_1, x_2) = F_{X(0),X(t_2-t_1)}(x_1, x_2) \qquad \text{for all } t_1 \text{ and } t_2 \qquad (4.60)$$

That is, *the second-order distribution function of a stationary random process depends only on the time difference between the observation times* and not on the particular times at which the random process is observed.

These two properties have profound implications for the statistical parameterization of a stationary random process; this issue is discussed in Section 4.8.

EXAMPLE 6

Consider Fig. 4.8, depicting three spatial windows located at times t_1, t_2, t_3. We wish to evaluate the probability of obtaining a sample function $x(t)$ of a random process $X(t)$ that passes through this set of windows, that is, the probability of the joint event

$$A = \{a_i < X(t_i) \le b_i\}, \qquad i = 1, 2, 3$$

In terms of the joint distribution function, this probability equals

$$P(A) = F_{X(t_1),X(t_2),X(t_3)}(b_1,b_2,b_3) - F_{X(t_1),X(t_2),X(t_3)}(a_1,a_2,a_3)$$

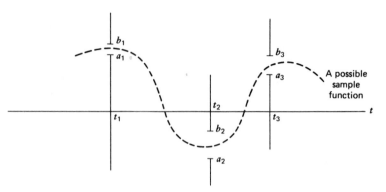

Figure 4.8 Illustrating the probability of a joint event.

Suppose now the random process $X(t)$ is known to be stationary. An implication of stationarity is that the probability of the set of sample functions of this process passing through the windows of Fig. 4.9a is equal to the probability of the set of sample functions passing through the corresponding time-shifted windows of Fig. 4.9b. Note, however, that it is not necessary that these two sets consist of the same sample functions.

4.8   MEAN, CORRELATION, AND COVARIANCE FUNCTIONS

Consider a stationary random process $X(t)$. We define the *mean* of the process $X(t)$ as the expectation of the random variable obtained by observing the process at some time t, as shown by

$$\mu_X(t) = E[X(t)]$$

$$= \int_{-\infty}^{\infty} x f_{X(t)}(x) \, dx \tag{4.61}$$

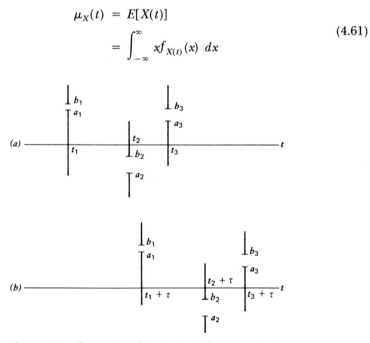

Figure 4.9 Illustrating the concept of stationarity in Example 6.

where $f_{X(t)}(x)$ is the first-order probability density function of the process. From Eq. (4.59) we deduce that for a stationary random process, $f_{X(t)}(x)$ is independent of time t. Consequently, *the mean of a stationary random process is a constant,* as shown by

$$\mu_X(t) = \mu_X \qquad \text{for all } t \tag{4.62}$$

We define the *autocorrelation function* of the process $X(t)$ as the expectation of the product of two random variables, $X(t_1)$ and $X(t_2)$, obtained by observing the process $X(t)$ at times t_1 and t_2, respectively. Specifically, we write

$$R_X(t_1,t_2) = E[X(t_1)X(t_2)]$$
$$= \int_{-\infty}^{\infty} \int_{-\infty}^{\infty} x_1 x_2 \, f_{X(t_1),X(t_2)}(x_1, x_2) \, dx_1 dx_2 \tag{4.63}$$

where $f_{X(t_1),X(t_2)}(x_1,x_2)$ is the second-order probability density function of the process. From Eq. (4.60), we deduce that for a stationary random process, $f_{X(t_1),X(t_2)}(x_1,x_2)$ depends only on the difference between the observation times t_1 and t_2. This, in turn, implies that *the autocorrelation function of a stationary random process depends only on the time difference $t_2 - t_1$,* as shown by

$$R_X(t_1,t_2) = R_X(t_2 - t_1) \qquad \text{for all } t_1 \text{ and } t_2 \tag{4.64}$$

Similarly, the *autocovariance function* of a stationary random process $X(t)$ is written as

$$C_X(t_1,t_2) = E[(X(t_1) - \mu_X)(X(t_2) - \mu_X)]$$
$$= R_X(t_2 - t_1) - \mu_X^2 \tag{4.65}$$

Equation (4.65) shows that, like the autocorrelation function, the autocovariance function of a stationary random process $X(t)$ depends only on the time difference $t_2 - t_1$. This equation also shows that if we know the mean and autocorrelation function of the process, we can readily determine the autocovariance function. The mean and autocorrelation function are therefore sufficient to describe the first two moments of the process.

However, there are two important points that should be carefully noted:

1. The mean and autocorrelation function only provide a *partial description* of the distribution of a random process $X(t)$.
2. The conditions of Eqs. (4.62) and (4.64) involving the mean and autocorrelation function, respectively, are *not* sufficient to guarantee that the random process $X(t)$ is stationary.

Nevertheless, practical considerations often dictate that we simply content ourselves with a partial description of the process given by the mean and autocorrelation function. A random process for which the conditions of Eqs. (4.62) and (4.64) hold is said to be *wide-sense stationary.*[2] Clearly, all strictly stationary processes are wide-sense stationary, but not all wide-sense stationary processes are strictly stationary.

Properties of the Autocorrelation Function

For convenience of notation, we redefine the autocorrelation function of a stationary process $X(t)$ as

$$R_X(\tau) = E[X(t + \tau)X(t)] \quad \text{for all } t \quad (4.66)$$

This autocorrelation function has several important properties:

1. The mean-square value of the process may be obtained from $R_X(\tau)$ simply by putting $\tau = 0$ in Eq. (4.66), as shown by

$$R_X(0) = E[X^2(t)] \quad (4.67)$$

2. The autocorrelation function $R_X(\tau)$ is an even function of τ, that is,

$$R_X(\tau) = R_X(-\tau) \quad (4.68)$$

This property follows directly from the defining equation (4.66). Accordingly, we may also define the autocorrelation function $R_X(\tau)$ as

$$R_X(\tau) = E[X(t)X(t - \tau)]$$

3. The autocorrelation function $R_X(\tau)$ has its maximum magnitude at $\tau = 0$, that is,

$$|R_X(\tau)| \leq R_X(0) \quad (4.69)$$

To probe this property, consider the nonnegative quantity

$$E[(X(t + \tau) \pm X(t))^2] \geq 0$$

Expanding terms and taking their individual expectations, we readily find that

$$E[X^2(t + \tau)] \pm 2E[X(t + \tau)X(t)] + E[X^2(t)] \geq 0$$

which, in light of Eqs. (4.66) and (4.67), reduces to

$$2R_X(0) \pm 2R_X(\tau) \geq 0$$

Equivalently, we may write

$$-R_X(0) \leq R_X(\tau) \leq R_X(0)$$

from which Eq. (4.69) follows directly.

The physical significance of the autocorrelation function $R_X(\tau)$ is that it provides a means of describing the "interdependence" of two random variables obtained by observing a random process $X(t)$ at times τ seconds apart. It is therefore apparent that the more rapidly the random process $X(t)$ changes with time, the more rapidly will the autocorrelation function $R_X(\tau)$ decrease from its maximum $R_X(0)$ as τ increases, as illustrated in Fig. 4.10. This decrease may be characterized by a *decorrelation time* τ_0, such that for $\tau > \tau_0$, the magnitude of the autocorrelation function $R_X(\tau)$ remains below some prescribed value. We may thus define the decorrelation time τ_0 of a wide-sense stationary random process $X(t)$ of zero mean as the time taken for the magnitude of the autocorrelation function $R_X(\tau)$ to decrease to 1 percent, say, of its maximum value $R_X(0)$.

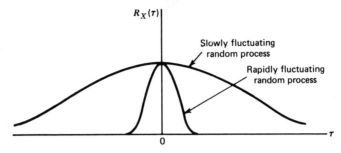

Figure 4.10 Illustrating the autocorrelation functions of slowly and rapidly fluctuating random processes.

EXAMPLE 7   Sinusoidal Wave with Random Phase

Consider a sinusoidal signal with random phase, defined by

$$X(t) = A \cos(2\pi f_c t + \Theta)$$

where A and f_c are constants and Θ is a random variable that is *uniformly distributed* over the interval $(-\pi, \pi)$, that is,

$$f_\Theta(\theta) = \begin{cases} \dfrac{1}{2\pi}, & -\pi \le \theta \le \pi \\ 0, & \text{elsewhere} \end{cases}$$

This means that the random variable Θ is equally likely to have any value in the interval $(-\pi, \pi)$. The autocorrelation function of $X(t)$ is

$$\begin{aligned} R_X(\tau) &= E[X(t + \tau)X(t)] \\ &= E[A^2 \cos(2\pi f_c t + 2\pi f_c \tau + \Theta)\cos(2\pi f_c t + \Theta)] \\ &= \frac{A^2}{2}E[\cos(4\pi f_c t + 2\pi f_c \tau + 2\Theta)] + \frac{A^2}{2} E[\cos(2\pi f_c \tau)] \\ &= \frac{A^2}{2} \int_{-\pi}^{\pi} \frac{1}{2\pi} \cos(4\pi f_c t + 2\pi f_c \tau + 2\theta) \, d\theta + \frac{A^2}{2} \cos(2\pi f_c \tau) \end{aligned}$$

The first term integrates to zero, and so we get

$$R_X(\tau) = \frac{A^2}{2} \cos(2\pi f_c \tau) \tag{4.70}$$

which is shown plotted in Fig. 4.11. We see therefore that the autocorrelation function of a sinusoidal wave with random phase is another sinusoid at the same frequency in the "τ domain" rather than the original time domain.

EXAMPLE 8   Random Binary Wave

Figure 4.12 shows the sample function $x(t)$ of a process $X(t)$ consisting of a random sequence of *binary symbols* 1 and 0. The following assumptions are made:

1.   The symbols 1 and 0 are represented by pulses of amplitude $+A$ and $-A$ volts, respectively, and duration T seconds.

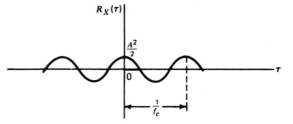

Figure 4.11 Autocorrelation function of a sine wave with random phase.

2. The pulses are not synchronized, so that the starting time t_d of the first complete pulse for positive time is equally likely to lie anywhere between zero and T seconds. That is, t_d is the sample value of a uniformly distributed random variable T_d, with its probability density function defined by

$$f_{T_d}(t_d) = \begin{cases} \dfrac{1}{T}, & 0 \leq t_d \leq T \\ 0, & \text{elsewhere} \end{cases}$$

3. During any time interval $(n-1)T < t - t_d < nT$, where n is an integer, the presence of a 1 or a 0 is determined by tossing a fair coin; specifically, if the outcome is "heads," we have a 1 and if the outcome is "tails," we have a 0. These two symbols are thus equally likely, and the presence of a 1 or 0 in any one interval is independent of all other intervals.

Since the amplitude levels $-A$ and $+A$ occur with equal probability, it follows immediately that $E[X(t)] = 0$ for all t, and the mean of the process is therefore zero.

To find the autocorrelation function $R_X(t_k, t_i)$, we have to evaluate $E[X(t_k)X(t_i)]$, where $X(t_k)$ and $X(t_i)$ are random variables obtained by observing the random process $X(t)$ at times t_k and t_i, respectively.

Consider first the case when $|t_k - t_i| > T$. Then the random variables $X(t_k)$ and $X(t_i)$ occur in different pulse intervals and are therefore independent. We thus have

$$E[X(t_k)X(t_i)] = E[X(t_k)]E[X(t_i)] = 0, \qquad |t_k - t_i| > T$$

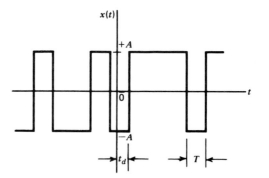

Figure 4.12 Sample function of random binary wave.

Consider next the case when $|t_k - t_i| < T$, with $t_k = 0$ and $t_i < t_k$. In such a situation we observe from Fig. 4.12 that the random variables $X(t_k)$ and $X(t_i)$ occur in the same pulse interval if and only if the delay t_d satisfies the condition $t_d < T - |t_k - t_i|$. We thus obtain the *conditional expectation*:

$$E[X(t_k)X(t_i)|t_d] = \begin{cases} A^2, & t_d < T - |t_k - t_i| \\ 0, & \text{elsewhere} \end{cases}$$

Averaging this result over all possible values of t_d, we get

$$E[X(t_k)X(t_i)] = \int_0^{T-|t_k-t_i|} A^2 f_{T_d}(t_d)\, dt_d$$

$$= \int_0^{T-|t_k-t_i|} \frac{A^2}{T}\, dt_d$$

$$= A^2\left(1 - \frac{|t_k - t_i|}{T}\right), \qquad |t_k - t_i| < T$$

By similar reasoning for any other value of t_k, we conclude that the autocorrelation function of a random binary wave, represented by the sample function shown in Fig. 4.12 is only a function of the time difference $\tau = t_k - t_i$, as shown by

$$R_X(\tau) = \begin{cases} A^2\left(1 - \dfrac{|\tau|}{T}\right), & |\tau| < T \\ 0, & |\tau| \ge T \end{cases} \qquad (4.71)$$

This result is shown plotted in Fig. 4.13.

Cross-Correlation Functions

Consider next the more general case of two random processes $X(t)$ and $Y(t)$ with autocorrelation functions $R_X(t,u)$ and $R_Y(t,u)$, respectively. The two *cross-correlation functions* of $X(t)$ and $Y(t)$ are defined by

$$R_{XY}(t,u) = E[X(t)Y(u)] \qquad (4.72)$$

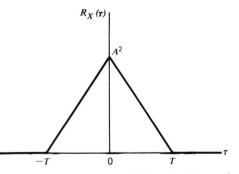

Figure 4.13 Autocorrelation function of random binary wave.

and

$$R_{YX}(t,u) = E[Y(t)X(u)] \tag{4.73}$$

where t and u denote two values of time at which the processes are observed. In this case, the correlation properties of the two random processes $X(t)$ and $Y(t)$ may be displayed conveniently in matrix form as follows:

$$\mathbf{R}(t,u) = \begin{bmatrix} R_X(t,u) & R_{XY}(t,u) \\ R_{YX}(t,u) & R_Y(t,u) \end{bmatrix} \tag{4.74}$$

which is called the *correlation matrix* of the random processes $X(t)$ and $Y(t)$. If the random processes $X(t)$ and $Y(t)$ are each wide-sense stationary and, in addition, they are jointly wide-sense stationary, then the correlation matrix can be written as

$$\mathbf{R}(\tau) = \begin{bmatrix} R_X(\tau) & R_{XY}(\tau) \\ R_{YX}(\tau) & R_Y(\tau) \end{bmatrix} \tag{4.75}$$

where $\tau = t - u$.

The cross-correlation function is not generally an even function of τ as was true for the autocorrelation function, nor does it have a maximum at the origin. However, it does obey a certain symmetry relationship as follows (see Problem 4.12):

$$R_{XY}(\tau) = R_{YX}(-\tau) \tag{4.76}$$

EXAMPLE 9   Quadrature-Modulated Processes

Consider a pair of *quadrature-modulated processes* $X_1(t)$ and $X_2(t)$ that are related to a wide-sense stationary process $X(t)$ as follows:

$$X_1(t) = X(t)\cos(2\pi f_c t + \Theta)$$
$$X_2(t) = X(t)\sin(2\pi f_c t + \Theta)$$

where f_c is a carrier frequency, and the random variable Θ is uniformly distributed over the interval $(0,2\pi)$. Moreover, Θ is independent of $X(t)$. One cross-correlation function of $X_1(t)$ and $X_2(t)$ is given by

$$\begin{aligned} R_{12}(\tau) &= E[X_1(t)X_2(t - \tau)] \\ &= E[X(t)X(t - \tau)\cos(2\pi f_c t + \Theta)\sin(2\pi f_c t - 2\pi f_c \tau + \Theta)] \\ &= E[X(t)X(t - \tau)]E[\cos(2\pi f_c t + \Theta)\sin(2\pi f_c t - 2\pi f_c \tau + \Theta)] \quad (4.77) \\ &= \tfrac{1}{2}R_X(\tau)E[\sin(4\pi f_c t - 2\pi f_c \tau + 2\Theta) - \sin(2\pi f_c \tau)] \\ &= -\tfrac{1}{2}R_X(\tau)\sin(2\pi f_c \tau) \end{aligned}$$

where, in the last line, we have made use of the uniform distribution of the random variable Θ representing phase. Note that at $\tau = 0$, the factor $\sin(2\pi f_c \tau)$ is zero and therefore

$$R_{12}(0) = E[X_1(t)X_2(t)]$$
$$= 0$$

This shows that the random variables obtained by simultaneously observing the quadrature-modulated processes $X_1(t)$ and $X_2(t)$ at some fixed value of time t are orthogonal to each other.

4.9 ERGODICITY

The *expectations* or *ensemble averages* of a stochastic process $X(t)$ are averages "across the process." For example, the mean of a stochastic process $X(t)$ at some fixed time t_k is the expectation of the random variable $X(t_k)$ that describes *all possible values* of the sample functions of the process observed at time $t = t_k$. Naturally, we may also define *long-term sample averages,* or *time averages* that are averages "along the process." We are therefore interested in relating ensemble averages to time averages, for time averages represent a practical means available to us for the *estimation* of ensemble averages of a random process. The key question, of course, is: When can we substitute time averages for ensemble averages? To explore this issue, consider the sample function $x(t)$ of a wide-sense stationary process $X(t)$, with the observation interval defined as $-T \le t \le T$. The *dc value* of $x(t)$ is defined by the time average

$$\mu_x(T) = \frac{1}{2T} \int_{-T}^{T} x(t) \, dt \qquad (4.78)$$

Clearly, the time average $\mu_x(T)$ is a random variable, as its value depends on the observation interval and which particular sample function of the random process $X(t)$ is picked for use in Eq. (4.78). Since the process $X(t)$ is assumed to be wide-sense stationary, the mean of the time average $\mu_x(T)$ is given by (after interchanging the operation of expectation and integration):

$$E[\mu_x(T)] = \frac{1}{2T} \int_{-T}^{T} E[x(t)] \, dt$$

$$= \frac{1}{2T} \int_{-T}^{T} \mu_X \, dt \qquad (4.79)$$

$$= \mu_X$$

where μ_X is the mean of the process $X(t)$. Accordingly, the time average $\mu_x(T)$ represents an *unbiased estimate* of the ensemble-averaged mean μ_X. We say that the process $X(t)$ is *ergodic in the mean* if two conditions are satisfied:

- The time average $\mu_x(T)$ approaches the ensemble average μ_X in the limit as the observation interval T approaches infinity; that is,

$$\lim_{T \to \infty} \mu_x(T) = \mu_X$$

- The variance of $\mu_x(T)$, treated as a random variable, approaches zero in the limit as the observation interval T approaches infinity; that is,

$$\lim_{T \to \infty} \text{var}[\mu_x(T)] = 0$$

The other time average of particular interest is the autocorrelation function $R_x(\tau, T)$ defined in terms of the sample function $x(t)$ observed over the interval

$-T \leq t \leq T$. Following Eq. (4.78), we may formally define the *time-averaged autocorrelation function* of a sample function $x(t)$ as follows:

$$R_x(\tau, T) = \frac{1}{2T} \int_{-T}^{T} x(t + \tau) x(t) \, dt \qquad (4.80)$$

This second time-average should also be viewed as a random variable with a mean and variance of its own. In a manner similar to ergodicity of the mean, we say that the process $x(t)$ is *ergodic in the autocorrelation function* if the following two limiting conditions are satisfied:

$$\lim_{T \to \infty} R_x(\tau, T) = R_X(\tau)$$

$$\lim_{T \to \infty} \text{var}[R_x(\tau, T)] = 0$$

We could, of course, go on in a similar way to define ergodicity in the most general sense by considering higher-order statistics of the process $X(t)$. In practice, however, ergodicity in the mean and ergodicity in the autocorrelation function, as described here, are often (but not always) considered to be adequate.[3] Note also that the use of Eqs. (4.79) and (4.80) to compute the time averages $\mu_x(T)$ and $R_x(t,T)$ requires that the process $X(t)$ be wide-sense stationary. In other words, for a random process to be ergodic, it has to be wide-sense stationary; however, the converse is not necessarily true.

4.10 TRANSMISSION OF A RANDOM PROCESS THROUGH A LINEAR FILTER

Suppose that a random process $X(t)$ is applied as input to a linear time-invariant filter of impulse response $h(t)$, producing a new random process $Y(t)$ at the filter output, as in Fig. 4.14. In general, it is difficult to describe the probability distribution of the output random process $Y(t)$, even when the probability distribution of the input random process $X(t)$ is completely specified for $-\infty < t < \infty$.

In this section, we wish to determine the time-domain form of the input–output relations of the filter for defining the mean and autocorrelation functions of the output random process $Y(t)$ in terms of those of the input $X(t)$, assuming that $X(t)$ is a wide-sense stationary random process.

Consider first the mean of the output random process $Y(t)$. By definition, we have

$$\mu_Y(t) = E[Y(t)] = E\left[\int_{-\infty}^{\infty} h(\tau_1) X(t - \tau_1) \, d\tau_1 \right] \qquad (4.81)$$

where τ_1 is a dummy variable. Provided that the expectation $E[X(t)]$ is finite for all t and the system is stable, we may interchange the order of the expectation and the integration with respect to τ_1 in Eq. (4.81), and so we write

$$\mu_Y(t) = \int_{-\infty}^{\infty} h(\tau_1) E[X(t - \tau_1)] \, d\tau_1$$

$$= \int_{-\infty}^{\infty} h(\tau_1) \mu_X(t - \tau_1) \, d\tau_1 \tag{4.82}$$

When the input random process $X(t)$ is wide-sense stationary, the mean $\mu_X(t)$ is a constant μ_X, so that we may simplify Eq. (4.82) as follows:

$$\mu_Y = \mu_X \int_{-\infty}^{\infty} h(\tau_1) \, d\tau_1$$

$$= \mu_X H(0) \tag{4.83}$$

where $H(0)$ is the zero-frequency (dc) response of the system. Equation (4.83) states that the mean of the random process $Y(t)$ produced at the output of a linear time-invariant system in response to $X(t)$ acting as the input process is equal to the mean of $X(t)$ multiplied by the dc response of the system, which is intuitively satisfying.

Consider next the autocorrelation function of the output random process $Y(t)$. By definition, we have

$$R_Y(t,u) = E[Y(t)Y(u)]$$

where t and u denote two values of the time at which the output process is observed. We may therefore use the convolution integral to write

$$R_Y(t,u) = E\left[\int_{-\infty}^{\infty} h(\tau_1) X(t - \tau_1) \, d\tau_1 \int_{-\infty}^{\infty} h(\tau_2) X(u - \tau_2) \, d\tau_2 \right] \tag{4.84}$$

Here again, provided that the mean-square value $E[X^2(t)]$ is finite for all t and the system is stable, we may interchange the order of the expectation and the integrations with respect to τ_1 and τ_2 in Eq. (4.84), obtaining

$$R_Y(t,u) = \int_{-\infty}^{\infty} d\tau_1 h(\tau_1) \int_{-\infty}^{\infty} d\tau_2 h(\tau_2) E[X(t - \tau_1) X(u - \tau_2)]$$

$$= \int_{-\infty}^{\infty} d\tau_1 h(\tau_1) \int_{-\infty}^{\infty} d\tau_2 h(\tau_2) R_X(t - \tau_1, u - \tau_2) \tag{4.85}$$

When the input $X(t)$ is a wide-sense stationary random process, the autocorrelation function of $X(t)$ is only a function of the difference between the obser-

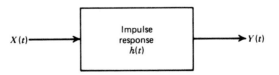

Figure 4.14   Transmission of a random process through a linear filter.

vation times $t - \tau_1$ and $u - \tau_2$. Thus, putting $\tau = t - u$ in Eq. (4.85), we may write

$$R_Y(\tau) = \int_{-\infty}^{\infty} \int_{-\infty}^{\infty} h(\tau_1) h(\tau_2) R_X(\tau - \tau_1 + \tau_2) \, d\tau_1 \, d\tau_2 \qquad (4.86)$$

On combining this result with that involving the mean μ_Y, we see that if the input to a stable linear time-invariant filter is a wide-sense stationary random process, then the output of the filter is also a wide-sense stationary random process.

Since $R_Y(0) = E[Y^2(t)]$, it follows that the mean-square value of the output random process $Y(t)$ is obtained by putting $\tau = 0$ in Eq. (4.86). We thus get the result

$$E[Y^2(t)] = \int_{-\infty}^{\infty} \int_{-\infty}^{\infty} h(\tau_1) h(\tau_2) R_X(\tau_2 - \tau_1) \, d\tau_1 \, d\tau_2 \qquad (4.87)$$

which is a constant.

4.11   POWER SPECTRAL DENSITY

Thus far we have considered the characterization of wide-sense stationary random processes in linear systems in the time domain. We turn next to the characterization of random processes in linear systems by using frequency-domain ideas. In particular, we wish to derive the frequency-domain equivalent to the result of Eq. (4.87) defining the mean-square value of the filter output.

By definition, the impulse response of a linear time-invariant filter is equal to the inverse Fourier transform of the transfer function of the system. We may thus write

$$h(\tau_1) = \int_{-\infty}^{\infty} H(f) \exp(j2\pi f \tau_1) \, df \qquad (4.88)$$

Substituting this expression for $h(\tau_1)$ in Eq. (4.87), we get

$$E[Y^2(t)] = \int_{-\infty}^{\infty} \int_{-\infty}^{\infty} \left[\int_{-\infty}^{\infty} H(f) \exp(j2\pi f \tau_1) \, df \right] h(\tau_2) R_X(\tau_2 - \tau_1) \, d\tau_1 \, d\tau_2$$

$$= \int_{-\infty}^{\infty} df H(f) \int_{-\infty}^{\infty} d\tau_2 h(\tau_2) \int_{-\infty}^{\infty} R_X(\tau_2 - \tau_1) \exp(j2\pi f \tau_1) \, d\tau_1 \qquad (4.89)$$

In the last integral on the right-hand side of Eq. (4.89), define a new variable

$$\tau = \tau_2 - \tau_1$$

Then we may rewrite Eq. (4.89) in the form

$$E[Y^2(t)] = \int_{-\infty}^{\infty} df H(f) \int_{-\infty}^{\infty} d\tau_2 h(\tau_2) \exp(j2\pi f \tau_2) \int_{-\infty}^{\infty} R_X(\tau) \exp(-j2\pi f \tau) \, d\tau$$

$$(4.90)$$

However, the middle integral on the right-hand side in Eq. (4.90) is simply the complex conjugate $H*(f)$ of the transfer function $H(f)$ of the filter, and so we may simplify this equation as

$$E[Y^2(t)] = \int_{-\infty}^{\infty} df \, |H(f)|^2 \int_{-\infty}^{\infty} R_X(\tau) \exp(-j2\pi f\tau) \, d\tau \qquad (4.91)$$

We may further simplify Eq. (4.91) by recognizing that the last integral is simply the Fourier transform of the autocorrelation function $R_X(\tau)$ of the input random process $X(t)$. Specifically, we introduce the definition of a new parameter

$$S_X(f) = \int_{-\infty}^{\infty} R_X(\tau) \exp(-j2\pi f\tau) \, d\tau \qquad (4.92)$$

The function $S_X(f)$ is called the *power spectral density* or *power spectrum* of the wide-sense stationary random process $X(t)$. Thus substituting Eq. (4.92) in (4.91), we obtain the desired relation:

$$E[Y^2(t)] = \int_{-\infty}^{\infty} |H(f)|^2 S_X(f) \, df \qquad (4.93)$$

Equation (4.93) states that *the mean-square value of the output of a stable linear time-invariant filter in response to a wide-sense stationary process is equal to the integral over all frequencies of the power spectral density of the input random process multiplied by the squared magnitude of the transfer function of the filter.* This is the desired frequency-domain equivalent to the time-domain relation of Eq. (4.87).

To investigate the physical significance of the power spectral density, suppose that the random process $X(t)$ is passed through an ideal narrow-band filter with an amplitude response centered about the frequency f_c, as shown in Fig. 4.15; that is,

$$|H(f)| = \begin{cases} 1, & |f \pm f_c| < \tfrac{1}{2}\Delta f \\ 0, & |f \pm f_c| > \tfrac{1}{2}\Delta f \end{cases} \qquad (4.94)$$

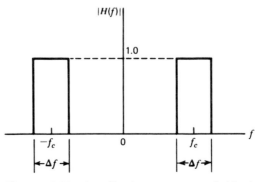

Figure 4.15 Amplitude response of ideal narrow-band filter.

where Δf is the bandwidth of the filter. Then from Eq. (4.93) we find that if the filter bandwidth Δf is sufficiently small compared to the mid-band frequency f_c and $S_X(f)$ is a continuous function, the mean-square value of the filter output is approximately

$$E[Y^2(t)] \simeq (2\Delta f) S_X(f_c) \tag{4.95}$$

The filter, however, passes only those frequency components of the input random process $X(t)$ that lie inside a narrow frequency band of width Δf centered about the frequency $\pm f_c$. Thus $S_X(f_c)$ represents the frequency density of the average power in the random process $X(t)$, evaluated at the frequency $f = f_c$. The dimensions of the power spectral density are therefore in watts per Hertz.

Properties of the Power Spectral Density

The power spectral density $S_X(f)$ and the autocorrelation function $R_X(\tau)$ of a wide-sense stationary random process $X(t)$ form a Fourier-transform pair with τ and f as the variables of interest, as shown by the pair of relations

$$S_X(f) = \int_{-\infty}^{\infty} R_X(\tau) \exp(-j2\pi f\tau) \ d\tau \tag{4.96}$$

$$R_X(\tau) = \int_{-\infty}^{\infty} S_X(f) \exp(j2\pi f\tau) \ df \tag{4.97}$$

Equations (4.96) and (4.97) are basic relations in the theory of spectral analysis of random processes, and together they constitute what are usually called the *Einstein–Wiener–Khintchine relations.*[4]

The Einstein–Wiener–Khintchine relations show that if either the autocorrelation function or power spectral density of a random process is known, the other can be found exactly. But these functions display different aspects of the correlation information about the process. It is commonly accepted that for practical purposes, the power spectral density is the more useful "parameter."

We now wish to use this pair of relations to derive some general properties of the power spectral density of a wide-sense stationary process.

PROPERTY 1

The zero-frequency value of the power spectral density of a wide-sense stationary random process equals the total area under the graph of the autocorrelation function; that is,

$$S_X(0) = \int_{-\infty}^{\infty} R_X(\tau) \ d\tau \tag{4.98}$$

This property follows directly from Eq. (4.96) by putting $f = 0$.

PROPERTY 2

The mean-square value of a wide-sense stationary random process equals the total area under the graph of the power spectral density; that is,

$$E[X^2(t)] = \int_{-\infty}^{\infty} S_X(f) \ df \tag{4.99}$$

This property follows directly from Eq. (4.97) by putting $\tau = 0$ and noting that $R_X(0) = E[X^2(t)]$.

PROPERTY 3

The power spectral density of a wide-sense stationary random process is always nonnegative; that is,

$$S_X(f) \geq 0 \qquad \text{for all } f \qquad (4.100)$$

This property is an immediate consequence of the fact that, in Eq. (4.95), the mean-square value $E[Y^2(t)]$ must always be nonnegative.

PROPERTY 4

The power spectral density of a real-valued random process is an even function of frequency; that is,

$$S_X(-f) = S_X(f) \qquad (4.101)$$

This property is readily obtained by substituting $-f$ for f in Eq. (4.96):

$$S_X(-f) = \int_{-\infty}^{\infty} R_X(\tau) \exp(j2\pi f\tau)\ d\tau$$

Next, substituting $-\tau$ for τ, and recognizing that $R_X(-\tau) = R_X(\tau)$, we get

$$S_X(-f) = \int_{-\infty}^{\infty} R_X(\tau) \exp(-j2\pi f\tau)\ d\tau = S_X(f)$$

which is the desired result.

PROPERTY 5

The power spectral density, appropriately normalized, has the properties usually associated with a probability density function.

The normalization we have in mind here is with respect to the total area under the graph of the power spectral density (i.e., the mean-square value of the process). Consider then the function

$$p_X(f) = \frac{S_X(f)}{\displaystyle\int_{-\infty}^{\infty} S_X(f)\ df} \qquad (4.102)$$

In light of Properties 2 and 3, we note that $p_X(f) \geq 0$ for all f. Moreover, the total area under the function $p_X(f)$ is unity. Hence, the normalized form of the power spectral density, as defined in Eq. (4.102), behaves similar to a probability density function.

As a useful application of Property 5, we may define the *root mean square (rms) bandwidth* for a wide-sense stationary process $X(t)$ as follows:

$$W_{rms} = \left(\int_{-\infty}^{\infty} f^2 p_X(f) \, df \right)^{1/2}$$

$$= \left\{ \frac{\int_{-\infty}^{\infty} f^2 S_X(f) \, df}{\int_{-\infty}^{\infty} S_X(f) \, df} \right\}^{1/2} \qquad (4.103)$$

As mentioned in Chapter 2, the rms bandwidth is of particular interest from a theoretical viewpoint, but it is not easily measurable in the laboratory.

EXAMPLE 10   Sinusoidal Wave with Random Phase (continued)

Consider the random process $X(t) = A \cos(2\pi f_c t + \Theta)$, where Θ is a uniformly distributed random variable over the interval $(-\pi, \pi)$. The autocorrelation function of this random process is given by Eq. (4.70), which is reproduced here for convenience:

$$R_X(\tau) = \frac{A^2}{2} \cos(2\pi f_c \tau)$$

Taking the Fourier transform of both sides of this relation, we find that the power spectral density of the sinusoidal process $X(t)$ is

$$S_X(f) = \frac{A^2}{4} [\delta(f - f_c) + \delta(f + f_c)] \qquad (4.104)$$

which consists of a pair of delta functions weighted by the factor $A^2/4$ and located at $\pm f_c$, as in Fig. 4.16. We note that the total area under a delta function is one. Hence, the total area under $S_X(f)$ is equal to $A^2/2$, as expected.

EXAMPLE 11   Random Binary Wave (continued)

Consider again a random binary wave consisting of a sequence of 1s and 0s represented by the values $+A$ and $-A$, respectively. In Example 8 we showed that the autocorrelation function of this random process has a triangular wave-

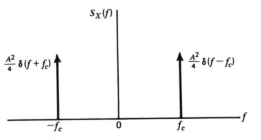

Figure 4.16 Power spectral density of sine wave with random phase.

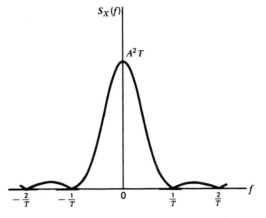

Figure 4.17 Power spectral density of random binary wave.

form, as shown by

$$R_X(\tau) = \begin{cases} A^2\left(1 - \dfrac{|\tau|}{T}\right), & |\tau| < T \\ 0, & |\tau| \geq T \end{cases}$$

The power spectral density of the process is therefore

$$S_X(f) = \int_{-T}^{T} A^2\left(1 - \frac{|\tau|}{T}\right)\exp(-j2\pi f\tau)\; d\tau$$

Using the Fourier transform of a triangular function evaluated in Example 7 of Chapter 2, we obtain

$$S_X(f) = A^2 T \operatorname{sinc}^2(fT) \qquad (4.105)$$

which is shown plotted in Fig. 4.17. Here again we see that the power spectral density is nonnegative for all f and that it is an even function of f. Noting that $R_X(0) = A^2$ and using Property 3, we find that the total area under $S_X(f)$, or the average power of the random binary wave described here, is A^2.

The result of Eq. (4.105) may be generalized as follows. We note that the energy spectral density of a rectangular pulse $g(t)$ of amplitude A and duration T is given by

$$\mathscr{E}_g(f) = A^2 T^2 \operatorname{sinc}^2(fT) \qquad (4.106)$$

We may therefore rewrite Eq. (4.105) in terms of $\mathscr{E}_g(f)$ as

$$S_X(f) = \frac{\mathscr{E}_g(f)}{T} \qquad (4.107)$$

Equation (4.107) states that for a random binary wave in which binary symbols 1 and 0 are represented by pulses $g(t)$ and $-g(t)$, respectively, the power spectral density $S_X(f)$ is equal to the energy spectral density $\mathscr{E}_g(f)$ of the *symbol shaping pulse* $g(t)$, divided by the *symbol duration T*.

EXAMPLE 12   Mixing of a Random Process with a Sinusoidal Process

A situation that often arises in practice is that of *mixing* (i.e., multiplication) of a wide-sense stationary random process $X(t)$ with a sinusoidal wave $\cos(2\pi f_c t + \Theta)$, where the phase Θ is a random variable that is uniformly distributed over the interval $(0, 2\pi)$. The addition of the random phase Θ in this manner merely recognizes the fact that the time origin is arbitrarily chosen when $X(t)$ and $\cos(2\pi f_c t + \Theta)$ come from physically independent sources, as is usually the case. We are interested in determining the power spectral density of the random process $Y(t)$ defined by

$$Y(t) = X(t)\cos(2\pi f_c t + \Theta) \qquad (4.108)$$

Using the definition of autocorrelation function of a wide-sense stationary process, and noting that the random variable Θ is independent of $X(t)$, we find that the autocorrelation function of $Y(t)$ is given by

$$
\begin{aligned}
R_Y(\tau) &= E[Y(t + \tau)Y(t)] \\
&= E[X(t + \tau)\cos(2\pi f_c t + 2\pi f_c \tau + \Theta)X(t)\cos(2\pi f_c t + \Theta)] \\
&= E[X(t + \tau)X(t)]E[\cos(2\pi f_c t + 2\pi f_c \tau + \Theta)\cos(2\pi f_c t + \Theta)] \quad (4.109) \\
&= \tfrac{1}{2}R_X(\tau)E[\cos(2\pi f_c \tau) + \cos(4\pi f_c t + 2\pi f_c \tau + 2\Theta)] \\
&= \tfrac{1}{2}R_X(\tau)\cos(2\pi f_c \tau)
\end{aligned}
$$

Because the power spectral density is the Fourier transform of the autocorrelation function, we find that the power spectral densities of the random process $X(t)$ and $Y(t)$ are related as follows:

$$S_Y(f) = \tfrac{1}{4}[S_X(f - f_c) + S_X(f + f_c)] \qquad (4.110)$$

According to Eq. (4.110), the power spectral density of the random process $Y(t)$ defined in Eq. (4.108) is obtained as follows: We shift the given power spectral density $S_X(f)$ of random process $X(t)$ to the right by f_c, shift it to the left by f_c, add the two shifted power spectra, and divide the result by 4.

Relation Among the Power Spectral Densities of the Input and Output Random Processes

Let $S_Y(f)$ denote the power spectral density of the output random process $Y(t)$ obtained by passing the random process $X(t)$ through a linear filter of transfer function $H(f)$. Then, recognizing by definition that the power spectral density of a random process is equal to the Fourier transform of its autocorrelation function and using Eq. (4.86), we obtain

$$
\begin{aligned}
S_Y(f) &= \int_{-\infty}^{\infty} R_Y(\tau)\exp(-j2\pi f\tau) \, d\tau \\
&= \int_{-\infty}^{\infty}\int_{-\infty}^{\infty}\int_{-\infty}^{\infty} h(\tau_1)h(\tau_2)R_X(\tau - \tau_1 + \tau_2)\exp(-j2\pi f\tau) \, d\tau_1 \, d\tau_2 \, d\tau
\end{aligned}
$$
$$(4.111)$$

Let $\tau - \tau_1 + \tau_2 = \tau_0$, or, equivalently, $\tau = \tau_0 + \tau_1 - \tau_2$. Then, by making this substitution in Eq. (4.111), we find that $S_Y(f)$ may be expressed as the product of three terms: the transfer function $H(f)$ of the filter, the complex conjugate of $H(f)$, and the power spectral density $S_X(f)$ of the input random process $X(t)$. We may thus simplify Eq. (4.111) as

$$S_Y(f) = H(f)H^*(f)S_X(f) \qquad (4.112)$$

Finally, since $|H(f)|^2 = H(f)H^*(f)$, we find that the relationship among the power spectral densities of the input and output random processes is expressed in the frequency domain by writing

$$S_Y(f) = |H(f)|^2 S_X(f) \qquad (4.113)$$

Equation (4.113) states that *the power spectral density of the output process $Y(t)$ equals the power spectral density of the input process $X(t)$ multiplied by the squared magnitude of the transfer function $H(f)$ of the filter*. By using this relation, we can therefore determine the effect of passing a random process through a stable, linear, time-invariant, filter. In computational terms, Eq. (4.113) is usually easier to handle than its time-domain counterpart of Eq. (4.86), involving the autocorrelation function.

EXAMPLE 13   Comb Filter

Consider the filter of Fig. 4.18a consisting of a delay line and a summing device. We wish to evaluate the power spectral density of the filter output $Y(t)$, given that the power spectral density of the filter input $X(t)$ is $S_X(f)$.

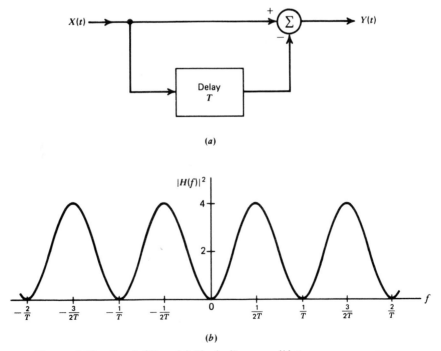

(a)

(b)

Figure 4.18 Comb filter. (a) Block diagram. (b) Frequency response.

The transfer function of this filter is

$$H(f) = 1 - \exp(-j2\pi fT)$$
$$= 1 - \cos(2\pi fT) + j\sin(2\pi fT)$$

The squared magnitude of $H(f)$ is

$$|H(f)|^2 = [1 - \cos(2\pi fT)]^2 + \sin^2(2\pi fT)$$
$$= 2[1 - \cos(2\pi fT)]$$
$$= 4\sin^2(\pi fT)$$

which is shown plotted in Fig. 4.18b. Because of the periodic form of this frequency response, the filter of Fig. 4.18a is sometimes referred to as a *comb filter*. The power spectral density of the filter output is therefore

$$S_Y(f) = 4\sin^2(\pi fT)S_X(f)$$

For values of frequency f that are small compared to $1/T$, we have

$$\sin(\pi fT) \simeq \pi fT$$

Under this condition, we may approximate the expression for $S_Y(f)$ as follows:

$$S_Y(f) \simeq 4\pi^2 f^2 T^2 S_X(f) \tag{4.114}$$

Because differentiation in the time domain corresponds to multiplication by $j2\pi f$ in the frequency domain, we see from Eq. (4.114) that the comb filter of Fig. 4.18a acts as a differentiator for low-frequency inputs.

Relation Among the Power Spectral Density and the Amplitude Spectrum of a Sample Function

We now wish to relate the power spectral density $S_X(f)$ directly to the spectral properties of a sample function $x(t)$ of a wide-sense stationary process $X(t)$ that is ergodic. For the sample function $x(t)$ to be Fourier transformable, however, it must be absolutely integrable; that is

$$\int_{-\infty}^{\infty} |x(t)|\, dt < \infty$$

This condition can never be satisfied by any sample function $x(t)$ of infinite duration. In order to use the Fourier transform technique, we consider a truncated segment of $x(t)$, defined over the observation interval $-T \le t \le T$, say. Thus, using $X(f,T)$ to denote the Fourier transform of the truncated sample function so defined, we may write

$$X(f,T) = \int_{-T}^{T} x(t)\exp(-j2\pi ft)\, dt \tag{4.115}$$

Assuming that the wide-sense stationary random process is also ergodic, we may evaluate the autocorrelation function $R_X(\tau)$ of the random process $X(t)$ using the time-average formula (see Section 4.9)

$$R_X(\tau) = \lim_{T\to\infty} \frac{1}{2T} \int_{-T}^{T} x(t + \tau)\, x(t)\, dt \qquad (4.116)$$

Viewing $x(t)$ as a power signal, we may formulate the following Fourier-transform pair:

$$\frac{1}{2T} \int_{-T}^{T} x(t + \tau)\, x(t)\, dt \rightleftharpoons \frac{1}{2T} |X(f,T)|^2 \qquad (4.117)$$

The parameter on the left-hand side is a time-averaged autocorrelation function. The parameter on the right-hand side is called the *periodogram*, whose dimensions are the same as those of the power spectral density. This terminology is a misnomer, however, since the periodogram is a function of frequency, not period. Nevertheless, it has wide usage. The quantity was first used by statisticians to look for periodicities such as seasonal trends in data.

Using the formula for the inverse Fourier transform in the Fourier-transform pair of Eq. (4.117), we may express the time-averaged autocorrelation function of the sample function $x(t)$ in terms of the periodogram as

$$\frac{1}{2T} \int_{-T}^{T} x(t + \tau)\, x(t)\, dt = \int_{-\infty}^{\infty} \frac{1}{2T} |X(f,T)|^2 \exp(j2\pi f T)\, df \quad (4.118)$$

Hence, substituting Eq. (4.118) in (4.116), we get

$$R_X(\tau) = \lim_{T\to\infty} \int_{-\infty}^{\infty} \frac{1}{2T} |X(f,T)|^2 \exp(j2\pi f\tau)\, df \qquad (4.119)$$

For a fixed value of the frequency f, the periodogram is a random variable in that its value varies in a random manner from one sample function of the random process to another. Thus, for a given sample function $x(t)$, the periodogram does not converge in any statistical sense to a limiting value as T tends to infinity. As such, it would be incorrect to interchange the order of the integration and limiting operations in Eq. (4.119). Suppose, however, that we take the expectation of both sides of Eq. (4.119) over the ensemble of all sample functions of the random process and recognize that for an ergodic process the autocorrelation function $R_X(\tau)$ is unchanged by such an operation. Then, since each sample function of an ergodic process eventually takes on nearly all the modes of behavior of each other sample function, we may write

$$R_X(\tau) = \lim_{T\to\infty} \int_{-\infty}^{\infty} \frac{1}{2T} E[|X(f,T)|^2]\exp(j2\pi f\tau)\, df \qquad (4.120)$$

Now we may interchange the order of the integration and limiting operations and so obtain

$$R_X(\tau) = \int_{-\infty}^{\infty} \left\{ \lim_{T\to\infty} \frac{1}{2T} E[|X(f,T)|^2] \right\} \exp(j2\pi f\tau)\, df \qquad (4.121)$$

Hence, comparing Eqs. (4.121) and (4.97), we obtain the desired relation between the power spectral density $S_X(f)$ of an ergodic process and the amplitude spectrum $|X(f,T)|$ of a truncated sample function of the process:

$$S_X(f) = \lim_{T \to \infty} \frac{1}{2T} E[|X(f,T)|^2]$$

$$= \lim_{T \to \infty} \frac{1}{2T} E\left[\left|\int_{-T}^{T} x(t) \exp(-j2\pi ft) \, dt\right|^2\right]$$

(4.122)

It is important to note that in Eq. (4.122) it is not possible to let $T \to \infty$ before taking the expectation. Equation (4.122) provides the mathematical basis for *estimating*[5] the power spectral density of an ergodic random process, given a sample function $x(t)$ of the process observed over the interval $(-T,T)$.

Cross-Spectral Densities

Just as the power spectral density provides a measure of the frequency distribution of a single random process, cross-spectral densities provide a measure of the frequency interrelationship between two random processes. In particular, let $X(t)$ and $Y(t)$ be two jointly wide-sense stationary random processes with their cross-correlation functions denoted by $R_{XY}(\tau)$ and $R_{YX}(\tau)$. We then define the *cross-spectral densities* $S_{XY}(f)$ and $S_{YX}(f)$ of this pair of random processes to be the Fourier transforms of their respective cross-correlation functions, as shown by

$$S_{XY}(f) = \int_{-\infty}^{\infty} R_{XY}(\tau) \exp(-j2\pi f\tau) \, d\tau$$

(4.123)

and

$$S_{YX}(f) = \int_{-\infty}^{\infty} R_{YX}(\tau) \exp(-j2\pi f\tau) \, d\tau$$

(4.124)

The cross-correlation functions and cross-spectral densities thus form Fourier-transform pairs. Accordingly, using the formula for inverse Fourier transformation we may also write

$$R_{XY}(\tau) = \int_{-\infty}^{\infty} S_{XY}(f) \exp(j2\pi f\tau) \, df$$

(4.125)

and

$$R_{YX}(\tau) = \int_{-\infty}^{\infty} S_{YX}(f) \exp(j2\pi f\tau) \, df$$

(4.126)

The cross-spectral densities $S_{XY}(f)$ and $S_{YX}(f)$ are not necessarily real functions of the frequency f. However, substituting the relationship

$$R_{XY}(\tau) = R_{YX}(-\tau)$$

in Eq. (4.123) and then using Eq. (4.124) we find that $S_{XY}(f)$ and $S_{YX}(f)$ are related by

$$S_{XY}(f) = S_{YX}(-f) = S_{YX}^*(f) \tag{4.127}$$

EXAMPLE 14

Suppose that the random processes $X(t)$ and $Y(t)$ have zero mean, and they are individually stationary in the wide sense. Consider the sum random process

$$Z(t) = X(t) + Y(t)$$

The problem is to determine the power spectral density of $Z(t)$.

The autocorrelation function of $Z(t)$ is given by

$$
\begin{aligned}
R_Z(t,u) &= E[Z(t)\,Z(u)] \\
&= E[(X(t) + Y(t))(X(u) + Y(u))] \\
&= E[X(t)X(u)] + E[X(t)Y(u)] + E[Y(t)X(u)] + E[Y(t)Y(u)] \\
&= R_X(t,u) + R_{XY}(t,u) + R_{YX}(t,u) + R_Y(t,u)
\end{aligned}
$$

Defining $\tau = t - u$, we may therefore write

$$R_Z(\tau) = R_X(\tau) + R_{XY}(\tau) + R_{YX}(\tau) + R_Y(\tau) \tag{4.128}$$

when the random processes $X(t)$ and $Y(t)$ are also stationary in the wide sense. Accordingly, taking the Fourier transform of both sides of Eq. (4.128), we get

$$S_Z(f) = S_X(f) + S_{XY}(f) + S_{YX}(f) + S_Y(f) \tag{4.129}$$

We thus see that the cross-spectral densities $S_{XY}(f)$ and $S_{YX}(f)$ represent the spectral components that must be added to the individual power spectral densities of a pair of correlated random processes in order to obtain the power spectral density of their sum.

When the wide-sense stationary random processes $X(t)$ and $Y(t)$ are uncorrelated, the cross-spectral densities $S_{XY}(f)$ and $S_{YX}(f)$ are zero, and so Eq. (4.129) reduces as follows:

$$S_Z(f) = S_X(f) + S_Y(f) \tag{4.130}$$

We may generalize this latter result by stating that when there is a multiplicity of zero-mean wide-sense stationary random processes that are uncorrelated with each other, the power spectral density of their sum is equal to the sum of their individual power spectral densities.

EXAMPLE 15

Consider next the problem of passing two jointly wide-sense stationary random processes through a pair of separate, stable, linear, time-invariant filters, as shown in Fig. 4.19. In particular, suppose that the random process $X(t)$ is the input to the filter of impulse response $h_1(t)$ and that the random process $Y(t)$ is the input to the filter of impulse response $h_2(t)$. Let $V(t)$ and $Z(t)$ denote the random processes at the respective filter outputs. The cross-correlation function of $V(t)$

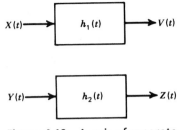

Figure 4.19 A pair of separate filters.

and $Z(t)$ is therefore

$$
\begin{aligned}
R_{VZ}(t,u) &= E[V(t)Z(u)] \\
&= E\left[\int_{-\infty}^{\infty} h_1(\tau_1)X(t-\tau_1)\,d\tau_1 \int_{-\infty}^{\infty} h_2(\tau_2)Y(u-\tau_2)\,d\tau_2\right] \\
&= \int_{-\infty}^{\infty}\int_{-\infty}^{\infty} h_1(\tau_1)h_2(\tau_2)E[X(t-\tau_1)Y(u-\tau_2)]\,d\tau_1\,d\tau_2 \\
&= \int_{-\infty}^{\infty}\int_{-\infty}^{\infty} h_1(\tau_1)h_2(\tau_2)R_{XY}(t-\tau_1,\,u-\tau_2)\,d\tau_1\,d\tau_2
\end{aligned}
\tag{4.131}
$$

where $R_{XY}(t,u)$ is the cross-correlation function of $X(t)$ and $Y(t)$. Because the input random processes are jointly wide-sense stationary (by hypothesis), we may set $\tau = t - u$ and so rewrite Eq. (4.131) as follows:

$$
R_{VZ}(\tau) = \int_{-\infty}^{\infty}\int_{-\infty}^{\infty} h_1(\tau_1)h_2(\tau_2)R_{XY}(\tau - \tau_1 + \tau_2)\,d\tau_1\,d\tau_2 \tag{4.132}
$$

Taking the Fourier transform of both sides of Eq. (4.132) and using a procedure similar to that which led to the development of Eq. (4.93), we finally get

$$
S_{VZ}(f) = H_1(f)H_2^*(f)S_{XY}(f) \tag{4.133}
$$

where $H_1(f)$ and $H_2(f)$ are the transfer functions of the respective filters in Fig. 4.19, and $H_2^*(f)$ is the complex conjugate of $H_2(f)$. This is the desired relationship between the cross-spectral density of the output processes and that of the input processes.

4.12 GAUSSIAN PROCESS

The material we have presented on random processes up to this point in the discussion has been of a fairly general nature. In this section, we consider an important family of random processes known as Gaussian processes.[6]

Let us suppose that we observe a random process $X(t)$ for an interval that starts at time $t = 0$ and lasts until $t = T$. Suppose also that we weight the random process $X(t)$ by some function $g(t)$ and then integrate the product $g(t)X(t)$ over this observation interval, thereby obtaining a random variable Y defined by

$$Y = \int_0^T g(t)X(t)\ dt \qquad (4.134)$$

We refer to Y as a *linear functional* of $X(t)$. The distinction between a function and a functional should be carefully noted. For example, the sum $Y = \Sigma_{i=1}^{N} a_i X_i$, where the a_i are constants and the X_i are random variables, is a linear *function* of the X_i; for each observed set of values for the random variables X_i, we have a corresponding value for the random variable Y. On the other hand, in Eq. (4.134) the value of the random variable Y depends on the course of the *argument function* $g(t)X(t)$ over the entire observation interval from 0 to T. Thus a functional is a quantity that depends on the entire course of one or more functions rather than on a number of discrete variables. In other words, the domain of a functional is a set or space of admissible functions rather than a region of a coordinate space.

If in Eq. (4.134) the weighting function $g(t)$ is such that the mean-square value of the random variable Y is finite, and if the random variable Y is a *Gaussian-distributed* random variable for every $g(t)$ in this class of functions, then the process $X(t)$ is said to be a *Gaussian process*. In other words, the process $X(t)$ is a Gaussian process if every linear functional of $X(t)$ is a Gaussian random variable.

In Example 4 we described the characterization of a Gaussian random variable. We say that the random variable Y has a Gaussian distribution if its probability density function has the form

$$f_Y(y) = \frac{1}{\sqrt{2\pi}\sigma_Y}\exp\left[-\frac{(y-\mu_Y)^2}{2\sigma_Y^2}\right] \qquad (4.135)$$

where μ_Y is the mean and σ_Y^2 is the variance of the random variable Y. A plot of this probability density function is given in Fig. 4.20 for the special case when the Gaussian random variable Y is *normalized* to have a mean μ_Y of zero and a variance σ_Y^2 of one, as shown by

$$f_Y(y) = \frac{1}{\sqrt{2\pi}}\exp\left(-\frac{y^2}{2}\right)$$

Such a normalized Gaussian distribution is commonly written as $\mathcal{N}(0,1)$.

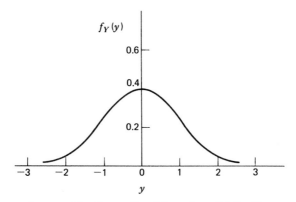

Figure 4.20 Normalized Gaussian distribution.

A Gaussian process has two main virtues. First, the Gaussian process has many properties that make analytic results possible; we will discuss these properties later in the section. Second, the random processes produced by physical phenomena are often such that a Gaussian model is appropriate. Furthermore, the use of a Gaussian model to describe such physical phenomena is usually confirmed by experiments. Thus the widespread occurrence of physical phenomena for which a Gaussian model is appropriate, together with the ease with which a Gaussian process is handled mathematically, make the Gaussian process very important in the study of communication systems.

Central Limit Theorem

The *central limit theorem* provides the mathematical justification for using a Gaussian process as a model for a large number of different physical phenomena in which the observed random variable, at a particular instant of time, is the result of a large number of individual random events. To formulate this important theorem, let X_i, $i = 1, 2, \ldots, N$, be a set of random variables that satisfies the following requirements:

1. The X_i are statistically independent.
2. The X_i have the same probability distribution with mean μ_X and variance σ_X^2.

The X_i so described are said to constitute a set of *independently and identically distributed* (i.i.d.) random variables. Let these random variables be *normalized* as follows:

$$Y_i = \frac{1}{\sigma_X} (X_i - \mu_X), \qquad i = 1, 2, \ldots, N$$

so that we have

$$E[Y_i] = 0$$

and

$$\mathrm{var}[Y_i] = 1$$

Define the random variable

$$V_N = \frac{1}{\sqrt{N}} \sum_{i=1}^{N} Y_i$$

The central limit theorem states that the probability distribution of V_N approaches a normalized Gaussian distribution $\mathcal{N}(0,1)$ in the limit as N approaches infinity.

It is important to realize, however, that the central limit theorem gives only the "limiting" form of the probability distribution of the normalized random variable V_N as N approaches infinity. When N is finite, it is sometimes found that the Gaussian limit gives a relatively poor approximation for the actual probability distribution of V_N even though N may be quite large.

Properties of a Gaussian Process

A Gaussian process has some useful properties that are described in the sequel.

PROPERTY 1

If a Gaussian process $X(t)$ is applied to a stable linear filter, then the random process $Y(t)$ developed at the output of the filter is also Gaussian.

This property is readily derived by using the definition of a Gaussian process based on Eq. (4.134). Consider the situation depicted in Fig. 4.14, where we have a linear time-invariant filter of impulse response $h(t)$, with the random process $X(t)$ as input and the random process $Y(t)$ as output. We assume that $X(t)$ is a Gaussian process. The random processes $Y(t)$ and $X(t)$ are related by the convolution integral

$$Y(t) = \int_0^T h(t - \tau) X(\tau) \ d\tau, \qquad 0 \leq t < \infty \qquad (4.136)$$

We assume that the impulse response $h(t)$ is such that the mean-square value of the output random process $Y(t)$ is finite for all t in the range $0 \leq t < \infty$ for which $Y(t)$ is defined. To demonstrate that the output process $Y(t)$ is Gaussian, we must show that any linear functional of it is a Gaussian random variable. That is, if we define the random variable

$$Z = \int_0^\infty g_Y(t) \int_0^T h(t - \tau) X(\tau) \ d\tau \ dt \qquad (4.137)$$

then Z must be a Gaussian random variable for every function $g_Y(t)$, such that the mean-square value of Z is finite. Interchanging the order of integration in Eq. (4.137), we get

$$Z = \int_0^T g(\tau) X(\tau) \ d\tau \qquad (4.138)$$

where

$$g(\tau) = \int_0^\infty g_Y(t) h(t - \tau) \ dt \qquad (4.139)$$

Since $X(t)$ is a Gaussian process by hypothesis, it follows from Eq. (4.138) that Z must be a Gaussian random variable. We have thus shown that if the input $X(t)$ to a linear filter is a Gaussian process, then the output $Y(t)$ is also a Gaussian process. Note, however, that although our proof was carried out assuming a time-invariant linear filter, this property is true for any arbitrary stable linear system.

PROPERTY 2

Consider the set of random variables or samples $X(t_1)$, $X(t_2)$, . . . , $X(t_n)$, obtained by observing a random process $X(t)$ at times $t_1, t_2, \ldots, t_n$. If the process $X(t)$ is Gaussian, *then this set of random variables are jointly Gaussian for any n, with their n-fold joint probability density function being completely determined by specifying the set of means*

$$\mu_{X(t_i)} = E[X(t_i)], \qquad i = 1, 2, \ldots, n$$

and the set of autocovariance functions

$$C_X(t_k,t_i) = E[(X(t_k) - \mu_{X(t_k)})(X(t_i) - \mu_{X(t_i)})], \qquad k, i = 1, 2, \ldots, n$$

Property 2 is frequently used as the definition of a Gaussian process.[7] However, this definition is more difficult to use than that based on Eq. (4.134) for evaluating the effects of filtering on a Gaussian process.

We may extend Property 2 to two (or more) random processes as follows. Consider the composite set of random variables $X(t_1)$, $X(t_2), \ldots, X(t_n)$, $Y(u_1)$, $Y(u_2), \ldots, Y(u_m)$ obtained by observing a random process $X(t)$ at times $\{t_i, i = 1, 2, \ldots, n\}$, and a second random process $Y(t)$ at times $\{u_k, k = 1, 2, \ldots, m\}$. We say that the processes $X(t)$ and $Y(t)$ are *jointly Gaussian* if this composite set of random variables are jointly Gaussian for any n and m. Note that in addition to the mean and correlation functions of the random processes $X(t)$ and $Y(t)$ individually, we must also know the cross-covariance function

$$E[(X(t_i) - \mu_{X(t_i)})(X(u_k) - \mu_{X(u_k)})] = R_{XY}(t_i, u_k) - \mu_{X(t_i)}\mu_{Y(u_k)}$$

for any pair of observation instants (t_i, u_k). This additional knowledge is embodied in the cross-correlation function, $R_{XY}(t_i, u_k)$, of the two processes $X(t)$ and $Y(t)$.

PROPERTY 3

If a Gaussian process is wide-sense stationary, then the process is also stationary in the strict sense.

This follows directly from Property 2.

PROPERTY 4

If the random variables $X(t_1)$, $X(t_2), \ldots, X(t_n)$, obtained by sampling a Gaussian process $X(t)$ at times $t_1, t_2, \ldots, t_n$, are uncorrelated, that is,

$$E[(X(t_k) - \mu_{X(t_k)})(X(t_i) - \mu_{X(t_i)})] = 0, \qquad i \neq k$$

then these random variables are statistically independent.

The implication of this property is that the joint probability density function of the set of random variables $X(t_1)$, $X(t_2), \ldots, X(t_n)$ can be expressed as the product of the probability density functions of the individual random variables in the set.

4.13   NOISE

The term *noise* is used customarily to designate unwanted waves that tend to disturb the transmission and processing of signals in communication systems and over which we have incomplete control. In practice, we find that there are many potential sources of noise in a communication system. The sources of noise may be external to the system (e.g., atmospheric noise, galactic noise, man-made noise), or internal to the system. The second category includes an important type of noise that arises from *spontaneous fluctuations* of current or voltage in electrical circuits.[8] This type of noise represents a basic limitation on the transmission or detection of signals in communication systems involving the use of electronic devices. The two most common examples of spontaneous fluctuations in electrical circuits are *shot noise* and *thermal noise*.

Shot Noise

Shot noise arises in electronic devices such as diodes and transistors because of the discrete nature of current flow in these devices. For example, in a *photodetector* circuit a current pulse is generated every time an electron is emitted by the cathode due to incident light from a source of constant intensity. The electrons are naturally emitted at random times denoted by τ_k, where $-\infty < k < \infty$. It is assumed that the random emissions of electrons have been going on for a long time. Thus, the total current flowing through the photodetector may be modeled as an infinite sum of current pulses, as shown by

$$X(t) = \sum_{k=-\infty}^{\infty} h(t - \tau_k) \tag{4.140}$$

where $h(t - \tau_k)$ is the current pulse generated at time τ_k. The process $X(t)$ defined by Eq. (4.140) is a stationary process, called *shot noise*.[9]

The number of electrons, $N(t)$, emitted in the time interval $(0, t)$ constitutes a discrete stochastic process, the value of which increases by one each time an electron is emitted. Figure 4.21 shows a sample function of such a process. Let the mean value of the number of electrons, ν, emitted between times t and $t + t_0$ be defined by

$$E[\nu] = \lambda t_0 \tag{4.141}$$

The parameter λ is a constant called the *rate* of the process. The total number of electrons emitted in the interval $(t, t + t_0)$, that is,

$$\nu = N(t + t_0) - N(t)$$

follows a *Poisson distribution* with a mean value equal to λt_0. In particular, the probability that k electrons are emitted in the interval $(t, t + t_0)$ is defined by

$$P(\nu = k) = \frac{(\lambda t_0)^k}{k!} e^{-\lambda t_0} \qquad k = 0, 1, \ldots \tag{4.142}$$

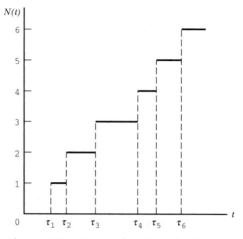

Figure 4.21   Sample function of a Poisson counting process.

Unfortunately, a detailed statistical characterization of the shot-noise process $X(t)$ defined in Eq. (4.140) is a difficult mathematical task. Here we simply quote the results pertaining to the first two moments of the process:

- The mean of $X(t)$ is

$$\mu_X = \lambda \int_{-\infty}^{\infty} h(t) \ dt \qquad (4.143)$$

where λ is the rate of the process and $h(t)$ is the waveform of a current pulse.
- The autocovariance function of $X(t)$ is

$$C_X(\tau) = \lambda \int_{-\infty}^{\infty} h(t) h(t + \tau) \ dt \qquad (4.144)$$

This latter result is known as *Campbell's theorem.*

For the special case of a waveform $h(t)$ consisting of a rectangular pulse of amplitude A and duration T, the mean of the shot-noise process $X(t)$ is λAT, and its autocovariance function is

$$C_X(\tau) = \begin{cases} \lambda A^2(T - |\tau|), & |\tau| < T \\ 0, & |\tau| \geq T \end{cases}$$

which has a triangular form similar to that shown in Fig. 4.13.

Thermal Noise

Thermal noise[10] is the name given to the electrical noise arising from the random motion of electrons in a conductor. The mean-square value of the thermal noise voltage V_{TN} appearing across the terminals of a resistor, measured in a bandwidth of Δf Hertz, is, for all practical purposes, given by

$$E[V_{TN}^2] = 4kTR \ \Delta f \ \text{volts}^2 \qquad (4.145)$$

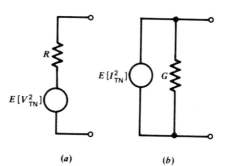

Figure 4.22 Models of a noisy resistor.
(a) Thévenin equivalent circuit. (b) Norton equivalent circuit.

where k is *Boltzmann's constant* equal to 1.38×10^{-23} joules per degree Kelvin, T is the *absolute temperature* in degrees Kelvin, and R is the resistance in ohms. We may thus model a noisy resistor by the *Thévenin equivalent circuit* consisting of a noise voltage generator of mean-square value $E[V_{TN}^2]$ in series with a noiseless resistor, as in Fig. 4.22*a*. Alternatively, we may use the *Norton equivalent circuit* consisting of a noise current generator in parallel with a noiseless conductance, as in Fig. 4.22*b*. The mean-square value of the noise current generator is

$$E[I_{TN}^2] = \frac{1}{R^2} E[V_{TN}^2]$$
$$= 4kTG\,\Delta f \text{ amps}^2 \tag{4.146}$$

where $G = 1/R$ is the conductance. It is also of interest to note that because the number of electrons in a resistor is very large and their random motions inside the resistor are statistically independent of each other, the central limit theorem indicates that thermal noise is Gaussian distributed with zero mean.

Noise calculations involve the transfer of power, and so we find that the use of the *maximum-power transfer theorem* is applicable to such calculations. This theorem states that the maximum possible power is transferred from a source of internal resistance R to a load of resistance R_l when $R_l = R$. Under this *matched condition*, the power produced by the source is divided equally between the internal resistance of the source and the load resistance, and the power delivered to the load is referred to as the *available power*. Applying the maximum-power transfer theorem to the Thévenin equivalent circuit of Fig. 4.22*a* or the Norton equivalent circuit of Fig. 4.22*b*, we find that a noisy resistor produces an *available noise power* equal to $kT\,\Delta f$ watts.

White Noise

The noise analysis of communication systems is customarily based on an idealized form of noise called *white noise*, the power spectral density of which is independent of the operating frequency. The adjective *white* is used in the sense that white light contains equal amounts of all frequencies within the visible band of electromagnetic radiation. We express the power spectral density of white noise, with a sample function denoted by $w(t)$, as

$$S_W(f) = \frac{N_0}{2} \qquad\qquad (4.147)$$

which is illustrated in Fig. 4.23a. The dimensions of N_0 are in watts per Hertz. The parameter N_0 is usually referenced to the input stage of the receiver of a communication system. It may be expressed as

$$N_0 = kT_e \qquad\qquad (4.148)$$

where k is Boltzmann's constant and T_e is the *equivalent noise temperature* of the receiver.[11] *The equivalent noise temperature of a system is defined as the temperature at which a noisy resistor has to be maintained such that, by connecting the resistor to the input of a noiseless version of the system, it produces the same available noise power at the output of the system as that produced by all the sources of noise in the actual system.* The important feature of the equivalent noise temperature is that it depends only on the parameters of the system.

Since the autocorrelation function is the inverse Fourier transform of the power spectral density, it follows that for white noise

$$R_W(\tau) = \frac{N_0}{2}\,\delta(\tau) \qquad\qquad (4.149)$$

That is, the autocorrelation function of white noise consists of a delta function weighted by the factor $N_0/2$ and occurring at $\tau = 0$, as in Fig. 4.23b. We note that $R_W(\tau)$ is zero for $\tau \neq 0$. Accordingly, any two different samples of white noise, no matter how closely together in time they are taken, are uncorrelated. If the white noise $w(t)$ is also Gaussian, then the two samples are statistically

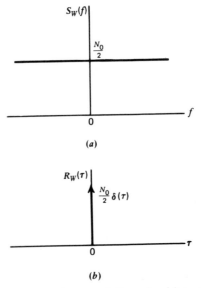

(a)

(b)

Figure 4.23 Characteristics of white noise. (a) Power spectral density. (b) Autocorrelation function.

independent. In a sense, white Gaussian noise represents the ultimate in "randomness."

Strictly speaking, white noise has infinite average power and, as such, it is not physically realizable. Nevertheless, white noise has simple mathematical properties exemplified by Eqs. (4.147) and (4.149), which make it useful in statistical system analysis.

The utility of a white noise process is parallel to that of an impulse function or delta function in the analysis of linear systems. Just as we may observe the effect of an impulse only after it has been passed through a system with a finite bandwidth, so it is with white noise whose effect is observed only after passing through a similar system. We may state, therefore, that as long as the bandwidth of a noise process at the input of a system is appreciably larger than that of the system itself, then we may model the noise process as white noise.

EXAMPLE 16   Ideal Low-Pass Filtered White Noise

Suppose that a white Gaussian noise $w(t)$ of zero mean and power spectral density $N_0/2$ is applied to an ideal low-pass filter of bandwidth B and passband amplitude response of one. The power spectral density of the noise $n(t)$ appearing at the filter output is therefore (see Fig. 4.24a)

$$S_N(f) = \begin{cases} \dfrac{N_0}{2}, & -B < f < B \\ 0, & |f| > B \end{cases} \qquad (4.150)$$

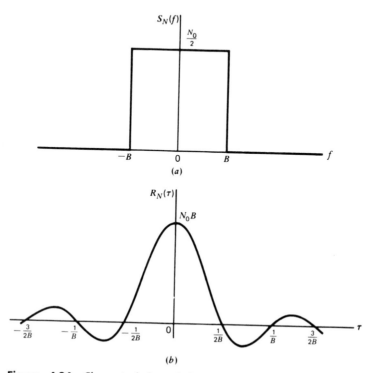

Figure 4.24 Characteristics of low-pass filtered white noise. (a) Power spectral density. (b) Autocorrelation function.

The autocorrelation function of $n(t)$ is the inverse Fourier transform of the power spectral density shown in Fig. 4.24a:

$$R_N(\tau) = \int_{-B}^{B} \frac{N_0}{2} \exp(j2\pi f\tau) \ df$$

$$= N_0 B \ \mathrm{sinc}(2B\tau)$$

(4.151)

This autocorrelation function is shown plotted in Fig. 4.24b. We see that $R_N(\tau)$ has its maximum value of $N_0 B$ at the origin, and it passes through zero at $\tau = \pm k/2B$, where $k = 1, 2, 3, \ldots$.

Since the input noise $w(t)$ is Gaussian (by hypothesis), it follows that the band-limited noise $n(t)$ at the filter output is also Gaussian. Suppose now that $n(t)$ is sampled at the rate of $2B$ times per second. From 4.24b, we see that the resulting noise samples are uncorrelated and, being Gaussian, they are statistically independent. Accordingly, the joint probability density function of a set of noise samples obtained in this way is equal to the product of the individual probability density functions. Note that each such noise sample has a mean of zero and variance of $N_0 B$.

EXAMPLE 17 RC Low-Pass Filtered White Noise

Consider next a white Gaussian noise $w(t)$ of zero mean and power spectral density $N_0/2$ applied to a low-pass RC filter, as in Fig. 4.25a. The transfer function of the filter is

$$H(f) = \frac{1}{1 + j2\pi f RC}$$

The power spectral density of the noise $n(t)$ appearing at the low-pass RC filter output is therefore (see Fig. 4.25b)

$$S_N(f) = \frac{N_0/2}{1 + (2\pi f RC)^2}$$

From Example 3 of Chapter 2, we recall the following Fourier-transform pair (using τ in place of t as the time variable to suit the problem at hand):

$$\exp(-a|\tau|) \rightleftharpoons \frac{2a}{a^2 + (2\pi f)^2}$$

(4.152)

where a is a constant. Therefore, setting $a = 1/RC$, we find that the autocorrelation function of the filtered noise $n(t)$ is

$$R_N(\tau) = \frac{N_0}{4RC} \exp\left(-\frac{|\tau|}{RC}\right)$$

(4.153)

which is shown plotted in Fig. 4.25c. The decorrelation time τ_0 for which $R_N(\tau)$ drops to 1 percent, say, of its maximum value of $N_0/4RC$ is equal to $4.61RC$. Thus, if the noise appearing at the filter output is sampled at a rate equal to or less than $0.217/RC$ samples per second, the resulting samples are essentially uncorrelated and, being Gaussian, they are statistically independent.

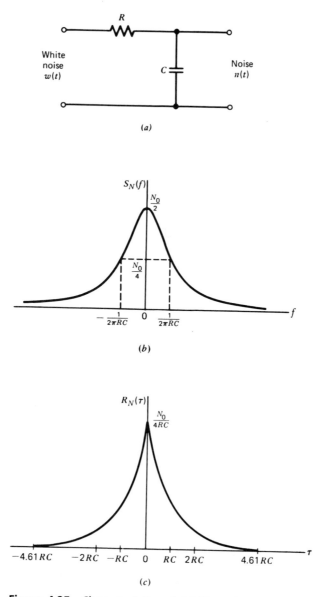

Figure 4.25 Characteristics of RC-filtered white noise. (a) Low-pass RC filter. (b) Power spectral density of filter output $n(t)$. (c) Autocorrelation function of $n(t)$.

COMPUTER EXPERIMENT **Autocorrelation of a Sinusoidal Wave Plus White Gaussian Noise**

In this computer experiment, we study the statistical characterization of a random process $X(t)$ consisting of a sinusoidal wave component $A \cos(2\pi f_c t + \Theta)$ and a white Gaussian noise process $W(t)$ of zero mean and power spectral density $N_0/2$. That is, we have

$$X(t) = A\cos(2\pi f_c t + \Theta) + W(t) \tag{4.154}$$

where Θ is a uniformly distributed random variable over the interval $(-\pi, \pi)$. Clearly, the two components of the process $X(t)$ are independent. The autocorrelation function of $X(t)$ is therefore the sum of the individual autocorrelation functions of the signal (sinusoidal wave) component and the noise component, as shown by

$$R_X(\tau) = \frac{A^2}{2} \cos(2\pi f_c \tau) + \frac{N_0}{2} \delta(\tau) \qquad (4.155)$$

This equation shows that for $|\tau| > 0$, the autocorrelation function $R_X(\tau)$ has the same sinusoidal waveform as the signal component. We may generalize this result by stating that the presence of a periodic signal component corrupted by additive white noise can be detected by computing the autocorrelation function of the composite process $X(t)$.

The purpose of the experiment described here is to perform this computation using two different methods: (1) ensemble averaging, and (2) time averaging. The top trace of Fig. 4.26a shows a sinusoidal signal of frequency $f_c = 0.002$ and phase $\theta = -\pi/2$, truncated to a finite duration $T = 1000$; the amplitude A of the sinusoidal signal is set to $\sqrt{2}$ to give unit average power. The bottom trace of Fig. 4.26a shows a particular realization $x(t)$ of the random process $X(t)$ consisting of this sinusoidal signal and additive white Gaussian noise; the power spectral density of the noise for this realization is $(N_0/2) = 1000$. The original sinusoidal is barely recognizable in $x(t)$.

For *ensemble-averaged computation* of the autocorrelation function, we may proceed as follows:

- Compute the product $x(t + \tau) x(t)$ for some fixed time t and specified time shift τ, where $x(t)$ is a particular realization of the random process $X(t)$.
- Repeat the computation of the product $x(t + \tau) x(t)$ for M independent realizations (i.e., sample functions) of the random process $X(t)$.
- Compute the average of these computations over M.
- Repeat this sequence of computations for different values of τ.

The results of this computation are plotted in Fig. 4.26b for $M = 50$ realizations. The picture portrayed here is in perfect agreement with theory defined by Eq. (4.155). The important point to note here is that the ensemble-averaging process yields a clean estimate of the true autocorrelation function $R_X(\tau)$ of the random process $X(t)$. Moreover, the presence of the sinusoidal signal is clearly visible in the plot of $R_X(\tau)$ versus τ.

For the *time-averaged estimation* of the autocorrelation function of the process $X(t)$, we invoke ergodicity and use the formula [see Eq. (4.80)]

$$R_X(\tau) = \lim_{T \to \infty} R_x(\tau, T) \qquad (4.156)$$

where $R_x(\tau, T)$ is the time-averaged autocorrelation function:

$$R_x(\tau, T) = \frac{1}{2T} \int_{-T}^{T} x(t + \tau) x(t) \, dt \qquad (4.157)$$

The $x(t)$ in Eq. (4.157) is a particular realization of the process $X(t)$, and $2T$ is the total observation interval. Define the *time-windowed function*

$$x_T(t) = \begin{cases} x(t), & -T \le t \le T \\ 0, & \text{otherwise} \end{cases} \qquad (4.158)$$

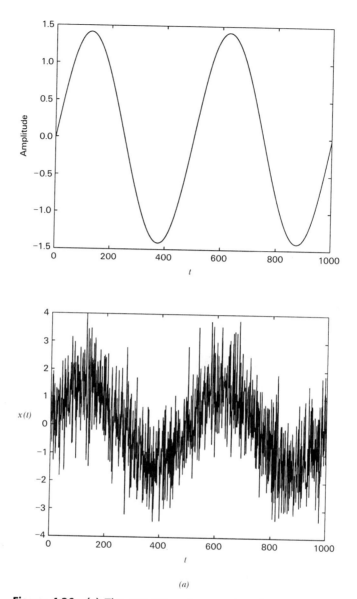

(a)

Figure 4.26 (a) The top trace shows original truncated si-nusoid $A \cos(2\pi f_c t)$, for $A = \sqrt{2}$, $f_c = 2/N = 0.002$, $t = 0, 1, \ldots, N - 1$; and bottom trace shows $x(t)$, a noisy version of the sinusoidal signal. (b) Ensemble-averaged autocorrela-tion function $R_x(\tau)$ for SNR = 0 dB, $N = 1000$, $M = 50$. (c) Periodogram; top trace shows a plot of the periodogram for frequency "samples" 0 to 1000, and bottom trace shows a magnified portion of the plot for frequency samples 490 to 510. (d) Autocorrelation function $\hat{R}_x(\tau)$ computed as the in-verse Fourier transform of the periodogram in Figure 4.26c. (e) Plots of $\hat{R}_x(\tau)$ for varying SNR: -10, 0, $+10$ dB, proceeding from top to bottom.

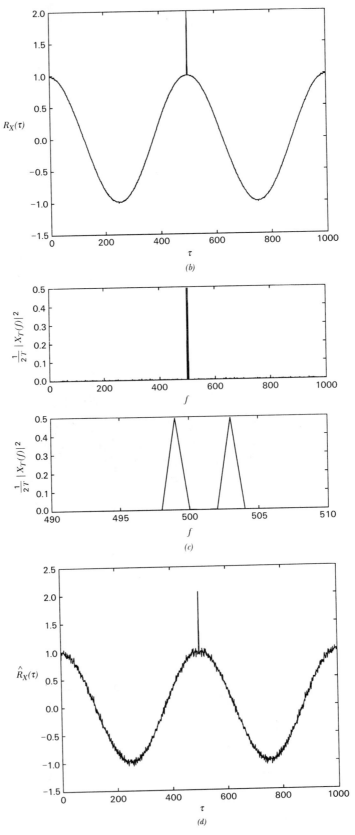

Figure 4.26 (*continued*)

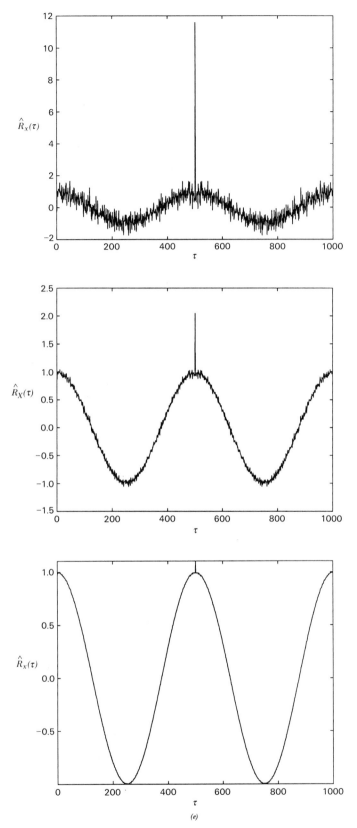

Figure 4.26 (*continued*)

We may then rewrite Eq. (4.157) as

$$R_x(\tau,T) = \frac{1}{2T} \int_{-\infty}^{\infty} x_T(t+\tau) x_T(t) \, dt \qquad (4.159)$$

For a specified time shift τ, we may compute $R_x(\tau,T)$ directly using Eq. (4.159). However, from a computational viewpoint, it is more efficient to use an *indirect* method based on Fourier transformation. First, we note from Eq. (4.159) that the time-averaged autocorrelation function $R_x(\tau,T)$ may be viewed as a scaled form of *convolution in the τ-domain* as follows:

$$R_x(\tau,T) = \frac{1}{2T} x_T(\tau) \bigstar x_T(-\tau) \qquad (4.160)$$

where the star denotes convolution and $x_T(\tau)$ is simply the time-windowed function $x_T(t)$ with t replaced by τ. Let $X_T(f)$ denote the Fourier transform of $x_T(\tau)$; note that $X_T(f)$ is the same as the Fourier transform $X(f,T)$ defined in Eq. (4.115). Since convolution in the τ-domain is transformed into multiplication in the frequency domain, we have the Fourier-transform pair:

$$R_x(\tau,T) \rightleftharpoons \frac{1}{2T} |X_T(f)|^2 \qquad (4.161)$$

The parameter $|X_T(f)|^2/2T$ is recognized as the periodogram of the process $X(t)$. Equation (4.161) is a mathematical description of the *correlation theorem*, which may be formally stated as follows: *The time-averaged autocorrelation function of a sample function pertaining to a random process and its periodogram, based on that sample function, constitute a Fourier-transform pair.*

We are now ready to describe the indirect method for computing the time-averaged autocorrelation function $R_x(t,\tau)$:

- Compute the Fourier transform $X_T(f)$ of time-windowed function $x_T(\tau)$.
- Compute the periodogram $|X_T(f)|^2/2T$.
- Compute the inverse Fourier transform of $|X_T(f)|^2/2T$.

To perform these calculations on a digital computer, the customary procedure is to use the fast Fourier transform (FFT) algorithm; the FFT algorithm was described in Section 2.15. With $x_T(\tau)$ uniformly sampled, the computational procedure described herein yields the desired values of $R_x(\tau,T)$ for $\tau = 0, \Delta, 2\Delta, \ldots, (N-1)\Delta$, where Δ is the sampling period and N is the total number of samples used in the computation. Figures 4.26c and 4.26d present the results obtained in the time-averaging approach of "estimating" the autocorrelation function $R_X(\tau)$ using the indirect method for the same set of parameters as those used for the ensemble-averaged results of Fig. 4.26b. The symbol $\hat{R}_X(\tau)$ is used in Fig. 4.26d to emphasize the fact that the computation described here results in an "estimate" of the autocorrelation function $R_X(\tau)$.

Figure 4.26e presents the estimation of $R_X(\tau)$ using the time-averaging approach for three different signal-to-noise ratios, namely, -10, 0, and $+10$ dB. The *signal-to-noise ratio* is defined by

$$\begin{aligned} \text{SNR} &= \frac{A^2/2}{N_0/(2T)} \\ &= \frac{A^2 T}{N_0} \end{aligned} \qquad (4.162)$$

On the basis of the results presented in Fig. 4.26, we may make the following observations:

- The ensemble-averaging and time-averaging approaches yield similar results for the autocorrelation function $R_X(\tau)$, signifying the fact that the random process $X(t)$ described herein is indeed ergodic.
- The indirect time-averaging approach, based on the FFT algorithm, provides an efficient method for the estimation of $R_X(\tau)$ using a digital computer.
- As the SNR is increased, the numerical accuracy of the estimation is improved, which is intuitively satisfying.

Noise Equivalent Bandwidth

In Example 16 we observe that when a source of white noise of zero mean and power spectral density $N_0/2$ is connected across the input of an ideal low-pass filter of bandwidth B and passband amplitude response of one, the average output noise power [or equivalently $R_N(0)$] is equal to $N_0 B$. In Example 17 we observe that when such a noise source is connected to the input of the simple RC low-pass filter of Fig. 4.25a, the corresponding value of the average output noise power is equal to $N_0/(4RC)$. For this filter, the half-power or 3-dB bandwidth is equal to $1/(2\pi RC)$. Here again we find that the average output noise power of the filter is proportional to the bandwidth.

We may generalize this statement to include all kinds of low-pass filters by defining a noise equivalent bandwidth as follows. Suppose that we have a source of white noise of zero mean and power spectral density $N_0/2$ connected to the input of an arbitrary low-pass filter of transfer function $H(f)$. The resulting average output noise power is therefore

$$
\begin{aligned}
N_{\text{out}} &= \frac{N_0}{2} \int_{-\infty}^{\infty} |H(f)|^2 \, df \\
&= N_0 \int_{0}^{\infty} |H(f)|^2 \, df
\end{aligned}
\tag{4.163}
$$

where, in the last line, we have made use of the fact that the amplitude response $|H(f)|$ is an even function of frequency.

Consider next the same source of white noise connected to the input of an *ideal* low-pass filter of zero-frequency response $H(0)$ and bandwidth B. In this case, the average output noise power is

$$
N_{\text{out}} = N_0 B H^2(0)
\tag{4.164}
$$

Therefore, equating this average output noise power to that in Eq. (4.163), we may define the *noise equivalent bandwidth* as

$$
B = \frac{\int_0^{\infty} |H(f)|^2 \, df}{H^2(0)}
\tag{4.165}
$$

Thus, the procedure for calculating the noise equivalent bandwidth consists of replacing the arbitrary low-pass filter of transfer function $H(f)$ by an equivalent

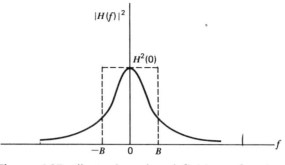

Figure 4.27 Illustrating the definition of noise-equivalent bandwidth.

ideal low-pass filter of zero-frequency response $H(0)$ and bandwidth B, as illustrated in Fig. 4.27. In a similar way, we may define a noise equivalent bandwidth for band-pass filters.

4.14 NARROW-BAND NOISE

The receiver of a communication system usually includes some provision for *preprocessing* the received signal. The preprocessing may take the form of a narrow-band filter whose bandwidth is just large enough to pass the signals of interest essentially undistorted but not so large as to admit excessive noise through the receiver. The noise process appearing at the output of such a filter is called *narrow-band noise*. With the spectral components of narrow-band noise concentrated about some mid-band frequency $\pm f_c$ as in Fig. 4.28a, we find that a sample function $n(t)$ of such a process appears somewhat similar to a sine wave of frequency f_c, which undulates slowly in both amplitude and phase, as in Fig. 4.28b.

Consider then the noise $n(t)$ produced at the output of a narrow-band filter in response to the sample function $w(t)$ of a white Gaussian noise process of zero mean and unit power spectral density applied to the filter input; $w(t)$ and $n(t)$ are sample functions of the respective processes $W(t)$ and $N(t)$. Let $H(f)$ denote the transfer function of this filter. Accordingly, we may express the power spectral density $S_N(f)$ of the noise $n(t)$ in terms of $H(f)$ as

$$S_N(f) \; = \; |H(f)|^2 \tag{4.166}$$

Indeed, any narrow-band noise encountered in practice may be modeled by applying a white noise to a suitable filter in the manner described here (see Problem 4.28).

In this section we wish to represent the narrow-band noise $n(t)$ in terms of its in-phase and quadrature components in a manner similar to that described for a narrow-band signal in Section 2.12. The derivation presented here is based on the idea of a pre-envelope and related concepts, which were discussed in Chapter 2.

Let $n_+(t)$ and $\tilde{n}(t)$ denote the pre-envelope and complex envelope of the narrow-band noise $n(t)$, respectively. We assume that the power spectrum of $n(t)$

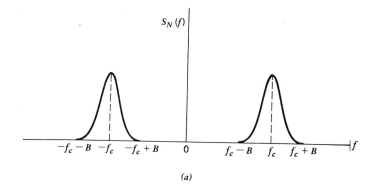

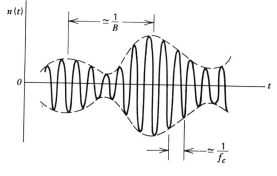

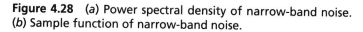

Figure 4.28 (a) Power spectral density of narrow-band noise. (b) Sample function of narrow-band noise.

is centered about the frequency f_c. Then we may write

$$n_+(t) = n(t) + j\hat{n}(t) \tag{4.167}$$

and

$$\tilde{n}(t) = n_+(t)\exp(-j2\pi f_c t) \tag{4.168}$$

where $\hat{n}(t)$ is the Hilbert transform of $n(t)$. The complex envelope $\tilde{n}(t)$ may itself be expressed as

$$\tilde{n}(t) = n_I(t) + jn_Q(t) \tag{4.169}$$

Hence, combining Eqs. (4.167) through (4.169), we find that the *in-phase component* $n_I(t)$ and the *quadrature component* $n_Q(t)$ of the narrow-band noise $n(t)$ are as follows, respectively:

$$n_I(t) = n(t)\cos(2\pi f_c t) + \hat{n}(t)\sin(2\pi f_c t) \tag{4.170}$$

and

$$n_Q(t) = \hat{n}(t)\cos(2\pi f_c t) - n(t)\sin(2\pi f_c t) \qquad (4.171)$$

Eliminating $\hat{n}(t)$ between Eqs. (4.170) and (4.171), we get the desired *canonical form* for representing the narrow-band noise $n(t)$, as shown by

$$n(t) = n_I(t)\cos(2\pi f_c t) - n_Q(t)\sin(2\pi f_c t) \qquad (4.172)$$

Using Eqs. (4.170) to (4.172), we may derive some important properties of the quadrature components of a narrow-band noise, as described next.

PROPERTY 1

The in-phase component $n_I(t)$ and quadrature component $n_Q(t)$ of narrow-band noise $n(t)$ have zero mean.

To prove this property, we first observe that the noise $\hat{n}(t)$ is obtained by passing $n(t)$ through a linear filter (i.e., Hilbert transformer). Accordingly, $\hat{n}(t)$ will have zero mean, because $n(t)$ has zero mean by virtue of its narrow-band nature. Furthermore, from Eqs. (4.170) and (4.171), we see that $n_I(t)$ and $n_Q(t)$ are weighted sums of $n(t)$ and $\hat{n}(t)$. It follows therefore that the quadrature components $n_I(t)$ and $n_Q(t)$ both have zero mean.

PROPERTY 2

If the narrow-band noise $n(t)$ is Gaussian, then its in-phase component $n_I(t)$ and quadrature component $n_Q(t)$ are jointly Gaussian.

To prove this property, we observe that $\hat{n}(t)$ is derived from $n(t)$ by a linear filtering operation. Then, if $n(t)$ is Gaussian, $\hat{n}(t)$ is Gaussian, and $n(t)$ and $\hat{n}(t)$ are jointly Gaussian. Therefore, the quadrature components $n_I(t)$ and $n_Q(t)$ are jointly Gaussian, since they are weighted sums of jointly Gaussian processes.

PROPERTY 3

If the narrow-band noise $n(t)$ is wide-sense stationary, then its in-phase component $n_I(t)$ and quadrature component $n_Q(t)$ are jointly wide-sense stationary.

If $n(t)$ is wide-sense stationary, so is its Hilbert transform $\hat{n}(t)$. However, since the quadrature components $n_I(t)$ and $n_Q(t)$ are weighted sums of $n(t)$ and $\hat{n}(t)$, and the weighting functions, namely, $\cos(2\pi f_c t)$, and $\sin(2\pi f_c t)$, vary with time, we cannot directly assert that $n_I(t)$ and $n_Q(t)$ are wide-sense stationary. To prove this, we have to evaluate their correlation functions.

Using Eqs. (4.170) and (4.171), we find that the in-phase and quadrature components $n_I(t)$ and $n_Q(t)$ of a narrow-band noise $n(t)$ have the same auto-correlation function, as shown by (see Problem 4.30)

$$R_{N_I}(\tau) = R_{N_Q}(\tau) = R_N(\tau)\cos(2\pi f_c \tau) + \hat{R}_N(\tau)\sin(2\pi f_c \tau) \qquad (4.173)$$

and their cross-correlation functions are given by

$$R_{N_I N_Q}(\tau) = -R_{N_Q N_I}(\tau) = R_N \sin(2\pi f_c \tau) - \hat{R}_N(\tau)\cos(2\pi f_c \tau) \quad (4.174)$$

where $R_N(\tau)$ is the autocorrelation function of $n(t)$, and $\hat{R}_N(\tau)$ is the Hilbert transform of $R_N(\tau)$. From Eqs. (4.173) and (4.174), we see that the correlation functions $R_{N_I}(\tau)$, $R_{N_Q}(\tau)$, and $R_{N_I N_Q}(\tau)$ of the quadrature components $n_I(t)$ and $n_Q(t)$ depend only on the time shift τ. This, in conjunction with Property 1, proves that $n_I(t)$ and $n_Q(t)$ are wide-sense stationary if the original narrow-band noise $n(t)$ is wide-sense stationary.

PROPERTY 4

Both the in-phase noise $n_I(t)$ and quadrature noise $n_Q(t)$ have the same power spectral density, which is related to the power spectral density $S_N(f)$ of the original narrow-band noise $n(t)$ as follows:

$$S_{N_I}(f) = S_{N_Q}(f) = \begin{cases} S_N(f - f_c) + S_N(f + f_c), & -B \leq f \leq B \\ 0, & \text{elsewhere} \end{cases} \quad (4.175)$$

where it is assumed that $S_N(f)$ occupies the frequency interval $f_c - B \leq |f| \leq f_c + B$ and $f_c > B$.

To prove this property, we take the Fourier transforms of both sides of Eq. (4.173), and use the fact that

$$\begin{aligned} F[\hat{R}_N(\tau)] &= -j\operatorname{sgn}(f)F[R_N(\tau)] \\ &= -j\operatorname{sgn}(f)S_N(f) \end{aligned} \quad (4.176)$$

We thus obtain the result

$$\begin{aligned} S_{N_I}(f) &= S_{N_Q}(f) \\ &= \tfrac{1}{2}[S_N(f - f_c) + S_N(f + f_c)] \\ &\quad - \tfrac{1}{2}[S_N(f - f_c)\operatorname{sgn}(f - f_c) - S_N(f + f_c)\operatorname{sgn}(f + f_c)] \\ &= \tfrac{1}{2}S_N(f - f_c)[1 - \operatorname{sgn}(f - f_c)] \\ &\quad + \tfrac{1}{2}S_N(f + f_c)[1 + \operatorname{sgn}(f + f_c)] \end{aligned} \quad (4.177)$$

Now, with the power spectral density $S_N(f)$ of the original narrow-band noise $n(t)$ occupying the frequency interval $f_c - B \leq |f| \leq f_c + B$, where $f_c > B$, as illustrated in Fig. 4.29a, we find that the corresponding shapes of $S_N(f - f_c)$ and $S_N(f + f_c)$ are as in Figs. 4.29b and 4.29c, respectively. Figures 4.29d, 4.29e, and 4.29f show the shapes of $\operatorname{sgn}(f)$, $\operatorname{sgn}(f - f_c)$, and $\operatorname{sgn}(f + f_c)$, respectively. Accordingly, we may make the following observations:

1. For frequencies defined by $-B \leq f \leq B$, we have

$$\operatorname{sgn}(f - f_c) = -1$$

and

$$\operatorname{sgn}(f + f_c) = +1$$

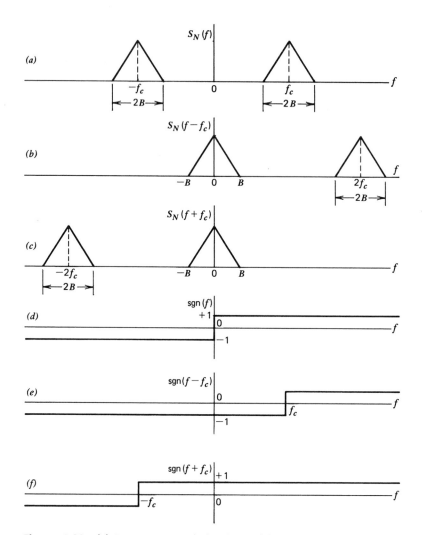

Figure 4.29 (a) Power spectral density $S_N(f)$ pertaining to narrow-band noise $n(t)$. Parts (b) and (c) represent frequency-shifted versions of $S_N(f)$ in opposite directions. (d) Signum function $\text{sgn}(f)$. Parts (e) and (f) represent frequency-shifted versions of $\text{sgn}(f)$ in opposite directions.

Hence, substituting these results in Eq. (4.177), we obtain

$$S_{N_I}(f) = S_{N_Q}(f)$$
$$= S_N(f - f_c) + S_N(f + f_c), \qquad -B \le f \le B$$

2.   For $2f_c - B \le f \le 2f_c + B$, we have

$$\text{sgn}(f - f_c) = 1$$

and

$$S_N(f + f_c) = 0$$

with the result that $S_{N_I}(f)$ and $S_{N_Q}(f)$ are both zero.

3. For $-2f_c - B \le f \le -2f_c + B$, we have

$$S_N(f - f_c) = 0$$

and

$$\text{sgn}(f + f_c) = -1$$

with the result that, here too, $S_{N_I}(f)$ and $S_{N_Q}(f)$ are both zero.

4. Outside the frequency intervals defined in 1, 2, and 3, both $S_N(f - f_c)$ and $S_N(f + f_c)$ are zero, and in a corresponding way, $S_{N_I}(f)$ and $S_{N_Q}(f)$ are also zero.

Combining these results, we obtain the relationship defined in Eq. (4.175).

As a consequence of this property, we may extract the in-phase component $n_I(t)$ and quadrature component $n_Q(t)$, except for scaling factors, from the narrow-band noise $n(t)$ by using the scheme shown in Fig. 4.30a, where both low-pass filters have a cutoff frequency at B. The scheme shown in Fig. 4.30a may be viewed as an *analyzer*. Given the in-phase component $n_I(t)$ and the quadrature component $n_Q(t)$, we may generate the narrow-band noise $n(t)$ using the scheme shown in Fig. 4.30b, which may be viewed as a *synthesizer*.

PROPERTY 5

The quadrature components $n_I(t)$ and $n_Q(t)$ have the same variance as the narrow-band noise $n(t)$.

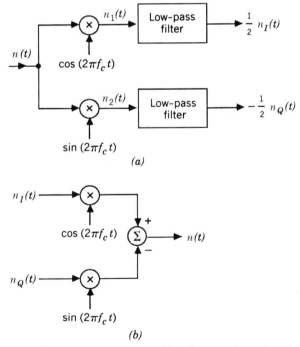

Figure 4.30 (a) Extraction of in-phase and quadrature components of a narrow-band process. (b) Generation of a narrow-band process from its in-phase and quadrature components.

This property follows directly from Eq. (4.175), according to which the total area under the power spectral density curve of $n_I(t)$ or $n_Q(t)$ is the same as the total area under the power spectral density curve $n(t)$. Hence, $n_I(t)$ and $n_Q(t)$ have the same mean-square value as $n(t)$. Earlier we showed that since $n(t)$ has zero mean, then $n_I(t)$ and $n_Q(t)$ have zero mean too. It follows therefore that $n_I(t)$ and $n_Q(t)$ have the same variance as the narrow-band noise $n(t)$.

PROPERTY 6

The cross-spectral densities of the quadrature components of a narrow-band noise are purely imaginary, as shown by

$$S_{N_I N_Q}(f) = -S_{N_Q N_I}(f)$$
$$= \begin{cases} j[S_N(f + f_c) - S_N(f - f_c)], & -B \leq f \leq B \\ 0, & \text{elsewhere} \end{cases} \quad (4.178)$$

To prove this property, we take the Fourier transforms of both sides of Eq. (4.174), and use the relation of Eq. (4.176), obtaining

$$S_{N_I N_Q}(f) = -S_{N_Q N_I}(f)$$

$$= -\frac{j}{2}[S_N(f - f_c) - S_N(f + f_c)]$$

$$+ \frac{j}{2}[S_N(f - f_c)\text{sgn}(f - f_c) + S_N(f + f_c)\text{sgn}(f + f_c)]$$

$$= \frac{j}{2}S_N(f + f_c)[1 + \text{sgn}(f + f_c)]$$

$$- \frac{j}{2}S_N(f - f_c)[1 - \text{sgn}(f - f_c)]$$

$$\quad (4.179)$$

Following a procedure similar to that described for proving Property 4, we may show that Eq. (4.179) reduces to the form shown in Eq. (4.178).

PROPERTY 7

If a narrow-band noise $n(t)$ is Gaussian with zero mean and a power spectral density $S_N(f)$ that is locally symmetric about the mid-band frequency $\pm f_c$, then the in-phase noise $n_I(t)$ and the quadrature noise $n_Q(t)$ are statistically independent.

To prove this property, we observe that if $S_N(f)$ is locally symmetric about $\pm f_c$, then

$$S_N(f - f_c) = S_N(f + f_c), \quad -B \leq f \leq B \quad (4.180)$$

Consequently, we find from Eq. (4.178) that the cross-spectral densities of the quadrature components $n_I(t)$ and $n_Q(t)$ are zero for all frequencies. This, in turn, means that the cross-correlation functions $R_{N_I N_Q}(\tau)$ and $R_{N_Q N_I}(\tau)$ are zero for all τ, as shown by

$$E[N_I(t_k + \tau) N_Q(t_k)] = 0 \qquad (4.181)$$

which implies that the random variables $N_I(t_k + \tau)$ and $N_Q(t_k)$ (obtained by observing the in-phase noise at time $t_k + \tau$ and the qudrature noise at time t_k, respectively) are orthogonal for all τ.

The narrow-band noise $n(t)$ is assumed to be Gaussian with zero mean; hence, from Properties 1 and 2 it follows that both $N_I(t_k + \tau)$ and $N_Q(t_k)$ are also Gaussian with zero mean. We thus conclude that because $N_I(t_k + \tau)$ and $N_Q(t_k)$ are orthogonal and have zero mean, they are uncorrelated, and being Gaussian, they are statistically independent for all τ. In other words, the in-phase noise $n_I(t)$ and the quadrature noise $n_Q(t)$ are statistically independent.

In accordance with Property 7, we may express the joint-probability density function of the random variables $N_I(t_k + \tau)$ and $N_Q(t_k)$ (for any delay τ) as the product of their individual probability density functions, as shown by

$$f_{N_I(t_k+\tau), N_Q(t_k)}(n_I, n_Q) = f_{N_I(t_k+\tau)}(n_I) f_{N_Q(t_k)}(n_Q)$$

$$= \frac{1}{\sqrt{2\pi}\sigma} \exp\left(-\frac{n_I^2}{2\sigma^2}\right) \frac{1}{\sqrt{2\pi}\sigma} \exp\left(-\frac{n_Q^2}{2\sigma^2}\right) \qquad (4.182)$$

$$= \frac{1}{2\pi\sigma^2} \exp\left(-\frac{n_I^2 + n_Q^2}{2\sigma^2}\right)$$

where σ^2 is the variance of the original narrow-band noise $n(t)$. Equation (4.182) holds only if the spectral density $S_N(f)$ of $n(t)$ is locally symmetric about $\pm f_c$. Otherwise, this relation holds only for $\tau = 0$ or those values of τ for which $n_I(t)$ and $n_Q(t)$ are uncorrelated.

To sum up, if the narrow-band noise $n(t)$ is zero-mean, stationary, and Gaussian, then its quadrature components $n_I(t)$ and $n_Q(t)$ are both zero-mean, jointly stationary, and jointly Gaussian. To evaluate the power spectral density of $n_I(t)$ or $n_Q(t)$, we may proceed as follows:

1. We shift the positive frequency portion of the power spectral density $S_N(f)$ of the original narrow-band noise $n(t)$ to the left by f_c.
2. We shift the negative frequency portion of $S_N(f)$ to the right by f_c.
3. We add these two shifted spectra to obtain the desired $S_{N_I}(f)$ or $S_{N_Q}(f)$.

EXAMPLE 18   Ideal Band-Pass Filtered White Noise

Consider a white Gaussian noise of zero mean and power spectral density $N_0/2$, which is passed through an ideal band-pass filter of passband amplitude response equal to one, mid-band frequency f_c, and bandwidth $2B$. The power spectral density characteristic of the filtered noise $n(t)$ will therefore be as shown in Fig. 4.31a. The problem is to determine the autocorrelation functions of $n(t)$ and its in-phase and quadrature components.

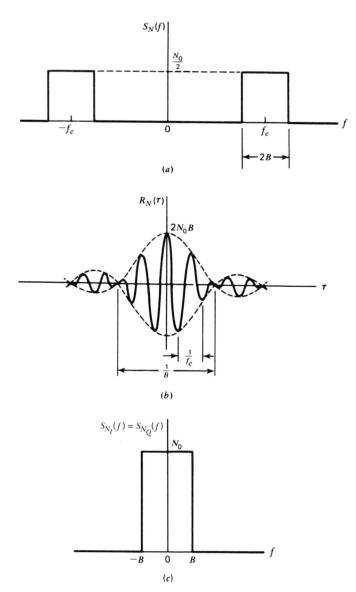

Figure 4.31 Characteristics of ideal band-pass filtered white noise. (a) Power spectral density. (b) Autocorrelation function. (c) Power spectral density of in-phase and quadrature components.

The autocorrelation function of $n(t)$ is the inverse Fourier transform of the power spectral density characteristic shown in Fig. 4.31a:

$$R_N(\tau) = \int_{-f_c-B}^{-f_c+B} \frac{N_0}{2}\exp(j2\pi f\tau)\ df + \int_{f_c-B}^{f_c+B} \frac{N_0}{2}\exp(j2\pi f\tau)\ df$$

$$= N_0 B\ \mathrm{sinc}(2B\tau)[\exp(-j2\pi f_c\tau) + \exp(j2\pi f_c\tau)] \qquad (4.183)$$

$$= 2N_0 B\ \mathrm{sinc}(2B\tau)\cos(2\pi f_c\tau)$$

which is shown plotted in Fig. 4.31b.

The spectral density characteristic of Fig. 4.31*a* is symmetric about $\pm f_c$. Therefore, we find that the corresponding spectral density characteristic of the in-phase noise component $n_I(t)$ or the quadrature noise component $n_Q(t)$ is as shown in Fig. 4.31*c*. The autocorrelation function of $n_I(t)$ or $n_Q(t)$ is therefore (see Example 16):

$$R_{N_I}(\tau) = R_{N_Q}(\tau) = 2N_0 B \operatorname{sinc}(2B\tau) \qquad (4.184)$$

EXAMPLE 19   Transmission of White Noise Through a High-Q Tuned Filter

Consider next the band-pass *LCR* filter of Fig. 4.32*a*. The transfer function of this filter, relating the output voltage to the input voltage, is given by

$$H(f) = \frac{R}{R + j2\pi fL + (1/j2\pi fC)} \qquad (4.185)$$

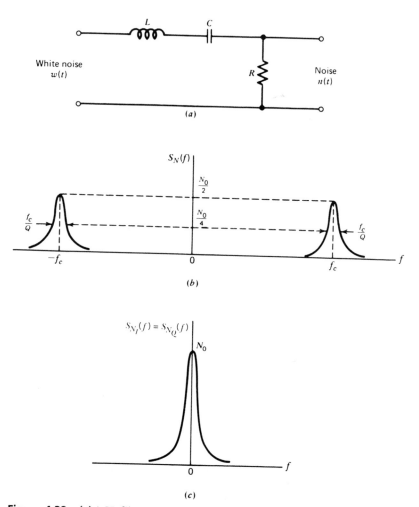

Figure 4.32   (a) LCR filter (b) Power spectral density of filter output $n(t)$ produced by white noise input. (c) Power spectral density of in-phase and quadrature components of $n(t)$.

Let the resonant frequency of the filter be defined by

$$f_c = \frac{1}{2\pi\sqrt{LC}} \tag{4.186}$$

and its Q-factor be defined by

$$Q = \frac{1}{R}\sqrt{\frac{L}{C}} \tag{4.187}$$

We may then rewrite Eq. (4.185) as

$$H(f) = \frac{1}{1 + jQ[(f/f_c) - (f_c/f)]} \tag{4.188}$$

If the Q-factor of the filter is high compared with unity, we may approximate Eq. (4.188) as follows:

$$H(f) \simeq \begin{cases} \dfrac{1}{1 + j2Q(f - f_c)/f_c}, & f > 0 \\[3mm] \dfrac{1}{1 + j2Q(f + f_c)f_c}, & f < 0 \end{cases} \tag{4.189}$$

Suppose that we connect, across the input terminals of this filter, a voltage source generating white Gaussian noise of zero mean and power spectral density $N_0/2$. The power spectral density of the resulting noise $n(t)$ at the filter output is therefore

$$S_N(f) \simeq \begin{cases} \dfrac{N_0/2}{1 + 4Q^2(f - f_c)^2/f_c^2}, & f > 0 \\[3mm] \dfrac{N_0/2}{1 + 4Q^2(f + f_c)^2/f_c^2}, & f < 0 \end{cases} \tag{4.190}$$

which is shown plotted in Fig. 4.32b. Thus the corresponding power spectral density of the in-phase noise component $n_I(t)$ or quadrature noise component $n_Q(t)$ is approximately given by

$$S_{N_I}(f) = S_{N_Q}(f) \simeq \frac{N_0}{1 + (2Qf/f_c)^2} \tag{4.191}$$

which is shown plotted in Fig. 4.32c. Comparing this last relation with the power spectral density of the RC low-filtered noise of Example 17, we see that they are both basically of a similar form. This means that the in-phase or quadrature component of the noise $n(t)$ at the output of the narrow-band filter of Fig. 4.32a has effectively the same characteristics as a noise process produced by passing a white noise through a corresponding low-pass RC filter.

Representation of Narrow-Band Noise in Terms of Envelope and Phase Components

Thus far we have considered the representation of a narrow-band noise $n(t)$ in terms of its in-phase and quadrature components. We may also represent the noise $n(t)$ in terms of its envelope and phase components as follows:

$$n(t) = r(t)\cos[2\pi f_c t + \psi(t)] \tag{4.192}$$

where

$$r(t) = [n_I^2(t) + n_Q^2(t)]^{1/2} \tag{4.193}$$

and

$$\psi(t) = \tan^{-1}\left[\frac{n_Q(t)}{n_I(t)}\right] \tag{4.194}$$

The function $r(t)$ is called the *envelope* of $n(t)$, and the function $\psi(t)$ is called the *phase* of $n(t)$.

The probability distributions of $r(t)$ and $\psi(t)$ may be obtained from those of $n_I(t)$ and $n_Q(t)$ as follows. Let N_I and N_Q represent the random variables obtained by observing (at some fixed time) the random processes represented by the sample functions $n_I(t)$ and $n_Q(t)$, respectively. We note that N_I and N_Q are independent Gaussian random variables of zero mean and variance σ^2, and so we may express their joint probability density function by

$$f_{N_I, N_Q}(n_I, n_Q) = \frac{1}{2\pi\sigma^2}\exp\left(-\frac{n_I^2 + n_Q^2}{\sigma^2}\right) \tag{4.195}$$

Accordingly, the probability of the joint event that N_I lies between n_I and $n_I + dn_I$ and that N_Q lies between n_Q and $n_Q + dn_Q$ (i.e., the pair of random variables N_I and N_Q lies jointly inside the shaded area of Fig. 4.33a) is given by

$$f_{N_I, N_Q}(n_I, n_Q)\, dn_I\, dn_Q = \frac{1}{2\pi\sigma^2}\exp\left(-\frac{n_I^2 + n_Q^2}{2\sigma^2}\right) dn_I\, dn_Q \tag{4.196}$$

Define the transformation (see Fig. 4.33a)

$$n_I = r\cos\psi \tag{4.197}$$

$$n_Q = r\sin\psi \tag{4.198}$$

In a limiting sense, we may equate the two incremental areas shown shaded in Figs. 4.33a and 4.33b and thus write

$$dn_I\, dn_Q = r\, dr\, d\psi \tag{4.199}$$

Now, let R and Ψ denote the random variables obtained by observing (at some time t) the random processes represented by the envelope $r(t)$ and phase $\psi(t)$, respectively. Then, substituting Eqs. (4.197) to (4.199) in (4.196), we find that the probability of the random variables R and Ψ lying jointly inside the shaded area of Fig. 4.33b is equal to

$$\frac{r}{2\pi\sigma^2}\exp\left(-\frac{r^2}{2\sigma^2}\right) dr\, d\psi$$

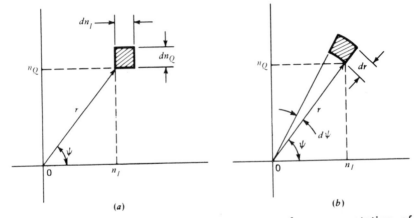

Figure 4.33 Illustrating the coordinate system for representation of narrow-band noise: (a) in terms of in-phase and quadrature components, and (b) in terms of envelope and phase.

That is, the joint probability density function of R and Ψ is

$$f_{R,\Psi}(r,\psi) = \frac{r}{2\pi\sigma^2}\exp\left(-\frac{r^2}{2\sigma^2}\right) \tag{4.200}$$

This probability density function is independent of the angle ψ, which means that the random variables R and Ψ are statistically independent. We may thus express $f_{R,\Psi}(r,\psi)$ as the product of $f_R(r)$ and $f_\Psi(\psi)$. In particular, the random variable Ψ representing phase is *uniformly distributed* inside the range 0 to 2π, as shown by

$$f_\Psi(\psi) = \begin{cases} \dfrac{1}{2\pi}, & 0 \le \psi \le 2\pi \\ 0, & \text{elsewhere} \end{cases} \tag{4.201}$$

This leaves the probability density function of the random variable R as

$$f_R(r) = \begin{cases} \dfrac{r}{\sigma^2}\exp\left(-\dfrac{r^2}{2\sigma^2}\right), & r \ge 0 \\ 0, & \text{elsewhere} \end{cases} \tag{4.202}$$

where σ^2 is the variance of the original narrow-band noise $n(t)$. A random variable having the probability density function of Eq. (4.202) is said to be *Rayleigh-distributed*.[12]

For convenience of graphical presentation, let

$$v = \frac{r}{\sigma} \tag{4.203}$$

$$f_V(v) = \sigma f_R(r) \tag{4.204}$$

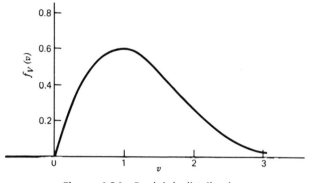

Figure 4.34   Rayleigh distribution.

Then we may rewrite the Rayleigh distribution of Eq. (4.202) in the normalized form

$$f_V(v) = \begin{cases} v \exp\left(-\dfrac{v^2}{2}\right), & v \geq 0 \\[2ex] 0, & \text{elsewhere} \end{cases} \tag{4.205}$$

Equation (4.205) is shown plotted in Fig. 4.34. The peak value of the distribution $f_V(v)$ occurs at $v = 1$ and is equal to 0.607. Note also that, unlike the Gaussian distribution, the Rayleigh distribution is zero for negative values of v. This is because the envelope $r(t)$ can assume only positive values.

4.15   SINE WAVE PLUS NARROW-BAND NOISE

Suppose next that we add the sinusoidal wave $A \cos(2\pi f_c t)$ to the narrow-band noise $n(t)$, where A and f_c are both constants. We assume that the frequency of the sinusoidal wave is the same as the nominal carrier frequency for the noise. A sample function of the sinusoidal wave plus noise is then expressed by

$$x(t) = A \cos(2\pi f_c t) + n(t) \tag{4.206}$$

Representing the narrow-band noise $n(t)$ in terms of its in-phase and quadrature components, we may write

$$x(t) = n_I'(t)\cos(2\pi f_c t) - n_Q(t)\sin(2\pi f_c t) \tag{4.207}$$

where

$$n_I'(t) = A + n_I(t) \tag{4.208}$$

We assume that $n(t)$ is Gaussian with zero mean and variance σ^2. Accordingly, we may state that

1. Both $n_I'(t)$ and $n_Q(t)$ are Gaussian and statistically independent.
2. The mean of $n_I'(t)$ is A and that of $n_Q(t)$ is zero.
3. The variance of both $n_I'(t)$ and $n_Q(t)$ is σ^2.

We may therefore express the joint probability density function of the random variables N_I' and N_Q corresponding to $n_I'(t)$ and $n_Q(t)$, as follows:

$$f_{N_I',N_Q}(n_I',n_Q) = \frac{1}{2\pi\sigma^2}\exp\left[-\frac{(n_I' - A)^2 + n_Q^2}{2\sigma^2}\right] \tag{4.209}$$

Let $r(t)$ denote the envelope of $x(t)$ and $\psi(t)$ denote its phase. From Eq. (4.207), we thus find that

$$r(t) = \{[n_I'(t)]^2 + n_Q^2(t)\}^{1/2} \tag{4.210}$$

$$\psi(t) = \tan^{-1}\left[\frac{n_Q(t)}{n_I'(t)}\right] \tag{4.211}$$

Following a procedure similar to that described in the previous section for the derivation of the Rayleigh distribution, we find that the joint probability density function of the random variables R and Ψ, corresponding to $r(t)$ and $\psi(t)$ for some fixed time t, is given by

$$f_{R,\Psi}(r,\psi) = \frac{r}{2\pi\sigma^2}\exp\left(-\frac{r^2 + A^2 - 2Ar\cos\psi}{2\sigma^2}\right) \tag{4.212}$$

We see that in this case, however, we cannot express the joint probability density function $f_{R,\Psi}(r,\psi)$ as a product $f_R(r)f_\Psi(\psi)$. This is because we now have a term involving the values of both random variables multiplied together as $r\cos\psi$. Hence, R and Ψ are dependent random variables for nonzero values of the amplitude A of the sinusoidal wave component.

We are interested, in particular, in the probability density function of R. To determine this probability density function, we integrate Eq. (4.212) over all possible values of ψ obtaining the marginal density

$$\begin{aligned}
f_R(r) &= \int_0^{2\pi} f_{R,\Psi}(r,\psi)\, d\psi \\
&= \frac{r}{2\pi\sigma^2}\exp\left(-\frac{r^2 + A^2}{2\sigma^2}\right)\int_0^{2\pi}\exp\left(\frac{Ar}{\sigma^2}\cos\psi\right)d\psi
\end{aligned} \tag{4.213}$$

The integral in the right-hand side of Eq. (4.213) can be identified in terms of the defining integral for the *modified Bessel function of the first kind of zero order* (see Appendix 4); that is

$$I_0(x) = \frac{1}{2\pi}\int_0^{2\pi}\exp(x\cos\psi)\, d\psi \tag{4.214}$$

Thus, letting $x = Ar/\sigma^2$, we may rewrite Eq. (4.213) in the compact form:

$$f_R(r) = \frac{r}{\sigma^2}\exp\left(-\frac{r^2 + A^2}{2\sigma^2}\right) I_0\left(\frac{Ar}{\sigma^2}\right) \tag{4.215}$$

This relation is called the *Rician distribution.*[13]

As with the Rayleigh distribution, the graphical presentation of the Rician distribution is simplified by putting

$$v = \frac{r}{\sigma} \tag{4.216}$$

$$a = \frac{A}{\sigma} \tag{4.217}$$

$$f_V(v) = \sigma f_R(r) \tag{4.218}$$

Then we may express the Rician distribution of Eq. (4.215) in the *normalized* form

$$f_V(v) = v\,\exp\left(-\frac{v^2 + a^2}{2}\right) I_0(av) \tag{4.219}$$

which is shown plotted in Fig. 4.35 for the values 0, 1, 2, 3, 5, of the parameter a. Based on these curves, we observe that

1. When a is zero, the Rician distribution reduces to the Rayleigh distribution.
2. The envelope distribution is approximately Gaussian in the vicinity of $v = a$ when a is large, that is, when the sine-wave amplitude A is large compared with σ, the square root of the average power of the noise $n(t)$.

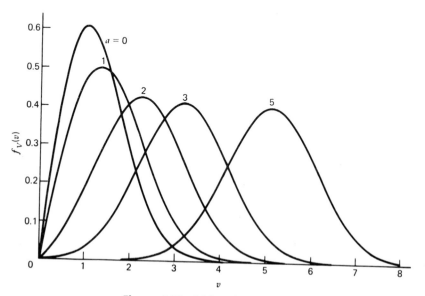

Figure 4.35 Rician distribution.

Table 4.1 Graphical Summary of Autocorrelation Functions and Power Spectral Densities of Random Processes of Zero Mean and Unit Variance

Type of Process, $X(t)$	Autocorrelation Function, $R_X(\tau)$	Power Spectral Density, $S_X(f)$
Sinusoidal process of unit frequency and random phase		
Random binary wave of unit symbol-duration		
RC low-pass filtered white noise		
Ideal low-pass filtered white noise		
Ideal band-pass filtered white noise		
RLC-filtered white noise		

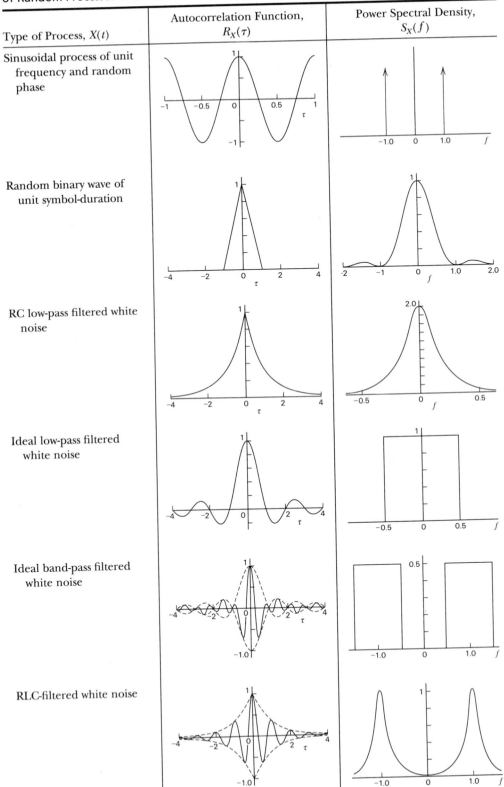

4.16 SUMMARY

Much of the material presented in this chapter has dealt with the characterization of a particular class of random processes known to be wide-sense stationary and ergodic. The implication of wide-sense stationarity is that we may develop a partial description of a random process in terms of two ensemble-averaged parameters: (1) a mean that is independent of time, and (2) an autocorrelation function that depends only on the difference between the times at which two observations of the process are made.[14] Ergodicity enables us to use time averages as "estimates" of these parameters. The time averages are computed using a sample function (i.e., single realization) of the random process.

Another important parameter of a random process is the power spectral density. The autocorrelation function and the power spectral density constitute a Fourier-transform pair. The formulas that define the power spectral density in terms of the autocorrelation function and vice versa are known as the Einstein–Wiener–Khintchine relations.

In Table 4.1 we present a graphical summary of the autocorrelation functions and power spectral densities of some important random processes studied in the chapter. All the processes described in this table are assumed to have zero mean and unit variance. This table should give the reader a feeling for (1) the interplay between the autocorrelation function and power spectral density of a random process, and (2) the role of linear filtering in shaping the autocorrelation function or, equivalently, the power spectral density of a white noise process.

The latter part of the chapter dealt with noise processes that are Gaussian and narrow band, which is the kind of filtered noise encountered at the front end of an idealized form of communication receiver. Gaussianity means that the random variable obtained by observing the output of the filter at some fixed time has a Gaussian distribution. The narrow-band nature of the noise means that it may be represented in terms of an in-phase and a quadrature component. These two components are both low-pass, Gaussian processes, each with zero mean and a variance equal to that of the original narrow-band noise. Alternatively, a Gaussian narrow-band noise may be represented in terms of a Rayleigh-distributed envelope and a uniformly distributed phase. Each of these representations has its own specific area of application, as shown in subsequent chapters of the book.

The material presented in this chapter has been confined entirely to *real* random processes. It may be generalized for *complex* random processes. A commonly encountered complex random process is a complex Gaussian low-pass process, which arises in the equivalent representation of a Gaussian narrow-band noise $n(t)$. From Section 4.14 we note that $n(t)$ is uniquely defined in terms of the in-phase component $n_I(t)$ and the quadrature component $n_Q(t)$. Equivalently, we may represent the narrow-band noise $n(t)$ in terms of the complex envelope $\tilde{n}(t)$ defined as $n_I(t) + jn_Q(t)$. For a statistical characterization of the complex envelope $\tilde{n}(t)$, the reader is referred to Appendix 7.

NOTES AND REFERENCES

1. For introductory treatment of probability theory by itself, see Hamming (1991). For introductory treatment of probability and random processes with an engineering emphasis, see Leon-Garcia (1989), and Helstrom (1990); the first chapter of Leon-Garcia's book discusses some interesting probability models. For an advanced treat-

ment of the subject, see Gray and Davisson (1986), Papoulis (1984), Wong (1983), Feller (1968), and Doob (1953). For foundations of probability theory, see Fine (1973), Jeffreys (1957), and Kolomogorov (1956).

2. There is another important class of random processes commonly encountered in practice, the mean and autocorrelation function of which exhibit *periodicity*, as in

$$\mu_X(t_1 + T) = \mu_X(t_1)$$

$$R_X(t_1 + T, t_2 + T) = R_X(t_1, t_2)$$

for all t_1 and t_2. A random process $X(t)$ satisfying this pair of conditions is said to be *cyclostationary* (in the wide sense). Modeling the process $X(t)$ as cyclostationary adds a new dimension, namely, period T to the partial description of the process. Examples of cyclostationary processes include a television signal obtained by raster-scanning a random video field, and a modulated process obtained by varying the amplitude, phase, or frequency of a sinusoidal carrier. For detailed discussion of cyclostationary processes, see Franks (1969), pp. 204–214, and the paper by Gardner and Franks (1975).

3. For a more detailed treatment of ergodicity, see Gray and Davisson (1986).

4. Traditionally, Eqs. (4.96) and (4.97) have been referred to in the literature as the Wiener–Khintchine relations in recognition of pioneering work done by Norbert Wiener and A. I. Khintchine; for their original papers, see Wiener (1930) and Khintchine (1934). A discovery of a forgotten paper by Albert Einstein on time-series analysis (delivered at the Swiss Physical Society's February 1914 meeting in Basel) reveals that Einstein had discussed the autocorrelation function and its relationship to the spectral content of a time series many years before Wiener and Khintchine. An English translation of Einstein's paper is reproduced in the *IEEE ASSP Magazine*, vol. 4, October 1987. This particular issue also contains articles by W. A. Gardner and A. M. Yaglom, which elaborate on Einstein's original work.

5. For further details of power spectrum estimation, see Blackman and Tukey (1958), Box and Jenkins (1976), Marple (1987), and Kay (1988).

6. The Gaussian distribution and associated Gaussian process are named after the great mathematician C. F. Gauss. At age 18, Gauss invented *the method of least squares* for finding the best value of a sequence of measurements of some quantity. Gauss later used the method of least squares in fitting orbits of planets to data measurements, a procedure that was published in 1809 in his book entitled *Theory of Motion of the Heavenly Bodies*. In connection with the error of observation, he developed the *Gaussian distribution*. This distribution is also known as the *normal distribution*. Partly for historical reasons, mathematicians commonly use the term normal, while engineers and physicists commonly use the term Gaussian.

7. The joint probability density function of the n-by-1 Gaussian vector

$$\mathbf{X} = \begin{bmatrix} X(t_1) \\ X(t_2) \\ \vdots \\ X(t_n) \end{bmatrix}$$

is defined by

$$f_{\mathbf{X}}(\mathbf{x}) = \frac{1}{(2\pi)^{n/2}|\mathbf{C}_X|^{1/2}} \exp\left[-\frac{1}{2}(\mathbf{x} - \boldsymbol{\mu}_X)^T \mathbf{C}_X^{-1}(\mathbf{x} - \boldsymbol{\mu}_X) \right]$$

where the superscript T denotes transposition, and

$$\boldsymbol{\mu}_X = \text{mean vector of } \mathbf{X}$$

$$= E[\mathbf{X}]$$

$$\mathbf{C}_X = \text{covariance matrix of } \mathbf{X}$$
$$= E[(\mathbf{X} - \boldsymbol{\mu}_X)(\mathbf{X} - \boldsymbol{\mu}_X)^T]$$
$$\mathbf{C}_X^{-1} = \text{inverse of the covariance matrix } \mathbf{C}_X$$
$$|\mathbf{C}_X| = \text{determinant of the covariance matrix } \mathbf{C}_X$$

For a derivation of this function see Thomas (1969, pp. 128–144), Davenport and Root (1958, pp. 147–154), and Sakrison (1968, pp. 87–97).

8. For a detailed treatment of electrical noise, see Van der Ziel (1970) and the collection of papers edited by Gupta (1977).

9. An introductory treatment of shot noise is presented in Helstrom (1990). For a more detailed treatment, see the paper by Yue, Luganani, and Rice (1978).

10. Thermal noise was first studied experimentally by J. B. Johnson in 1928, and for this reason it is sometimes referred to as the "Johnson noise." Johnson's experiments were confirmed theoretically by Nyquist (1928).

11. The noisiness of a receiver may also be measured in terms of the so-called *noise figure*. The relationship between the noise figure and the equivalent noise temperature is developed in Appendix 6 at the end of the book.

12. The Rayleigh distribution is named after the English physicist J. W. Strutt, Lord Rayleigh.

13. The Rician distribution is named in honor of Stephen O. Rice for the original contribution reported in a pair of papers published in 1944 and 1945.

14. The statistical characterization of communication systems presented in this book is confined to the first two moments, mean and autocorrelation function (equivalently, autocovariance function) of the pertinent random process. However, when a random process is transmitted through a nonlinear system, valuable information is contained in higher-order moments of the resulting output process. The parameters used to characterize higher-order moments in the time domain are called *cumulants*; their multidimensional Fourier transforms are called *polyspectra*. For a discussion of higher-order cumulants and polyspectra and their estimation, see the papers by Brillinger (1965) and Nikias and Raghuveer (1987).

PROBLEMS

Problem 4.1

(a) Show that the characteristic function of a Gaussian random variable X of mean μ_X and variance σ_X^2 is

$$\phi_X(v) = \exp(jv\mu_X - \tfrac{1}{2}v^2\sigma_X^2)$$

(b) Using the result of part (a), show that the nth central moment of this Gaussian random variable is

$$E[(X - \mu_X)^n] = \begin{cases} 1 \times 3 \times 5 \ldots (n-1)\sigma_X^n & \text{for } n \text{ even} \\ 0 & \text{for } n \text{ odd} \end{cases}$$

Problem 4.2 A Gaussian-distributed random variable X of zero mean and variance σ_X^2 is transformed by a piecewise-linear rectifier characterized by the input–output relation (see Fig. P4.1):

$$Y = \begin{cases} X, & X \geq 0 \\ 0, & X < 0. \end{cases}$$

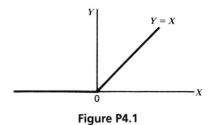

Figure P4.1

The probability density function of the new random variable Y is described by

$$f_Y(y) = \begin{cases} 0, & y < 0 \\ k\delta(y), & y = 0 \\ \dfrac{1}{\sqrt{2\pi}\sigma_X}\exp\left(-\dfrac{y^2}{2\sigma_X^2}\right), & y > 0 \end{cases}$$

(a) Explain the physical reasons for the functional form of this result.
(b) Determine the value of the constant k by which the delta function $\delta(y)$ is weighted.

Problem 4.3   A Gaussian-distributed random variable X of zero mean and variance σ_X^2 is transformed by a square-law device defined by (see Fig. P4.2)

$$Y = X^2$$

Show that the probability density function of the new random variable Y is defined by the *chi-distribution with two degrees of freedom:*

$$f_Y(y) = \begin{cases} \dfrac{1}{\sqrt{2\pi y}\sigma_X}\exp\left(-\dfrac{y}{2\sigma_X^2}\right), & y \geq 0 \\ 0, & y < 0 \end{cases}$$

Hint: Evaluate the probability $P(Y \leq y)$ for the two intervals: $y < 0$, and $y \geq 0$.

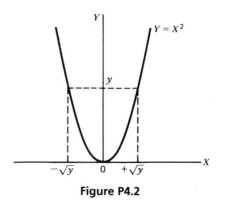

Figure P4.2

Problem 4.4   Consider a random process $X(t)$ defined by

$$X(t) = \sin(2\pi f t)$$

in which the frequency f is a random variable uniformly distributed over the interval $(0, W)$. Show that $X(t)$ is nonstationary. *Hint:* Examine specific sample functions of the random process $X(t)$ for the frequency $f = W/4$, $W/2$, and W, say.

Problem 4.5 Consider the sinusoidal process

$$X(t) = A \cos(2\pi f_c t)$$

where the frequency f_c is constant and the amplitude A is uniformly distributed:

$$f_A(a) = \begin{cases} 1, & 0 \le a \le 1 \\ 0, & \text{otherwise} \end{cases}$$

Determine whether or not this process is stationary in the strict sense.

Problem 4.6 A random process $X(t)$ is defined by

$$X(t) = A \cos(2\pi f_c t)$$

where A is a Gaussian-distributed random variable of zero mean and variance σ_A^2. This random process is applied to an ideal integrator, producing the output

$$Y(t) = \int_0^t X(\tau) \, d\tau$$

(a) Determine the probability density function of the output $Y(t)$ at a particular time t_k.

(b) Determine whether or not $Y(t)$ is stationary.

(c) Determine whether or not $Y(t)$ is ergodic.

Problem 4.7 Let X and Y be statistically independent Gaussian-distributed random variables, each with zero mean and unit variance. Define the Gaussian process

$$Z(t) = X \cos(2\pi t) + Y \sin(2\pi t)$$

(a) Determine the joint probability density function of the random variables $Z(t_1)$ and $Z(t_2)$ obtained by observing $Z(t)$ at times t_1 and t_2, respectively.

(b) Is the process $Z(t)$ stationary? Why?

Problem 4.8 Prove the following two properties of the autocorrelation function $R_X(\tau)$ of a random process $X(t)$:

(a) If $X(t)$ contains a dc component equal to A, then $R_X(\tau)$ will contain a constant component equal to A^2.

(b) If $X(t)$ contains a sinusoidal component, then $R_X(\tau)$ will also contain a sinusoidal component of the same frequency.

Problem 4.9 The square wave $x(t)$ of Fig. P4.3 of constant amplitude A, period T_0, and delay t_d, represents the sample function of a random process $X(t)$. The delay is random, described by the probability density function

$$f_{T_d}(t_d) = \begin{cases} \dfrac{1}{T_0}, & -\dfrac{1}{2}T_0 \le t_d \le \dfrac{1}{2}T_0 \\ 0, & \text{otherwise} \end{cases}$$

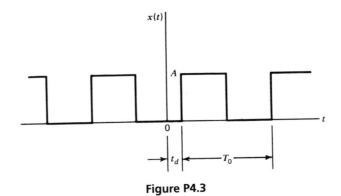

Figure P4.3

(a) Determine the probability density function of the random variable $X(t_k)$ obtained by observing the random process $X(t)$ at time t_k.

(b) Determine the mean and autocorrelation function of $X(t)$ using ensemble-averaging.

(c) Determine the mean and autocorrelation function of $X(t)$ using time-averaging.

(d) Establish whether or not $X(t)$ is wide-sense stationary. In what sense is it ergodic?

Problem 4.10   A binary wave consists of a random sequence of symbols 1 and 0, similar to that described in Example 8, with one basic difference: symbol 1 is now represented by a pulse of amplitude A volts and symbol 0 is represented by zero volts. All other parameters are the same as before. Show that for this new random binary wave $X(t)$:

(a) The autocorrelation function is

$$R_X(\tau) = \begin{cases} \dfrac{A^2}{4} + \dfrac{A^2}{4}\left(1 - \dfrac{|\tau|}{T}\right), & |\tau| < T \\[4mm] \dfrac{A^2}{4}, & |\tau| \geq T \end{cases}$$

(b) The power spectral density is

$$S_X(f) = \frac{A^2}{4}\,\delta(f) + \frac{A^2 T}{4}\,\text{sinc}^2(fT)$$

What is the percentage power contained in the dc component of the binary wave?

Problem 4.11   A random process $Y(t)$ consists of a dc component of $\sqrt{3/2}$ volts, a periodic component $g(t)$, and a random component $X(t)$. The autocorrelation function of $Y(t)$ is shown in Fig. P4.4.

(a) What is the average power of the periodic component $g(t)$?

(b) What is the average power of the random component $X(t)$?

Problem 4.12   Consider a pair of wide-sense stationary random processes $X(t)$ and $Y(t)$. Show that the cross-correlations $R_{XY}(\tau)$ and $R_{YX}(\tau)$ of these processes have the following properties:

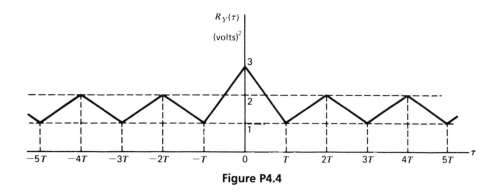

Figure P4.4

(a) $R_{XY}(\tau) = R_{YX}(-\tau)$

(b) $|R_{XY}(\tau)| \leq \frac{1}{2}[R_X(0) + R_Y(0)]$

where $R_X(\tau)$ and $R_Y(\tau)$ are the autocorrelation functions of $X(t)$ and $Y(t)$, respectively.

Problem 4.13 Consider two linear filters connected in cascade as in Fig. P4.5. Let $X(t)$ be a wide-sense stationary process with autocorrelation function $R_X(\tau)$. The random process appearing at the first filter output is $V(t)$ and that at the second filter output is $Y(t)$.

(a) Find the autocorrelation function of $Y(t)$.

(b) Find the cross-correlation function $R_{VY}(\tau)$ of $V(t)$ and $Y(t)$.

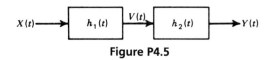

Figure P4.5

Problem 4.14 A wide-sense stationary random process $X(t)$ is applied to a linear time-invariant filter of impulse response $h(t)$, producing an output $Y(t)$.

(a) Show that the cross-correlation function $R_{YX}(\tau)$ of the output $Y(t)$ and the input $X(t)$ is equal to the impulse response $h(\tau)$ convolved with the auto-correlation function $R_X(\tau)$ of the input, as shown by

$$R_{YX}(\tau) = \int_{-\infty}^{\infty} h(u) R_X(\tau - u) \, du$$

Show that the second cross-correlation function $R_{XY}(\tau)$ equals

$$R_{XY}(\tau) = \int_{-\infty}^{\infty} h(-u) R_X(\tau - u) \, du$$

(b) Find the cross-spectral densities $S_{YX}(f)$ and $S_{XY}(f)$.

(c) Assuming that $X(t)$ is a white noise process with zero mean and power spectral density $N_0/2$, show that

$$R_{YX}(\tau) = \frac{N_0}{2} h(\tau)$$

Comment on the practical significance of this result.

Problem 4.15 The power spectral density of a random process $X(t)$ is shown in Fig. P4.6.

(a) Determine and sketch the autocorrelation function $R_X(\tau)$ of $X(t)$.

(b) What is the dc power contained in $X(t)$?

(c) What is the ac power contained in $X(t)$?

(d) What sampling rates will give uncorrelated samples of $X(t)$? Are the samples statistically independent?

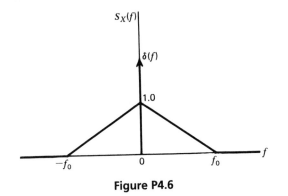

Figure P4.6

Problem 4.16 A pair of noise processes $n_1(t)$ and $n_2(t)$ are related by

$$n_2(t) = n_1(t)\cos(2\pi f_c t + \theta) - n_1(t)\sin(2\pi f_c t + \theta)$$

where f_c is a constant, and θ is the value of a random variable Θ whose probability density function is defined by

$$f_\Theta(\theta) = \begin{cases} \dfrac{1}{2\pi}, & 0 \le \theta \le 2\pi \\ 0, & \text{otherwise} \end{cases}$$

The noise process $n_1(t)$ is stationary and its power spectral density is as shown in Fig. P4.7. Find and plot the corresponding power spectral density of $n_2(t)$.

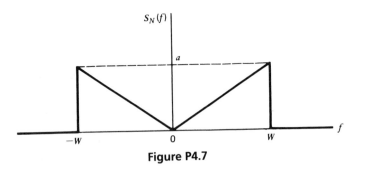

Figure P4.7

Problem 4.17 A *random telegraph signal* $X(t)$, characterized by the autocorrelation function

$$R_X(\tau) = \exp(-2v|\tau|)$$

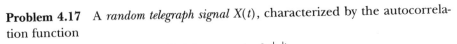

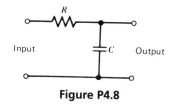

Figure P4.8

where v is a constant, is applied to the low-pass RC filter of Fig. P4.8. Determine the power spectral density and autocorrelation function of the random process at the filter output.

Problem 4.18 The output of an oscillator is described by

$$X(t) = A \cos(2\pi f t - \Theta),$$

where A is a constant, and f and Θ are independent random variables. The probability density function of Θ is defined by

$$f_\Theta(\theta) = \begin{cases} \dfrac{1}{2\pi}, & 0 \le \theta \le 2\pi \\ 0, & \text{otherwise} \end{cases}$$

Find the power spectral density of $X(t)$ in terms of the probability density function of the frequency f. What happens to this power spectral density when the frequency f assumes a constant value?

Problem 4.19 A stationary, Gaussian process $X(t)$ has zero mean and power spectral density $S_X(f)$. Determine the probability density function of a random variable obtained by observing the process $X(t)$ at some time t_k.

Problem 4.20 A Gaussian process $X(t)$ of zero mean and variance σ_X^2 is passed through a full-wave rectifier, which is described by the input–output relation of Fig. P4.9. Show that the probability density function of the random variable $Y(t_k)$, obtained by observing the random process $Y(t)$ at the rectifier output at time t_k, is as follows:

$$f_{Y(t_k)}(y) = \begin{cases} \sqrt{\dfrac{2}{\pi}}\, \dfrac{1}{\sigma_X} \exp\left(-\dfrac{y^2}{2\sigma_X^2}\right), & y \ge 0 \\ 0, & y < 0 \end{cases}$$

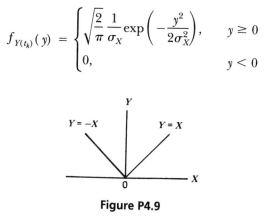

Figure P4.9

Problem 4.21 Let $X(t)$ be a zero-mean, stationary, Gaussian process with autocorrelation function $R_X(\tau)$. This process is applied to a square-law device defined by the input–output relation (see Fig. P4.2)

$$Y(t) = X^2(t)$$

where $Y(t)$ is the output.

(a) Using the result of Problem 4.3, show that the mean of $Y(t)$ is $R_X(0)$.

(b) Show that the autocovariance function of $Y(t)$ is $2R_X^2(\tau)$.

Problem 4.22 A stationary, Gaussian process $X(t)$ with mean μ_X and variance σ_X^2 is passed through two linear filters with impulse responses $h_1(t)$ and $h_2(t)$, yielding processes $Y(t)$ and $Z(t)$, as shown in Fig. P4.10.

(a) Determine the joint probability density function of the random variables $Y(t_1)$ and $Z(t_2)$.

(b) What conditions are necessary and sufficient to ensure that $Y(t_1)$ and $Z(t_2)$ are statistically independent?

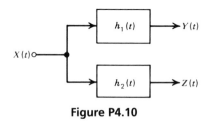

Figure P4.10

Problem 4.23 A stationary, Gaussian process $X(t)$ with zero mean and power spectral density $S_X(f)$ is applied to a linear filter whose impulse response $h(t)$ is shown in Fig. P4.11. A sample Y is taken of the random process at the filter output at time T.

(a) Determine the mean and variance of Y.

(b) What is the probability density function of Y?

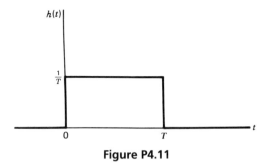

Figure P4.11

Problem 4.24 Consider a white Gaussian noise process of zero mean and power spectral density $N_0/2$ that is applied to the input of the high-pass RL filter shown in Fig. P4.12.

(a) Find the autocorrelation function and power spectral density of the random process at the output of the filter.

(b) What are the mean and variance of this output?

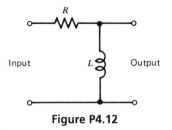

Figure P4.12

Problem 4.25   A white noise $w(t)$ of power spectral density $N_0/2$ is applied to a *Butterworth* low-pass filter of order n, whose amplitude response is defined by

$$|H(f)| = \frac{1}{[1 + (f/f_0)^{2n}]^{1/2}}$$

(a)   Determine the noise equivalent bandwidth for this low-pass filter.

(b)   What is the limiting value of the noise equivalent bandwidth as n approaches infinity?

Problem 4.26   The shot-noise process $X(t)$ defined by Eq. (4.140) is stationary. Why?

Problem 4.27   White Gaussian noise of zero mean and power spectral density $N_0/2$ is applied to the filtering scheme shown in Fig. P4.13. The noise at the low-pass filter output is denoted by $n(t)$.

(a)   Find the power spectral density and the autocorrelation function of $n(t)$.

(b)   Find the mean and variance of $n(t)$.

(c)   What is the rate at which $n(t)$ can be sampled so that the resulting samples are essentially uncorrelated?

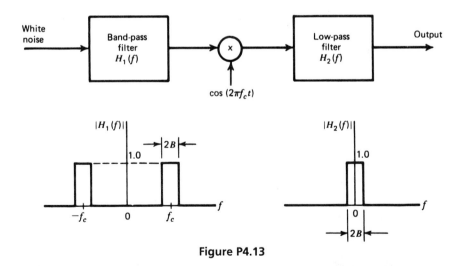

Figure P4.13

Problem 4.28   Let $X(t)$ be a stationary process with zero mean, autocorrelation function $R_X(\tau)$, and power spectral density $S_X(f)$. We are required to find a linear filter with impulse response $h(t)$, such that the filter output is $X(t)$ when the input is white noise of power spectral density $N_0/2$.

(a) Determine the condition that the impulse response $h(t)$ must satisfy in order to achieve this requirement.

(b) What is the corresponding condition on the transfer function $H(f)$ of the filter?

(c) Using the Paley–Wiener criterion (see Section 2.9), find the requirement on $S_X(f)$ for the filter to be causal.

Problem 4.29 Consider a narrow-band noise $n(t)$ with its Hilbert transform denoted by $\hat{n}(t)$.

(a) Show that the cross-correlation functions of $n(t)$ and $\hat{n}(t)$ are given by

$$R_{N\hat{N}}(\tau) = -\hat{R}_N(\tau)$$

and

$$R_{\hat{N}N}(\tau) = \hat{R}_N(\tau)$$

where $\hat{R}_N(\tau)$ is the Hilbert transform of the autocorrelation function $R_N(\tau)$ of $n(t)$. *Hint:* Use the formula

$$\hat{n}(t) = \frac{1}{\pi} \int_{-\infty}^{\infty} \frac{n(\lambda)}{t - \lambda} \, d\lambda$$

(b) Show that $R_{N\hat{N}}(0) = 0$.

Problem 4.30 A narrow-band noise $n(t)$ has zero mean and autocorrelation function $R_N(\tau)$. Its power spectral density $S_N(f)$ is centered about $\pm f_c$. The quadrature components $n_I(t)$ and $n_Q(t)$ of $n(t)$ are defined by the weighted sums

$$n_I(t) = n(t)\cos(2\pi f_c t) + \hat{n}(t)\sin(2\pi f_c t)$$

and

$$n_Q(t) = \hat{n}(t)\cos(2\pi f_c t) - n(t)\sin(2\pi f_c t)$$

Using the result of part (a) of Problem 4.29, show that $n_I(t)$ and $n_Q(t)$ have the correlation functions

$$R_{N_I}(\tau) = R_{N_Q}(\tau) = R_N(\tau)\cos(2\pi f_c \tau) + \hat{R}_N(\tau)\sin(2\pi f_c \tau)$$

and

$$R_{N_I N_Q}(\tau) = -R_{N_Q N_I}(\tau) = R_N(\tau)\sin(2\pi f_c \tau) - \hat{R}_N(\tau)\cos(2\pi f_c \tau)$$

Problem 4.31 The power spectral density of a narrow-band noise $n(t)$ is as shown in Fig. P4.14. The carrier frequency is 5 Hz.

(a) Find the power spectral densities of the in-phase and quadrature components of $n(t)$.

(b) Find their cross-spectral densities.

Problem 4.32 Consider a Gaussian noise $n(t)$ with zero mean and the power spectral density $S_N(f)$ shown in Fig. P4.15.

(a) Find the probability density function of the envelope of $n(t)$.

(b) What are the mean and variance of this envelope?

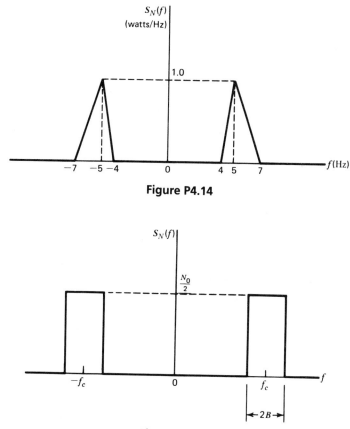

Figure P4.14

Figure P4.15

Problem 4.33 Consider the problem of propagating signals through so-called *random* or *fading communication channels*. Examples of such channels include the *ionosphere* from which short-wave (high-frequency) signals are reflected back to the earth producing long-range radio transmission, and *underwater communications*. A simple model of such a channel is shown in Fig. P4.16, which consists of

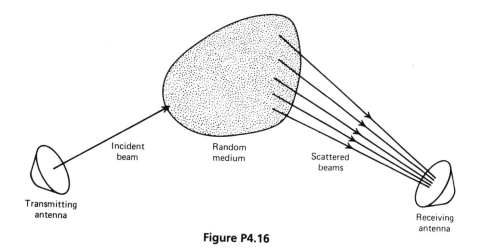

Figure P4.16

a large collection of *random scatterers*, with the result that a single incident beam is converted into a correspondingly large number of scattered beams at the receiver. The transmitted signal is equal to $A \exp(j2\pi f_c t)$. Assume that all scattered beams travel at the same mean velocity. However, each scattered beam differs in amplitude and phase from the incident beam, so that the kth scattered beam is given by $A_k \exp(j2\pi f_c t + j\Theta_k)$, where the amplitude A_k and phase Θ_k vary slowly and randomly with time. In particular, assume that the Θ_k are all independent of one another and uniformly distributed random variables.

(a) With the received signal denoted by

$$x(t) = r(t)\exp[j2\pi f_c t + \psi(t)]$$

show that the random variable R, obtained by observing the envelope of the received signal at time t, is Rayleigh-distributed, and that the random variable Ψ, obtained by observing the phase at some fixed time, is uniformly distributed.

(b) Assuming that the channel includes a line-of-sight path, so that the received signal contains a sinusoidal component of frequency f_c, show that in this case the envelope of the received signal is Rician distributed.

CHAPTER 5

Noise in CW Modulation Systems

5.1   INTRODUCTION

In Chapter 3, we studied the characterization of continuous-wave (CW) modulation [i.e., amplitude modulation (AM) and frequency modulation (FM)] techniques, entirely from a deterministic perspective. Then in Chapter 4, we equipped ourselves with the mathematical tools needed for the statistical characterization of random signals and noise. We are now ready to resume the study of CW modulation systems by evaluating the effects of noise on their performance, and thereby develop a deeper understanding of analog communications.

To undertake an analysis of noise in CW modulation systems, we need to do a number of things. First and foremost, however, we must have a *receiver model*. In formulating such a model, the customary practice is to model the receiver noise (channel noise) as *additive, white,* and *Gaussian*. These simplifying assumptions enable us to obtain a basic understanding of the way in which noise affects the performance of the receiver. Moreover, it provides a framework for the comparison of the noise performances of different CW modulation–demodulation schemes.

The material of the chapter is organized as follows. In Section 5.2 we describe a receiver model and define some related quantitative measures of noise per-

313

formance. This is followed by three sections on noise in AM receivers, namely, double sideband–suppressed carrier, single sideband, and standard amplitude modulation receivers. Next, we discuss noise in FM receivers, the analysis of which is a more difficult task. The chapter concludes with a comparative evaluation of the noise performance of AM and FM systems.

5.2   RECEIVER MODEL

The idea of *modeling* is fundamental to the study of all physical systems, including communication systems. Through modeling, we improve our understanding of the capabilities and limitations of a system. In formulating a receiver model for the study of noise in CW modulation systems, we need to keep the following points in mind:

- The model provides an adequate description of the form of receiver noise that is of common concern.
- The model accounts for the inherent filtering and modulation characteristics of the system.
- The model is simple enough for a statistical analysis of the system to be possible.

For the situation at hand, we propose to use the *receiver model* of Fig. 5.1, shown in its most basic form. In this figure, $s(t)$ denotes the incoming *modulated signal* and $w(t)$ denotes *front-end receiver noise*. The *received signal* is therefore made up of the sum of $s(t)$ and $w(t)$; this is the signal that the receiver has to work on. The *band-pass filter* in the model of Fig. 5.1 represents the combined filtering action of the tuned amplifiers used in the actual receiver for the purpose of signal amplification prior to demodulation. The bandwidth of this band-pass filter is just wide enough to pass the modulated signal $s(t)$ without distortion. As for the *demodulator* in the model of Fig. 5.1, its details naturally depend on the type of modulation used.

In performing the noise analysis of a communication system, the customary practice is to assume that the noise $w(t)$ is *additive, white,* and *Gaussian*. We thus let the power spectral density of the noise $w(t)$ be denoted by $N_0/2$, defined for both positive and negative frequencies; that is, N_0 *is the average noise power per unit bandwidth measured at the front end of the receiver.* We also assume that the band-pass filter in the receiver model of Fig. 5.1 is ideal, having a bandwidth equal to the transmission bandwidth B_T of the modulated signal $s(t)$ and a mid-band frequency equal to the carrier frequency f_c. The latter assumption is justified for double sideband–suppressed carrier (DSB-SC) modulation, full amplitude modulation (AM), and frequency modulation (FM); the cases of single sideband

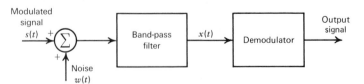

Figure 5.1   Noisy receiver model.

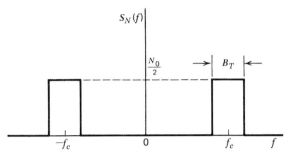

Figure 5.2 Idealized characteristic of band-pass filtered noise.

(SSB) modulation and vestigial sideband (VSB) modulation require special considerations. Taking the mid-band frequency of the band-pass filter to be the same as the carrier frequency f_c, we may model the power spectral density $S_N(f)$ of the noise $n(t)$, resulting from the passage of the white noise $w(t)$ through the filter, as shown in Fig. 5.2. Typically, the carrier frequency f_c is large compared to the transmission bandwidth B_T. We may therefore treat the *filtered noise n(t)* as a narrow-band noise represented in the canonical form

$$n(t) = n_I(t)\cos(2\pi f_c t) - n_Q(t)\sin(2\pi f_c t) \tag{5.1}$$

where $n_I(t)$ is the *in-phase noise component* and $n_Q(t)$ is the *quadrature noise component*, both measured with respect to the carrier wave $A_c\cos(2\pi f_c t)$. The filtered signal $x(t)$ available for demodulation is defined by

$$x(t) = s(t) + n(t) \tag{5.2}$$

The details of $s(t)$ depend on the type of modulation used. In any event, the average noise power at the demodulator input is equal to the total area under the curve of the power spectral density $S_N(f)$. From Fig. 5.2 we readily see that this average noise power is equal to $N_0 B_T$. Given the format of $s(t)$, we may also determine the average signal power at the demodulator input. With the modulated signal $s(t)$ and the filtered noise $n(t)$ appearing additively at the demodulator input in accordance with Eq. (5.2), we may go on to define an *input signal-to-noise ratio*, $(\text{SNR})_I$, *as the ratio of the average power of the modulated signal s(t) to the average power of the filtered noise n(t).*

A more useful measure of noise performance, however, is the *output signal-to-noise ratio*, $(\text{SNR})_O$, *defined as the ratio of the average power of the demodulated message signal to the average power of the noise, both measured at the receiver output.* The output signal-to-noise ratio provides an intuitive measure for describing the fidelity with which the demodulation process in the receiver recovers the message signal from the modulated signal in the presence of additive noise. For such a criterion to be well defined, the recovered message signal and the corruptive noise component must appear *additively* at the demodulator output. This condition is perfectly valid in the case of a receiver using coherent detection. On the other hand, when the receiver uses envelope detection as in full AM or frequency discrimination as in FM, we have to assume that the average power of the filtered noise

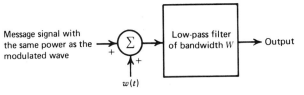

Figure 5.3 The baseband transmission model, assuming a message signal of bandwidth *W*, used for calculating channel signal-to-noise ratio.

$n(t)$ is relatively low to justify the use of output signal-to-noise ratio as a measure of receiver performance.

The output signal-to-noise ratio depends, among other factors, on the type of modulation used in the transmitter and the type of demodulation used in the receiver. Thus it is informative to compare the output signal-to-noise ratios for different modulation–demodulation systems. However, for this comparison to be of meaningful value, it must be made on an equal basis as described here:

- The modulated signal $s(t)$ transmitted by each system has the same average power.
- The front-end receiver noise $w(t)$ has the same average power measured in the message bandwidth *W*.

Accordingly, as a frame of reference we define the *channel signal-to-noise ratio,* $(SNR)_C$, as *the ratio of the average power of the modulated signal to the average power of noise in the message bandwidth, both measured at the receiver input.* This ratio may be viewed as the signal-to-noise ratio that results from *baseband (direct) transmission* of the message signal $m(t)$ without modulation, as modeled in Fig. 5.3. Here it is assumed that (1) the message power at the low-pass filter input is adjusted to be the same as the average power of the modulated signal, and (2) the low-pass filter passes the message signal and rejects out-of-band noise.

For the purpose of comparing different continuous-wave (CW) modulation systems, we *normalize* the receiver performance by dividing the output signal-to-noise ratio by the channel signal-to-noise ratio. We thus define a *figure of merit* for the receiver as follows:

$$\text{Figure of merit} = \frac{(SNR)_O}{(SNR)_C} \tag{5.3}$$

Clearly, the higher the value of the figure of merit, the better will the noise performance of the receiver be. The figure of merit may equal one, be less than one, or be greater than one, depending on the type of modulation used.

In the next four sections, we use the ideas described herein to perform a noise analysis of (1) DSB-SC receivers using coherent detection, (2) SSB receivers using coherent detection, (3) AM receivers using envelope detection, and (4) FM receivers using frequency discrimination; we also consider related issues that arise under high noise levels. These receivers pertain to typical examples of CW modulation systems, which exhibit different noise behavior.

5.3   NOISE IN DSB-SC RECEIVERS

The noise analysis of a DSB-SC receiver using coherent detection is the simplest of the above-mentioned cases. Figure 5.4 shows the model of a DSB-SC receiver using a coherent detector. The use of coherent detection requires multiplication of the filtered signal $x(t)$ by a locally generated sinusoidal wave cos $(2\pi f_c t)$ and then low-pass filtering the product. To simplify the analysis, we assume that the amplitude of the locally generated sinusoidal wave is unity. For this demodulation scheme to operate satisfactorily, however, it is necessary that the local oscillator be synchronized both in phase and in frequency with the oscillator generating the carrier wave in the transmitter. We assume that this synchronization has been achieved.

The DSB-SC component of the filtered signal $x(t)$ is expressed as

$$s(t) \ = \ CA_c \cos(2\pi f_c t)\, m(t) \tag{5.4}$$

where $A_c \cos(2\pi f_c t)$ is the sinusoidal carrier wave and $m(t)$ is the message signal. In the expression for $s(t)$ in Eq. (5.4) we have included a *system-dependent scaling factor C*, the purpose of which is to ensure that the signal component $s(t)$ is measured in the same units as the additive noise component $n(t)$. We assume that $m(t)$ is the sample function of a stationary process of zero mean, whose power spectral density $S_M(f)$ is limited to a maximum frequency W; that is, W is the *message bandwidth*. The average power P of the message signal is the total area under the curve of power spectral density, as shown by

$$P = \int_{-W}^{W} S_M(f)\ df \tag{5.5}$$

The carrier wave is statistically independent of the message signal. To emphasize this independence, the carrier should include a random phase that is uniformly distributed over 2π radians. In the defining equation for $s(t)$ this random phase angle has been omitted for convenience of presentation. Using the result of Example 12 of Chapter 4 on a modulated random process, we may express the average power of the DSB-SC modulated signal component $s(t)$ as $C^2 A_c^2 P/2$. With a noise spectral density of $N_0/2$, the average noise power in the message bandwidth W is equal to WN_0. The channel signal-to-noise ratio of the DSB-SC mod-

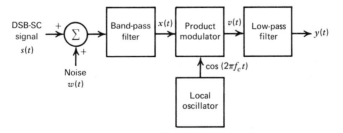

Figure 5.4 Model of DSB-SC receiver using coherent detection.

ulation system is therefore

$$(\text{SNR})_{C,\text{DSB}} = \frac{C^2 A_c^2 P}{2WN_0} \tag{5.6}$$

where the constant C^2 in the numerator ensures that this ratio is dimensionless.

Next, we wish to determine the output signal-to-noise ratio of the system. Using the narrow-band representation of the filtered noise $n(t)$, the total signal at the coherent detector input may be expressed as

$$\begin{aligned}
x(t) &= s(t) + n(t) \\
&= CA_c\cos(2\pi f_c t)\,m(t) + n_I(t)\cos(2\pi f_c t) - n_Q(t)\sin(2\pi f_c t)
\end{aligned} \tag{5.7}$$

where $n_I(t)$ and $n_Q(t)$ are the in-phase and quadrature components of $n(t)$ with respect to the carrier. The output of the product–modulator component of the coherent detector is therefore

$$\begin{aligned}
v(t) &= x(t)\cos(2\pi f_c t) \\
&= \tfrac{1}{2}CA_c m(t) + \tfrac{1}{2}n_I(t) \\
&\quad + \tfrac{1}{2}[CA_c m(t) + n_I(t)]\cos(4\pi f_c t) - \tfrac{1}{2}A_c n_Q(t)\sin(4\pi f_c t)
\end{aligned}$$

The low-pass filter in the coherent detector removes the high frequency components of $v(t)$, yielding a receiver output

$$y(t) = \tfrac{1}{2}CA_c m(t) + \tfrac{1}{2}n_I(t) \tag{5.8}$$

Equation (5.8) indicates the following:

1. The message signal $m(t)$ and in-phase noise component $n_I(t)$ of the filtered noise $n(t)$ appear additively at the receiver output.
2. The quadrature component $n_Q(t)$ of the noise $n(t)$ is completely rejected by the coherent detector.

These two results are independent of the input signal-to-noise ratio. Thus, coherent detection distinguishes itself from other demodulation techniques in the important property: The output message component is unmutilated and the noise component always appears additively with the message, irrespective of the input signal-to-noise ratio.

The message signal component at the receiver output is $CA_c m(t)/2$. Therefore, the average power of this component may be expressed as $C^2 A_c^2 P/4$, where P is the average power of the original message signal $m(t)$ and C is the system-dependent scaling factor referred to earlier.

In the case of DSB-SC modulation, the band-pass filter in Fig. 5.4 has a bandwidth B_T equal to $2W$ in order to accommodate the upper and lower sidebands of the modulated signal $s(t)$. It follows therefore that the average power of the filtered noise $n(t)$ is $2WN_0$. From Property 5 of narrow-band noise described in Section 4.14, we know that the average power of the (low-pass) in-phase noise component $n_I(t)$ is the same as that of the (band-pass) filtered noise $n(t)$. Since from Eq. (5.8) the noise component at the receiver output is $n_I(t)/2$, it follows that the average power of the noise at the receiver output is

$$\left(\tfrac{1}{2}\right)^2 2WN_0 = \tfrac{1}{2}WN_0$$

The output signal-to-noise for a DSB-SC receiver using coherent detection is therefore

$$
\begin{aligned}
(\text{SNR})_O &= \frac{C^2 A_c^2 P/4}{WN_0/2} \\
&= \frac{C^2 A_c^2 P}{2WN_0}
\end{aligned}
\tag{5.9}
$$

Using Eqs. (5.6) and (5.9), we obtain the figure of merit

$$\left.\frac{(\text{SNR})_O}{(\text{SNR})_C}\right|_{\text{DSB-SC}} = 1 \tag{5.10}$$

Note that the factor C^2 is common to both the output and channel signal-to-noise ratios, and therefore cancels out in evaluating the figure of merit.

Note also that at the coherent detector output in the receiver of Fig. 5.4 using DSB-SC modulation, the translated signal sidebands sum coherently, whereas the translated noise sidebands sum incoherently. This means that the output signal-to-noise ratio in this receiver is twice the signal-to-noise ratio at the coherent detector input.

5.4   NOISE IN SSB RECEIVERS

Consider next the case of a receiver using coherent detection, with an incoming single-sideband (SSB) modulated wave. We assume that only the lower sideband is transmitted, so that we may express the modulated wave as

$$s(t) = \tfrac{1}{2}CA_c \cos(2\pi f_c t) m(t) + \tfrac{1}{2}CA_c \sin(2\pi f_c t) \hat{m}(t) \tag{5.11}$$

where $\hat{m}(t)$ is the Hilbert transform of the message signal $m(t)$. As before, the system-dependent scaling factor C is included to make the signal component $s(t)$ have the same units as the noise component $n(t)$. We may make the following observations concerning the in-phase and quadrature components of $s(t)$ in Eq. (5.11):

1. The two components $m(t)$ and $\hat{m}(t)$ are orthogonal to each other. Therefore, with the message signal $m(t)$ assumed to have zero mean, which is a reasonable assumption to make, it follows that $m(t)$ and $\hat{m}(t)$ are uncorrelated; hence, their power spectral densities are additive.
2. The Hilbert transform $\hat{m}(t)$ is obtained by passing $m(t)$ through a linear filter with a transfer function $-j\,\text{sgn}(f)$. The squared magnitude of this transfer function is equal to one for all f. Accordingly, we find that both $m(t)$ and $\hat{m}(t)$ have the same power spectral density.

Thus, using a procedure similar to that in Section 5.3, we find that the in-phase and quadrature components of the modulated signal $s(t)$ contribute an average power of $C^2 A_c^2 P/8$ each, where P is the average power of the message signal $m(t)$. The average power of $s(t)$ is therefore $C^2 A_c^2 P/4$. This result is half that in the DSB-SC receiver, which is intuitively satisfying.

The average noise power in the message bandwidth W is WN_0, as in the DSB-SC receiver. Thus the channel signal-to-noise ratio of a coherent receiver with SSB modulation is

$$(\text{SNR})_{C,\text{SSB}} = \frac{C^2 A_c^2 P}{4 W N_0} \qquad (5.12)$$

As illustrated in Fig. 5.5a, in an SSB system the transmission bandwidth B_T is W and the mid-band frequency of the power spectral density $S_N(f)$ of the narrow-band noise $n(t)$ is offset from the carrier frequency f_c by $W/2$. Therefore, we may express $n(t)$ as

$$n(t) = n_I(t) \cos\left[2\pi\left(f_c - \frac{W}{2}\right)t\right] - n_Q(t) \sin\left[2\pi\left(f_c - \frac{W}{2}\right)t\right] \qquad (5.13)$$

The output of the coherent detector, due to the combined influence of the modulated signal $s(t)$ and noise $n(t)$, is thus given by

$$y(t) = \tfrac{1}{4} C A_c m(t) + \tfrac{1}{2} n_I(t) \cos(\pi W t) + \tfrac{1}{2} n_Q(t) \sin(\pi W t) \qquad (5.14)$$

As expected, we see that the quadrature component $\hat{m}(t)$ of the modulated message signal $s(t)$ has been eliminated from the detector output, but unlike the case of DSB-SC modulation, the quadrature component of the narrow-band noise $n(t)$ now appears at the receiver output.

The message component in the receiver output is $C A_c m(t)/4$, and so we may express the average power of the recovered message signal as $C^2 A_c^2 P/16$. The noise component in the receiver output is $[n_I(t)\cos(\pi W t) + n_Q(t)\sin(\pi W t)]/2$. To determine the average power of the output noise, we note the following:

1. The power spectral density of both $n_I(t)$ and $n_Q(t)$ is as shown in Fig. 5.5b.
2. The sinusoidal wave $\cos(\pi W t)$ is independent of both $n_I(t)$ and $n_Q(t)$. Hence, the power spectral density of $n_I'(t) = n_I(t)\cos(\pi W t)$ is obtained by shifting $S_{N_I}(f)$ to the left by $W/2$, shifting it to the right by $W/2$, adding the shifted spectra, and dividing the result by 4, in accordance with Example 12 of Chapter 4. The power spectral density of $n_Q'(t) = n_Q(t)\sin(\pi W t)$ is obtained in a similar way. The power spectral density of both $n_I'(t)$ and $n_Q'(t)$, obtained in this manner, is shown sketched in Fig. 5.5c.

From Fig. 5.5c we see that the average power of the noise component $n_I'(t)$ or $n_Q'(t)$ is $WN_0/2$. Therefore, from Eq. (5.14), the average output noise power is $WN_0/4$. We thus find that the output signal-to-noise ratio of a system, using SSB modulation in the transmitter and coherent detection in the receiver, is given by

$$(\text{SNR})_{O,\text{SSB}} = \frac{C^2 A_c^2 P}{4 W N_0} \qquad (5.15)$$

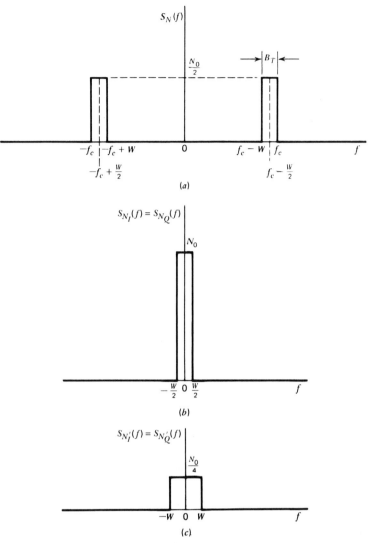

Figure 5.5 Noise analysis of SSB modulation system using coherent detection. (a) Power spectral density of narrow-band noise $n(t)$ at the coherent detector input. (b) Power spectral density of in-phase and quadrature components of $n(t)$ with respect to $f_c - W/2$. (c) Power spectral density of $n_{I'}(t) = n_I(t)\cos(\pi Wt)$ and $n_{Q'}(t) = n_Q(t)\sin(\pi Wt)$.

Hence, from Eqs. (5.12) and (5.15), the figure of merit of such a system is

$$\left.\frac{(\text{SNR})_O}{(\text{SNR})_C}\right|_{\text{SSB}} = 1 \tag{5.16}$$

where again we see that the factor C^2 cancels out.

Comparing Eqs. (5.10) and (5.16), we conclude that *for the same average transmitted (or modulated message) signal power and the same average noise power in the message bandwidth, an SSB receiver will have exactly the same output signal-to-noise ratio as a DSB-SC receiver, when both receivers use coherent detection for the recovery of the message signal.* Furthermore, in both cases, the noise performance of the receiver is the same as that obtained by simply transmitting the message signal itself in the presence of the same noise. The only effect of the modulation process is to translate the message signal to a different frequency band to facilitate its transmission over a band-pass channel.

5.5   NOISE IN AM RECEIVERS

The next noise analysis we perform is for an amplitude modulation (AM) system using an envelope detector in the receiver, as shown in the model of Fig. 5.6. In a full AM signal, both sidebands and the carrier wave are transmitted as shown by

$$s(t) = A_c[1 + k_a m(t)] \cos(2\pi f_c t) \tag{5.17}$$

where $A_c \cos(2\pi f_c t)$ is the carrier wave, $m(t)$ is the message signal, and k_a is a constant that determines the percentage modulation. In the expression for the amplitude-modulated signal component $s(t)$ given in Eq. (5.17) we see no need for the use of a scaling factor, because it is reasonable to assume that the carrier amplitude A_c has the same units as the additive noise component.

As with the DSB-SC receiver, we perform the noise analysis of the AM receiver by first determining the channel signal-to-noise ratio, and then the output signal-to-noise ratio.

The average power of the carrier component in the AM signal $s(t)$ is $A_c^2/2$. The average power of the information-bearing component $A_c k_a m(t) \cos(2\pi f_c t)$ is $A_c^2 k_a^2 P/2$, where P is the average power of the message signal $m(t)$. The average power of the full AM signal $s(t)$ is therefore equal to $A_c^2(1 + k_a^2 P)/2$. As for the DSB-SC system, the average power of noise in the message bandwidth is WN_0. The channel signal-to-noise ratio for AM is therefore

$$(\text{SNR})_{C,\text{AM}} = \frac{A_c^2(1 + k_a^2 P)}{2WN_0} \tag{5.18}$$

To evaluate the output signal-to-noise ratio, we first represent the filtered noise $n(t)$ in terms of its in-phase and quadrature components. We may therefore define the filtered signal $x(t)$ applied to the envelope detector in the receiver model of Fig. 5.6 as follows:

$$\begin{aligned} x(t) &= s(t) + n(t) \\ &= [A_c + A_c k_a m(t) + n_I(t)]\cos(2\pi f_c t) - n_Q(t)\sin(2\pi f_c t) \end{aligned} \tag{5.19}$$

It is informative to represent the components that comprise the signal $x(t)$ by means of phasors, as in Fig. 5.7a. From this phasor diagram, the receiver output is readily obtained as

$$\begin{aligned} y(t) &= \text{envelope of } x(t) \\ &= \{[A_c + A_c k_a m(t) + n_I(t)]^2 + n_Q^2(t)\}^{1/2} \end{aligned} \tag{5.20}$$

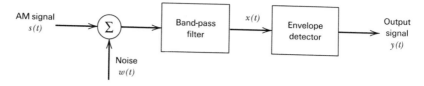

Figure 5.6 Noisy model of AM receiver.

The signal $y(t)$ defines the output of an ideal envelope detector. The phase of $x(t)$ is of no interest to us, because an ideal envelope detector is totally insensitive to variations in the phase of $x(t)$.

The expression defining $y(t)$ is somewhat complex and needs to be simplified in some manner in order to permit the derivation of insightful results. Specifically, we would like to approximate the output $y(t)$ as the sum of a message term plus a term due to noise. In general, this is quite difficult to achieve. However, when the average carrier power is large compared with the average noise power, so that the receiver is operating satisfactorily, then the signal term $A_c[1 + k_a m(t)]$ will be large compared with the noise terms $n_I(t)$ and $n_Q(t)$, at least most of the time. Then we may approximate the output $y(t)$ as (see Problem 5.8):

$$y(t) \simeq A_c + A_c k_a m(t) + n_I(t) \tag{5.21}$$

The presence of the dc or constant term A_c in the envelope detector output $y(t)$ of Eq. (5.21) is due to demodulation of the transmitted carrier wave. We may ignore this term, however, because it bears no relation whatsoever to the message signal $m(t)$. In any case, it may be removed simply by means of a blocking capacitor. Thus if we neglect the dc term A_c in Eq. (5.21), we find that the remainder has, except for scaling factors, a form similar to the output of a DSB-SC receiver using coherent detection. Accordingly, the output signal-to-noise

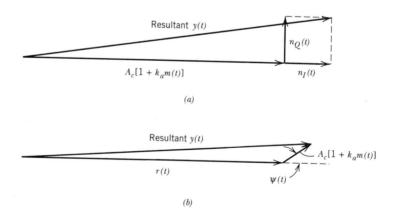

Figure 5.7 (a) Phasor diagram for AM wave plus narrow-band noise for the case of high carrier-to-noise ratio. (b) Phasor diagram for AM wave plus narrow-band noise for the case of low carrier-to-noise ratio.

ratio of an AM receiver using an envelope detector is approximately

$$(\text{SNR})_{O,\text{AM}} \simeq \frac{A_c^2 k_a^2 P}{2WN_0} \tag{5.22}$$

Equation (5.22) is, however, valid only if the following two conditions are satisfied:

1. The average noise power is small compared to the average carrier power at the envelope detector input.
2. The amplitude sensitivity k_a is adjusted for a percentage modulation less than or equal to 100 percent.

Using Eqs. (5.18) and (5.22), we obtain the following figure of merit for amplitude modulation:

$$\left.\frac{(\text{SNR})_O}{(\text{SNR})_C}\right|_{\text{AM}} \simeq \frac{k_a^2 P}{1 + k_a^2 P} \tag{5.23}$$

Thus, whereas the figure of merit of a DSB-SC receiver or that of an SSB receiver using coherent detection is always unity, the corresponding figure of merit of an AM receiver using envelope detection is always less than unity. In other words, *the noise performance of an AM receiver is always inferior to that of a DSB-SC receiver.* This is due to the wastage of transmitter power, which results from transmitting the carrier as a component of the AM wave.

EXAMPLE 1   Single-Tone Modulation

Consider the special case of a sinusoidal wave of frequency f_m and amplitude A_m as the modulating wave, as shown by

$$m(t) = A_m \cos(2\pi f_m t)$$

The corresponding AM wave is

$$s(t) = A_c[1 + \mu \cos(2\pi f_m t)]\cos(2\pi f_c t)$$

where $\mu = k_a A_m$ is the modulation factor. The average power of the modulating wave $m(t)$ is (assuming a load resistor of 1 ohm)

$$P = \tfrac{1}{2}A_m^2$$

Therefore, using Eq. (5.23), we get

$$\left.\frac{(\text{SNR})_O}{(\text{SNR})_C}\right|_{\text{AM}} = \frac{\dfrac{1}{2}k_a^2 A_m^2}{1 + \dfrac{1}{2}k_a^2 A_m^2} \tag{5.24}$$

$$= \frac{\mu^2}{2 + \mu^2}$$

When $\mu = 1$, which corresponds to 100 percent modulation, we get a figure of merit equal to 1/3. This means that, other factors being equal, an AM system (using envelope detection) must transmit three times as much average power as a suppressed-carrier system (using coherent detection) in order to achieve the same quality of noise performance.

Threshold Effect

When the carrier-to-noise ratio is small compared with unity, the noise term dominates and the performance of the envelope detector changes completely from that described above. In this case it is more convenient to represent the narrow-band noise $n(t)$ in terms of its envelope $r(t)$ and phase $\psi(t)$, as shown by

$$n(t) = r(t)\cos[2\pi f_c t) + \psi(t)] \qquad (5.25)$$

The corresponding phasor diagram for the detector input $x(t) = s(t) + n(t)$ is shown in Fig. 5.7b, where we have used the noise as reference, because it is now the dominant term. To the noise phasor $r(t)$ we have added a phasor representing the signal term $A_c[1 + k_a m(t)]$, with the angle between them being equal to the phase $\psi(t)$ of the noise $n(t)$. In Fig. 5.7b it is assumed that the carrier-to-noise ratio is so low that the carrier amplitude A_c is small compared with the noise envelope $r(t)$, at least most of the time. Then we may neglect the quadrature component of the signal with respect to the noise, and thus find directly from Fig. 5.7b that the envelope detector output is approximately

$$y(t) \simeq r(t) + A_c\cos[\psi(t)] + A_c k_a m(t)\cos[\psi(t)] \qquad (5.26)$$

This relation reveals that when the carrier-to-noise ratio is low, the detector output has no component strictly proportional to the message signal $m(t)$. The last term of the expression defining $y(t)$ contains the message signal $m(t)$ multiplied by noise in the form of $\cos[\psi(t)]$. From Section 4.14 we recall that the phase $\psi(t)$ of the narrow-band noise $n(t)$ is uniformly distributed over 2π radians. It follows therefore that we have a complete loss of information in that the detector output does not contain the message signal $m(t)$ at all. The loss of a message in an envelope detector that operates at a low carrier-to-noise ratio is referred to as the *threshold effect*.[1] By threshold we mean *a value of the carrier-to-noise ratio below which the noise performance of a detector deteriorates much more rapidly than proportionately to the carrier-to-noise ratio.* It is important to recognize that every nonlinear detector (e.g., envelope detector) exhibits a threshold effect. On the other hand, such an effect does *not* arise in a coherent detector.

A detailed analysis of the threshold effect in envelope detectors is complicated, and certainly beyond the level of treatment in this book. We may develop some insight into the threshold effect, however, by using the following qualitative approach. Let R denote the random variable obtained by observing the envelope process, with sample function $r(t)$, at some fixed time. Intuitively, an envelope detector is expected to be operating well into the threshold region if the probability that the random variable R exceeds the carrier amplitude A_c is, say, 0.5. On the other hand, if this same probability is only 0.01, the envelope detector is expected to be relatively free of loss of message and the threshold effect. The evaluation of the carrier-to-noise ratios, corresponding to these probabilities, is best illustrated by way of an example.

EXAMPLE 2

From Section 4.14 we recall that the envelope $r(t)$ of the narrow-band noise $n(t)$ is Rayleigh distributed; that is

$$f_R(r) = \frac{r}{\sigma_N^2}\exp\left(-\frac{r^2}{2\sigma_N^2}\right)$$

where σ_N^2 is the variance of the noise $n(t)$. For an AM system, the variance σ_N^2 is $2WN_0$. Therefore, the probability of the event that the envelope R of the narrow-band noise $n(t)$ is large compared to the carrier amplitude A_c is defined by

$$
\begin{aligned}
P(R \geq A_c) &= \int_{A_c}^{\infty} f_R(r)\ dr \\
&= \int_{A_c}^{\infty} \frac{r}{2WN_0} \exp\left(-\frac{r^2}{4WN_0}\right) dr \\
&= \exp\left(-\frac{A_c^2}{4WN_0}\right)
\end{aligned}
\tag{5.27}
$$

Define the *carrier-to-noise ratio* as

$$
\rho = \frac{\text{average carrier power}}{\text{average noise power in bandwidth of the modulated message signal}}
\tag{5.28}
$$

Since the bandwidth of the AM signal is $2W$, the average noise power in this bandwidth is $2WN_0$. The average power of the carrier is $A_c^2/2$. The carrier-to-noise ratio is therefore

$$
\rho = \frac{A_c^2}{4WN_0}
\tag{5.29}
$$

Then we may use this definition to rewrite Eq. (5.27) in the compact form

$$
P(R \geq A_c) = \exp(-\rho)
\tag{5.30}
$$

Solving $P(R \geq A_c) = 0.5$ for ρ, we get

$$
\rho = \ln 2 = 0.69
$$

Similarly, for $P(R \geq A_c) = 0.01$, we get

$$
\rho = \ln 100 = 4.6
$$

Thus with a carrier-to-noise ratio $10 \log_{10} 0.69 = -1.6$ dB, the envelope detector is expected to be well into the threshold region, whereas with a carrier-to-noise ratio $10 \log_{10} 4.6 = 6.6$ dB, the detector is expected to be operating satisfactorily. We ordinarily need a signal-to-noise ratio considerably greater than 6.6 dB for satisfactory intelligibility, and therefore threshold effects are seldom of great importance in AM receivers using envelope detection.

5.6 NOISE IN FM RECEIVERS

We finally turn our attention to the noise analysis of a frequency modulation (FM) system, for which we use the receiver model shown in Fig. 5.8. As before, the noise $w(t)$ is modeled as white Gaussian noise of zero mean and power spectral density $N_0/2$. The received FM signal $s(t)$ has a carrier frequency f_c and transmission bandwidth B_T, such that only a negligible amount of power lies outside the frequency band $f_c \pm B_T/2$ for positive frequencies.

As in the AM case, the band-pass filter has a mid-band frequency f_c and bandwidth B_T and therefore passes the FM signal essentially without distortion.

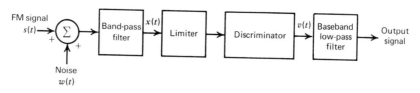

Figure 5.8 Noisy model of an FM receiver.

Ordinarily, B_T is small compared with the mid-band frequency f_c, so that we may use the narrow-band representation for $n(t)$, the filtered version of receiver noise $w(t)$, in terms of its in-phase and quadrature components.

In an FM system, the message information is transmitted by variations of the instantaneous frequency of a sinusoidal carrier wave, and its amplitude is maintained constant. Therefore, any variations of the carrier amplitude at the receiver input must result from noise or interference. The amplitude *limiter*, following the bandpass filter in the receiver model of Fig. 5.8, is used to remove amplitude variations by clipping the modulated wave at the filter output almost to the zero axis. The resulting rectangular wave is rounded off by another bandpass filter that is an integral part of the limiter, thereby suppressing harmonics of the carrier frequency. Thus, the filter output is again sinusoidal, with an amplitude that is practically independent of the carrier amplitude at the receiver input.

The discriminator in the model of Fig. 5.8 consists of two components:

1. A *slope network* or *differentiator* with a purely imaginary transfer function that varies linearly with frequency. It produces a hybrid-modulated wave in which both amplitude and frequency vary in accordance with the message signal.
2. An envelope detector that recovers the amplitude variation and thus reproduces the message signal.

The slope network and envelope detector are usually implemented as integral parts of a single physical unit.

The *postdetection filter,* labeled "baseband low-pass filter" in Fig. 5.8, has a bandwidth that is just large enough to accommodate the highest frequency component of the message signal. This filter removes the out-of-band components of the noise at the discriminator output and thereby keeps the effect of the output noise to a minimum.

The filtered noise $n(t)$ at the band-pass filter output in Fig. 5.8 is defined in terms of its in-phase and quadrature components by

$$n(t) = n_I(t)\cos(2\pi f_c t) - n_Q(t)\sin(2\pi f_c t)$$

Equivalently, we may express $n(t)$ in terms of its envelope and phase as

$$n(t) = r(t)\cos[2\pi f_c t) + \psi(t)] \tag{5.31}$$

where the envelope is

$$r(t) = [n_I^2(t) + n_Q^2(t)]^{1/2} \tag{5.32}$$

and the phase is

$$\psi(t) = \tan^{-1}\left[\frac{n_Q(t)}{n_I(t)}\right] \tag{5.33}$$

The envelope $r(t)$ is Rayleigh distributed, and the phase $\psi(t)$ is uniformly distributed over 2π radians (see Section 4.14).

The incoming FM signal $s(t)$ is defined by

$$s(t) = A_c \cos\left[2\pi f_c t + 2\pi k_f \int_0^t m(t)\,dt\right] \tag{5.34}$$

where A_c is the carrier amplitude, f_c is the carrier frequency, k_f is the frequency sensitivity, and $m(t)$ is the message signal. Note that, as with the standard AM, in FM there is no need to introduce a scaling factor in the definition of the modulated signal $s(t)$, since it is reasonable to assume that its amplitude A_c has the same units as the additive noise component $n(t)$. To proceed, we define

$$\phi(t) = 2\pi k_f \int_0^t m(t)\,dt \tag{5.35}$$

We may thus express $s(t)$ in the simple form

$$s(t) = A_c \cos[2\pi f_c t + \phi(t)] \tag{5.36}$$

The noisy signal at the band-pass filter output is therefore

$$\begin{aligned} x(t) &= s(t) + n(t) \\ &= A_c \cos[2\pi f_c t + \phi(t)] + r(t)\cos[2\pi f_c t + \psi(t)] \end{aligned} \tag{5.37}$$

It is informative to represent $x(t)$ by means of a phasor diagram, as in Fig. 5.9. In this diagram we have used the signal term as reference. The phase $\theta(t)$ of the resultant phasor representing $x(t)$ is obtained directly from Fig. 5.9 as

$$\theta(t) = \phi(t) + \tan^{-1}\left\{\frac{r(t)\sin[\psi(t) - \phi(t)]}{A_c + r(t)\cos[\psi(t) - \phi(t)]}\right\} \tag{5.38}$$

The envelope of $x(t)$ is of no interest to us, because any envelope variations at the bandpass output are removed by the limiter.

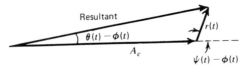

Figure 5.9 Phasor diagram for FM wave plus narrow-band noise for the case of high carrier-to-noise ratio.

Our motivation is to determine the error in the instantaneous frequency of the carrier wave caused by the presence of the filtered noise $n(t)$. With the discriminator assumed ideal, its output is proportional to $\theta'(t)/2\pi$ where $\theta'(t)$ is the derivative of $\theta(t)$ with respect to time. In view of the complexity of the expression defining $\theta(t)$, however, we need to make certain simplifying approximations, so that our analysis may yield useful results.

We assume that the carrier-to-noise ratio measured at the discriminator input is large compared with unity. Let R denote the random variable obtained by observing (at some fixed time) the envelope process with sample function $r(t)$ [due to the noise $n(t)$]. Then, at least most of the time, the random variable R is small compared with the carrier amplitude A_c, and so the expression for the phase $\theta(t)$ simplifies considerably as follows:

$$\theta(t) \simeq \phi(t) + \frac{r(t)}{A_c} \sin[\psi(t) - \phi(t)] \tag{5.39}$$

or, using the expression for $\phi(t)$ given in Eq. (5.35),

$$\theta(t) \simeq 2\pi k_f \int_0^t m(t)\ dt + \frac{r(t)}{A_c} \sin[\psi(t) - \phi(t)] \tag{5.40}$$

The discriminator output is therefore

$$\begin{aligned} v(t) &= \frac{1}{2\pi} \frac{d\theta(t)}{dt} \\ &\simeq k_f m(t) + n_d(t) \end{aligned} \tag{5.41}$$

where the noise term $n_d(t)$ is defined by

$$n_d(t) = \frac{1}{2\pi A_c} \frac{d}{dt}\{r(t)\sin[\psi(t) - \phi(t)]\} \tag{5.42}$$

We thus see that provided the carrier-to-noise ratio is high, the discriminator output $v(t)$ consists of the original message or modulating wave $m(t)$ multiplied by the constant factor k_f, plus an additive noise component $n_d(t)$. Accordingly, we may use the output signal-to-noise ratio as previously defined to assess the quality of performance of the FM receiver. Before doing this, however, it is instructive to see if we can simplify the expression defining the noise $n_d(t)$.

From the phasor diagram of Fig. 5.9, we note that the effect of variations in the phase $\psi(t)$ of the narrow-band noise appear referred to the signal term $\phi(t)$. We know that the phase $\psi(t)$ is uniformly distributed over 2π radians. It would therefore be tempting to assume that the phase difference $\psi(t) - \phi(t)$ is also uniformly distributed over 2π radians. If such an assumption were true, then the noise $n_d(t)$ at the discriminator output would be independent of the modulating signal and would depend only on the characteristics of the carrier and narrow-band noise. Theoretical considerations show that this assumption is justified provided that the carrier-to-noise ratio is high.[2] Then we may simplify Eq. (5.42) as:

$$n_d(t) \simeq \frac{1}{2\pi A_c} \frac{d}{dt}\{r(t)\sin[\psi(t)]\} \tag{5.43}$$

However, from the defining equations for $r(t)$ and $\psi(t)$, we note that the quadrature component $n_Q(t)$ of the filtered noise $n(t)$ is

$$n_Q(t) = r(t)\sin[\psi(t)] \tag{5.44}$$

Therefore, we may rewrite Eq. (5.43) as

$$n_d(t) \simeq \frac{1}{2\pi A_c} \frac{dn_Q(t)}{dt} \tag{5.45}$$

This means that *the additive noise $n_d(t)$ appearing at the discriminator output is determined effectively by the carrier amplitude A_c and the quadrature component $n_Q(t)$ of the narrow-band noise $n(t)$.*

The output signal-to-noise ratio is defined as the ratio of the average output signal power to the average output noise power. From Eq. (5.41), we see that the message component in the discriminator output, and therefore the low-pass filter output, is $k_f m(t)$. Hence, the average output signal power is equal to $k_f^2 P$, where P is the average power of the message signal $m(t)$.

To determine the average output noise power, we note that the noise $n_d(t)$ at the discriminator output is proportional to the time derivative of the quadrature noise component $n_Q(t)$. Since the differentiation of a function with respect to time corresponds to multiplication of its Fourier transform by $j2\pi f$, it follows that we may obtain the noise process $n_d(t)$ by passing $n_Q(t)$ through a linear filter with a transfer function equal to

$$\frac{j2\pi f}{2\pi A_c} = \frac{jf}{A_c}$$

This means that the power spectral density $S_{N_d}(f)$ of the noise $n_d(t)$ is related to the power spectral density $S_{N_Q}(f)$ of the quadrature noise component $n_Q(t)$ as follows:

$$S_{N_d}(f) = \frac{f^2}{A_c^2} S_{N_Q}(f) \tag{5.46}$$

With the band-pass filter in the receiver model of Fig. 5.8 having an ideal frequency response characterized by bandwidth B_T and mid-band frequency f_c, it follows that the narrow-band noise $n(t)$ will have a power spectral density characteristic that is similarly shaped. This means that the quadrature component $n_Q(t)$ of the narrow-band noise $n(t)$ will have the ideal low-pass characteristic shown in Fig. 5.10a. The corresponding power spectral density of the noise $n_d(t)$ is shown in Fig. 5.10b; that is,

$$S_{N_d}(f) = \begin{cases} \dfrac{N_0 f^2}{A_c^2}, & |f| \le \dfrac{B_T}{2} \\ 0, & \text{otherwise} \end{cases} \tag{5.47}$$

In the receiver model of Fig. 5.8, the discriminator output is followed by a low-pass filter with a bandwidth equal to the message bandwidth W. For wide-band

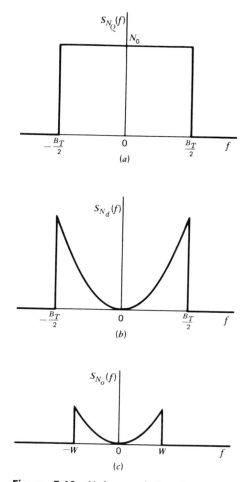

Figure 5.10 Noise analysis of FM receiver. (a) Power spectral density of quadrature component $n_Q(t)$ of narrowband noise $n(t)$. (b) Power spectral density of noise $n_d(t)$ at discriminator output. (c) Power spectral density of noise $n_o(t)$ at receiver output.

FM, we usually find that W is smaller than $B_T/2$, where B_T is the transmission bandwidth of the FM signal. This means that the out-of-band components of noise $n_d(t)$ will be rejected. Therefore, the power spectral density $S_{N_o}(f)$ of the noise $n_o(t)$ appearing at the receiver output is defined by

$$S_{N_o}(f) = \begin{cases} \dfrac{N_0 f^2}{A_c^2}, & |f| \leq W \\ \\ 0, & \text{otherwise} \end{cases} \tag{5.48}$$

as shown in Fig. 5.10c. The average output noise power is determined by integrating the power spectral density $S_{N_o}(f)$ from $-W$ to W. We thus get the result:

$$\text{Average power of output noise} = \frac{N_0}{A_c^2} \int_{-W}^{W} f^2 \, df$$

$$= \frac{2N_0 W^3}{3A_c^2} \tag{5.49}$$

Note that the average output noise power is inversely proportional to the average carrier power $A_c^2/2$. Accordingly, in an FM system, increasing the carrier power has a *noise-quieting effect*.

Earlier we determined the average output signal power as $k_f^2 P$. Therefore, provided the carrier-to-noise ratio is high, we may divide this average output signal power by the average output noise power of Eq. (5.49) to obtain the output signal-to-noise ratio

$$(\text{SNR})_{O,\text{FM}} = \frac{3A_c^2 k_f^2 P}{2N_0 W^3} \tag{5.50}$$

The average power in the modulated signal $s(t)$ is $A_c^2/2$, and the average noise power in the message bandwidth is WN_0. Thus the channel signal-to-noise ratio is

$$(\text{SNR})_{C,\text{FM}} = \frac{A_c^2}{2WN_0} \tag{5.51}$$

Dividing the output signal-to-noise ratio by the channel signal-to-noise ratio, we get the following figure of merit for frequency modulation:

$$\left. \frac{(\text{SNR})_O}{(\text{SNR})_C} \right|_{\text{FM}} = \frac{3k_f^2 P}{W^2} \tag{5.52}$$

From Section 3.10 we recall that the frequency deviation Δf is proportional to the frequency sensitivity k_f of the modulator. Also, by definition, the deviation ratio D is equal to the frequency deviation Δf divided by the message bandwidth W. In other words, the deviation ratio D is proportional to the ratio $k_f P^{1/2}/W$. It follows therefore from Eq. (5.52) that the figure of merit of a wide-band FM system is a quadratic function of the deviation ratio. Now, in wide-band FM, the transmission bandwidth B_T is approximately proportional to the deviation ratio D. Accordingly, we may state that *when the carrier-to-noise ratio is high, an increase in the transmission bandwidth B_T provides a corresponding quadratic increase in the output signal-to-noise ratio or figure of merit of the FM system.* The important point to note from this statement is that, unlike amplitude modulation, the use of frequency modulation does provide a useful mechanism for the exchange of increased transmission bandwidth for improved noise performance.

EXAMPLE 3 Single-Tone Modulation

Consider the case of a sinusoidal wave of frequency f_m as the modulating signal, and assume a peak frequency deviation Δf. The modulated FM signal is thus defined by

$$s(t) = A_c \cos\left[2\pi f_c t + \frac{\Delta f}{f_m} \sin(2\pi f_m t) \right]$$

Therefore, we may write

$$2\pi k_f \int_0^t m(t) \; dt = \frac{\Delta f}{f_m} \sin(2\pi f_m t)$$

Differentiating both sides with respect to time and solving for $m(t)$, we get

$$m(t) = \frac{\Delta f}{k_f} \cos(2\pi f_m t)$$

Hence, the average power of the message signal $m(t)$, developed across a 1-ohm load, is

$$P = \frac{(\Delta f)^2}{2k_f^2}$$

Substituting this result into the formula for the output signal-to-noise ratio given in Eq. (5.50), we get

$$(\text{SNR})_{O,\text{FM}} = \frac{3A_c^2(\Delta f)^2}{4N_0 W^3}$$

$$= \frac{3A_c^2 \beta^2}{4N_0 W}$$

where $\beta = \Delta f / W$ is the modulation index. Using Eq. (5.52) to evaluate the corresponding figure of merit, we get

$$\left. \frac{(\text{SNR})_O}{(\text{SNR})_C} \right|_{\text{FM}} = \frac{3}{2} \left(\frac{\Delta f}{W} \right)^2$$

$$= \frac{3}{2}\beta^2 \qquad (5.53)$$

It is important to note that the modulation index $\beta = \Delta f / W$ is determined by the bandwidth W of the postdetection low-pass filter and is not related to the sinusoidal message frequency f_m, except insofar as this filter is usually chosen so as to pass the spectrum of the desired message; this is merely a matter of consistent design. For a specified system bandwidth W, the sinusoidal message frequency f_m may lie anywhere between 0 and W and would yield the same output signal-to-noise ratio.

It is of particular interest to compare the noise performances of AM and FM systems. An insightful way of making this comparison is to consider the figures of merit of the two systems based on a sinusoidal modulating signal. For an AM system operating with a sinusoidal modulating signal and 100 percent modulation, we have (from Example 1):

$$\left. \frac{(\text{SNR})_O}{(\text{SNR})_C} \right|_{\text{AM}} = \frac{1}{3}$$

Comparing this figure of merit with the corresponding result described in Eq. (5.53) for an FM system, we see that the use of frequency modulation offers the possibility of improved noise performance over amplitude modulation when

$$\tfrac{3}{2}\beta^2 > \tfrac{1}{3}$$

that is,

$$\beta > \frac{\sqrt{2}}{3} = 0.471$$

We may therefore consider $\beta = 0.5$ as defining roughly *the transition between narrow-band FM and wide-band FM*. This statement, based on noise considerations, further confirms a similar observation that was made in Section 3.10 when considering the bandwidth of FM waves.

Capture Effect

The inherent ability of an FM system to minimize the effects of unwanted signals (e.g., noise, as discussed above) also applies to *interference* produced by another frequency-modulated signal whose frequency content is close to the carrier frequency of the desired FM wave. However, interference suppression in an FM receiver works well only when the interference is weaker than the desired FM input. When the interference is the stronger one of the two, the receiver locks onto the stronger signal and thereby suppresses the desired FM input. When they are of nearly equal strength, the receiver fluctuates back and forth between them. This phenomenon is known as the *capture effect*, which describes another distinctive characteristic of frequency modulation.

FM Threshold Effect

The formula of Eq. (5.50) defining the output signal-to-noise ratio of an FM receiver, is valid only if the carrier-to-noise ratio, measured at the discriminator input, is high compared with unity. It is found experimentally that as the input noise power is increased so that the carrier-to-noise ratio is decreased, the FM receiver *breaks*. At first, individual clicks are heard in the receiver output, and as the carrier-to-noise ratio decreases still further, the clicks rapidly merge into a *crackling* or *sputtering sound*. Near the breaking point, Eq. (5.50) begins to fail by predicting values of output signal-to-noise ratio larger than the actual ones. This phenomenon is known as the *threshold effect*.[3] The threshold is defined as the minimum carrier-to-noise ratio yielding an FM improvement that is not significantly deteriorated from the value predicted by the usual signal-to-noise formula assuming a small noise power.

For a qualitative discussion of the FM threshold effect, consider first the case when there is no signal present, so that the carrier wave is unmodulated. Then the composite signal at the frequency discriminator input is

$$x(t) = [A_c + n_I(t)]\cos(2\pi f_c t) - n_Q(t)\sin(2\pi f_c t) \qquad (5.54)$$

where $n_I(t)$ and $n_Q(t)$ are the in-phase and quadrature components of the narrow-band noise $n(t)$ with respect to the carrier wave. The phasor diagram of Fig. 5.11 displays the phase relations between the various components of $x(t)$ in Eq. (5.54). As the amplitudes and phases of $n_I(t)$ and $n_Q(t)$ change with time in a random manner, the point P_1 [the tip of the phasor representing $x(t)$] wanders around the point P_2 (the tip of the phasor representing the carrier). When the carrier-to-noise ratio is large, $n_I(t)$ and $n_Q(t)$ are usually much smaller than the carrier amplitude A_c, and so the wandering point P_1 in Fig. 5.11 spends most of

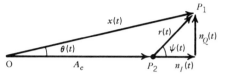

Figure 5.11 A phasor diagram interpretation of Eq. (5.54).

its time near point P_2. Thus the angle $\theta(t)$ is approximately $n_Q(t)/A_c$ to within a multiple of 2π. When the carrier-to-noise ratio is low, on the other hand, the wandering point P_1 occasionally sweeps around the origin and $\theta(t)$ increases or decreases by 2π radians. Figure 5.12 illustrates how in a rough way the excursions in $\theta(t)$, depicted in Fig. 5.12a, produce impulselike components in $\theta'(t) = d\theta/dt$. The discriminator output $v(t)$ is equal to $\theta'(t)/2\pi$. These impulselike components have different heights depending on how close the wandering point P_1 comes to the origin O, but all have areas nearly equal to $\pm 2\pi$ radians, as illustrated in Fig. 5.12b. When the signal shown in Fig. 5.12b is passed through the postdetection low-pass filter, corresponding but wider impulselike components are excited in the receiver output and are heard as clicks. The clicks are produced only when $\theta(t)$ changes by $\pm 2\pi$ radians.

From the phasor diagram of Fig. 5.11, we may deduce the conditions re-

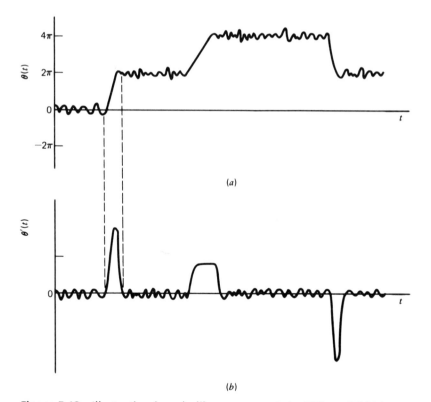

Figure 5.12 Illustrating impulselike components in $\theta'(t) = d\theta(t)/dt$ produced by changes of 2π in $\theta(t)$; (a) and (b) are graphs of $\theta(t)$ and $\theta'(t)$, respectively.

quired for clicks to occur. A positive-going click occurs when the envelope $r(t)$ and phase $\psi(t)$ of the narrow-band noise $n(t)$ satisfy the following conditions:

$$r(t) > A_c$$

$$\psi(t) < \pi < \psi(t) + d\psi(t)$$

$$\frac{d\psi(t)}{dt} > 0$$

These conditions ensure that the phase $\theta(t)$ of the resultant phasor $x(t)$ changes by 2π radians in the time increment dt, during which the phase of the narrow-band noise increases by the incremental amount $d\psi(t)$. Similarly, the conditions for a negative-going click to occur are as follows:

$$r(t) > A_c$$

$$\psi(t) > -\pi > \psi(t) + d\psi(t)$$

$$\frac{d\psi(t)}{dt} < 0$$

These conditions ensure that $\theta(t)$ changes by -2π radians during the time increment dt.

The *carrier-to-noise ratio* is defined by

$$\rho = \frac{A_c^2}{2B_T N_0} \tag{5.55}$$

As ρ is decreased, the average number of clicks per unit time increases. When this number becomes appreciably large, the threshold is said to occur.

The output signal-to-noise ratio is calculated as follows:

1. The output signal is taken as the receiver output measured in the absence of noise. The average output signal power is calculated assuming a sinusoidal modulation that produces a frequency deviation Δf equal to $B_T/2$, so that the carrier swings back and forth across the entire input frequency band.

2. The average output noise power is calculated when there is no signal present; that is, the carrier is unmodulated, with no restriction imposed on the value of the carrier-to-noise ratio ρ.

On this basis, theory[4] yields Curve I of Fig. 5.13 presenting a plot of the output signal-to-noise ratio versus the carrier-to-noise ratio when the ratio $B_T/2W$ is equal to 5. This curve shows that the output signal-to-noise ratio deviates appreciably from a linear function of the carrier-to-noise ratio ρ when ρ is less than about 10 dB. Curve II of Fig. 5.13 shows the effect of modulation on the output signal-to-noise ratio when the modulating signal (assumed sinusoidal) and the noise are present at the same time. The average output signal power pertaining to curve II may be taken to be effectively the same as for curve I. The average output noise power, however, is strongly dependent on the presence of the modulating signal, which accounts for the noticeable deviation of curve II from curve I. In particular, we find that as ρ decreases from infinity, the output signal-

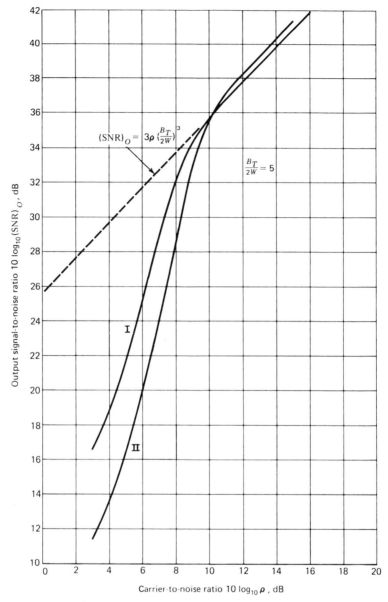

Figure 5.13 Dependence of output signal-to-noise ratio on input carrier-to-noise ratio. In curve I, the average output noise power is calculated assuming an unmodulated carrier. In curve II, the average output noise power is calculated assuming a sinusoidally modulated carrier. Both curves I and II are calculated from theory.

to-noise ratio deviates appreciably from a linear function of ρ when ρ is about 11 dB. Also when the signal is present, the resulting modulation of the carrier tends to increase the average number of clicks per second. Experimentally, it is found that occasional clicks are heard in the receiver output at a carrier-to-noise ratio of about 13 dB, which appears to be only slightly higher than what theory indicates. Also it is of interest to note that the enhanced increase in the average number of clicks per second tends to cause the output signal-to-noise ratio to fall off somewhat more sharply just below the threshold level in the presence of modulation.

From the foregoing discussion we may conclude that threshold effects in FM receivers may be avoided in most practical cases of interest if the carrier-to-noise ratio ρ is equal to or greater than 20 or, equivalently, 13 dB. Thus, using Eq. (5.55) we find that the loss of message at the discriminator output is negligible if

$$\frac{A_c^2}{2B_T N_0} \geq 20$$

or, equivalently, if the average transmitted power $A_c^2/2$ satisfies the condition

$$\frac{A_c^2}{2} \geq 20 B_T N_0 \tag{5.56}$$

To use this formula, we may proceed as follows:

1. For a specified modulation index β and message bandwidth W, we determine the transmission bandwidth of the FM wave, B_T, using the universal curve of Fig. 3.36 or Carson's rule.
2. For a specified average noise power per unit bandwidth, N_0, we use Eq. (5.56) to determine the minimum value of the average transmitted power $A_c^2/2$ that is necessary to operate above threshold.

FM Threshold Reduction

In certain applications such as space communications using frequency modulation, there is particular interest in reducing the noise threshold in an FM receiver so as to satisfactorily operate the receiver with the minimum signal power possible. Threshold reduction in FM receivers may be achieved by using an FM demodulator with negative feedback[5] (commonly referred to as an *FMFB demodulator*), or by using a *phase-locked loop demodulator*. Such devices are referred to as *extended-threshold demodulators*, the idea of which is illustrated in Fig. 5.14. The threshold extension shown in this figure is measured with respect to the standard frequency discriminator (i.e., one without feedback).

The block diagram of an FMFB demodulator[6] is shown in Fig. 5.15. We see that the local oscillator of the conventional FM receiver has been replaced by a voltage-controlled oscillator (VCO) whose instantaneous output frequency is controlled by the demodulated signal. In order to understand the operation of this receiver, suppose for the moment that the VCO is removed from the circuit and the feedback loop is left open. Assume that a wide-band FM signal is applied to the receiver input, and a second FM signal, from the same source but whose modulation index is a fraction smaller, is applied to the VCO terminal of the mixer. The output of the mixer would consist of the difference frequency component, because the sum frequency component is removed by the band-pass filter. The frequency deviation of the mixer output would be small, although the frequency deviation of both input FM waves is large, since the difference between their instantaneous deviations is small. Hence, the modulation indices would

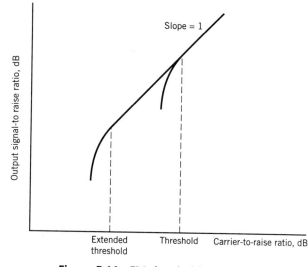

Figure 5.14 FM threshold extension.

subtract and the resulting FM wave at the mixer output would have a smaller modulation index. The FM wave with reduced modulation index may be passed through a band-pass filter, whose bandwidth need only be a fraction of that required for either wide-band FM, and then frequency demodulated. It is now apparent that the second wide-band FM signal applied to the mixer may be obtained by feeding the output of the frequency discriminator back to the VCO.

It will now be shown that the signal-to-noise ratio of an FMFB receiver is the same as that of a conventional FM receiver with the same input signal and noise power if the carrier-to-noise ratio is sufficiently large. Assume for the moment that there is no feedback around the demodulator. In the combined presence of an unmodulated carrier $A_c \cos(2\pi f_c t)$ and narrow-band noise

$$n(t) = n_I(t)\cos(2\pi f_c t) - n_Q(t)\sin(2\pi f_c t)$$

the phase of the composite signal $x(t)$ at the limiter-discriminator input is approximately equal to $n_Q(t)/A_c$, assuming that the carrier-to-noise ratio is high. The envelope of $x(t)$ is of no interest to us, because the limiter removes all variations in the envelope. Thus the composite signal at the frequency discriminator input consists of a small index phase-modulated wave with the modulation

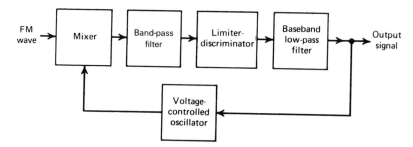

Figure 5.15 FMFB demodulator.

derived from the component $n_Q(t)$ of noise that is in phase quadrature with the carrier. When feedback is applied, the VCO generates a frequency-modulated signal that reduces the phase-modulation index of the wave in the band-pass filter output, that is, the quadrature component $n_Q(t)$ of noise. Thus we see that as long as the carrier-to-noise ratio is sufficiently large, the FMFB receiver does not respond to the in-phase noise component $n_I(t)$, but that it would demodulate the quadrature noise component $n_Q(t)$ in exactly the same fashion as it would demodulate signal modulation. Signal and quadrature noise are reduced in the same proportion by the applied feedback, with the result that the baseband signal-to-noise ratio is independent of feedback. For large carrier-to-noise ratios, the baseband signal-to-noise ratio of an FMFB receiver is then the same as that of a conventional FM receiver.

The reason that an FMFB receiver is able to extend the threshold is that, unlike a conventional FM receiver, it uses a very important piece of *a priori* information, namely, that even though the carrier frequency of the incoming FM wave will usually have large frequency deviations, its rate of change will be at the baseband rate. An FMFB demodulator is essentially a *tracking filter* that can track only the slowly varying frequency of a wide-band FM signal, and consequently it responds only to a narrow band of noise centered about the instantaneous carrier frequency. The bandwidth of noise to which the FMFB receiver responds is precisely the band of noise that the VCO tracks. The end result is that an FMFB receiver is capable of realizing a threshold extension on the order of 5–7 dB, which represents a significant improvement in the design of minimum power FM systems.

Like the FMFB demodulator, the phase-locked loop (discussed in Chapter 3) is also a tracking filter and, as such, the noise bandwidth to which it responds is precisely the band of noise tracked by the VCO. Indeed, the phase-locked loop demodulator[7] offers a threshold extension capability with a relatively simple circuit. Unfortunately, the amount of threshold extension is not predictable by any existing theory, and it depends on signal parameters. Roughly speaking, improvement by a few (on the order of 2 to 3) decibels is achieved in typical applications, which is not as good as an FMFB demodulator.

5.7 PRE-EMPHASIS AND DE-EMPHASIS IN FM

In Section 5.6 we showed that the power spectral density of the noise at the output of an FM receiver has a square-law dependence on the operating frequency; this is illustrated in Fig. 5.16a. In Fig. 5.16b, we have included the power spectral density of a typical message source; audio and video signals typically have spectra of this form. In particular, we see that the power spectral density of the message usually falls off appreciably at higher frequencies. On the other hand, the power spectral density of the output noise increases rapidly with frequency. Thus, around $f = \pm W$, the relative spectral density of the message is quite low, whereas that of the output noise is quite high in comparison. Clearly, the message is not utilizing the frequency band allotted to it in an efficient manner. It may appear that one way of improving the noise performance of the system is to slightly reduce the bandwidth of the post-detection low-pass filter so as to reject a large amount of noise power while losing only a small amount of message power. Such an approach, however, is usually not satisfactory because

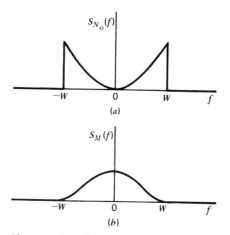

Figure 5.16 (a) Power spectral density of noise at FM receiver output. (b) Power spectral density of a typical message signal.

the distortion of the message caused by the reduced filter bandwidth, even though slight, may not be tolerable. For example, in the case of music, we find that although the high-frequency notes contribute only a very small fraction of the total power, nonetheless, they contribute a great deal from an esthetic viewpoint.

A more satisfactory approach to the efficient utilization of the allowed frequency band is based on the use of *pre-emphasis* in the transmitter and *de-emphasis* in the receiver, as illustrated in Fig. 5.17. In this method, we artificially emphasize the high-frequency components of the message signal prior to modulation in the transmitter, and therefore before the noise is introduced in the receiver. In effect, the low-frequency and high-frequency portions of the power spectral density of the message are equalized in such a way that the message fully occupies the frequency band allotted to it. Then, at the discriminator output in the receiver, we perform the inverse operation by de-emphasizing the high-frequency components, so as to restore the original signal-power distribution of the message. In such a process, the high-frequency components of the noise at the discriminator output are also reduced, thereby effectively increasing the output signal-to-noise ratio of the system. Such a pre-emphasis and de-emphasis process is widely used in commercial FM radio transmission and reception.

In order to produce an undistorted version of the original message at the receiver output, the pre-emphasis filter in the transmitter and the de-emphasis filter in the receiver must ideally have transfer functions that are the inverse of

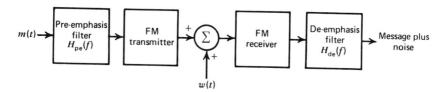

Figure 5.17   Use of pre-emphasis and de-emphasis in an FM system.

each other. That is, if $H_{\text{pe}}(f)$ designates the transfer function of the pre-emphasis filter, then the transfer function $H_{\text{de}}(f)$ of the de-emphasis filter must ideally be (ignoring transmission delay)

$$H_{\text{de}}(f) = \frac{1}{H_{\text{pe}}(f)}, \qquad -W \le f \le W \tag{5.57}$$

This choice of transfer functions makes the average message power at the receiver output independent of the pre-emphasis and de-emphasis procedure.

From our previous noise analysis in FM systems, assuming a high carrier-to-noise ratio, the power spectral density of the noise $n_d(t)$ at the discriminator output is

$$S_{N_d}(f) = \begin{cases} \dfrac{N_0 f^2}{A_c^2}, & |f| \le \dfrac{B_T}{2} \\[2mm] 0, & \text{otherwise} \end{cases} \tag{5.58}$$

Therefore, the modified power spectral density of the noise at the de-emphasis filter output is equal to $|H_{\text{de}}(f)|^2 S_{N_d}(f)$. Recognizing, as before, that the post-detection low-pass filter has a bandwidth W that is, in general, less than $B_T/2$, we find that the average power of the modified noise at the receiver output is as follows:

$$\begin{pmatrix} \text{Average output noise} \\ \text{power with de-emphasis} \end{pmatrix} = \frac{N_0}{A_c^2} \int_{-W}^{W} f^2 |H_{\text{de}}(f)|^2 \, df \tag{5.59}$$

Because the average message power at the receiver output is ideally unaffected by the pre-emphasis and de-emphasis procedure, it follows that the improvement in output signal-to-noise ratio produced by the use of pre-emphasis in the transmitter and de-emphasis in the receiver is defined by

$$I = \frac{\text{average output noise power without pre-emphasis and de-emphasis}}{\text{average output noise power with pre-emphasis and de-emphasis}} \tag{5.60}$$

Earlier we showed that the average output noise power without pre-emphasis and de-emphasis is equal to $(2N_0 W^3/3A_c^2)$. Therefore, after cancellation of common terms, we may express the improvement factor I as

$$I = \frac{2W^3}{3 \displaystyle\int_{-W}^{W} f^2 |H_{\text{de}}(f)|^2 \, df} \tag{5.61}$$

It must be emphasized that this improvement factor assumes the use of a high carrier-to-noise ratio at the discriminator input in the receiver.

EXAMPLE 4

A simple pre-emphasis filter that emphasizes high frequencies and is commonly used in practice is defined by the transfer function

$$H_{\text{pe}}(f) = 1 + \frac{jf}{f_0}$$

This transfer function is closely realized by the RC-amplifier network shown in Fig. 5.18a, provided that $R \ll r$ and $2\pi fCr \ll 1$ inside the frequency band of interest. The amplifier in Fig. 5.18a is intended to make up for the attenuation introduced by the RC network at low frequencies. The frequency parameter f_0 is $1/(2\pi Cr)$.

The corresponding de-emphasis filter in the receiver is defined by the transfer function

$$H_{\text{de}}(f) = \frac{1}{1 + jf/f_0}$$

which can be realized using the simple RC network of Fig. 5.18b.

The improvement in output signal-to-noise ratio of the FM receiver, resulting from use of the pre-emphasis and de-emphasis filters of Fig. 5.18, is therefore

$$I = \frac{2W^3}{3\displaystyle\int_{-W}^{W} \frac{f^2\,df}{1 + (f/f_0)^2}}$$

$$= \frac{(W/f_0)^3}{3[(W/f_0) - \tan^{-1}(W/f_0)]}$$

(5.62)

In commercial FM broadcasting, we typically have $f_0 = 2.1$ kHz, and we may reasonably assume $W = 15$ kHz. This set of values yields $I = 22$, which corresponds to an improvement of 13 dB in the output signal-to-noise ratio of the receiver. The output signal-to-noise ratio of an FM receiver without pre-emhasis and de-emphasis is typically 40–50 dB. We see, therefore, that by using the simple pre-emphasis and de-emphasis filters shown in Fig. 5.18, we can realize a significant improvement in the noise performance of the receiver.

The use of the simple *linear* pre-emphasis and de-emphasis filters described above is an example of how the performance of an FM system may be improved

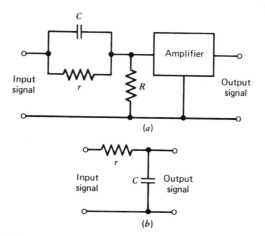

Figure 5.18 (a) Pre-emphasis filter. (b) De-emphasis filter.

by utilizing the differences between characteristics of signals and noise in the system. These simple filters also find application in audio tape-recording. Specifically, *nonlinear* pre-emphasis and de-emphasis techniques have been applied successfully to tape-recording. These techniques[8] (known as *Dolby-A*, *Dolby-B*, and *DBX* systems) use a combination of filtering and dynamic range compression to reduce the effects of noise, particularly when the signal level is low.

5.8   SUMMARY AND DISCUSSION

We conclude the noise analysis of CW modulation systems by presenting a comparison of the relative merits of the different modulation techniques. For this comparison, we assume that the modulation is produced by a sinusoidal wave. For the comparison to be meaningful, we also assume that all the different modulation systems operate with exactly the same channel signal-to-noise ratio. In making the comparison, it is informative to keep in mind the transmission bandwidth requirement of the modulation system in question. In this regard, we use a *normalized transmission bandwidth* defined by

$$B_n = \frac{B_T}{W} \tag{5.63}$$

where B_T is the transmission bandwidth of the modulated signal and W is the message bandwidth. We may thus make the following observations:

1. In a full AM system using envelope detection, the output signal-to-noise ratio, assuming sinusoidal modulation, is given by [see Eq. (5.24)]

$$(\text{SNR})_O = \frac{\mu^2}{2 + \mu^2} (\text{SNR})_C$$

    This relation is shown plotted as curve I in Fig. 5.19, assuming $\mu = 1$. In this curve we have also included the AM threshold effect. Since in a full AM system both sidebands are transmitted, the normalized transmission bandwidth B_n equals 2.

2. In the case of a DSB-SC or SSB modulation system using coherent detection, the output signal-to-noise ratio is given by [see Eqs. (5.10) and (5.16)]

$$(\text{SNR})_O = (\text{SNR})_C$$

    This relation is shown plotted as curve II in Fig. 5.19. We see, therefore, that the noise performance of a DSB-SC or SSB system, using coherent detection, is superior to that of a full AM system using envelope detection by 4.8 dB. It should also be noted that neither the DSB-SC nor the SSB system exhibits a threshold effect. With regard to transmission bandwidth requirement, we have $B_n = 2$ for the DSB-SC system and $B_n = 1$ for the SSB system. Thus, among the family of AM systems, SSB modulation is optimum with regard to noise performance as well as bandwidth conservation.

3. In an FM system using a conventional discriminator, the output signal-to-noise ratio, assuming sinusoidal modulation, is given by [see Eq. (5.53)]

$$(\text{SNR})_O = \tfrac{3}{2}\beta^2(\text{SNR})_C$$

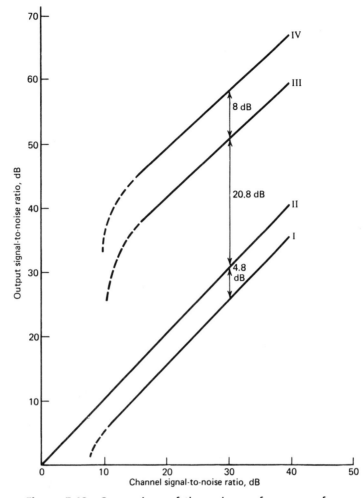

Figure 5.19 Comparison of the noise performance of various CW modulation systems. Curve I: Full AM, $\mu = 1$. Curve II: DSB-SC, SSB. Curve III: FM, $\beta = 2$. Curve IV: FM, $\beta = 5$. (Curves III and IV include 13-dB pre-emphasis, de-emphasis improvement.)

where β is the modulation index. This relation is shown plotted as curves III and IV in Fig. 5.19, corresponding to $\beta = 2$ and $\beta = 5$, respectively. In each case, we have included a 13-dB improvement, which is typically obtained by using pre-emphasis in the transmitter and de-emphasis in the receiver as described in Section 5.7. To determine the transmission bandwidth requirement, we use the universal curve of Fig. 3.36 and so find that

$$B_n = 8 \qquad \text{for } \beta = 2$$
$$B_n = 16 \qquad \text{for } \beta = 5$$

Thus we see that, compared with the SSB system, which is the optimum form of linear modulation, by using wide-band FM, we obtain an improvement in output signal-to-noise ratio equal to 20.8 dB for a normalized bandwidth

$B_n = 8$, and an improvement of 28.8 dB for $B_n = 16$. This clearly illustrates
the improvement in noise performance that is achievable by using wide-band
FM. However, the price that we have to pay for this improvement is excessive
transmission bandwidth. It is, of course, assumed that the FM system operates
above threshold for the noise improvement to be realizable.

An important point to conclude the discussion is that, unlike amplitude
modulation, frequency modulation offers the capability to trade off transmission
bandwidth for improved noise performance. The trade-off follows a square law,
which is the best that we can do with CW modulation (i.e., analog communications). In the next chapter we describe pulse-code modulation, which is basic to
the transmission of analog information-bearing signals by a digital communication system, and which can indeed do better.

NOTES AND REFERENCES

1. For a detailed analysis of the threshold effect in AM receivers see Middleton (1960,
   pp. 563–574). For a qualitative discussion of this effect see Downing (1964, p. 71).
2. For a justification of the critical assumption on which the simplification presented in
   Eq. (5.43) rests, see Rice (1963).
3. For a detailed discussion of the threshold effect in FM receivers, see Rice (1963) and
   Schwartz, Bennett, and Stein (1966, pp. 129–163).
4. Figure 5.13 is adapted from Rice (1963). The validity of the theoretical curve II in
   this figure has been confirmed experimentally; see Schwartz, Bennett, and Stein
   (1966, p. 153). For some earlier experimental work on the threshold phenomenon
   in FM, see Crosby (1937).
5. The idea of using feedback around an FM demodulator was originally proposed by
   Chaffee (1939), long before the advent of space communications.
6. The treatment of the FMFB demodulator presented in Section 5.6 is based on the
   paper by Enloe (1962); see also Roberts (1977, pp. 166–181).
7. For a full discussion of threshold effects in phase-locked loops, see Gardner (1979,
   pp. 178–196) and Roberts (1977, pp. 200–202).
8. For a detailed description of Dolby systems mentioned in the latter part of Section
   5.7, see Stremler (1990, pp. 732–734).

PROBLEMS

Problem 5.1 The sample function

$$x(t) = A_c \cos(2\pi f_c t) + w(t)$$

is applied to the low-pass RC filter of Fig. P5.1. The amplitude A_c and frequency
f_c of the sinusoidal component are constants, and $w(t)$ is a white Gaussian noise
of zero mean and power spectral density $N_0/2$. Find an expression for the output
signal-to-noise ratio with the sinusoidal component of $x(t)$ regarded as the signal
of interest.

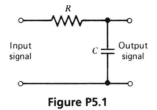

Figure P5.1

Problem 5.2 Suppose next the sample function $x(t)$ of Problem 5.1 is applied to the band-pass LCR filter of Fig. P5.2, which is tuned to the frequency f_c of the sinusoidal component. Assume that the Q-factor of the filter is high compared with unity. Find an expression for the output signal-to-noise ratio by treating the sinusoidal component of $x(t)$ as the signal of interest.

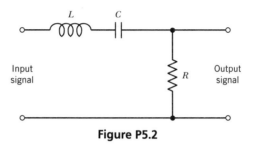

Figure P5.2

Problem 5.3 A DSB-SC modulated signal is transmitted over a noisy channel, with the power spectral density of the noise being as shown in Fig. P5.3. The message bandwidth is 4 kHz and the carrier frequency is 200 kHz. Assuming that the average power of the modulated wave is 10 watts, determine the output signal-to-noise ratio of the receiver.

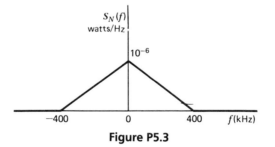

Figure P5.3

Problem 5.4 Evaluate the autocorrelation functions and cross-correlation functions of the in-phase and quadrature components of the narrow-band noise at the coherent detector input for (a) the DSB-SC system, (b) an SSB system using the lower sideband, and (c) an SSB system using the upper sideband.

Problem 5.5 In a receiver using coherent detection, the sinusoidal wave generated by the local oscillator suffers from a phase error $\theta(t)$ with respect to the carrier wave $\cos(2\pi f_c t)$. Assuming that $\theta(t)$ is a sample function of a zero-mean

Gaussian process of variance σ_Θ^2, and that most of the time the maximum value of $\theta(t)$ is small compared with unity, find the mean-square error of the receiver output for (a) DSB-SC modulation and (b) SSB modulation. The mean-square error is defined as the expected value of the squared difference between the receiver output and the message signal component of the receiver output.

Problem 5.6   Let a message signal $m(t)$ be transmitted using single-sideband modulation. The power spectral density of $m(t)$ is

$$S_M(f) = \begin{cases} a\dfrac{|f|}{W}, & |f| \leq W \\ 0, & \text{otherwise} \end{cases}$$

where a and W are constants. White Gaussian noise of zero mean and power spectral density $N_0/2$ is added to the SSB modulated wave at the receiver input. Find an expression for the output signal-to-noise ratio of the receiver.

Problem 5.7   The average noise power per unit bandwidth measured at the front end of an AM receiver is 10^{-3} watt per Hertz. The modulating wave is sinusoidal, with a carrier power of 80 kilowatts, and a sideband power of 10 kilowatts per sideband. The message bandwidth is 4 kHz. Assuming the use of an envelope detector in the receiver, determine the output signal-to-noise ratio of the system. By how many decibels is this system inferior to a DSB-SC modulation system?

Problem 5.8   Consider the output of an envelope detector defined by Eq. (5.20), which is reproduced here for convenience

$$y(t) = \{[A_c + A_c k_a m(t) + n_I(t)]^2 + n_Q^2(t)\}^{1/2}$$

(a)   Assume that the probability of the event

$$|n_Q(t)| > \varepsilon A_c |1 + k_a m(t)|$$

is equal to or less than δ_1, where $\varepsilon \ll 1$. What is the probability that the effect of the quadrature component $n_Q(t)$ is negligible?

(b)   Suppose that k_a is adjusted relative to the message signal $m(t)$ such that the probability of the event

$$A_c[1 + k_a m(t)] + n_I(t) < 0$$

is equal to δ_2. What is the probability that the approximation

$$y(t) \simeq A_c[1 + k_a m(t)] + n_I(t)$$

is valid?

(c)   Comment on the significance of the result in part (b) for the case when δ_1 and δ_2 are both small compared with unity.

Problem 5.9   An unmodulated carrier of amplitude A_c and frequency f_c and band-limited white noise are summed and then passed through an ideal envelope detector. Assume the noise spectral density to be of height $N_0/2$ and bandwidth $2W$, centered about the carrier frequency f_c. Determine the output signal-to-noise ratio for the case when the carrier-to-noise ratio is high.

Problem 5.10   An AM receiver, operating with a sinusoidal modulating signal and 80 percent modulation, has an output signal-to-noise ratio of 30 dB.

(a) What is the corresponding carrier-to-noise ratio?

(b) By how many decibels can we decrease the carrier-to-noise ratio so that the system is operating just above threshold?

Problem 5.11   Evaluate the output signal-to-noise ratio of a vestigial sideband system, the receiver of which uses coherent detection. The additive noise at the detector input is narrow band.

Problem 5.12   Consider a phase modulation (PM) system, with the modulated wave defined by

$$s(t) = A_c \cos[2\pi f_c t + k_p m(t)]$$

where k_p is a constant and $m(t)$ is the message signal. The additive noise $n(t)$ at the phase detector input is

$$n(t) = n_I(t)\cos(2\pi f_c t) - n_Q(t)\sin(2\pi f_c t)$$

Assuming that the carrier-to-noise ratio at the detector input is high compared with unity, determine (a) the output signal-to-noise ratio and (b) the figure of merit of the system. Compare your results with the FM system for the case of sinusoidal modulation.

Problem 5.13   An FDM system uses single-sideband modulation to combine 12 independent voice signals and then uses frequency modulation to transmit the composite baseband signal. Each voice signal has an average power P and occupies the frequency band 0.3–3.4 kHz; the system allocates it a bandwidth of 4 kHz. For each voice signal, only the lower sideband is transmitted. The subcarrier waves used for the first stage of modulation are defined by

$$c_k(t) = A_k \cos(2\pi k f_0 t), \qquad 0 \le k \le 11$$

The received signal consists of the transmitted FM signal plus white Gaussian noise of zero mean and power spectral density $N_0/2$.

(a) Sketch the power spectral density of the signal produced at the frequency discriminator output, showing both the signal and noise components.

(b) Find the relationship between the subcarrier amplitudes A_k so that the modulated voice signals have equal signal-to-noise ratios.

Problem 5.14   In the discussion on FM threshold effect presented in Section 5.6, we described the conditions for positive-going and negative-going clicks in terms of the envelope $r(t)$ and phase $\psi(t)$ of the narrow-band noise $n(t)$. Reformulate these conditions in terms of the in-phase component $n_I(t)$ and quadrature component $n_Q(t)$ of $n(t)$.

Problem 5.15   By using the pre-emphasis filter shown in Fig. 5.18a and with a voice signal as the modulating wave, an FM transmitter produces a signal that is essentially frequency-modulated by the lower audio frequencies and phase-modulated by the higher audio frequencies. Explain the reasons for this phenomenon.

Problem 5.16   Suppose that the transfer functions of the pre-emphasis and de-emphasis filters of an FM system are scaled as follows:

$$H_{\mathrm{pe}}(f) = k\left(1 + \frac{jf}{f_0}\right)$$

and

$$H_{de}(f) = \frac{1}{k}\left(\frac{1}{1 + jf/f_0}\right)$$

The scaling factor k is to be chosen so that the average power of the emphasized message signal is the same as that of the original message signal $m(t)$.

(a) Find the value of k that satisfies this requirement for the case when the power spectral density of the message signal $m(t)$ is

$$S_M(f) = \begin{cases} \dfrac{S_0}{1 + (f/f_0)^2}, & -W \le f \le W \\ 0, & \text{elsewhere} \end{cases}$$

(b) What is the corresponding value of the improvement factor I produced by using this pair of pre-emphasis and de-emphasis filters? Compare this ratio with that obtained in Example 4. The improvement factor I is defined by Eq. (5.60).

Problem 5.17 A phase modulation (PM) system uses a pair of pre-emphasis and de-emphasis filters defined by the transfer functions

$$H_{pe}(f) = 1 + \frac{jf}{f_0}$$

and

$$H_{de}(f) = \frac{1}{1 + (jf/f_0)}$$

Show that the improvement in output signal-to-noise ratio produced by using this pair of filters is

$$I = \frac{W/f_0}{\tan^{-1}(W/f_0)}$$

where W is the message bandwidth. Evaluate this improvement for the case when $W = 15$ kHz and $f_0 = 2.1$ kHz, and compare your result with the corresponding value for an FM system.

CHAPTER 6

Pulse
Modulation

6.1 INTRODUCTION

In *continuous-wave (CW) modulation,* which we studied in Chapters 3 and 5, some parameter of a sinusoidal carrier wave is varied continuously in accordance with the message signal. This is in direct contrast to pulse modulation, which we study in the present chapter. In *pulse modulation,*[1] some parameter of a pulse train is varied in accordance with the message signal. We may distinguish two families of pulse modulation: *analog pulse modulation* and *digital pulse modulation.* In analog pulse modulation, a periodic pulse train is used as the carrier wave, and some characteristic feature of each pulse (e.g., amplitude, duration, or position) is varied in a continuous manner in accordance with the corresponding *sample* value of the message signal. Thus, in analog pulse modulation, information is transmitted basically in analog form, but the transmission takes place at discrete times. In digital pulse modulation, on the other hand, the message signal is represented in a form that is discrete in both time and amplitude, thereby permitting its transmission in digital form as a sequence of *coded pulses;* this form of signal transmission has *no* CW counterpart.

The use of coded pulses for the transmission of analog information-bearing signals represents a basic ingredient in the application of digital communications. This chapter may therefore be viewed as a transition from analog to digital communications in our study of the principles of communication systems.

We begin the chapter by describing the sampling process, which is basic to all pulse modulation systems. This is followed by a discussion of pulse-amplitude modulation, which is the simplest form of analog pulse modulation. Next, we describe time-division multiplexing as a method for the common use of a channel by a multiplicity of message signals, which is a natural extension of pulse-amplitude modulation. We complete the discussion of analog pulse modulation by considering pulse-position modulation, which is another important method of pulse modulation.

The remaining sections of the chapter are devoted to a discussion of the different forms of digital pulse modulation, their practical virtues, limitations, and modifications.

6.2   THE SAMPLING PROCESS

Much of the material on the representation of signals and systems covered up to this stage in the book has been devoted to signals and systems that are continuous in both time and frequency. At various points in Chapter 2, however, we did consider the representation of periodic signals. In particular, referring to Eq. (2.88), we see that the Fourier transform of a periodic signal with period T_0 consists of an infinite sequence of delta functions occurring at integer multiples of the fundamental frequency $f_0 = 1/T_0$. On the basis of this observation, we may state that making a signal periodic in the time domain has the effect of sampling the spectrum of the signal in the frequency domain. We may go one step further by invoking the duality property of the Fourier transform, and thus make the observation that sampling a signal in the time domain has the effect of making the spectrum of the signal periodic in the frequency domain. This latter issue is the subject of this section.

The *sampling process* is usually described in the time domain. As such, it is an operation that is basic to digital signal processing and digital communications. Through use of the sampling process, an analog signal is converted into a corresponding sequence of samples that are usually spaced uniformly in time. Clearly, for such a procedure to have practical utility, it is necessary that we choose the sampling rate properly, so that the sequence of samples uniquely defines the original analog signal. This is the essence of the sampling theorem, which is derived in what follows.

Consider an arbitrary signal $g(t)$ of finite energy, which is specified for all time. A segment of the signal $g(t)$ is shown in Fig. 6.1a. Suppose that we sample the signal $g(t)$ instantaneously and at a uniform rate, once every T_s seconds. Consequently, we obtain an infinite sequence of samples spaced T_s seconds apart and denoted by $\{g(nT_s)\}$, where n takes on all possible integer values. We refer to T_s as the *sampling period*, and to its reciprocal $f_s = 1/T_s$ as the *sampling rate*. This ideal form of sampling is called *instantaneous sampling*.

Let $g_\delta(t)$ denote the signal obtained by individually weighting the elements of a periodic sequence of Dirac delta functions spaced T_s seconds apart by the sequence of numbers $\{g(nT_s)\}$, as shown by (see Fig. 6.1b)

$$g_\delta(t) = \sum_{n=-\infty}^{\infty} g(nT_s)\delta(t - nT_s) \qquad (6.1)$$

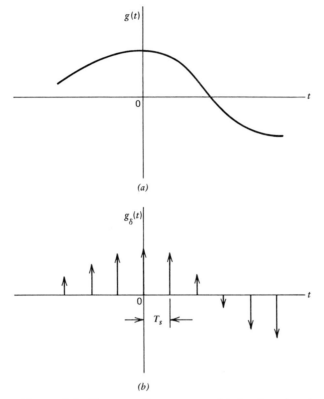

Figure 6.1   The sampling process. (*a*) Analog signal.
(*b*) Instantaneously sampled version of the signal.

We refer to $g_\delta(t)$ as the *ideal sampled signal.* The term $\delta(t - nT_s)$ represents a delta function positioned at time $t = nT_s$. From the definition of the delta function presented in Chapter 2, we recall that such an idealized function has unit area. We may therefore view the multiplying factor $g(nT_s)$ in Eq. (6.1) as a "mass" assigned to the delta function $\delta(t - nT_s)$. A delta function weighted in this manner is closely approximated by a rectanglar pulse of duration Δt and amplitude $g(nT_s)/\Delta t$; the smaller we make Δt the better will be the approximation.

The ideal sampled signal $g_\delta(t)$ has a mathematical form similar to that of the Fourier transform of a periodic signal. This is readily established by comparing Eq. (6.1) for $g_\delta(t)$ with the Fourier transform of a periodic signal given in Eq. (2.88). This correspondence suggests that we may determine the Fourier transform of the ideal sampled signal $g_\delta(t)$ by applying the duality property of the Fourier transform to the transform pair of Eq. (2.88). By so doing, and using the fact that a delta function is an even function of time, we get the desired result:

$$g_\delta(t) \rightleftharpoons f_s \sum_{m=-\infty}^{\infty} G(f - mf_s) \qquad (6.2)$$

where $G(f)$ is the Fourier transform of the original signal $g(t)$, and f_s is the sampling rate. Equation (6.2) states that *the process of uniformly sampling a contin-*

uous-time signal of finite energy results in a periodic spectrum with a period equal to the sampling rate.

Another useful expression for the Fourier transform of the ideal sampled signal $g_\delta(t)$ may be obtained by taking the Fourier transform of both sides of Eq. (6.1) and noting that the Fourier transform of the delta function $\delta(t - nT_s)$ is equal to $\exp(-j2\pi f T_s)$. Let $G_\delta(f)$ denote the Fourier transform of $g_\delta(t)$. We may therefore write

$$G_\delta(f) = \sum_{n=-\infty}^{\infty} g(nT_s)\exp(-j2\pi nf T_s) \qquad (6.3)$$

This relation is called the *discrete-time Fourier transform.*[2] It may be viewed as a complex Fourier series representation of the periodic frequency function $G_\delta(f)$, with the sequence of samples $\{g(nT_s)\}$ defining the coefficients of the expansion.

The relations, as derived here, apply to any continuous-time signal $g(t)$ of finite energy and infinite duration. Suppose, however, that the signal $g(t)$ is *strictly band-limited*, with no frequency components higher than W Hertz. That is, the Fourier transform $G(f)$ of the signal $g(t)$ has the property that $G(f)$ is zero for $|f| \geq W$, as illustrated in Fig. 6.2a; the shape of the spectrum shown in this figure is intended for the purpose of illustration only. Suppose also that we choose the sampling period $T_s = 1/2W$. Then the corresponding spectrum $G_\delta(f)$ of the sampled signal $g_\delta(t)$ is as shown in Fig. 6.2b. Putting $T_s = 1/2W$ in Eq. (6.3) yields

$$G_\delta(f) = \sum_{n=-\infty}^{\infty} g\left(\frac{n}{2W}\right)\exp\left(-\frac{j\pi nf}{W}\right) \qquad (6.4)$$

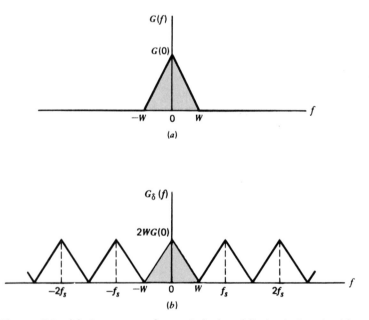

Figure 6.2 (a) Spectrum of a strictly band-limited signal g(t). (b) Spectrum of sampled version of g(t) for a sampling period $T_s = 1/2W$.

From Eq. (6.2), we readily see that the Fourier transform of $g_\delta(t)$ may also be expressed as

$$G_\delta(f) = f_s G(f) + f_s \sum_{\substack{m=-\infty \\ m \neq 0}}^{\infty} G(f - mf_s) \qquad (6.5)$$

Hence, under the following two conditions:

1. $G(f) = 0$ for $|f| \geq W$
2. $f_s = 2W$

we find from Eq. (6.5) that

$$G(f) = \frac{1}{2W} G_\delta(f), \qquad -W < f < W \qquad (6.6)$$

Substituting Eq. (6.4) in Eq. (6.6), we may also write

$$G(f) = \frac{1}{2W} \sum_{n=-\infty}^{\infty} g\left(\frac{n}{2W}\right) \exp\left(-\frac{j\pi n f}{W}\right), \qquad -W < f < W \qquad (6.7)$$

Therefore, if the sample values $g(n/2W)$ of a signal $g(t)$ are specified for all time, then the Fourier transform $G(f)$ of the signal is uniquely determined by using the discrete-time Fourier transform of Eq. (6.7). Because $g(t)$ is related to $G(f)$ by the inverse Fourier transform, it follows that the signal $g(t)$ is itself uniquely determined by the sample values $g(n/2W)$ for $-\infty < n < \infty$. In other words, the sequence $\{g(n/2W)\}$ has all the information contained in $g(t)$.

Consider next the problem of reconstructing the signal $g(t)$ from the sequence of sample values $\{g(n/2W)\}$. Substituting Eq. (6.7) in the formula for the inverse Fourier transform defining $g(t)$ in terms of $G(f)$, we get

$$g(t) = \int_{-\infty}^{\infty} G(f) \exp(j2\pi ft) \, df$$

$$= \int_{-W}^{W} \frac{1}{2W} \sum_{n=-\infty}^{\infty} g\left(\frac{n}{2W}\right) \exp\left(-\frac{j\pi n f}{W}\right) \exp(j2\pi ft) \, df$$

Interchanging the order of summation and integration:

$$g(t) = \sum_{n=-\infty}^{\infty} g\left(\frac{n}{2W}\right) \frac{1}{2W} \int_{-W}^{W} \exp\left[j2\pi f\left(t - \frac{n}{2W}\right)\right] df \qquad (6.8)$$

The integral term in Eq. (6.8) is readily evaluated, yielding the final result

$$g(t) = \sum_{n=-\infty}^{\infty} g\left(\frac{n}{2W}\right) \frac{\sin(2\pi Wt - n\pi)}{(2\pi Wt - n\pi)}$$

$$= \sum_{n=-\infty}^{\infty} g\left(\frac{n}{2W}\right) \text{sinc}(2Wt - n), \qquad -\infty < t < \infty \qquad (6.9)$$

Equation (6.9) provides an *interpolation formula* for reconstructing the original signal $g(t)$ from the sequence of sample values $\{g(n/2W)\}$, with the sinc function $\text{sinc}(2Wt)$ playing the role of an *interpolation function*. Each sample is multiplied by a delayed version of the interpolation function, and all the resulting waveforms are added to obtain $g(t)$.

We may now state the *sampling theorem* for strictly band-limited signals of finite energy in two equivalent parts:

1. *A band-limited signal of finite energy, which has no frequency components higher than W Hertz, is completely described by specifying the values of the signal at instants of time separated by 1/2W seconds.*

2. *A band-limited signal of finite energy, which has no frequency components higher than W Hertz, may be completely recovered from a knowledge of its samples taken at the rate of 2W samples per second.*

The sampling rate of $2W$ samples per second, for a signal bandwidth of W Hertz, is called the *Nyquist rate*; its reciprocal $1/2W$ (measured in seconds) is called the *Nyquist interval*.

The derivation of the sampling theorem, as described herein, is based on the assumption that the signal $g(t)$ is strictly band limited. In practice, however, an information-bearing signal is *not* strictly band limited, with the result that some degree of undersampling is encountered. Consequently, some *aliasing* is produced by the sampling process. Aliasing refers to the phenomenon of a high-frequency component in the spectrum of the signal seemingly taking on the identity of a lower frequency in the spectrum of its sampled version, as illustrated in Fig. 6.3. The aliased spectrum shown by the solid curve in Fig. 6.3b pertains to an "undersampled" version of the message signal represented by the spec-

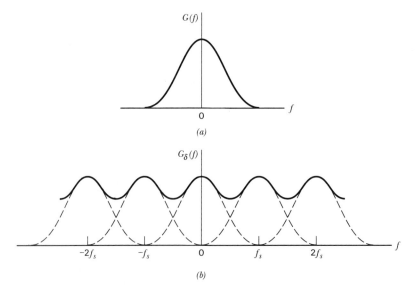

Figure 6.3 (*a*) Spectrum of a signal. (*b*) Spectrum of an undersampled version of the signal exhibiting the aliasing phenomenon.

trum of Fig. 6.3a. To combat the effects of aliasing in practice, we may use two corrective measures, as described here:

1. Prior to sampling, a low-pass *pre-alias filter* is used to attenuate those high-frequency components of the signal that are not essential to the information being conveyed by the signal.
2. The filtered signal is sampled at a rate slightly higher than the Nyquist rate.

The use of a sampling rate higher than the Nyquist rate also has the beneficial effect of easing the design of the *reconstruction filter* used to recover the original signal from its sampled version. Consider the example of a message signal that has been pre-alias (low-pass) filtered, resulting in the spectrum shown in Fig. 6.4a. The corresponding spectrum of the instantaneously sampled version of the signal is shown in Fig. 6.4b, assuming a sampling rate higher than the Nyquist rate. According to Fig. 6.4b, we readily see that the design of the reconstruction filter may be specified as follows (see Fig. 6.4c):

• The reconstruction filter is low-pass with a passband extending from $-W$ to W, which is itself determined by the pre-alias filter.
• The filter has a transition band extending (for positive frequencies) from W to $f_s - W$, where f_s is the sampling rate.

The fact that the reconstruction filter has a well-defined transition band means that it is physically realizable.

6.3 PULSE-AMPLITUDE MODULATION

Now that we understand the essence of the sampling process, we are ready to formally define pulse-amplitude modulation, which is the simplest and most basic form of analog pulse modulation. In *pulse-amplitude modulation* (PAM), *the amplitudes of regularly spaced pulses are varied in proportion to the corresponding sample values of a continuous message signal*; the pulses can be of a rectangular form or some other appropriate shape. Pulse-amplitude modulation as defined here is somewhat similar to natural sampling, where the message signal is multiplied by a periodic train of rectangular pulses. However, in natural sampling the top of each modulated rectangular pulse varies with the message signal, whereas in PAM it is maintained flat; natural sampling is explored further in Problem 6.1.

The waveform of a PAM signal is illustrated in Fig. 6.5. The dashed curve in this figure depicts the waveform of a message signal $m(t)$, and the sequence of amplitude-modulated rectangular pulses shown as solid lines represents the corresponding PAM signal $s(t)$. There are two operations involved in the generation of the PAM signal:

1. *Instantaneous sampling* of the message signal $m(t)$ every T_s seconds, where the sampling rate $f_s = 1/T_s$ is chosen in accordance with the sampling theorem.
2. *Lengthening* the duration of each sample so obtained to some constant value T.

In digital circuit technology, these two operations are jointly referred to as "sample and hold." One important reason for intentionally lengthening the duration of each sample is to avoid the use of an excessive channel bandwidth, since

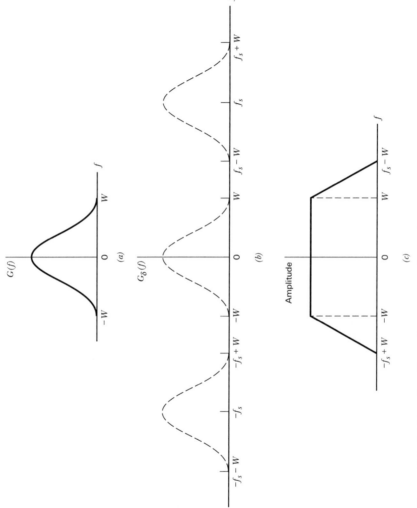

Figure 6.4 (a) Pre-alias filtered spectrum of an information-bearing signal. (b) Spectrum of instantaneously sampled version of the signal, assuming the use of a sampling rate greater than the Nyquist rate. (c) Amplitude response of reconstruction filter.

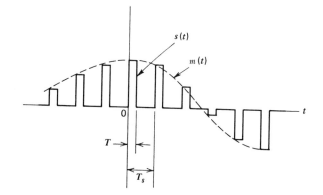

Figure 6.5 Flat-top samples.

bandwidth is inversely proportional to pulse duration. However, care has to be exercised in how long we make the sample duration T, as the following analysis reveals.

Let $s(t)$ denote the sequence of flat-top pulses generated in the manner described in Fig. 6.5. Hence, we may express the PAM signal as

$$s(t) = \sum_{n=-\infty}^{\infty} m(nT_s) h(t - nT_s) \tag{6.10}$$

where T_s is the *sampling period* and $m(nT_s)$ is the sample value of $m(t)$ obtained at time $t = nT_s$. The $h(t)$ is a standard rectangular pulse of unit amplitude and duration T, defined as follows (see Fig. 6.6a):

$$h(t) = \begin{cases} 1, & 0 < t < T \\ \frac{1}{2}, & t = 0, t = T \\ 0, & \text{otherwise} \end{cases} \tag{6.11}$$

By definition, the instantaneously sampled version of $m(t)$ is given by

$$m_\delta(t) = \sum_{n=-\infty}^{\infty} m(nT_s) \delta(t - nT_s) \tag{6.12}$$

where $\delta(t - nT_s)$ is a time-shifted delta function. Therefore, convolving $m_\delta(t)$ with the pulse $h(t)$, we get

$$m_\delta(t) \bigstar h(t) = \int_{-\infty}^{\infty} m_\delta(\tau) h(t - \tau) \, d\tau$$

$$= \int_{-\infty}^{\infty} \sum_{n=-\infty}^{\infty} m(nT_s) \delta(\tau - nT_s) h(t - \tau) \, d\tau \tag{6.13}$$

$$= \sum_{n=-\infty}^{\infty} m(nT_s) \int_{-\infty}^{\infty} \delta(\tau - nT_s) h(t - \tau) \, d\tau$$

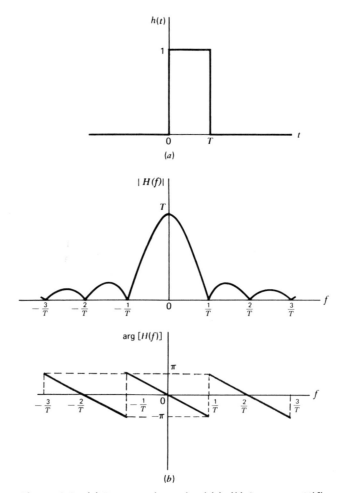

Figure 6.6   (a) Rectangular pulse $h(t)$. (b) Spectrum $H(f)$.

Using the sifting property of the delta function, we thus obtain

$$m_\delta(t) \star h(t) = \sum_{n=-\infty}^{\infty} m(nT_s) h(t - nT_s) \tag{6.14}$$

From Eqs. (6.10) and (6.14) it thus follows that the PAM signal $s(t)$ is mathematically equivalent to the convolution of $m_\delta(t)$, the instantaneously sampled version of $m(t)$, and the pulse $h(t)$, as shown by

$$s(t) = m_\delta(t) \star h(t) \tag{6.15}$$

Taking the Fourier transform of both sides of Eq. (6.15) and recognizing that the convolution of two time functions is transformed into the multiplication of their respective Fourier transforms, we get

$$S(f) = M_\delta(f) H(f) \tag{6.16}$$

where $S(f) = F[s(t)], M_\delta(f) = F[m_\delta(t)]$, and $H(f) = F[h(t)]$. From Eq. (6.2) we note that the Fourier transform $M_\delta(f)$ is related to the Fourier transform $M(f)$ of the original message signal $m(t)$ as follows:

$$M_\delta(f) = f_s \sum_{k=-\infty}^{\infty} M(f - kf_s) \qquad (6.17)$$

where $f_s = 1/T_s$ is the sampling rate. Therefore, substitution of Eq. (6.17) in (6.16) yields

$$S(f) = f_s \sum_{k=-\infty}^{\infty} M(f - kf_s) H((f) \qquad (6.18)$$

Given a PAM signal $s(t)$ whose Fourier transform $S(f)$ is as defined in Eq. (6.18), how do we recover the original message signal $m(t)$? As a first step in this reconstruction, we may pass $s(t)$ through a low-pass filter whose frequency response is defined in Fig. 6.4c; here it is assumed that the message is limited to bandwidth W and the sampling rate f_s is larger than the Nyquist rate $2W$. Then, from Eq. (6.18) we find that the spectrum of the resulting filter output is equal to $M(f)H(f)$. This output is equivalent to passing the original message signal $m(t)$ through another low-pass filter of transfer function $H(f)$.

From Eq. (6.11) we note that the Fourier transform of the rectangular pulse $h(t)$ is given by

$$H(f) = T \operatorname{sinc}(fT)\exp(-j\pi fT) \qquad (6.19)$$

which is shown plotted in Fig. 6.6b. We see therefore that by using flat-top samples to generate a PAM signal, we have introduced *amplitude distortion* as well as a *delay* of $T/2$. This effect is rather similar to the variation in transmission with frequency that is caused by the finite size of the scanning aperture in television. Accordingly, the distortion caused by the use of pulse-amplitude modulation to transmit an analog information-bearing signal is referred to as the *aperture effect*.

This distortion may be corrected by connecting an *equalizer* in cascade with the low-pass reconstruction filter, as shown in Fig. 6.7. The equalizer has the effect of decreasing the in-band loss of the reconstruction filter as the frequency increases in such a manner as to compensate for the aperture effect. Ideally, the amplitude response of the equalizer is given by

$$\frac{1}{|H(f)|} = \frac{1}{T \operatorname{sinc}(fT)} = \frac{\pi f}{\sin(\pi fT)}$$

The amount of equalization needed in practice is usually small. Indeed, for a duty cycle $T/T_s \leq 0.1$, the amplitude distortion is less than 0.5 percent, in which case the need for equalization may be omitted altogether.

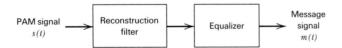

Figure 6.7 Recovering $m(t)$ from PAM signal $s(t)$.

The transmission of a PAM signal imposes rather stringent requirements on the amplitude and phase responses of the channel, because of the relatively short duration of the transmitted pulses. Furthermore, the noise performance of a PAM system can never be better than baseband-signal transmission. Accordingly, we find that for transmission over long distances, PAM would be used only as a means of message processing for time-division multiplexing, from which conversion to some other form of pulse modulation is subsequently made. The concept of time-division multiplexing is discussed in the next section.

6.4 TIME-DIVISION MULTIPLEXING

The sampling theorem provides the basis for transmitting the information contained in a band-limited message signal $m(t)$ as a sequence of samples of $m(t)$ taken uniformly at a rate that is usually slightly higher than the Nyquist rate. An important feature of the sampling process is a *conservation of time*. That is, the transmission of the message samples engages the communication channel for only a fraction of the sampling interval on a periodic basis, and in this way some of the time interval between adjacent samples is cleared for use by other independent message sources on a time-shared basis. We thereby obtain a *time-division multiplex* (TDM) *system*, which enables the joint utilization of a common communication channel by a plurality of independent message sources without mutual interference among them.

The concept of TDM is illustrated by the block diagram shown in Fig. 6.8. Each input message signal is first restricted in bandwidth by a low-pass pre-alias filter to remove the frequencies that are nonessential to an adequate signal representation. The low-pass filter outputs are then applied to a *commutator*, which is usually implemented using electronic switching circuitry. The function of the commutator is twofold: (1) to take a narrow sample of each of the N input messages at a rate f_s that is slightly higher than $2W$, where W is the cutoff frequency of the pre-alias filter, and (2) to sequentially interleave these N samples inside the sampling interval T_s. Indeed, this latter function is the essence of the time-division multiplexing operation. Following the commutation process, the multiplexed signal is applied to a *pulse modulator*, the purpose of which is to transform the multiplexed signal into a form suitable for transmission over the common channel. It is clear that the use of time-division multiplexing introduces a bandwidth expansion factor N, because the scheme must squeeze N samples derived from N independent message sources into a time slot equal to one sampling interval. At the receiving end of the system, the received signal is applied to a *pulse demodulator*, which performs the reverse operation of the pulse modulator. The narrow samples produced at the pulse demodulator output are distributed to the appropriate low-pass reconstruction filters by means of a *decommutator*, which operates in *synchronism* with the commutator in the transmitter. This synchronization is essential for a satisfactory operation of the system. The way this synchronization is implemented depends naturally on the method of pulse modulation used to transmit the multiplexed sequence of samples.

The TDM system is highly sensitive to dispersion in the common channel, that is, to variations of amplitude with frequency or lack of proportionality of phase with frequency. Accordingly, accurate equalization of both amplitude and phase responses of the channel is necessary to ensure a satisfactory operation of

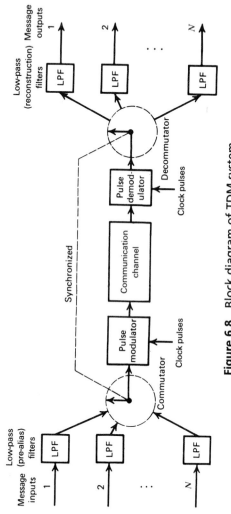

Figure 6.8 Block diagram of TDM system.

the system. This issue is discussed in detail in Chapter 7. To a first approximation, however, TDM is immune to amplitude nonlinearities in the channel as a source of cross-talk, because the different message signals are not simultaneously impressed on the channel.

6.5   PULSE-POSITION MODULATION

In a pulse modulation system, we may use the increased bandwidth consumed by pulses to obtain an improvement in noise performance by representing the sample values of the message signal by some property of the pulse other than amplitude. *In pulse-duration modulation (PDM), the samples of the message signal are used to vary the duration of the individual pulses.* This form of modulation is also referred to as *pulse-width modulation* or *pulse-length modulation.* The modulating signal may vary the time of occurrence of the leading edge, the trailing edge, or both edges of the pulse. In Fig. 6.9*c* the trailing edge of each pulse is varied in

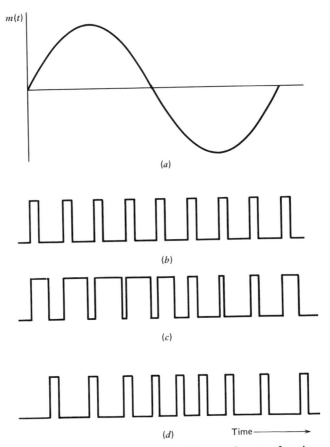

Figure 6.9 Illustrating two different forms of pulse-time modulation for the case of a sinusoidal modulating wave. (*a*) Modulating wave. (*b*) Pulse carrier. (*c*) PDM wave. (*d*) PPM wave.

accordance with the message signal, assumed to be sinusoidal as shown in Fig. 6.9a. The periodic pulse carrier is shown in Fig. 6.9b.

In PDM, long pulses expend considerable power during the pulse while bearing no additional information. If this unused power is subtracted from PDM, so that only time transitions are preserved, we obtain a more efficient type of pulse modulation known as *pulse-position modulation* (PPM). In *PPM, the position of a pulse relative to its unmodulated time of occurrence is varied in accordance with the message signal,* as illustrated in Fig. 6.9d for the case of sinusoidal modulation.

Let T_s denote the sample duration. Using the sample $m(nT_s)$ of a message signal $m(t)$ to modulate the position of the nth pulse, we obtain the PPM signal

$$s(t) = \sum_{n=-\infty}^{\infty} g(t - nT_s - k_p m(nT_s)) \tag{6.20}$$

where k_p is the *sensitivity* of the pulse-position modulator and $g(t)$ denotes a standard pulse of interest. Clearly, the different pulses constituting the PPM signal $s(t)$ must be *strictly nonoverlapping*; a sufficient condition for this requirement to be satisfied is to have

$$g(t) = 0, \qquad |t| > \frac{T_s}{2} - k_p |m(t)|_{\max} \tag{6.21}$$

which, in turn, requires that

$$k_p |m(t)|_{\max} < \frac{T_s}{2} \tag{6.22}$$

The closer $k_p |m(t)|_{\max}$ is to one half the sampling duration T_s, the narrower must the standard pulse $g(t)$ be in order to ensure that the individual pulses of the PPM signal $s(t)$ do not interfere with each other, and the wider will the bandwidth occupied by the PPM signal be. Assuming that Eq. (6.21) is satisfied, and that there is no interference between adjacent pulses of the PPM signal $s(t)$, then the signal samples $m(nT_s)$ can be recovered perfectly. Furthermore, if the message signal $m(t)$ is strictly band limited, it follows from the sampling theorem that the original message signal $m(t)$ can be recovered from the PPM signal $s(t)$ without distortion.

Generation of PPM Waves

The PPM signal described by Eq. (6.20) may be generated using the system described in Fig. 6.10. The message signal $m(t)$ is first converted into a PAM signal by means of a sample-and-hold circuit, generating a staircase waveform $u(t)$; note that the pulse duration T of the sample-and-hold circuit is the same as the sampling duration T_s. This operation is illustrated in Fig. 6.11b for the message signal $m(t)$ shown in Fig. 6.11a. Next, the signal $u(t)$ is added to a sawtooth wave (shown in Fig. 6.11c), yielding the combined signal $v(t)$ shown in Fig. 6.11d. The combined signal $v(t)$ is applied to a *threshold detector* that produces a very narrow pulse (approximating an impulse) each time $v(t)$ passes through

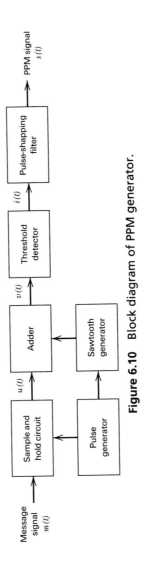

Figure 6.10 Block diagram of PPM generator.

366

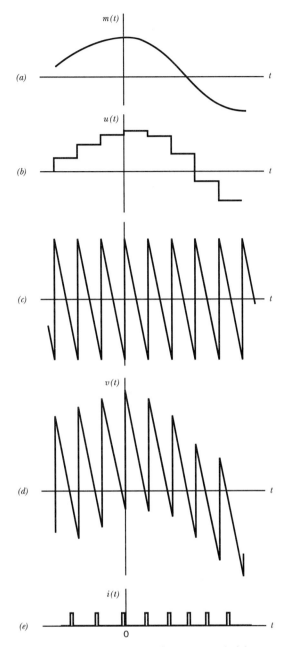

Figure 6.11 Generation of PPM signal. (a) Message signal. (b) Staircase approximation of the message signal. (c) Sawtooth wave. (d) Composite wave obtained by adding (b) and (c). (e) Sequence of "impulses" used to generate the PPM signal.

a zero-crossing in the negative-going direction. The resulting sequence of "impulses" $i(t)$ is shown in Fig. 6.11e. Finally, the PPM signal $s(t)$ is generated by using this sequence of impulses to excite a filter whose impulse response is defined by the standard pulse $g(t)$.

Detection of PPM Waves

Consider a PPM wave $s(t)$ with uniform sampling, as defined by Eqs. (6.20) and (6.21), and assume that the message (modulating) signal $m(t)$ is strictly band-limited. The operation of one type of PPM receiver may proceed as follows:

- Convert the received PPM wave into a PDM wave with the same modulation.
- Integrate this PDM wave using a device with a finite integration time, thereby computing the area under each pulse of the PDM wave.
- Sample the output of the integrator at a uniform rate to produce a PAM wave, whose pulse amplitudes are proportional to the signal samples $m(nT_s)$ of the original PPM wave $s(t)$.
- Finally, demodulate the PAM wave to recover the message signal $m(t)$.

All the operations described here are linear. In addition, a practical PPM receiver includes a nonlinear device called a *slicer* at its input end. The input–output characteristic of an ideal slicer is shown in Fig. 6.12, where the slicing level is normally set at approximately half the peak pulse amplitude of the received PPM wave. The function of the slicer is to preserve the positions of the edges of the received pulses (as modified by noise) and remove everything else. It does so by producing almost "rectangular" pulses with fairly sharp leading and trailing edges at the same instants as the corresponding edges of the received pulses. Thus, in a loose sense, the slicer acts as a "noise cleaning device" in that the final noise level at the output of the receiver is greatly reduced by eliminating all the noise in the received PPM wave except in the neighborhood of the leading and trailing edges.

The output of the slicer is differentiated and then half-wave rectified, yielding a very short pulse (approximating an impulse) each time the amplitude of a pulse in the received PPM wave passes through the slicing level. Figure 6.13a shows the nth pulse of a PPM wave, and Fig. 6.13b shows the short pulse produced (by the operations described herein) as the pulse passes through the slicing level. In Fig. 6.13c an appropriate delay is applied to the short pulse, and the corresponding PDM pulse is shown in Fig. 6.13d.

Having converted the received (noisy) PPM wave into a PDM wave with the same modulation, the receiver then proceeds to reconstruct the original baseband signal $m(t)$ in the manner described above.

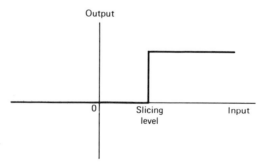

Figure 6.12 Input–output relation of slicer.

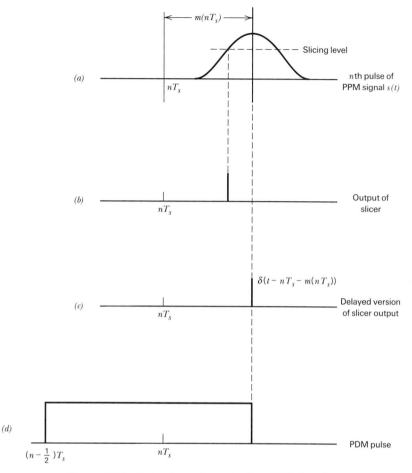

Figure 6.13 Detection of a noiseless PPM signal.

Noise in Pulse-Position Modulation

In a PPM system, the transmitted information is contained in the relative positions of the modulated pulses. The presence of additive noise affects the performance of such a system by falsifying the time at which the modulated pulses are judged to occur. Immunity to noise can be established by making the pulse build up so rapidly that the time interval during which noise can exert any perturbation is very short. Indeed, additive noise would have no effect on the pulse positions if the received pulses were perfectly rectangular, because the presence of noise introduces only vertical perturbations. However, the reception of perfectly rectangular pulses implies an infinite channel bandwidth, which is of course impractical. Thus, with a finite channel bandwidth in practice, we find that the received pulses have a finite rise time, and so the performance of the PPM receiver is affected by noise.

As with a CW modulation system, the noise performance of a PPM system may be described in terms of the output signal-to-noise ratio. Also, to find the noise improvement produced by PPM over baseband transmission of a message signal, we may use the figure of merit defined as the output signal-to-noise ratio of the PPM system divided by the channel signal-to-noise ratio. We illustrate this evaluation by considering the example of a PPM system using a raised cosine pulse and sinusoidal modulation.

EXAMPLE 1   Signal-to-Noise Ratios of a PPM System Using Sinusoidal Modulation

Consider a PPM system whose pulse train, in the absence of modulation is as shown in Fig. 6.14a. The standard pulse of the carrier is assumed to be a *raised cosine pulse*, which is a convenient type of pulse for analysis. This pulse, centered at $t = 0$ and denoted as $g(t)$, is defined by

$$g(t) = \frac{A}{2}[1 + \cos(\pi B_T t)], \quad -T \le t \le T \tag{6.23}$$

where $B_T = 1/T$. The time of occurrence of such a pulse may be determined by applying the pulse to an ideal slicer with the input–output amplitude characteristic shown in Fig. 6.12 and then observing the slicer output. We assume that the slicing level is set at half the peak pulse amplitude, namely, $A/2$, as in Fig. 6.14a. For inputs below the slicing level the output is zero, and for inputs above the slicing level the output is constant.

The Fourier transform of the pulse $g(t)$ is given by

$$G(f) = \frac{A \sin(2\pi f/B_T)}{2\pi f(1 - 4f^2/B_T^2)}$$

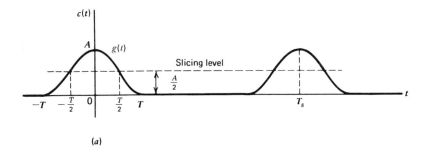

(a)

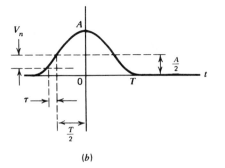

(b)

Figure 6.14 Noise analysis of PPM system. (a) Unmodulated pulse train. (b) Illustrating the effect of noise on pulse detection time.

As indicated in Fig. 6.15, this transform has its first nulls at $f = \pm B_T$ and is small outside this interval, so that the transmission bandwidth required to pass such a pulse may be taken as essentially equal to B_T.

Let the peak-to-peak swing in the position of a pulse be denoted by T_s. Then, in response to a full-load sinusoidal modulating wave, the peak-to-peak amplitude of the receiver output will be KT_s, where K is a constant determined by the receiver circuitry. The root-mean-square (rms) value of the receiver output is $KT_s/2\sqrt{2}$, and the corresponding average signal power at the receiver output (assuming a 1-ohm load) is given by

$$\left(\frac{KT_s}{2\sqrt{2}}\right)^2 = \frac{K^2 T_s^2}{8}$$

In the presence of additive noise, both amplitude and position of the pulse will be perturbed. Random variations in the pulse amplitude are removed by the slicer. Random variations in the pulse position due to noise will remain, however, thereby contributing to noise at the receiver output. We assume that, at the receiver input, the noise power is small compared with the peak pulse power. Then, if at a particular instant of time the noise amplitude is V_n, the time of pulse detection will be replaced by a small amount τ as depicted in Fig. 6.14b. To a first order of approximation, V_n/τ is equal to the slope of the pulse $g(t)$ at time $t = -T/2$. Thus, using Eq. (6.23), we get

$$\frac{V_n}{\tau} = \left.\frac{dg(t)}{dt}\right|_{t=-T/2}$$

$$= \frac{\pi B_T A}{2}$$

Solving this equation for τ, we have

$$\tau = \frac{2V_n}{\pi B_T A} \tag{6.24}$$

The error τ in the position of the pulse $g(t)$ will produce an average noise power at the receiver output equal to $K^2 E[\tau^2]$, where E is the statistical expectation operator. Assuming that the noise at the front end of the receiver has a power

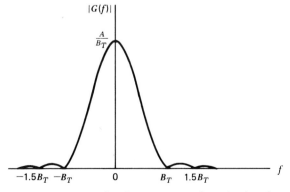

Figure 6.15 Amplitude spectrum of a raised cosine pulse.

spectral density $N_0/2$, we find that the mean-square value of V_n in a bandwidth B_T is given by

$$E[V_n^2] = B_T N_0 \tag{6.25}$$

Using Eqs. (6.24) and (6.25), we obtain the following result:

$$\text{Average power of output noise} = K^2 E[\tau^2]$$
$$= \frac{4K^2 N_0}{\pi^2 B_T A^2} \tag{6.26}$$

The output signal-to-noise ratio, assuming a full-load sinusoidal modulation, is therefore

$$(\text{SNR})_O = \frac{K^2 T_s^2/8}{4K^2 N_0/\pi^2 B_T A^2}$$
$$= \frac{\pi^2 B_T T_s^2 A^2}{32 N_0} \tag{6.27}$$

The average transmitted power P in a PPM system is independent of the applied modulation. Accordingly, we may determine P by averaging the power in a single pulse of the PPM wave over the sampling period T_s, as shown by

$$P = \frac{1}{T_s} \int_{-T_s/2}^{T_s/2} g^2(t) \, dt$$
$$= \frac{3A^2}{4T_s B_T} \tag{6.28}$$

The average noise power in a message bandwidth W is equal to WN_0. The channel signal-to-noise ratio is therefore

$$(\text{SNR})_C = \frac{3A^2/4T_s B_T}{WN_0}$$
$$= \frac{3A^2}{4T_s B_T WN_0} \tag{6.29}$$

Thus the figure of merit of a PPM system using a raised cosine pulse is as follows:

$$\text{Figure of merit} = \frac{(\text{SNR})_O}{(\text{SNR})_C}$$
$$= \frac{\pi^2}{24} B_T^2 T_s^3 W \tag{6.30}$$

Assuming that the message signal is sampled at its Nyquist rate, we have $T_s = 1/2W$. Then, we find from Eq. (6.30) that the corresponding value of the figure of merit is $(\pi^2/192)(B_T/W)^2$, which is greater than unity if $B_T > 4.41W$. We also see that the figure of merit of a PPM system is proportional to the square of the normalized transmission bandwidth B_T/W.

In the noise analysis presented here for PPM, we assumed that the average power of the additive noise at the front end of the receiver is small compared with the peak pulse power. In particular, it is assumed that there are two crossings of the slicing level for each pulse, one for the leading edge and one for the trailing edge. A Gaussian noise will have occasional peaks that produce additional crossings of the slicing level, however, and so the occasional noise peaks are mistaken for message pulses. The analysis neglects the *false pulses* produced by high noise peaks. It is apparent that these false pulses have a finite though small probability of occurrence when the noise is Gaussian, no matter how small its standard deviation is compared with the peak amplitude of the pulses. As the transmission bandwidth is increased indefinitely, the accompanying increase in average noise power eventually causes the false pulses to occur often enough, thereby causing loss of the wanted message signal at the receiver output. We thus find, in practice, that a PPM system suffers from a *threshold effect.*

6.6 BANDWIDTH–NOISE TRADE-OFF

In the context of noise performance, a PPM system represents the optimum form of analog pulse modulation. The noise analysis of a PPM system presented in the preceding example reveals that pulse-position modulation (PPM) and frequency modulation (FM) systems exhibit a similar noise performance, as summarized here:

1. Both systems have a figure of merit proportional to the square of the transmission bandwidth normalized with respect to the message bandwidth.
2. Both systems exhibit a threshold effect as the signal-to-noise ratio is reduced.

The practical implication of point 1 is that, in terms of a trade-off of increased transmission bandwidth for improved noise performance, the best that we can do with continuous-wave (CW) modulation and analog pulse modulation systems is to follow a *square law*. A question that arises at this point in the discussion is: Can we produce a trade-off better than a square law? The answer is an emphatic yes, and *digital pulse modulation* is the way to do it. The use of such a method is a radical departure from CW modulation.

Specifically, in a basic form of digital pulse modulation known as *pulse-code modulation* (PCM)[3] a message signal is represented in discrete form in both time and amplitude. This form of signal representation permits the transmission of the message signal as a sequence of *coded binary pulses*. Given such a sequence, the effect of receiver noise at the final system output can be reduced to a negligible level simply by making the average power of the transmitted binary PCM wave large enough compared to the average power of the noise.

There are two fundamental processes involved in the generation of a binary PCM wave: *sampling* and *quantization*. The sampling process takes care of the discrete-time representation of the message signal; for its proper application, we have to follow the sampling theorem described in Section 6.2. The quantization process takes care of the discrete-amplitude representation of the message signal; quantization is a new process, the details of which are described in the next section. For now it suffices to say that the combined use of sampling and quantization permits the transmission of a message signal in coded form. This, in turn, makes it possible to realize an *exponential law* for the bandwidth-noise trade-off, which is also demonstrated in the next section.

6.7   THE QUANTIZATION PROCESS[4]

A continuous signal, such as voice, has a continuous range of amplitudes and therefore its samples have a continuous amplitude range. In other words, within the finite amplitude range of the signal, we find an infinite number of amplitude levels. It is not necessary in fact to transmit the exact amplitudes of the samples. Any human sense (the ear or the eye), as ultimate receiver, can detect only finite intensity differences. This means that the original continuous signal may be *approximated* by a signal constructed of discrete amplitudes selected on a minimum error basis from an available set. The existence of a finite number of discrete amplitude levels is a basic condition of pulse-code modulation. Clearly, if we assign the discrete amplitude levels with sufficiently close spacing, we may make the approximated signal practically indistinguishable from the original continuous signal.

Amplitude *quantization* is defined as the process of transforming the sample amplitude $m(nT_s)$ of a message signal $m(t)$ at time $t = nT_s$ into a discrete amplitude $v(nT_s)$ taken from a *finite* set of possible amplitudes. In this book, we assume that the quantization process is *memoryless* and *instantaneous*, which means that the transformation at time $t = nT_s$ is not affected by earlier or later samples of the message signal. This simple form of quantization, though not optimum, is commonly used in practice.

When dealing with a memoryless quantizer, we may simplify the notation by dropping the time index. We may thus use the symbol m in place of $m(nT_s)$, as indicated in the block diagram of a quantizer shown in Fig. 6.16a. Then, as shown in Fig. 6.16b, the signal amplitude m is specified by the index k if it lies inside the interval

$$\mathcal{I}_k : \{m_k < m \le m_{k+1}\}, \qquad k = 1, 2, \ldots, L \tag{6.31}$$

where L is the total number of amplitude levels used in the quantizer. The amplitudes m_k, $k = 1, 2, \ldots, L$, are called *decision levels* or *decision thresholds*. At the quantizer output, the index k is transformed into an amplitude v_k that represents all amplitudes of the interval $\mathcal{I}_k$; the amplitudes v_k, $k = 1, 2, \ldots, L$, are called *representation levels* or *reconstruction levels*, and the spacing between two adjacent representation levels is called a *quantum* or *step-size*. Thus, the quantizer output v equals v_k if the input signal sample m belongs to the interval $\mathcal{I}_k$. The mapping (see Fig. 6.16a)

$$v = g(m) \tag{6.32}$$

is the *quantizer characteristic*, which is a staircase function by definition.

Quantizers can be of a *uniform* or *nonuniform* type. In a uniform quantizer,

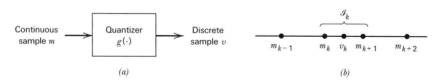

Figure 6.16   Description of a memoryless quantizer.

the representation levels are uniformly spaced; otherwise, the quantizer is non-uniform. In this section, we consider only uniform quantizers; nonuniform quantizers are considered in the next section. The quantizer characteristic can also be of *midtread* or *midrise type*. Figure 6.17*a* shows the input–output characteristic of a uniform quantizer of the midtread type, which is so called because the origin lies in the middle of a tread of the staircaselike graph. Figure 6.17*b* shows the corresponding input–output characteristic of a uniform quantizer of the midrise type, in which the origin lies in the middle of a rising part of the staircaselike graph. Note that both the midtread and midrise types of uniform quantizers illustrated in Fig. 6.17 are *symmetric* about the origin.

Quantization Noise

The use of quantization introduces an error defined as the difference between the input signal m and the output signal v. This error is called *quantization noise*. Figure 6.18 illustrates a typical variation of the quantization noise as a function of time, assuming the use of a uniform quantizer of the midtread type.

Let the quantizer input m be the sample value of a zero-mean random variable M. (If the input has a nonzero mean, we can always remove it by subtracting the mean from the input and then adding it back after quantization.) A quantizer $g(\cdot)$ maps the input random variable M of continuous amplitude into a discrete random variable V; their respective sample values m and v are related by Eq. (6.32). Let the quantization error be denoted by the random variable Q of sample value q. We may thus write

$$q = m - v \qquad (6.33)$$

or, correspondingly,

$$Q = M - V \qquad (6.34)$$

With the input M having zero mean, and the quantizer assumed to be symmetric as in Fig. 6.17, it follows that the quantizer output V and therefore the quantization error Q, will also have zero mean. Thus, for a partial statistical character-

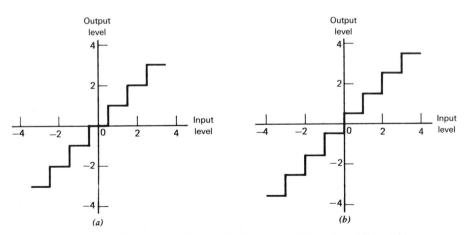

Figure 6.17 Two types of quantization: (a) midtread and (b) midrise.

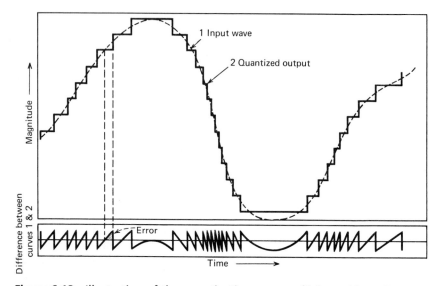

Figure 6.18 Illustration of the quantization process. (Adapted from Bennett, 1948, with permission of AT&T.)

ization of the quantizer in terms of output signal-to-noise ratio, we need to find only the mean-square value of the quantization error Q.

Consider then an input m of continuous amplitude in the range $(-m_{max}, m_{max})$. Assuming a uniform quantizer of the midrise type illustrated in Fig. 6.17b, we find that the step-size of the quantizer is given by

$$\Delta = \frac{2m_{max}}{L} \tag{6.35}$$

where L is the total number of representation levels. For a uniform quantizer, the quantization error Q will have its sample values bounded by $-\Delta/2 \leq q \leq \Delta/2$. If the step-size Δ is sufficiently small (i.e., the number of representation levels L is sufficiently large), it is reasonable to assume that the quantization error Q is a *uniformly distributed* random variable, and the interfering effect of the quantization noise on the quantizer input is similar to that of thermal noise. We may thus express the probability density function of the quantization error Q as follows:

$$f_Q(q) = \begin{cases} \dfrac{1}{\Delta}, & -\dfrac{\Delta}{2} < q \leq \dfrac{\Delta}{2} \\ 0, & \text{otherwise} \end{cases} \tag{6.36}$$

For this to be true, however, we must ensure that the incoming signal does *not* overload the quantizer. Then, with the mean of the quantization error being zero, its variance σ_Q^2 is the same as the mean-square value

$$\sigma_Q^2 = \int_{-\Delta/2}^{\Delta/2} q^2 f_Q(q) \; dq = E[Q^2] \tag{6.37}$$

Substituting Eq. (6.36) in (6.37), we get

$$\sigma_Q^2 = \frac{1}{\Delta} \int_{-\Delta/2}^{\Delta/2} q^2 \, dq$$
$$= \frac{\Delta^2}{12}$$

(6.38)

Typically, the L-ary number k, denoting the kth representation level of the quantizer, is transmitted to the receiver in binary form. Let R denote the number of *bits per sample* used in the construction of the binary code. We may then write

$$L = 2^R \tag{6.39}$$

or, equivalently,

$$R = \log_2 L \tag{6.40}$$

Hence, substituting Eq. (6.39) in (6.35), we get the step size

$$\Delta = \frac{2m_{\text{max}}}{2^R} \tag{6.41}$$

Thus, the use of Eq. (6.41) in (6.38) yields

$$\sigma_Q^2 = \tfrac{1}{3} m_{\text{max}}^2 2^{-2R} \tag{6.42}$$

Let P denote the average power of the message signal $m(t)$. We may then express the *output signal-to-noise ratio* of a uniform quantizer as

$$(\text{SNR})_O = \frac{P}{\sigma_Q^2}$$
$$= \left(\frac{3P}{m_{\text{max}}^2}\right) 2^{2R}$$

(6.43)

Equation (6.43) shows that the output signal-to-noise ratio of the quantizer increases *exponentially* with increasing number of bits per sample, R. Recognizing that an increase in R requires a proportionate increase in the channel (transmission) bandwidth B_T, we thus see that the use of a binary code for the representation of a message signal (as in pulse-code modulation) provides a more efficient method than either frequency modulation (FM) or pulse-position modulation (PPM) for the trade-off of increased channel bandwidth for improved noise performance. In making this statement, we presume that the FM and PPM systems are limited by receiver noise, whereas the binary-coded modulation system is limited by quantization noise. We have more to say on the latter issue in Section 6.9.

EXAMPLE 2   Sinusoidal Modulating Signal

Consider the special case of a full-load sinusoidal modulating signal of amplitude A_m, which utilizes all the representation levels provided. The average signal

Table 6.1 Signal-to-Quantization Noise Ratio for
Varying Number of Representation Levels

Number of Representation Levels, L	Number of Bits/Sample, R	Signal-to-Noise Ratio (dB)
32	5	31.8
64	6	37.8
128	7	43.8
256	8	49.8

power is (assuming a load of 1 ohm)

$$P = \frac{A_m^2}{2}$$

The total range of the quantizer input is $2A_m$, because the modulating signal swings between $-A_m$ and A_m. We may therefore set $m_{\max} = A_m$, in which case the use of Eq. (6.42) yields the average power (variance) of the quantization noise as

$$\sigma_Q^2 = \tfrac{1}{3}A_m^2 2^{-2R}$$

Thus the output signal-to-noise ratio of a uniform quantizer, for a full-load test tone, is

$$(\text{SNR})_O = \frac{A_m^2/2}{A_m^2 2^{-2R}/3} = \frac{3}{2}(2^{2R}) \qquad (6.44)$$

Expressing the signal-to-noise ratio in decibels, we get

$$10\log_{10}(\text{SNR})_O = 1.8 + 6R \qquad (6.45)$$

For various values of L and R, the corresponding values of signal-to-noise ratio are as given in Table 6.1. From Table 6.1 we can make a quick estimate of the number of bits per sample required for a desired output signal-to-noise ratio.

6.8   PULSE-CODE MODULATION

With the sampling and quantization processes at our disposal, we are now ready to describe pulse-code modulation which, as mentioned previously, is the most basic form of digital pulse modulation. In *pulse-code modulation (PCM) a message signal is represented by a sequence of coded pulses, which is accomplished by representing the signal in discrete form in both time and amplitude.* The basic operations performed in the transmitter of a PCM system are *sampling*, *quantizing*, and *encoding*, as shown in Fig. 6.19a; the low-pass filter prior to sampling is included to prevent aliasing of the message signal. The quantizing and encoding operations are usually performed in the same circuit, which is called an *analog-to-digital converter.* The basic operations in the receiver are *regeneration* of impaired signals, *decoding*, and *reconstruction* of the train of quantized samples, as shown in Fig. 6.19c. Regeneration also occurs at intermediate points along the transmission path as necessary,

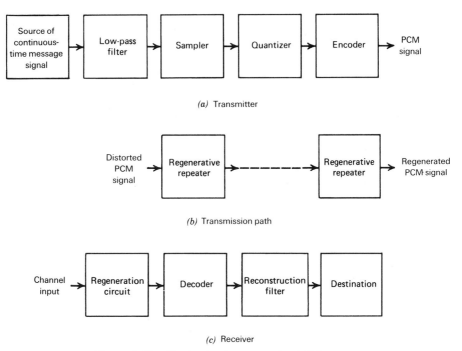

(a) Transmitter

(b) Transmission path

(c) Receiver

Figure 6.19 The basic elements of a PCM system.

as indicated in Fig. 6.19*b*. When time-division multiplexing is used, it becomes necessary to synchronize the receiver to the transmitter for the overall system to operate satisfactorily. In what follows we describe the various operations that constitute a PCM system.

Sampling

The incoming message signal is sampled with a train of narrow rectangular pulses so as to closely approximate the instantaneous sampling process. In order to ensure perfect reconstruction of the message signal at the receiver, the sampling rate must be greater than twice the highest frequency component W of the message signal in accordance with the sampling theorem. In practice, a pre-alias (low-pass) filter is used at the front end of the sampler in order to exclude frequencies greater than W before sampling. Thus the application of sampling permits the reduction of the continuously varying message signal (of some finite duration) to a limited number of discrete values per second.

Quantization

The sampled version of the message signal is then quantized, thereby providing a new representation of the signal that is discrete in both time and amplitude. The quantization process may follow a uniform law as described in Section 6.7. In certain applications, however, it is preferable to use a variable separation between the representation levels. For example, the range of voltages covered by voice signals, from the peaks of loud talk to the weak passages of weak talk, is on the order of 1000 to 1. By using a *nonuniform quantizer* with the feature that the step-size increases as the separation from the origin of the input–output amplitude characteristic is increased, the large end step of the quantizer can take care of possible excursions of the voice signal into the large amplitude ranges that occur relatively infrequently. In other words, the weak passages, which need more protection, are favored at the expense of the loud passages. In this way, a nearly uniform percentage precision is achieved throughout the greater part of the amplitude range of the input signal, with the result that fewer steps are needed than would be the case if a uniform quantizer were used.

The use of a nonuniform quantizer is equivalent to passing the baseband signal through a *compressor* and then applying the compressed signal to a uniform quantizer. A particular form of compression law that is used in practice is the so called μ-*law*[5] defined by

$$|v| = \frac{\log(1 + \mu|m|)}{\log(1 + \mu)} \tag{6.46}$$

where m and v are the normalized input and output voltages, and μ is a positive constant. In Fig. 6.20a, we have plotted the μ-law for varying μ. The case of uniform quantization corresponds to $\mu = 0$. For a given value of μ, the reciprocal slope of the compression curve, which defines the quantum steps, is given by the derivative of $|m|$ with respect to $|v|$; that is,

$$\frac{d|m|}{d|v|} = \frac{\log(1 + \mu)}{\mu}(1 + \mu|m|) \tag{6.47}$$

We see therefore that the μ-law is neither strictly linear nor strictly logarithmic, but it is approximately linear at low input levels corresponding to $\mu|m| \ll 1$, and approximately logarithmic at high input levels corresponding to $\mu|m| \gg 1$.

Another compression law that is used in practice is the so-called A-*law* defined by

$$|v| = \begin{cases} \dfrac{A|m|}{1 + \log A}, & 0 \le |m| \le \dfrac{1}{A} \\[3mm] \dfrac{1 + \log(A|m|)}{1 + \log A}, & \dfrac{1}{A} \le |m| \le 1 \end{cases} \tag{6.48}$$

which is shown plotted in Fig. 6.20b. Practical values of A(as of μ in the μ-law) tend to be in the vicinity of 100. The case of uniform quantization corresponds to $A = 1$. The reciprocal slope of this compression curve is given by the derivative of $|m|$ with respect to $|v|$, as shown by

$$\frac{d|m|}{d|v|} = \begin{cases} \dfrac{1 + \log A}{A}, & 0 \le |m| \le \dfrac{1}{A} \\[3mm] (1 + \log A)|m|, & \dfrac{1}{A} \le |m| \le 1 \end{cases} \tag{6.49}$$

Thus the quantum steps over the central linear segment, which have the dominant effect on small signals, are diminished by the factor $A/(1 + \log A)$. This is typically about 25 dB in practice, as compared with uniform quantization.

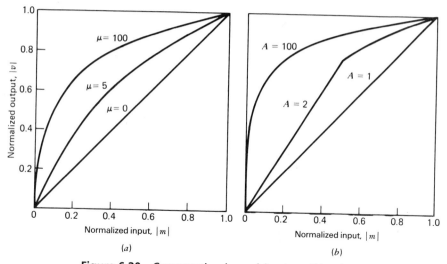

Figure 6.20 Compression laws. (a) μ-law. (b) A-law.

In order to restore the signal samples to their correct relative level, we must, of course, use a device in the receiver with a characteristic complementary to the compressor. Such a device is called an *expander*. Ideally, the compression and expansion laws are exactly inverse so that, except for the effect of quantization, the expander output is equal to the compressor input. The combination of a *comp*ressor and an exp*ander* is called a *compander*.

In actual PCM systems, the companding circuitry does not produce an exact replica of the nonlinear compression curves shown in Fig. 6.20. Rather, it provides a *piecewise linear* approximation to the desired curve. By using a large enough number of linear segments, the approximation can approach the true compression curve very closely. This form of approximation is illustrated in Example 3 at the end of the section.

Encoding

In combining the processes of sampling and quantizing, the specification of a continuous message (baseband) signal becomes limited to a discrete set of values, but not in the form best suited to transmission over a line or radio path. To exploit the advantages of sampling and quantizing for the purpose of making the transmitted signal more robust to noise, interference and other channel degradations, we require the use of an *encoding process* to translate the discrete set of sample values to a more appropriate form of signal. Any plan for representing each of this discrete set of values as a particular arrangement of discrete events is called a *code*. One of the discrete events in a code is called a *code element* or *symbol*. For example, the presence or absence of a pulse is a symbol. A particular arrangement of symbols used in a code to represent a single value of the discrete set is called a *code word* or *character*.

In a *binary code*, each symbol may be either of two distinct values or kinds, such as the presence or absence of a pulse. The two symbols of a binary code are customarily denoted as 0 and 1. In a *ternary code*, each symbol may be one of three distinct values or kinds, and so on for other codes. However, *the maximum advantage over the effects of noise in a transmission medium is obtained by using a binary code, because a binary symbol withstands a relatively high level of noise and is easy to regenerate.* Suppose that, in a binary code, each code word consists of R bits: the bit is an acronym for *binary digit*; thus R denotes the number of *bits per sample*. Then, using such a code, we may represent a total of 2^R distinct numbers. For example, a sample quantized into one of 256 levels may be represented by an 8-bit code word.

There are several ways of establishing a one-to-one correspondence between representation levels and code words. A convenient method is to express the ordinal number of the representation level as a binary number. In the binary number system, each digit has a place-value that is a power of 2, as illustrated in Table 6.2 for the case of four bits per sample (i.e., $R = 4$).

There are several *line codes* that can be used for the electrical representation of binary symbols 1 and 0, as described here:

1. *On–off signaling*, in which symbol 1 is represented by transmitting a pulse of constant amplitude for the duration of the symbol, and symbol 0 is represented by switching off the pulse, as in Fig. 6.21*a*.

2. *Nonreturn-to-zero (NRZ) signaling*, in which symbols 1 and 0 are represented by pulses of equal positive and negative amplitudes, as illustrated in Fig. 6.21*b*.

3. *Return-to-zero (RZ) signaling*, in which symbol 1 is represented by a positive rectangular pulse of half-symbol width, and symbol 0 is represented by transmitting *no* pulse, as illustrated in Fig. 6.21*c*.

Table 6.2 Binary Number System for $R = 4$.

Ordinal Number of Representation Level	Level Number Expressed as Sum of Powers of 2	Binary Number
0		0000
1	2^0	0001
2	2^1	0010
3	$2^1 + 2^0$	0011
4	2^2	0100
5	$2^2 \quad\;\; + 2^0$	0101
6	$2^2 + 2^1$	0110
7	$2^2 + 2^1 + 2^0$	0111
8	2^3	1000
9	$2^3 \qquad\;\;\; + 2^0$	1001
10	$2^3 \quad\; + 2^1$	1010
11	$2^3 \quad\; + 2^1 + 2^0$	1011
12	$2^3 + 2^2$	1100
13	$2^3 + 2^2 \quad\;\; + 2^0$	1101
14	$2^3 + 2^2 + 2^1$	1110
15	$2^3 + 2^2 + 2^1 + 2^0$	1111

Binary data

0   1   1   0   1   0   0   1

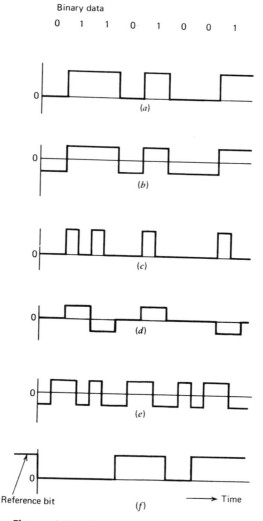

Figure 6.21 Electrical representations of binary data. (*a*) On–off signaling. (*b*) Nonreturn-to-zero signaling. (*c*) Return-to-zero signaling. (*d*) Bipolar signaling. (*e*) Split phase or Manchester code. (*f*) Differential encoding.

4. *Bipolar return-to-zero (BRZ) signaling*, which uses three amplitude levels as indicated in Fig. 6.21*d*. Specifically, positive and negative pulses of equal amplitude are used alternately for symbol 1, and no pulse is always used for symbol 0. A useful property of the BRZ signaling is that the power spectrum of the transmitted signal has no dc component and relatively insignificant low-frequency components for the case when symbols 1 and 0 occur with equal probability.

5. *Split-phase (Manchester code)*, which is illustrated in Fig. 6.21*e*. In this method of signaling, symbol 1 is represented by a positive pulse followed by a negative pulse, with both pulses being of equal amplitude and half-symbol width. For

symbol 0, the polarities of these two pulses are reversed. The Manchester code suppresses the dc component and has relatively insignificant low-frequency components, regardless of the signal statistics. This property is essential in some applications.

6. *Differential encoding*, in which the information is encoded in terms of signal transitions, as illustrated in Fig. 6.21*f*. In the example of the binary PCM signal shown here, a transition is used to designate symbol 0, while no transition is used to designate symbol 1. It is apparent that a differentially encoded signal may be inverted without affecting its interpretation. The original binary information is recovered by comparing the polarity of adjacent symbols to establish whether or not a transition has occurred.

The waveforms shown in Figs. 6.21*a* to 6.21*f* are for the binary data stream 01101001. Note that no pulse shaping is used in these waveforms. The issue of pulse shaping and its beneficial effect for the baseband transmission of PCM signals are discussed in Chapter 7.

Regeneration

The most important feature of PCM systems lies in the ability to control the effects of distortion and noise produced by transmitting a PCM signal through a channel. This capability is accomplished by reconstructing the PCM signal by means of a chain of *regenerative repeaters* located at sufficiently close spacing along the transmission route. As illustrated in Fig. 6.22, three basic functions are performed by a regenerative repeater: *equalization, timing,* and *decision making.* The equalizer shapes the received pulses so as to compensate for the effects of amplitude and phase distortions produced by the transmission characteristics of the channel. The timing circuitry provides a periodic pulse train, derived from the received pulses, for sampling the equalized pulses at the instants of time where the signal-to-noise ratio is a maximum. The sample so extracted is compared to a predetermined *threshold* in the decision-making device. In each bit interval a decision is then made whether the received symbol is a 1 or a 0 on the basis of whether the threshold is exceeded or not. If the threshold is exceeded, a clean new pulse representing symbol 1 is transmitted to the next repeater. Otherwise, another clean new pulse representing symbol 0 is transmitted. In this way, the accumulation of distortion and noise in a repeater span is completely removed, provided that the disturbance is not too large to cause an error in the decision-making process. Ideally, except for delay, the regenerated signal is exactly the same as the signal originally transmitted. In practice, however, the regenerated signal departs from the original signal for two main reasons:

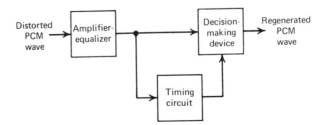

Figure 6.22 Block diagram of a regenerative repeater.

1. The unavoidable presence of channel noise and interference causes the repeater to make wrong decisions occasionally, thereby introducing *bit errors* into the regenerated signal.
2. If the spacing between received pulses deviates from its assigned value, a *jitter* is introduced into the regenerated pulse position, thereby causing distortion.

Decoding

The first operation in the receiver is to regenerate (i.e., reshape and clean up) the received pulses one last time. These clean pulses are then regrouped into code words and decoded (i.e., mapped back) into a quantized PAM signal. The *decoding* process involves generating a pulse the amplitude of which is the linear sum of all the pulses in the code word, with each pulse being weighted by its place value $(2^0, 2^1, 2^2, 2^3, \ldots, 2^{R-1})$ in the code, where R is the number of bits per sample.

Filtering

The final operation in the receiver is to recover the message signal wave by passing the decoder output through a low-pass reconstruction filter whose cutoff frequency is equal to the message bandwidth W. Assuming that the transmission path is error free, the recovered signal includes no noise with the exception of the initial distortion introduced by the quantization process.

Multiplexing

In applications using PCM, it is natural to multiplex different messages sources by time division, whereby each source keeps its individuality throughout the journey from the transmitter to the receiver. This individuality accounts for the comparative ease with which message sources may be dropped or reinserted in a time-division multiplex system. As the number of independent message sources is increased, the time interval that may be allotted to each source has to be reduced, since all of them must be accommodated into a time interval equal to the reciprocal of the sampling rate. This in turn means that the allowable duration of a code word representing a single sample is reduced. However, pulses tend to become more difficult to generate and to transmit as their duration is reduced. Furthermore, if the pulses become too short, impairments in the transmission medium begin to interfere with the proper operation of the system. Accordingly, in practice, it is necessary to restrict the number of independent message sources that can be included within a time-division group.

Synchronization

For a PCM system with time-division multiplexing to operate satisfactorily, it is necessary that the timing operations at the receiver, except for the time lost in transmission and regenerative repeating, follow closely the corresponding operations at the transmitter. In a general way, this amounts to requiring a local clock at the receiver to keep the same time as a distant standard clock at the transmitter, except that the local clock is somewhat slower by an amount corresponding to the time required to transport the message signals from the transmitter to the receiver. One possible procedure to synchronize the transmitter and receiver clocks is to set aside a code element or pulse at the end of a *frame* (consisting of a code word derived from each of the independent message sources in succession) and to transmit this pulse every other frame only. In such a case, the receiver includes a circuit that would search for the pattern of 1s and 0s alternating at half the frame rate, and thereby establish synchronization between the transmitter and receiver.

When the transmission path is interrupted, it is highly unlikely that transmitter and receiver clocks will continue to indicate the same time for long. Accordingly, in carrying out a synchronization process, we must set up an orderly procedure for detecting the synchronizing pulse. The procedure consists of observing the code elements one by one until the synchronizing pulse is detected. That is, after observing a particular code element long enough to establish the absence of the synchronizing pulse, the receiver clock is set back by one code element and the next code element is observed. This *searching process* is repeated until the synchronizing pulse is detected. Clearly, the time required for synchronization depends on the epoch at which proper transmission is reestablished.

EXAMPLE 3 The T1 System

In this example, we describe the important characteristics of a PCM system known as the *T1 carrier system,*[6] which is designed to accommodate 24 voice channels, primarily for short-distance, heavy usage in metropolitan areas. The T1 system was pioneered by the Bell System in the United States in the early 1960s, and with its introduction the shift to digital communication facilities started. The T1 system has been adopted for use throughout the United States, Canada, and Japan. It forms the basis for a complete hierarchy of higher-order multiplexed systems that are used for either long-distance transmission or transmission in heavily populated urban centers.

A voice signal (male or female) is essentially limited to a band from 300 to 3100 Hz in that frequencies outside this band do not contribute much to articulation efficiency. Indeed, telephone circuits that respond to this range of frequencies give quite satisfactory service. Accordingly, it is customary to pass the voice signal through a low-pass filter with a cutoff frequency of about 3.1 kHz prior to sampling. Hence, with $W = 3.1$ kHz, the nominal value of the Nyquist rate is 6.2 kHz. The filtered voice signal is usually sampled at a slightly higher rate, namely, 8 kHz, which is the *standard* sampling rate in telephone systems.

For companding, the T1 system uses a *piecewise-linear* characteristic (consisting of 15 linear segments) to approximate the logarithmic μ-law of Eq. (6.46) with the constant $\mu = 255$. This approximation is constructed in such a way that the segment end points lie on the compression curve computed from Eq. (6.46), and their projections onto the vertical axis are spaced uniformly. Table 6.3 gives the projections of the segment end points onto the horizontal axis and the step-sizes of the individual segments. The table is normalized to 8159, so that all values are represented as integer numbers. Segment 0 of the approximation is a colinear segment, passing through the origin; it contains a total of 30 uniform decision levels. Linear segments $1a, 2a, \ldots, 7a$ lie above the horizontal axis, whereas linear segments $1b, 2b, \ldots, 7b$ lie below the horizontal axis; each of these 14 segments contains 16 uniform decision levels. For colinear segment 0 the decision levels at the quantizer input are $\pm 1, \pm 3, \ldots, \pm 31$, and the corresponding representation levels at the quantizer output are $0, \pm 1, \ldots, \pm 15$.

Table 6.3 The 15-Segment Companding Characteristic ($\mu = 255$)

Linear Segment Number	Step-Size	Projections of Segment End Point onto the Horizontal Axis
0	2	
		± 31
$1a, 1b$	4	
		± 95
$2a, 2b$	8	
		± 223
$3a, 3b$	16	
		± 479
$4a, 4b$	32	
		± 991
$5a, 5b$	64	
		± 2015
$6a, 6b$	128	
		± 4063
$7a, 7b$	256	
		± 8159

For linear segments $1a$ and $1b$, the decision levels at the quantizer input are ± 31, $\pm 35, \ldots, \pm 95$, and the corresponding representation levels at the quantizer output are $\pm 16, \pm 17, \ldots, \pm 31$, and so on for the other linear segments.

There are a total of $31 + (14 \times 16) = 255$ representation levels associated with the 15-segment companding characteristic described above. To accommodate this number of representation levels, each of the 24 voice channels uses a binary code with an 8-bit word. The first bit indicates whether the input voice sample is positive or negative; this bit is a 1 if positive and a 0 if negative. The next three bits of the code word identify the particular segment inside which the amplitude of the input voice sample lies, and the last four bits identify the actual representation level inside that segment.

With a sampling rate of 8 kHz, each frame of the multiplexed signal occupies a period of 125 μs. In particular, it consists of twenty-four 8-bit words, plus a single bit that is added at the end of the frame for the purpose of synchronization. Hence, each frame consists of a total of $(24 \times 8) + 1 = 193$ bits. Correspondingly, the duration of each bit equals 0.647 μs, and the resultant transmission rate is 1.544 megabits per second (Mb/s).

In addition to the voice signal, a telephone system must also pass special supervisory signals to the far end. This *signaling information* is needed to transmit dial pulses, as well as telephone off-hook/on-hook signals. In the T1 system, this requirement is accomplished as follows. Every sixth frame, the least significant (that is, the eighth) bit of each voice channel is deleted and a *signaling bit* is inserted in its place, thereby yielding an average $7\frac{5}{6}$-bit operation for each voice input. The sequence of signaling bits is thus transmitted at a rate equal to the sampling rate divided by six, that is, 1.333 kilobits per second (kb/s).

6.9 NOISE CONSIDERATIONS IN PCM SYSTEMS

The performance of a PCM system is influenced by two major sources of noise:

1. *Channel noise*, which is introduced anywhere between the transmitter output and the receiver input. Channel noise is always present, once the equipment is switched on.

2. *Quantization noise*, which is introduced in the transmitter and is carried all the way along to the receiver output. Unlike channel noise, quantization noise is *signal-dependent* in the sense that it disappears when the message signal is switched off.

Naturally, these two sources of noise appear simultaneously once the PCM system is in operation. However, the traditional practice is to consider them separately, so that we may develop insight into their individual effects on the system performance.

The main effect of channel noise is to introduce *bit errors* into the received signal. In the case of a binary PCM system, the presence of a bit error causes symbol 1 to be mistaken for symbol 0, or vice versa. Clearly, the more frequently bit errors occur, the more dissimilar the receiver output becomes compared to the original message signal. The fidelity of information transmission by PCM in the presence of channel noise may be measured in terms of the *average probability of symbol error*, which is defined as the probability that the reconstructed symbol at the receiver output differs from the transmitted binary symbol, on the average. The average probability of symbol error, also referred to as the *error rate*, assumes that all the bits in the received binary wave are of equal importance. When, however, there is more interest in reconstructuring the analog waveform of the original message signal, different symbol errors may need to be *weighted* differently; for example, an error in the most significant bit in a code word (representing a quantized sample of the message signal) is more harmful than an error in the least significant bit.

To optimize system performance in the presence of channel noise, we need to minimize the average probability of symbol error. For this evaluation, it is customary to model the channel noise, originating at the front end of the receiver, as additive, white, and Gaussian. The effect of channel noise can be made practically negligible by ensuring the use of an adequate signal energy-to-noise density ratio through the provision of proper spacing between the regenerative repeaters in the PCM system. In such a situation, the performance of the PCM system is essentially limited by quantization noise acting alone.

From the discussion of quantization noise presented in Sections 6.7 and 6.8, we recognize that quantization noise is essentially under the designer's control. It can be made negligibly small through the use of an adequate number of representation levels in the quantizer and the selection of a companding strategy matched to the characteristics of the type of message signal being transmitted. We thus find that the use of PCM offers the possibility of building a communication system that is *rugged* with respect to channel noise on a scale that is beyond the capability of any CW modulation or analog pulse modulation system.

Error Threshold

The underlying theory of error rate calculation in a PCM system is deferred to the next chapter. For the present, it suffices to say that the average probability of symbol error in a binary encoded PCM receiver due to additive white Gaussian noise depends solely on E_b/N_0, *the ratio of the transmitted signal energy per bit, E_b, to the noise spectral density, N_0*. Note that the ratio E_b/N_0 is dimensionless even though the quantities E_b and N_0 have different physical meaning. In Table 6.4 we present a summary of this dependence for the case of a binary PCM system using nonre-

Table 6.4 Influence of E_b/N_0 on the Probability of Error

E_b/N_0	Probablity of Error P_e	For a Bit Rate of 10^5 b/s This Is About One Error Every	
4.3 dB	10^{-2}	10^{-3}	second
8.4	10^{-4}	10^{-1}	second
10.6	10^{-6}	10	seconds
12.0	10^{-8}	20	minutes
13.0	10^{-10}	1	day
14.0	10^{-12}	3	months

turn-to-zero signaling. The results presented in the last column of the table assume a bit rate of 10^5 b/s.

From Table 6.4 it is clear that there is an *error threshold* (at about 11 dB). For E_b/N_0 below the error threshold the receiver performance involves significant numbers of errors, and above it the effect of channel noise is practically negligible. In other words, provided that the ratio E_b/N_0 exceeds the error threshold, channel noise has virtually no effect on the receiver performance, which is precisely the goal of PCM. When, however, E_b/N_0 drops below the error threshold, there is a sharp increase in the rate at which errors occur in the receiver. Because decision errors result in the construction of incorrect code words, we find that when the errors are frequent, the reconstructed message at the receiver output bears little resemblance to the original message.

Comparing the figure of 11 dB for the error threshold in a PCM system using NRZ signaling with the 60–70 dB required for high-quality transmission of speech using amplitude modulation, we see that PCM requires much less power, even though the average noise power in the PCM system is increased by the R-fold increase in bandwidth, where R is the number of bits in a code word (i.e., bits per sample).

In most transmission systems, the effects of noise and distortion from the individual links accumulate. For a given quality of overall transmission, the longer the physical separation between the transmitter and the receiver, the more severe are the requirements on each link in the system. In a PCM system, however, because the signal can be regenerated as often as necessary, the effects of amplitude, phase, and nonlinear distortions in one link (if not too severe) have practically no effect on the regenerated input signal to the next link. We have also seen that the effect of channel noise can be made practically negligible by using a ratio E_b/N_0 above threshold. For all practical purposes, then, the transmission requirements for a PCM link are almost independent of the physical length of the communication channel.

Another important characteristic of a PCM system is its *ruggedness to interference*, caused by stray impulses or cross-talk. The combined presence of channel noise and interference causes the error threshold necessary for satisfactory operation of the PCM system to increase. If an adequate margin over the error threshold is provided in the first place, however, the system can withstand the presence of relatively large amounts of interference. In other words, a PCM system is quite *rugged*.

6.10   VIRTUES, LIMITATIONS, AND MODIFICATIONS OF PCM

In a generic sense, pulse-code modulation (PCM) has emerged as the most favored modulation scheme for the transmission of analog information-bearing signals such as voice and video signals. The advantages of PCM may all be traced to the use of *coded pulses for the digital representation of analog signals*, a feature that distinguishes it from all other analog methods of modulation. We may summarize the important advantages of PCM as follows:

1. *Ruggedness* to channel noise and interference.
2. Efficient *regeneration* of the coded signal along the transmission path.
3. Efficient *exchange* of increase channel bandwidth for improved signal-to-noise ratio, obeying an exponential law.
4. A *uniform format* for the transmission of different kinds of baseband signals; hence their integration with other forms of digital data in a common network.
5. Comparative *ease* with which message sources may be dropped or reinserted in a time-division multiplex system.
6. *Secure* communication through the use of special modulation schemes or encryption.

These advantages, however, are attained at the cost of increased system complexity and increased channel bandwidth. These two issues are considered in the sequel in turn.

Although the use of PCM involves many complex operations, today they can all be implemented in a cost-effective fashion using commercially available and/or custom-made *very-large-scale integrated* (VLSI) chips. In other words, the requisite device technology for the implementation of a PCM system is already in place. Moreover, with continuing improvements in VLSI technology, we are likely to see an ever-expanding use of PCM for the transmission of analog signals.

If, however, the simplicity of implementation is a necessary requirement, then we may use delta modulation as an alternative to pulse-code modulation. In delta modulation, the baseband signal is intentionally "oversampled" in order to permit the use of a simple quantizing strategy for constructing the encoded signal; delta modulation is discussed in Section 6.11.

Turning next to the issue of bandwidth, we do recognize that the increased bandwidth requirement of PCM may have been a reason for justifiable concern in the past. Today, however, it is of no real concern for two different reasons. First, the increasing availability of *wide-band communication channels* means that bandwidth is no longer a system constraint in the traditional way it used to be. Liberation from the bandwidth constraint has been made possible by the deployment of communication satellites for broadcasting and the ever-increasing use of fiber optics for networking; a discussion of these communication channel concepts is presented in the final chapter of the book.

The second reason is that through the use of sophisticated *data compression* techniques, it is indeed possible to remove the redundancy inherently present in a PCM signal and thereby reduce the bit rate of the transmitted data without serious degradation in system performance. In effect, increased processing complexity (and therefore increased cost of implementation) is traded off for a reduced bit rate and therefore reduced bandwidth requirement. A major motivation for bit reduction is for secure communication over radio channels that are inherently of low capacity. The issue of coding speech at low bit rates is discussed in Section 6.13, where the treatment is from a signal processing perspective. We revisit the idea of data compression in a more general setting using information-theoretic ideas in Chapter 10.

6.11 DELTA MODULATION

In *delta modulation*[7] (DM), an incoming message signal is oversampled (i.e., at a rate much higher than the Nyquist rate) to purposely increase the correlation between adjacent samples of the signal. This is done to permit the use of a simple quantizing strategy for constructing the encoded signal.

In its basic form, DM provides a *staircase approximation* to the oversampled version of the message signal, as illustrated in Fig. 6.23a. The difference between the input and the approximation is quantized into only two levels, namely, $\pm\Delta$, corresponding to positive and negative differences, respectively. Thus, if the approximation falls below the signal at any sampling epoch, it is increased by Δ. If, on the other hand, the approximation lies above the signal, it is diminished by Δ. Provided that the signal does not change too rapidly from sample to sample, we find that the staircase approximation remains within $\pm\Delta$ of the input signal.

Denoting the input signal as $m(t)$, and its staircase approximation as $m_q(t)$, the basic principle of delta modulation may be formalized in the following set of discrete-time relations:

$$e(nT_s) = m(nT_s) - m_q(nT_s - T_s) \tag{6.50}$$

$$e_q(nT_s) = \Delta \, \mathrm{sgn}[e(nT_s)] \tag{6.51}$$

$$m_q(nT_s) = m_q(nT_s - T_s) + e_q(nT_s) \tag{6.52}$$

where T_s is the sampling period; $e(nT_s)$ is an error signal representing the difference between the present sample value $m(nT_s)$ of the input signal and the latest

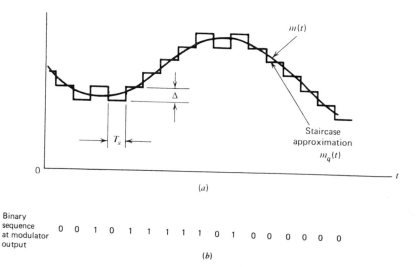

(a)

Binary
sequence
at modulator   0   0   1   0   1   1   1   1   1   0   1   0   0   0   0   0   0
output

(b)

Figure 6.23 Illustration of delta modulation.

approximation to it, that is, $m(nT_s) - m_q(nT_s - T_s)$; and $e_q(nT_s)$ is the quantized version of $e(nT_s)$. The quantizer output $e_q(nT_s)$ is finally coded to produce the desired DM signal.

Figure 6.23a illustrates the way in which the staircase approximation $m_q(t)$ follows variations in the input signal $m(t)$ in accordance with Eqs.(6.50) to (6.52), and Fig. 6.23b displays the corresponding binary sequence at the delta modulator output. It is apparent that in a delta modulation system the rate of information transmission is simply equal to the sampling rate $f_s = 1/T_s$.

The principal virtue of delta modulation is its simplicity. It may be generated by applying the sampled version of the incoming message signal to a modulator that involves a *comparator, quantizer,* and *accumulator* interconnected as shown in Fig. 6.24a. Details of the modulator follow directly from Eqs. (6.50) to (6.52). The comparator computes the difference between its two inputs. The quantizer consists of a *hard limiter* with an input–output relation that is a scaled version of the signum function. The quantizer output is then applied to an accumulator, producing the result

$$m_q(nT_s) = \Delta \sum_{i=1}^{n} \text{sgn}[e(iT_s)]$$

$$= \sum_{i=1}^{n} e_q(iT_s)$$

(6.53)

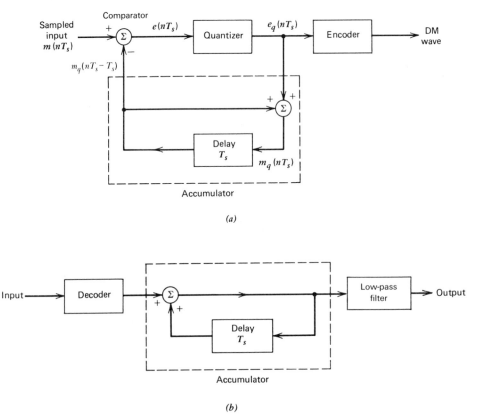

Figure 6.24   DM system. (a) Transmitter. (b) Receiver.

which is obtained by solving Eqs. (6.51) and (6.52) for $m_q(nT_s)$. Thus, at the sampling instant nT_s, the accumulator increments the approximation by a step Δ in a positive or negative direction, depending on the algebraic sign of the error signal $e(nT_s)$. If the input signal $m(nT_s)$ is greater than the most recent approximation $m_q(nT_s)$, a positive increment $+\Delta$ is applied to the approximation. If, on the other hand, the input signal is smaller, a negative increment $-\Delta$ is applied to the approximation. In this way, the accumulator does the best it can to track the input samples by one step (of amplitude $+\Delta$ or $-\Delta$) at a time. In the receiver shown in Fig. 6.24b, the staircase approximation $m_q(t)$ is reconstructed by passing the sequence of positive and negative pulses, produced at the decoder output, through an accumulator in a manner similar to that used in the transmitter. The out-of-band quantization noise in the high-frequency staircase waveform $m_q(t)$ is rejected by passing it through a low-pass filter, as in Fig. 6.24b, with a bandwidth equal to the original message bandwidth.

Delta modulation is subject to two types of quantization error: (1) slope overload distortion and (2) granular noise. We first discuss the cause of slope overload distortion, and then granular noise.

We observe that Eq. (6.52) is the digital equivalent of integration in the sense that it represents the accumulation of positive and negative increments of magnitude Δ. Also, denoting the quantization error by $q(nT_s)$, as shown by,

$$m_q(nT_s) = m(nT_s) + q(nT_s) \tag{6.54}$$

we observe from Eq. (6.50) that the input to the quantizer is

$$e(nT_s) = m(nT_s) - m(nT_s - T_s) - q(nT_s - T_s) \tag{6.55}$$

Thus, except for the quantization error $q(nT_s - T_s)$, the quantizer input is a *first backward difference* of the input signal, which may be viewed as a digital approximation to the derivative of the input signal or, equivalently, as the inverse of the digital integration process. If we consider the maximum slope of the original input waveform $m(t)$, it is clear that in order for the sequence of samples $\{m_q(nT_s)\}$ to increase as fast as the input sequence of samples $\{m(nT_s)\}$ in a region of maximum slope of $m(t)$, we require that the condition

$$\frac{\Delta}{T_s} \geq \max\left|\frac{dm(t)}{dt}\right| \tag{6.56}$$

be satisfied. Otherwise, we find that the step-size Δ is too small for the staircase approximation $m_q(t)$ to follow a steep segment of the input waveform $m(t)$, with the result that $m_q(t)$ falls behind $m(t)$, as illustrated in Fig. 6.25. This condition is called *slope overload*, and the resulting quantization error is called *slope-overload distortion* (noise). Note that since the maximum slope of the staircase approximation $m_q(t)$ is fixed by the step-size Δ, increases and decreases in $m_q(t)$ tend to occur along straight lines. For this reason, a delta modulator using a fixed step-size is often referred to as a *linear delta modulator*.

In contrast to slope-overload distortion, *granular noise* occurs when the step-size Δ is too large relative to the local slope characteristics of the input waveform $m(t)$, thereby causing the staircase approximation $m_q(t)$ to hunt around a rela-

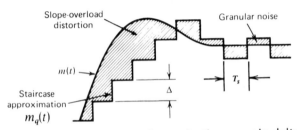

Figure 6.25 Illustration of quantization error in delta modulation.

tively flat segment of the input waveform; this phenomenon is also illustrated in Fig. 6.25. Granular noise is analogous to quantization noise in a PCM system.

We thus see that there is a need to have a large step-size to accommodate a wide dynamic range, whereas a small step-size is required for the accurate representation of relatively low-level signals. It is therefore clear that the choice of the optimum step-size that minimizes the mean-square value of the quantization error in a linear delta modulator will be the result of a compromise between slope overload distortion and granular noise. To satisfy such a requirement, we need to make the delta modulator "adaptive," in the sense that the step-size is made to vary in accordance with the input signal.[8]

Delta-Sigma Modulation

As mentioned previously, the quantizer input in the conventional form of delta modulation may be viewed as an approximation to the *derivative* of the incoming message signal. This behavior leads to a drawback of delta modulation in that transmission disturbances such as noise result in an accumulative error in the demodulated signal. This drawback can be overcome by *integrating* the message signal prior to delta modulation. The use of integration in the manner described here has also the following beneficial effects:

- The low-frequency content of the input signal is pre-emphasized.
- Correlation between adjacent samples of the delta modulator input is increased, which tends to improve overall system performance by reducing the variance of the error signal at the quantizer input.
- Design of the receiver is simplified.

A delta modulation scheme that incorporates integration at its input is called *delta-sigma modulation*[9] (D-ΣM). To be more precise, however, it should be called *sigma-delta modulation*, because the integration is in fact performed before the delta modulation. Nevertheless, the former terminology is the one commonly used in the literature.

Figure 6.26a shows the block diagram of a delta-sigma modulation system. In this diagram, the message signal $m(t)$ is defined in its continuous-time form, which means that the pulse modulator now consists of a hard-limiter followed by a multiplier; the latter component is also fed from an external pulse generator (clock) to produce a 1-bit encoded signal. The use of integration at the transmitter input clearly requires an inverse signal emphasis, namely, differentiation, at the receiver. The need for this differentiation is, however, eliminated because

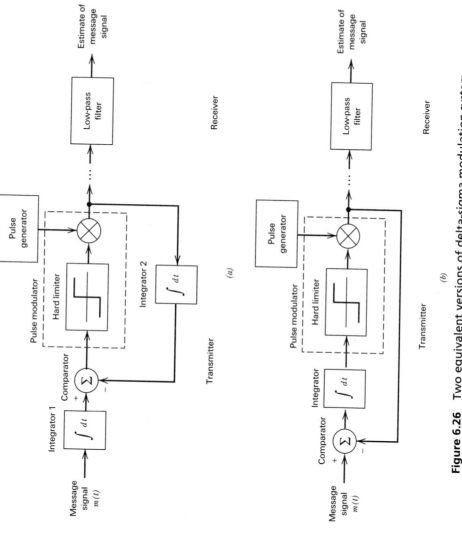

Figure 6.26 Two equivalent versions of delta-sigma modulation system.

of its cancellation by integration in the conventional DM receiver. Thus, the receiver of a delta-sigma modulation system consists simply of a low-pass filter, as indicated in Fig. 6.26a.

Moreover, we note that integration is basically a linear operation. Accordingly, we may simplify the design of the transmitter by combining the two integrators 1 and 2 of Fig. 6.26a into a single integrator placed after the comparator, as shown in Fig. 6.26b. This latter form of the delta-sigma modulation system is not only simpler than that of Fig. 6.26a, but it also provides an interesting interpretation of delta-sigma modulation as a "smoothed" version of 1-bit pulse-code modulation: The term *smoothness* refers to the fact that the comparator output is integrated prior to quantization, and the term *1-bit* merely restates that the quantizer consists of a hard-limiter with only two representation levels.

6.12   DIFFERENTIAL PULSE-CODE MODULATION

When a voice or video signal is sampled at a rate slightly higher than the Nyquist rate, the resulting sampled signal is found to exhibit a high correlation between adjacent samples. The meaning of this high correlation is that, in an average sense, the signal does not change rapidly from one sample to the next, with the result that the difference between adjacent samples has a variance that is smaller than the variance of the signal itself. When these highly correlated samples are encoded, as in a standard PCM system, the resulting encoded signal contains *redundant information*. This means that symbols that are not absolutely essential to the transmission of information are generated as a result of the encoding process. By removing this redundancy before encoding, we obtain a more efficient coded signal.

Now, if we know a sufficient part of a redundant signal, we may infer the rest, or at least make the most probable estimate. In particular, if we know the past behavior of a signal up to a certain point in time, it is possible to make some inference about its future values; such a process is commonly called *prediction*. Suppose then a baseband signal $m(t)$ is sampled at the rate $f_s = 1/T_s$ to produce a sequence of correlated samples T_s seconds apart; this sequence is denoted by $\{m(nT_s)\}$. The fact that it is possible to predict future values of the signal $m(t)$ provides motivation for the *differential quantization* scheme shown in Fig. 6.27a. In this scheme the input signal to the quantizer is defined by

$$e(nT_s) = m(nT_s) - \hat{m}(nT_s) \tag{6.57}$$

which is the difference between the unquantized input sample $m(nT_s)$ and a prediction of it, denoted by $\hat{m}(nT_s)$. This predicted value is produced by using a *prediction filter* whose input, as we will see, consists of a quantized version of the input sample $m(nT_s)$. The difference signal $e(nT_s)$ is called the *prediction error*, since it is the amount by which the prediction filter fails to predict the input exactly. A simple and yet effective approach to implement the prediction filter is to use a *tapped-delay-line filter*, with the basic delay set equal to the sampling period. The block diagram of this filter is shown in Fig. 6.28, according to which the prediction $\hat{m}(nT_s)$ is modeled as a linear combination of p past sample values of the quantized input, where p is the *prediction order*.

By encoding the quantizer output, as in Fig. 6.27a, we obtain a variation of

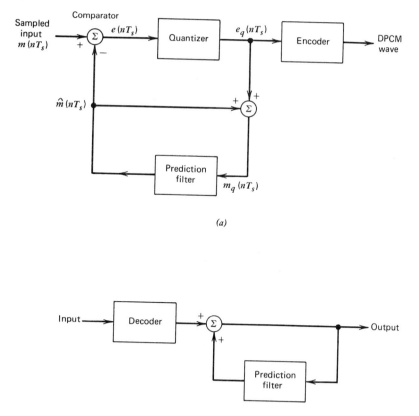

(a)

(b)

Figure 6.27 DPCM system. (*a*) Transmitter. (*b*) Receiver.

PCM, which is known as *differential pulse-code modulation*[10] (DPCM). It is this encoded signal that is used for transmission.

The quantizer output may be expressed as

$$e_q(nT_s) = e(nT_s) + q(nT_s) \tag{6.58}$$

where $q(nT_s)$ is the quantization error. According to Fig. 6.27a, the quantizer output $e_q(nT_s)$ is added to the predicted value $\hat{m}(nT_s)$ to produce the prediction-

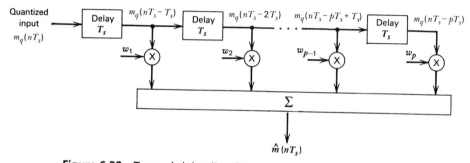

Figure 6.28 Tapped-delay-line filter used as a prediction filter.

filter input

$$m_q(nT_s) \; = \; \hat{m}(nT_s) \; + \; e_q(nT_s) \tag{6.59}$$

Substituting Eq. (6.58) in (6.59), we get

$$m_q(nT_s) \; = \; \hat{m}(nT_s) \; + \; e(nT_s) \; + \; q(nT_s) \tag{6.60}$$

However, from Eq. (6.57) we observe that the sum term $\hat{m}(nT_s) + e(nT_s)$ is equal to the input signal $m(nT_s)$. Therefore, we may rewrite Eq. (6.60) as

$$m_q(nT_s) \; = \; m(nT_s) \; + \; q(nT_s) \tag{6.61}$$

which represents a quantized version of the input signal $m(nT_s)$. That is, irrespective of the properties of the prediction filter, the quantized signal $m_q(nT_s)$ at the prediction filter input differs from the original input signal $m(nT_s)$ by the quantization error $q(nT_s)$. Accordingly, if the prediction is good, the variance of the prediction error $e(nT_s)$ will be smaller than the variance of $m(nT_s)$, so that a quantizer with a given number of levels can be adjusted to produce a quantization error with a smaller variance than would be possible if the input signal $m(nT_s)$ were quantized directly as in a standard PCM system.

The receiver for reconstructing the quantized version of the input is shown in Fig. 6.27b. It consists of a decoder to reconstruct the quantized error signal. The quantized version of the original input is reconstructed from the decoder output using the same prediction filter used in the transmitter of Fig. 6.27a. In the absence of channel noise, we find that the encoded signal at the receiver input is identical to the encoded signal at the transmitter output. Accordingly, the corresponding receiver output is equal to $m_q(nT_s)$, which differs from the original input $m(nT_s)$ only by the quantization error $q(nT_s)$ incurred as a result of quantizing the prediction error $e(nT_s)$.

From the foregoing analysis we observe that, in a noise-free environment, the prediction filters in the transmitter and receiver operate on the same sequence of samples, $m_q(nT_s)$. It is with this purpose in mind that a feedback path is added to the quantizer in the transmitter, as shown in Fig. 6.27a.

Differential pulse-code modulation includes delta modulation as a special case. In particular, comparing the DPCM system of Fig. 6.27 with the DM system of Fig. 6.24, we see that they are basically similar, except for two important differences: the use of a one-bit (two-level) quantizer in the delta modulator, and the replacement of the prediction filter by a single delay element (i.e., zero prediction order). Simply put, DM is the 1-bit version of DPCM. Note that unlike a standard PCM system, the transmitters of both the DPCM and DM involve the use of *feedback*.

DPCM, like DM, is subject to slope-overload distortion whenever the input signal changes too rapidly for the prediction filter to track it. Also like PCM, DPCM suffers from quantization noise.

Processing Gain

The output signal-to-noise ratio of the DPCM system shown in Fig. 6.27 is, by definition,

$$(SNR)_O = \frac{\sigma_M^2}{\sigma_Q^2} \qquad (6.62)$$

where σ_M^2 is the variance of the original input $m(nT_s)$, assumed to be of zero mean, and σ_Q^2 is the variance of the quantization error $q(nT_s)$. We may rewrite Eq. (6.62) as the product of two factors as follows:

$$(SNR)_O = \left(\frac{\sigma_M^2}{\sigma_E^2}\right)\left(\frac{\sigma_E^2}{\sigma_Q^2}\right)$$
$$= G_P(SNR)_Q \qquad (6.63)$$

where σ_E^2 is the variance of the prediction error. The factor $(SNR)_Q$ is the *signal-to-quantization noise ratio*, defined by

$$(SNR)_Q = \frac{\sigma_E^2}{\sigma_Q^2} \qquad (6.64)$$

The other factor G_P is the *processing gain* produced by the differential quantization scheme; it is defined by

$$G_P = \frac{\sigma_M^2}{\sigma_E^2} \qquad (6.65)$$

The quantity G_P, when greater than unity, represents the gain in signal-to-noise ratio that is due to the differential quantization scheme of Fig. 6.27. Now, for a given baseband (message) signal, the variance σ_M^2 is fixed, so that G_P is maximized by minimizing the variance σ_E^2 of the prediction error $e(nT_s)$. Accordingly, our objective should be to design the prediction filter so as to minimize σ_E^2.

In the case of voice signals, it is found that the optimum signal-to-quantization noise advantage of DPCM over standard PCM is in the neighborhood of 4–11 dB. The greatest improvement occurs in going from no prediction to first-order prediction, with some additional gain resulting from increasing the order of the prediction filter up to 4 or 5, after which little additional gain is obtained. Since 6 dB of quantization noise is equivalent to 1 bit per sample, by virtue of Eq. (6.45), the advantage of DPCM may also be expressed in terms of bit rate. For a constant signal-to-quantization noise ratio, and assuming a sampling rate of 8 kHz, the use of DPCM may provide a saving of about 8–16 kb/s (i.e., 1–2 bits per sample) over standard PCM.

6.13   CODING SPEECH AT LOW BIT RATES

The use of PCM at the standard rate of 64 kb/s demands a high channel bandwidth for its transmission. In certain applications, however, such as secure transmission over radio channels that are inherently of low capacity, channel bandwidth is at a premium. In application of this kind, there is a definite need for *speech coding at low bit rates, while maintaining acceptable fidelity or quality of reproduction.*[11]

For coding speech at low bit rates, a waveform coder of prescribed configuration is optimized by exploiting both *statistical characterization of speech waveforms
and properties of hearing.* In particular, the design philosophy has two aims in mind:

1. To remove redundancies from the speech signal as far as possible.
2. To assign the available bits to code the nonredundant parts of the speech
    signal in a perceptually efficient manner.

As we strive to reduce the bit rate from 64 kb/s (used in standard PCM) to 32,
16, 8, and 4 kb/s, the schemes used for redundancy removal and bit assignment
become increasingly more sophisticated. As a rule of thumb, in the 64 to 8 kb/s
range, the computational complexity (measured in terms of multiply–add operations) required to code speech increases by an order of magnitude when the
bit rate is halved, for approximately equal speech quality.

In the sequel, we briefly describe two schemes for coding speech, one at 32
kb/s and the other at 16 kb/s.

Adaptive Differential Pulse-Code Modulation

Reduction in the number of bits per sample from 8 (as used in standard PCM)
to 4 involves the combined use of *adaptive quantization* and *adaptive prediction.* In
this context, the term "adaptive" means being responsive to changing level and
spectrum of the input speech signal. The variation of performance with speakers
and speech material, together with variations in signal level inherent in the
speech communication process, make the combined use of adaptive quantization
and adaptive prediction necessary to achieve best performance over a wide range
of speakers and speaking situations. A digital coding scheme that uses both adaptive quantization and adaptive prediction is called *adaptive differential pulse-code
modulation* (ADPCM).

Adaptive quantization refers to a quantizer that operates with a *time-varying*
step-size $\Delta(nT_s)$, where T_s is the sampling period. At any given time identified by
the index n, the adaptive quantizer is assumed to have a uniform transfer characteristic. The step-size $\Delta(nT_s)$ is varied so as to match the variance σ_M^2 of the
input signal $m(nT_s)$. In particular, we write

$$\Delta(nT_s) \;=\; \phi\hat{\sigma}_M(nT_s) \tag{6.66}$$

where ϕ is a constant, and $\hat{\sigma}_M(nT_s)$ is an *estimate* of the standard deviation
$\sigma_M(nT_s)$ (i.e., square root of the variance σ_M^2). For a nonstationary input,
$\sigma_M(nT_s)$ is time varying. The problem of adaptive quantization, according to
Eq. (6.66) is, therefore, one of computing the estimate $\hat{\sigma}_M(nT_s)$ continuously.

The implementation of Eq. (6.66) may proceed in one of two ways:

1. *Adaptive quantization with forward estimation* (AQF), in which unquantized
    samples of the input signal are used to derive forward estimates of $\sigma_M(nT_s)$.
2. *Adaptive quantization with backward estimation* (AQB), in which samples of the
    quantizer output are used to derive backward estimates of $\sigma_M(nT_s)$.

The AQF scheme requires the use of a *buffer* to store unquantized samples of the
input speech signal needed for the learning period. It also requires the explicit
transmission of level information (typically, about 5–6 bits per step-size sample)

to a remote decoder, thereby burdening the system with additional *side information* that has to be transmitted to the receiver. Moreover, a processing *delay* (on the order of 16 ms for speech) in the encoding operation results from the use of AQF, which is unacceptable in some applications. The problems of level transmission, buffering, and delay intrinsic to AQF are all avoided in AQB. In the latter scheme the recent history of the quantizer output is used to extract information for the computation of the step-size $\Delta(nT_s)$. In practice, AQB is therefore usually preferred over AQF.

Figure 6.29 shows the block diagram of an adaptive quantizer with backward estimation. It represents a nonlinear feedback system; hence, it is not obvious that the system will be stable. However, provided that the quantizer input $m(nT_s)$ is *bounded*, then so are the backward estimate $\hat{\sigma}_M(nT_s)$ and the corresponding step-size $\Delta(nT_s)$; under such a condition, the system is indeed stable.

The use of adaptive prediction in ADPCM is justified because speech signals are inherently *nonstationary*, a phenomenon that manifests itself in the fact that the autocorrelation function and power spectral density of speech signals are time-varying functions of their respective arguments. This implies that the design of predictors for such inputs should likewise be time varying, that is, adaptive. As with adaptive quantization, there are two schemes for performing adaptive prediction:

1. *Adaptive prediction with forward estimation* (APF), in which unquantized samples of the input signal are used to derive estimates of the predictor coefficients.

2. *Adaptive prediction with backward estimation* (APB), in which samples of the quantizer output and the prediction error are used to derive estimates of the predictor coefficients.

However, APF suffers from the same intrinsic disadvantages (side information, buffering, and delay) as AQF. These disadvantages are eliminated by using the APB scheme shown in Fig. 6.30, where the box labeled "logic for adaptive prediction" is intended to represent the mechanism for updating the predictor coefficients. In the latter scheme, the optimum predictor coefficients are estimated on the basis of quantized and transmitted data; they can therefore be updated as frequently as desired, from sample to sample, say. Accordingly, APB is the preferred method of prediction for ADPCM.

A signal processing scheme known as the *least-mean-square (LMS) algorithm* for the predictor and an adaptive scheme for the quantizer have been combined in a synchronous fashion for the design of both the encoder and decoder. The

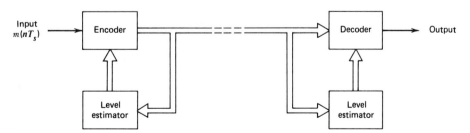

Figure 6.29 Adaptive quantization with backward estimation (AQB).

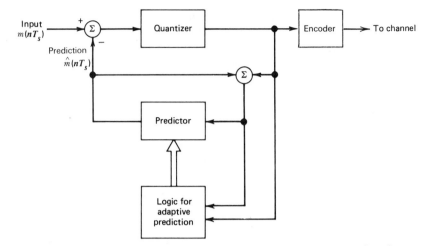

Figure 6.30 Adaptive prediction with backward estimation (APB).

performance of this combination is so impressive at 32 kb/s that ADPCM is now accepted internationally as a standard coding technique for voice signals, along with 64 kb/s using standard PCM. A description of the LMS algorithm is presented in Chapter 7.

Adaptive Subband Coding

PCM and ADPCM are both *time-domain coders* in the sense that the speech signal is processed in the time domain as a single-full-band signal. We next describe a *frequency-domain coder*, in which the speech signal is divided into a number of subbands and each one is encoded separately. The coder is capable of digitizing speech at a rate of 16 kb/s with a quality comparable to that of 64 kb/s PCM. To accomplish this performance, it exploits the quasi-periodic nature of *voiced speech* and a characteristic of the hearing mechanism known as *noise masking*. A voiced speech sound is generated from quasi-periodic vocal-chord sound, whereas an unvoiced speech sound is generated from random sound produced by turbulent airflows. A more detailed description of voiced sounds and their distinction from unvoiced sounds is presented in Appendix 1.

Periodicity of voiced speech manifests itself in the fact that people speak with a characteristic *pitch frequency*. This periodicity permits pitch prediction, and therefore a further reduction in the level of the prediction error that requires quantization, compared to differential pulse-code modulation without pitch prediction. The number of bits per sample that needs to be transmitted is thereby greatly reduced, without a serious degradation in speech quality.

The number of bits per sample can be reduced further by making use of the *noise-masking phenomenon* in perception. Specifically, the human ear does not perceive noise in a given frequency band if the noise is about 15 dB below the signal level in that band. This means that a relatively large coding error (the equivalent of noise) can be tolerated near *formants*, and that the coding rate can be correspondingly reduced. In the context of speech production, the *formant frequencies* (or simply formants) are the resonance frequencies of the vocal tract tube. The formants depend on the shape and dimensions of the vocal tract.

In *adaptive subband coding* (ASBC), noise shaping is accomplished by *adaptive bit assignment.* In particular, the number of bits used to encode each subband is varied dynamically and shared with other subbands, such that the encoding accuracy is always placed where it is needed in the frequency-domain characterization of the speech signal. Indeed, subbands with little or no energy may not be encoded at all.

A block diagram of the adaptive subband coding scheme is shown in Fig. 6.31. Specifically, the speech band is divided into a number of contiguous bands by a bank of band-pass filters (typically four to eight). The output of each band-pass filter is frequency-translated to assume a low-pass form by a modulation process equivalent to single-sideband modulation. It is then sampled (or resampled) at a rate slightly higher than its Nyquist rate (i.e., twice the width of the pertinent subband), and then digitally encoded by using an ADPCM with fixed (typically, first-order) prediction. A specific coding strategy is employed for each subband in accordance with perceptual criteria tailored to that band. Bit assignment information is transmitted to the receiver, enabling it to decode the subband signals individually and modulate them back to their original locations in

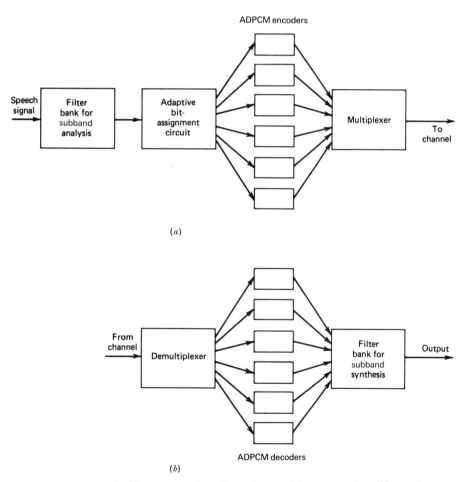

Figure 6.31 Adaptive subband coding scheme. (*a*) Transmitter. (*b*) Receiver.

the frequency band. Finally, they are summed to produce an output signal that provides a close replica of the original speech signal.

The complexity of a 16 kb/s adaptive subband coder is typically 100 times that of a 64 kb/s PCM coder for about the same reproduction quality. As a result of the large number of arithmetic operations involved in designing the adaptive subband coder, there is a processing delay of 25 ms; such a delay is not encountered in the PCM coder. Note, however, that delay is of no concern in applications that involve voice storage, as in "voice mail," for example.

6.14   SUMMARY AND DISCUSSION

In this chapter we introduced two fundamental and complementary processes:

- *Sampling*, which operates in the time domain; the sampling process is the link between an analog waveform and its discrete-time representation.
- *Quantization*, which operates in the amplitude domain; the quantization process is the link between an analog waveform and its discrete-amplitude representation.

The sampling process builds on the *sampling theorem*, which states that a strictly band-limited signal with no frequency components higher than W Hz is represented uniquely by a sequence of samples taken at a uniform rate equal to or greater than the Nyquist rate of $2W$ samples per second. The quantization process exploits the fact that any human sense, as ultimate receiver, can only detect finite intensity differences.

The sampling process is basic to the operation of all pulse modulation systems, which may be classified into analog pulse modulation and digital pulse modulation. The distinguishing feature between them is that analog pulse modulation systems maintain a continuous amplitude representation of the message signal, whereas digital pulse modulation systems also employ quantization to provide a representation of the message signal that is discrete in both time and amplitude.

Analog pulse modulation results from varying some parameter of the transmitted pulses, such as amplitude, duration, or position, in which case we speak of pulse-amplitude modulation (PAM), pulse-duration modulation (PDM), or pulse-position modulation (PPM), respectively. In time-division multiplexing (TDM) of several channels, signal processing usually begins with PAM. To use PDM or PPM in such an application, we have to ensure that full-scale modulation will not cause a pulse from one message signal to enter a time slot belonging to another message signal. This restriction results in a wasteful use of time space in telephone systems that are characterized by high peak factors, which is one reason for not using PDM[12] or PPM in telephony. Also, despite the fact that PPM is more efficient than PDM, they both fall short of the ideal system for exchanging transmission bandwidth for improved noise performance.

Digital pulse modulation systems transmit analog message signals as a sequence of coded pulses, which is made possible through the combined use of sampling and quantization. Pulse-code modulation is an important form of digital pulse modulation that is endowed with some unique system advantages, which in turn have made it the preferred method of modulation for the transmission of such analog signals as voice and video signals. The advantages of pulse-code modulation include robustness to noise and interference, efficient regeneration of the coded pulses along the transmission path, and a uniform format for different kinds of baseband signals.

Delta modulation and differential pulse-code modulation are two other useful forms of digital pulse modulation. The principal advantage of delta modulation is the simplicity of its circuitry. In contrast, differential pulse-code modulation employs increased circuit complexity to improve system performance. The improvement is achieved by using the idea of prediction to remove redundant symbols from an incoming data stream. A further improvement in the operation of differential pulse-code modulation can be made through the use of adaptivity to account for statistical variations in the input data. By so doing, the bandwidth requirement of pulse-code modulation is reduced significantly without serious degradation in system performance.

At this point in the discussion, it is informative to take a critical look at the different forms of pulse modulation that we have described in this chapter. In a strict sense, the term "pulse modulation" is a misnomer in that all of its different forms, be they analog or digital, are in fact *source coding* techniques. We say this for the simple reason that a message signal remains to be a baseband signal after undergoing all the changes involved in a pulse modulation process. The baseband nature of a pulse-modulated signal is exemplified by the fact that, irrespective of its exact description, it can be transmitted over a baseband channel of adequate bandwidth. Indeed, the material presented in the next chapter is devoted to the baseband transmission of data represented by a sequence of pulses.

It is also important to recognize that pulse modulation techniques are *lossy* in the sense that some information is lost as a result of the signal representation that they perform. For example, in pulse-amplitude modulation, the customary practice is to use pre-alias (low-pass) filtering prior to sampling; in so doing, information is lost by virtue of the fact that high-frequency components considered to be unessential are removed by the filter. The lossy nature of pulse modulation is most vividly seen in pulse-code modulation that is characterized by the generation of quantization noise (i.e., distortion); the transmitted sequence of encoded pulses does not have the infinite precision needed to represent continuous samples exactly. Nevertheless, the loss of information incurred by the use of a pulse modulation process is *under the designer's control* in that it can be made small enough for it to be nondiscernible by the end user.

The material presented in this chapter on pulse modulation has been from a signal processing perspective. We will revisit pulse-code modulation and its variant, differential pulse-code modulation, in Chapter 10, which is devoted to information-theoretic considerations of communication systems. In so doing we will develop deeper insight into their operation as source coding techniques.

NOTES AND REFERENCES

1. The classic book on pulse modulation is Black (1953). A more detailed treatment of the subject is presented in the book by Rowe (1965).

2. By sampling the discrete-time Fourier transform of Section 6.2 in the frequency domain, we obtain the discrete Fourier transform (DFT) that was discussed in Chapter 2.

3. Pulse-code modulation was invented by Reeves in 1937. For a historical account of this invention, see the paper by Reeves (1975). The book by Jayant and Noll (1984) presents the most complete treatment of pulse-code modulation, differential pulse-code modulation, delta modulation, and their variants. The book edited by Jayant (1976) provides a collection of important papers written on waveform quantization and coding.

4. For a detailed discussion of quantization noise in PCM systems, see the paper by Bennett (1948) and also the book by Rowe (1965, pp. 311–321).

5. The μ-law used for signal compression is described in Smith (1957). The μ-law is used in the United States, Canada, and Japan. In Europe the A-law is used for signal compression; this compression law is described in Cattermole (1969, pp. 133–140); for a discussion of the μ-law and A-law, see also the paper by Kaneko (1970).

6. For a description of the original version of the T1 PCM system, see the paper by Fultz and Penick (1965). The description given in Example 3 is based on an updated version of this system; see Henning and Pan (1972).

7. For the original papers on delta modulation, see Schouten, DeJager, and Greefkes (1952) and DeJager (1952). For a review paper on delta modulation, see the paper by Schindler (1970).

8. For a discussion of adaptive delta modulation, see the paper by Abate (1967); see also Jayant and Noll (1984).

9. Delta-sigma modulation is described in the book by Jayant and Noll (1984, pp. 399–400); see also the paper by Inose, Yasuda, and Murakami (1962).

10. Differential pulse-code modulation was invented by Cutler; the invention is described in a patent issued in 1952.

    For a comparison of the noise performances of PCM and DPCM, see the paper by Jayant (1974); see also Rabiner and Schafer (1978, Chapter 5).

11. For a discussion of coding speech at low bit rates, see the paper by Jayant (1986); see also Jayant and Noll (1984, pp. 188–210, 290–311), and Flanagan et al. (1979). Much of the material presented in Section 6.13 is based on these three references.

12. Pulse-duration (width) modulation finds a useful application in *computer control*, where it is used to convey digital information from the computer to a physical plant. In this application, the pulse widths take on a finite set of possible values, making the controller's behavior similar to that of a relay. For more details, see Vanlandingham (1985, pp. 176–178).

PROBLEMS

Problem 6.1 In *natural sampling* an analog signal $g(t)$ is multiplied by a periodic train of rectangular pulses $c(t)$. Given that the pulse repetition frequency of this periodic train is f_s and the duration of each rectangular pulse is T (with $f_s T \gg 1$), do the following:

(a) Find the spectrum of the signal $s(t)$ that results from the use of natural sampling; you may assume that time $t = 0$ corresponds to the midpoint of a rectangular pulse in $c(t)$.

(b) Show that the original signal $m(t)$ may be recovered exactly from its naturally sampled version, provided that the conditions embodied in the sampling theorem are satisfied.

Problem 6.2 Specify the Nyquist rate and the Nyquist interval for each of the following signals:

(a) $g(t) = \text{sinc}(200t)$

(b) $g(t) = \text{sinc}^2(200t)$

(c) $g(t) = \text{sinc}(200t) + \text{sinc}^2(200t)$

Problem 6.3

(a) Plot the spectrum of a PAM wave produced by the modulating signal

$$m(t) = A_m \cos(2\pi f_m t)$$

assuming a modulation frequency $f_m = 0.25$ Hz, sampling period $T_s = 1$ s, and pulse duration $T = 0.45$ s.

(b) Using an ideal reconstruction filter, plot the spectrum of the filter output. Compare this result with the output that would be obtained if there were no aperture effect.

Problem 6.4 In this problem we evaluate the equalization needed for the aperture effect in a PAM system. The operating frequency $f = f_s/2$, which corresponds to the highest frequency component of the message signal for a sampling rate equal to the Nyquist rate. Plot $1/\text{sinc}(0.5T/T_s)$ versus T/T_s, and hence find the equalization needed when $T/T_s = 0.1$.

Problem 6.5 Consider a PAM wave transmitted through a channel with white Gaussian noise and minimum bandwidth $B_T = 1/2T_s$, where T_s is the sampling period. The noise is of zero mean and power spectral density $N_0/2$. The PAM signal uses a standard pulse $g(t)$ with its Fourier transform defined by

$$G(f) = \begin{cases} \dfrac{1}{2B_T}, & |f| < B_T \\ 0, & |f| > B_T \end{cases}$$

By considering a full-load sinusoidal modulating wave, show that PAM and baseband-signal transmission have equal signal-to-noise ratios for the same average transmitted power.

Problem 6.6 Twenty-four voice signals are sampled uniformly and then time-division multiplexed, The sampling operation uses flat-top samples with 1 μs duration. The multiplexing operation includes provision for synchronization by adding an extra pulse of sufficient amplitude and also 1 μs duration. The highest frequency component of each voice signal is 3.4 kHz.

(a) Assuming a sampling rate of 8 kHz, calculate the spacing between successive pulses of the multiplexed signal.

(b) Repeat your calculation assuming the use of Nyquist rate sampling.

Problem 6.7 Twelve different message signals, each with a bandwidth of 10 kHz, are to be multiplexed and transmitted. Determine the minimum bandwidth required for each method if the multiplexing/modulation method used is

(a) FDM, SSB.

(b) TDM, PAM.

Problem 6.8 A PAM *telemetry* system involves the multiplexing of four input signals: $s_i(t)$, $i = 1, 2, 3, 4$. Two of the signals $s_1(t)$ and $s_2(t)$ have bandwidths of

80 Hz each, whereas the remaining two signals $s_3(t)$ and $s_4(t)$ have bandwidths of 1 kHz each. The signals $s_3(t)$ and $s_4(t)$ are each sampled at the rate of 2400 samples per second. This sampling rate is divided by 2^R (i.e., an integer power of 2) in order to derive the sampling rate for $s_1(t)$ and $s_2(t)$.

(a) Find the maximum value of R.

(b) Using the value of R found in part (a), design a multiplexing system that first multiplexes $s_1(t)$ and $s_2(t)$ into a new sequence, $s_5(t)$, and then multiplexes $s_3(t)$, $s_4(t)$, and $s_5(t)$.

Problem 6.9 The unmodulated pulse train in a PPM system is as shown in Fig. P6.1. The slicing level in the receiver is set at $A/2$.

(a) Assuming a full-load sinusoidal modulating wave and front-end receiver noise of zero mean and power spectral density $N_0/2$, determine the output signal-to-noise ratio and figure of merit of the system. Assume a high peak pulse-to-noise ratio.

(b) For the case when the message signal is sampled at its Nyquist rate, find the value of the transmission bandwidth for which the figure of merit of the system is greater than unity.

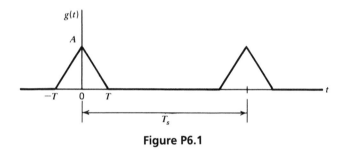

Figure P6.1

Problem 6.10 Consider Fig. P6.2, which shows the unmodulated pulse train for PDM. The PDM pulse consists of a rectangular pulse of duration D, which is preceded and followed by leading and trailing segments that are identical to the corresponding halves of the PPM pulse shown in Fig. 6.14. The slicer in the receiver is set at half the peak pulse amplitude, removing all noise effects except for the displacement of edge detection time by a small amount τ similar to that evaluated for the PPM system in Example 1. Assume that one edge of the duration-modulated pulse is fixed by means of a noise-free reference.

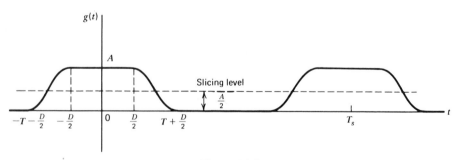

Figure P6.2

(a) Find the output signal-to-noise ratio of the PDM system.

(b) Find its channel signal-to-noise ratio.

(c) Compare the figure of merit for the PDM system to that of the corresponding PPM system.

Problem 6.11

(a) A sinusoidal signal, with an amplitude of 3.25 volts, is applied to a uniform quantizer of the midtread type whose output takes on the values $0, \pm 1, \pm 2, \pm 3$ volts. Sketch the waveform of the resulting quantizer output for one complete cycle of the input.

(b) Repeat this evaluation for the case when the quantizer is of the midrise type whose output takes on the values $\pm 0.5, \pm 1.5, \pm 2.5, \pm 3.5$ volts.

Problem 6.12 Consider the following sequences of 1s and 0s:

(a) An alternating sequence of 1s and 0s.

(b) A long sequence of 1s followed by a long sequence of 0s.

(c) A long sequence of 1s followed by a single 0 and then a long sequence of 1s.

Sketch the waveform for each of these sequences using the following methods of representing symbols 1 and 0:

(a) On–off signaling.

(b) Bipolar return-to-zero signaling.

Problem 6.13 This problem is intended to show that the power spectral density of a PCM wave depends on the signaling format used to represent the 1s and 0s. Assuming that the 1s and 0s occur with equal probability, and the symbols in adjacent time slots are statistically independent, determine and plot the power spectral density of the PCM wave for each of the signaling formats:

(a) On–off signaling.

(b) Manchester code.

Problem 6.14 The signal

$$m(t) = 6 \sin(2\pi t) \text{ volts}$$

is transmitted using a 4-bit binary PCM system. The quantizer is of the midrise type, with a step-size of 1 volt. Sketch the resulting PCM wave for one complete cycle of the input. Assume a sampling rate of four samples per second, with samples taken at $t = \pm 1/8, \pm 3/8, \pm 5/8, \ldots$, seconds.

Problem 6.15 Figure P6.3 shows a PCM signal in which the amplitude levels of $+1$ volt and -1 volt are used to represent binary symbols 1 and 0, respectively. The code word used consists of three bits. Find the sampled version of an analog signal from which this PCM signal is derived.

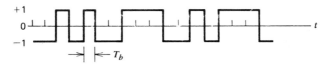

Figure P6.3

Problem 6.16 Consider a uniform quantizer characterized by the input–output relation illustrated in Fig. 6.17a. Assume that a Gaussian-distributed random variable with zero mean and unit variance is applied to this quantizer input.

(a) What is the probability that the amplitude of the input lies outside the range -4 to $+4$?

(b) Using the result of part (a), show that the output signal-to-noise ratio of the quantizer is given by

$$(\text{SNR})_O = 6R - 7.2 \text{ dB}$$

where R is the number of bits per sample. Specifically, you may assume that the quantizer input extends from -4 to $+4$. Compare the result of part (b) with that obtained in Example 2.

Problem 6.17 A PCM system uses a uniform quantizer followed by a 7-bit binary encoder. The bit rate of the system is equal to 50×10^6 b/s.

(a) What is the maximum message bandwidth for which the system operates satisfactorily?

(b) Determine the output signal-to-quantization noise ratio when a full-load sinusoidal modulating wave of frequency 1 MHz is applied to the input.

Problem 6.18 Show that, with a nonuniform quantizer, the mean-square value of the quantization error is approximately equal to $(1/12)\Sigma_i\Delta_i^2 p_i$, where Δ_i is the ith step size and p_i is the probability that the input signal amplitude lies within the ith interval. Assume that the step-size Δ_i is small compared with the excursion of the input signal.

Problem 6.19 Consider a chain of $(n - 1)$ regenerative repeaters, with a total of n sequential decisions made on a binary PCM wave, including the final decision made at the receiver. Assume that any binary symbol transmitted through the system has an independent probability p_1 of being inverted by any repeater. Let p_n represent the probability that a binary symbol is in error after transmission through the complete system.

(a) Show that

$$p_n = \tfrac{1}{2}[1 - (1 - 2p_1)^n]$$

(b) If p_1 is very small and n is not too large, what is the corresponding value of p_n?

Problem 6.20 Consider a sine wave of frequency f_m and amplitude A_m, applied to a delta modulator of step-size Δ. Show that slope-overload distortion will occur if

$$A_m > \frac{\Delta}{2\pi f_m T_s}$$

where T_s is the sampling period. What is the maximum power that may be transmitted without slope-overload distortion?

Problem 6.21 In the DPCM system depicted in Fig. P6.4, show that in the absence of channel noise, the transmitting and receiving prediction filters operate on slightly different input signals.

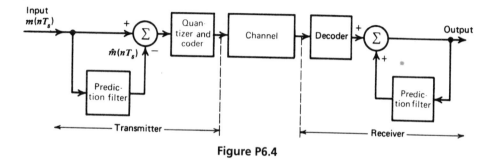

Figure P6.4

Problem 6.22 Consider the first-order prediction defined by

$$\hat{m}(nT_s) = w\, m(nT_s - T_s)$$

where $m(nT_s)$ is the sample of a stationary signal of zero mean, T_s is the sampling period, and w is a constant.

(a) Show that the prediction error

$$e(nT_s) = m(nT_s) - \hat{m}(nT_s)$$

has variance

$$\sigma_E^2 = \sigma_M^2 \left[1 + w^2 - \frac{2wR_M(T_s)}{\sigma_M^2} \right]$$

where σ_M^2 is the variance of the input signal, and $R_M(T_s)$ is its autocorrelation function for a lag of T_s.

(b) Show that the variance of the prediction error is minimized for

$$w_{opt} = \frac{R_M(T_s)}{\sigma_M^2}$$

and that the corresponding value of σ_E^2 is

$$\sigma_{E,min}^2 = \sigma_M^2 - \frac{R_M^2(T_s)}{\sigma_M^2}$$

(c) What is the condition for which σ_E^2 is less than σ_M^2?

Baseband Pulse Transmission

7.1 INTRODUCTION

In the previous chapter we described techniques for converting an analog information-bearing signal into digital form. There is another way in which digital data can arise in practice: the data may represent the output of a source of information that is inherently discrete in nature (e.g., a digital computer). In this chapter we study the transmission of digital data (of whatever origin) over a *baseband channel*.[1] Data transmission over a band-pass channel using modulation is covered in the next chapter.

Digital data have a broad spectrum with a significant low-frequency content. Baseband transmission of digital data therefore requires the use of a low-pass channel with a bandwidth large enough to accommodate the essential frequency content of the data stream. Typically, however, the channel is *dispersive* in that its frequency response deviates from that of an ideal low-pass filter. The result of data transmission over such a channel is that each received pulse is affected somewhat by adjacent pulses, thereby giving rise to a common form of interference called *intersymbol interference* (ISI). Intersymbol interference is a major source of bit errors in the reconstructed data stream at the receiver. To correct for it, control has to be exercised over the pulse shape in the overall system.

Thus much of the material covered in this chapter is devoted to *pulse shaping* in one form or another.

Another source of bit errors in a baseband data transmission system is the ubiquitous *receiver noise* (channel noise). Naturally, noise and ISI arise in the system simultaneously. However, to understand how they affect the performance of the system, we propose to consider them separately. We thus begin the chapter by describing a fundamental result in communication theory, which deals with the *detection* of a pulse signal of known waveform that is immersed in additive white noise. The device for the optimum detection of such a pulse involves the use of a linear-time-invariant filter known as a *matched filter*,[2] which is so called because its impulse response is matched to the pulse signal.

7.2 MATCHED FILTER

A basic problem that often arises in the study of communication systems is that of *detecting* a pulse transmitted over a channel that is corrupted by additive noise at the front end of the receiver. For the purpose of the discussion presented in this section, we assume that the major source of system limitation is the noise.

Consider then the receiver model shown in Fig. 7.1, involving a linear time-invariant filter of impulse response $h(t)$. The filter input $x(t)$ consists of a pulse signal $g(t)$ corrupted by additive noise $w(t)$, as shown by

$$x(t) = g(t) + w(t) \qquad 0 \le t \le T \tag{7.1}$$

where T is an arbitrary observation interval. The pulse signal $g(t)$ may represent a binary symbol 1 or 0 in a digital communication system. The $w(t)$ is the sample function of a white noise process of zero mean and power spectral density $N_0/2$. It is assumed that the receiver has knowledge of the waveform of the pulse signal $g(t)$. The source of uncertainty lies in the noise $w(t)$. The function of the receiver is to detect the pulse signal $g(t)$ in an optimum manner, given the received signal $x(t)$. To satisfy this requirement, we have to optimize the design of the filter so as to minimize the effects of noise at the filter output, and thereby enhance the detection of the pulse signal $g(t)$.

Since the filter is linear, the resulting output $y(t)$ may be expressed as

$$y(t) = g_o(t) + n(t) \tag{7.2}$$

where $g_o(t)$ and $n(t)$ are produced by the signal and noise components of the input $x(t)$, respectively. A simple way of describing the requirement that the

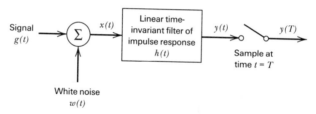

Figure 7.1 Linear receiver.

output signal component $g_o(t)$ be considerably greater than the output noise component $n(t)$ is to have the filter make the instantaneous power in the output signal $g_o(t)$, measured at time $t = T$, as large as possible compared with the average power of the output noise $n(t)$. This is equivalent to maximizing the *peak pulse signal-to-noise ratio*, defined as

$$\eta = \frac{|g_o(T)|^2}{E[n^2(t)]} \tag{7.3}$$

where $|g_o(T)|^2$ is the instantaneous power in the output signal, E is the statistical expectation operator, and $E[n^2(t)]$ is a measure of the average output noise power. The requirement is to specify the impulse response $h(t)$ of the filter such that the output signal-to-noise ratio in Eq. (7.3) is maximized.

Let $G(f)$ denote the Fourier transform of the known signal $g(t)$, and $H(f)$ denote the transfer function of the filter. Then the Fourier transform of the output signal $g_o(t)$ is equal to $H(f)G(f)$, and $g_o(t)$ is itself given by the inverse Fourier transform

$$g_o(t) = \int_{-\infty}^{\infty} H(f)G(f)\exp(j2\pi ft)\ df \tag{7.4}$$

Hence, when the filter output is sampled at time $t = T$, we have (in the absence of receiver noise)

$$|g_o(T)|^2 = \left| \int_{-\infty}^{\infty} H(f)G(f)\exp(j2\pi fT)\ df \right|^2 \tag{7.5}$$

Consider next the effect on the filter output due to the noise $w(t)$ acting alone. The power spectral density $S_N(f)$ of the output noise $n(t)$ is equal to the power spectral density of the input noise $w(t)$ times the squared magnitude of the transfer function $H(f)$ (see Section 4.10). Since $w(t)$ is white with constant power spectral density $N_0/2$, it follows that

$$S_N(f) = \frac{N_0}{2}|H(f)|^2 \tag{7.6}$$

The average power of the output noise $n(t)$ is therefore

$$\begin{aligned} E[n^2(t)] &= \int_{-\infty}^{\infty} S_N(f)\ df \\ &= \frac{N_0}{2}\int_{-\infty}^{\infty} |H(f)|^2\ df \end{aligned} \tag{7.7}$$

Thus, substituting Eqs. (7.5) and (7.7) into (7.3), we may rewrite the expression for the peak pulse signal-to-noise ratio as

$$\eta = \frac{\left| \int_{-\infty}^{\infty} H(f)G(f)\exp(j2\pi T)\ df \right|^2}{\dfrac{N_0}{2}\int_{-\infty}^{\infty} |H(f)|^2\ df} \tag{7.8}$$

Our problem is to find, for a given $G(f)$, the particular form of the transfer function $H(f)$ of the filter that makes η a maximum. To find the solution to this optimization problem, we apply a mathematical result known as Schwarz's inequality to the numerator of Eq. (7.8).

A derivation of *Schwarz's inequality* is given in Appendix 5, where it is shown that if we have two complex functions $\phi_1(x)$ and $\phi_2(x)$ in the real variable x, satisfying the conditions

$$\int_{-\infty}^{\infty} |\phi_1(x)|^2 \, dx < \infty$$

and

$$\int_{-\infty}^{\infty} |\phi_2(x)|^2 \, dx < \infty$$

then we may write

$$\left| \int_{-\infty}^{\infty} \phi_1(x) \phi_2(x) \, dx \right|^2 \leq \int_{-\infty}^{\infty} |\phi_1(x)|^2 \, dx \int_{-\infty}^{\infty} |\phi_2(x)|^2 \, dx \tag{7.9}$$

The equality in (7.9) holds if, and only if, we have

$$\phi_1(x) = k\phi_2^*(x) \tag{7.10}$$

where k is an arbitrary constant, and the asterisk denotes complex conjugation.

Returning to the problem at hand, we readily see that by invoking Schwarz's inequality (7.9), and setting $\phi_1(x) = H(f)$ and $\phi_2(x) = G(f)\exp(j\pi fT)$, the numerator in Eq. (7.8) may be rewritten as

$$\left| \int_{-\infty}^{\infty} H(f) G(f) \exp(j2\pi fT) \, df \right|^2 \leq \int_{-\infty}^{\infty} |H(f)|^2 \, df \int_{-\infty}^{\infty} |G(f)|^2 \, df \tag{7.11}$$

Using this relation in Eq. (7.8), we may redefine the peak pulse signal-to-noise ratio as

$$\eta \leq \frac{2}{N_0} \int_{-\infty}^{\infty} |G(f)|^2 \, df \tag{7.12}$$

The right-hand side of this relation does not depend on the transfer function $H(f)$ of the filter but only on the signal energy and the noise power spectral density. Consequently, the peak pulse signal-to-noise ratio η will be a maximum when $H(f)$ is chosen so that the equality holds; that is,

$$\eta_{\text{max}} = \frac{2}{N_0} \int_{-\infty}^{\infty} |G(f)|^2 \, df \tag{7.13}$$

Correspondingly, $H(f)$ assumes its optimum value denoted by $H_{\text{opt}}(f)$. To find this optimum value we use Eq. (7.10), which, for the situation at hand, yields

$$H_{\text{opt}}(f) = kG^*(f)\exp(-j2\pi fT) \tag{7.14}$$

where $G^*(f)$ is the complex conjugate of the Fourier transform of the input signal $g(t)$, and k is a scaling factor of appropriate dimensions. This relation states that, except for the factor $k \exp(-j2\pi fT)$, the transfer function of the optimum filter is the same as the complex conjugate of the spectrum of the input signal.

Equation (7.14) specifies the optimum filter in the frequency domain. To characterize it in the time domain, we take the inverse Fourier transform of $H_{\text{opt}}(f)$ in Eq. (7.14) to obtain the impulse response of the optimum filter as

$$h_{\text{opt}}(t) = k \int_{-\infty}^{\infty} G^*(f) \exp[-j2\pi f(T - t)] \, df \qquad (7.15)$$

Since for a real signal $g(t)$ we have $G^*(f) = G(-f)$, we may rewrite Eq. (7.15) as

$$
\begin{aligned}
h_{\text{opt}}(t) &= k \int_{-\infty}^{\infty} G(-f) \exp[-j2\pi f(T - t)] \, df \\
&= kg(T - t)
\end{aligned}
\qquad (7.16)
$$

Equation (7.16) shows that the impulse response of the optimum filter, except for the scaling factor k, is a time-reversed and delayed version of the input signal $g(t)$; that is, it is "matched" to the input signal. A linear time-invariant filter defined in this way is called a matched filter. Note that in deriving the matched filter the only assumption we have made about the input noise $w(t)$ is that it is stationary and white with zero mean and power spectral density $N_0/2$.

Properties of Matched Filters

We note that a filter, which is matched to a pulse signal $g(t)$ of duration T, is characterized by an impulse response that is a time-reversed and delayed version of the input $g(t)$, as shown by

$$h_{\text{opt}}(t) = kg(T - t)$$

In other words, the impulse response $h_{\text{opt}}(t)$ is uniquely defined, except for the delay T and the scaling factor k, by the waveform of the pulse signal $g(t)$ to which the filter is matched. In the frequency domain, the matched filter is characterized by a transfer function that is, except for a delay factor, the complex conjugate of the Fourier transform of the input $g(t)$, as shown by

$$H_{\text{opt}}(f) = kG^*(f) \exp(-j2\pi fT)$$

The most important result in the calculation of the performance of signal processing systems using matched filters is perhaps the following:

• *The peak pulse signal-to-noise ratio of a matched filter depends only on the ratio of the signal energy to the power spectral density of the white noise at the filter input.*

To demonstrate this property, consider a filter matched to a known signal $g(t)$. The Fourier transform of the resulting matched filter output $g_o(t)$ is

$$
\begin{aligned}
G_o(f) &= H_{\mathrm{opt}}(f)\,G(f) \\
&= kG^*(f)\,G(f)\exp(-j2\pi fT) \\
&= k|G(f)|^2\exp(-j2\pi fT)
\end{aligned}
\tag{7.17}
$$

Using Eq. (7.17) in the formula for the inverse Fourier transform, we find that the matched filter output at time $t = T$ is

$$
\begin{aligned}
g_o(T) &= k\int_{-\infty}^{\infty} G_o(f)\exp(j2\pi fT)\ df \\
&= k\int_{-\infty}^{\infty} |G(f)|^2\ df
\end{aligned}
$$

Using Rayleigh's energy theorem, this result reduces to

$$
g_o(T) = kE
\tag{7.18}
$$

where E is the energy of the pulse signal $g(t)$. Next, substituting Eq. (7.14) in (7.7), we find that the average output noise power is

$$
\begin{aligned}
E[n^2(t)] &= \frac{k^2 N_0}{2}\int_{-\infty}^{\infty} |G(f)|^2\ df \\
&= k^2 N_0 E/2
\end{aligned}
\tag{7.19}
$$

where again we have made use of Rayleigh's energy theorem. Therefore, the peak pulse signal-to-noise ratio has the maximum value

$$
\eta_{\max} = \frac{(kE)^2}{(k^2 N_0 E/2)} = \frac{2E}{N_0}
\tag{7.20}
$$

From Eq. (7.20) we see that dependence on the waveform of the input $g(t)$ has been completely removed by the matched filter. Accordingly, in evaluating the ability of a matched-filter receiver to combat additive white Gaussian noise, we find that all signals that have the same energy are equally effective. Note that the signal energy E is in joules and the noise spectral density $N_0/2$ is in watts per hertz, so that the ratio $2E/N_0$ is dimensionless; however, the two quantities have different physical meaning. We refer to E/N_0 as the *signal energy-to-noise spectral density ratio*.

EXAMPLE 1   Matched Filter for Rectangular Pulse

Consider a signal $g(t)$ in the form of a rectangular pulse of amplitude A and duration T, as shown in Fig. 7.2a. In this example, the impulse response $h(t)$ of the matched filter has exactly the same waveform as the signal itself. The output signal $g_o(t)$ of the matched filter produced in response to the input signal $g(t)$ has a triangular waveform, as shown in Fig. 7.2b.

The maximum value of the output signal $g_o(t)$ is equal to kA^2T, which is the energy of the input signal $g(t)$ scaled by the factor k; this maximum value occurs at $t = T$, as indicated in Fig. 7.2b.

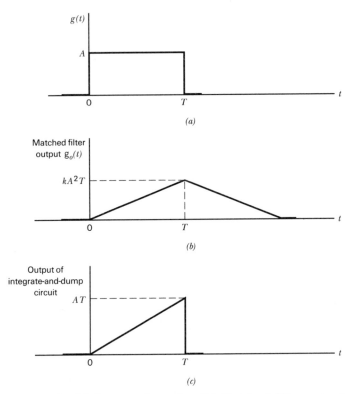

Figure 7.2 (a) Rectangular pulse. (b) Matched filter output. (c) Integrator output.

For the special case of a rectangular pulse, the matched filter may be implemented using a circuit known as the *integrate-and-dump circuit*, a block diagram of which is shown in Fig. 7.3. The integrator computes the area under the rectangular pulse, and the resulting output is then sampled at time $t = T$, where T is the duration of the pulse. Immediately after $t = T$, the integrator is restored to its initial condition; hence the name of the circuit. Figure 7.2c shows the output waveform of the integrate-and-dump circuit for the rectangular pulse of Fig. 7.2a. We see that for $0 \leq t \leq T$, the output of this circuit has the *same waveform* as that appearing at the output of the matched filter.

7.3 ERROR RATE DUE TO NOISE

In Section 6.9 we presented a qualitative discussion of the effect of noise on the performance of a binary PCM system. Now that we are equipped with the matched filter as the optimum detector of a known pulse in additive white noise, we are ready to derive a formula for the error rate in such a system due to noise.

To proceed with the analysis, consider a binary PCM system based on *nonre-turn-to-zero (NRZ) signaling*. In this form of signaling, symbols 1 and 0 are represented by positive and negative rectangular pulses of equal amplitude and equal duration. The noise is modeled as *additive white Gaussian noise $w(t)$* of zero mean and power spectral density $N_0/2$; the Gaussian assumption is needed for later

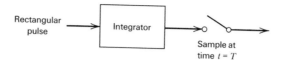

Figure 7.3 Integrate-and-dump circuit.

calculations. In the signaling interval $0 \leq t \leq T_b$, the received signal is thus written as follows:

$$x(t) = \begin{cases} +A + w(t), & \text{symbol 1 was sent} \\ -A + w(t), & \text{symbol 0 was sent} \end{cases} \qquad (7.21)$$

where T_b is the *bit duration*, and A is the *transmitted pulse amplitude*. It is assumed that the receiver has acquired knowledge of the starting and ending times of each transmitted pulse; in other words, the receiver has prior knowledge of the pulse shape, but not its polarity. Given the noisy signal $x(t)$, the receiver is required to make a decision in each signaling interval as to whether the transmitted symbol is a 1 or a 0.

The structure of the receiver used to perform this decision-making process is shown in Fig. 7.4. It consists of a matched filter followed by a sampler, and then finally a decision device. The filter is matched to a rectangular pulse of amplitude A and duration T_b, exploiting the bit-timing information available to the receiver. The resulting matched filter output is sampled at the end of each signaling interval. The presence of receiver noise $w(t)$ adds randomness to the matched filter output.

Let y denote the sample value obtained at the end of a signaling interval. The sample value y is compared to a preset *threshold* λ in the decision device. If the threshold is exceeded, the receiver makes a decision in favor of symbol 1; if not, a decision is made in favor of symbol 0. We adopt the convention that when the sample value y is exactly equal to the threshold λ, the receiver just makes a guess as to which symbol was transmitted; such a decision is the same as obtained by flipping a fair coin, the outcome of which will not alter the average probability of error.

There are two possible kinds of error to be considered:

1. Symbol 1 is chosen when a 0 was actually transmitted; we refer to this error as an *error of the first kind*.

2. Symbol 0 is chosen when a 1 was actually transmitted; we refer to this error as an *error of the second kind*.

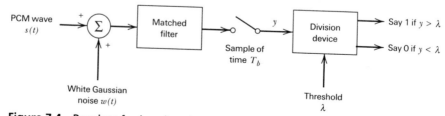

Figure 7.4 Receiver for baseband transmission of binary-encoded PCM wave using NRZ signaling.

To determine the average probability of error, we consider these two situations separately.

Suppose that symbol 0 was sent. Then, according to Eq. (7.21) the received signal is

$$x(t) = -A + w(t), \qquad 0 \le t \le T_b \tag{7.22}$$

Correspondingly, the matched filter output, sampled at time $t = T_b$, is given by (in light of Example 1 with kAT_b set equal to unity for convenience of presentation)

$$y = \int_0^{T_b} x(t) \; dt \tag{7.23}$$

$$= -A + \frac{1}{T_b} \int_0^{T_b} w(t) \; dt$$

which represents the sample value of a random variable Y. By virtue of the fact that the noise $w(t)$ is white and Gaussian, we may characterize the random variable Y as follows:

- The random variable Y is Gaussian distributed with a mean of $-A$.
- The variance of the random variable Y is

$$\sigma_Y^2 = E[(Y + A)^2]$$

$$= \frac{1}{T_b^2} E\left[\int_0^{T_b} \int_0^{T_b} w(t) w(u) \; dt \; du \right]$$

$$= \frac{1}{T_b^2} \int_0^{T_b} \int_0^{T_b} E[w(t) w(u)] \; dt \; du \tag{7.24}$$

$$= \frac{1}{T_b^2} \int_0^{T_b} \int_0^{T_b} R_W(t,u) \; dt \; du$$

where $R_W(t,u)$ is the autocorrelation function of the white noise $w(t)$. Since $w(t)$ is white with a power spectral density $N_0/2$, we have

$$R_W(t,u) = \frac{N_0}{2} \delta(t - u) \tag{7.25}$$

where $\delta(t - u)$ is a time-shifted Dirac delta function. Hence, substituting Eq. (7.25) in (7.24) yields

$$\sigma_Y^2 = \frac{1}{T_b^2} \int_0^{T_b} \int_0^{T_b} \frac{N_0}{2} \delta(t - u) \; dt \; du \tag{7.26}$$

$$= \frac{N_0}{2T_b}$$

The probability density function of the random variable Y, given that symbol 0 was sent, is therefore

$$f_Y(y|0) = \frac{1}{\sqrt{\pi N_0/T_b}} \exp\left(-\frac{(y + A)^2}{N_0/T_b}\right) \quad (7.27)$$

This function is shown plotted in Fig. 7.5(a). Let P_{e0} denote the *conditional probability of error, given that symbol 0 was sent*. This probability is defined by the shaded area under the curve of $f_Y(y|0)$ from the threshold λ to infinity, which corresponds to the range of values assumed by y for a decision in favor of symbol 1. In the absence of noise, the matched filter output y sampled at time $t = T_b$ is equal to $-A$. When noise is present, y occasionally assumes a value greater than λ, in which case an error is made. The probability of this error, conditional on sending symbol 0, is defined by

$$P_{e0} = P(y > \lambda|\text{symbol 0 was sent})$$

$$= \int_\lambda^\infty f_Y(y|0)\, dy \quad (7.28)$$

$$= \frac{1}{\sqrt{\pi N_0/T_b}} \int_\lambda^\infty \exp\left(-\frac{(y + A)^2}{N_0/T_b}\right) dy$$

To proceed further we need to assign an appropriate value to the threshold λ. Such an assignment requires knowledge of the *a priori probabilities* of binary symbols 0 and 1, denoted by p_0 and p_1, respectively. It is clear that we must always have

$$p_0 + p_1 = 1 \quad (7.29)$$

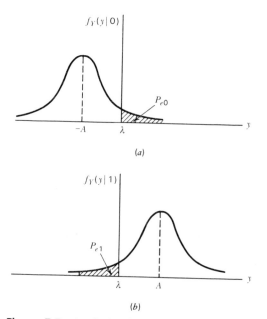

(a)

(b)

Figure 7.5 Analysis of the effect of channel noise on a PCM system. (a) Probability density function of random variable Y at matched filter output when a 0 is transmitted. (b) Probability density function of Y when a 1 is transmitted.

In what follows we assume that symbols 0 and 1 occur with equal probability, in which case we have

$$p_0 = p_1 = \tfrac{1}{2} \tag{7.30}$$

Moreover, in the absence of noise, we note that the sampled value of the matched filter output is $-A$ when symbol 0 is sent, and $+A$ when symbol 1 is sent. Thus, in light of Eq. (7.30) it is reasonable to set the threshold halfway between these two values, that is,

$$\lambda = 0 \tag{7.31}$$

Accordingly, Eq. (7.28) for the conditional probability of error of the first kind takes the form

$$P_{e0} = \frac{1}{\sqrt{\pi N_0 / T_b}} \int_0^{\infty} \exp\left(-\frac{(y + A)^2}{N_0 / T_b}\right) dy \tag{7.32}$$

Define a new variable

$$z = \frac{y + A}{\sqrt{N_0 / T_b}} \tag{7.33}$$

Then we may reformulate Eq. (7.32) as

$$P_{e0} = \frac{1}{\sqrt{\pi}} \int_{\sqrt{E_b/N_0}}^{\infty} \exp(-z^2) \, dz \tag{7.34}$$

where E_b is the *transmitted signal energy per bit*, defined by

$$E_b = A^2 T_b \tag{7.35}$$

At this point we find it convenient to introduce the definition of the so-called *complementary error function*:

$$\text{erfc}(u) = \frac{2}{\sqrt{\pi}} \int_u^{\infty} \exp(-z^2) \, dz \tag{7.36}$$

As discussed in Appendix 7, the complementary error function is closely related to the Gaussian distribution.

We may finally reformulate the conditional probability of error P_{e0} in terms of the complementary error function as follows:

$$P_{e0} = \frac{1}{2}\text{erfc}\left(\sqrt{\frac{E_b}{N_0}}\right) \tag{7.37}$$

Assume next that symbol 1 was transmitted. This time the Gaussian random variable Y represented by the sample value y of the matched filter output has a mean $+A$ and variance $N_0/2T_b$. Note that, compared to the situation when symbol 0 was sent, the mean of the random variable Y has changed but its variance is exactly the same as before. The conditional probability density function of Y, given that symbol 1 was sent, is therefore

$$f_Y(y|1) = \frac{1}{\sqrt{\pi N_0/T_b}} \exp\left(-\frac{(y-A)^2}{N_0/T_b}\right) \tag{7.38}$$

which is plotted in Fig. 7.5b. Let P_{e1} denote the *conditional probability of error, given that symbol 1 was sent*. This probability is defined by the shaded area under the curve of $f_Y(y|1)$ extending from $-\infty$ to the threshold λ, which corresponds to the range of values assumed by y for a decision in favor of symbol 0. In the absence of noise, the matched filter output y sampled at time $t = T_b$ is equal to $+A$. When noise is present, y occasionally assumes a value less than λ, and an error is then made. The probability of this error, conditional on sending symbol 1, is defined by

$$P_{e1} = P(y < \lambda | \text{symbol 1 was sent})$$

$$= \int_{-\infty}^{\lambda} f_Y(y|1)\, dy \tag{7.39}$$

$$= \frac{2}{\sqrt{\pi N_0/T_b}} \int_{-\infty}^{\lambda} \exp\left(-\frac{(y-A)^2}{N_0/T_b}\right) dy$$

Setting the threshold $\lambda = 0$, as before, and putting

$$\frac{y - A}{\sqrt{N_0/T_b}} = -z$$

we readily find that $P_{e1} = P_{e0}$. This result is indeed a consequence of setting the threshold at the midpoint between $-A$ and $+A$, which was justified earlier on the assumption that symbol 0 and 1 are equiprobable. A channel for which the conditional error probabilities P_{e1} and P_{e0} are equal is said to be *binary symmetric.*

To determine the average probability of symbol error in the receiver, we note that the two possible kinds of error considered above are mutually exclusive events in that if the receiver, at a particular sampling instant, chooses symbol 1, then symbol 0 is excluded from appearing, and vice versa. Furthermore, P_{e0} and P_{e1} are conditional probabilities with P_{e0} assuming that symbol 0 was sent and P_{e1} assuming that symbol 1 was sent. Thus, the *average probability of symbol error P_e* in the receiver is given by

$$P_e = p_0 P_{e0} + p_1 P_{e1} \tag{7.40}$$

where p_0 and p_1 are the *a priori* probabilities of binary symbols 0 and 1, respectively. Since $P_{e1} = P_{e0}$, and $p_0 = p_1 = \frac{1}{2}$ in accordance with Eq. (7.30), we obtain

$$P_e = P_{e1} = P_{e0}$$

or

$$P_e = \frac{1}{2}\text{erfc}\left(\sqrt{\frac{E_b}{N_0}}\right) \tag{7.41}$$

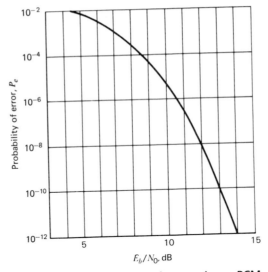

Figure 7.6 Probability of error in a PCM receiver.

We have thus shown that *the average probability of symbol error in a binary-encoded PCM receiver depends solely on E_b/N_0, the ratio of the transmitted signal energy per bit to the noise spectral density.*

In Fig. 7.6 we have used the formula of Eq. (7.41) to plot the average probability of symbol error P_e versus the dimensionless ratio E_b/N_0. This figure shows that the error probability P_e decreases very rapidly as the ratio E_b/N_0 is increased, so that eventually a very "small increase" in transmitted signal energy will make the reception of binary pulses almost error free, as discussed previously in Section 6.9. Note however, that in practical terms the increase in signal energy has to be viewed in the context of what the bias level is; for example, a 3-dB increase in E_b/N_0 is much easier to implement when E_b has a small value than when its value is orders of magnitude larger.

7.4   INTERSYMBOL INTERFERENCE

The next source of bit errors in a baseband-pulse transmission system that we wish to study is intersymbol interference (ISI), which arises when the communication channel is *dispersive*. First of all, however, we need to address a key question: Given a pulse shape of interest, how do we use it to transmit data in *M*-ary form? The answer lies in the use of *discrete pulse modulation*, in which the amplitude, duration, or position of the transmitted pulses is varied in a discrete manner in accordance with the given data stream. However, for the baseband transmission of digital data, the use of *discrete pulse-amplitude modulation* (PAM) is the most efficient one in terms of power and bandwidth utilization. Accordingly, we confine our attention to discrete PAM systems. We begin the study by first considering the case of binary data; later in the chapter, we consider the more general case of *M*-ary data.

Consider then a *baseband binary PAM system*, a generic form of which is shown in Fig. 7.7. The incoming binary sequence $\{b_k\}$ consists of symbols 1 and 0, each

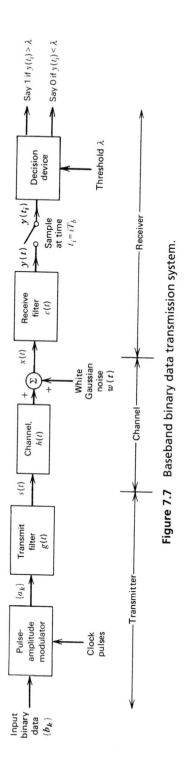

Figure 7.7 Baseband binary data transmission system.

of duration T_b. The *pulse-amplitude modulator* transforms this binary sequence into a new sequence of short pulses (approximating a unit impulse), whose amplitude a_k is represented in the polar form

$$a_k = \begin{cases} +1 & \text{if symbol } b_k \text{ is } 1 \\ -1 & \text{if symbol } b_k \text{ is } 0 \end{cases} \tag{7.42}$$

The sequence of short pulses so produced is applied to a *transmit filter* of impulse response $g(t)$, producing the transmitted signal

$$s(t) = \sum_k a_k g(t - kT_b) \tag{7.43}$$

The signal $s(t)$ is modified as a result of transmission through the *channel* of impulse response $h(t)$. In addition, the channel adds random noise to the signal at the receiver input. The noisy signal $x(t)$ is then passed through a *receive filter* of impulse response $c(t)$. The resulting filter output $y(t)$ is sampled *synchronously* with the transmitter, with the sampling instants being determined by a *clock* or *timing signal* that is usually extracted from the receive filter output. Finally, the sequence of samples thus obtained is used to reconstruct the original data sequence by means of a *decision device*. Specifically, the amplitude of each sample is compared to a *threshold* λ. If the threshold λ is exceeded, a decision is made in favor of symbol 1. If the threshold λ is not exceeded, a decision is made in favor of symbol 0. If the sample amplitude equals the threshold exactly, the flip of a fair coin will determine which symbol was transmitted (i.e., the receiver simply makes a guess).

The receive filter output is written as

$$y(t) = \mu \sum_k a_k p(t - kT_b) + n(t) \tag{7.44}$$

where μ is a scaling factor, and the pulse $p(t)$ is to be defined. To be precise, an arbitrary time delay t_0 should be included in the argument of the pulse $p(t - kT_b)$ in Eq. (7.44) to represent the effect of transmission delay through the system. To simplify the exposition, we have put this delay equal to zero in Eq. (7.44) without loss of generality.

The scaled pulse $\mu p(t)$ is obtained by a double convolution involving the impulse response $g(t)$ of the transmit filter, the impulse response $h(t)$ of the channel, and the impulse response $c(t)$ of the receive filter, as shown by

$$\mu p(t) = g(t) \star h(t) \star c(t) \tag{7.45}$$

where the star denotes convolution. We assume that the pulse $p(t)$ is *normalized* by setting

$$p(0) = 1 \tag{7.46}$$

which justifies the use of μ as a scaling factor to account for amplitude changes incurred in the course of signal transmission through the system.

Since the convolution in the time domain is transformed into multiplication in the frequency domain, we may use the Fourier transform to change Eq. (7.45) into the equivalent form

$$\mu P(f) = G(f)H(f)C(f) \tag{7.47}$$

where $P(f)$, $G(f)$, $H(f)$, and $C(f)$ are the Fourier transforms of $p(t)$, $g(t)$, $h(t)$, and $c(t)$, respectively.

Finally, the term $n(t)$ in Eq. (7.44) is the noise produced at the output of the receive filter due to the additive noise $w(t)$ at the receiver input. It is customary to model $w(t)$ as a white Gaussian noise of zero mean.

The receive filter output $y(t)$ is sampled at time $t_i = iT_b$ (with i taking on integer values), yielding [in light of Eq. (7.46)]

$$
\begin{aligned}
y(t_i) &= \mu \sum_{k=-\infty}^{\infty} a_k p[(i-k)T_b)] + n(t_i) \\
&= \mu a_i + \mu \sum_{\substack{k=-\infty \\ k \neq i}}^{\infty} a_k p[(i-k)T_b)] + n(t_i)
\end{aligned} \tag{7.48}
$$

In Eq. (7.48), the first term μa_i represents the contribution of the ith transmitted bit. The second term represents the residual effect of all other transmitted bits on the decoding of the ith bit; this residual effect due to the occurrence of pulses before and after the sampling instant t_i is called intersymbol interference (ISI). The last term $n(t_i)$ represents the noise sample at time t_i.

In the absence of both ISI and noise, we observe from Eq. (7.48) that

$$y(t_i) = \mu a_i$$

which shows that, under these ideal conditions, the ith transmitted bit is decoded correctly. The unavoidable presence of ISI and noise in the system, however, introduces errors in the decision device at the receiver output. Therefore, in the design of the transmit and receive filters, the objective is to minimize the effects of noise and ISI and thereby deliver the digital data to its destination with the smallest error rate possible.

When the signal-to-noise ratio is high, as is the case in a telephone system, for example, the operation of the system is largely limited by ISI rather than noise; in other words, we may ignore $n(t_i)$. In the next couple of sections, we assume that this condition holds so that we may focus our attention on ISI and the techniques for its control. In particular, the issue we wish to consider is to determine the pulse waveform $p(t)$ for which the ISI is completely eliminated.

7.5   NYQUIST'S CRITERION FOR DISTORTIONLESS BASEBAND BINARY TRANSMISSION

Typically, the transfer function of the channel and the transmitted pulse shape are specified, and the problem is to determine the transfer functions of the transmit and receive filters so as to reconstruct the original binary data sequence $\{b_k\}$. The receiver does this by *extracting* and then *decoding* the corresponding sequence of coefficients, $\{a_k\}$, from the output $y(t)$. The *extraction* involves sampling the output $y(t)$ at time $t = iT_b$. The *decoding* requires that the weighted pulse contribution $a_k p(iT_b - kT_b)$ for $k = i$ be *free* from ISI due to the overlapping tails of all other weighted pulse contributions represented by $k \neq i$. This, in turn, requires that we *control* the overall pulse $p(t)$, as shown by

$$p(iT_b - kT_b) = \begin{cases} 1, & i = k \\ 0, & i \neq k \end{cases} \tag{7.49}$$

where $p(0) = 1$, by normalization. If $p(t)$ satisfies the condition of Eq. (7.49), the receiver output $y(t_i)$ given in Eq. (7.48) simplifies to (ignoring the noise term)

$$y(t_i) = \mu a_i \qquad \text{for all } i$$

which implies zero intersymbol interference. Hence, the condition of Eq. (7.49) ensures *perfect reception in the absence of noise.*

From a design point of view, it is informative to transform the condition of Eq. (7.49) into the frequency domain. Consider then the sequence of samples $\{p(nT_b)\}$, where $n = 0, \pm 1, \pm 2, \ldots$. From the discussion presented in Chapter 6 on the sampling process, we recall that sampling in the time domain produces periodicity in the frequency domain. In particular, we may write

$$P_\delta(f) = R_b \sum_{n=-\infty}^{\infty} P(f - nR_b) \tag{7.50}$$

where $R_b = 1/T_b$ is the *bit rate* in bits per second (b/s); $P_\delta(f)$ is the Fourier transform of an infinite periodic sequence of delta functions of period T_b, whose areas are weighted by the respective sample values of $p(t)$. That is, $P_\delta(f)$ is given by

$$P_\delta(f) = \int_{-\infty}^{\infty} \sum_{m=-\infty}^{\infty} [p(mT_b)\,\delta(t - mT_b)]\exp(-j2\pi ft)\,dt \tag{7.51}$$

Let the integer $m = i - k$. Then, $i = k$ corresponds to $m = 0$, and likewise $i \neq k$ corresponds to $m \neq 0$. Accordingly, imposing the condition of Eq. (7.49) on the sample values of $p(t)$ in the integral of Eq. (7.51), we get

$$P_\delta(f) = \int_{-\infty}^{\infty} p(0)\,\delta(t)\exp(-j2\pi ft)\,dt$$
$$= p(0) \tag{7.52}$$

where we have made use of the sifting property of the delta function. Since from Eq. (7.46) we have $p(0) = 1$, it follows from Eqs. (7.50) and (7.52) that the condition for zero intersymbol interference is satisfied if

$$\sum_{n=-\infty}^{\infty} P(f - nR_b) = T_b \tag{7.53}$$

We may now state the *Nyquist criterion*[3] *for distortionless baseband transmission in the absence of noise: The frequency function P(f) eliminates intersymbol interference for samples taken at intervals T_b provided that it satisfies Eq. (7.53).* Note that $P(f)$ refers to the overall system, incorporating the transmit filter, the channel, and the receive filter in accordance with Eq. (7.47).

Ideal Nyquist Channel

The simplest way of satisfying Eq. (7.53) is to specify the frequency function $P(f)$ to be in the form of a *rectangular function*, as shown by

$$P(f) = \begin{cases} \dfrac{1}{2W}, & -W < f < W \\ 0, & |f| > W \end{cases}$$

$$= \frac{1}{2W} \text{rect}\left(\frac{f}{2W}\right) \tag{7.54}$$

where the overall system bandwidth W is defined by

$$W = \frac{R_b}{2} = \frac{1}{2T_b} \tag{7.55}$$

According to the solution described by Eqs. (7.54) and (7.55), no frequencies of absolute value exceeding half the bit rate are needed. Hence, one signal waveform that produces zero intersymbol interference is defined by the *sinc function*:

$$p(t) = \frac{\sin(2\pi Wt)}{2\pi Wt}$$

$$= \text{sinc}(2Wt) \tag{7.56}$$

The special value of the bit rate $R_b = 2W$ is called the *Nyquist rate*, and W is itself called the *Nyquist bandwidth*. Correspondingly, the ideal baseband pulse transmission system described by Eq. (7.54) in the frequency domain or, equivalently, Eq. (7.56) in the time domain is called the *ideal Nyquist channel*.

Figures 7.8a and 7.8b show plots of $P(f)$ and $p(t)$, respectively. In Fig. 7.8a, the normalized form of the frequency function $P(f)$ is shown plotted for positive and negative frequencies. In Fig. 7.8b, we have also included the signaling intervals and the corresponding centered sampling instants. The function $p(t)$ can be regarded as the impulse response of an ideal low-pass filter with passband amplitude response $1/2W$ and bandwidth W. The function $p(t)$ has its peak value at the origin and goes through zero at integer multiples of the bit duration T_b. It is apparent that if the received waveform $y(t)$ is sampled at the instants of time $t = 0, \pm T_b, \pm 2T_b, \ldots$, then the pulses defined by $\mu p(t - iT_b)$ with arbitrary amplitude μ and index $i = 0, \pm 1, \pm 2, \ldots$, will not interfere with each other. This condition is illustrated in Fig. 7.9 for the binary sequence 1011010.

Although the use of the ideal Nyquist channel does indeed achieve economy in bandwidth in that it solves the problem of zero intersymbol interference with

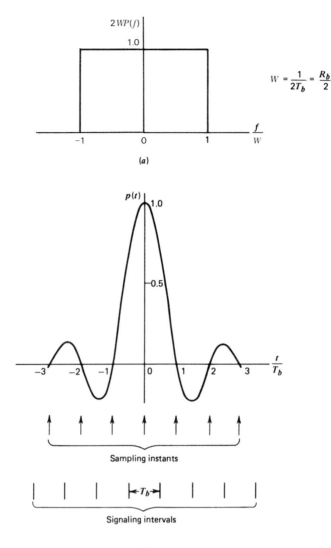

Figure 7.8   (a) Ideal amplitude response. (b) Ideal basic pulse shape.

the minimum bandwidth possible, there are two practical difficulties that make it an undesirable objective for system design:

1. It requires that the amplitude characteristic of $P(f)$ be flat from $-W$ to W, and zero elsewhere. This is physically unrealizable because of the abrupt transitions at the band edges $\pm W$.

2. The function $p(t)$ decreases as $1/|t|$ for large $|t|$, resulting in a slow rate of decay. This is also caused by the discontinuity of $P(f)$ at $\pm W$. Accordingly, there is practically no margin of error in sampling times in the receiver.

   To evaluate the effect of this *timing error*, consider the sample of $y(t)$ at $t = \Delta t$, where Δt is the timing error. To simplify the exposition, we may put the

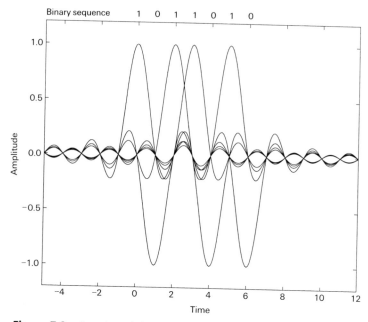

Figure 7.9   A series of sinc pulses corresponding to the sequence 1011010.

correct sampling time t_i equal to zero. In the absence of noise, we thus have

$$y(\Delta t) = \mu \sum_k a_k p(\Delta t - k T_b)$$

$$= \mu \sum_k a_k \frac{\sin[2\pi W(\Delta t - k T_b)]}{2\pi W(\Delta t - k T_b)} \tag{7.57}$$

Since $2W T_b = 1$, by definition, we may rewrite Eq. (7.57) as

$$y(\Delta t) = \mu a_0 \operatorname{sinc}(2W\Delta t) + \frac{\mu \sin(2\pi W\Delta t)}{\pi} \sum_{\substack{k \\ k \neq 0}} \frac{(-1)^k a_k}{(2W\Delta t - k)} \tag{7.58}$$

The first term on the right-hand side of Eq. (7.58) defines the desired symbol, whereas the remaining series represents the intersymbol interference caused by the timing error Δt in sampling the output $y(t)$. Unfortunately, it is possible for this series to diverge, thereby causing erroneous decisions in the receiver.

Raised Cosine Spectrum

We may overcome the practical difficulties encountered with the ideal Nyquist channel by extending the bandwidth from the minimum value $W = R_b/2$ to an adjustable value between W and $2W$. We now specify the frequency function $P(f)$ to satisfy a condition more elaborate than that for the ideal Nyquist channel; specifically, we retain three terms of Eq. (7.53) and restrict the frequency band of interest to $[-W, W]$, as shown by

$$P(f) + P(f - 2W) + P(f + 2W) = \frac{1}{2W}, \qquad -W \le f \le W \quad (7.59)$$

We may devise several band-limited functions that satisfy Eq. (7.59). A particular form of $P(f)$ that embodies many desirable features is provided by a *raised cosine spectrum*. This frequency characteristic consists of a flat portion and a *rolloff* portion that has a sinusoidal form, as follows:

$$P(f) = \begin{cases} \dfrac{1}{2W}, & 0 \le |f| < f_1 \\[2ex] \dfrac{1}{4W}\left\{1 - \sin\left[\dfrac{\pi(|f| - W)}{2W - 2f_1}\right]\right\}, & f_1 \le |f| < 2W - f_1 \\[2ex] 0, & |f| \ge 2W - f_1 \end{cases} \quad (7.60)$$

The frequency parameter f_1 and bandwidth W are related by

$$\alpha = 1 - \frac{f_1}{W} \quad (7.61)$$

The parameter α is called the *rolloff factor*; it indicates the *excess bandwidth* over the ideal solution, W. Specifically, the transmission bandwidth B_T is defined by $2W - f_1 = W(1 + \alpha)$.

The frequency response $P(f)$, normalized by multiplying it by $2W$, is shown plotted in Fig. 7.10a for three values of α, namely, 0, 0.5, and 1. We see that for $\alpha = 0.5$ or 1, the function $P(f)$ cuts off gradually as compared with the ideal Nyquist channel (i.e., $\alpha = 0$) and is therefore easier to implement in practice. Also the function $P(f)$ exhibits odd symmetry with respect to the Nyquist bandwidth W, making it possible to satisfy the condition of Eq. (7.59).

The time response $p(t)$ is the inverse Fourier transform of the function $P(f)$. Hence, using the $P(f)$ defined in Eq. (7.60), we obtain the result (see Problem 7.9)

$$p(t) = \left(\text{sinc}(2Wt)\right)\left(\frac{\cos(2\pi\alpha Wt)}{1 - 16\alpha^2 W^2 t^2}\right) \quad (7.62)$$

which is shown plotted in Fig. 7.10b for $\alpha = 0$, 0.5, and 1.

The function $p(t)$ consists of the product of two factors: the factor $\text{sinc}(2Wt)$ characterizing the ideal Nyquist channel and a second factor that decreases as $1/|t|^2$ for large $|t|$. The first factor ensures zero crossings of $p(t)$ at the desired sampling instants of time $t = iT$ with i an integer (positive and negative). The second factor reduces the tails of the pulse considerably below that obtained from the ideal Nyquist channel, so that the transmission of binary waves using such pulses is relatively insensitive to sampling time errors. In fact, for $\alpha = 1$ we have the most gradual rolloff in that the amplitudes of the oscillatory tails of $p(t)$ are smallest. Thus, the amount of intersymbol interference resulting from timing error decreases as the rolloff factor α is increased from zero to unity.

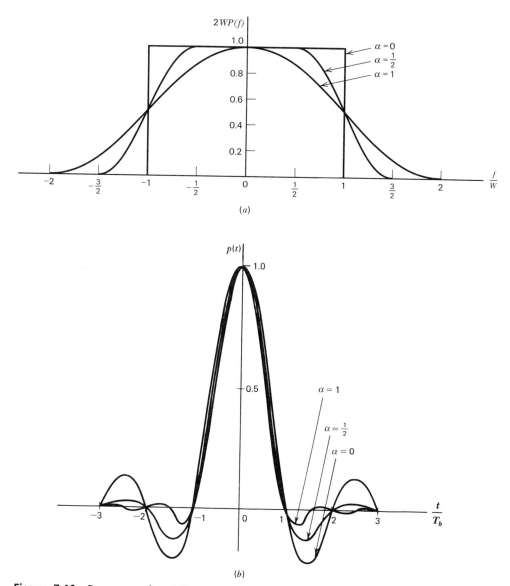

Figure 7.10 Responses for different rolloff factors. (*a*) Frequency response. (*b*) Time response.

The special case with $\alpha = 1$ (i.e., $f_1 = 0$) is known as the *full-cosine rolloff* characteristic, for which the frequency response of Eq. (7.60) simplifies to

$$P(f) = \begin{cases} \dfrac{1}{4W}\left[1 + \cos\left(\dfrac{\pi f}{2W}\right)\right], & 0 < |f| < 2W \\ 0, & |f| \geq 2W \end{cases} \tag{7.63}$$

Correspondingly, the time response $p(t)$ simplifies to

$$p(t) = \frac{\text{sinc}(4Wt)}{1 - 16W^2 t^2} \tag{7.64}$$

This time response exhibits two interesting properties:

1. At $t = \pm T_b/2 = \pm 1/4W$, we have $p(t) = 0.5$; that is, the pulse width measured at half amplitude is exactly equal to the bit duration T_b.
2. There are zero crossings at $t = \pm 3T_b/2, \pm 5T_b/2, \ldots$ in addition to the usual zero crossings at the sampling times $t = \pm T_b, \pm 2T_b, \ldots$.

These two properties are extremely useful in extracting a timing signal from the received signal for the purpose of synchronization. However, the price paid for this desirable property is the use of a channel bandwidth double that required for the ideal Nyquist channel corresponding to $\alpha = 0$.

EXAMPLE 2   Bandwidth Requirement of the T1 System

In Example 3 of Chapter 6, we described the signal format for the T1 carrier system that is used to multiplex 24 independent voice inputs, based on an 8-bit PCM word. It was shown that the bit duration of the resulting time-division multiplexed signal (including a framing bit) is

$$T_b = 0.647 \ \mu s$$

Assuming the use of an ideal Nyquist channel, it follows that the minimum transmission bandwidth B_T of the T1 system is (for $\alpha = 0$)

$$B_T = W = \frac{1}{2T_b} = 772 \text{ kHz}$$

However, a more realistic value for the necessary transmission bandwidth is obtained by using a full-cosine rolloff characteristic with $\alpha = 1$. In this case, we find that

$$B_T = W(1 + \alpha) = 2W = \frac{1}{T_b} = 1.544 \text{ MHz}$$

It is interesting to compare the transmission bandwidth requirement of the T1 system with the minimum bandwidth requirement of a corresponding frequency-division multiplexing (FDM) system. We recall from Chapter 3 that of all the CW modulation techniques, the use of single sideband (SSB) modulation requires the minimum bandwidth possible. Thus, to accommodate an FDM system using SSB modulation to transmit 24 independent voice inputs, and assuming a bandwidth of 4 kHz for each voice input, the channel must provide the transmission bandwidth

$$B_T = 24 \times 4 = 96 \text{ kHz}$$

This is more than an order of magnitude smaller than the minimum bandwidth requirement of the T1 system.

7.6   CORRELATIVE-LEVEL CODING

Thus far we have treated intersymbol interference as an undesirable phenomenon that produces a degradation in system performance. Indeed, its very name connotes a nuisance effect. Nevertheless, by adding intersymbol interference to the transmitted signal in a controlled manner, it is possible to achieve a signaling rate equal to the Nyquist rate of $2W$ symbols per second in a channel of band-width W Hertz. Such schemes are called *correlative-level coding* or *partial-response signaling* schemes.[4] The design of these schemes is based on the following prem-ise: Since intersymbol interference introduced into the transmitted signal is known, its effect can be interpreted at the receiver in a deterministic way. Thus correlative-level coding may be regarded as a practical method of achieving the theoretical maximum signaling rate of $2W$ symbols per second in a bandwidth of W Hertz, as postulated by Nyquist, using realizable and perturbation-tolerant filters.

Duobinary Signaling

The basic idea of correlative-level coding will now be illustrated by considering the specific example of *duobinary signaling*, where "duo" implies doubling of the transmission capacity of a straight binary system. This particular form of corre-lative-level coding is also called *class I partial response*.

Consider a binary input sequence $\{b_k\}$ consisting of uncorrelated binary sym-bols 1 and 0, each having duration T_b. As before, this sequence is applied to a pulse-amplitude modulator producing a two-level sequence of short pulses (ap-proximating a unit impulse), whose amplitude a_k is defined by

$$a_k = \begin{cases} +1 & \text{if symbol } b_k \text{ is 1} \\ -1 & \text{if symbol } b_k \text{ is 0} \end{cases} \qquad (7.65)$$

When this sequence is applied to a *duobinary encoder*, it is converted into a *three-level output*, namely, $-2, 0$, and $+2$. To produce this transformation, we may use the scheme shown in Fig. 7.11. The two-level sequence $\{a_k\}$ is first passed through a simple filter involving a single delay element and summer. For every unit impulse applied to the input of this filter, we get two unit impulses spaced T_b seconds apart at the filter output. We may therefore express the duobinary coder output c_k as the sum of the present input pulse a_k and its previous value a_{k-1}, as shown by

$$c_k = a_k + a_{k-1} \qquad (7.66)$$

One of the effects of the transformation described by Eq. (7.66) is to change the

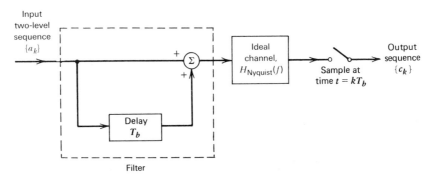

Figure 7.11 Duobinary signaling scheme.

input sequence $\{a_k\}$ of uncorrelated two-level pulses into a sequence $\{c_k\}$ of correlated three-level pulses. This correlation between the adjacent pulses may be viewed as introducing intersymbol interference into the transmitted signal in an artificial manner. However, the intersymbol interference so introduced is under the designer's control, which is the basis of correlative coding.

An ideal delay element, producing a delay of T_b seconds, has the transfer function $\exp(-j2\pi f T_b)$, so that the transfer function of the simple delay-line filter in Fig. 7.11 is $1 + \exp(-j2\pi f T_b)$. Hence, the overall transfer function of this filter connected in cascade with an ideal Nyquist channel is

$$
\begin{aligned}
H_I(f) &= H_{\text{Nyquist}}(f)[1 + \exp(-j2\pi f T_b)] \\
&= H_{\text{Nyquist}}(f)[\exp(j\pi f T_b) + \exp(-j\pi f T_b)]\exp(-j\pi f T_b) \quad (7.67) \\
&= 2H_{\text{Nyquist}}(f)\cos(\pi f T_b)\exp(-j\pi f T_b)
\end{aligned}
$$

where the subscript I in $H_I(f)$ indicates the pertinent class of partial response. For an ideal Nyquist channel of bandwidth $W = 1/2T_b$, we have (ignoring the scaling factor T_b)

$$
H_{\text{Nyquist}}(f) = \begin{cases} 1, & |f| \le 1/2T_b \\ 0, & \text{otherwise} \end{cases} \quad (7.68)
$$

Thus the overall frequency response of the duobinary signaling scheme has the form of a half-cycle cosine function, as shown by

$$
H_I(f) = \begin{cases} 2\cos(\pi f T_b)\exp(-j\pi f T_b), & |f| \le 1/2T_b \\ 0, & \text{otherwise} \end{cases} \quad (7.69)
$$

for which the amplitude response and phase response are as shown in Figs. 7.12a and 7.12b, respectively. An advantage of this frequency response is that it can be easily approximated, in practice, by virtue of the fact that there is continuity at the band edges.

From the first line of Eq. (7.67) and the definition of $H_{\text{Nyquist}}(f)$ in Eq. (7.68), we find that the impulse response corresponding to the transfer function $H_I(f)$ consists of two sinc (Nyquist) pulses that are time-displaced by T_b seconds with respect to each other, as shown by (except for a scaling factor)

$$
\begin{aligned}
h_I(t) &= \frac{\sin(\pi t/T_b)}{\pi t/T_b} + \frac{\sin[\pi(t-T_b)/T_b]}{\pi(t-T_b)/T_b} \\
&= \frac{\sin(\pi t/T_b)}{\pi t/T_b} - \frac{\sin(\pi t/T_b)}{\pi(t-T_b)/T_b} \quad (7.70) \\
&= \frac{T_b^2 \sin(\pi t/T_b)}{\pi t(T_b - t)}
\end{aligned}
$$

The impulse response $h_I(t)$ is shown plotted in Fig. 7.13, where we see that it has only *two* distinguishable values at the sampling instants. The form of $h_I(t)$ shown here explains why we also refer to this type of correlative coding as partial-response signaling. The response to an input pulse is spread over more than one

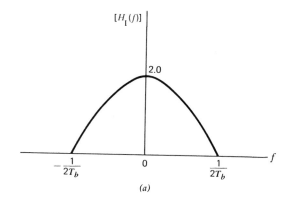

(a)

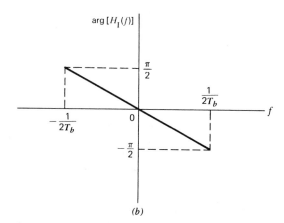

(b)

Figure 7.12 Frequency response of the duo-binary conversion filter. (a) Amplitude response. (b) Phase response.

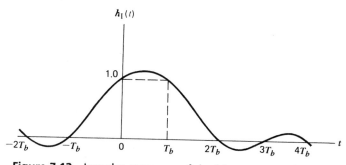

Figure 7.13 Impulse response of duobinary conversion filter.

signaling interval; stated in another way, the response in any signaling interval is "partial." Note also that the tails of $h_1(t)$ decay as $1/|t|^2$, which is a faster rate of decay than the $1/|t|$ encountered in the ideal Nyquist channel.

The original two-level sequence $\{a_k\}$ may be detected from the duobinary-coded sequence $\{c_k\}$ by invoking the use of Eq. (7.66). Specifically, let $\hat{a}_k$ represent the *estimate* of the original pulse a_k as conceived by the receiver at time $t = kT_b$. Then, subtracting the previous estimate $\hat{a}_{k-1}$ from c_k, we get

$$\hat{a}_k = c_k - \hat{a}_{k-1} \qquad (7.71)$$

It is apparent that if c_k is received without error and if also the previous estimate $\hat{a}_{k-1}$ at time $t = (k-1)T_b$ corresponds to a correct decision, then the current estimate $\hat{a}_k$ will be correct too. The technique of using a stored estimate of the previous symbol is called *decision feedback.*

We observe that the detection procedure just described is essentially an inverse of the operation of the simple delay-line filter at the transmitter. However, a major drawback of this detection procedure is that once errors are made, they tend to *propagate* through the output. This is due to the fact that a decision on the current input a_k depends on the correctness of the decision made on the previous input a_{k-1}.

A practical means of avoiding the error propagation phenomenon is to use *precoding* before the duobinary coding, as shown in Fig. 7.14. The precoding operation performed on the binary data sequence $\{b_k\}$ converts it into another binary sequence $\{d_k\}$ defined by

$$d_k = b_k \oplus d_{k-1} \qquad (7.72)$$

where the symbol $\oplus$ denotes *modulo-two addition* of the binary digits b_k and d_{k-1}. This addition is equivalent to a two-input EXCLUSIVE OR operation, which is performed as follows:

$$d_k = \begin{cases} \text{symbol 1} & \text{if either symbol } b_k \text{ or symbol } d_{k-1} \text{ is 1} \\ \text{symbol 0} & \text{otherwise} \end{cases} \qquad (7.73)$$

The precoded binary sequence $\{d_k\}$ is applied to a pulse-amplitude modulator, producing a corresponding two-level sequence of short pulses $\{a_k\}$, where $a_k = \pm 1$ as before. This sequence of short pulses is next applied to the duobinary coder, thereby producing the sequence $\{c_k\}$ that is related to $\{a_k\}$ as follows:

$$c_k = a_k + a_{k-1} \qquad (7.74)$$

Note that unlike the linear operation of duobinary coding, the precoding described by Eq. (7.72) is a *nonlinear* operation.

The combined use of Eqs. (7.72) and (7.74) yields

$$c_k = \begin{cases} 0 & \text{if data symbol } b_k \text{ is 1} \\ \pm 2 & \text{if data symbol } b_k \text{ is 0} \end{cases} \qquad (7.75)$$

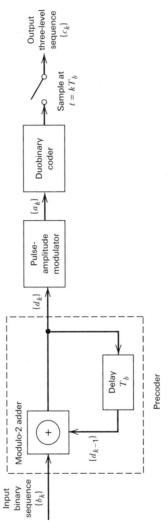

Figure 7.14 A precoded duobinary scheme; details of the duobinary coder are given in Figure 7.11.

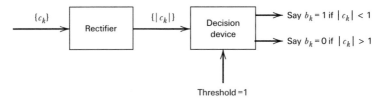

Figure 7.15 Detector for recovering original binary sequence from the precoded duobinary coder output.

which is illustrated in Example 3 below. From Eq. (7.75) we deduce the following decision rule for detecting the original binary sequence $\{b_k\}$ from $\{c_k\}$:

$$
\begin{aligned}
&\text{If } |c_k| < 1, &&\text{say symbol } b_k \text{ is } 1 \\
&\text{If } |c_k| > 1, &&\text{say symbol } b_k \text{ is } 0
\end{aligned}
\tag{7.76}
$$

When $|c_k| = 1$, the receiver simply makes a random guess in favor of symbol 1 or 0. According to this decision rule, the detector consists of a rectifier, the output of which is compared in a decision device to a threshold of 1. A block diagram of the detector is shown in Fig. 7.15. A useful feature of this detector is that no knowledge of any input sample other than the present one is required. Hence, error propagation cannot occur in the detector of Fig. 7.15.

EXAMPLE 3   Duobinary Coding with Precoding

Consider the binary data sequence 0010110. To proceed with the precoding of this sequence, which involves feeding the precoder output back to the input, we add an extra bit to the precoder output. This extra bit is chosen arbitrarily to be 1. Hence, using Eq. (7.73), we find that the sequence $\{d_k\}$ at the precoder output is as shown in row 2 of Table 7.1. The polar representation of the precoded sequence $\{d_k\}$ is shown in row 3 of Table 7.1. Finally, using Eq. (7.74), we find that the duobinary coder output has the amplitude levels given in row 4 of Table 7.1.

To detect the original binary sequence, we apply the decision rule of Eq. (7.76), and so obtain the binary sequence given in row 5 of Table 7.1. This latter result shows that, in the absence of noise, the original binary sequence is detected correctly.

Table 7.1   Illustrating Example 3 on Duobinary Coding

Binary sequence $\{b_k\}$		0	0	1	0	1	1	0
Precoded sequence $\{d_k\}$	1	1	1	0	0	1	0	0
Two-level sequence $\{a_k\}$	+1	+1	+1	−1	−1	+1	−1	−1
Duobinary coder output $\{c_k\}$		+2	+2	0	−2	0	0	−2
Binary sequence obtained by applying decision rule of Eq. (7.76)		0	0	1	0	1	1	0

Modified Duobinary Signaling

In the duobinary signaling technique the transfer function $H(f)$, and consequently the power spectral density of the transmitted pulse, is nonzero at the origin. This is considered to be an undesirable feature in some applications, since many communication channels cannot transmit a dc component. We may correct for this deficiency by using the *class IV partial response* or *modified duobinary* technique, which involves a correlation span of two binary digits. This special form of correlation is achieved by subtracting amplitude-modulated pulses spaced $2T_b$ seconds apart, as indicated in the block diagram of Fig. 7.16. The precoder involves a delay of $2T_b$ seconds. The output of the modified duobinary conversion filter is related to the input two-level sequence $\{a_k\}$ at the pulse-amplitude modulator output as follows:

$$c_k = a_k - a_{k-2} \tag{7.77}$$

Here, again, we find that a three-level signal is generated. With $a_k = \pm 1$, we find that c_k takes on one of three values: $+2$, 0, and -2.

The overall transfer function of the delay-line filter connected in cascade with an ideal Nyquist channel, as in Fig. 7.16, is given by

$$
\begin{aligned}
H_{\text{IV}}(f) &= H_{\text{Nyquist}}(f)[1 - \exp(-j4\pi f T_b)] \\
&= 2j H_{\text{Nyquist}}(f)\sin(2\pi f T_b)\exp(-j2\pi f T_b)
\end{aligned}
\tag{7.78}
$$

where the subscript IV in $H_{\text{IV}}(f)$ indicates the pertinent class of partial response and $H_{\text{Nyquist}}(f)$ is as defined in Eq. (7.68). We therefore have an overall frequency response in the form of a half-cycle sine function, as shown by

$$
H_{\text{IV}}(f) = \begin{cases} 2j\sin(2\pi f T_b)\exp(-j2\pi f T_b), & |f| \le 1/2T_b \\ 0, & \text{elsewhere} \end{cases}
\tag{7.79}
$$

The corresponding amplitude response and phase response of the modified duobinary coder are shown in Figs. 7.17a and 7.17b, respectively. A useful feature of the modified duobinary coder is the fact that its output has no dc component. Owing to this property, modified duobinary signaling is well suited for use in conjunction with single sideband transmission. Note also that this second form of correlative-level coding exhibits the same continuity at the band edges as in duobinary signaling.

From the first line of Eq. (7.78) and the definition of $H_{\text{Nyquist}}(f)$ in Eq. (7.68), we find that the impulse response of the modified duobinary coder consists of two sinc (Nyquist) pulses that are time-displaced by $2T_b$ seconds with respect to each other, as shown by (except for a scaling factor)

$$
\begin{aligned}
h_{\text{IV}}(t) &= \frac{\sin(\pi t/T_b)}{\pi t/T_b} - \frac{\sin[\pi(t - 2T_b)/T_b]}{\pi(t - 2T_b)/T_b} \\
&= \frac{\sin(\pi t/T_b)}{\pi t/T_b} - \frac{\sin(\pi t/T_b)}{\pi(t - 2T_b)/T_b} \\
&= \frac{2T_b^2 \sin(\pi t/T_b)}{\pi t(2T_b - t)}
\end{aligned}
\tag{7.80}
$$

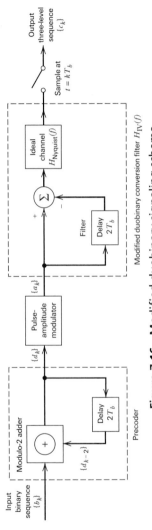

Figure 7.16 Modified duobinary signaling scheme.

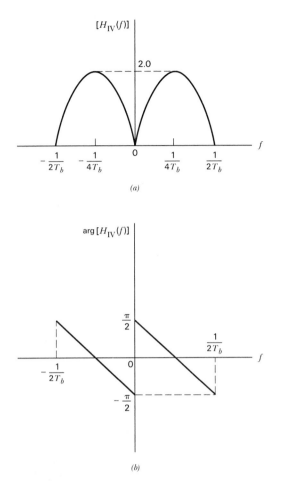

Figure 7.17 Frequency response of modified duobinary conversion filter. (*a*) Amplitude response. (*b*) Phase response.

This impulse response is plotted in Fig. 7.18, which shows that it has *three* distinguishable levels at the sampling instants. Note also that, as with duobinary signaling, the tails of $h_{IV}(t)$ for the modified duobinary signaling decay as $1/|t|^2$.

To eliminate the possibility of error propagation in the modified duobinary system, we use a precoding procedure similar to that used for the duobinary case. Specifically, prior to the generation of the modified duobinary signal, a modulo-two logical addition is used on signals $2T_b$ seconds apart, as shown by (see the front end of Fig. 7.16)

$$d_k = b_k \oplus d_{k-2} \tag{7.81}$$
$$= \begin{cases} \text{symbol } 1 & \text{if either symbol } b_k \text{ or symbol } d_{k-2} \text{ is } 1 \\ \text{symbol } 0 & \text{otherwise} \end{cases}$$

where $\{b_k\}$ is the incoming binary data sequence and $\{d_k\}$ is the sequence at the precoder output. The precoded sequence $\{d_k\}$ thus produced is then applied to

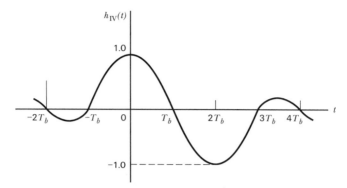

Figure 7.18 Impulse response of the modified duobinary conversion filter.

a pulse-amplitude modulator and then to the modified duobinary conversion filter.

In Fig. 7.16, the output digit c_k equals -2, 0, or $+2$, assuming that the pulse-amplitude modulator uses a polar representation for the precoded sequence $\{d_k\}$. Also we find that the detected digit $\hat{b}_k$ at the receiver output may be extracted from c_k by disregarding the polarity of c_k. Specifically, we may formulate the following decision rule

$$
\begin{aligned}
&\text{If } |c_k| > 1, \qquad \text{say symbol } b_k \text{ is } 1 \\
&\text{If } |c_k| < 1, \qquad \text{say symbol } b_k \text{ is } 0
\end{aligned}
\tag{7.82}
$$

When $|c_k| = 1$, the receiver makes a random guess in favor of symbol 1 or 0. As with the duobinary signaling, we may note the following:

- In the absence of channel noise, the detected binary sequence $\{\hat{b}_k\}$ is exactly the same as the original binary sequence $\{b_k\}$ at the transmitter input.
- The use of Eq. (7.81) requires the addition of two extra bits to the precoded sequence $\{a_k\}$. The composition of the decoded sequence $\{\hat{b}_k\}$ using Eq. (7.82) is invariant to the selection made for these two bits.

Generalized Form of Correlative-Level Coding (Partial-Response Signaling)

The duobinary and modified duobinary techniques have correlation spans of 1 binary digit and 2 binary digits, respectively. It is a straightforward matter to generalize these two techniques to other schemes, which are known collectively as *correlative-level coding* or *partial-response signaling* schemes. This generalization is shown in Fig. 7.19, where $H_{\text{Nyquist}}(f)$ is defined in Eq. (7.68). It involves the use of a tapped-delay-line filter with tap-weights $w_0, w_1, \ldots, w_{N-1}$. Specifically, different classes of partial-response signaling schemes may be achieved by using a weighted linear combination of N ideal Nyquist (sinc) pulses, as shown by

$$
h(t) = \sum_{n=0}^{N-1} w_n \operatorname{sinc}\left(\frac{t}{T_b} - n\right)
\tag{7.83}
$$

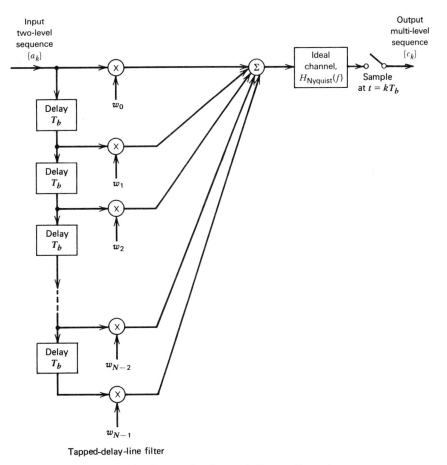

Tapped-delay-line filter

Figure 7.19 Generalized correlative coding scheme.

An appropriate choice of the tap-weights in Eq. (7.83) results in a variety of spectral shapes designed to suit individual applications. Table 7.2 presents the specific details of 5 different classes of partial-response signaling schemes. For example, in the duobinary case (class I partial response), we have

$$w_0 = +1$$
$$w_1 = +1$$

Table 7.2 Different Classes of Partial-Response Signaling Schemes

Type of Class	N	w_0	w_1	w_2	w_3	w_4	Comments
I	2	1	1				Duobinary coding
II	3	1	2	1			
III	3	2	1	-1			
IV	3	1	0	-1			Modified duobinary coding
V	5	-1	0	2	0	-1	

and $w_n = 0$ for $n \geq 2$. In the modified duobinary case (class IV partial response), we have

$$w_0 = +1$$

$$w_1 = 0$$

$$w_2 = -1$$

and $w_n = 0$ for $n \geq 3$.

The useful characteristics of partial-response signaling schemes may now be summarized as follows:

- Binary data transmission over a physical baseband channel can be accomplished at a rate close to the Nyquist rate, using realizable filters with gradual cutoff characteristics.
- Different spectral shapes can be produced, appropriate for the application at hand.

However, these desirable characteristics are achieved at a price: A larger signal-to-noise ratio is required to yield the same average probability of symbol error in the presence of noise as in the corresponding binary PAM systems, because of an increase in the number of signal levels used.

7.7 BASEBAND *M*-ARY PAM TRANSMISSION

In the baseband binary PAM system of Fig. 7.7, the pulse-amplitude modulator produces binary pulses, that is, pulses with one of two possible amplitude levels. On the other hand, in a *baseband M-ary PAM system*, the pulse-amplitude modulator produces one of M possible amplitude levels, with $M > 2$, as illustrated in Fig. 7.20*a* for the case of a *quaternary* ($M = 4$) system and the binary data sequence 0010110111. The corresponding electrical representation for each of the four possible *dibits* (pairs of bits) is shown in Fig. 7.20*b*. In an *M*-ary system, the information source emits a sequence of symbols from an alphabet that consists of M symbols. Each amplitude level at the pulse-amplitude modulator output corresponds to a distinct symbol, so that there are M distinct amplitude levels to be transmitted. Consider then an *M*-ary PAM system with a signal alphabet that contains M equally likely and statistically independent symbols, with the symbol duration denoted by T seconds. We refer to $1/T$ as the *signaling rate* of the system, which is expressed in *symbols per second* or *bauds*. It is informative to relate the signaling rate of this system to that of an equivalent binary PAM system for which the value of M is 2 and the successive binary symbols 1 and 0 are equally likely and statistically independent, with the duration of either symbol denoted by T_b seconds. Under the conditions described here, the binary PAM system produces information at the rate of $1/T_b$ bits per seconds. We also observe that in the case of a quaternary PAM system, for example, the four possible symbols may be identified with the dibits 00, 01, 10, and 11. We thus see that each symbol represents 2 bits of information, and 1 baud is equal to 2 bits per second. We may generalize this result by stating that in an *M*-ary PAM system, 1 baud is equal to $\log_2 M$ bits per second, and the symbol duration T of the *M*-ary PAM system is

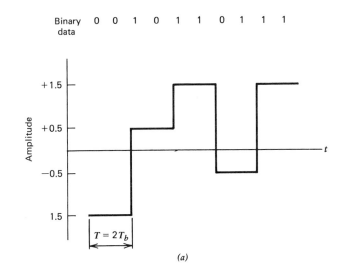

Figure 7.20 Output of a quaternary system. (*a*) Waveform. (*b*) Representation of the 4 possible dibits.

related to the bit duration T_b of the equivalent binary PAM system as

$$T = T_b \log_2 M \qquad (7.84)$$

Therefore, in a given channel bandwidth, we find that by using an *M*-ary PAM system, we are able to transmit information at a rate that is $\log_2 M$ faster than the corresponding binary PAM system. However, to realize the same average probability of symbol error, an *M*-ary PAM system requires more transmitted power. Specifically, we find that for *M* much larger than 2 and an average probability of symbol error small compared to 1, the transmitted power must be increased by a factor of $M^2/\log_2 M$, compared to a binary PAM system.

In a baseband *M*-ary system, first of all, the sequence of symbols emitted by the information source is converted into an *M*-level PAM pulse train by a pulse-amplitude modulator at the transmitter input. Next, as with the binary PAM system, this pulse train is shaped by a transmit filter and then transmitted over the communication channel, which corrupts the signal waveform with both noise and distortion. The received signal is passed through a receive filter and then sampled at an appropriate rate in synchronism with the transmitter. Each sample is compared with preset *threshold* values (also called *slicing* levels), and a decision is made as to which symbol was transmitted. We therefore find that the designs of the pulse-amplitude modulator and the decision-making device in an *M*-ary PAM are more complex than those in a binary PAM system. Intersymbol interference, noise, and imperfect synchronization cause errors to appear at the receive output. The transmit and receive filters are designed to minimize these errors. Procedures used for the design of these filters are similar to those discussed in Sections 7.4 and 7.5 for baseband binary PAM systems.

7.8   TAPPED-DELAY-LINE EQUALIZATION

A communication channel that readily comes to mind for the transmission of digital data (e.g., computer data) is a *telephone channel*, which is characterized by a high signal-to-noise ratio. However, the telephone channel is *bandwidth-limited*, as illustrated in Fig. 7.21 for a typical toll connection. Figure 7.21*a* shows the insertion loss of the channel plotted versus frequency; *insertion loss* (in dB) is defined as $10 \log_{10} (P_0/P_2)$, where P_2 is the power delivered to a load by the channel, and P_0 is the power delivered to the same load when it is connected directly to the source (i.e., the channel is removed). Figure 7.21*b* shows the corresponding plots of the phase response and envelope (group) delay versus frequency; for the definition of envelope delay, see Section 2.14. Figure 7.21 clearly illustrates the dispersive nature of the telephone channel. An efficient approach to *high-speed transmission* of digital data over such a channel uses a combination of two basic signal-processing operations:

- Discrete PAM, which involves encoding the amplitudes of successive pulses in a periodic pulse train with a discrete set of possible amplitude levels.
- A linear modulation scheme, which offers bandwidth conservation to transmit the encoded pulse train over the telephone channel.

At the receiving end of the system, the received signal is demodulated and synchronously sampled, and then decisions are made as to which particular symbols were transmitted. As a result of dispersion of the pulse shape by the telephone channel, we find that the number of detectable amplitude levels is often limited by intersymbol interference rather than by additive noise. In principle, if the channel is known precisely, it is virtually always possible to make the intersymbol interference at the sampling instants arbitrarily small by using a suitable pair of transmit and receive filters, so as to control the overall pulse shape in the manner described previously. The transmit filter is placed directly before the modulator, whereas the receive filter is placed directly after the demodulator. Thus, insofar as intersymbol interference is concerned, we may consider the data transmission over the telephone channel as being baseband.

   In practice, however, we seldom have prior knowledge of the exact channel characteristics. Also, there is the unavoidable problem of imprecision that arises in the physical implementation of the transmit and receive filters. The net result of all these effects is that there will be some residual distortion for ISI to be a limiting factor on the data rate of the system. To compensate for the intrinsic residual distortion, we may use a process known as *equalization*. The filter used to perform such a process is called an *equalizer*.

   A device well-suited for the design of a linear equalizer is the tapped-delay-line filter, as depicted in Fig. 7.22. For symmetry, the total number of taps is

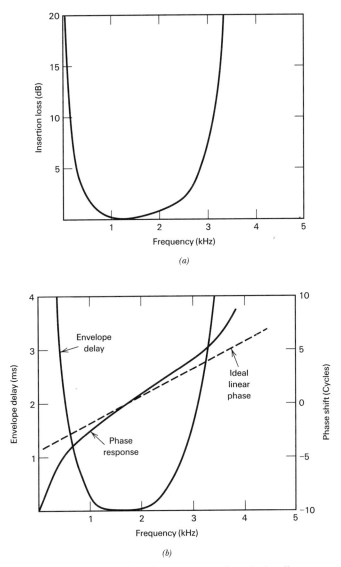

Figure 7.21 (a) Amplitude response of typical toll connection. (b) Envelope delay and phase response of typical toll connection. (Bellamy, 1982.)

chosen to be $(2N + 1)$, with the weights denoted by $w_{-N}, \ldots, w_{-1}, w_0, w_1, \ldots,$ w_N. The impulse response of the tapped-delay-line equalizer is therefore

$$h(t) = \sum_{k=-N}^{N} w_k \, \delta(t - kT) \tag{7.85}$$

where $\delta(t)$ is the Dirac delta function, and the delay T is chosen equal to the symbol duration.

Suppose that the tapped-delay-line-equalizer is connected in cascade with a linear system whose impulse response is $c(t)$, as depicted in Fig. 7.23. Let $p(t)$

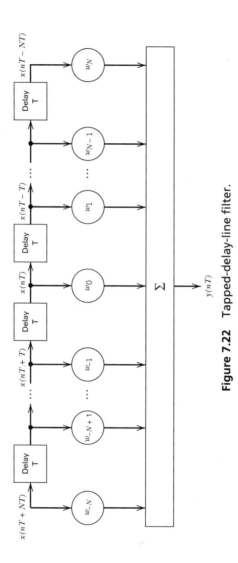

Figure 7.22 Tapped-delay-line filter.

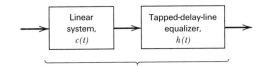

Impulse response, $p(t)$

Figure 7.23 Cascade connection of linear system and tapped-delay-line equalizer.

denote the impulse response of the equalized system. Then $p(t)$ is equal to the convolution of $c(t)$ and $h(t)$, as shown by

$$p(t) = c(t) \bigstar h(t)$$

$$= c(t) \bigstar \sum_{k=-N}^{N} w_k \, \delta(t - kT)$$

Interchanging the order of summation and convolution:

$$p(t) = \sum_{k=-N}^{N} w_k c(t) \bigstar \delta(t - kT)$$

$$= \sum_{k=-N}^{N} w_k c(t - kT) \tag{7.86}$$

where we have made use of the sifting property of the delta function. Evaluating Eq. (7.86) at the sampling times $t = nT$, we get the *discrete convolution sum*

$$p(nT) = \sum_{k=-N}^{N} w_k \, c((n - k)T) \tag{7.87}$$

Note that the sequence $\{p(nT)\}$ is longer than $\{c(nT)\}$.

To eliminate intersymbol interference completely, we must satisfy the Nyquist criterion for distortionless transmission described in Eq. (7.49), with T used in place of T_b. It is assumed that the $p(t)$ is defined in such a way that the normalized condition $p(0) = 1$ is satisfied in accordance with Eq. (7.46). Thus, for no intersymbol interference we require that

$$p(nT) = \begin{cases} 1, & n = 0 \\ 0, & n \neq 0 \end{cases}$$

But from Eq. (7.87) we note that there are only $(2N + 1)$ adjustable coefficients at our disposal. Hence, this ideal condition can only be satisfied approximately as follows:

$$p(nT) = \begin{cases} 1, & n = 0 \\ 0, & n = \pm 1, \pm 2, \ldots, \pm N \end{cases} \tag{7.88}$$

To simplify the notation, we let the nth sample of the impulse response $c(t)$ be written as

$$c_n = c(nT) \qquad (7.89)$$

Then, imposing the condition of Eq. (7.88) on the discrete convolution sum of Eq. (7.87), we obtain a set of $(2N + 1)$ simultaneous equations:

$$\sum_{k=-N}^{N} w_k\, c_{n-k} = \begin{cases} 1, & n = 0 \\ 0, & n = \pm 1, \pm 2, \ldots, \pm N \end{cases} \qquad (7.90)$$

Equivalently, in matrix form we may write

$$
\begin{bmatrix}
c_0 & \cdots & c_{-N+1} & c_{-N} & c_{-N-1} & \cdots & c_{-2N} \\
\vdots & & \vdots & & & & \\
c_{N-1} & \cdots & c_0 & c_{-1} & c_{-2} & \cdots & c_{-N-1} \\
c_N & \cdots & c_1 & c_0 & c_{-1} & \cdots & c_{-N} \\
c_{N+1} & \cdots & c_2 & c_1 & c_0 & \cdots & c_{-N+1} \\
\vdots & & \vdots & \vdots & \vdots & & \vdots \\
c_{2N} & \cdots & c_{N+1} & c_N & c_{N-1} & \cdots & c_0
\end{bmatrix}
\begin{bmatrix}
w_{-N} \\ \vdots \\ w_{-1} \\ w_0 \\ w_1 \\ \vdots \\ w_N
\end{bmatrix}
=
\begin{bmatrix}
0 \\ \vdots \\ 0 \\ 1 \\ 0 \\ \vdots \\ 0
\end{bmatrix}
\qquad (7.91)
$$

A tapped-delay-line equalizer described by Eq. (7.90) or, equivalently Eq. (7.91), is referred to as a *zero-forcing equalizer*. Such an equalizer is optimum in the sense that it minimizes the peak distortion (intersymbol interference). It also has the nice feature of being relatively simple to implement. In theory, the longer we make the equalizer (i.e., permit N to approach infinity), the more closely will the equalized system approach the ideal condition specified by the Nyquist criterion for distortionless transmission.

7.9   ADAPTIVE EQUALIZATION

The zero-forcing strategy described above works well in the laboratory, where we have access to the system to be equalized, in which case we know the system coefficients $c_{-N}, \ldots, c_{-1}, c_0, c_1, \ldots, c_N$ that are needed for the solution of Eq. (7.91). In a telecommunications environment, however, the channel is usually time varying. For example, in a switched telephone network, we find that two factors contribute to the distribution of pulse distortion on different link connections:

• Differences in the transmission characteristics of the individual links that may be switched together.
• Differences in the number of links in a connection.

The result is that the telephone channel is random in the sense of being one of an ensemble of possible physical realizations. Consequently, the use of a fixed equalizer designed on the basis of average channel characteristics may not adequately reduce intersymbol interference. To realize the full transmission capability of a telephone channel, there is need for *adaptive equalization*.[5] The process of equalization is said to be adaptive when the equalizer adjusts itself continuously and automatically by operating on the input signal.

Among the philosophies for adaptive equalization of data transmission systems, we have *prechannel equalization* at the transmitter and *postchannel equalization* at the receiver. Because the first approach requires a feedback channel, we consider only adaptive equalization at the receiving end of the system. This equalization can be achieved, prior to data transmission, by training the filter with the guidance of a suitable *training sequence* transmitted through the channel so as to adjust the filter parameters to optimum values. The typical telephone channel changes little during an average data call, so that precall equalization with a training sequence is sufficient in most cases encountered in practice. As mentioned previously, the equalizer is positioned after the receive filter in the receiver.

In this section we study an adaptive equalizer based on the tapped-delay-line filter, which is *synchronous* in the sense that the tap spacing of the equalizer is the same as the symbol duration T of the transmitted signal (i.e., the reciprocal of the signaling rate). This equalizer is not only simple to implement but is also capable of realizing a satisfactory performance.

Least-Mean-Square Algorithm

Consider a tapped-delay-line equalizer, whose tap-weights are adjustable as indicated in Fig. 7.24. The input sequence $\{x(nT)\}$ applied to this equalizer is produced by the transmission of a binary sequence through an unknown channel that is both dispersive and noisy. It is assumed that some form of pulse shaping is included in the design of the transmission system. The requirement is to correct for the combined effects of residual distortion and noise in the system through the use of an adaptive equalizer.

To simplify notational matters, we let

$$x_n = x(nT) \tag{7.92}$$

$$y_n = y(nT) \tag{7.93}$$

Then, the output y_n of the tapped-delay-line equalizer in response to the input sequence $\{x_n\}$ is defined by the discrete convolution sum (see Fig. 7.24)

$$y_n = \sum_{k=-N}^{N} w_k x_{n-k} \tag{7.94}$$

where w_k is the weight at the kth tap, and $2N + 1$ is the total number of taps. The tap-weights constitute the adaptive filter coefficients. We assume that the input sequence $\{x_n\}$ has finite energy.

The adaptation may be achieved by observing the error between the desired pulse shape and the actual pulse shape at the filter output, measured at the sampling instants, and then using this error to estimate the direction in which the tap-weights of the filter should be changed so as to approach an optimum set of values. For the adaptation, we may use a *peak distortion criterion* that minimizes the peak distortion, defined as the worst-case intersymbol interference at

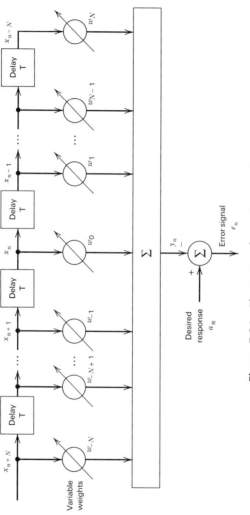

Figure 7.24 Elements of an adaptive filter.

the output of the equalizer. The development of an adaptive equalizer using such a criterion builds on the zero-forcing concept described in the preceding section. However, the equalizer is optimum only when the peak distortion at its input is less than 100 percent (i.e., the intersymbol interference is not too severe). Alternatively, we may use a mean-square error criterion, which is more general in application; also an adaptive equalizer based on the mean-square error criterion appears to be less sensitive to timing perturbations than one based on the peak distortion criterion. Accordingly, in what follows we will use the mean-square error criterion for the development of the adaptive equalizer.

Let a_n denote the *desired response* defined as the polar representation of the nth transmitted binary symbol. Let e_n denote the *error signal* defined as the difference between the desired response a_n and the actual response y_n of the equalizer, as shown by

$$e_n = a_n - y_n \tag{7.95}$$

Then, a criterion commonly used in practice (because of its mathematical tractability) is the *mean-square error*, defined by the cost function

$$\mathscr{E} = E[e_n^2] \tag{7.96}$$

where E is the statistical expectation operator. Using Eqs. (7.94) to (7.96), the gradient of the mean-square error $\mathscr{E}$ with respect to the kth tap-weight w_k may be expressed as

$$
\begin{aligned}
\frac{\partial \mathscr{E}}{\partial w_k} &= 2\, E\left[e_n \frac{\partial e_n}{\partial w_k} \right] \\
&= -2\, E\left[e_n \frac{\partial y_n}{\partial w_k} \right] \\
&= -2\, E[e_n x_{n-k}]
\end{aligned}
\tag{7.97}
$$

The expectation on the right-hand side of Eq. (7.97) is the ensemble-averaged cross-correlation between the error signal e_n and the input signal x_n for a lag of k samples; that is,

$$R_{ex}(k) = E[e_n x_{n-k}] \tag{7.98}$$

We may thus simplify Eq. (7.97) to

$$\frac{\partial \mathscr{E}}{\partial w_k} = -2\, R_{ex}(k) \tag{7.99}$$

The optimality condition for minimum mean-square error may now be expressed simply as

$$\frac{\partial \mathscr{E}}{\partial w_k} = 0 \qquad \text{for } k = 0, \pm 1, \ldots, \pm N \tag{7.100}$$

In light of Eq. (7.99), this condition is equivalent to the requirement that

$$R_{ex}(k) = 0 \qquad \text{for } k = 0, \pm 1, \ldots, \pm N \qquad (7.101)$$

That is, *for minimum mean-square error, the cross-correlation between the output error sequence $\{e_n\}$ and the input sequence $\{x_n\}$ must have zeros for the $(2N + 1)$ components with integer lags corresponding to the index values of the available tap-weights of the filter.* This important result is known as the *principle of orthogonality.*

Substituting Eqs. (7.94) and (7.95) in (7.96) and expanding terms, we find that the mean-square error $\mathscr{E}$ is precisely a second-order function of the tap-weights $w_{-N}, \ldots, w_{-1}, w_0, w_1, \ldots, w_N$. The mean-square error performance of the equalizer may therefore be visualized as a multidimensional bowl-shaped surface that is a parabolic function of the tap-weights. The adaptive process, through successive adjustments of the tap-weights, has the task of continually seeking the *bottom of the bowl*; at this unique point, the mean-square error $\mathscr{E}$ attains its *minimum value* $\mathscr{E}_{\min}$. It is therefore intuitively reasonable that successive adjustments to the tap-weights be in the direction of steepest descent of the error surface (i.e., in a direction opposite to the vector of gradients $\partial \mathscr{E}/\partial w_k$, $-N \leq k \leq N$), which should lead to the minimum mean-square error $\mathscr{E}_{\min}$. This is the basic idea of the *steepest descent algorithm*, described by the recursive formula

$$w_k(n + 1) = w_k(n) - \frac{1}{2}\mu \frac{\partial \mathscr{E}}{\partial w_k}, \qquad k = 0, \pm 1, \ldots, \pm N \qquad (7.102)$$

where μ is a small positive constant called the *step-size parameter*, and the factor $1/2$ has been introduced to cancel the factor 2 in the defining equation for $\partial \mathscr{E}/\partial w_k$. The index n is the iteration number. Thus the use of Eq. (7.99) in (7.102) yields

$$w_k(n + 1) = w_k(n) + \mu R_{ex}(k), \qquad k = 0, \pm 1, \ldots, \pm N \qquad (7.103)$$

The use of the steepest-descent algorithm requires knowledge of the cross-correlation function $R_{ex}(k)$. However, this knowledge is not available when operating in an unknown environment. We may overcome this difficulty by using an *instantaneous estimate* for the cross-correlation function $R_{ex}(k)$. Specifically, on the basis of the defining equation (7.98), we may use the following estimate:

$$\hat{R}_{ex}(k) = e_n x_{n-k}, \qquad k = 0, \pm 1, \ldots, \pm N \qquad (7.104)$$

In a corresponding fashion, we use the estimate $\hat{w}_k(n)$ in place of the tap-weight $w_k(n)$. Naturally, the use of these estimates in Eq. (7.103) results in an *approximation* to the steepest-descent algorithm. We may express the new recursive formula for updating the tap-weights of the equalizer as follows:

$$\hat{w}_k(n + 1) = \hat{w}_k(n) + \mu e_n x_{n-k}, \qquad k = 0, \pm 1, \ldots, \pm N \qquad (7.105)$$

This algorithm is known as the *least-mean-square (LMS) algorithm.*[6] Viewing n as index for the previous iteration, $\hat{w}_k(n)$ is the "old value" of the kth tap-weight, and $\mu e_n x_{n-k}$ is the "correction" applied to it to compute the "updated value" $\hat{w}_k(n + 1)$.

The LMS algorithm is an example of a feedback system, as illustrated in the block diagram of Fig. 7.25. It is therefore possible for the algorithm to diverge (i.e., for the adaptive equalizer to become unstable). Unfortunately, the convergence behavior of the LMS algorithm is difficult to analyze. Nevertheless, provided that the step-size parameter μ is assigned a small value, we find that after a large number of iterations the behavior of the LMS algorithm is roughly similar to that of the steepest-descent algorithm, which uses the actual gradient rather than a noisy estimate for the computation of the tap-weights.

We may simplify the formulation of the LMS algorithm using matrix notation. Let the $(2N + 1)$-by-1 vector $\mathbf{x}_n$ denote the tap-inputs of the equalizer:

$$\mathbf{x}_n = [x_{n+N}, \ldots, x_{n+1}, x_n, x_{n-1}, \ldots, x_{n-N}]^T \tag{7.106}$$

where the superscript T denotes matrix transposition. Correspondingly, let the $(2N + 1)$-by-1 vector $\hat{\mathbf{w}}_n$ denote the tap-weights of the equalizer:

$$\hat{\mathbf{w}}_n = [\hat{w}_{-N}(n), \ldots, \hat{w}_{-1}(n), \hat{w}_0(n), \hat{w}_1(n), \ldots, \hat{w}_N(n)]^T \tag{7.107}$$

We may then use matrix notation to recast the convolution sum of Eq. (7.94) in the compact form

$$y_n = \mathbf{x}_n^T \hat{\mathbf{w}}_n \tag{7.108}$$

where $\mathbf{x}_n^T \hat{\mathbf{w}}_n$ is referred to as the *inner product* of the vectors $\mathbf{x}_n$ and $\hat{\mathbf{w}}_n$. We may now summarize the LMS algorithm as follows:

1. Initialize the algorithm by setting $\hat{\mathbf{w}}_1 = \mathbf{0}$ (i.e., set all the tap-weights of the equalizer to zero at $n = 1$, which corresponds to time $t = T$).

2. For $n = 1, 2, \ldots,$ compute

$$y_n = \mathbf{x}_n^T \hat{\mathbf{w}}_n$$

$$e_n = a_n - y_n$$

$$\hat{\mathbf{w}}_{n+1} = \hat{\mathbf{w}}_n + \mu e_n \mathbf{x}_n$$

where μ is the step-size parameter.

3. Continue the computation until steady-state conditions are reached.

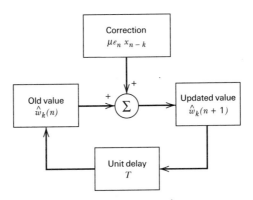

Figure 7.25 Signal-flow graph representation of the LMS algorithm.

Operation of the Equalizer

There are two modes of operation for an adaptive equalizer, namely, the training mode and decision-directed mode, as shown in Fig. 7.26. During the *training mode*, a known sequence is transmitted and a synchronized version of this signal is generated in the receiver, where (after a time shift equal to the transmission delay) it is applied to the adaptive equalizer as the desired response; the tap-weights of the equalizer are thereby adjusted in accordance with the LMS algorithm. A training sequence commonly used in practice is the so-called pseudo-noise (PN) sequence, which consists of a deterministic sequence with noiselike characteristics; a full discussion of this sequence is presented in Chapter 9.

When the training process is completed, the adaptive equalizer is switched to its second mode of operation: the *decision-directed mode*. In this mode of operation, the error signal is defined by

$$e_n = \hat{a}_n - y_n \tag{7.109}$$

where y_n is the equalizer output at time $t = nT$, and $\hat{a}_n$ is the final (not necessarily) correct estimate of the transmitted symbol a_n. Now, in normal operation the decisions made by the receiver are correct with high probability. This means that the error estimates are correct most of the time, thereby permitting the adaptive equalizer to operate satisfactorily. Furthermore, an adaptive equalizer operating in a decision-directed mode is able to *track* relatively slow variations in channel characteristics.

It turns out that the larger the step-size parameter μ, the faster the tracking capability of the adaptive equalizer. However, a large step-size parameter μ may result in an unacceptably high *excess mean-square error,* defined as that part of the mean-square value of the error signal in excess of the minimum attainable value $\mathscr{E}_{min}$ (which results when the tap-weights are at their optimum settings). We therefore find that in practice the choice of a suitable value for the step-size parameter μ involves making a compromise between fast tracking and reducing the excess mean-square error.

Implementation Approaches

An important advantage of the LMS algorithm is that it is simple to implement. The methods of implementing adaptive equalizers may be divided into three

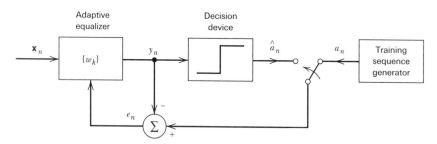

Figure 7.26   Illustrating the two modes of operation of an adaptive equalizer.

broad categories: *analog, hardwired digital,* and *programmable digital,* as described here:

1. The analog approach is primarily based on the use of *charge-coupled device* (CCD) technology. The basic circuit realization of the CCD is a row of field-effect transistors with drains and sources connected in series, and the drains capacitively coupled to gates. The set of adjustable tap-weights are stored in digital memory locations, and the multiplications of the analog sample values by the digitized tap-weights take place in analog fashion. This approach has significant potential in applications where the symbol rate is too high for digital implementation.

2. In hardwired digital implementation of an adaptive equalizer, the equalizer input is first sampled and then quantized into a form suitable for storage in shift registers. The set of adjustable tap-weights is also stored in shift registers. Logic circuits are used to perform the required digital arithmetic (e.g., multiply and accumulate). In this second approach, the circuitry is hardwired for the sole purpose of performing equalization. It is the most widely used method of building adaptive equalizers and lends itself to implementation in very-large-scale integrated (VLSI) circuit form.

3. The use of a programmable digital processor in the form of a *microprocessor,* for example, offers flexibility in that the adaptive equalization is performed as a series of steps or instructions in the microprocessor. An important advantage of this approach is that the same hardware may be time shared to perform a multiplicity of signal-processing functions such as filtering, modulation, and demodulation in a modem (*mo*dulator–*dem*odulator) used to transmit digital data over a telephone channel.

Decision-Feedback Equalization

To develop further insight into adaptive equalization, consider a baseband channel with impulse response denoted in its sampled form by the sequence $\{h_n\}$ where $h_n = h(nT)$. The response of this channel to an input sequence $\{x_n\}$, in the absence of noise, is given by the discrete convolution sum

$$
\begin{aligned}
y_n &= \sum_k h_k x_{n-k} \\
&= h_0 x_n + \sum_{k<0} h_k x_{n-k} + \sum_{k>0} h_k x_{n-k}
\end{aligned}
\tag{7.110}
$$

The first term of Eq. (7.110) represents the desired data symbol. The second term is due to the *precursors* of the channel impulse response that occur before the main sample h_0 associated with the desired data symbol. The third term is due to the *postcursors* of the channel impulse response that occur after the main sample h_0. The precursors and postcursors of a channel impulse response are illustrated in Fig. 7.27. The idea of *decision-feedback equalization*[7] is to use data decisions made on the basis of precursors of the channel impulse response to take care of the postcursors; for the idea to work, however, the decisions would obviously have to be correct. Provided that this condition is satisfied, a decision-

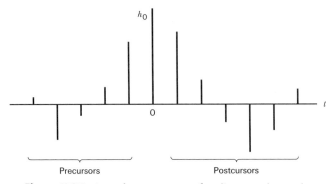

Figure 7.27 Impulse response of a discrete channel.

feedback equalizer is able to provide an improvement over the performance of the tapped-delay-line equalizer.

A *decision-feedback equalizer* consists of a feedforward section, a feedback section, and a decision device connected together as shown in Fig. 7.28. The feedforward section consists of a tapped-delay-line filter whose taps are spaced at the reciprocal of the signaling rate. The data sequence to be equalized is applied to this section. The feedback section consists of another tapped-delay-line filter whose taps are also spaced at the reciprocal of the signaling rate. The input applied to the feedback section consists of the decisions made on previously detected symbols of the input sequence. The function of the feedback section is to subtract out that portion of the intersymbol interference produced by previously detected symbols from the estimates of future samples.

Note that the inclusion of the decision device in the feedback loop makes the equalizer intrinsically *nonlinear* and therefore more difficult to analyze than an ordinary tapped-delay-line equalizer. Nevertheless, the mean-square error criterion can be used to obtain a mathematically tractable optimization of a decision-feedback equalizer. Indeed, the LMS algorithm can be used to jointly adapt both the feedforward tap-weights and the feedback tap-weights based on a *common* error signal. To be specific, let the augmented vector $\mathbf{c}_n$ denote the combination of the feedforward and feedback tap-weights, as shown by

$$\mathbf{c}_n = \begin{bmatrix} \hat{\mathbf{w}}_n^{(1)} \\ \hat{\mathbf{w}}_n^{(2)} \end{bmatrix} \qquad (7.111)$$

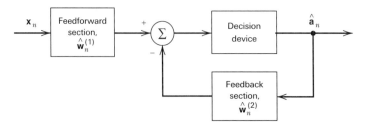

Figure 7.28 Block diagram of decision-feedback equalizer.

where the vector $\hat{\mathbf{w}}_n^{(1)}$ denotes the tap-weights of the feedforward section, and $\hat{\mathbf{w}}_n^{(2)}$ denotes the tap-weights of the feedback section. Let the augmented vector $\mathbf{v}_n$ denote the combination of input samples for both sections:

$$\mathbf{v}_n = \begin{bmatrix} \mathbf{x}_n \\ \hat{\mathbf{a}}_n \end{bmatrix} \tag{7.112}$$

where $\mathbf{x}_n$ is the vector of tap-inputs in the feedforward section, and $\hat{\mathbf{a}}_n$ is the vector of tap-inputs (i.e., present and past decisions) in the feedback section. The common error signal is defined by

$$e_n = a_n - \mathbf{c}_n^T \mathbf{v}_n \tag{7.113}$$

where the superscript T denotes matrix transposition, and a_n is the polar representation of the nth transmitted binary symbol. The LMS algorithm for the decision-feedback equalizer is described by the update equations:

$$\hat{\mathbf{w}}_{n+1}^{(1)} = \hat{\mathbf{w}}_n^{(1)} + \mu_1 e_n \mathbf{x}_n \tag{7.114}$$

$$\hat{\mathbf{w}}_{n+1}^{(2)} = \hat{\mathbf{w}}_n^{(2)} + \mu_2 e_n \hat{\mathbf{a}}_n \tag{7.115}$$

where μ_1 and μ_2 are the step-size parameters for the feedforward and feedback sections, respectively.

A decision-feedback equalizer yields good performance in the presence of moderate to severe intersymbol interference as experienced in a fading radio channel, for example.

7.10 EYE PATTERN

In previous sections of this chapter we have discussed various techniques for dealing with the effects of receiver noise and intersymbol interference on the performance of a baseband pulse-transmission system. In the final analysis, what really matters is how to evaluate the combined effect of these impairments on overall system performance in an operational environment. An experimental tool for such an evaluation in an insightful manner is the so-called *eye pattern*, which is defined as the synchronized superposition of all possible realizations of the signal of interest (e.g., received signal, receiver output) viewed within a particular signaling interval. The eye pattern derives its name from the fact that it resembles the human eye for binary waves. The interior region of the eye pattern is called the *eye opening*.

An eye pattern provides a great deal of useful information about the performance of a data transmission system, as described in Fig. 7.29. Specifically, we may make the following statements:

- The width of the eye opening defines the *time interval over which the received signal can be sampled without error from intersymbol interference*; it is apparent that the preferred time for sampling is the instant of time at which the eye is open the widest.

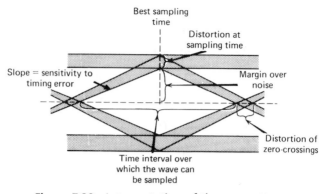

Figure 7.29   Interpretation of the eye pattern.

- The *sensitivity of the system to timing errors* is determined by the rate of closure of the eye as the sampling time is varied.
- The height of the eye opening, at a specified sampling time, defines the *noise margin* of the system.

When the effect of intersymbol interference is severe, traces from the upper portion of the eye pattern cross traces from the lower portion, with the result that the eye is completely closed. In such a situation, it is impossible to avoid errors due to the combined presence of intersymbol interference and noise in the system.

In the case of an *M*-ary system, the eye pattern contains $(M - 1)$ eye openings stacked up vertically one on the other, where M is the number of discrete amplitude levels used to construct the transmitted signal. In a strictly linear system with truly random data, all these eye openings would be identical. In practice, however, it is often possible to discern asymmetries in the eye pattern, which are caused by nonlinearities in the communication channel.

COMPUTER EXPERIMENT   Eye Diagrams for Binary and Quaternary Systems

Figures 7.30*a* and 7.30*b* show the eye diagrams for a baseband PAM transmission system using $M = 2$ and $M = 4$, respectively. The channel has no bandwidth limitation, and the source symbols used are randomly generated on a computer. A raised cosine pulse is used in both cases. The system parameters used for the generation of these eye diagrams are as follows: Nyquist bandwidth $W = 0.5$ Hz, and rolloff factor $\alpha = 0.5$. For the binary case of $M = 2$, the symbol duration T and the bit duration T_b are the same, with $T_b = 1$ s. For the case of $M = 4$, we have $T = T_b \log_2 M = 2T_b$. In both cases, we see that the *eyes* are open, indicating reliable operation of the system.

Figures 7.31*a* and 7.31*b* show the eye diagrams for these baseband-pulse transmission systems using the same system parameters as before, but this time under a bandwidth-limited condition. Specifically, the channel is now modeled by a low-pass *Buttworth filter*, whose frequency response is defined by

$$|H(f)| = \frac{1}{1 + (f/f_0)^{2N}}$$

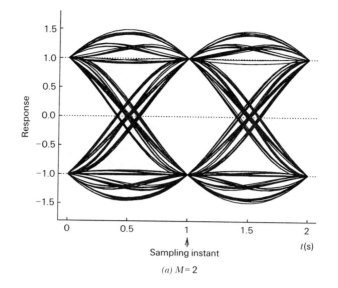

(a) $M = 2$

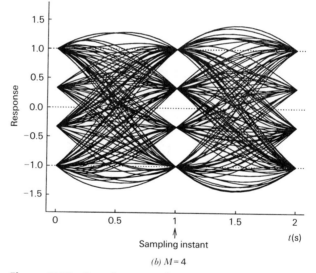

(b) $M = 4$

Figure 7.30 Eye diagram of received signal with no bandwidth limitation.

where N is the *order* of the filter, and f_0 is its 3-dB cutoff frequency. For the computer experiment described in Fig. 7.31, the following values are used:

$$N = 25 \quad \text{and} \quad f_0 = 0.975 \text{ Hz}$$

The bandwidth required by the PAM transmission system is computed to be

$$B_T = W(1 + \alpha) = 0.75 \text{ Hz}$$

Although the channel bandwidth cutoff frequency is greater than absolutely necessary, its effect on the passband is observed in a decrease in the size of the eye opening. Instead of the distinct values at time $t = 1$s (as shown in the previous figures), now there is a blurred region. If the channel bandwidth were reduced

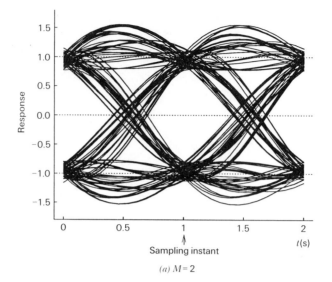

(a) $M = 2$

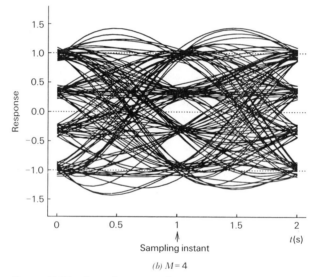

(b) $M = 4$

Figure 7.31 Eye diagram of received signal, using a
bandwidth-limited channel response.

further, the eye would close even more until finally no distinct eye opening would
be recognizable.

7.11 SUMMARY AND DISCUSSION

In this chapter, we studied the effects of noise and intersymbol interference on
the performance of baseband-pulse transmission systems. Intersymbol interference (ISI) is different from noise in that it is a *signal-dependent* form of interference that arises because of deviations in the frequency response of a channel

from the ideal low-pass filter (Nyquist channel); it disappears when the transmitted signal is switched off. The result of these deviations is that the received pulse corresponding to a particular data symbol is affected by (1) the tail ends of the pulses representing the previous symbols and (2) the front ends of the pulses representing the subsequent symbols.

Depending on the received signal-to-noise ratio, we may distinguish three different situations that can arise in baseband-pulse transmission systems for channels with fixed characteristics:

1. *The effect of ISI is negligible in comparison to that of the noise.* The proper procedure in this case is to use a matched filter, which is the optimum linear time-invariant filter for maximizing the peak pulse signal-to-noise ratio.

2. *The received signal-to-noise ratio is high enough to ignore the effect of noise.* In this case, we need to guard against the effects of ISI on the reconstruction of the transmitted data at the receiver. In particular, control must be exercised over the shape of the received pulse. This design objective can be achieved in one of two different ways:

   - Using a raised cosine spectrum for the overall frequency response of the baseband-pulse transmission system.

   - Using correlative-level coding or partial-response signaling that adds ISI to the transmitted signal in a controlled manner.

3. *The ISI and noise are both significant.* The solution for this third situation requires the joint optimization of the transmit and receive filters. In brief, a suitable pulse shape is first used to reduce the ISI to zero at the sampling instants, and then Schwarz's inequality is invoked to maximize the output signal-to-noise ratio at the sampling instants.[8]

When, however, the channel is random in the sense of being one of an ensemble of possible physical realizations, which is frequently the case in a telecommunications environment, the use of fixed filter designs based on average channel characteristics may not be adequate. In situations of this kind, the preferred approach is to use an adaptive equalizer, positioned after the receive filter in the receiver. The purpose of the adaptive equalizer is to compensate for variations in the frequency response of the channel automatically during the course of data transmission. The combined use of a tapped-delay-line filter and the least-mean-square (LMS) algorithm for adjusting the tap-weights provides the basis of a simple and yet highly effective method for implementing the adaptive equalizer. Such a device is capable of dealing with the combined effects of ISI and receiver noise in a nonstationary environment. Its practical value lies in the fact that almost every modem (modulator–demodulator) in commercial use today for the transmission of digital data over a voice-grade telephone channel uses an adaptive equalizer as an integral part.

NOTES AND REFERENCES

1. The classic books on baseband-pulse transmission are Lucky, Salz, and Weldon (1968) and Sunde (1969). For detailed treatment of different aspects of the subject, see Gitlin, Hayes, and Weinstein (1992), Proakis (1989), and Benedetto, Biglieri, and Castellani (1987).

2. The characterization of a matched filter was first derived by North in a classified report (RCA Laboratories Report PTR-6C, June 1943), which was published 20 years later: see the paper by North (1963). A similar result was obtained independently by Van Vleck and Middleton, who coined the term "matched filter": see the paper by Van Vleck and Middleton (1946). For review material on the matched filter and its properties, see the papers by Turin (1960, 1976).

3. The criterion described by Eq. (7.49) or Eq. (7.53) was first formulated by Nyquist in the study of telegraph transmission theory; the 1928 paper by Nyquist is a classic. In the literature, this criterion is referred to as *Nyquist's first criterion*. In his 1928 paper, Nyquist described another method, referred to in the literature as *Nyquist's second criterion*. The second method makes use of the instants of transition between unlike symbols in the received signal rather than centered samples. A discussion of the first and second criteria is presented in Bennett (1970, pp. 78–92) and in the paper by Gibby and Smith (1965). A third criterion attributed to Nyquist is discussed in Sunde (1969); see also the papers by Pasupathy (1974) and Sayar and Pasupathy (1987).

4. Correlative-level coding and partial-response signaling are synonymous; both terms are used in the literature. The idea of correlative coding was originated by Lender (1963). Lender's work was generalized for binary data transmission by Kretzmer (1966). For further details on correlative coding techniques, see the book by Gitlin, Hayes, and Weinstein (1992); see also the papers by Pasupathy (1977), Kabal and Pasupathy (1975), and Sousa and Pasupathy (1983).

5. Adaptive equalization of telephone channels was pioneered by Lucky (1965, 1966). Since that time, numerous adaptive equalization schemes have been published in the literature, which provide equalization for specific synchronous data-transmission systems. For review papers on adaptive equalization, see Proakis (1975) and Qureshi (1982, 1985). Adaptive equalization is also discussed in detail in the books by Gitlin, Hayes, and Weinstein (1992, Chapter 8) and Proakis (1989, Chapter 6).

6. The LMS algorithm was originated by Widrow and Hoff, Jr. (1960). For a detailed convergence analysis of the LMS algorithm, see Haykin (1991, pp. 314–336), and Widrow and Stearns (1985, Chapter 6).

7. Decision-feedback equalization was first described in a report by Austin (1967). The optimization of the decision-feedback equalizer for minimum mean-squared error was first accomplished by Monsen (1971). A readable account of decision-feedback equalization is presented in the book by Gitlin, Hayes, and Weinstein (1992, pp. 500–510).

8. The joint optimization of transmit and receive filters, in the combined presence of intersymbol interference and noise, is discussed in Shanmugam (1979), pp. 197–201.

PROBLEMS

Problem 7.1 Consider the signal $s(t)$ shown in Fig. P7.1.

(a) Determine the impulse response of a filter matched to this signal and sketch it as a function of time.

(b) Plot the matched filter output as a function of time.

(c) What is the peak value of the output?

Problem 7.2 It is proposed to implement a matched filter in the form of a tapped-delay-line filter with a set of tap-weights $\{w_k, k = 0, 1, \ldots, K\}$. Given a signal $s(t)$ of duration T seconds to which the filter is matched, find the value of w_k. Assume that the signal is uniformly sampled.

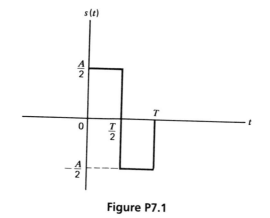

Figure P7.1

Problem 7.3 Consider a rectangular pulse defined by

$$g(t) = \begin{cases} A, & 0 \leq t \leq T \\ 0, & \text{otherwise} \end{cases}$$

It is proposed to approximate the matched filter for $g(t)$ by an ideal low-pass filter of bandwidth B; maximization of the peak pulse signal-to-noise ratio is the primary objective.

(a) Determine the optimum value of B for which the ideal low-pass filter provides the best approximation to the matched filter.

(b) By how many decibels is the ideal low-pass filter worse off than the matched filter?

Problem 7.4 A binary PCM wave uses on–off signaling to transmit symbols 1 and 0; symbol 1 is represented by a rectangular pulse of amplitude A and duration T_b. The additive noise at the receiver input is white and Gaussian, with zero mean and power spectral density $N_0/2$. Assuming that symbols 1 and 0 occur with equal probability, find an expression for the average probability of error at the receiver output, using a matched filter as described in Section 7.3.

Problem 7.5 A binary PCM system, using NRZ signaling, operates just above the error threshold with an average probability of error equal to 10^{-6}. Suppose that the signaling rate is doubled. Find the new value of the average probability of error. You may use Table 1 of Appendix 7 to evaluate the complementary error function.

Problem 7.6 A continuous-time signal is sampled and then transmitted as a PCM signal. The random variable at the input of the decision device in the receiver has a variance of 0.01 volts^2.

(a) Assuming the use of NRZ signaling, determine the pulse amplitude that must be transmitted for the average error rate not to exceed 1 bit in 10^8 bits.

(b) If the added presence of interference causes the error rate to increase to 1 bit in 10^6 bits, what is the variance of the interference?

Problem 7.7 In a binary PCM system, symbols 0 and 1 have *a priori probabilities* p_0 and p_1, respectively. The conditional probability density function of the random variable Y (with sample value y) obtained by sampling the matched filter output in the receiver of Fig. 7.4 at the end of a signaling interval, given that symbol 0 was transmitted, is denoted by $f_Y(y|0)$. Similarly, $f_Y(y|1)$ denotes the conditional probability density function of Y, given that symbol 1 was transmitted. Let λ denote the threshold used in the receiver, so that if the sample value y exceeds λ, the receiver decides in favor of symbol 1; otherwise, it decides in favor of symbol 0. Show that the optimum threshold λ_{opt}, for which the average probability of error is a minimum, is given by the solution of

$$\frac{f_Y(\lambda_{opt}|1)}{f_Y(\lambda_{opt}|0)} = \frac{p_0}{p_1}$$

Problem 7.8 The overall pulse shape $p(t)$ of a baseband binary PAM system is defined by

$$p(t) = \text{sinc}\left(\frac{t}{T_b}\right)$$

where T_b is the bit duration of the input binary data. The amplitude levels at the pulse modulator output are $+1$ or -1, depending on whether the binary symbol at the input is 1 or 0, respectively. Sketch the waveform at the output of the receive filter in response to the input data 001101001.

Problem 7.9 Determine the inverse Fourier transform of the frequency function $P(f)$ defined in Eq. (7.60).

Problem 7.10 An analog signal is sampled, quantized, and encoded into a binary PCM wave. The specifications of the PCM system include the following:

Sampling rate = 8 kHz
Number of representation levels = 64

The PCM wave is transmitted over a baseband channel using discrete pulse-amplitude modulation. Determine the minimum bandwidth required for transmitting the PCM wave if each pulse is allowed to take on the following number of amplitude levels: 2, 4, or 8.

Problem 7.11 Consider a baseband binary PAM system that is designed to have a raised-cosine spectrum $P(f)$. The resulting pulse $p(t)$ is defined in Eq. (7.62). How would this pulse be modified if the system was designed to have a linear phase response?

Problem 7.12 A computer puts out binary data at the rate of 56 kb/s. The computer output is transmitted using a baseband binary PAM system that is designed to have a raised-cosine spectrum. Determine the transmission bandwidth required for each of the following rolloff factors: $\alpha = 0.25, 0.5, 0.75, 1.0$.

Problem 7.13 Repeat Problem 7.12, given that each set of three successive binary digits in the computer output are coded into one of eight possible amplitude levels, and the resulting signal is transmitted using an eight-level PAM system designed to have a raised-cosine spectrum.

Problem 7.14 An analog signal is sampled, quantized, and encoded into a binary PCM wave. The number of representation levels used is 128. A synchronizing pulse is added at the end of each code word representing a sample of the analog signal. The resulting PCM wave is transmitted over a channel of bandwidth 12 kHz using a quaternary PAM system with raised-cosine spectrum. The rolloff factor is unity.

(a) Find the rate (b/s) at which information is transmitted through the channel.
(b) Find the rate at which the analog signal is sampled. What is the maximum possible value for the highest frequency component of the analog signal?

Problem 7.15 A binary PAM wave is to be transmitted over a baseband channel with an absolute maximum bandwidth of 75 kHz. The bit duration is 10 μs. Find a raised-cosine spectrum that satisfies these requirements.

Problem 7.16 The duobinary, ternary, and bipolar signaling techniques have one common feature: They all employ three amplitude levels. In what way does the duobinary technique differ from the other two?

Problem 7.17 The binary data stream 001101001 is applied to the input of a duobinary system.

(a) Construct the duobinary coder output and corresponding receiver output, without a precoder.
(b) Suppose that owing to error during transmission, the level at the receiver input produced by the second digit is reduced to zero. Construct the new receiver output.

Problem 7.18 Repeat Problem 7.17, assuming the use of a precoder in the transmitter.

Problem 7.19 The scheme shown in Fig. P7.2 may be viewed as a differential encoder (consisting of the modulo-2 adder and the 1-unit delay element) connected in cascade with a special form of correlative coder (consisting of the 1-unit delay element and summer). A single delay element is shown in Fig. P7.2 since it is common to both the differential encoder and the correlative coder. In this differential encoder, a transition is represented by symbol 0 and no transition by symbol 1.

(a) Find the frequency response and impulse response of the correlative coder part of the scheme shown in Fig. P7.2.
(b) Show that this scheme may be used to convert the on–off representation of a binary sequence (applied to the input) into the bipolar representation of the sequence at the output. You may illustrate this conversion by considering the sequence 010001101.

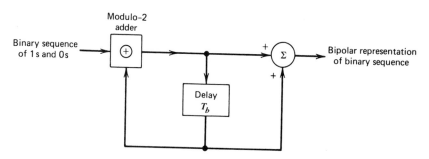

Figure P7.2

For descriptions of on–off, bipolar, and differential encoding of binary sequences, see Section 6.8.

Problem 7.20 Consider a random binary wave $x(t)$ in which the 1s and 0s occur with equal probability, the symbols in adjacent time slots are statistically independent, and symbol 1 is represented by A volts and symbol 0 by zero volts. This on–off binary wave is applied to the circuit of Fig. P7.2.

(a) Using the result of Problem 7.19, show that the power spectral density of the bipolar wave $y(t)$ appearing at the output of the circuit equals

$$S_Y(f) = T_b A^2 \sin^2(\pi f T_b)\operatorname{sinc}^2(\pi f T_b)$$

(b) Plot the power spectral densities of the on–off and bipolar binary waves, and compare them.

Problem 7.21 The binary data stream 011100101 is applied to the input of a modified duobinary system.

(a) Construct the modified duobinary coder output and corresponding receiver output, without a precoder.

(b) Suppose that due to error during transmission, the level produced by the third digit is reduced to zero. Construct the new receiver output.

Problem 7.22 Repeat Problem 7.21 assuming the use of a precoder in the transmitter.

Problem 7.23 Consider a baseband *M*-ary system using M discrete amplitude levels. The receiver model is as shown in Fig. P7.3, the operation of which is governed by the following assumptions:

(a) The signal component in the received wave is

$$m(t) = \sum_n a_n \operatorname{sinc}\left(\frac{t}{T} - n\right)$$

where $1/T$ is the signaling rate in bauds.

(b) The amplitude levels are $a_n = \pm A/2, \pm 3A/2, \ldots, \pm(M-1)A/2$ if M is even, and $a_n = 0, \pm A, \ldots, \pm(M-1)A/2$ if M is odd.

(c) The M levels are equiprobable, and the symbols transmitted in adjacent time slots are statistically independent.

(d) The noise $w(t)$ at the receiver input is white and Gaussian with zero mean and power spectral density $N_0/2$.

(e) The low-pass filter is ideal with bandwidth $B = 1/2T$.

(f) The threshold levels used in the decision device are $0, \pm A, \ldots, \pm(M-3)A/2$ if M is even, and $\pm A/2, \pm 3A/2, \ldots, \pm(M-3)A/2$ if M is odd.

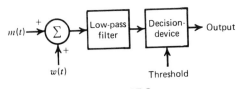

Figure P7.3

The average probability of symbol error in this system is defined by

$$P_e = \left(1 - \frac{1}{M}\right)\text{erfc}\left(\frac{A}{2\sqrt{2}\sigma}\right)$$

where σ is the standard deviation of the noise at the input of the decision device. Demonstrate the validity of this general formula by determining P_e for the following three cases: $M = 2, 3, 4$.

Problem 7.24 Suppose that in a baseband M-ary PAM system with M equally likely amplitude levels, as described in Problem 7.23, the average probability of symbol error P_e is less than 10^{-6} so as to make the occurrence of decoding errors negligible. Show that the minimum value of received signal-to-noise ratio in such a system is approximately given by

$$(\text{SNR})_{R,\min} \simeq 7.8(M^2 - 1)$$

Problem 7.25 Some radio systems suffer from *multipath distortion*, which is caused by the existence of more than one propagation path between the transmitter and the receiver. Consider a channel the output of which, in response to a signal $s(t)$, is defined by (in the absence of noise)

$$x(t) = K_1 s(t - t_{01}) + K_2 s(t - t_{02})$$

where K_1 and K_2 are constants, and t_{01} and t_{02} represent transmission delays. It is proposed to use the three-tap delay-line-filter of Fig. P7.4 to equalize the multipath distortion produced by this channel.

(a) Evaluate the transfer function of the channel.
(b) Evaluate the parameters of the tapped-delay-line filter in terms of K_1, K_2, t_{01}, and t_{02}, assuming that $K_2 \ll K_1$ and $t_{02} > t_{01}$.

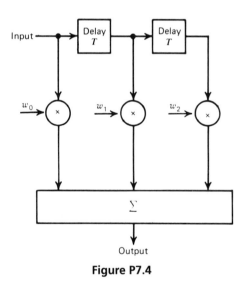

Figure P7.4

Problem 7.26 Let the sequence $\{x(nT)\}$ denote the input applied to a tapped-delay-line equalizer. Show that intersymbol interference is eliminated completely by the equalizer provided that its transfer function satisfies the condition

$$H(f) = \frac{T}{\sum_k X(f - k/T)}$$

where T is the symbol duration.

As the number of taps in the equalizer approaches infinity, the transfer function of the equalizer becomes a Fourier series with real coefficients and can therefore approximate any function in the interval $(-1/2T, 1/2T)$. Demonstrate this property of the equalizer.

CHAPTER 8

Digital Passband Transmission

8.1 INTRODUCTION

In *baseband pulse transmission*, which we studied in the previous chapter, a data stream represented in the form of a discrete pulse-amplitude modulated (PAM) signal is transmitted directly over a low-pass channel. An issue of particular concern in baseband pulse transmission is that of pulse shaping designed to bring the intersymbol interference (ISI) problem under control. In *digital passband transmission*, on the other hand, the incoming data stream is modulated onto a carrier (usually sinusoidal) with fixed frequency limits imposed by a band-pass channel of interest; digital passband transmission is studied in the present chapter. The major issue of concern here is the optimum design of the receiver so as to minimize the average probability of symbol error in the presence of noise. This does not mean, of course, that noise is of no concern in baseband pulse transmission, nor does it mean that ISI is of no concern in digital passband transmission; it merely points out the issues that are of high priority in these two different domains of data transmission.

The communication channel used for passband data transmission may be a microwave radio link, a satellite channel, or the like. In any event, the modulation process making the transmission possible involves switching (keying) the amplitude, frequency, or phase of a sinusoidal carrier in some fashion in accordance with the incoming data. Thus there are three basic signaling schemes known as *amplitude-shift keying* (ASK), *frequency-shift keying* (FSK), and *phase-shift keying* (PSK), which may be viewed as special cases of amplitude modulation, frequency modulation, and phase modulation, respectively. A distinguishing feature of FSK and PSK signals is that ideally they both have a constant envelope. This feature makes them impervious to amplitude nonlinearities, commonly encountered in microwave radio links and satellite channels. It is for this reason that we find that, in practice, FSK and PSK signals are preferred to ASK signals for digital passband transmission over nonlinear channels.

In this chapter we study digital passband transmission techniques with emphasis on the following issues: (1) *optimum design* of the receiver in the sense that it will make fewer errors in the long run than any other receiver, (2) calculation of the *average probability of symbol error* of the receiver, and (3) *spectral properties* of the modulated signals. Two different cases are considered in the study: *coherent receivers* and *noncoherent receivers*. In a coherent receiver the receiver is *phase locked* to the transmitter, whereas in a noncoherent receiver there is *no* phase synchronization between the local oscillator used in the receiver for demodulation and the oscillator supplying the sinusoidal carrier in the transmitter for modulation.

8.2   PASSBAND TRANSMISSION MODEL

We may model a digital passband transmission system as shown in Fig. 8.1. First, there is assumed to exist a *message source* that emits one *symbol* every T seconds, with the symbols belonging to an alphabet of M symbols, which we denote by $m_1, m_2, \ldots, m_M$. Consider, for example, the remote connection of two digital computers, with one computer acting as an information source that calculates digital outputs based on observations and inputs fed into it. The resulting computer output is expressed as a sequence of 0s and 1s, which are transmitted to a second computer. In this example, the alphabet consists simply of the two binary symbols 0 and 1. A second example is that of a quaternary PCM encoder with an alphabet consisting of four possible symbols: 00, 01, 10, and 11. In any event, the *a priori probabilities* $P(m_1), P(m_2), \ldots, P(m_M)$ specify the message source output. In the absence of prior information to the contrary, we assume that the M symbols of the alphabet are *equally likely*. Then we may write

$$p_i = P(m_i)$$
$$= \frac{1}{M} \quad \text{for all } i \tag{8.1}$$

The M-ary output of the message source is presented to a *signal transmission encoder*, producing a corresponding vector $\mathbf{s}_i$ made up of N real elements, one such set for each of the M symbols of the source alphabet; the dimension N is less than or equal to M. With the vector $\mathbf{s}_i$ as input, the *modulator* then constructs a *distinct* signal $s_i(t)$ of duration T seconds as the representation of the symbol m_i generated by the message source. The signal $s_i(t)$ is necessarily of finite energy, as shown by

$$E_i = \int_0^T s_i^2(t)\, dt, \quad i = 1, 2, \ldots, M \tag{8.2}$$

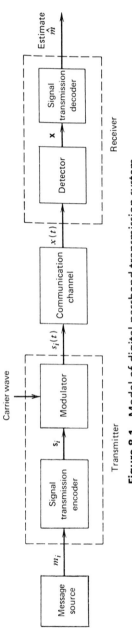

Figure 8.1 Model of digital passband transmission system.

Note that $s_i(t)$ is real valued. One such signal is transmitted every T seconds. The particular signal chosen for transmission depends in some fashion on the incoming message and possibly on the signals transmitted in preceding time slots. With a sinusoidal carrier, the feature that is used by the modulator to distinguish one signal from another is a *step* change in the amplitude, frequency, or phase of the carrier. (Sometimes, a hybrid form of modulation is used, combining changes in both amplitude and phase or amplitude and frequency.) The result of the modulation process is amplitude-shift keying (ASK), frequency-shift keying (FSK), or phase-shift keying (PSK), respectively, as illustrated in Fig. 8.2 for the special case of a source of binary data for which the *symbol duration T* is the same as the *bit duration T_b*. It is of interest to note that although in general it is not easy to distinguish between frequency-modulated and phase-modulated signals (on an oscilloscope, say), this is not so in the case of FSK and PSK signals; for example, compare the waveforms in Figs. 8.2*b* and 8.2*c*.

Returning to the model of Fig. 8.1, the bandpass communication channel, coupling the transmitter to the receiver, is assumed to have two characteristics:

1. The channel is linear, with a bandwidth that is wide enough to accommodate the transmission of the modulated signal $s_i(t)$ with negligible or no distortion.

2. The transmitted signal $s_i(t)$ is perturbed by an *additive, zero-mean, stationary, white, Gaussian noise* process, a sample function of which is denoted by $w(t)$. The reasons for this assumption are that it makes calculations tractable, and also it is a reasonable description of the type of noise present in many practical communication systems.

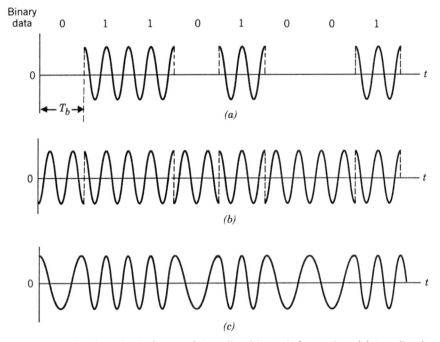

Figure 8.2   The three basic forms of signaling binary information. (*a*) Amplitude-shift keying. (*b*) Phase-shift keying. (*c*) Frequency-shift keying with continuous phase.

We refer to such a channel as an *additive white Gaussian noise* (AWGN) *channel.* Accordingly, we may express the *received signal x(t)* as

$$x(t) = s_i(t) + w(t), \qquad \begin{cases} 0 \le t \le T \\ i = 1, 2, \ldots, M \end{cases} \tag{8.3}$$

We may thus model the channel as in Fig. 8.3.

The receiver has the task of observing the received signal $x(t)$ for a duration of T seconds and making a best *estimate* of the transmitted signal $s_i(t)$ or, equivalently, the symbol m_i. This task is accomplished in two stages. The first stage is a *detector* that operates on the received signal $x(t)$ to produce an observation vector **x**. By using the observation vector **x**, prior knowledge of the modulation format used in the transmitter, and the a priori probabilities $P(m_i)$, the *signal transmission decoder* constituting the second stage of the receiver produces an estimate $\hat{m}$. However, owing to the presence of additive noise at the receiver input, this decision-making process is statistical in nature, with the result that the receiver will make occasional errors. The requirement is to design the receiver so as to minimize the *average probability of symbol error* defined as

$$P_e = \sum_{i=1}^{M} P(\hat{m} \ne m_i) P(m_i) \tag{8.4}$$

where m_i is the transmitted symbol, $\hat{m}$ is the estimate produced by the decision device, and $P(\hat{m} \ne m)$ is the conditional error probability given that the ith symbol was sent. The resulting receiver is said to be *optimum in the minimum probability of error* sense.

It is customary to assume that the receiver is *time synchronized* with the transmitter, which means that the receiver knows the instants of time when the modulation changes state. Sometimes, it is also assumed that the receiver is *phase locked* to the transmitter. In such a case, we speak of *coherent detection*, and we refer to the receiver as a *coherent receiver*. On the other hand, there may be no phase synchronism between transmitter and receiver. In this second case, we speak of *noncoherent detection*, and we refer to the receiver as a *noncoherent receiver*. In this chapter, we assume the existence of time synchronism; however, we shall distinguish between coherent and noncoherent detection.

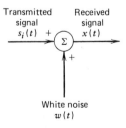

Figure 8.3 Model of additive white Gaussian noise channel.

The model described above provides a basis for the design of the optimum receiver, for which we will use *geometric representation* of the known set of transmitted signals, $\{s_i(t)\}$. This method provides a great deal of insight, with considerable simplification of detail.

8.3 GRAM–SCHMIDT ORTHOGONALIZATION PROCEDURE

According to the model of Fig. 8.1, the task of transforming an incoming message m_i, $i = 1, 2, \ldots, M$, into a modulated wave $s_i(t)$ may be divided into separate discrete-time and continuous-time operations. The justification for this separation lies in the *Gram–Schmidt orthogonalization procedure*, which permits the representation of any set of M energy signals, $\{s_i(t)\}$, as linear combinations of N *orthonormal basis functions*, where $N \leq M$. That is to say, we may represent the given set of real-valued energy signals $s_1(t), s_2(t), \ldots, s_M(t)$, each of duration T seconds, in the form

$$s_i(t) = \sum_{j=1}^{N} s_{ij} \phi_j(t), \qquad \begin{cases} 0 \leq t \leq T \\ i = 1, 2, \ldots, M \end{cases} \tag{8.5}$$

where the coefficients of the expansion are defined by

$$s_{ij} = \int_0^T s_i(t)\,\phi_j(t)\ dt, \qquad \begin{cases} i = 1, 2, \ldots, M \\ j = 1, 2, \ldots, N \end{cases} \tag{8.6}$$

The real-valued basis functions $\phi_1(t), \phi_2(t), \ldots, \phi_N(t)$ are *orthonormal*, by which we mean

$$\int_0^T \phi_i(t)\phi_j(t)\ dt = \begin{cases} 1 & \text{if } i = j \\ 0 & \text{if } i \neq j \end{cases} \tag{8.7}$$

The first condition of Eq. (8.7) states that each basis function is *normalized* to have unit energy. The second condition states that the basis functions $\phi_1(t)$, $\phi_2(t), \ldots, \phi_N(t)$ are *orthogonal* with respect to each other over the interval $0 \leq t \leq T$.

The coefficient s_{ij} may be viewed as the jth element of the N-dimensional vector $\mathbf{s}_i$ in Fig. 8.1. Given the N elements of the vector $\mathbf{s}_i$, that is, $s_{i1}, s_{i2}, \ldots, s_{iN}$, operating as input, we may use the scheme shown in Fig. 8.4a to generate the signal $s_i(t)$, which follows directly from Eq. (8.5). It consists of a bank of N multipliers, with each multiplier supplied with its own basis function, followed by a summer. This scheme may be viewed as performing a similar role to that of the second stage or modulator in the transmitter of Fig. 8.1. Conversely, given the signals $s_i(t)$, $i = 1, 2, \ldots, M$, operating as input, we may use the scheme shown in Fig. 8.4b to calculate the coefficients $s_{i1}, s_{i2}, \ldots, s_{iN}$ which follows directly from Eq. (8.7). This second scheme consists of a bank of N *product-integrators* or *correlators* with a common input, and with each one supplied with its own basis function. Later, we show that such a bank of correlators may be used as the first stage or detector in the receiver of Fig. 8.1.

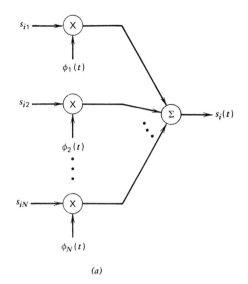

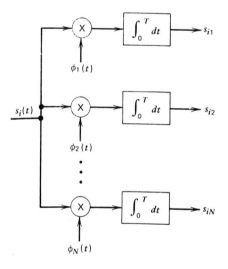

Figure 8.4 (a) Scheme for generating the signal $s_i(t)$. (b) Scheme for generating the set of coefficients $\{s_i\}$.

To prove the Gram–Schmidt orthogonalization procedure, we may proceed by defining the first basis function as

$$\phi_1(t) = \frac{s_1(t)}{\sqrt{E_1}} \tag{8.8}$$

where E_1 is the energy of the signal $s_1(t)$. Then, clearly, we have

$$
\begin{aligned}
s_1(t) &= \sqrt{E_1}\,\phi_1(t) \\
&= s_{11}\phi_1(t)
\end{aligned}
\tag{8.9}
$$

where the coefficient $s_{11} = \sqrt{E_1}$ and $\phi_1(t)$ has unit energy, as required.

Next, using the signal $s_2(t)$, we define the coefficient s_{21} as

$$
s_{21} = \int_0^T s_2(t)\,\phi_1(t)\ dt
\tag{8.10}
$$

We may thus introduce a new intermediate function

$$
g_2(t) = s_2(t) - s_{21}\phi_1(t)
\tag{8.11}
$$

which is orthogonal to $\phi_1(t)$ over the interval $0 \leq t \leq T$. Now, we are ready to define the second basis function as

$$
\phi_2(t) = \frac{g_2(t)}{\sqrt{\int_0^T g_2^2(t)\ dt}}
\tag{8.12}
$$

Substituting Eq. (8.11) in (8.12) and simplifying, we get the desired result

$$
\phi_2(t) = \frac{s_2(t) - s_{21}\phi_1(t)}{\sqrt{E_2 - s_{21}^2}}
\tag{8.13}
$$

where E_2 is the energy of the signal $s_2(t)$. It is clear from Eq. (8.12) that

$$
\int_0^T \phi_2^2(t)\ dt = 1
$$

and from Eq. (8.13) that

$$
\int_0^T \phi_1(t)\phi_2(t)\ dt = 0
$$

That is to say, $\phi_1(t)$ and $\phi_2(t)$ form an orthonormal set, as required.

Continuing in this fashion, we may in general define

$$
g_i(t) = s_i(t) - \sum_{j=1}^{i-1} s_{ij}\phi_j(t)
\tag{8.14}
$$

where the coefficients s_{ij} are themselves defined by

$$
s_{ij} = \int_0^T s_i(t)\,\phi_j(t)\ dt, \qquad j = 1, 2, \ldots, i-1
\tag{8.15}
$$

Given the $g_i(t)$, we may now define the set of basis functions

$$\phi_i(t) = \frac{g_i(t)}{\sqrt{\int_0^T g_i^2(t)\ dt}}, \qquad i = 1, 2, \ldots, N \tag{8.16}$$

which form an orthonormal set. The dimension N is less than or equal to the number of given signals, M, depending on one of two possibilities:

- The signals $s_1(t), s_2(t), \ldots, s_M(t)$ form a *linearly independent set*, in which case $N = M$.
- The signals $s_1(t), s_2(t), \ldots, s_M(t)$ are *not* linearly independent, in which case $N < M$, and the intermediate function $g_i(t)$ is zero for $i > N$.

Note that the conventional Fourier series expansion of a periodic signal is an example of a particular expansion of this type. Also, the representation of a band-limited signal in terms of its samples taken at the Nyquist rate may be viewed as another example of a particular expansion of this type. There are, however, two important distinctions that should be made:

1. The form of the basis functions $\phi_1(t), \phi_2(t), \ldots, \phi_N(t)$ has not been specified. That is to say, unlike the Fourier series expansion of a periodic signal or the sampled representation of a band-limited signal, we have not restricted the Gram–Schmidt orthogonalization procedure to be in terms of sinusoidal functions or sinc functions of time.

2. The expansion of the signal $s_i(t)$ in terms of a finite number of terms is not an approximation wherein only the first N terms are significant but rather an *exact* expression where N and only N terms are significant.

EXAMPLE 1

Consider the signals $s_1(t), s_2(t), s_3(t)$, and $s_4(t)$ shown in Fig. 8.5a. We wish to use the Gram–Schmidt orthogonalization procedure to find an orthonormal basis for this set of signals.

Step 1 We note that the energy of signal $s_1(t)$ is

$$E_1 = \int_0^T s_1^2(t)\ dt$$

$$= \int_0^{T/3} (1)^2\ dt$$

$$= \frac{T}{3}$$

The first basis function $\phi_1(t)$ is therefore [see Eq. (8.8)]

$$\phi_1(t) = \frac{s_1(t)}{\sqrt{E_1}}$$

$$= \begin{cases} \sqrt{3/T}, & 0 \le t \le T/3 \\ 0, & \text{elsewhere} \end{cases}$$

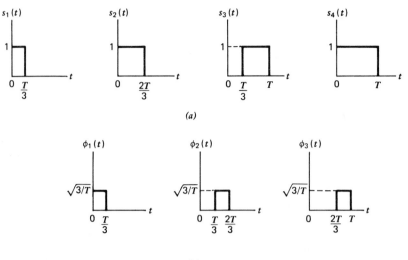

Figure 8.5 (a) Set of signals to be orthonormalized. (b) The resulting set of orthonormal functions.

Step 2 From Eq. (8.10), we find that

$$s_{21} = \int_0^T s_2(t)\,\phi_1(t)\ dt$$

$$= \int_0^{T/3} (1)\left(\sqrt{\frac{3}{T}}\right) dt$$

$$= \sqrt{\frac{T}{3}}$$

The energy of signal $s_2(t)$ is

$$E_2 = \int_0^T s_2^2(t)$$

$$= \int_0^{2T/3} (1)^2\ dt$$

$$= \frac{2T}{3}$$

The second basis function $\phi_2(t)$ is therefore [see Eq. (8.13)]

$$\phi_2(t) = \frac{s_2(t) - s_{21}\phi_1(t)}{\sqrt{E_2 - s_{21}^2}}$$

$$= \begin{cases} \sqrt{3/T}, & T/3 \le t \le 2T/3 \\ 0, & \text{elsewhere} \end{cases}$$

Step 3 Using Eq. (8.15), we find that

$$s_{31} = \int_0^T s_3(t)\,\phi_1(t)\ dt$$

$$= 0$$

and the coefficient s_{32} equals

$$s_{32} = \int_0^T s_3(t)\,\phi_2(t)\ dt$$

$$= \int_{T/3}^{2T/3} (1)\left(\sqrt{\frac{3}{T}}\right)\ dt$$

$$= \sqrt{\frac{T}{3}}$$

The corresponding value of the intermediate function $g_i(t)$, with $i = 3$, is therefore [see Eq. (8.14)]

$$g_3(t) = s_3(t) - s_{31}\phi_1(t) - s_{32}\phi_2(t)$$

$$= \begin{cases} 1, & 2T/3 \le t \le T \\ 0, & \text{elsewhere} \end{cases}$$

Using Eq. (8.16), we find that the third basis function $\phi_3(t)$ is

$$\phi_3(t) = \frac{g_3(t)}{\sqrt{\displaystyle\int_0^T g_3^2(t)\ dt}}$$

$$= \begin{cases} \sqrt{3/T}, & 2T/3 \le t \le T \\ 0, & \text{elsewhere} \end{cases}$$

Finally, using Eq. (8.14) with $i = 4$, we find that $g_4(t) = 0$ and the orthogonalization process is completed.

The three basis functions $\phi_1(t)$, $\phi_2(t)$, and $\phi_3(t)$ form an orthonormal set, as shown in Fig. 8.5b. In this example, we thus have $M = 4$ and $N = 3$, which means that the four signals $s_1(t)$, $s_2(t)$, $s_3(t)$, and $s_4(t)$ described in Fig. 8.5a do not form a linearly independent set. This is readily confirmed by noting that $s_4(t) = s_1(t) + s_3(t)$. Moreover, we note that any of these four signals can be expressed as a linear combination of the three basis functions, which is the essence of the Gram–Schmidt orthogonalization procedure.

8.4 GEOMETRIC INTERPRETATION OF SIGNALS

Once we have adopted a convenient set of orthonormal basis functions $\{\phi_j(t)\,|\,j = 1, 2, \ldots, N\}$, then each signal in the set $\{s_i(t)\,|\,i = 1, 2, \ldots, M\}$ may be expanded as in Eq. (8.5), reproduced here for convenience:

$$s_i(t) = \sum_{j=1}^{N} s_{ij}\phi_j(t), \qquad \begin{cases} 0 \leq t \leq T \\ i = 1, 2, \ldots, M \end{cases} \tag{8.17}$$

The coefficients of the expansion s_{ij} are themselves defined by Eq. (8.6), also reproduced here for convenience:

$$s_{ij} = \int_0^T s_i(t)\,\phi_j(t)\ dt, \qquad \begin{cases} i = 1, 2, \ldots, M \\ j = 1, 2, \ldots, N \end{cases} \tag{8.18}$$

Accordingly, we may state that each signal in the set $\{s_i(t)\}$ is completely determined by the *vector* of its coefficients

$$\mathbf{s}_i = \begin{bmatrix} s_{i1} \\ s_{i2} \\ \cdot \\ \cdot \\ \cdot \\ s_{iN} \end{bmatrix}, \qquad i = 1, 2, \ldots, M \tag{8.19}$$

The vector $\mathbf{s}_i$ is called the *signal vector*. Furthermore, if we conceptually extend our conventional notion of two- and three-dimensional Euclidean spaces to an *N-dimensional Euclidean space*, we may visualize the set of signal vectors $\{\mathbf{s}_i | i = 1, 2, \ldots, M\}$ as defining a corresponding set of M points in an N-dimensional Euclidean space, with N mutually perpendicular axes labeled $\phi_1, \phi_2, \ldots, \phi_N$. This N-dimensional Euclidean space is called the *signal space*.

The idea of visualizing a set of energy signals geometrically, as described above, is of profound importance. It provides the mathematical basis for the geometric representation of energy signals, thereby paving the way for the noise analysis of digital passband transmission systems in a conceptually satisfying manner. This form of representation is illustrated in Fig. 8.6 for the case of a two-dimensional signal space with three signals, that is, $N = 2$ and $M = 3$.

In an N-dimensional Euclidean space, we may define *lengths* of vectors and *angles* between vectors. It is customary to denote the length (also called the *absolute value* or *norm*) of a signal vector $\mathbf{s}_i$ by the symbol $\|\mathbf{s}_i\|$. The squared-length of any signal vector $\mathbf{s}_i$ is defined to be the *inner product* or *dot product* of $\mathbf{s}_i$ with itself, as shown by

$$\begin{aligned} \|\mathbf{s}_i\|^2 &= \mathbf{s}_i^T \mathbf{s}_i \\ &= \sum_{j=1}^{N} s_{ij}^2 \end{aligned} \tag{8.20}$$

where s_{ij} is the jth element of $\mathbf{s}_i$, and the superscript T denotes matrix transposition.

The *cosine of the angle* between two vectors is defined as the inner product of the two vectors divided by the product of their individual norms. Let θ_{ij} denote

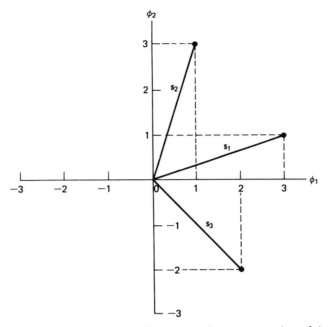

Figure 8.6 Illustrating the geometric representation of signals for the case when $N = 2$ and $M = 3$.

the angle between the vectors $\mathbf{s}_i$ and $\mathbf{s}_j$. The cosine of this angle is defined by

$$\cos\theta_{ij} = \frac{\mathbf{s}_i^T \mathbf{s}_j}{\|\mathbf{s}_i\|\,\|\mathbf{s}_j\|} \tag{8.21}$$

The two vectors $\mathbf{s}_i$ and $\mathbf{s}_j$ are thus *orthogonal* or *perpendicular* to each other if their inner product is zero, in which case $\theta_{ij} = 90$ degrees.

There is an interesting relationship between the energy content of a signal and its representation as a vector. By definition, the energy of a signal $s_i(t)$ of duration T seconds is equal to

$$E_i = \int_0^T s_i^2(t)\ dt \tag{8.22}$$

Therefore, substituting Eq. (8.17) in (8.22), we get

$$E_i = \int_0^T \left[\sum_{j=1}^N s_{ij}\phi_j(t)\right]\left[\sum_{k=1}^N s_{ik}\phi_k(t)\right] dt$$

Interchanging the order of summation and integration, and rearranging terms:

$$E_i = \sum_{j=1}^N \sum_{k=1}^N s_{ij}s_{ik} \int_0^T \phi_j(t)\phi_k(t)\ dt \tag{8.23}$$

But, since the $\phi_j(t)$ form an orthonormal set, then, in accordance with the two conditions of Eq. (8.7), we find that Eq. (8.23) reduces simply to

$$E_i = \sum_{j=1}^{N} s_{ij}^2 \tag{8.24}$$

Thus Eqs. (8.20) and (8.24) show that the energy of a signal $s_i(t)$ is equal to the squared-length of the signal vector $\mathbf{s}_i$ representing it.

In the case of a pair of signals $s_i(t)$ and $s_k(t)$, represented by the signal vectors $\mathbf{s}_i$ and $\mathbf{s}_k$, respectively, we may similarly show that

$$
\begin{aligned}
\|\mathbf{s}_i - \mathbf{s}_k\|^2 &= \sum_{j=1}^{N} (s_{ij} - s_{kj})^2 \\
&= \int_0^T [s_i(t) - s_k(t)]^2 \, dt
\end{aligned}
\tag{8.25}
$$

where $\|\mathbf{s}_i - \mathbf{s}_k\|$ is the *Euclidean distance* d_{ik} between the points represented by the signal vectors $\mathbf{s}_i$ and $\mathbf{s}_k$.

8.5   RESPONSE OF BANK OF CORRELATORS TO NOISY INPUT

Suppose that the input to the bank of N product integrators or correlators in Fig. 8.4b is not the transmitted signal $s_i(t)$ but rather the received signal $x(t)$ defined in accordance with the idealized AWGN channel of Fig. 8.3. That is to say,

$$x(t) = s_i(t) + w(t), \qquad \begin{cases} 0 \le t \le T \\ i = 1, 2, \ldots, M \end{cases} \tag{8.26}$$

where $w(t)$ is a sample function of a white Gaussian noise process $W(t)$ of zero mean and power spectral density $N_0/2$. Correspondingly, we find that the output of correlator j, say, is the sample value of a random variable X_j, as shown by

$$
\begin{aligned}
x_j &= \int_0^T x(t)\phi_j(t) \, dt \\
&= s_{ij} + w_j, \qquad\qquad j = 1, 2, \ldots, N
\end{aligned}
\tag{8.27}
$$

The first component, s_{ij}, is a deterministic quantity contributed by the transmitted signal $s_i(t)$; it is defined by

$$s_{ij} = \int_0^T s_i(t)\phi_j(t) \, dt \tag{8.28}$$

The second component, w_j, is the sample of a random variable W_j that arises because of the presence of noise at the receiver input; it is defined by

$$w_j = \int_0^T w(t)\,\phi_j(t)\ dt \tag{8.29}$$

Consider next a new random process $X'(t)$ whose sample function $x'(t)$ is related to the received signal $x(t)$ as follows:

$$x'(t) = x(t) - \sum_{j=1}^N x_j\phi_j(t) \tag{8.30}$$

Substituting Eqs. (8.26) and (8.27) in (8.30), and then using the expansion of Eq. (8.17), we get

$$\begin{aligned}
x'(t) &= s_i(t) + w(t) - \sum_{j=1}^N (s_{ij} + w_j)\phi_j(t) \\
&= w(t) - \sum_{j=1}^N w_j\phi_j(t) \tag{8.31} \\
&= w'(t)
\end{aligned}$$

The sample function $x'(t)$ therefore depends only on the noise $w(t)$ at the front end of the receiver, but not at all on the transmitted signal $s_i(t)$. On the basis of Eqs. (8.30) and (8.31), we may thus express the received signal as

$$\begin{aligned}
x(t) &= \sum_{j=1}^N x_j\phi_j(t) + x'(t) \\
&= \sum_{j=1}^N x_j\phi_j(t) + w'(t)
\end{aligned} \tag{8.32}$$

Accordingly, we may view $w'(t)$ as a sort of remainder term that must be included on the right in order to preserve the equality in Eq. (8.32). It is informative to contrast the expansion of the received signal $x(t)$ given in Eq. (8.32) with the corresponding expansion of the transmitted signal $s_i(t)$ given in Eq. (8.17): The latter expansion is entirely deterministic, whereas that of Eq. (8.32) is entirely random (stochastic).

Statistical Characterization of the Correlator Outputs

We now wish to develop a statistical characterization of the set of N correlator outputs. Let $X(t)$ denote the random process a sample function of which is represented by the received signal $x(t)$. Correspondingly, let X_j denote the random variable whose sample value is represented by the correlator output x_j, $j = 1, 2, \ldots, N$. According to the AWGN model of Fig. 8.3, the random process $X(t)$ is a Gaussian process. It follows therefore that X_j is a Gaussian random variable for all j (see Property 1 of a Gaussian process, Section 4.12). Hence, X_j is characterized completely by its mean and variance, which are determined next.

Let W_j denote the random variable represented by the sample value w_j produced by the jth correlator in response to the white Gaussian noise component

$w(t)$. The random variable W_j has zero mean, because the noise process $W(t)$ represented by $w(t)$ in the AWGN model of Fig. 8.3 has zero mean by definition. Consequently, the mean of X_j depends only on s_{ij}, as shown by

$$
\begin{aligned}
\mu_{X_j} &= E[X_j] \\
&= E[s_{ij} + W_j] \\
&= s_{ij} + E[W_j] \\
&= s_{ij}
\end{aligned}
\tag{8.33}
$$

To find the variance of X_j, we note that

$$
\begin{aligned}
\sigma_{X_j}^2 &= \text{var}[X_j] \\
&= E[(X_j - s_{ij})^2] \\
&= E[W_j^2]
\end{aligned}
\tag{8.34}
$$

According to Eq. (8.29), the random variable W_j is defined by

$$
W_j = \int_0^T W(t)\phi_j(t)\ dt
$$

We may therefore expand Eq. (8.34) as follows:

$$
\begin{aligned}
\sigma_{X_j}^2 &= E\left[\int_0^T W(t)\phi_j(t)\ dt \int_0^T W(u)\phi_j(u)\ du\right] \\
&= E\left[\int_0^T \int_0^T \phi_j(t)\phi_j(u)W(t)W(u)\ dt\ du\right]
\end{aligned}
\tag{8.35}
$$

Interchanging the order of integration and expectation:

$$
\begin{aligned}
\sigma_{X_j}^2 &= \int_0^T \int_0^T \phi_j(t)\phi_j(u)E[W(t)W(u)]\ dt\ du \\
&= \int_0^T \int_0^T \phi_j(t)\phi_j(u)R_W(t,u)\ dt\ du
\end{aligned}
\tag{8.36}
$$

where $R_W(t,u)$ is the autocorrelation function of the noise process $W(t)$. Since this noise is stationary, $R_W(t,u)$ depends only on the time difference $t - u$. Furthermore, since the noise $W(t)$ is white with a constant power spectral density $N_0/2$, we may express $R_W(t,u)$ as follows [see Eq. (4.150)]:

$$
R_W(t,u) = \frac{N_0}{2}\delta(t - u)
\tag{8.37}
$$

Therefore, substituting Eq. (8.37) in (8.36), and then using the sifting property of the Dirac delta function $\delta(t)$, we get

$$\sigma_{X_j}^2 = \frac{N_0}{2} \int_0^T \int_0^T \phi_j(t)\,\phi_j(u)\,\delta(t - u)\; dt\; du$$

$$= \frac{N_0}{2} \int_0^T \phi_j^2(t)\; dt \tag{8.38}$$

Since the $\phi_j(t)$ have unit energy, by definition, we finally get the simple result

$$\sigma_{X_j}^2 = \frac{N_0}{2} \qquad \text{for all } j \tag{8.39}$$

This important result shows that all the correlator outputs denoted by X_j with $j = 1, 2, \ldots, N$, have a variance equal to the power spectral density $N_0/2$ of the noise process $W(t)$.

Moreover, since the $\phi_j(t)$ form an orthogonal set, we find that the X_j are mutually uncorrelated, as shown by

$$\begin{aligned}
\text{cov}[X_j X_k] &= E[(X_j - \mu_{X_j})(X_k - \mu_{X_k})]\\
&= E[(X_j - s_{ij})(X_k - s_{ik})]\\
&= E[W_j W_k]\\
&= E\left[\int_0^T W(t)\phi_j(t)\; dt \int_0^T W(u)\phi_k(u)\; du\right]\\
&= \int_0^T \int_0^T \phi_j(t)\phi_k(u)\,R_W(t,u)\; dt\; du \qquad (8.40)\\
&= \frac{N_0}{2} \int_0^T \int_0^T \phi_j(t)\phi_k(u)\;\delta(t - u)\; dt\; du\\
&= \frac{N_0}{2} \int_0^T \phi_j(t)\phi_k(t)\; dt\\
&= 0, \qquad j \neq k
\end{aligned}$$

Since the X_j are Gaussian random variables, Eq. (8.40) implies that they are also statistically independent (see Property 4 of a Gaussian Process, Section 4.12).

Define the vector of N random variables

$$\mathbf{X} = \begin{bmatrix} X_1 \\ X_2 \\ \vdots \\ X_N \end{bmatrix} \tag{8.41}$$

whose elements are independent Gaussian random variables with mean values equal to s_{ij} and variances equal to $N_0/2$. Since the elements of the vector $\mathbf{X}$ are statistically independent, we may express the conditional probability density function of the vector $\mathbf{X}$, given that the signal $s_i(t)$ or correspondingly the symbol

m_i was transmitted, as the product of the conditional probability density functions of its individual elements, as shown by

$$f_{\mathbf{X}}(\mathbf{x}|m_i) = \prod_{j=1}^{N} f_{X_j}(x_j|m_i), \qquad i = 1, 2, \ldots, M \tag{8.42}$$

where the vector $\mathbf{x}$ and scalar x_j are sample values of the random vector $\mathbf{X}$ and random variable X_j, respectively. The vector $\mathbf{x}$ is called the *observation vector,* corresponding, x_j is called an *observable element.* The conditional probability density functions, $f_{\mathbf{X}}(\mathbf{x}|m_i)$, for each transmitted message m_i, $i = 1, 2, \ldots, M$ are called *likelihood functions.* These likelihood functions, which are in fact the channel characterization, are also called *channel transition probabilities.* Any channel whose likelihood functions satisfy Eq. (8.42) is called a *memoryless* channel.

Since each X_j is a Gaussian random variable with mean s_{ij} and variance $N_0/2$, we have

$$f_{X_j}(x_j|m_i) = \frac{1}{\sqrt{\pi N_0}} \exp\left[-\frac{1}{N_0} (x_j - s_{ij})^2 \right], \qquad \begin{matrix} j = 1, 2, \ldots, N \\ i = 1, 2, \ldots, M \end{matrix} \tag{8.43}$$

Therefore, substituting Eq. (8.43) in (8.42), we find that the likelihood functions of an AWGN channel are defined by

$$f_{\mathbf{X}}(\mathbf{x}|m_i) = (\pi N_0)^{-N/2} \exp\left[-\frac{1}{N_0} \sum_{j=1}^{N} (x_j - s_{ij})^2 \right], \qquad i = 1, 2, \ldots, M \tag{8.44}$$

It is now clear that the elements of the random vector $\mathbf{X}$ completely characterize the summation term $\Sigma X_j \phi_j(t)$, whose sample value is represented by the first term in Eq. (8.32). However, there remains the noise term $w'(t)$ in this equation, which depends only on the original noise $w(t)$. Since the noise process $W(t)$ represented by $w(t)$ is Gaussian with zero mean, it follows that the noise process $W'(t)$ represented by the sample function $w'(t)$ is also a zero-mean Gaussian process. Finally, we note that any random variable $W'(t_k)$, say, derived from the noise process $W'(t)$ by sampling it at time t_k, is in fact statistically independent of the set of random variables $\{X_j\}$; that is to say (see Problem 8.4)

$$E[X_j W'(t_k)] = 0, \qquad \begin{cases} j = 1, 2, \ldots, N \\ 0 \le t_k \le T \end{cases} \tag{8.45}$$

Since any random variable based on the remainder noise process $W'(t)$ is independent of the set of random variables $\{X_j\}$ and the set of transmitted signals $\{s_i(t)\}$, we conclude that such a random variable is *irrelevant* to the decision as to which signal was transmitted. In other words, the correlator outputs determined by the received signal $x(t)$ are the only data that are useful for the decision-making process, and hence, represent *sufficient statistics* for the problem at hand. By definition, sufficient statistics summarize the whole of the relevant information supplied by an observation vector.

8.6   COHERENT DETECTION OF SIGNALS IN NOISE

Assume that, in each time slot of duration T seconds, one of the M possible signals $s_1(t), s_2(t), \ldots, s_M(t)$ is transmitted with equal probability, namely $1/M$. Then, for the AWGN channel model of Fig. 8.3, the received signal $x(t)$ is defined by Eq. (8.26), reproduced here for convenience of presentation

$$x(t) = s_i(t) + w(t), \qquad \begin{cases} 0 \le t \le T \\ i = 1, 2, \ldots, M \end{cases} \tag{8.46}$$

where $w(t)$ is a sample function of a white Gaussian noise process of zero mean and power spectral density $N_0/2$. Given the received signal $x(t)$, the receiver has to make a "best estimate" of the transmitted signal $s_i(t)$ or equivalently the symbol m_i.

We note that when the transmitted signal $s_i(t)$, $i = 1, 2, \ldots, M$, is applied to a bank of correlators, with a common input and supplied with an appropriate set of N orthonormal basis functions, the resulting correlator outputs define the *signal vector* $\mathbf{s}_i$ [see Eq. (8.19)]. Since knowledge of the signal vector $\mathbf{s}_i$ is as good as knowing the transmitted signal $s_i(t)$ itself, and vice versa, we may represent $s_i(t)$ by a point in a Euclidean space of dimension $N \le M$. We refer to this point as the *transmitted signal point* or *message point*. The set of message points corresponding to the set of transmitted signals $\{s_i(t) | i = 1, 2, \ldots, M\}$ is called a *signal constellation*.

However, the representation of the received signal $x(t)$ is complicated by the presence of the additive noise $w(t)$. We note that when the received signal $x(t)$ is applied to the bank of N correlators, the correlator outputs define the observation vector $\mathbf{x}$. The vector $\mathbf{x}$ differs from the signal vector $\mathbf{s}_i$ by the *noise vector* $\mathbf{w}$ whose orientation is completely random. In particular, in light of Eq. (8.27) we have

$$\mathbf{x} = \mathbf{s}_i + \mathbf{w}, \qquad i = 1, 2, \ldots, M \tag{8.47}$$

which may be viewed as the vector counterpart to Eq. (8.46). The noise vector $\mathbf{w}$ is completely characterized by the noise $w(t)$; the converse of this statement, however, is not true. The noise vector $\mathbf{w}$ represents that portion of the noise $w(t)$ that will interfere with the detection process; the remaining portion of this noise, denoted by $w'(t)$, is tuned out by the bank of correlators.

Now, based on the observation vector $\mathbf{x}$, we may represent the received signal $x(t)$ by a point in the same Euclidean space used to represent the transmitted signal. We refer to this second point as the *received signal point*. The received signal point wanders about the message point in a completely random fashion, in the sense that it may lie anywhere inside a Gaussian-distributed "cloud" centered on the message point. This is illustrated in Fig. 8.7a for the case of a three-dimensional signal space. For a particular realization of the noise vector $\mathbf{w}$ (i.e., a particular point inside the random cloud of Fig. 8.7a), the relationship between the observation vector $\mathbf{x}$ and the signal vector $\mathbf{s}_i$ is as illustrated in Fig. 8.7b.

We are now ready to state the detection problem: *Given the observation vector* $\mathbf{x}$, *we have to perform a mapping from* $\mathbf{x}$ *to an estimate* $\hat{m}$ *of the transmitted symbol,* m_i, *in a way that would minimize the probability of error in the decision-making process.*

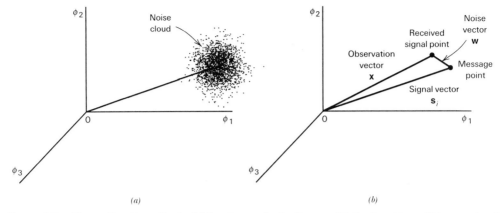

Figure 8.7   Illustrating the effect of (a) noise perturbation on (b) the location of the received signal point.

Assuming that all the M transmitted symbols are equally likely, the maximum-likelihood decoder, discussed next, provides the solution to this basic signal processing problem.

Maximum-Likelihood Decoder

Suppose that given the observation vector $\mathbf{x}$, we make the decision $\hat{m} = m_i$. The probability of error in this decision, which we denote by $P_e(m_i, \mathbf{x})$, is simply

$$
\begin{aligned}
P_e(m_i, \mathbf{x}) &= P(m_i \text{ not sent } |\mathbf{x}) \\
&= 1 - P(m_i \text{ sent } |\mathbf{x})
\end{aligned}
\tag{8.48}
$$

The decision-making criterion is to minimize the probability of error in mapping each given observation vector $\mathbf{x}$ into a decision. On the basis of Eq. (8.48), we may therefore state the *optimum decision rule*:

$$
\begin{aligned}
&\text{Set } \hat{m} = m_i \text{ if} \\
&P(m_i \text{ sent } |\mathbf{x}) \geq P(m_k \text{ sent } |\mathbf{x}) \qquad \text{for all } k \neq i
\end{aligned}
\tag{8.49}
$$

where $k = 1, 2, \ldots, M$. This decision rule is referred to as the *maximum a posteriori probability (MAP) rule*.

The condition of Eq. (8.49) may be expressed more explicitly in terms of the a priori probabilities of the transmitted signals and in terms of the likelihood functions. Using Bayes' rule in Eq. (8.49), and for the moment ignoring possible ties in the decision-making process, we may restate the MAP rule as follows:

$$
\begin{aligned}
&\text{Set } \hat{m} = m_i \text{ if} \\
&\frac{p_k f_{\mathbf{X}}(\mathbf{x}|m_k)}{f_{\mathbf{X}}(\mathbf{x})} \text{ is maximum for } k = i
\end{aligned}
\tag{8.50}
$$

where p_k is the a priori probability of occurrence of symbol m_k, $f_{\mathbf{X}}(\mathbf{x}|m_k)$ is the likelihood function that results when symbol m_k is transmitted, and $f_{\mathbf{X}}(\mathbf{x})$ is the unconditional joint probability density function of the random vector $\mathbf{X}$. How-ever, in Eq. (8.50), we note that (1) the denominator term $f_{\mathbf{X}}(\mathbf{x})$ is independent of the transmitted signal and (2) the a priori probability $p_k = p_i$ when all the signals are transmitted with equal probability. Therefore, we may simplify the decision rule of Eq. (8.50) as follows:

$$
\text{Set } \hat{m} = m_i \text{ if}
$$
$$
f_{\mathbf{X}}(\mathbf{x}|m_k) \text{ is maximum for } k = i \tag{8.51}
$$

Ordinarily, we find it more convenient to work with the natural logarithm of the likelihood function rather than the likelihood function itself. For a mem-oryless channel, the logarithm of the likelihood function is commonly called the *metric*. Since the likelihood function $f_{\mathbf{X}}(\mathbf{x}|m_k)$ is always nonnegative, and since the logarithmic function is a monotone increasing function of its argument, we may restate the decision rule of Eq. (8.51) in terms of the metric as follows:

$$
\text{Set } \hat{m} = m_i \text{ if}
$$
$$
\ln[f_{\mathbf{X}}(\mathbf{x}|m_k)] \text{ is maximum for } k = i \tag{8.52}
$$

where ln denotes the natural logarithm. This decision rule is referred to as the *maximum-likelihood rule*, and the device for its implementation is correspond-ingly referred to as the *maximum-likelihood decoder*. According to Eq. (8.52), a maximum-likelihood decoder computes the metrics for all the transmitted messages, compares them, and then decides in favor of the maximum.

It is useful to have a graphical interpretation of the maximum-likelihood decision rule. Let Z denote the N-dimensional space of all possible observation vectors $\mathbf{x}$. We refer to this space as the *observation space*. Because we have assumed that the decision rule must say $\hat{m} = m_i$, where $i = 1, 2, \ldots, M$, the total obser-vation space Z is correspondingly partitioned into M *decision regions*, denoted by $Z_1, Z_2, \ldots, Z_M$. Accordingly, we may restate the decision rule of Eq. (8.52) as follows:

$$
\text{Observation vector } \mathbf{x} \text{ lies in region } Z_i \text{ if}
$$
$$
\ln[f_{\mathbf{X}}(\mathbf{x}|m_k)] \text{ is maximum for } k = i \tag{8.53}
$$

Aside from the boundaries between the decision regions $Z_1, Z_2, \ldots, Z_M$, it is clear that this set of regions covers the entire space of possible observation vectors $\mathbf{x}$. We adopt the convention that all ties are resolved at random, that is, the receiver simply makes a guess. Specifically, if the observation vector $\mathbf{x}$ falls on the boundary between any two decision regions, Z_i and Z_k, say, the choice between the two possible decisions $\hat{m} = m_i$ and $\hat{m} = m_k$ is resolved a priori by the flip of a fair coin. Clearly, the outcome of such an event does not affect the ultimate value of the probability of error since, on this boundary, the condition of Eq. (8.49) is satisfied with the equality sign.

To illustrate the maximum likelihood decision rule described by Eq. (8.53), consider an AWGN channel for which the likelihood function is defined by Eq. (8.44). We may thus write

$$
f_{\mathbf{X}}(\mathbf{x}|m_k) = (\pi N_0)^{-N/2} \exp\left[-\frac{1}{N_0} \sum_{j=1}^{N} (x_j - s_{kj})^2\right], \quad k = 1, 2, \ldots, M \tag{8.54}
$$

The corresponding value of the metric is therefore

$$\ln[f_{\mathbf{X}}(\mathbf{x}|m_k)] = -\frac{N}{2}\ln(\pi N_0) - \frac{1}{N_0}\sum_{j=1}^{N}(x_j - s_{kj})^2, \quad k = 1, 2, \ldots, M \quad (8.55)$$

The constant term $-(N/2)\ln(\pi N_0)$ on the right-hand side of Eq. (8.55) is of no consequence insofar as application of the decision rule is concerned. Therefore, ignoring this term, and then substituting the remainder of Eq. (8.55) into (8.53), we may formulate the maximum-likelihood decision rule for an AWGN channel as follows:

> Observation vector $\mathbf{x}$ lies in region Z_i if
>
> $$-\frac{1}{N_0}\sum_{j=1}^{N}(x_j - s_{kj})^2 \text{ is maximum for } k = i$$

$$(8.56)$$

Equivalently, we may state

> Observation vector $\mathbf{x}$ lies in region Z_i if
>
> $$\sum_{j=1}^{N}(x_j - s_{kj})^2 \text{ is minimum for } k = i$$

$$(8.57)$$

Next, we note from our earlier discussion that

$$\sum_{j=1}^{N}(x_j - s_{kj})^2 = \|\mathbf{x} - \mathbf{s}_k\|^2 \quad (8.58)$$

where $\|\mathbf{x} - \mathbf{s}_k\|$ is the Euclidean distance between the received signal point and message point, represented by the vectors $\mathbf{x}$ and $\mathbf{s}_k$, respectively. Accordingly, we may rewrite the decision rule of Eq. (8.57) as follows:

> Observation vector $\mathbf{x}$ lies in region Z_i if
>
> the Euclidean distance $\|\mathbf{x} - \mathbf{s}_k\|$ is minimum for $k = i$

$$(8.59)$$

Equation (8.59) states that *the maximum likelihood decision rule is simply to choose the message point closest to the received signal point*, which is intuitively satisfying.

In practice the need for squarers, as in the decision rule of Eq. (8.59), is avoided by recognizing that

$$\sum_{j=1}^{N}(x_j - s_{kj})^2 = \sum_{j=1}^{N}x_j^2 - 2\sum_{j=1}^{N}x_j s_{kj} + \sum_{j=1}^{N}s_{kj}^2 \quad (8.60)$$

The first summation term of this expansion is independent of the index k and may therefore be ignored. The second summation term is the inner product of the observation vector $\mathbf{x}$ and signal vector $\mathbf{s}_k$. The third summation term is the energy of the transmitted signal $s_k(t)$. Accordingly, we may formulate a decision rule equivalent to that of Eq. (8.59) as follows:

> Observation vector $\mathbf{x}$ lies in region Z_i if
>
> $$\sum_{j=1}^{N}x_j s_{kj} - \frac{1}{2}E_k \text{ is maximum for } k = i$$

$$(8.61)$$

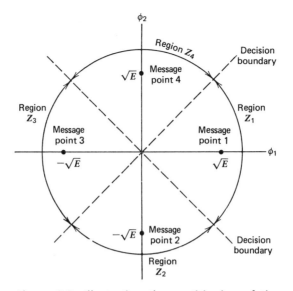

Figure 8.8 Illustrating the partitioning of the observation space into decision regions for the case when $N = 2$ and $M = 4$; it is assumed that the M transmitted symbols are equally likely.

where E_k is the energy of the transmitted signal $s_k(t)$:

$$E_k = \sum_{j=1}^{N} s_{kj}^2 \tag{8.62}$$

From Eq. (8.61) we deduce that, for an AWGN channel, the decision regions are regions of the N-dimensional observation space Z, bounded by linear [$(N - 1)$-dimensional hyperplane] boundaries. Figure 8.8 shows the example of decision regions for $M = 4$ signals and $N = 2$ dimensions, assuming that the signals are transmitted with equal energy, E, and equal probability.

8.7 PROBABILITY OF ERROR

Suppose that the observation space Z is partitioned, in accordance with the maximum-likelihood decision rule, into a set of M regions $\{Z_i | i = 1, 2, \ldots, M\}$. Suppose also that symbol m_i (or, equivalently, signal vector $\mathbf{s}_i$) is transmitted, and an observation vector $\mathbf{x}$ is received. Then an error occurs whenever the received signal point represented by $\mathbf{x}$ does not fall inside region Z_i associated with the message point represented by $\mathbf{s}_i$. Averaging over all possible transmitted symbols, we readily see that the *average probability of symbol error*, P_e, is

$$P_e = \sum_{i=1}^{M} P(\mathbf{x} \text{ does not lie in } Z_i | m_i \text{ sent}) \ P(m_i \text{ sent})$$

$$= \frac{1}{M} \sum_{i=1}^{M} P(\mathbf{x} \text{ does not lie in } Z_i | m_i \text{ sent}) \tag{8.63}$$

$$= 1 - \frac{1}{M} \sum_{i=1}^{M} P(\mathbf{x} \text{ lies in } Z_i | m_i \text{ sent})$$

where we have used standard notation to denote the probability of an event and the conditional probability of an event (see Section 4.2). Since $\mathbf{x}$ is the sample value of random vector $\mathbf{X}$, we may rewrite Eq. (8.63) in terms of the likelihood function (when m_i is sent) as follows:

$$P_e = 1 - \frac{1}{M} \sum_{i=1}^{M} \int_{Z_i} f_{\mathbf{X}}(\mathbf{x}|m_i) \ d\mathbf{x} \tag{8.64}$$

For an N-dimensional observation vector, the integral in Eq. (8.64) is likewise N-dimensional.

Union Bound on the Probability of Error[2]

For AWGN channels, the formulation of the average probability of symbol error, P_e, is conceptually straightforward. We simply write P_e in integral form by substituting Eq. (8.44) in Eq. (8.64). Unfortunately, however, numerical computation of the integral is impractical, except in a few simple (but important) cases. To overcome this computational difficulty, we may resort to the use of *bounds*, which are usually adequate to predict the signal-to-noise ratio (within a decibel or so) required to maintain a prescribed error rate. The approximation to the integral defining P_e is made by simplifying the integral or simplifying the region of integration. In the sequel, we use the latter procedure to develop a simple and yet useful upper bound called the *union bound* as an approximation to the average probability of symbol error for a set of M equally likely signals (symbols) in an AWGN channel.

Let A_{ik}, with $i, k = 1, 2, \ldots, M$, denote the event that the observation vector $\mathbf{x}$ is closer to the signal vector $\mathbf{s}_k$ than to $\mathbf{s}_i$, when the symbol m_i (vector $\mathbf{s}_i$) is sent. The conditional probability of symbol error when symbol m_i is sent, $P_e(m_i)$, is equal to the probability of the *union of events*, $A_{i1}, A_{i2}, \ldots, A_{i,i-1}, A_{i,i+1}, \ldots, A_{i,M}$. From probability theory we know that *the probability of a finite union of events is overbounded by the sum of the probabilities of the constituent events*. We may therefore write

$$P_e(m_i) \leq \sum_{\substack{k=1 \\ k \neq 1}}^{M} P(A_{ik}), \qquad i = 1, 2, \ldots, M \tag{8.65}$$

This relationship is illustrated in Fig. 8.9 for the case of $M = 4$. In Fig. 8.9a, we show the four message points and associated decision regions, with the point $\mathbf{s}_1$ assumed to represent the transmitted symbol. In Fig. 8.9b, we show the three constituent signal-space descriptions where, in each case, the transmitted message point $\mathbf{s}_1$ and one other message point are retained. According to Fig. 8.9a, the conditional probability of symbol error, $P_e(m_i)$, is equal to the probability that the observation vector $\mathbf{x}$ lies in the shaded region of the two-dimensional signal-space diagram. Clearly, this probability is less than the sum of the probabilities of the three individual events that $\mathbf{x}$ lies in the shaded regions of the three constituent signal spaces depicted in Fig. 8.9b.

It is important to note that, in general, the probability $P(A_{ik})$ is different from the probability $P(\hat{m} = m_k | m_i)$. The latter is the probability that the observation vector $\mathbf{x}$ is closer to the signal vector $\mathbf{s}_k$ than every other, when $\mathbf{s}_i$ (or m_i)

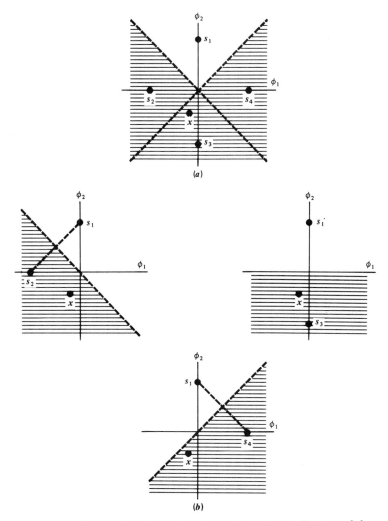

Figure 8.9 Illustrating the union bound. (a) Constellation of four message points. (b) Three constellations with a common message point and one other message point retained.

is sent. On the other hand, the probability $P(A_{ik})$ depends on only two signal vectors, $\mathbf{s}_i$ and $\mathbf{s}_k$. To emphasize this difference, we rewrite Eq. (8.65) by adopting $P_2(\mathbf{s}_i, \mathbf{s}_k)$ in place of $P(A_{ik})$. We thus have

$$P_e(m_i) \leq \sum_{\substack{k=1 \\ k \neq i}}^{M} P_2(\mathbf{s}_i, \mathbf{s}_k), \qquad i = 1, 2, \ldots, M \qquad (8.66)$$

Consider then a simplified digital communication system that involves the use of two equally likely messages represented by the vectors $\mathbf{s}_i$ and $\mathbf{s}_k$. Since white Gaussian noise is identically distributed along any set of orthogonal axes, we may temporarily choose the first axis in such a set as one that passes through the points $\mathbf{s}_i$ and $\mathbf{s}_k$; for examples, see Fig. 8.9b. The corresponding decision boundary is represented by the bisector that is perpendicular to the line joining the points $\mathbf{s}_i$ and $\mathbf{s}_k$. Accordingly, when the symbol m_i (vector $\mathbf{s}_i$) is sent, and if the observation vector $\mathbf{x}$ lies on the side of the bisector where $\mathbf{s}_k$ lies, an error is made. The probability of this event is given by

$$P_2(\mathbf{s}_i,\mathbf{s}_k) = P(\mathbf{x} \text{ is closer to } \mathbf{s}_k \text{ than } \mathbf{s}_i, \text{ when } \mathbf{s}_i \text{ is sent})$$

$$= \int_{d_{ik}/2}^{\infty} \frac{1}{\sqrt{\pi N_0}} \exp\left(-\frac{u^2}{N_0}\right) du \tag{8.67}$$

where d_{ik} is the Euclidean distance between $\mathbf{s}_i$ and $\mathbf{s}_k$; that is

$$d_{ik} = \|\mathbf{s}_i - \mathbf{s}_k\| \tag{8.68}$$

From the definition of the complementary error function, we have

$$\text{erfc}(u) = \frac{2}{\sqrt{\pi}} \int_u^{\infty} \exp(-z^2)\, dz$$

Thus, in terms of this function, with z set equal to $u/\sqrt{N_0}$, we find that Eq. (8.67) takes on the compact form

$$P_2(\mathbf{s}_i,\mathbf{s}_k) = \frac{1}{2} \text{erfc}\left(\frac{d_{ik}}{2\sqrt{N_0}}\right) \tag{8.69}$$

The complementary error function is a monotone decreasing function of its argument. We may therefore state that as the distance separating the message points $\mathbf{s}_i$ and $\mathbf{s}_k$ is increased, the probability of error is reduced, a result that is intuitively satisfying.

Substituting Eq. (8.69) into Eq. (8.66), we get

$$P_e(m_i) \leq \frac{1}{2} \sum_{\substack{k=1 \\ k \neq i}}^{M} \text{erfc}\left(\frac{d_{ik}}{2\sqrt{N_0}}\right), \qquad i = 1, 2, \ldots, M \tag{8.70}$$

Finally, with the M transmitted messages assumed equally likely, we find that the average probability of symbol error is overbounded as follows:

$$P_e = \frac{1}{M} \sum_{i=1}^{M} P_e(m_i)$$

$$\leq \frac{1}{2M} \sum_{i=1}^{M} \sum_{\substack{k=1 \\ k \neq i}}^{M} \text{erfc}\left(\frac{d_{ik}}{2\sqrt{N_0}}\right) \tag{8.71}$$

where the distance d_{ik} is defined in Eq. (8.68). The second line of Eq. (8.71) defines the union bound on the average probability of symbol error for any set of M equally likely signals in an AWGN channel.

Equation (8.71) is the most general formulation of the union bound. It is particularly useful for the special case of a signal set that has a *symmetric geometry*, which is of common occurrence in practice. In such a case, the conditional error probability $P_e(m_i)$ is the same for all i, and so we may simplify Eq. (8.71) as

$$P_e = P_e(m_i)$$

$$\leq \frac{1}{2} \sum_{\substack{k=1 \\ k \neq i}}^{M} \text{erfc}\left(\frac{d_{ik}}{2\sqrt{N_0}}\right) \qquad \text{for all } i \tag{8.72}$$

The complementary error function may be upper-bounded as follows (see Appendix 7)

$$\text{erfc}\left(\frac{d_{ik}}{2\sqrt{N_0}}\right) \leq \frac{1}{\sqrt{\pi}} \exp\left(-\frac{d_{ik}^2}{2N_0}\right)$$

Hence, we may rewrite Eq. (8.72) as

$$P_e \leq \frac{1}{2\sqrt{\pi}} \sum_{\substack{k=1 \\ k \neq i}}^{M} \exp\left(-\frac{d_{ik}^2}{2N_0}\right) \qquad \text{for all } i \tag{8.73}$$

Provided that the transmitted signal energy is high enough compared to the noise spectral density N_0, the exponential term with the smallest distance d_{ik} will dominate the summation in Eq. (8.73). Accordingly, we may approximate the bound as

$$P_e \leq \frac{M_{\min}}{2\sqrt{\pi}} \exp\left[-\min_{\substack{i,k \\ i \neq k}}\left(\frac{d_{ik}^2}{2N_0}\right)\right] \tag{8.74}$$

where $M_{\min}$ is the number of transmitted signals that attain the minimum Euclidean distance for each m_i. Equation (8.74) describes a simplified form of the union bound for a symmetric signal set, which is easy to calculate.

Bit Versus Symbol Error Probabilities

Thus far, the only figure of merit we have used to assess the noise performance of a digital passband transmission system has been the average probability of symbol error. This figure of merit is the natural choice when messages of length $m = \log_2 M$ are transmitted, such as alphanumeric symbols. However, when the requirement is to transmit binary data such as digital computer data, it is often more meaningful to use another figure of merit called the *probability of bit error* or *bit error rate* (BER). Although, in general, there are no unique relationships between these two figures of merit, it is fortunate that such relationships can be derived for two cases of practical interest, as discussed next.

CASE 1

In the first case, we assume that it is possible to perform the mapping from binary to *M*-ary symbols in such a way that the two binary *M*-tuples corresponding to any pair of adjacent symbols in the *M*-ary modulation scheme differ in only one bit position. This mapping constraint is satisfied by using a *Gray code*. When the probability of symbol error P_e is acceptably small, we find that the probability of mistaking one symbol for either of the two "nearest" symbols is much greater than any other kind of symbol error. Moreover, given a symbol error, the most probable number of bit errors is one, subject to the aforementioned mapping constraint. Since there are $\log_2 M$ bits per symbol, it follows that the average probability of symbol error is related to the bit error rate as follows

$$
\begin{aligned}
P_e &= P\left(\bigcup_{i=1}^{\log_2 M} \{i\text{th bit is in error}\} \right) \\
&\leq \sum_{i=1}^{\log_2 M} P(i\text{th bit is in error}) \\
&= \log_2 M \cdot (\text{BER})
\end{aligned}
\tag{8.75}
$$

We also note that

$$
P_e \geq P(i\text{th bit is in error}) = \text{BER}
\tag{8.76}
$$

It follows therefore that the bit error rate is bounded as follows:

$$
\frac{P_e}{\log_2 M} \leq \text{BER} \leq P_e
\tag{8.77}
$$

CASE 2

Let $M = 2^K$, where K is an integer. We assume that all symbol errors are equally likely and occur with probability

$$
\frac{P_e}{M-1} = \frac{P_e}{2^K - 1}
$$

where P_e is the average probability of symbol error. What is the probability that the ith bit in a symbol is in error? Well, there are 2^{K-1} cases of symbol error in which this particular bit is changed, and there are $2^{K-1} - 1$ cases in which it is not changed. Hence, the bit error rate is

$$
\text{BER} = \left(\frac{2^{K-1}}{2^K - 1} \right) P_e
\tag{8.78}
$$

or, equivalently,

$$
\text{BER} = \left(\frac{M/2}{M-1} \right) P_e
\tag{8.79}
$$

Note that for large M, the bit error rate approaches the limiting value of $P_e/2$. The same idea described here also shows that bit errors are not independent, since we have

$$
P(i\text{th and }j\text{th bits are in error}) = \frac{2^{K-2}}{2^K - 1} P_e \neq (\text{BER})^2
$$

8.8   CORRELATION RECEIVER

We note that for an AWGN channel and for the case when the transmitted signals $s_1(t)$, $s_2(t), \ldots, s_M(t)$ are equally likely, the optimum receiver consists of two subsystems, detailed in Fig. 8.10 and described below:

1. The *detector* part of the receiver is shown in Fig. 8.10*a*. It consists of a bank of M *product-integrators* or *correlators*, supplied with a corresponding set of coherent reference signals or orthonormal basis functions $\phi_1(t)$, $\phi_2(t), \ldots,$ $\phi_N(t)$ that are generated locally. This bank of correlators operates on the received signal $x(t)$, $0 \leq t \leq T$, to produce the observation vector **x**.

2. The second part of the receiver, namely, the *signal transmission decoder* is shown in Fig. 8.10*b*. It is implemented in the form of a maximum-likelihood decoder that operates on the observation vector **x** to produce an estimate, $\hat{m}$, of the transmitted symbol m_i, $i = 1, 2, \ldots, M$, in a way that would minimize the average probability of symbol error. In accordance with Eq. (8.61), the N elements of the observation vector **x** are first multiplied by the corresponding N elements of each of the M signal vectors $\mathbf{s}_1$, $\mathbf{s}_2, \ldots, \mathbf{s}_M$, and the resulting products are successively summed in accumulators to form the corresponding set of inner products $\{\mathbf{x}^T\mathbf{s}_k | k = 1, 2, \ldots, M\}$. Next, the inner products are corrected for the fact that the transmitted signal energies may be unequal. Finally, the largest in the resulting set of numbers is selected, and an appropriate decision on the transmitted message is made.

The optimum receiver of Fig. 8.10 is commonly referred to as a *correlation receiver*.

Equivalence of Correlation and Matched Filter Receivers

The detector shown in Fig. 8.10*a* involves a set of correlators. Alternatively, we may use a corresponding set of *matched filters* to build the detector; the matched filter and its properties were considered in Section 7.2. To demonstrate the equivalence of a correlator and a matched filter, consider a linear time-invariant filter with impulse response $h_j(t)$. With the received signal $x(t)$ used as the filter input, the resulting filter output, $y_j(t)$, is defined by the convolution integral:

$$y_j(t) = \int_{-\infty}^{\infty} x(\tau) h_j(t - \tau) \, d\tau \qquad (8.80)$$

From the definition of a matched filter presented in Section 7.2, we recall that the impulse response $h_j(t)$ of a linear time-invariant filter matched to an input signal $\phi_j(t)$ is a time-reversed and delayed version of the input $\phi_j(t)$. Suppose that we set the impulse response

$$h_j(t) = \phi_j(T - t) \qquad (8.81)$$

Then the resulting filter output is

$$y_j(t) = \int_{-\infty}^{\infty} x(\tau) \phi_j(T - t + \tau) \, d\tau \qquad (8.82)$$

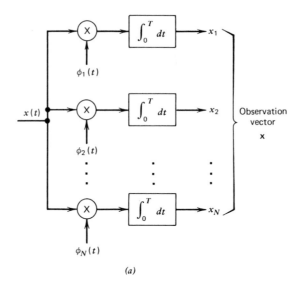

(a)

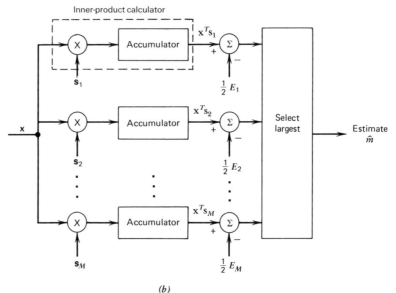

(b)

Figure 8.10 (a) Detector or demodulator. (b) Signal transmission decoder.

Sampling this output at time $t = T$, we get

$$y_j(T) = \int_{-\infty}^{\infty} x(\tau)\phi_j(\tau)\,d\tau$$

Since, by definition, $\phi_j(T)$ is zero outside the interval $0 \leq t \leq T$, we find that $y_j(T)$ is in actual fact the jth correlator output x_j produced by the received signal $x(t)$ in Fig. 8.10a, as shown by

$$y_j(T) = \int_0^T x(\tau)\phi_j(\tau)\,d\tau \tag{8.83}$$

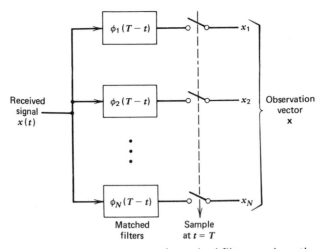

Figure 8.11 Detector part of matched filter receiver; the signal transmission decoder is as shown in Fig. 8.10(b).

Accordingly, the detector part of the optimum receiver may also be implemented using a bank of matched filters, as shown in Fig. 8.11. It is important to note, however, that the output of each correlator in Fig. 8.10a is equivalent to the output of a corresponding matched filter in Fig. 8.11 only when that output is sampled at time $t = T$.

8.9 DETECTION OF SIGNALS WITH UNKNOWN PHASE

Up to this point in our discussion, we have assumed that the information-bearing signal is completely known at the receiver. In practice, however, it is often found that in addition to the uncertainty due to receiver noise, there is an additional uncertainty due to the randomness of certain signal parameters. The usual cause of this uncertainty is distortion in the transmission medium. Perhaps the most common random signal parameter is the phase, which is especially true for narrow-band signals. For example, transmission over a multiplicity of paths of different and variable lengths, or rapidly varying delays in the propagating medium from transmitter to receiver, may cause the phase of the received signal to change in a way that the receiver cannot follow. Synchronization with the phase of the transmitted carrier may then be too costly, and the designer may simply choose to disregard the phase information in the received signal at the expense of some degradation in system performance.

Consider then a digital communication system in which the transmitted signal equals

$$s_i(t) = \sqrt{\frac{2E}{T}} \cos(2\pi f_i t), \qquad 0 \le t \le T \tag{8.84}$$

where E is the signal energy, T is the duration of the signaling interval, and the frequency f_i is an integral multiple of $1/2T$. When no provision is made to phase synchronize the receiver with the transmitter, the received signal may, for an

AWGN channel, be written in the form

$$x(t) = \sqrt{\frac{2E}{T}} \cos(2\pi f_i t + \theta) + w(t), \qquad 0 \le t \le T \qquad (8.85)$$

where $w(t)$ is the sample function of a white Gaussian noise process of zero mean and power spectral density $N_0/2$. The phase θ is unknown and is usually considered to be the sample value of a random variable uniformly distributed between 0 and 2π radians; this assumption implies a complete lack of knowledge of the phase. A digital communication system characterized in this way is said to be *noncoherent*.

The detection theory presented previously is inadequate for dealing with noncoherent systems, because if the received signal has the form described by Eq. (8.85), the output of the associated correlator in the receiver will be a function of the unknown phase θ. We now discuss, in a rather intuitive manner, the necessary modifications that may be introduced into the receiver in order to deal with this new situation.

Using a well-known trigonometric identity, we may rewrite Eq. (8.85) in the expanded form

$$x(t) = \sqrt{\frac{2E}{T}} \left(\cos\theta \cos(2\pi f_i t) - \sin\theta \sin(2\pi f_i t) \right) + w(t), \, 0 \le t \le T \quad (8.86)$$

Suppose that the received signal $x(t)$ is applied to a pair of correlators; one correlator is supplied with the reference signal $\sqrt{2/T} \cos(2\pi f_i t)$, and the other is supplied with the reference signal $\sqrt{2/T} \sin(2\pi f_i t)$. For both correlators, the observation interval is $0 \le t \le T$. Then in the absence of noise, we find that the first correlator output equals $\sqrt{E} \cos\theta$ and the second correlator output equals $-\sqrt{E} \sin\theta$. The dependence on the unknown phase θ may be removed by summing the squares of the two correlator outputs and then taking the square root of the sum. Thus, when the noise $w(t)$ is zero, the result of these operations is simply $\sqrt{E}$, which is independent of the unknown phase θ. This suggests that for the detection of a sinusoidal signal of arbitrary phase, one which is corrupted by an additive white Gaussian noise, as in the model of Eq. (8.85), we may use the so-called *quadrature receiver* shown in Fig. 8.12a. Indeed, this receiver is optimum in the sense that it realizes this detection with the minimum probability of error.[3]

We next derive two equivalent forms of the quadrature receiver. The first form is obtained easily by replacing each correlator in Fig. 8.12a with a corresponding equivalent matched filter. We thus obtain the alternative form of quadrature receiver shown in Fig. 8.12b. In one branch of this receiver, we have a filter matched to the signal $\sqrt{2/T} \cos(2\pi f_i t)$, and in the other branch we have a filter matched to $\sqrt{2/T} \sin(2\pi f_i t)$, both defined for the time interval $0 \le t \le T$. The filter outputs are sampled at time $t = T$, squared, and then added together.

To obtain the second equivalent form of the quadrature receiver, suppose we have a filter that is matched to $s(t) = \sqrt{2/T} \cos(2\pi f_i t + \theta)$ for $0 \le t \le T$. The envelope of the matched filter output is obviously unaffected by the value of phase θ. Therefore, for convenience, we may simply choose a matched filter with impulse response $\sqrt{2/T} \cos[2\pi f_i(T - t)]$, corresponding to $\theta = 0$. The

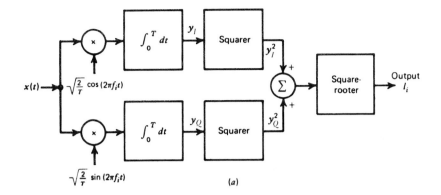

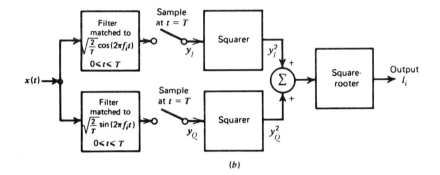

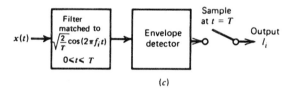

Figure 8.12   Noncoherent receivers. (a) Quadrature receiver using correlators. (b) Quadrature receiver using matched filters. (c) Noncoherent
matched filter.

output of such a filter in response to the received signal $x(t)$ is given by

$$
\begin{aligned}
y(t) &= \sqrt{\frac{2}{T}} \int_0^T x(\tau)\cos[2\pi f_i t (T - t + \tau)]\ d\tau \\
&= \sqrt{\frac{2}{T}} \cos[2\pi f_i (T - t)] \int_0^T x(\tau)\cos(2\pi f_i \tau)\ d\tau \\
&\quad - \sqrt{\frac{2}{T}} \sin[2\pi f_i (T - t)] \int_0^T x(\tau)\sin(2\pi f_i \tau)\ d\tau
\end{aligned}
\tag{8.87}
$$

The envelope of the matched filter output is proportional to the square root of
the sum of the squares of the integrals in Eq. (8.87). The envelope, evaluated at

time $t = T$, is therefore

$$l_i = \left\{ \left[\int_0^T x(\tau) \sqrt{\frac{2}{T}} \cos(2\pi f_i \tau) \; d\tau \right]^2 + \left[\int_0^T x(\tau) \sqrt{\frac{2}{T}} \sin(2\pi f_i \tau) \; d\tau \right]^2 \right\}^{1/2}$$

(8.88)

But this is just the output of the quadrature receiver. Therefore, the output (at time T) of a filter matched to the signal $\sqrt{2/T} \cos(2\pi f_i t + \theta)$, of arbitrary phase θ, followed by an envelope detector is the same as the corresponding output of the quadrature receiver of Fig. 8.12a. This form of receiver is shown in Fig. 8.12c. The combination of matched filter and envelope detector shown in Fig. 8.12c, is called a *noncoherent matched filter*.

The need for an envelope detector following the matched filter in Fig. 8.12c may also be justified intuitively as follows. The output of a filter matched to a rectangular RF wave reaches a positive peak at the sampling instant $t = T$. If, however, the phase of the filter is not matched to that of the signal, the peak may occur at a time different from the sampling instant. In actual fact, if the phases differ by 180 degrees, we get a negative peak at the sampling instant. Figure 8.13 illustrates the matched filter output for the two limiting conditions: $\theta = 0$ and $\theta = 180$ degrees. To avoid poor sampling that arises in the absence of prior information about the phase θ, it is reasonable to retain only the envelope

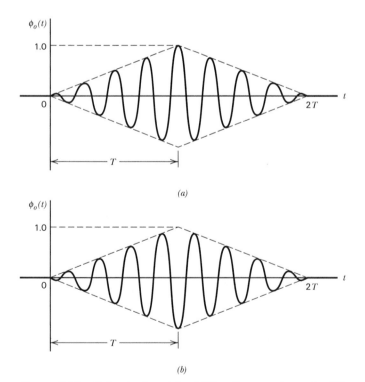

Figure 8.13 Output of matched filter for a rectangular RF wave: (a) $\theta = 0$, and (b) $\theta = 180$ degrees.

of the matched filter output, since it is completely independent of the phase mismatch.

8.10 HIERARCHY OF DIGITAL MODULATION TECHNIQUES

Now that we have a basic understanding of the optimum receiver design problem for both coherent and noncoherent forms of reception, we are ready to study the noise analysis of digital passband transmission systems. Before doing so, however, it is befitting that we first outline the *hierarchy of digital modulation techniques* used for data transmission.

Digital modulation techniques[4] may be classified into *coherent* and *noncoherent* techniques. Each of these two classes may be subdivided into *binary* and *M-ary* techniques. Naturally, an *M*-ary technique includes the corresponding binary one as a special case with $M = 2$. Traditionally, however, binary and *M*-ary techniques have been considered separately.

Under coherent binary modulation techniques, we have *amplitude-shift keying* (ASK), *phase-shift keying* (PSK), and *frequency-shift keying* (FSK), which involve applying two-level changes to the amplitude, phase, and frequency of a sinusoidal carrier wave, respectively. However, as mentioned in the introductory section, PSK and FSK signals are much more widely used in practice because of their robustness with respect to changes in their amplitude that may be caused by transmission over a nonlinear channel. Accordingly, we focus our attention on coherent binary PSK and FSK signals, details of which are presented in Sections 8.11 and 8.12, respectively; the case of coherent binary ASK signals is presented as a problem to the reader (see Problem 8.14). A special form of coherent binary FSK known as *minimum-shift keying* is studied in Section 8.14.

Under coherent *M*-ary modulation techniques, we have *M-ary ASK, M-ary PSK,* and *M-ary FSK,* which represent extensions of their respective binary counterparts. A popular form of coherent *M*-ary PSK known as *quadriphase-shift keying* (QPSK), for which $M = 4$, is discussed in Section 8.13. Coherent *M*-ary modulation techniques, in a general context, are discussed in Section 8.19.

Another way of generating *M*-ary signals is to combine different methods of modulation into a hybrid form. For example, we may combine discrete changes in both the amplitude and phase of a carrier to produce *M-ary amplitude-phase keying* (APK). A special form of this hybrid modulation, called *M-ary QAM,* has some attractive properties. *M*-ary QAM is also discussed under *M*-ary modulation techniques in Section 8.19.

Turning next to noncoherent systems, we have *noncoherent ASK* and *noncoherent FSK.* In the case of phase-shift keying, we cannot have "noncoherent PSK," because noncoherent means doing without phase information. Instead, we have a "pseudo PSK" technique known as *differential phase-shift keying* (DPSK), which (in a loose sense) may be viewed as the noncoherent form of binary PSK. Noncoherent FSK and DPSK are discussed in Sections 8.16 and 8.17, respectively. For the noise analysis of noncoherent binary ASK signals, the reader is referred to Problem 8.14.

Under noncoherent *M*-ary techniques, we have *M*-ary ASK, *M*-ary FSK, and *M*-ary DPSK. Unfortunately, a mathematical analysis of the effects of noise on these modulation schemes is beyond the scope of our present book; we will merely make a brief reference to noncoherent *M*-ary FSK in Section 8.19.

8.11   COHERENT BINARY PSK

In a coherent binary PSK system, the pair of signals $s_1(t)$ and $s_2(t)$ used to represent binary symbols 1 and 0, respectively, are defined by

$$s_1(t) = \sqrt{\frac{2E_b}{T_b}} \cos(2\pi f_c t) \tag{8.89}$$

$$s_2(t) = \sqrt{\frac{2E_b}{T_b}} \cos(2\pi f_c t + \pi) = -\sqrt{\frac{2E_b}{T_b}} \cos(2\pi f_c t) \tag{8.90}$$

where $0 \le t \le T_b$, and E_b is the *transmitted signal energy per bit*. To ensure that each transmitted bit contains an integral number of cycles of the carrier wave, the carrier frequency f_c is chosen equal to n_c/T_b for some fixed integer n_c. A pair of sinusoidal waves that differ only in a relative phase-shift of 180 degrees, as defined in Eqs. (8.89) and (8.90), are referred to as *antipodal signals*.

From this pair of equations it is clear that, in the case of PSK, there is only one basis function of unit energy, namely

$$\phi_1(t) = \sqrt{\frac{2}{T_b}} \cos(2\pi f_c t), \quad 0 \le t < T_b \tag{8.91}$$

Then we may expand the transmitted signals $s_1(t)$ and $s_2(t)$ in terms of $\phi_1(t)$ as follows:

$$s_1(t) = \sqrt{E_b}\phi_1(t), \quad 0 \le t < T_b \tag{8.92}$$

and

$$s_2(t) = -\sqrt{E_b}\phi_1(t), \quad 0 \le t < T_b \tag{8.93}$$

A coherent binary PSK system is therefore characterized by having a signal space that is one-dimensional (i.e., $N = 1$), with a signal constellation consisting of two message points (i.e., $M = 2$) as shown in Fig. 8.14. The coordinates of the message points are

$$s_{11} = \int_0^{T_b} s_1(t)\phi_1(t)\ dt$$
$$= +\sqrt{E_b} \tag{8.94}$$

and

$$s_{21} = \int_0^{T_b} s_2(t)\phi_1(t)\ dt$$
$$= -\sqrt{E_b} \tag{8.95}$$

The message point corresponding to $s_1(t)$ is located at $s_{11} = +\sqrt{E_b}$, and the message point corresponding to $s_2(t)$ is located at $s_{21} = -\sqrt{E_b}$.

To realize a *rule for making a decision* in favor of symbol 1 or symbol 0, we apply Eq. (8.59). Specifically, we partition the signal space of Fig. 8.14 into two regions:

- The set of points closest to message point 1 at $+\sqrt{E_b}$.
- The set of points closest to message point 2 at $-\sqrt{E_b}$.

This is accomplished by constructing the midpoint of the line joining these two message points, and then marking off the appropriate decision regions. In Fig. 8.14 these decision regions are marked Z_1 and Z_2, according to the message point around which they are constructed.

The decision rule is now simply to decide that signal $s_1(t)$ (i.e., binary symbol 1) was transmitted if the received signal point falls in region Z_1, and decide that signal $s_2(t)$ (i.e., binary symbol 0) was transmitted if the received signal point falls in region Z_2. Two kinds of erroneous decisions may, however, be made. Signal $s_2(t)$ is transmitted, but the noise is such that the received signal point falls inside region Z_1 and so the receiver decides in favor of signal $s_1(t)$. Alternatively, signal $s_1(t)$ is transmitted, but the noise is such that the received signal point falls inside region Z_2 and so the receiver decides in favor of signal $s_2(t)$.

To calculate the probability of making an error of the first kind, we note from Fig. 8.14 that the decision region associated with symbol 1 or signal $s_1(t)$ is described by

$$Z_1: \qquad 0 < x_1 < \infty \qquad\qquad (8.96)$$

where the observable element x_1 is related to the received signal $x(t)$ by

$$x_1 = \int_0^{T_b} x(t)\,\phi_1(t)\ dt \qquad\qquad (8.97)$$

From Eq. (8.43) we find that the likelihood function when symbol 0 [i.e., signal

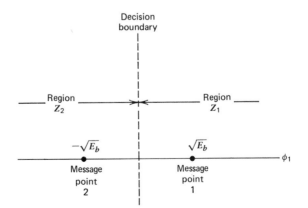

Figure 8.14 Signal-space diagram for coherent binary PSK system.

$s_2(t)$] is transmitted is defined by

$$
\begin{aligned}
f_{X_1}(x_1|0) &= \frac{1}{\sqrt{\pi N_0}} \exp\left[-\frac{1}{N_0} (x_1 - s_{21})^2 \right] \\
&= \frac{1}{\sqrt{\pi N_0}} \exp\left[-\frac{1}{N_0} (x_1 + \sqrt{E_b})^2 \right]
\end{aligned}
\tag{8.98}
$$

The conditional probability of the receiver deciding in favor of symbol 1, given that symbol 0 was transmitted, is therefore

$$
\begin{aligned}
P_{e0} &= \int_0^\infty f_{X_1}(x_1|0)\, dx_1 \\
&= \frac{1}{\sqrt{\pi N_0}} \int_0^\infty \exp\left[-\frac{1}{N_0} (x_1 + \sqrt{E_b})^2 \right] dx_1
\end{aligned}
\tag{8.99}
$$

Putting

$$
z = \frac{1}{\sqrt{N_0}} (x_1 + \sqrt{E_b})
\tag{8.100}
$$

and changing the variable of integration from x_1 to z, we may rewrite Eq. (8.99) in the compact form

$$
\begin{aligned}
P_{e0} &= \frac{1}{\sqrt{\pi}} \int_{\sqrt{E_b/N_0}}^\infty \exp(-z^2)\, dz \\
&= \frac{1}{2} \operatorname{erfc}\left(\sqrt{\frac{E_b}{N_0}} \right)
\end{aligned}
\tag{8.101}
$$

where $\operatorname{erfc}(\cdot)$ is the complementary error function.

Consider next an error of the second kind. We note that the signal space of Fig. 8.14 is symmetric with respect to the origin. It follows therefore that P_{e1}, the conditional probability of the receiver deciding in favor of symbol 0 given that symbol 1 was transmitted, also has the same value as in Eq. (8.101).

Thus, averaging the conditional error probabilities P_{e0} and P_{e1}, we find that the *average probability of symbol error* or, equivalently, the *bit error rate for coherent binary PSK* is

$$
P_e = \frac{1}{2} \operatorname{erfc}\left(\sqrt{\frac{E_b}{N_0}} \right)
\tag{8.102}
$$

As we increase the transmitted signal energy per bit, E_b, for a specified noise spectral density N_0, the message points corresponding to symbols 1 and 0 move further apart, and the average probability of error P_e is correspondingly reduced in accordance with Eq. (8.102).

Generation and Detection of Coherent Binary PSK Signals

To generate a binary PSK signal, we see from Eqs. (8.89) to (8.91) that we have to represent the input binary sequence in polar form with symbols 1 and 0 represented by constant amplitude levels of $+\sqrt{E_b}$ and $-\sqrt{E_b}$, respectively. This signal transmission encoding is performed by a nonreturn-to-zero (NRZ) level

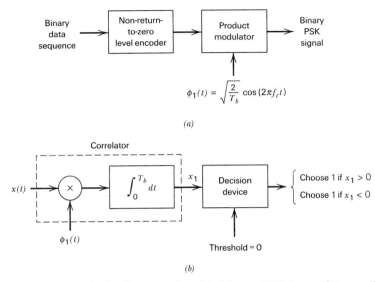

Figure 8.15 Block diagrams for (a) binary PSK transmitter and (b) coherent binary PSK receiver.

encoder. The resulting binary wave (in polar form) and a sinusoidal carrier $\phi_1(t)$ [whose frequency $f_c = (n_c/T_b)$ for some fixed integer n_c] are applied to a product modulator, as in Fig. 8.15a. The carrier and the timing pulses used to generate the binary wave are usually extracted from a common master clock. The desired PSK wave is obtained at the modulator output.

To detect the original binary sequence of 1s and 0s, we apply the noisy PSK signal $x(t)$ (at the channel output) to a correlator, which is also supplied with a locally generated coherent reference signal $\phi_1(t)$, as in Fig. 8.15b. The correlator output, x_1, is compared with a threshold of zero volts. If $x_1 > 0$, the receiver decides in favor of symbol 1. On the other hand, if $x_1 < 0$, it decides in favor of symbol 0. If x_1 is exactly zero, the receiver makes a random guess in favor of 0 or 1.

8.12   COHERENT BINARY FSK

In a binary FSK system, symbols 1 and 0 are distinguished from each other by transmitting one of two sinusoidal waves that differ in frequency by a fixed amount. A typical pair of sinusoidal waves is described by

$$s_i(t) = \begin{cases} \sqrt{\dfrac{2E_b}{T_b}} \cos(2\pi f_i t), & 0 \le t \le T_b \\ 0, & \text{elsewhere} \end{cases} \qquad (8.103)$$

where $i = 1, 2$, and E_b is the transmitted signal energy per bit; the transmitted frequency is

$$f_i = \frac{n_c + i}{T_b} \qquad \text{for some fixed integer } n_c \text{ and } i = 1, 2 \qquad (8.104)$$

Thus symbol 1 is represented by $s_1(t)$, and symbol 0 by $s_2(t)$. The FSK signal described here is known as *Sunde's FSK*. It is a *continuous-phase signal* in the sense that phase continuity is always maintained, including the inter-bit switching times. This form of digital modulation is an example of *continuous-phase frequency-shift keying* (CPFSK), on which we have more to say in Section 8.14.

From Eqs. (8.103) and (8.104), we observe directly that the signals $s_1(t)$ and $s_2(t)$ are orthogonal, but not normalized to have unit energy. We therefore deduce that the most useful form for the set of orthonormal basis functions is

$$\phi_i(t) = \begin{cases} \sqrt{\dfrac{2}{T_b}} \cos(2\pi f_i t), & 0 \le t \le T_b \\ 0, & \text{elsewhere} \end{cases} \tag{8.105}$$

where $i = 1, 2$. Correspondingly, the coefficient s_{ij} for $i = 1, 2$, and $j = 1, 2$, is defined by

$$\begin{aligned} s_{ij} &= \int_0^{T_b} s_i(t)\phi_j(t)\ dt \\ &= \int_0^{T_b} \sqrt{\frac{2E_b}{T_b}} \cos(2\pi f_i t) \sqrt{\frac{2}{T_b}} \cos(2\pi f_j t)\ dt \\ &= \begin{cases} \sqrt{E_b}, & i = j \\ 0, & i \ne j \end{cases} \end{aligned} \tag{8.106}$$

Thus, unlike coherent binary PSK, a coherent binary FSK system is characterized by having a signal space that is two-dimensional (i.e., $N = 2$) with two message points (i.e., $M = 2$), as in Fig. 8.16. The two message points are defined by the signal vectors:

$$\mathbf{s}_1 = \begin{bmatrix} \sqrt{E_b} \\ 0 \end{bmatrix} \tag{8.107}$$

and

$$\mathbf{s}_2 = \begin{bmatrix} 0 \\ \sqrt{E_b} \end{bmatrix} \tag{8.108}$$

Note that the Euclidean distance between the two message points is equal to $\sqrt{2E_b}$.

The observation vector $\mathbf{x}$ has has two elements x_1 and x_2 that are defined by, respectively,

$$x_1 = \int_0^{T_b} x(t)\phi_1(t)\ dt \tag{8.109}$$

and

$$x_2 = \int_0^{T_b} x(t)\phi_2(t)\ dt \tag{8.110}$$

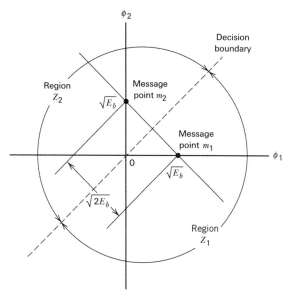

Figure 8.16 Signal-space diagram for coherent binary FSK system.

where $x(t)$ is the received signal, the form of which depends on which symbol was transmitted. Given that symbol 1 was transmitted, $x(t)$ equals $s_1(t) + w(t)$, where $w(t)$ is the sample function of a white Gaussian noise process of zero mean and power spectral density $N_0/2$. If, on the other hand, symbol 0 was transmitted, $x(t)$ equals $s_2(t) + w(t)$.

Now, applying the decision rule of Eq. (8.59), we find that the observation space is partitioned into two decision regions, labeled Z_1 and Z_2 in Fig. 8.16. The decision boundary, separating region Z_1 from region Z_2 is the perpendicular bisector of the line joining the two message points. The receiver decides in favor of symbol 1 if the received signal point represented by the observation vector $\mathbf{x}$ falls inside region Z_1. This occurs when $x_1 > x_2$. If, on the other hand, we have $x_1 < x_2$, the received signal point falls inside region Z_2, and the receiver decides in favor of symbol 0. On the decision boundary, we have $x_1 = x_2$, in which case the receiver makes a random guess in favor of symbol 1 or 0.

Define a new Gaussian random variable L whose sample value l is equal to the difference between x_1 and x_2; that is,

$$l = x_1 - x_2 \tag{8.111}$$

The mean value of the random variable L depends on which binary symbol was transmitted. Given that symbol 1 was transmitted, the Gaussian random variables X_1 and X_2, whose sample values are denoted by x_1 and x_2, have mean values equal to $\sqrt{E_b}$ and zero, respectively. Correspondingly, the conditional mean of the random variable L, given that symbol 1 was transmitted, is

$$
\begin{aligned}
E[L|1] &= E[X_1|1] - E[X_2|1] \\
&= +\sqrt{E_b}
\end{aligned}
\tag{8.112}
$$

On the other hand, given that symbol 0 was transmitted, the random variables X_1 and X_2 have mean values equal to zero and $\sqrt{E_b}$, respectively. Correspondingly, the conditional mean of the random variable L, given that symbol 0 was transmitted, is

$$
\begin{aligned}
E[L|0] &= E[X_1|0] - E[X_2|0] \\
&= -\sqrt{E_b}
\end{aligned}
\tag{8.113}
$$

The variance of the random variable L is independent of which binary symbol was transmitted. Since the random variables X_1 and X_2 are statistically independent, each with a variance equal to $N_0/2$, it follows that

$$
\begin{aligned}
\text{var}[L] &= \text{var}[X_1] + \text{var}[X_2] \\
&= N_0
\end{aligned}
\tag{8.114}
$$

Suppose we know that symbol 0 was transmitted. The conditional probability density function of the random variable L is then given by

$$
f_L(l|0) = \frac{1}{\sqrt{2\pi N_0}} \exp\left[-\frac{(l + \sqrt{E_b})^2}{2N_0} \right]
\tag{8.115}
$$

Since the condition $x_1 > x_2$, or equivalently, $l > 0$, corresponds to the receiver making a decision in favor of symbol 1, we deduce that the conditional probability of error, given that symbol 0 was transmitted, is

$$
\begin{aligned}
P_{e0} &= P(l > 0|\text{symbol 0 was sent}) \\
&= \int_0^\infty f_L(l|0)\ dl \\
&= \frac{1}{\sqrt{2\pi N_0}} \int_0^\infty \exp\left[-\frac{(l + \sqrt{E_b})^2}{2N_0} \right] dl
\end{aligned}
\tag{8.116}
$$

Put

$$
\frac{l + \sqrt{E_b}}{\sqrt{2N_0}} = z
\tag{8.117}
$$

Then, changing the variable of integration from l to z, we may rewrite Eq. (8.116) as follows:

$$
\begin{aligned}
P_{e0} &= \frac{1}{\sqrt{\pi}} \int_{\sqrt{E_b/2N_0}}^\infty \exp(-z^2)\ dz \\
&= \frac{1}{2} \text{erfc}\left(\sqrt{\frac{E_b}{2N_0}} \right)
\end{aligned}
\tag{8.118}
$$

Similarly, we may show that P_{e1}, the conditional probability of error, given that symbol 1 was transmitted, has the same value as in Eq. (8.118). Accordingly, averaging P_{e0} and P_{e1}, we find that the *average probability of symbol error* or, equivalently, the *bit error rate for coherent binary FSK* is

$$P_e = \frac{1}{2} \operatorname{erfc}\left(\sqrt{\frac{E_b}{2N_0}}\right) \qquad (8.119)$$

Comparing Eqs. (8.102) and (8.119), we see that, in a coherent binary FSK system, we have to double the *bit energy-to-noise density ratio, E_b/N_0,* in order to maintain the same bit error rate as in a coherent binary PSK system. This result is in perfect accord with the signal-space diagrams of Fig. 8.14 and 8.16, where we see that in a binary PSK system the distance between the two message points is equal to $2\sqrt{E_b}$, whereas in a binary FSK system the corresponding distance is $\sqrt{2E_b}$. This result clearly illustrates that, in an AWGN channel, the detection performance of equal energy binary signals depends only on the "distance" between the two pertinent message points in the signal space. In particular, the larger we make this distance, the smaller will be the average probability of error. This is intuitively appealing, since the larger the distance between the message points, the smaller will be the probability of mistaking one signal for the other.

Generation and Detection of Coherent FSK Signals

To generate a binary FSK signal, we may use the scheme shown in Fig. 8.17*a*. The incoming binary data sequence is first applied to an *on–off level encoder,* at the output of which symbol 1 is represented by a constant amplitude of $\sqrt{E_b}$ volts and symbol 0 is represented by zero volts. By using an *inverter* in the lower channel in Fig. 8.17*a*, we in effect make sure that when we have symbol 1 at the input, the oscillator with frequency f_1 in the upper channel is switched on while the oscillator with frequency f_2 in the lower channel is switched off, with the result that frequency f_1 is transmitted. Conversely, when we have symbol 0 at the input, the oscillator in the upper channel is switched off and the oscillator in the lower channel is switched on, with the result that frequency f_2 is transmitted. The two frequencies f_1 and f_2 are chosen to equal integer multiples of the bit rate $1/T_b$, as in Eq. (8.104).

In the transmitter of Fig. 8.17*a*, we assume that the two oscillators are synchronized, so that their outputs satisfy the requirements of the two orthonormal basis functions $\phi_1(t)$ and $\phi_2(t)$, as in Eq. (8.105). Alternatively, we may use a single keyed (voltage-controlled) oscillator. In either case, the frequency of the modulated wave is shifted with a continuous phase, in accordance with the input binary wave.

To detect the original binary sequence given the noisy received signal $x(t)$, we may use the receiver shown in Fig. 8.17*b*. It consists of two correlators with a common input, which are supplied with locally generated coherent reference signals $\phi_1(t)$ and $\phi_2(t)$. The correlator outputs are then subtracted, one from the other, and the resulting difference, l, is compared with a threshold of zero volts. If $l > 0$, the receiver decides in favor of 1. On the other hand, if $l < 0$, it decides in favor of 0. If l is exactly zero, the receiver makes a random guess in favor of 1 or 0.

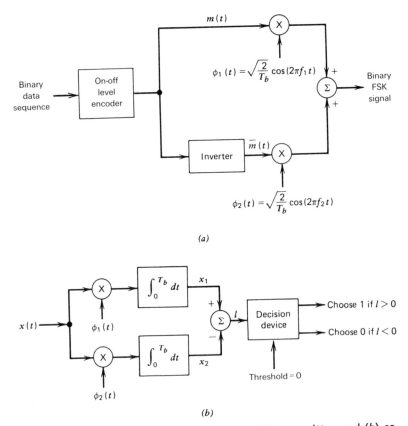

Figure 8.17 Block diagrams for (a) binary FSK transmitter and (b) coherent binary FSK receiver.

8.13 COHERENT QUADRIPHASE-SHIFT KEYING

The provision of reliable performance, exemplified by a very low probability of error, is one important goal in the design of a digital communication system. Another important goal is the efficient utilization of channel bandwidth. In this section, we study a bandwidth-conserving modulation scheme known as coherent quadriphase-shift keying, which is an example of *quadrature-carrier multiplexing*.

In *quadriphase-shift keying* (QPSK), as with binary PSK, information carried by the transmitted signal is contained in the phase. In particular, the phase of the carrier takes on one of four equally spaced values, such as $\pi/4$, $3\pi/4$, $5\pi/4$, and $7\pi/4$. For this set of values we may define the transmitted signal as

$$s_i(t) = \begin{cases} \sqrt{\dfrac{2E}{T}} \cos\left[2\pi f_c t + (2i - 1)\dfrac{\pi}{4}\right], & 0 \le t \le T \\ 0, & \text{elsewhere} \end{cases} \quad (8.120)$$

where $i = 1, 2, 3, 4$; E is the transmitted signal energy per symbol, and T is the symbol duration; the carrier frequency f_c equals n_c/T for some fixed integer n_c. Each possible value of the phase corresponds to a unique pair of bits called a *dibit.* Thus, for example, we may choose the foregoing set of phase values to represent the *Gray encoded* set of dibits: 10, 00, 01, and 11; here we see that only a single bit is changed from one dibit to the next.

Signal-Space Diagram

Using a well-known trigonometric identity, we may use Eq. (8.120) to redefine the transmitted signal $s_i(t)$ for the interval $0 \le t \le T$ in the equivalent form:

$$s_i(t) = \sqrt{\frac{2E}{T}} \cos\left[(2i - 1)\frac{\pi}{4}\right] \cos(2\pi f_c t) - \sqrt{\frac{2E}{T}} \sin\left[(2i - 1)\frac{\pi}{4}\right] \sin(2\pi f_c t)$$

$$(8.121)$$

where $i = 1, 2, 3, 4$. Based on this representation, we can make the following observations:

- There are only two orthonormal basis functions, $\phi_1(t)$ and $\phi_2(t)$, contained in the expansion of $s_i(t)$. Specifically, $\phi_1(t)$ and $\phi_2(t)$ are defined by a pair of *quadrature carriers*:

$$\phi_1(t) = \sqrt{\frac{2}{T}} \cos(2\pi f_c t), \qquad 0 \le t \le T \tag{8.122}$$

$$\phi_2(t) = \sqrt{\frac{2}{T}} \sin(\pi f_c t), \qquad 0 \le t \le T \tag{8.123}$$

- There are four message points, and the associated signal vectors are defined by

$$\mathbf{s}_i = \begin{bmatrix} \sqrt{E}\cos\left((2i - 1)\frac{\pi}{4}\right) \\ -\sqrt{E}\sin\left((2i - 1)\frac{\pi}{4}\right) \end{bmatrix}, \qquad i = 1, 2, 3, 4 \tag{8.124}$$

The elements of the signal vectors, namely, s_{i1} and s_{i2}, have their values summarized in Table 8.1. The first two columns of this table give the associated dibits and phase of the QPSK signal.

Accordingly, a QPSK signal is characterized by having a two-dimensional signal constellation (i.e., $N = 2$) and four message points (i.e., $M = 4$), as illustrated in Fig. 8.18.

Table 8.1 Signal-Space Characterization of QPSK

Input Dibit $0 \le t \le T$	Phase of QPSK Signal (radians)	Coordinates of Message Points	
		s_{i1}	s_{i2}
10	$\pi/4$	$+\sqrt{E/2}$	$-\sqrt{E/2}$
00	$3\pi/4$	$-\sqrt{E/2}$	$-\sqrt{E/2}$
01	$5\pi/4$	$-\sqrt{E/2}$	$+\sqrt{E/2}$
11	$7\pi/4$	$+\sqrt{E/2}$	$+\sqrt{E/2}$

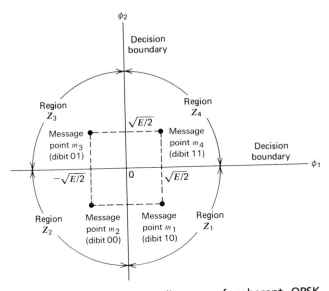

Figure 8.18 Signal-space diagram of coherent QPSK system.

EXAMPLE 2

Figure 8.19 illustrates the sequences and waveforms involved in the generation of a QPSK signal. The input binary sequence 01101000 is shown in Fig. 8.19a. This sequence is divided into two other sequences, consisting of odd- and even-numbered bits of the input sequence. These two sequences are shown in the top lines of Figs. 8.19b and 8.19c. The waveforms representing the in-phase and quadrature components of the QPSK signal are also shown in Figs. 8.19b and 8.19c, respectively. These two waveforms may individually be viewed as examples of a binary PSK signal. Adding them, we get the QPSK waveform shown in Fig. 8.19d.

To define the decision rule for the detection of the transmitted data sequence, we partition the signal space into four regions, in accordance with Eq. (8.59). The individual regions are defined by the set of points closest to the message point represented by signal vector s_1, s_2, s_3, and s_4. This is readily accomplished by constructing the perpendicular bisectors of the square formed by joining the four message points and then marking off the appropriate regions. We thus find that the decision regions are quadrants whose vertices coincide with the origin. These regions are marked Z_1, Z_2, Z_3, and Z_4, in Fig. 8.18, according to the message point around which they are constructed.

Average Probability of Error

In a coherent QPSK system, the received signal $x(t)$ is defined by

$$x(t) = s_i(t) + w(t), \qquad \begin{cases} 0 \le t \le T \\ i = 1, 2, 3, 4 \end{cases} \tag{8.125}$$

where $w(t)$ is the sample function of a white Gaussian noise process of zero mean

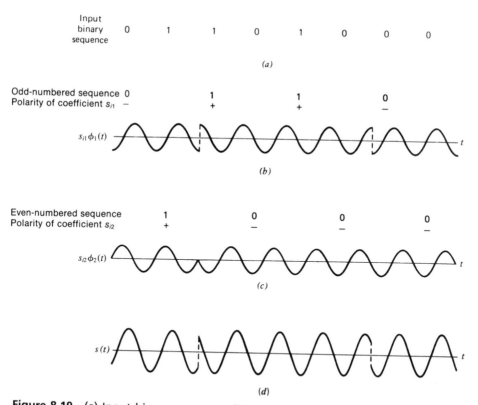

Input
binary   0   1   1   0   1   0   0   0
sequence

(a)

Odd-numbered sequence 0   1   1   0
Polarity of coefficient s_{i1} −   +   +   −

$s_{i1}\phi_1(t)$

(b)

Even-numbered sequence   1   0   0   0
Polarity of coefficient s_{i2} +   −   −   −

$s_{i2}\phi_2(t)$

(c)

$s(t)$

(d)

Figure 8.19 *(a)* Input binary sequence. *(b)* Odd-numbered bits of input sequence and associated binary PSK wave. *(c)* Even-numbered bits of input sequence and associated binary PSK wave. *(d)* QPSK waveform.

and power spectral density $N_0/2$. Correspondingly, the observation vector **x** has two elements, x_1 and x_2, defined by

$$x_1 = \int_0^T x(t)\phi_1(t) \, dt$$

$$= \sqrt{E}\cos\left[(2i-1)\frac{\pi}{4}\right] + w_1 \qquad (8.126)$$

$$= \pm\sqrt{\frac{E}{2}} + w_1$$

and

$$x_2 = \int_0^T x(t)\phi_2(t) \, dt$$

$$= -\sqrt{E}\sin\left[(2i-1)\frac{\pi}{4}\right] + w_2 \qquad (8.127)$$

$$= \mp\sqrt{\frac{E}{2}} + w_2$$

Thus, the observable elements x_1 and x_2 are sample values of independent Gaussian random variables with mean values equal to $\pm\sqrt{E/2}$, and $\mp\sqrt{E/2}$, respectively, and with a common variance equal to $N_0/2$.

The decision rule is now simply to decide that $s_1(t)$ was transmitted if the received signal point associated with the observation vector $\mathbf{x}$ falls inside region Z_1, decide that $s_2(t)$ was transmitted if the received signal point falls inside region Z_2, and so on. An erroneous decision will be made if, for example, signal $s_4(t)$ is transmitted but the noise $w(t)$ is such that the received signal point falls outside region Z_4.

To calculate the average probability of symbol error, we note from Eq. (8.121) that a coherent QPSK system is in fact equivalent to two coherent binary PSK systems working in parallel and using two carriers that are in phase quadrature; this is merely a statement of the quadrature-carrier multiplexing property of coherent QPSK. The in-phase channel output x_1 and the quadrature channel output x_2 (i.e., the two elements of the observation vector $\mathbf{x}$) may be viewed as the individual outputs of the two coherent binary PSK systems. Thus, according to Eqs. (8.126) and (8.127), these two binary PSK systems may be characterized as follows:

- The signal energy per bit is $\sqrt{E/2}$.

- The noise spectral density is $N_0/2$.

Hence, using Eq. (8.102) for the average probability of bit error of a coherent binary PSK system, we may now state that the average probability of bit error in *each* channel of the coherent QPSK system is

$$\begin{aligned}
P' &= \frac{1}{2}\text{erfc}\left(\sqrt{\frac{E/2}{N_0}}\right) \\
&= \frac{1}{2}\text{erfc}\left(\sqrt{\frac{E}{2N_0}}\right)
\end{aligned} \tag{8.128}$$

Another important point to note is that the bit errors in the in-phase and quadrature channels of the coherent QPSK system are statistically independent. The in-phase channel makes a decision on one of the two bits constituting a symbol (dibit) of the QPSK signal, and the quadrature channel takes care of the other bit. Accordingly, the *average probability of a correct decision* resulting from the combined action of the two channels working together is

$$\begin{aligned}
P_c &= (1 - P')^2 \\
&= \left[1 - \frac{1}{2}\text{erfc}\left(\sqrt{\frac{E}{2N_0}}\right)\right]^2 \\
&= 1 - \text{erfc}\left(\sqrt{\frac{E}{2N_0}}\right) + \frac{1}{4}\text{erfc}^2\left(\sqrt{\frac{E}{2N_0}}\right)
\end{aligned} \tag{8.129}$$

The average probability of symbol error for coherent QPSK is therefore

$$\begin{aligned}
P_e &= 1 - P_c \\
&= \text{erfc}\left(\sqrt{\frac{E}{2N_0}}\right) - \frac{1}{4}\text{erfc}^2\left(\sqrt{\frac{E}{2N_0}}\right)
\end{aligned} \tag{8.130}$$

In the region where $(E/2N_0) \gg 1$, we may ignore the second term on the right-hand side of Eq. (8.130), and so approximate the formula for the average probability of symbol error for coherent QPSK as

$$P_e \simeq \text{erfc}\left(\sqrt{\frac{E}{2N_0}}\right) \tag{8.131}$$

The formula of Eq. (8.131) may also be derived in another insightful way, using the signal-space diagram of Fig. 8.18. Since the four message points of this diagram are completely symmetric, we may apply Eq. (8.72), reproduced here in the form

$$P_e \leq \frac{1}{2} \sum_{\substack{k=1 \\ k \neq i}}^{4} \text{erfc}\left(\frac{d_{ik}}{2\sqrt{N_0}}\right) \tag{8.132}$$

where i refers to the message point m_i sent. Consider, for example, message point m_1 (corresponding to dibit 10) chosen as m_i. The message points m_2 and m_4 (corresponding to dibits 00 and 11) are the *closest* to m_1. From Fig. 8.18 we readily find that m_1 is equidistant from m_2 and m_4 in a Euclidean sense, as shown by

$$d_{12} = d_{14} = \sqrt{2E}$$

Assuming that E/N_0 is large enough to ignore the contribution of the most distant message point m_3 (corresponding to dibit 01) relative to m_1, we find that the use of Eq. (8.132) yields an approximate expression for P_e that is the same as Eq. (8.131). Note that in mistaking either m_2 or m_4 for m_1, a single bit error is made; on the other hand, in mistaking m_3 for m_1, two bit errors are made. For a high enough E/N_0, the likelihood of both bits of a symbol being in error is much less than a single bit, which is a further justification for ignoring m_3 in calculating P_e when m_1 is sent.

In a QPSK system, we note that since there are two bits per symbol, the transmitted signal energy per symbol is twice the signal energy per bit, as shown by

$$E = 2E_b \tag{8.133}$$

Thus, expressing the average probability of symbol error in terms of the ratio E_b/N_0, we may write

$$P_e \simeq \text{erfc}\left(\sqrt{\frac{E_b}{N_0}}\right) \tag{8.134}$$

With Gray encoding used for the incoming dibits (symbols), we find from Eqs. (8.128) and (8.133) that the bit error rate of QPSK is exactly

$$\text{BER} = \frac{1}{2}\text{erfc}\left(\sqrt{\frac{E_b}{N_0}}\right) \tag{8.135}$$

We may therefore state that a coherent QPSK system achieves the same average probability of bit error as a coherent binary PSK system for the same bit rate and the same E_b/N_0, but uses only half the channel bandwidth. Stated in a different way, for the same E_b/N_0 and therefore the same average probability of bit error, a coherent QPSK system transmits information at twice the bit rate of a coherent binary PSK system for the same channel bandwidth.

Generation and Detection of Coherent QPSK Signals

Consider next the generation and detection of QPSK signals. Figure 8.20a shows a block diagram of a typical QPSK transmitter. The incoming binary data sequence is first transformed into polar form by a *nonreturn-to-zero level* encoder. Thus, symbols 1 and 0 are represented by $+\sqrt{E_b}$ and $-\sqrt{E_b}$, respectively. This binary wave is next divided by means of a *demultiplexer* into two separate binary waves consisting of the odd- and even-numbered input bits. These two binary waves are denoted by $a_1(t)$ and $a_2(t)$. We note that in any signaling interval, the amplitudes of $a_1(t)$ and $a_2(t)$ equal s_{i1} and s_{i2}, respectively, depending on the particular dibit that is being transmitted. The two binary waves $a_1(t)$ and $a_2(t)$

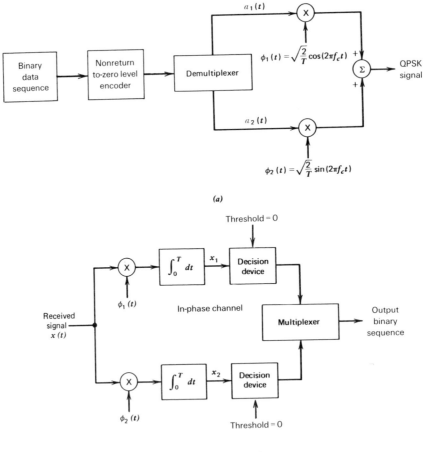

Figure 8.20 Block diagrams for (a) QPSK transmitter and (b) QPSK receiver.

are used to modulate a pair of quadrature carriers or orthonormal basis functions: $\phi_1(t)$ equal to $\sqrt{2/T}\cos(2\pi f_c t)$ and $\phi_2(t)$ equal to $\sqrt{2/T}\sin(2\pi f_c t)$. The result is a pair of binary PSK signals, which may be detected independently due to the orthogonality of $\phi_1(t)$ and $\phi_2(t)$. Finally, the two binary PSK signals are added to produce the desired QPSK signal.

The QPSK receiver consists of a pair of correlators with a common input and supplied with a locally generated pair of coherent reference signals $\phi_1(t)$ and $\phi_2(t)$, as in Fig. 8.20b. The correlator outputs x_1 and x_2, produced in response to the received signal $x(t)$, are each compared with a threshold of zero. If $x_1 > 0$, a decision is made in favor of symbol 1 for the in-phase channel output, but if $x_1 < 0$, a decision is made in favor of symbol 0. Similarly, if $x_2 > 0$, a decision is made in favor of symbol 1 for the quadrature channel output, but if $x_2 < 0$, a decision is made in favor of symbol 0. Finally, these two binary sequences at the in-phase and quadrature channel outputs are combined in a *multiplexer* to reproduce the original binary sequence at the transmitter input with the minimum probability of symbol error in an AWGN channel.

8.14 COHERENT MINIMUM SHIFT KEYING

In the coherent detection of binary FSK signal described in Section 8.12, the phase information contained in the received signal was not fully exploited, other than to provide for synchronization of the receiver to the transmitter. We now show that by proper utilization of the phase when performing detection, it is possible to improve the noise performance of the receiver significantly. This improvement is, however, achieved at the expense of increased receiver complexity.

Consider a continuous-phase frequency-shift keying (CPFSK) signal, which is defined for the interval $0 \leq t \leq T_b$ as follows:

$$s(t) = \begin{cases} \sqrt{\dfrac{2E_b}{T_b}}\cos[2\pi f_1 t + \theta(0)] & \text{for symbol 1} \\[2em] \sqrt{\dfrac{2E_b}{T_b}}\cos[2\pi f_2 t + \theta(0)] & \text{for symbol 0} \end{cases} \tag{8.136}$$

where E_b is the transmitted signal energy per bit, and T_b is the bit duration. The phase $\theta(0)$, denoting the value of the phase at time $t = 0$, depends on the past history of the modulation process. The frequencies f_1 and f_2 are sent in response to binary symbols 1 and 0 appearing at the modulator input, respectively.

Another useful way of representing the CPFSK signal $s(t)$ is to express it in the conventional form of an *angle-modulated signal* as follows

$$s(t) = \sqrt{\dfrac{2E_b}{T_b}}\cos[2\pi f_c t + \theta(t)] \tag{8.137}$$

where $\theta(t)$ is the phase of $s(t)$. When the phase $\theta(t)$ is a continuous function of time, we find that the modulated signal $s(t)$ itself is also continuous at all times, including the inter-bit switching times. The phase $\theta(t)$ of a CPFSK signal increases or decreases linearly with time during each bit duration of T_b seconds, as shown by

$$\theta(t) = \theta(0) \pm \frac{\pi h}{T_b}t, \qquad 0 \leq t \leq T_b \tag{8.138}$$

where the plus sign corresponds to sending symbol 1, and the minus sign corresponds to sending symbol 0. Substituting Eq. (8.138) in (8.137), and then comparing the angle of the cosine function with that of Eq. (8.136), we deduce the following pair of relations:

$$f_c + \frac{h}{2T_b} = f_1 \tag{8.139}$$

$$f_c - \frac{h}{2T_b} = f_2 \tag{8.140}$$

Solving Eqs. (8.139) and (8.140) for f_c and h, we thus get

$$f_c = \frac{1}{2}(f_1 + f_2) \tag{8.141}$$

and

$$h = T_b(f_1 - f_2) \tag{8.142}$$

The nominal carrier frequency f_c is therefore the arithmetic mean of the frequencies f_1 and f_2. The difference between the frequencies f_1 and f_2, normalized with respect to the bit rate $1/T_b$, defines the dimensionless parameter h, which is referred to as the *deviation ratio*.

Phase Trellis

From Eq. (8.138) we find that at time $t = T_b$,

$$\theta(T_b) - \theta(0) = \begin{cases} \pi h & \text{for symbol 1} \\ -\pi h & \text{for symbol 0} \end{cases} \tag{8.143}$$

That is to say, the sending of symbol 1 increases the phase of a CPFSK signal $s(t)$ by πh radians, whereas the sending of symbol 0 reduces it by an equal amount.

The variation of phase $\theta(t)$ with time t follows a path consisting of a sequence of straight lines, the slopes of which represent frequency changes. Figure 8.21 depicts possible paths starting from time $t = 0$. A plot like that shown in Fig. 8.21 is called a *phase tree*. The tree makes clear the transitions of phase across interval boundaries of the incoming sequence of data bits. Moreover, it is evident from Fig. 8.21 that the phase of a CPFSK signal is an odd or even multiple of πh radians at odd or even multiples of the bit duration T_b, respectively.

The phase tree described in Fig. 8.21 is a manifestation of phase continuity, which is an inherent characteristic of a CPFSK signal. To appreciate the notion of phase continuity, let us go back for a moment to Sunde's FSK, which is a

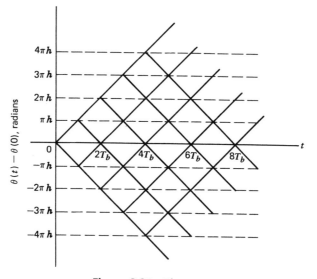

Figure 8.21 Phase tree.

CPFSK scheme as described in Section 8.12. In this case the deviation ratio h equals unity. Hence, according to Fig. 8.21 the phase change over one bit interval is $\pm\pi$ radians. But, a change of $+\pi$ radians is exactly the same as a change of $-\pi$ radians, module 2π. It follows therefore that in the case of Sunde's FSK there is *no* memory; that is, knowing which particular change occurred in the *previous* bit interval provides no help in the *current* bit interval.

In contrast, we have a completely different situation when the deviation ratio h is assigned the special value of $1/2$. We now find that the phase can take on only the two values $\pm\pi/2$ at odd multiples of T_b, and only the two values 0 and π at even multiples of T_b, as in Fig. 8.22. This second graph is called a *phase trellis*, since a "trellis" is a treelike structure with remerging branches. Each path from left to right through the trellis of Fig. 8.22 corresponds to a specific binary sequence input. For example, the path shown in boldface in Fig. 8.22 corresponds to the binary sequence 1101000 with $\theta(0) = 0$. Henceforth, we assume that $h = 1/2$.

With $h = 1/2$, we find from Eq. (8.142) that the frequency deviation (i.e., the difference between the two signaling frequencies f_1 and f_2) equals half the

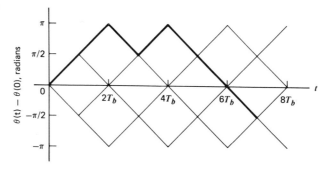

Figure 8.22 Phase trellis; boldfaced path represents the sequence 1101000.

bit rate. This is the minimum frequency spacing that allows the two FSK signals representing symbols 1 and 0, as in Eq. (8.136), to be coherently orthogonal in the sense that they do not interfere with one another in the process of detection. It is for this reason that a CPFSK signal with a deviation ratio of one-half is commonly referred to as *minimum shift keying* (MSK).[5]

Signal-Space Diagram

Using a well-known trigonometric identity in Eq. (8.137), we may express the CPFSK signal $s(t)$ in terms of its in-phase and quadrature components as follows:

$$s(t) = \sqrt{\frac{2E_b}{T_b}}\cos[\theta(t)]\cos(2\pi f_c t) - \sqrt{\frac{2E_b}{T_b}}\sin[\theta(t)]\sin(2\pi f_c t) \quad (8.144)$$

Consider first the in-phase component $\sqrt{2E_b/T_b}\cos[\theta(t)]$. With the deviation ratio $h = 1/2$, we have from Eq. (8.138) that

$$\theta(t) = \theta(0) \pm \frac{\pi}{2T_b}t, \qquad 0 \leq t \leq T_b \quad (8.145)$$

where the plus sign corresponds to symbol 1 and the minus sign corresponds to symbol 0. A similar result holds for $\theta(t)$ in the interval $-T_b \leq t \leq 0$, except that the algebraic sign is not necessarily the same in both intervals. Since the phase $\theta(0)$ is 0 or π, depending on the past history of the modulation process, we find that, in the interval $-T_b \leq t \leq T_b$, the polarity of $\cos[\theta(t)]$ depends only on $\theta(0)$, regardless of the sequence of 1s and 0s transmitted before or after $t = 0$. Thus, for this time interval, the in-phase component $s_I(t)$ consists of a *half-cycle cosine pulse* defined as follows:

$$
\begin{aligned}
s_I(t) &= \sqrt{\frac{2E_b}{T_b}}\cos[\theta(t)] \\
&= \sqrt{\frac{2E_b}{T_b}}\cos[\theta(0)]\cos\left(\frac{\pi}{2T_b}t\right) \\
&= \pm\sqrt{\frac{2E_b}{T_b}}\cos\left(\frac{\pi}{2T_b}t\right), \qquad -T_b \leq t \leq T_b
\end{aligned}
\quad (8.146)
$$

where the plus sign corresponds to $\theta(0) = 0$ and the minus sign corresponds to $\theta(0) = \pi$. In a similar way, we may show that, in the interval $0 \leq t \leq 2T_b$, the quadrature component $s_Q(t)$ consists of a *half-cycle sine pulse*, whose polarity depends only on $\theta(T_b)$, as shown by

$$
\begin{aligned}
s_Q(t) &= \sqrt{\frac{2E_b}{T_b}}\sin[\theta(t)] \\
&= \sqrt{\frac{2E_b}{T_b}}\sin[\theta(T_b)]\sin\left(\frac{\pi}{2T_b}t\right) \\
&= \pm\sqrt{\frac{2E_b}{T_b}}\sin\left(\frac{\pi}{2T_b}t\right), \qquad 0 \leq t \leq 2T_b
\end{aligned}
\quad (8.147)
$$

where the plus sign corresponds to $\theta(T_b) = \pi/2$ and the minus sign corresponds to $\theta(T_b) = -\pi/2$.

From the foregoing discussion we see that since the phase states $\theta(0)$ and $\theta(T_b)$ can each assume one of two possible values, any one of four possibilities can arise, as described here:

- The phase $\theta(0) = 0$ and $\theta(T_b) = \pi/2$, corresponding to the transmission of symbol 1.
- The phase $\theta(0) = \pi$ and $\theta(T_b) = \pi/2$, corresponding to the transmission of symbol 0.
- The phase $\theta(0) = \pi$ and $\theta(T_b) = -\pi/2$ (or, equivalently, $3\pi/2$, modulo 2π), corresponding to the transmission of symbol 1.
- The phase $\theta(0) = 0$ and $\theta(T_b) = -\pi/2$, corresponding to the transmission of symbol 0.

This, in turn, means that the MSK signal itself may assume any one of four possible forms, depending on the values of $\theta(0)$ and $\theta(T_b)$.

From the expansion of Eq. (8.144), we deduce that the orthonormal basis functions $\phi_1(t)$ and $\phi_2(t)$ for MSK are defined by a pair of sinusoidally modulated quadrature carriers:

$$\phi_1(t) = \sqrt{\frac{2}{T_b}} \cos\left(\frac{\pi}{2T_b}t\right) \cos(2\pi f_c t), \qquad 0 \leq t \leq T_b \qquad (8.148)$$

$$\phi_2(t) = \sqrt{\frac{2}{T_b}} \sin\left(\frac{\pi}{2T_b}t\right) \sin(2\pi f_c t), \qquad 0 \leq t \leq T_b \qquad (8.149)$$

Correspondingly, we may express the MSK signal in the form

$$s(t) = s_1 \phi_1(t) + s_2 \phi_2(t), \qquad 0 \leq t \leq T_b \qquad (8.150)$$

where the coefficients s_1 and s_2 are related to the phase states $\theta(0)$ and $\theta(T_b)$, respectively. To evaluate s_1, we integrate the product $s(t)\phi_1(t)$ between the limits $-T_b$ and T_b, as shown by

$$s_1 = \int_{-T_b}^{T_b} s(t)\phi_1(t)\ dt$$
$$= \sqrt{E_b}\cos[\theta(0)], \qquad -T_b \leq t \leq T_b \qquad (8.151)$$

Similarly, to evaluate s_2 we integrate the product $s(t)\phi_2(t)$ between the limits 0 and $2T_b$, as shown by

$$s_2 = \int_{0}^{2T_b} s(t)\phi_2(t)\ dt$$
$$= -\sqrt{E_b}\sin[\theta(T_b)], \qquad 0 \leq t \leq 2T_b \qquad (8.152)$$

Note that in Eqs. (8.151) and (8.152):

- Both integrals are evaluated for a time interval equal to twice the bit duration.
- Both the lower and upper limits of the product integration used to evaluate

the coefficient s_1 are shifted by the bit duration T_b with respect to those used to evaluate the coefficient s_2.

- The time interval $0 \le t \le T_b$, for which the phase states $\theta(0)$ and $\theta(T_b)$ are defined, is common to both integrals.

Accordingly, the signal constellation for an MSK signal is two-dimensional (i.e., $N = 2$), with four message points (i.e., $M = 4$), as illustrated in Fig. 8.23. The coordinates of the message points are as follows: $(+\sqrt{E_b}, -\sqrt{E_b})$, $(-\sqrt{E_b}, -\sqrt{E_b})$, $(-\sqrt{E_b}, +\sqrt{E_b})$, and $(+\sqrt{E_b}, +\sqrt{E_b})$. The possible values of $\theta(0)$ and $\theta(T_b)$, corresponding to these four message points, are also included in Fig. 8.23. The signal-space diagram of MSK is thus similar to that of QPSK in that both of them have four message points. However, they differ in a subtle way that should be carefully noted: In QPSK, the transmitted symbol can be represented by any one of the four message points, whereas in MSK, exactly two of the four message points can be used to represent the transmitted symbol at any one time.

Table 8.2 presents a summary of the values of $\theta(0)$ and $\theta(T_b)$, as well as the corresponding values of s_1 and s_2 that are calculated for the time intervals $-T_b \le t \le T_b$ and $0 \le t \le 2T_b$, respectively. The first column of this table indicates whether symbol 1 or symbol 0 was sent in the interval $0 \le t \le T_b$. Note that the coordinates of the message points, s_1 and s_2, have opposite signs when symbol 1 is sent in this interval, but the same sign when symbol 0 is sent. Ac-

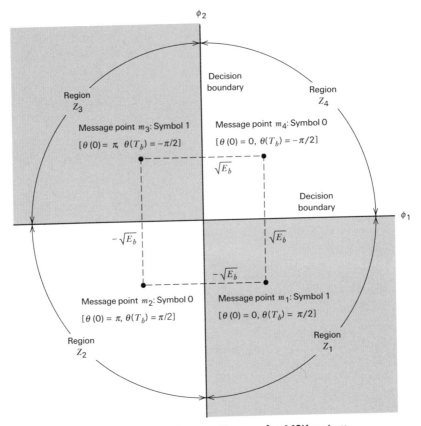

Figure 8.23 Signal-space diagram for MSK system.

Table 8.2 Signal-Space Characterization of MSK

Transmitted Binary Symbol, $0 \leq t \leq T_b$	Phase States (radians)		Coordinates of Message Points	
	$\theta(0)$	$\theta(T_b)$	s_1	s_2
1	0	$+\pi/2$	$+\sqrt{E_b}$	$-\sqrt{E_b}$
0	π	$+\pi/2$	$-\sqrt{E_b}$	$-\sqrt{E_b}$
1	π	$-\pi/2$	$-\sqrt{E_b}$	$+\sqrt{E_b}$
0	0	$-\pi/2$	$+\sqrt{E_b}$	$+\sqrt{E_b}$

cordingly, for a given input data sequence, we may use the entries of Table 8.2 to derive, on a bit-by-bit basis, the two sequences of coefficients required to scale $\phi_1(t)$ and $\phi_2(t)$, and thereby determine the MSK signal $s(t)$.

EXAMPLE 3

Figure 8.24 shows the sequences and waveforms involved in the generation of an MSK signal for the binary sequence 1101000. The input binary sequence is shown Fig. 8.24a. Assuming that, at time $t = 0$ the phase $\theta(0)$ is zero, the sequence of phase states is as shown in Fig. 8.22, modulo 2π. The polarities of the two sequences of factors used to scale the time functions $\phi_1(t)$ and $\phi_2(t)$ are shown in the top lines of Figs. 8.24b and 8.24c. Note that these two sequences are *offset* relative to each other by an interval equal to the bit duration T_b. The waveforms of the resulting in-phase and quadrature components are also shown in Figs. 8.24b and 8.24c. Adding these two modulated waveforms, we get the desired MSK signal shown in Fig. 8.24d.

Average Probability of Error

In the case of an AWGN channel, the received signal is given by

$$x(t) = s(t) + w(t)$$

where $s(t)$ is the transmitted MSK signal, and $w(t)$ is the sample function of a white Gaussian noise process of zero mean and power spectral density $N_0/2$. To decide whether symbol 1 or symbol 0 was transmitted in the interval $0 \leq t \leq T_b$, say, we have to establish a procedure for the use of $x(t)$ to detect the phase states $\theta(0)$ and $\theta(T_b)$. For the optimum detection of $\theta(0)$, we first determine the projection of the received signal $x(t)$ onto the reference signal $\phi_1(t)$ over the interval $-T_b \leq t \leq T_b$, obtaining

$$x_1 = \int_{-T_b}^{T_b} x(t)\phi_1(t) \, dt$$

$$= s_1 + w_1, \quad -T_b \leq t \leq T_b \tag{8.153}$$

where s_1 is as defined by Eq. (8.151) and w_1 is the sample value of a Gaussian random variable of zero mean and variance $N_0/2$. From the signal-space dia-

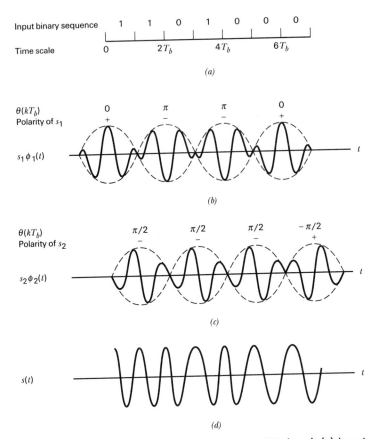

Figure 8.24 Sequence and waveforms for MSK signal. (a) Input binary sequence. (b) Scaled time function $s_1\phi_1(t)$. (c) Scaled time function $s_2\phi_2(t)$. (d) MSK signal $s(t)$ obtained by adding $s_1\phi_1(t)$ and $s_2\phi_2(t)$ and on a bit-by-bit basis.

gram of Fig. 8.23, we observe that if $x_1 > 0$, the receiver chooses the estimate $\hat{\theta}(0) = 0$. On the other hand, if $x_1 < 0$, it chooses the estimate $\hat{\theta}(0) = \pi$.

Similarly, for the optimum detection of $\theta(T_b)$, we determine the projection of the received signal $x(t)$ onto the second reference signal $\phi_2(t)$ over the interval $0 \leq t \leq 2T_b$, obtaining

$$x_2 = \int_0^{2T_b} x(t)\phi_2(t)\ dt \tag{8.154}$$

$$= s_2 + w_2, \qquad 0 \leq t \leq 2T_b$$

where s_2 is as defined by Eq. (8.152) and w_2 is the sample value of another independent Gaussian random variable of zero mean and variance $N_0/2$. Referring again to the signal space diagram of Fig. 8.23, we observe that if $x_2 > 0$, the receiver chooses the estimate $\hat{\theta}(T_b) = -\pi/2$. If, on the other hand, $x_2 < 0$, it chooses the estimate $\hat{\theta}(T_b) = \pi/2$.

To reconstruct the original binary sequence, we interleave the above two sets of phase decisions, as described next (see Table 8.2):

- If we have the estimates $\hat{\theta}(0) = 0$ and $\hat{\theta}(T_b) = -\pi/2$, or alternatively if we have the estimates $\hat{\theta}(0) = \pi$ and $\hat{\theta}(T_b) = \pi/2$, the receiver makes a decision in favor of symbol 0.

- If we have the estimates $\hat{\theta}(0) = \pi$ and $\hat{\theta}(T_b) = -\pi/2$, or alternatively if we have the estimates $\hat{\theta}(0) = 0$ and $\hat{\theta}(T_b) = \pi/2$, the receiver makes a decision in favor of symbol 1.

Thus, referring to the signal-space diagram of Fig. 8.23, we see that the decision made by the receiver is between the message points m_1 and m_3 corresponding to symbol 1, or between the message points m_2 and m_4 corresponding to symbol 0. The corresponding decisions whether $\theta(0)$ is 0 or π and whether $\theta(T_b)$ is $-\pi/2$ or $+\pi/2$ (i.e., the bit decisions) are made *alternately* in the I- and Q-channels of the receiver, with each channel looking at the input signal for $2T_b$ seconds. The signal from *other bits* does not interfere with the receiver's decision for a given bit in either channel. The receiver makes an error when the I-channel assigns the wrong value to $\theta(0)$ or the Q-channel assigns the wrong value to $\theta(T_b)$. Accordingly, using the statistical characterizations of the product-integrator outputs x_1 and x_2 of these two channels, defined by Eqs. (8.153) and (8.154), respectively, we readily find that the bit error rate for coherent MSK is given by

$$\text{BER} = \frac{1}{2}\text{erfc}\left(\sqrt{\frac{E_b}{N_0}}\right)$$

which is exactly the same as that for binary PSK and QPSK. It is important to note, however, that this good performance is the result of the detection of the MSK signal being performed in the receiver on the basis of observations over $2T_b$ seconds.

Generation and Detection of MSK Signals

Consider next the generation and demodulation of MSK. Figure 8.25 shows the block diagram of a typical MSK transmitter. The advantage of this method of generating MSK signals is that the signal coherence and deviation ratio are largely unaffected by variations in the input data rate. Two input sinusoidal waves, one of frequency $f_c = n_c/4T_b$ for some fixed integer n_c, and the other of frequency $1/4T_b$, are first applied to a product modulator. This produces two phase-coherent sinusoidal waves at frequencies f_1 and f_2, which are related to the carrier frequency f_c and the bit rate $1/T_b$ by Eqs. (8.139) and (8.140) for $h = 1/2$. These two sinusoidal waves are separated from each other by two narrow-band filters, one centered at f_1 and the other at f_2. The resulting filter outputs are next linearly combined to produce the pair of quadrature carriers or orthonormal basis functions $\phi_1(t)$ and $\phi_2(t)$. Finally, $\phi_1(t)$ and $\phi_2(t)$ are multiplied with two binary waves $a_1(t)$ and $a_2(t)$, both of which have a bit rate equal to $1/2T_b$. These two binary waves are extracted from the incoming binary sequence in the manner described in Example 3.

Figure 8.25b shows the block diagram of a typical MSK receiver. The received signal $x(t)$ is correlated with locally generated replicas of the coherent reference signals $\phi_1(t)$ and $\phi_2(t)$. Note that in both cases the integration interval is $2T_b$ seconds, and that the integration in the quadrature channel is delayed by T_b seconds with respect to that in the in-phase channel. The resulting in-phase and

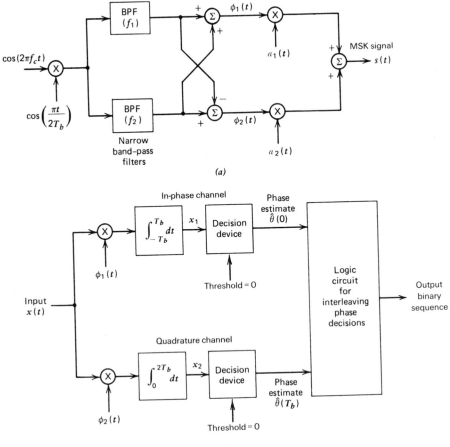

Figure 8.25 Block diagrams for (a) MSK transmitter and (b) MSK receiver.

quadrature channel correlator outputs, x_1 and x_2, are each compared with a
threshold of zero, and estimates of the phase $\theta(0)$ and $\theta(T_b)$ are derived in the
manner described previously. Finally, these phase decisions are interleaved so as
to reconstruct the original input binary sequence with the minimum average
probability of symbol error in an AWGN channel.

8.15 NONCOHERENT ORTHOGONAL MODULATION

Coherent detection exploits knowledge of the carrier wave's phase reference,
thereby providing the optimum error performance attainable with a digital modulation format of interest. When, however, it is impractical to have knowledge
of the carrier phase at the receiver, we resort to the use of *noncoherent detection*.
In this section, we study the noise performance of *noncoherent orthogonal modulation* that includes noncoherent binary frequency-shift keying and differential
phase-shift keying as special cases.[6]

Consider a binary signaling scheme that involves the use of two orthogonal
signals $s_1(t)$ and $s_2(t)$, which have equal energy. During the interval $0 \leq t \leq T$,

one of these two signals is sent over an imperfect channel that shifts the carrier phase by an unknown amount. Let $g_1(t)$ and $g_2(t)$ denote the phase-shifted versions of $s_1(t)$ and $s_2(t)$, respectively. It is assumed that the signals $g_1(t)$ and $g_2(t)$ remain orthogonal and have the same energy E, regardless of the unknown carrier phase. We refer to such a signaling scheme as *noncoherent orthogonal modulation*. Depending on how we define the orthogonal pair of signals $s_1(t)$ and $s_2(t)$, noncoherent binary FSK and DPSK may be treated as special cases of this modulation scheme.

The channel also introduces an additive white Gaussian noise $w(t)$ of zero mean and power spectral density $N_0/2$. We may thus express the received signal $x(t)$ as

$$x(t) = \begin{cases} g_1(t) + w(t), & 0 \le t \le T \\ g_2(t) + w(t), & 0 \le t \le T \end{cases} \qquad (8.155)$$

The requirement is to use $x(t)$ to discriminate between $s_1(t)$ and $s_2(t)$, regardless of the carrier phase.

For this purpose, we employ the receiver shown in Fig. 8.26a. The receiver consists of a pair of filters matched to the basis functions $\phi_1(t)$ and $\phi_2(t)$ that are scaled versions of the transmitted signals $s_1(t)$ and $s_2(t)$, respectively. Because the carrier phase is unknown, the receiver relies on amplitude as the only possible discriminant. Accordingly, the matched filter outputs are envelope detected, sampled, and then compared with each other. If the upper path in Fig. 8.26a has an output amplitude l_1 greater than the output amplitude l_2 of the lower path, the receiver makes a decision in favor of $s_1(t)$. If the converse is true, it decides in favor of $s_2(t)$. When they are equal, the decision may be made by flipping a fair coin. In any event, a decision error occurs when the matched filter that rejects the signal component of the received signal $x(t)$ has a larger output amplitude (due to noise alone) than the matched filter that passes it.

From the discussion presented in Section 8.9, we note that a noncoherent matched filter (constituting the upper or lower path in the receiver of Fig. 8.26a) may be viewed as being equivalent to a *quadrature receiver*. The quadrature receiver itself has two channels. One version of the quadrature receiver is shown in Fig. 8.26b. In the upper channel, called the *in-phase channel*, the received signal $x(t)$ is correlated with the basis function $\phi_i(t)$, representing a scaled version of the transmitted signal $s_1(t)$ or $s_2(t)$ with zero carrier phase. In the lower channel, called the *quadrature channel*, on the other hand, $x(t)$ is correlated with another basis function $\hat{\phi}_i(t)$, representing the version of $\phi_i(t)$ that results from shifting the carrier phase by $-90°$. Naturally, $\phi_i(t)$ and $\hat{\phi}_i(t)$ are orthogonal to each other.

The signal $\hat{\phi}_i(t)$ is in fact the *Hilbert transform* of $\phi_i(t)$. To illustrate the nature of this relationship, let

$$\phi_i(t) = m(t)\cos(2\pi f_i t)$$

where $m(t)$ is a band-limited message signal. Typically, the carrier frequency f_i is greater than the highest frequency component of $m(t)$. Then, the Hilbert transform of $\phi_i(t)$ is defined by (see Problem 2.26)

$$\hat{\phi}_i(t) = m(t)\sin(2\pi f_i t)$$

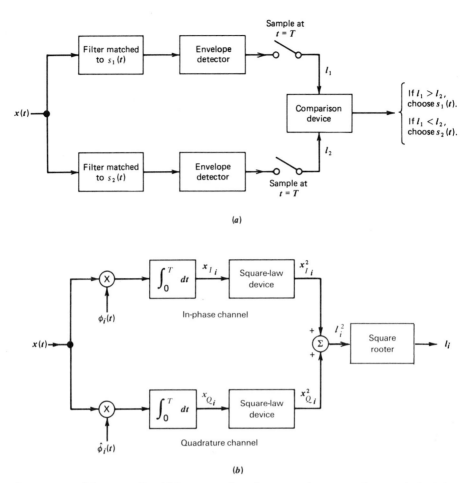

Figure 8.26 (a) Generalized binary receiver for noncoherent orthogonal modulation. (b) Quadrature receiver equivalent to either one of the two matched filters in part (a); the index $i = 1, 2$.

Since

$$\cos\left(2\pi f_i t - \frac{\pi}{2}\right) = \sin(2\pi f_i t),$$

we see that $\hat{\phi}_i(t)$ is indeed obtained from $\phi_i(t)$ by shifting the carrier $\cos(2\pi f_i t)$ by -90 degrees. An important property of Hilbert transformation is that a signal and its Hilbert transform are orthogonal to each other (see Chapter 2). Thus, $\phi_i(t)$ and $\hat{\phi}_i(t)$ are orthogonal to each other, as already stated.

The average probability of error for the noncoherent receiver of Fig. 8.26a is given by the simple formula

$$P_e = \frac{1}{2}\exp\left(\frac{E}{2N_0}\right) \tag{8.156}$$

where E is the signal energy per symbol, and N_0 is the noise spectral density.* To derive this formula, we make use of the equivalence depicted in Fig. 8.26b. In particular, we observe that since the carrier phase is unknown, noise at the output of each matched filter in Fig. 8.26a has *two degrees of freedom*, namely, in-phase and quadrature. Accordingly, the noncoherent receiver of Fig. 8.26a has a total of four noise parameters that are *conditionally independent* given the phase θ, and also *identically distributed*. These four noise parameters have sample values denoted by x_{I1}, x_{Q1}, x_{I2}, and x_{Q2}; the first two account for degrees of freedom associated with the upper path of Fig. 8.26a, and the latter two account for degrees of freedom associated with the lower path.

The receiver of Fig. 8.26a has a *symmetric* structure. Hence, the probability of choosing $s_2(t)$, given that $s_1(t)$ was transmitted, is the same as the probability of choosing $s_1(t)$, given that $s_2(t)$ was transmitted. This means that the average probability of error may be obtained by transmitting $s_1(t)$ and calculating the probability of choosing $s_2(t)$, or vice versa.

Suppose that signal $s_1(t)$ is transmitted for the interval $0 \leq t \leq T$. An error occurs if the receiver noise $w(t)$ is such that the output l_2 of the lower path in Fig. 8.26a is greater than the output l_1 of the upper path. Then the receiver makes a decision in favor of $s_2(t)$ rather than $s_1(t)$. To calculate the probability of error so made, we must have the probability density function of the random variable L_2 (represented by sample value l_2). Since the filter in the lower path is matched to $s_2(t)$, and $s_2(t)$ is orthogonal to the transmitted signal $s_1(t)$, it follows that the output of this matched filter is due to *noise alone*. Let x_{I2} and x_{Q2} denote the in-phase and quadrature components of the matched filter output in the lower path of Fig. 8.26a. Then, from the equivalence depicted in Fig. 8.26b, we see that (for $i = 2$)

$$l_2 = \sqrt{x_{I2}^2 + x_{Q2}^2} \tag{8.157}$$

Figure 8.27a shows a geometric interpretation of this relation. The channel noise $w(t)$ is both white (with power spectral density $N_0/2$) and Gaussian (with zero mean). Correspondingly, we find that the random variables X_{I2} and X_{Q2} (represented by sample values x_{I2} and x_{Q2}) are both Gaussian-distributed with zero mean and variance $N_0/2$, given the phase θ. Hence, we may write

$$f_{X_{I2}}(x_{I2}) = \frac{1}{\sqrt{\pi N_0}} \exp\left(-\frac{x_{I2}^2}{N_0}\right) \tag{8.158}$$

and

$$f_{X_{Q2}}(x_{Q2}) = \frac{1}{\sqrt{\pi N_0}} \exp\left(-\frac{x_{Q2}^2}{N_0}\right) \tag{8.159}$$

Next, we use a well-known result in probability theory, namely, the fact that the envelope of a Gaussian process is *Rayleigh-distributed* and independent of the

---

*A reader who is not interested in the formal derivation of Eq. (8.156) may at this point wish to move on to the treatment of noncoherent binary frequency-shift keying (in Section 8.16) and differential phase-shift keying (in Section 8.17) as special cases of noncoherent orthogonal modulation, without loss of continuity.

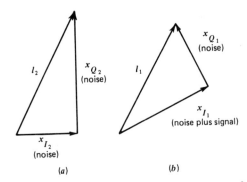

Figure 8.27 Geometric interpretations of the two path outputs l_1 and l_2 in the generalized noncoherent receiver.

phase θ (see Section 4.14). Specifically, for the situation at hand, we may state that the random variable L_2 [whose sample value l_2 is related to x_{I2} and x_{Q2} by Eq. (8.157)] has the following probability density function:

$$f_{L_2}(l_2) = \begin{cases} \dfrac{2l_2}{N_0}\exp\left(-\dfrac{l_2^2}{N_0}\right), & l_2 \geq 0 \\ 0, & \text{elsewhere} \end{cases} \tag{8.160}$$

Figure 8.28 shows a plot of this probability density function. The conditional probability that $l_2 > l_1$, given the sample value l_1, is defined by the shaded area in Fig. 8.28. Hence, we have

$$P(l_2 > l_1|l_1) = \int_{l_1}^{\infty} f_{L_2}(l_2)\, dl_2 \tag{8.161}$$

Substituting Eq. (8.160) in Eq. (8.161) and integrating, we get

$$P(l_2 > l_1|l_1) = \exp\left(-\dfrac{l_1^2}{N_0}\right) \tag{8.162}$$

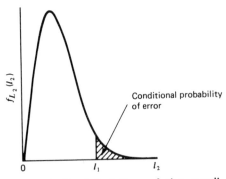

Figure 8.28 Calculation of the conditional probability that $l_2 > l_1$, given l_1.

Consider next the output amplitude l_1, pertaining to the upper path in Fig. 8.26a. Since the filter in this path is matched to $s_1(t)$, and it is assumed that $s_1(t)$ is transmitted, it follows that l_1 is due to *signal plus noise*. Let x_{I1} and x_{Q1} denote the components at the output of the matched filter (in the upper path of Fig. 8.26a) that are in phase and in quadrature with respect to the received signal, respectively. Then from the equivalence depicted in Fig. 8.26b, we see that (for $i = 1$)

$$l_1 = \sqrt{x_{I1}^2 + x_{Q1}^2} \tag{8.163}$$

Figure 8.27b presents a geometric interpretation of this relation. Since the Hilbert transform $\hat{s}_1(t)$ is orthogonal to $s_1(t)$, it follows that x_{I1} is due to signal plus noise, whereas x_{Q1} is due to noise alone. This means that (1) the random variable X_{I1} represented by the sample value x_{I1} is Gaussian distributed with mean $\sqrt{E}$ and variance $N_0/2$, where E is the signal energy per symbol, and (2) the random variable X_{Q1} represented by the sample value x_{Q1} is Gaussian distributed with zero mean and variance $N_0/2$. Hence, we may express the probability density functions of these two independent random variables as follows:

$$f_{X_{I1}}(x_{I1}) = \frac{1}{\sqrt{\pi N_0}} \exp\left(-\frac{(x_{I1} - \sqrt{E})^2}{N_0}\right) \tag{8.164}$$

and

$$f_{X_{Q1}}(x_{Q1}) = \frac{1}{\sqrt{\pi N_0}} \exp\left(-\frac{x_{Q1}^2}{N_0}\right) \tag{8.165}$$

Since the two random variables X_{I1} and X_{Q1} are independent, their joint probability density function is simply the product of the probability density functions given in Eqs. (8.164) and (8.165).

To find the average probability of error, we have to average the conditional probability of error given in Eq. (8.162) over all possible values of l_1. Naturally, this calculation requires knowledge of the probability density function of random variables L_1 represented by sample value l_1. The standard method is now to combine Eqs. (8.164) and (8.165) to find the probability density function of L_1 due to signal plus noise. However, this leads to rather complicated calculations involving the use of Bessel functions. This analytic difficulty may be circumvented by the following approach. Given x_{I1} and x_{Q1}, an error occurs when, in Fig. 8.26a, the lower path's output amplitude l_2 due to noise alone exceeds l_1 due to signal plus noise; from Eq. (8.163) we have

$$l_1^2 = x_{I1}^2 + x_{Q1}^2 \tag{8.166}$$

The probability of such an occurrence is obtained by substituting Eq. (8.166) in Eq. (8.162), as shown by

$$P(\text{error}|x_{I1}, x_{Q1}) = \exp\left(-\frac{x_{I1}^2 + x_{Q1}^2}{N_0}\right) \tag{8.167}$$

This is now a conditional probability of error, conditional on the output of the matched filter in the upper path taking on values X_{I1} and X_{Q1}. This conditional probability multiplied by the joint probability density function of X_{I1} and X_{Q1} is then the *error-density, given x_{I1} and x_{Q1}*. Since X_{I1} and X_{Q1} are statistically independent, their joint probability density function equals the product of their individual probability density functions. The resulting error-density is a complicated expression in x_{I1} and x_{Q1}. However, the average probability of error, which is the issue of interest, may be obtained in a relatively simple manner. We first use Eqs. (8.164), (8.165), and (8.167) to evaluate the desired error-density as

$$
\begin{aligned}
&P(\text{error}|x_{I1}, x_{Q1}) f_{X_{I1}}(x_{I1}) f_{X_{Q1}}(x_{Q1}) \\
&= \frac{1}{\pi N_0} \exp\left\{ -\frac{1}{N_0} [x_{I1}^2 + x_{Q1}^2 + (x_{I1} - \sqrt{E})^2 + x_{Q1}^2] \right\}
\end{aligned}
\tag{8.168}
$$

Completing the square in the exponent of Eq. (8.168), we may rewrite the exponent as

$$
x_{I1}^2 + x_{Q1}^2 + (x_{I1} - \sqrt{E})^2 + x_{Q1}^2 = 2\left(x_{I1} - \frac{\sqrt{E}}{2} \right)^2 + 2x_{Q1}^2 + \frac{E}{2}
\tag{8.169}
$$

Next, we substitute Eq. (8.169) in Eq. (8.168) and integrate the error-density over all x_{I1} and x_{Q1}. We thus evaluate the average probability of error as

$$
\begin{aligned}
P_e &= \int_{-\infty}^{\infty} \int_{-\infty}^{\infty} P(\text{error}|x_{I1}, x_{Q1}) f_{X_{I1}}(x_{I1}) f_{X_{Q1}}(x_{Q1}) \, dx_{I1} \, dx_{Q1} \\
&= \frac{1}{\pi N_0} \exp\left(-\frac{E}{2N_0} \right) \int_{-\infty}^{\infty} \exp\left[-\frac{2}{N_0} \left(x_{I1} - \frac{\sqrt{E}}{2} \right)^2 \right] dx_{I1} \\
&\quad \cdot \int_{-\infty}^{\infty} \exp\left(-\frac{2x_{Q1}^2}{N_0} \right) dx_{Q1}
\end{aligned}
\tag{8.170}
$$

We now use the following two identities:

$$
\int_{-\infty}^{\infty} \exp\left[-\frac{2}{N_0} \left(x_{I1} - \frac{\sqrt{E}}{2} \right)^2 \right] dx_{I1} = \sqrt{\frac{N_0 \pi}{2}}
\tag{8.171}
$$

and

$$
\int_{-\infty}^{\infty} \exp\left(-\frac{2x_{Q1}^2}{N_0} \right) dx_{Q1} = \sqrt{\frac{N_0 \pi}{2}}
\tag{8.172}
$$

The identity of Eq. (8.171) is obtained by considering a Gaussian-distributed variable with mean $\sqrt{E}/2$ and variance $N_0/4$, and recognizing that the total area under the curve of a random variable's probability density function equals unity; the identity of Eq. (8.172) follows as a special case of Eq. (8.171). Thus, in light of these two identities, Eq. (8.170) simplifies as follows:

$$P_e = \frac{1}{2}\exp\left(-\frac{E}{2N_0}\right)$$

which is the desired result.

With this formula at our disposal, we are ready to consider noncoherent binary FSK and DPSK as special cases, which we do in the next two sections, respectively.

8.16 NONCOHERENT BINARY FREQUENCY-SHIFT KEYING

In the binary FSK case, the transmitted signal is defined by

$$s_i(t) = \begin{cases} \sqrt{\dfrac{2E_b}{T_b}}\cos(2\pi f_i t), & 0 \le t \le T_b \\ 0, & \text{elsewhere} \end{cases} \qquad (8.173)$$

where the carrier frequency f_i equals one of two possible values f_1 and f_2; to ensure that the signals representing these two frequencies are orthogonal, we choose $f_i = n_i/T_b$, where n_i is an integer. The transmission of frequency f_1 represents symbol 1, and the transmission of frequency f_2 represents symbol 0. For the noncoherent detection of this frequency-modulated wave, the receiver consists of a pair of matched filters followed by envelope detectors, as in Fig. 8.29. The filter in the upper path of the receiver is matched to $\sqrt{2/T_b}\cos(2\pi f_1 t)$, and the filter in the lower path is matched to $\sqrt{2/T_b}\cos(2\pi f_2 t)$, and $0 \le t \le T_b$. The resulting envelope detector outputs are sampled at $t = T_b$, and their values are compared. The envelope samples of the upper and lower paths in Fig. 8.29 are shown as l_1 and l_2, respectively; then, if $l_1 > l_2$, the receiver decides in favor of

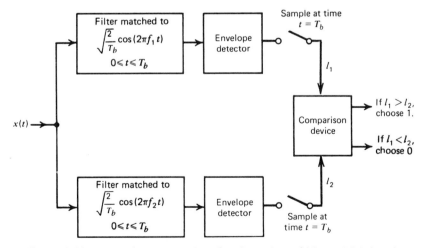

Figure 8.29 Noncoherent receiver for detection of binary FSK signals.

symbol 1, and if $l_1 < l_2$, it decides in favor of symbol 0. If $l_1 = l_2$, the receiver simply makes a guess in favor of symbol 1 or 0.

The noncoherent binary FSK described herein is a special case of noncoherent orthogonal modulation with

$$T = T_b \tag{8.174}$$

and

$$E = E_b \tag{8.175}$$

where T_b is the bit duration and E_b is the signal energy per bit. Hence, substituting Eq. (8.175) in Eq. (8.156), we find that the *average probability of error* or equivalently, the *bit error rate for noncoherent binary FSK* is

$$P_e = \frac{1}{2}\exp\left(-\frac{E_b}{2N_0}\right) \tag{8.176}$$

The formula of Eq. (8.176) is derived as a special case of noncoherent orthogonal modulation. In Problem 8.25 we address the same issue using a direct approach that invokes the application of Rayleigh and Rician distributions; these distributions pertain respectively to the random variables L_2 and L_1 whose sample values are defined by Eqs. (8.157) and (8.163).

8.17 DIFFERENTIAL PHASE-SHIFT KEYING

We may view *differential phase-shift keying* (DPSK) as the noncoherent version of PSK. It eliminates the need for a coherent reference signal at the receiver by combining two basic operations at the transmitter: (1) *differential encoding* of the input binary wave and (2) *phase-shift keying*—hence, the name, *differential phase-shift keying* (DPSK). In effect, to send symbol 0, we phase advance the current signal waveform by 180 degrees, and to send symbol 1 we leave the phase of the current signal waveform unchanged. The receiver is equipped with a *storage* capability, so that it can measure the *relative phase difference* between the waveforms received during two successive bit intervals. Provided that the unknown phase θ contained in the received wave varies slowly (that is, slow enough for it to be considered essentially constant over two bit intervals), the phase difference between waveforms received in two successive bit intervals will be independent of θ.

DPSK is another example of noncoherent orthogonal modulation, when it is considered over two bit intervals. Suppose the transmitted DPSK signal equals $\sqrt{E_b/2T_b}\cos(2\pi f_c t)$ for $0 \le t \le T_b$, where T_b is the bit duration and E_b is the signal energy per bit. Let $s_1(t)$ denote the transmitted DPSK signal for $0 \le t \le 2T_b$ for the case when we have binary symbol 1 at the transmitter input for the second part of this interval, namely, $T_b \le t \le 2T_b$. The transmission of symbol 1 leaves the carrier phase unchanged, and so we define $s_1(t)$ as

$$s_1(t) = \begin{cases} \sqrt{\dfrac{E_b}{2T_b}}\cos(2\pi f_c t), & 0 \le t \le T_b \\[2ex] \sqrt{\dfrac{E_b}{2T_b}}\cos(2\pi f_c t), & T_b \le t \le 2T_b \end{cases} \tag{8.177}$$

Let $s_2(t)$ denote the transmitted DPSK signal for $0 \le t \le 2T_b$ for the case when we have binary symbol 0 at the transmitter input for $T_b \le t \le 2T_b$. The transmission of 0 advances the carrier phase by 180 degrees, and so we define $s_2(t)$ as

$$s_2(t) = \begin{cases} \sqrt{\dfrac{E_b}{2T_b}}\cos(2\pi f_c t), & 0 \le t \le T_b \\[2ex] \sqrt{\dfrac{E_b}{2T_b}}\cos(2\pi f_c t + \pi), & T_b \le t \le 2T_b \end{cases} \tag{8.178}$$

We readily see from Eqs. (8.177) and (8.178) that $s_1(t)$ and $s_2(t)$ are indeed orthogonal over the two-bit interval $0 \le t \le 2T_b$. In other words, DPSK is a special case of noncoherent orthogonal modulation with

$$T = 2T_b \tag{8.179}$$

and

$$E = 2E_b \tag{8.180}$$

Hence, substituting Eq. (8.180) in Eq. (8.156), we find that the *average probability of error* or equivalently, the *bit error rate for DPSK* is given by

$$P_e = \frac{1}{2}\exp\left(-\frac{E_b}{N_0}\right) \tag{8.181}$$

which provides a gain of 3 dB over noncoherent FSK for the same E_b/N_0.

Generation of DPSK

The next issue to be considered is the generation of DPSK signals. The differential encoding process at the transmitter input starts with an arbitrary first bit, serving as reference. Let $\{d_k\}$ denote the differentially encoded sequence with this added reference bit. We now introduce the following definitions in the generation of this sequence:

- If the incoming binary symbol b_k is 1, leave the symbol d_k unchanged with respect to the previous bit.
- If the incoming binary symbol b_k is 0, change the symbol d_k with respect to the previous bit.

The differentially encoded sequence $\{d_k\}$ thus generated is used to phase-shift a carrier with phase angles 0 and π radians representing symbols 1 and 0, respec-

Table 8.3 Illustrating the Generation of DPSK Signal

$\{b_k\}$		1	0	0	1	0	0	1	1
$\{d_{k-1}\}$		1	1	0	1	1	0	1	1
Differentially encoded sequence $\{d_k\}$	1	1	0	1	1	0	1	1	1
Transmitted phase (*radians*)	0	0	π	0	0	π	0	0	0

tively. The differential-phase encoding process is illustrated in Table 8.3. Note that d_k is the complement of the modulo-2 sum of b_k and d_{k-1}.

The block diagram of a DPSK transmitter is shown in Fig. 8.30a. It consists, in part, of a logic network and a one-bit delay element interconnected so as to convert the raw binary sequence $\{b_k\}$ into a differentially encoded sequence $\{d_k\}$. This sequence is amplitude-level encoded and then used to modulate a carrier wave of frequency f_c, thereby producing the desired DPSK signal.

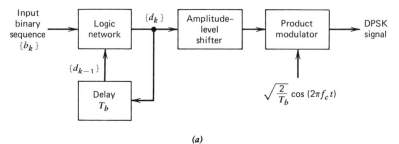

(a)

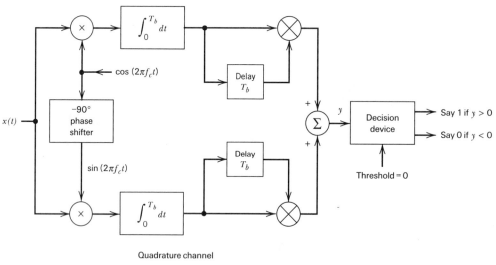

(b)

Figure 8.30 Block diagrams for (a) DPSK transmitter and (b) DPSK receiver.

Optimum Receiver

Consider that in differentially coherent detection of binary DPSK, the carrier phase is unknown. Then, in light of the receiver being equipped with an in-phase and a quadrature channel, we have a signal space diagram where the received signal points are $(A \cos\theta, A \sin\theta)$ and $(-A \cos\theta, -A \sin\theta)$, with θ denoting the unknown phase and A denoting amplitude. This geometry of possible signals is illustrated in Figure 8.31. The receiver measures the coordinates (x_{I_0}, x_{Q_0}) at time $t = T_b$ and (x_{I_1}, x_{Q_1}) at time $t = 2T_b$. The issue to be resolved is whether these two points map to the same signal point or different ones. Recognizing that the two vectors $\mathbf{x}_0$ and $\mathbf{x}_1$, with end points (x_{I_0}, x_{Q_0}) and (x_{I_1}, x_{Q_1}) are pointed roughly in the same direction if their inner product is positive, we may formulate the test as follows:

Is the inner product $\mathbf{x}_0^T\mathbf{x}_1$ positive or negative?

Accordingly, we may write

$$x_{I_0}x_{I_1} + x_{Q_0}x_{Q_1} \underset{\substack{< \\ \text{say } 0}}{\overset{\substack{\text{say } 1 \\ >}}{}} 0 \qquad (8.182)$$

This test is equivalent to the following

$$\tfrac{1}{4}[(x_{I_0} + x_{I_1})^2 + (x_{Q_0} + x_{Q_1})^2 - (x_{I_0} - x_{I_1})^2 - (x_{Q_0} - x_{Q_1})^2] \underset{\substack{< \\ \text{say } 0}}{\overset{\substack{\text{say } 1 \\ >}}{}} 0$$

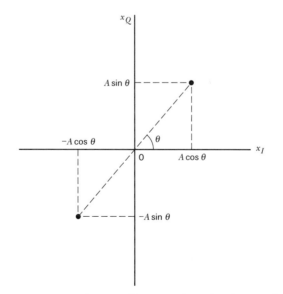

Figure 8.31 Signal-space diagram of received DPSK signal

That is,

$$(x_{I_0} + x_{I_1})^2 + (x_{Q_0} + x_{Q_1})^2 - (x_{I_0} - x_{I_1})^2 - (x_{Q_0} - x_{I_1})^2 \underset{\text{say } 0}{\overset{\text{say } 1}{\underset{<}{\overset{>}{}}}} 0$$

The decision-making process may therefore be thought of as testing whether the point (x_{I_0}, x_{Q_0}) is closer to (x_{I_1}, x_{Q_1}) or its image $(-x_{I_1}, -x_{Q_1})$.

Thus, the optimum receiver[7] for differentially coherent detection of binary DPSK is as shown in Fig. 8.30b, which follows directly from Eq. (8.182). This implementation merely requires that *sample* values be stored, thereby avoiding the need for fancy delay lines that may be needed otherwise. The equivalent receiver implementation that tests squared elements is more complicated, but its use makes the *analysis* easier to handle in that the two signals to be considered are orthogonal over the interval $(0,2T_b)$; hence, the noncoherent orthogonal demodulation analysis applies.

8.18 COMPARISON OF BINARY AND QUATERNARY MODULATION SCHEMES

In Table 8.4, we have summarized the expressions for the bit error rate (BER) for coherent binary PSK, conventional coherent binary FSK with one-bit decoding, DPSK, noncoherent binary FSK, coherent QPSK, and coherent MSK, when operating over an AWGN channel. In Fig. 8.31 we have used the expressions summarized in Table 8.4 to plot BER as a function of the signal energy per bit-to-noise spectral density ratio, E_b/N_0.

Based on the performance curves shown in Fig. 8.32, the summary of formulas given in Table 8.4, and the defining equations for the pertinent modulation formats, we can make the following statements:

1. The bit error rates for all the systems decrease monotonically with increasing values of E_b/N_0.
2. For any value of E_b/N_0, coherent binary PSK, QPSK, and MSK produce a smaller bit error rate than any of the other systems.

Table 8.4 Summary of Formulas for the Bit Error Rate of Different Digital Modulation Schemes

Signaling Scheme	Bit Error Rate
(a) Coherent binary PSK, Coherent QPSK, Coherent MSK	$\frac{1}{2}\text{erfc}(\sqrt{E_b/N_0})$
(b) Coherent binary FSK	$\frac{1}{2}\text{erfc}(\sqrt{E_b/2N_0})$
(c) DPSK	$\frac{1}{2}\exp(-E_b/N_0)$
(d) Noncoherent binary FSK	$\frac{1}{2}\exp(-E_b/2N_0)$

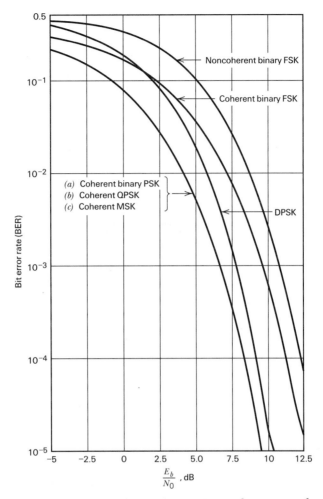

Figure 8.32 Comparison of the noise performances of different PSK and FSK schemes.

3. Coherent binary PSK and DPSK require an E_b/N_0 that is 3 dB less than the corresponding values for conventional coherent binary FSK and noncoherent binary FSK, respectively, to realize the same bit error rate.

4. At high values of E_b/N_0, DPSK and noncoherent binary FSK perform almost as well (to within about 1 dB) as coherent binary PSK and conventional coherent binary FSK, respectively, for the same bit rate and signal energy per bit.

5. In coherent QPSK, two orthogonal carriers $\sqrt{2/T}\cos(2\pi f_c t)$ and $\sqrt{2/T}\sin(2\pi f_c t)$ are used, where the carrier frequency f_c is an integral multiple of the symbol rate $1/T$, with the result that two independent bit streams can be transmitted simultaneously and subsequently detected in the receiver.

6. In coherent MSK, the two orthogonal carriers $\sqrt{2/T_b}\cos(2\pi f_c t)$ and $\sqrt{2/T_b}\sin(2\pi f_c t)$ are modulated by the two antipodal symbol shaping pulses $\cos(\pi t/2T_b)$ and $\sin(\pi t/2T_b)$, respectively, over $2T_b$ intervals, where T_b is the

bit duration. Correspondingly, the receiver uses a coherent phase decoding process over two successive bit intervals to recover the original bit stream.

7. The MSK scheme differs from the other signaling schemes in that its receiver has *memory*. In particular, the MSK receiver makes decisions based on observations over two successive bit intervals. Thus, although the transmitted signal has a binary format represented by the transmission of two distinct frequencies, the presence of memory in the receiver makes it assume a two-dimensional signal space diagram. There are four message points, depending on which binary symbol (0 or 1) was sent and the past phase history of the FSK signal.

8.19 *M*-ARY MODULATION TECHNIQUES

In an *M-ary signaling scheme*, we may send one of M possible signals, $s_1(t)$, $s_2(t), \ldots, s_M(t)$, during each signaling interval of duration T. For almost all applications, the number of possible signals $M = 2^n$, where n is an integer. The symbol duration $T = nT_b$, where T_b is the bit duration. These signals are generated by changing the amplitude, phase, or frequency of a carrier in M discrete steps. Thus, we have *M*-ary ASK, *M*-ary PSK, and *M*-ary FSK digital modulation schemes. The QPSK system considered in Section 8.13 is an example of *M*-ary PSK with $M = 4$.

Another way of generating *M*-ary signals is to combine different methods of modulation into a hybrid form. For example, we may combine discrete changes in both the amplitude and phase of a carrier to produce *M-ary amplitude-phase keying* (APK). A special form of this hybrid modulation, called *M-ary QAM*, has some attractive properties.

M-ary signaling schemes are preferred over binary signaling schemes for transmitting digital information over band-pass channels when the requirement is to conserve bandwidth at the expense of increased power. In practice, we rarely find a communication channel that has the exact bandwidth required for transmitting the output of an information source by means of binary signaling schemes. Thus, when the bandwidth of the channel is less than the required value, we may use *M*-ary signaling schemes so as to utilize the channel efficiently.

To illustrate the bandwidth-conservation capability of *M*-ary signaling schemes, consider the transmission of information consisting of a binary sequence with bit duration T_b. If we were to transmit this information by means of binary PSK, for example, we would require a bandwidth inversely proportional to T_b. However, if we take blocks of n bits and use an *M*-ary PSK scheme with $M = 2^n$ and symbol duration $T = nT_b$, the bandwidth required is proportional to $1/nT_b$. This shows that the use of *M*-ary PSK enables a reduction in transmission bandwidth by the factor $n = \log_2 M$ over binary PSK.

In this section we consider three different *M*-ary signaling schemes: *M*-ary PSK, *M*-ary QAM, and *M*-ary FSK, each of which offers virtues of its own.

M-ary PSK

In *M*-ary PSK, the phase of the carrier takes on one of M possible values, namely, $\theta_i = 2(i - 1)\pi/M$, where $i = 1, 2, \ldots, M$. Accordingly, during each signaling interval of duration T, one of the M possible signals

$$s_i(t) = \sqrt{\frac{2E}{T}} \cos\left(2\pi f_c t + \frac{2\pi}{M}(i-1)\right), \qquad i = 1, 2, \ldots, M \quad (8.183)$$

is sent, where E is the signal energy per symbol. The carrier frequency $f_c = n_c/T$ for some fixed integer n_c.

Each $s_i(t)$ may be expanded in terms of two basis functions $\phi_1(t)$ and $\phi_2(t)$ defined as

$$\phi_1(t) = \sqrt{\frac{2}{T}} \cos(2\pi f_c t), \qquad 0 \le t \le T \quad (8.184)$$

$$\phi_2(t) = \sqrt{\frac{2}{T}} \sin(2\pi f_c t), \qquad 0 \le t \le T \quad (8.185)$$

Both $\phi_1(t)$ and $\phi_2(t)$ have unit energy. The signal constellation of M-ary PSK is therefore two-dimensional. The M message points are equally spaced on a circle of radius $\sqrt{E}$ and center at the origin, as illustrated in Fig. 8.33a, for the case of *octaphase-shift-keying* (i.e., $M = 8$).

From Fig. 8.33a we note that the signal-space diagram is symmetric. We may therefore apply Eq. (8.72), based on the union bound, to develop an approximate formula for the average probability of symbol error for M-ary PSK. Suppose that the transmitted signal corresponds to the message point m_1 whose coordinates along the ϕ_1- and ϕ_2-axes are $+\sqrt{E}$ and 0, respectively. Suppose that the ratio E/N_0 is large enough to consider the nearest two message points, one on either side of m_1, for the application of Eq. (8.72); this is illustrated in Fig. 8.33b for the case of $M = 8$. The Euclidean distance of each of these two points from m_1 is (for $M = 8$)

$$d_{12} = d_{18} = 2\sqrt{E} \sin\left(\frac{\pi}{M}\right)$$

Hence, the use of Eq. (8.72) yields the average probability of symbol error for coherent M-ary PSK as[8]

$$P_e \simeq \operatorname{erfc}\left(\sqrt{\frac{E}{N_0}} \sin\left(\frac{\pi}{M}\right)\right) \quad (8.186)$$

where it is assumed that $M \ge 4$. The approximation becomes extremely tight, for fixed M, as E/N_0 is increased. For $M = 4$, Eq. (8.186) reduces to the same form given in Eq. (8.131) for QPSK.

Coherent M-ary PSK requires exact knowledge of the carrier frequency and phase for the receiver to be accurately synchronized to the transmitter. When carrier recovery at the receiver is impractical, we may use differential encoding based on the phase difference between successive symbols at the cost of some degradation in performance. If the incoming data are encoded by a phase-shift rather than by absolute phase, the receiver performs detection by comparing the phase of one symbol with that of the previous symbol, and the need for a co-

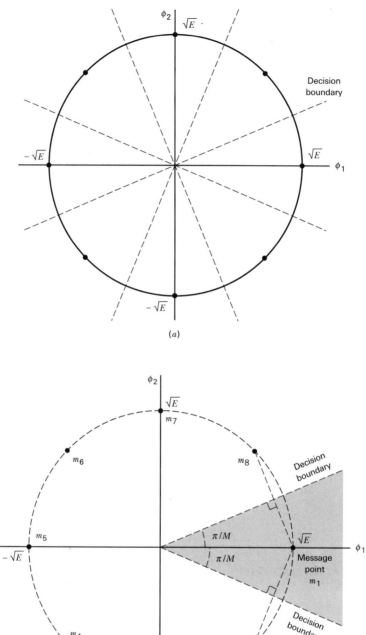

Figure 8.33 (a) Signal-space diagram for octaphase-shift keying (i.e., $M = 8$). The decision boundaries are shown as dashed lines. (b) Signal-space diagram illustrating the application of the union bound for octaphase-shift keying.

herent reference is thereby eliminated. This procedure is the same as that described for binary DPSK in Section 8.17. The exact calculation of probability of
symbol error for the differential detection of differential *M*-ary PSK (commonly
referred to as *M*-ary DPSK) is much too complicated for $M > 2$. However, for
large values of E/N_0 and $M \geq 4$, the probability of symbol error is approximately
given by

$$P_e \simeq \text{erfc}\left(\sqrt{\frac{2E}{N_0}} \sin\left(\frac{\pi}{2M}\right) \right), \qquad M \geq 4 \tag{8.187}$$

Comparing the approximate formulas of Eqs. (8.186) and (8.187), we see
that for $M \geq 4$ an *M*-ary DPSK system attains the same probability of symbol error
as the corresponding coherent *M*-ary PSK system provided that the transmitted
energy per symbol is increased by the following factor:

$$k(M) = \frac{\sin^2\left(\dfrac{\pi}{M}\right)}{2 \sin^2\left(\dfrac{\pi}{2M}\right)}, \qquad M \geq 4 \tag{8.188}$$

For example, $k(4) = 1.7$. That is, differential QPSK (which is noncoherent) is
approximately 2.3 dB poorer in performance than coherent QPSK.

M-ary QAM

In an *M*-ary PSK system, in-phase and quadrature components of the modulated
signal are interrelated in such a way that the envelope is constrained to remain
constant. This constraint manifests itself in a circular constellation for the message points. However, if this constraint is removed, and the in-phase and quadrature components are thereby permitted to be independent, we get a new
modulation scheme called *M*-ary *quadrature amplitude modulation* (QAM). In
this latter modulation scheme, the carrier experiences amplitude as well as phase
modulation.

The signal constellation for *M*-ary QAM consists of a *square lattice* of message
points, as illustrated in Fig. 8.34 for $M = 16$. The corresponding signal constellations for the in-phase and quadrature components of the amplitude-phase
modulated signal are shown in Figs. 8.35*a* and 8.35*b*, respectively. The basic
format of the signal constellations shown in the latter figures is recognized to be
that of a *polar L-ary ASK signal* with $L = 4$. Thus, in general, an *M*-ary QAM
scheme enables the transmission of $M = L^2$ independent symbols with the same
channel bandwidth as that required for one polar *L*-ary ASK scheme.

The general form of *M*-ary QAM is defined by the transmitted signal

$$s_i(t) = \sqrt{\frac{2E_0}{T}} a_i \cos(2\pi f_c t) + \sqrt{\frac{2E_0}{T}} b_i \sin(2\pi f_c t), \quad 0 \leq t \leq T \tag{8.189}$$

where E_0 is the energy of the signal with the lowest amplitude, and a_i and b_i are
a pair of independent integers chosen in accordance with the location of the

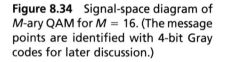

Figure 8.34 Signal-space diagram of *M*-ary QAM for *M* = 16. (The message points are identified with 4-bit Gray codes for later discussion.)

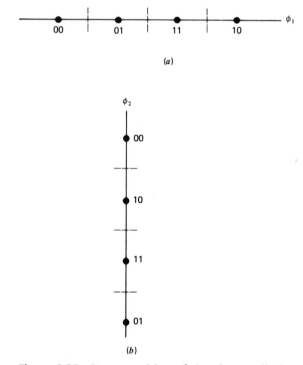

(a)

(b)

Figure 8.35 Decomposition of signal constellation of *M*-ary QAM (for *M* = 16) into two signal-space diagrams for (a) in-phase component $\phi_1(t)$ and (b) quadrature component $\phi_2(t)$. (The message points are identified by 2-bit Gray codes for later discussion.)

pertinent message point. The signal $s_i(t)$ consists of two phase-quadrature carriers, each of which is modulated by a set of discrete amplitudes; hence, the name "quadrature amplitude modulation."

The signal $s_i(t)$ can be expanded in terms of a pair of basis functions:

$$\phi_1(t) = \sqrt{\frac{2}{T}}\cos(2\pi f_c t), \qquad 0 \le t \le T \tag{8.190}$$

$$\phi_2(t) = \sqrt{\frac{2}{T}}\sin(2\pi f_c t), \qquad 0 \le t \le T \tag{8.191}$$

The coordinates of the ith message point are $a_i\sqrt{E_0}$ and $b_i\sqrt{E_0}$, where (a_i, b_i) is an element of the L-by-L matrix:

$$\{a_i, b_i\} = \begin{bmatrix} (-L+1, L-1) & (-L+3, L-1) & \cdots & (L-1, L-1) \\ (-L+1, L-3) & (-L+3, L-3) & \cdots & (L-1, L-3) \\ \vdots & \vdots & & \vdots \\ (-L+1, -L+1) & (-L+3, -L+1) & \cdots & (L-1, -L+1) \end{bmatrix} \tag{8.192}$$

where

$$L = \sqrt{M} \tag{8.193}$$

For example, for a 16-QAM whose signal constellation is depicted in Fig. 8.33, where $L = 4$, we have the matrix

$$\{a_i, b_i\} = \begin{bmatrix} (-3,3) & (-1,3) & (1,3) & (3,3) \\ (-3,1) & (-1,1) & (1,1) & (3,1) \\ (-3,-1) & (-1,-1) & (1,-1) & (3,-1) \\ (-3,-3) & (-1,-3) & (1,-3) & (3,-3) \end{bmatrix}$$

To calculate the probability of symbol error for *M*-ary QAM, we proceed as follows:

1. Since the in-phase and quadrature components of *M*-ary QAM are independent, the probability of correct detection for such a scheme may be written as

$$P_c = (1 - P_e')^2 \tag{8.194}$$

where P_e' is the probability of symbol error for either component.

2. The signal constellation for the in-phase or quadrature component has a geometry similar to that of *L*-ary pulse-amplitude modulation (PAM) with a corresponding number of amplitude levels; for this modulation scheme, we have (see Problem 8.27)

$$P_e' = \left(1 - \frac{1}{L}\right)\mathrm{erfc}\left(\sqrt{\frac{E_0}{N_0}}\right) \tag{8.195}$$

where $L = \sqrt{M}$.

3. The probability of symbol error for M-ary QAM is given by

$$P_e = 1 - P_c$$
$$= 1 - (1 - P'_e)^2 \qquad (8.196)$$
$$\simeq 2P'_e$$

where it is assumed that P'_e is small compared to unity.

Hence, using Eqs. (8.193) and (8.195) in Eq. (8.196), we find that the probability of symbol error for M-ary QAM is approximately given by

$$P_e \simeq 2\left(1 - \frac{1}{\sqrt{M}}\right) \text{erfc}\left(\sqrt{\frac{E_0}{N_0}}\right) \qquad (8.197)$$

The transmitted energy in M-ary QAM is variable in that its instantaneous value depends on the particular symbol transmitted. It is therefore logical to express P_e in terms of the *average* value of the transmitted energy rather than E_0. Assuming that the L amplitude levels of the in-phase or quadrature component are equally likely, we have

$$E_{\text{av}} = 2\left[\frac{2E_0}{L}\sum_{i=1}^{L/2}(2i - 1)^2\right] \qquad (8.198)$$

where the multiplying factor of 2 accounts for the equal contributions made by the in-phase and quadrature components. The limits of the summation take account of the symmetric nature of the pertinent amplitude levels around zero. Summing the series in Eq. (8.198), we get

$$E_{\text{av}} = \frac{2(L^2 - 1)E_0}{3}$$
$$= \frac{2(M - 1)E_0}{3} \qquad (8.199)$$

Accordingly, we may rewrite Eq. (8.197) in terms of E_{av} as

$$P_e \simeq 2\left(1 - \frac{1}{\sqrt{M}}\right)\text{erfc}\left(\sqrt{\frac{3E_{\text{av}}}{2(M - 1)N_0}}\right) \qquad (8.200)$$

which is the desired result.

The case of $M = 4$ is of special interest. The signal constellation for this value of M is shown in Fig. 8.36, which is recognized to be the same as that for QPSK. Indeed, putting $M = 4$ in Eq. (8.200) and noting that E_0 equals $E/2$, where E is the energy per symbol, we find that the resulting formula for the probability of symbol error becomes identical to that in Eq. (8.131), and so it should.

M-ary FSK

In an M-ary FSK scheme, the transmitted signals are defined by

$$s_i(t) = \sqrt{\frac{2E}{T}}\cos\left[\frac{\pi}{T}(n_c + i)t\right], \qquad 0 \le t \le T \qquad (8.201)$$

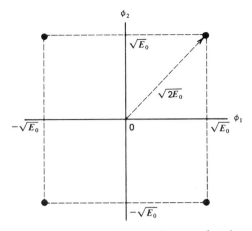

Figure 8.36 Signal-space diagram for the special case of *M*-ary QAM for *M* = 4.

where $i = 1, 2, \ldots, M$, and the carrier frequence $f_c = n_c/2T$ for some fixed integer n_c. The transmitted signals are of equal duration T and have equal energy E. Since the individual signal frequencies are separated by $1/2T$ Hertz, the signals in Eq. (8.201) are orthogonal; that is

$$\int_0^T s_i(t)\,s_j(t) \; dt = 0, \qquad i \neq j \tag{8.202}$$

For coherent *M*-ary FSK, the optimum receiver consists of a bank of *M* correlators or matched filters, with the signals in Eq. (8.201) providing the pertinent references. At the sampling times $t = kT$, the receiver makes decisions based on the largest matched filter output. An upper bound for the probability of symbol error may be obtained by applying Eq. (8.72) based on the union bound. The resulting bound is given by

$$P_e \leq \frac{1}{2}(M - 1)\,\mathrm{erfc}\left(\sqrt{\frac{E}{2N_0}}\right) \tag{8.203}$$

For fixed *M*, this bound becomes increasingly tight as E/N_0 is increased. Indeed, it becomes a good approximation to P_e for values of $P_e \leq 10^{-3}$. Moreover, for $M = 2$ (i.e., binary FSK), the bound of Eq. (8.203) becomes an equality.

Coherent detection of *M*-ary FSK requires the use of exact phase references, but providing these at the receiver can be costly and difficult to maintain. We may avoid the need for such a provision by using noncoherent detection, which results in a slightly inferior performance. In a noncoherent receiver, the individual matched filters are followed by envelope detectors that destroy the phase information. The probability of symbol error for the noncoherent detection of *M*-ary FSK is given by[9]

$$P_e = \sum_{k=1}^{M-1} \frac{(-1)^{k+1}}{k+1} \binom{M-1}{k} \exp\left(-\frac{kE}{(k+1)N_0}\right) \tag{8.204}$$

where $\binom{M-1}{k}$ is a binomial coefficient, defined by

$$\binom{M-1}{k} = \frac{(M-1)!}{(M-1-k)!\,k!} \tag{8.205}$$

The leading term of the series in Eq. (8.204) provides an upper bound on the probability of symbol error for the noncoherent detection of M-ary FSK:

$$P_e \leq \frac{M-1}{2}\exp\left(-\frac{E}{2N_0}\right) \tag{8.206}$$

For fixed M, this bound becomes increasingly close to the actual value of P_e as E/N_0 is increased. Indeed, for $M = 2$ (i.e., binary FSK), the bound of Eq. (8.206) becomes an equality.

Comparison of M-ary Digital Modulation Techniques

We conclude this section on M-ary digital modulation techniques by presenting some notes on the comparative performances and merits of the three M-ary modulation schemes considered here.

In Table 8.5, we have summarized typical values of power-bandwidth requirements for coherent binary and M-ary PSK schemes, assuming an average probability of symbol error equal to 10^{-4} and the systems operating in identical noise environments. This table shows that, among the family of M-ary PSK signals, QPSK (corresponding to $M = 4$) offers the best trade-off between power and bandwidth requirements. It is for this reason that we find QPSK is widely used in practice. For $M > 8$, power requirements become excessive; accordingly, M-ary PSK schemes with $M > 8$ are not as widely used in practice. Also, coherent M-ary PSK schemes require considerably more complex equipment than coherent binary PSK schemes for signal generation or detection, especially when $M > 8$.

Basically, M-ary PSK and M-ary QAM have similar spectral and bandwidth characteristics. For $M > 4$, however, the two schemes have different signal constellations. For M-ary PSK the signal constellation is circular, whereas for M-ary QAM it is rectangular. Moreover, a comparison of these two constellations reveals

Table 8.5[a] Comparison of Power-Bandwidth Requirements for M-ary PSK with Binary PSK. Probability of Symbol Error $= 10^{-4}$

Value of M	$\dfrac{(\text{Bandwidth})_{M\text{-ary}}}{(\text{Bandwidth})_{\text{Binary}}}$	$\dfrac{(\text{Average power})_{M\text{-ary}}}{(\text{Average power})_{\text{Binary}}}$
4	0.5	0.34 dB
8	0.333	3.91 dB
16	0.25	8.52 dB
32	0.2	13.52 dB

[a]This table is taken from Shanmugam (1979, p. 424).

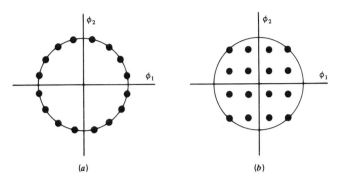

Figure 8.37 Signal constellations for (a) *M*-ary QPSK and (b) *M*-ary QAM, for *M* = 16.

that the distance between the message points of *M*-ary PSK is smaller than the distance between the message points of *M*-ary QAM, for a fixed peak transmitted power. This basic difference between the two schemes is illustrated in Fig. 8.37 for *M* = 16. Accordingly, in an AWGN channel, *M*-ary QAM outperforms the corresponding *M*-ary PSK in error performance for *M* > 4. However, the superior performance of *M*-ary QAM can be realized only if the channel is free of nonlinearities.

As for *M*-ary FSK, we find that for a fixed probability of error, increasing *M* results in a reduced power requirement. However, this reduction in transmitted power is achieved at the cost of increased channel bandwidth.

8.20 POWER SPECTRA

A study of digital passband transmission systems would be incomplete without considering the power spectra of the modulated signals of interest. Recognizing that ASK, PSK, and FSK signals are usually narrow band, we may use the complex notation of Section 2.12 to describe them in terms of the in-phase and quadrature components as

$$
\begin{aligned}
s(t) &= s_I(t)\cos(2\pi f_c t) - s_Q(t)\sin(2\pi f_c t) \\
&= \mathrm{Re}[\tilde{s}(t)\exp(j2\pi f_c t)]
\end{aligned}
\tag{8.207}
$$

where Re[·] is the real part of the expression contained inside the square brackets. We also have

$$
\tilde{s}(t) = s_I(t) + js_Q(t)
\tag{8.208}
$$

and

$$
\exp(j2\pi f_c t) = \cos(2\pi f_c t) + j\sin(2\pi f_c t)
\tag{8.209}
$$

The signal $\tilde{s}(t)$ is the *complex envelope* of the band-pass signal $s(t)$. The components $s_I(t)$ and $s_Q(t)$ and therefore $\tilde{s}(t)$ are all low-pass signals. They are uniquely defined in terms of the band-pass signal $s(t)$ and the carrier frequency f_c, provided that the half-bandwidth of $s(t)$ is less than the carrier frequency f_c.

Let $S_B(f)$ denote the power spectral density of the complex envelope $\tilde{s}(t)$. We refer to $S_B(f)$ as the *baseband power spectral density*. The power spectral density, $S_S(f)$, of the original band-pass signal $s(t)$ is a frequency-shifted version of $S_B(f)$, except for a scaling factor, as shown by

$$S_S(f) = \tfrac{1}{4}[S_B(f - f_c) + S_B(f + f_c)] \tag{8.210}$$

It is therefore sufficient to evaluate the baseband power spectral density $S_B(f)$. Since $\tilde{s}(t)$ is a low-pass signal, the calculation of $S_B(f)$ should be simpler than the calculation of $S_S(f)$. In the sequel, we calculate the baseband power spectral density for binary PSK, binary FSK, QPSK, and MSK signals.

Power Spectra of Binary PSK and FSK Signals

Consider first the case of a binary PSK signal. From the modulator of Fig. 8.15a, we see that the complex envelope of a binary PSK wave consists of an in-phase component only. Furthermore, depending on whether we have a symbol 1 or a symbol 0 at the modulator input during the signaling interval $0 \leq t \leq T_b$, we find that this in-phase component equals $+g(t)$ or $-g(t)$, respectively, where $g(t)$ is the *symbol shaping function* defined by

$$g(t) = \begin{cases} \sqrt{\dfrac{2E_b}{T_b}}, & 0 \leq t \leq T_b \\ 0, & \text{elsewhere} \end{cases} \tag{8.211}$$

We assume that the input binary wave is random, with symbols 1 and 0 equally likely and the symbols transmitted during the different time slots being statistically independent. In Example 11 of Chapter 4, it is shown that the power spectral density of a random binary wave so described is equal to the energy spectral density of the symbol shaping function divided by the symbol duration. The energy spectral density of a Fourier transformable signal $g(t)$ is defined as the squared magnitude of the signal's Fourier transform. Hence, the baseband power spectral density of a binary PSK signal equals

$$\begin{aligned} S_B(f) &= \frac{2E_b \sin^2(\pi T_b f)}{(\pi T_b f)^2} \\ &= 2E_b \, \text{sinc}^2(T_b f) \end{aligned} \tag{8.212}$$

This power spectrum falls off as the inverse square of frequency.

Consider next the case of Sunde's FSK, for which the two transmitted frequencies f_1 and f_2 differ by an amount equal to the bit rate $1/T_b$, and their arithmetic mean equals the nominal carrier frequency f_c as described in Section 8.12; phase continuity is always maintained, including inter-bit switching times. We may express this special binary FSK signal as follows:

$$s(t) = \sqrt{\frac{2E_b}{T_b}}\cos\left(2\pi f_c t \pm \frac{\pi t}{T_b}\right), \qquad 0 \leq t \leq T_b$$

and using a well-known trigonometric identity, we get

$$s(t) = \sqrt{\frac{2E_b}{T_b}}\cos\left(\pm\frac{\pi t}{T_b}\right)\cos(2\pi f_c t) - \sqrt{\frac{2E_b}{T_b}}\sin\left(\pm\frac{\pi t}{T_b}\right)\sin(2\pi f_c t)$$

$$= \sqrt{\frac{2E_b}{T_b}}\cos\left(\frac{\pi t}{T_b}\right)\cos(2\pi f_c t) \mp \sqrt{\frac{2E_b}{T_b}}\sin\left(\frac{\pi t}{T_b}\right)\sin(2\pi f_c t)$$

(8.213)

In the last line of Eq. (8.213), the plus sign corresponds to transmitting symbol 0, and the minus sign corresponds to transmitting symbol 1. As before, we assume that the symbols 1 and 0 in the random binary wave at the modulator input are equally likely, and that the symbols transmitted in adjacent time slots are statistically independent. Then, based on the representation of Eq. (8.213), we may make the following observations pertaining to the in-phase and quadrature components of a binary FSK signal with continuous phase:

1. The in-phase component is completely independent of the input binary wave. It equals $\sqrt{2E_b/T_b}\cos(\pi t/T_b)$ for all values of time t. The power spectral density of this component therefore consists of two delta functions, weighted by the factor $E_b/2T_b$, and occurring at $f = \pm 1/2T_b$.

2. The quadrature component is directly related to the input binary wave. During the signaling interval $0 \leq t \leq T_b$, it equals $-g(t)$ when we have symbol 1, and $+g(t)$ when we have symbol 0. The symbol shaping function $g(t)$ is defined by

$$g(t) = \begin{cases} \sqrt{\dfrac{2E_b}{T_b}}\sin\left(\dfrac{\pi t}{T_b}\right), & 0 \leq t \leq T_b \\ 0, & \text{elsewhere} \end{cases}$$

(8.214)

The energy spectral density of this symbol shaping function equals

$$\Psi_g(f) = \frac{8E_b T_b \cos^2(\pi T_b f)}{\pi^2(4T_b^2 f^2 - 1)^2}$$

(8.215)

The power spectral density of the quadrature component equals $\Psi_g(f)/T_b$. It is also apparent that the in-phase and quadrature components of the binary FSK signal are independent of each other. Accordingly, the baseband power spectral density of Sunde's FSK signal equals the sum of the power spectral densities of these two components, as shown by

$$S_B(f) = \frac{E_b}{2T_b}\left[\delta\left(f - \frac{1}{2T_b}\right) + \delta\left(f + \frac{1}{2T_b}\right)\right] + \frac{8E_b \cos^2(\pi T_b f)}{\pi^2(4T_b^2 f^2 - 1)^2}$$

(8.216)

Substituting Eq. (8.216) in Eq. (8.210), we find that the power spectrum of the binary FSK signal contains two discrete frequency components located at $(f_c + 1/2T_b) = f_1$ and $(f_c - 1/2T_b) = f_2$, with their average powers adding up to one-half the total power of the binary FSK signal. The presence of these two discrete frequency components provides a means of synchronizing the receiver with the transmitter.

Note also that the baseband power spectral density of a binary FSK signal with continuous phase ultimately falls off as the inverse fourth power of frequency. This is readily established by taking the limit in Eq. (8.216) as f approaches infinity. If, however, the FSK signal exhibits phase discontinuity at the inter-bit switching instants (this arises when the two oscillators supplying frequencies f_1 and f_2 operate independently of each other), the power spectral density ultimately falls off as the inverse square of frequency. Accordingly, an FSK signal with continuous phase does not produce as much interference outside the signal band of interest as an FSK signal with discontinuous phase.

In Fig. 8.38, we have plotted the baseband power spectra of Eqs. (8.212) and (8.216). (To simplify matters, we have only plotted the results for positive frequency.) In both cases, $S_B(f)$ is shown normalized with respect to $2E_b$, and the frequency is normalized with respect to the bit rate $R_b = 1/T_b$. The difference in the rates of falloff of these spectra can be explained on the basis of the pulse shape $g(t)$. The smoother the pulse, the faster the drop of spectral tails to zero. Thus, since binary FSK (with continuous phase) has a smoother pulse shape, it has lower sidelobes than binary PSK.

Power Spectra of QPSK and MSK Signals

Consider first the case of QPSK. We assume that the binary wave at the modulator input is random, with symbols 1 and 0 being equally likely, and with the symbols transmitted during adjacent time slots being statistically independent. We make the following observations pertaining to the in-phase and quadrature components of a QPSK signal:

1. Depending on the dibit sent during the signaling interval $-T_b \leq t \leq T_b$, the in-phase component equals $+g(t)$ or $-g(t)$, and similarly for the quadrature component. The $g(t)$ denotes the symbol shaping function, defined by

$$g(t) = \begin{cases} \sqrt{\dfrac{E}{T}}, & 0 \leq t \leq T \\ 0, & \text{elsewhere} \end{cases} \tag{8.217}$$

    Hence, the in-phase and quadrature components have a common power spectral density, namely, $E \operatorname{sinc}^2(Tf)$.

2. The in-phase and quadrature components are statistically independent. Accordingly, the baseband power spectral density of the QPSK signal equals the sum of the individual power spectral densities of the in-phase and quadrature components, and so we may write

$$\begin{aligned} S_B(f) &= 2E \operatorname{sinc}^2(Tf) \\ &= 4E_b \operatorname{sinc}^2(2T_b f) \end{aligned} \tag{8.218}$$

Consider next the MSK signal. Here again we assume that the input binary wave is random, with symbols 1 and 0 equally likely, and the symbols transmitted during different time slots being statistically independent. In this case, we make the following observations:

1. Depending on the value of phase state $\theta(0)$, the in-phase component equals $+g(t)$ or $-g(t)$, where

$$g(t) = \begin{cases} \sqrt{\dfrac{2E_b}{T_b}} \cos\left(\dfrac{\pi t}{2T_b}\right), & -T_b \leq t \leq T_b \\ 0, & \text{elsewhere} \end{cases} \tag{8.219}$$

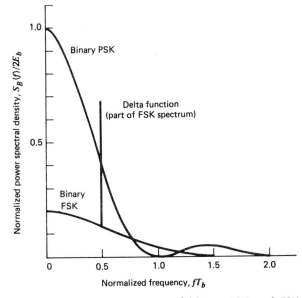

Figure 8.38 Power spectra of binary PSK and FSK signals.

The energy spectral density of this symbol-shaping function is

$$\Psi_g(f) = \frac{32E_bT_b}{\pi^2}\left[\frac{\cos(2\pi T_b f)}{16T_b^2 f^2 - 1}\right]^2 \tag{8.220}$$

Hence, the power spectral density of the in-phase component equals $\Psi_g(f)/2T_b$.

2. Depending on the value of the phase state $\theta(T_b)$, the quadrature component equals $+g(t)$ or $-g(t)$, where

$$g(t) = \begin{cases} \sqrt{\dfrac{2E_b}{T_b}}\sin\left(\dfrac{\pi t}{2T_b}\right), & 0 \le t \le 2T_b \\ 0, & \text{elsewhere} \end{cases} \tag{8.221}$$

The energy spectral density of this second symbol shaping function is also given by Eq. (8.220). Hence, the in-phase and quadrature components have the same power spectral density.

3. As with the QPSK signal, the in-phase and quadrature component of the MSK signal are also statistically independent. Hence, the baseband power spectral density of the MSK signal is given by

$$\begin{aligned} S_B(f) &= 2\left[\frac{\Psi_g(f)}{2T_b}\right] \\ &= \frac{32E_b}{\pi^2}\left[\frac{\cos(2\pi T_b f)}{16T_b^2 f^2 - 1}\right]^2 \end{aligned} \tag{8.222}$$

The baseband power spectra of Eqs. (8.218) and (8.222) for the QPSK and MSK signals, respectively, are shown plotted in Fig. 8.39. The power spectral

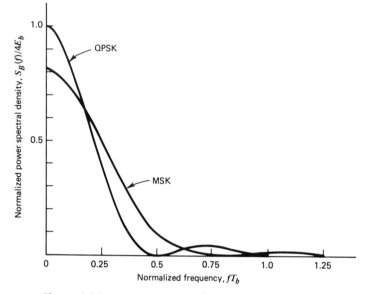

Figure 8.39 Power spectra of QPSK and MSK signals.

density is normalized with respect to $4E_b$, and the frequency is normalized with respect to the bit rate $1/T_b$. Note that for $f \gg 1/T_b$, the baseband power spectral density of the MSK signal falls off as the inverse fourth power of frequency, whereas in the case of the QPSK signal, it falls off as the inverse square of frequency. Accordingly, MSK does not produce as much interference outside the signal band of interest as QPSK. This is a desirable characteristic of MSK, especially when it operates with a bandwidth limitation.

Power Spectra of *M*-ary Signals

Binary PSK and QPSK are special cases of *M*-ary PSK signals. The symbol duration of *M*-ary PSK is defined by

$$T = T_b \log_2 M \tag{8.223}$$

where T_b is the bit duration. Proceeding in a manner similar to that described for a QPSK signal, we may show that the baseband power spectral density of an *M*-ary PSK signal is given by

$$
\begin{aligned}
S_B(f) &= 2E \operatorname{sinc}^2(Tf) \\
&= 2E_b \log_2 M \operatorname{sinc}^2(T_b f \log_2 M)
\end{aligned}
\tag{8.224}
$$

In Fig. 8.40, we show the normalized power spectral density $S_B(f)/2E_b$ plotted versus the normalized frequency fT_b for three different values of M, namely, $M = 2, 4, 8$.

The spectral analysis of *M*-ary FSK signals[10] is much more complicated than that of *M*-ary PSK signals. A case of particular interest occurs when the frequencies assigned to the multilevels make the frequency spacing uniform and the

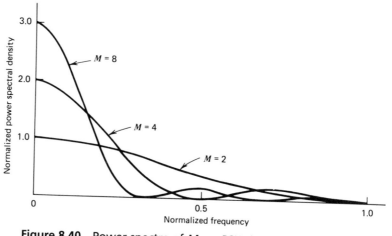

Figure 8.40 Power spectra of *M*-ary PSK signals for *M* = 2, 4, 8.

frequency deviation $k = 0.5$. That is, the M signal frequencies are separated by $1/2T$, where T is the symbol duration. For $k = 0.5$, the baseband power spectral density of M-ary FSK signals is shown plotted in Fig. 8.41 for $M = 2, 4, 8$.

8.21 BANDWIDTH EFFICIENCY

Channel bandwidth and *transmitted power* constitute two primary "communication resources," the efficient utilization of which provides the motivation for the search for *spectrally efficient* schemes. The primary objective of spectrally efficient modulation is to maximize the *bandwidth efficiency*, defined as the ratio of data

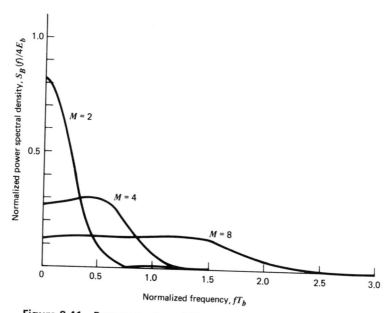

Figure 8.41 Power spectra of *M*-ary FSK signals for *M* = 2, 4, 8.

rate to channel bandwidth; it is measured in units of bits per second per Hertz. A secondary objective is to achieve this bandwidth efficiency at a minimum practical expenditure of average signal power or, equivalently, in a channel perturbed by additive white Gaussian noise, a minimum practical expenditure of average signal-to-noise ratio. Bandwidth efficiency is also referred to as *spectral efficiency*.

With the data rate denoted by R_b and the channel bandwidth by B, we may express the bandwidth efficiency, ρ, as

$$\rho = \frac{R_b}{B} \quad \text{b/s/Hz} \tag{8.225}$$

The data rate R_b is well defined. Unfortunately, however, there is no universally satisfying definition for the bandwidth B. This means that the bandwidth efficiency of a digital modulation scheme depends on the particular definition adopted for the bandwidth of the modulated signal. In the sequel, we consider the evaluation of the bandwidth efficiency of *M*-ary PSK signals, followed by that of *M*-ary FSK signals.

Bandwidth Efficiency of *M*-ary PSK Signals

The power spectra of *M*-ary PSK signals possess a *main lobe* bounded by well-defined *spectral nulls* (i.e., frequencies at which the power spectral density is zero). Accordingly, the spectral width of the main lobe provides a simple and popular measure for the bandwidth of *M*-ary PSK signals. This definition is referred to as the *null-to-null bandwidth*. With the null-to-null bandwidth encompassing the main lobe of the power spectrum of an *M*-ary signal, we find that it contains most of the signal power. This is readily seen by looking at the power spectral plots of Fig. 8.41.

The channel bandwidth required to pass *M*-ary PSK signals (more precisely, the main spectral lobe of *M*-ary PSK signals) is given by

$$B = \frac{2}{T} \tag{8.226}$$

where T is the symbol duration. But the symbol duration T is related to the bit duration T_b by Eq. (8.223). Moreover, the bit rate $R_b = 1/T_b$. Hence, we may redefine the channel bandwidth of Eq. (8.226) in terms of the bit rate R_b as

$$B = \frac{2R_b}{\log_2 M} \tag{8.227}$$

Equivalently, we may express the bandwidth efficiency of *M*-ary PSK signals as

$$\rho = \frac{R_b}{B}$$
$$= \frac{\log_2 M}{2} \tag{8.228}$$

Table 8.6 gives the values of ρ calculated from Eq. (8.228) for varying *M*.

Table 8.6 Bandwidth Efficiency of
M-ary PSK Signals

M	2	4	8	16	32	64
ρ (bits/s/Hz)	0.5	1	1.5	2	2.5	3

Bandwidth Efficiency of *M*-ary FSK Signals

Consider next an *M*-ary FSK signal that consists of an *orthogonal* set of *M* frequency-shifted signals. When the orthogonal signals are detected coherently, the adjacent signals need only be separated from each other by a frequency difference $1/2T$ so as to maintain orthogonality. Hence, we may define the channel bandwidth required to transmit *M*-ary FSK signals as

$$B = \frac{M}{2T} \tag{8.229}$$

For multilevels with frequency assignments that make the frequency spacing uniform and equal to $1/2T$, the bandwidth *B* of Eq. (8.229) contains a large fraction of the signal power. This is readily confirmed by looking at the baseband power spectral plots shown in Fig. 8.41. Hence, substituting Eq. (8.223) in Eq. (8.229) and using $R_b = 1/T_b$, we may redefine the channel bandwidth *B* for *M*-ary FSK signals as

$$B = \frac{R_b M}{2 \log_2 M} \tag{8.230}$$

This, in turn, means that the bandwidth efficiency of *M*-ary FSK signals is given by

$$\begin{aligned} \rho &= \frac{R_b}{B} \\ &= \frac{2 \log_2 M}{M} \end{aligned} \tag{8.231}$$

Table 8.7 gives the values of ρ calculated from Eq. (8.231) for varying *M*.

Comparing Tables 8.6 and 8.7, we see that increasing the number of levels *M* tends to increase the bandwidth efficiency of *M*-ary PSK signals, but tends to decrease the bandwidth efficiency of *M*-ary FSK signals. In other words, *M*-ary PSK signals are spectrally efficient, whereas *M*-ary FSK signals are spectrally inefficient.

Table 8.7 Bandwidth Efficiency
of *M*-ary FSK Signals

M	2	4	8	16	32	64
ρ (bits/s/Hz)	1	1	0.75	0.5	0.3125	0.1875

8.22   SYNCHRONIZATION

The coherent reception of a digitally modulated signal requires that the receiver be synchronous to the transmitter. We say that two sequences of events (representing a transmitter and a receiver) are *synchronous* relative to each other when the events in one sequence and the corresponding events in the other occur simultaneously. The process of making a situation synchronous, and maintaining it in this condition, is called *synchronization.*

From the discussion presented on the operation of digital modulation techniques, we recognize the need for two basic modes of synchronization:

1. When coherent detection is used, knowledge of both the frequency and phase of the carrier is necessary. The estimation of carrier phase and frequency is called *carrier recovery* or *carrier synchronization.*
2. To perform demodulation, the receiver has to know the instants of time at which the modulation can change its state. That is, it has to know the starting and finishing times of the individual symbols, so that it may determine when to sample and when to quench the product-integrators. The estimation of these times is called *clock recovery* or *symbol synchronization.*

These two modes of synchronization can be coincident with each other, or they can occur sequentially one after the other. Naturally, in a noncoherent system, carrier synchronization is of no concern.

Both modes of the synchronization problem may be formulated in statistical terms. In what follows, we present a qualitative discussion of the synchronization problem, and describe circuits for carrier and clock recovery.[11]

Carrier Synchronization

The most straightforward method of carrier synchronization is to modulate the data-bearing signal onto a carrier in such a way that the power spectrum of the modulated signal contains a discrete component at the carrier frequency. Then a narrow-band *phase-locked loop* (PLL) can be used to track this component, thereby providing the desired reference signal at the receiver; a phase-locked loop consists of a voltage-controlled oscillator (VCO), a loop filter, and a multiplier that are connected together in the form of a negative feedback system, as discussed in Section 3.12. The disadvantage of such an approach is that since the discrete component does not convey any information other than the frequency and phase of the carrier, its transmission represents a waste of power.

Accordingly, modulation techniques that conserve power are always of interest in practice. In particular, the modulation employed is often of such a form that, in the absence of a dc component in the power spectrum of the data-bearing signal, the receiver requires the use of a *suppressed carrier-tracking loop* for providing a coherent secondary carrier (subcarrier) reference. For example, Fig. 8.42 shows the block diagram of a carrier-recovery circuit for M-ary PSK. This circuit is called the *Mth-power loop.* For the special case of $M = 2$, the circuit is called a *squaring loop.* However, when the squaring loop or its generalization is used for carrier recovery, we encounter a *phase ambiguity* problem. Consider, for example, the simple case of binary PSK. Since a squaring loop contains a squaring device at its input end, it is clear that changing the sign of the input signal leaves the sign of the recovered carrier unaltered. In other words, the squaring loop ex-

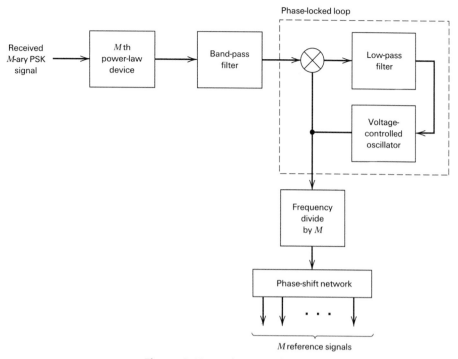

Figure 8.42 *M*th power loop.

hibits a 180 degree phase ambiguity. Correspondingly, the generalization of the squaring loop for *M*-ary PSK exhibits *M* phase ambiguities in the interval $(0, 2\pi)$.

Another method for carrier recovery involves the use of a *Costas loop*, which consists of two paths, one referred to as *in-phase* and the other referred to as *quadrature*; these two paths are coupled together via a common voltage-controlled oscillator (VCO) to form a negative feedback system, as discussed in Section 3.4. When synchronization is attained, the demodulated data waveform appears at the output of the in-phase path, and the corresponding output of the quadrature path is zero under ideal conditions. Analysis of the Costas loop shows that it too exhibits the same phase ambiguity problem as the squaring loop. Moreover, the Costas loop is equivalent to the squaring loop in terms of noise performance, provided that the two low-pass filters in the two paths of the Costas loop are the low-pass (baseband) equivalent of the band-pass filter in the squaring loop. The Costas loop may be generalized for *M*-ary PSK, in which case it also exhibits *M* phase ambiguities in the interval $(0, 2\pi)$. However, compared to the *M*th order loop, the *M*th order Costas loop has a practical disadvantage in that the amount of circuitry needed for its implementation becomes prohibitive for large *M*.

One method of resolving the phase ambiguity problem is to exploit differential encoding. Specifically, the incoming data sequence is differentially encoded before modulation at the transmitter, and differentially decoded after detection at the receiver, resulting in a small degradation in noise performance. This method is called the *coherent detection of differentially encoded M-ary PSK*. As such, it is different from *M*-ary DPSK. For the special case of coherent detection of differentially encoded binary PSK, the average probability of symbol error is given by[12]

$$P_e = \text{erfc}\left(\sqrt{\frac{E_b}{N_0}}\right) - \frac{1}{2}\text{erfc}^2\left(\sqrt{\frac{E_b}{N_0}}\right) \tag{8.232}$$

In the region where $(E_b/N_0) \gg 1$, the second term on the right-hand side of Eq. (8.232) has a negligible effect; hence, this modulation scheme has an average probability of symbol error practically the same as that for coherent QPSK or MSK. For the coherent detection of differentially encoded QPSK, the average probability of symbol error is given by

$$P_e = 2\text{erfc}\left(\sqrt{\frac{E_b}{N_0}}\right) - 2\text{erfc}^2\left(\sqrt{\frac{E_b}{N_0}}\right) + \text{erfc}^3\left(\sqrt{\frac{E_b}{N_0}}\right) - \frac{1}{4}\text{erfc}^4\left(\sqrt{\frac{E_b}{N_0}}\right) \tag{8.233}$$

For large E_b/N_0, this average probability of symbol error is approximately twice that of coherent QPSK.

Symbol Synchronization

As mentioned previously, clock recovery (i.e., symbol synchronization) can be processed alongside carrier recovery. Alternatively, clock recovery is accomplished first, followed by carrier recovery; sometimes, the reverse procedure is followed. The choice of one particular approach or the other is determined by the application of interest.

In one approach, the symbol synchronization problem is solved by transmitting a clock along with the data-bearing signal, in multiplexed form. Then, at the receiver, the clock is extracted by appropriate filtering of the modulated waveforms. Such an approach minimizes the time required for carrier/clock recovery. However, a disadvantage of the approach is that a fraction of the transmitted power is allocated to the transmission of the clock.

In another approach, a good method is, first, to use a noncoherent detector to extract the clock. Here, use is made of the fact that clock timing is usually much more stable than carrier phase. Then, the carrier is recovered by processing the noncoherent detector output in each clocked interval.

In yet another approach, when clock recovery follows carrier recovery, the clock is extracted by processing demodulated (not necessarily detected) baseband waveforms, thereby avoiding any wastage of transmitted power. In the sequel, we use heuristic arguments to develop a symbol synchronizer that satisfies this requirement.

Consider first a rectangular pulse defined by (see Fig. 8.43a)

$$g(t) = \begin{cases} a, & 0 \le t \le T \\ 0, & \text{otherwise} \end{cases}$$

The output of a filter matched to the pulse $g(t)$ is shown in Fig. 8.43b. We observe that the matched filter output attains its peak value at time $t = T$ and that it is symmetric about this point. Clearly, the proper time to sample the matched filter output is at $t = T$. Suppose, however, that the matched filter output is sampled

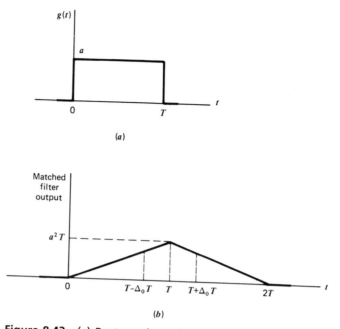

Figure 8.43 (a) Rectangular pulse g(t). (b) Output of filter matched to g(t).

early at $t = T - \Delta_0 T$ or *late* at $t = T + \Delta_0 T$. Then, the absolute values of the two samples so obtained will (on the average in the presence of additive noise) be equal, and smaller than the peak value at $t = T$. Moreover, the *error signal*, the difference between the absolute values of the two samples, is zero; and the proper sampling time is the midpoint between $t = T - \Delta_0 T$ and $t = T + \Delta_0 T$. This special condition may be viewed as an *equilibrium point* in that if we deviate from it the error signal becomes nonzero.

Figure 8.44 shows the block diagram of a symbol synchronizer, which exploits this notion. For obvious reasons, it is called an *early–late gate symbol synchronizer of the absolute value type*. In Fig. 8.44 correlators are used in place of equivalent matched filters. Both correlators integrate over a full symbol interval T, with one starting $\Delta_0 T$ early relative to the transition time estimate and the other starting $\Delta_0 T$ late. An error signal, $e(kT)$, is generated by taking the difference between the absolute values of the two correlator outputs at time $t = kT$. For a given offset λT between the actual transition times $\{t(kT)\}$ and their local estimates $\{\hat{t}(kT)\}$, the error signal is zero for zero offset; otherwise, it is linearly proportional to λ, irrespective of the polarity of λ. This polarity independence is the result of taking absolute values of the correlator outputs before evaluating the difference between them. The error signal is low-pass filtered and then applied to a voltage-controlled oscillator that controls (through a symbol waveform generator) the charging and discharging instants of the correlators. The closed loop is designed to be narrow band relative to the symbol rate $1/T$. The instantaneous frequency of the local clock is advanced or retarded in an iterative manner until the equilibrium point is reached, and symbol synchronization is thereby established.

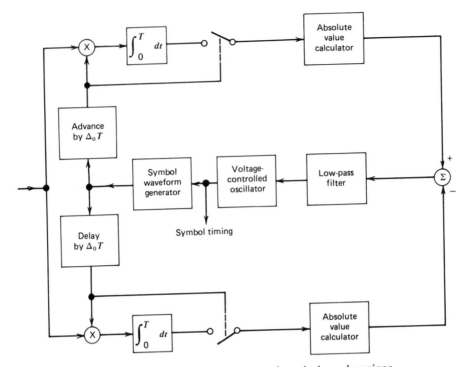

Figure 8.44 Early–late gate type of symbol synchronizer.

8.23 SUMMARY AND DISCUSSION

In this rather long chapter, we presented a systematic analysis of the effects of noise on the performance of passband data transmission systems. The analysis proceeded by first using the Gram–Schmidt orthogonalization procedure to develop a geometric interpretation of transmitted signals, and then extending this representation to include the effects of channel noise modeled as additive white Gaussian noise (AWGN). The important result emerging from this analysis was the idea of a correlation receiver or equivalently a matched filter receiver for the optimum detection of a known signal in an AWGN channel. The receiver is optimum in the sense that it minimizes the average probability of symbol error. We also developed a noncoherent matched filter receiver for the optimum detection of a signal in an AWGN channel when there is uncertainty in the carrier phase of the received signal, and it is therefore difficult to phase-lock the receiver to the transmitter.

With this basic background theory on optimum receivers at our disposal, we presented detailed derivations of formulas for the bit error rate error for some important digital modulation techniques in an AWGN channel:

1. Coherent modulation techniques:
    - Coherent binary phase-shift keying
    - Coherent binary frequency-shift keying
    - Coherent quadriphase-shift keying
    - Coherent minimum shift keying

2. Noncoherent binary modulation techniques:
   - Noncoherent binary frequency-shift keying
   - Differential phase-shift keying

This was followed by a brief discussion of coherent M-ary modulation techniques: M-ary phase-shift keying, M-ary quadrature amplitude modulation, and M-ary frequency-shift keying. We also evaluated the spectral properties of the bandpass signals resulting from the use of these digital modulation techniques and described practical synchronization circuits for their operation.

From the discussion presented in this chapter, we conclude that the performance analysis of passband data transmission systems in the presence of additive white Gaussian noise is well understood for both coherent and noncoherent reception. In practice, however, we find that because bandwidth occupancy is a major factor in the design of these systems, there is indeed a second source of intereference, namely, intersymbol interference (ISI), which must be accounted for in error rate calculations. As explained in Chapter 7, intersymbol interference is generated by the use of band-limiting filters at the transmitter output, in the transmission medium, and at the receiver input, or combinations thereof. When ISI is present, we find that (for the case of coherent reception) the correlation receiver or the matched filter receiver is no longer optimum, with the result that there is degradation in the actual error rate of the receiver. This also applies to noncoherent receivers.

A final comment is in order: When explicit performance analysis of a passband transmission system defies a satisfactory solution, particularly when the effect of ISI is noticeable, the use of *computer simulation*[13] provides the only alternative approach to actual hardware evaluation. The speed and flexibility usually associated with digital computers are compelling reasons for adopting this approach. The simulation procedure involves the formulation of a baseband equivalent model for the system using the complex notation described in Chapter 2.

NOTES AND REFERENCES

1. The geometric representation of signals was first developed by Kotel'nikov in 1947: V. A. Kotel'nikov, *The Theory of Optimum Noise Immunity* (Dover Publications, 1960), which is a translation of the original doctoral dissertation presented in January 1947 before the Academic Council of the Molotov Energy Institute in Moscow. In particular, see Part II of the book. This method was subsequently brought to fuller fruition in the classic book by Wozencraft and Jacobs (1965).

2. In Section 8.7 we derive the union bound on the average probability of symbol error; the classic reference for this bound is Wozencraft and Jacobs (1965). For the derivation of tighter bounds, see Viterbi and Omura (1979, pp. 58–59).

3. For mathematical details of the derivation of the quadrature receiver of Fig. 8.12, see Whalen (1971, pp. 196–205).

4. For a detailed tutorial review of different digital modulation techniques (ASK, FSK, and PSK) using a geometric viewpoint, see Arthurs and Dym, 1962. See also the following list of books:
   - Ziemer and Tranter (1990, Chapter 9)
   - Lafrance (1990, Chapter 6)

- Proakis (1989, Chapter 4)
- Gibson (1989, Chapter 11)
- Viterbi and Omura (1979, pp. 47–127)
- Franks (1969, pp. 1–65)
- Sakrison (1968, pp. 219–271)

5. The MSK signal was first described in Doelz and Heald (1961). For a tutorial review of MSK and comparison with QPSK, see Pasupathy (1979). Since the frequency spacing is only half as much as the conventional spacing of $1/T_b$ that is used in the coherent detection of binary FSK signals, this signaling scheme is also referred to as *fast* FSK; see deBuda (1972).

6. The standard method of deriving the bit error rate for noncoherent binary FSK, presented in Whalen (1971), and that for differential phase-shift keying presented in Arthurs and Dym (1962), involves the use of the Rician distribution. This distribution arises when the envelope of a sine wave plus additive Gaussian noise is of interest; see Chapter 4 for a discussion of the Rician distribution. The evaluations presented in Sections 8.16 and 8.17 avoid the complications encountered in the standard method.

7. The optimum receiver for differential phase-shift keying is discussed in Simon and Divsalar (1992).

8. For a table of values of P_e for varying E/N_0 and for $M = 2, 4, 8, 16, 32, 64$ for the case of coherent M-ary PSK, see Lindsey and Simon (1973, pp. 232–233).

9. For the derivation of Eq. (8.204) for the average probability of symbol error for noncoherent M-ary FSK, see Lindsey and Simon (1973, p. 489).

10. A detailed analysis of the spectra of M-ary FSK for an arbitrary value of frequency deviation is presented in the paper by Anderson and Salz (1965). The results shown plotted in Fig. 8.41 represent a special case of a formula derived in that paper for a frequency deviation of $k = 0.5$.

11. For a statistical analysis of the synchronization problems and descriptions of carrier and clock recovery circuits, see Stiffler (1971), Lindsey (1972), and Lindsey and Simon (1973, Chapters 2 and 9); see also the collection of papers edited by Lindsey and Simon (1978).

12. For the noise analysis of coherent detection of differentially encoded M-ary PSK, of which Eqs. (8.232) and (8.233) are special cases, see Lindsey and Simon (1973, pp. 242–246).

13. For a detailed treatment of computer simulation of digital communication systems, see Jeruchim, Balaban, and Shanmugan (1992).

PROBLEMS

Problem 8.1

(a) Using the Gram–Schmidt orthogonalization procedure, find a set of orthonormal basis functions to represent the three signals $s_1(t)$, $s_2(t)$, and $s_3(t)$ shown in Fig. P8.1.

(b) Express each of these signals in terms of the set of basis functions found in part (a).

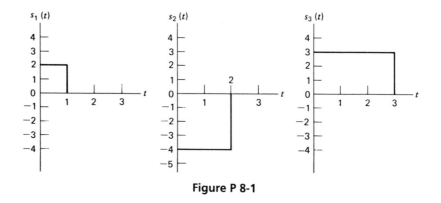

Figure P 8-1

Problem 8.2 Consider the set of signals

$$s_i(t) = \begin{cases} \sqrt{\dfrac{2E}{T}}\cos\left(2\pi f_c t + i\dfrac{\pi}{4}\right), & 0 \le t \le T \\ 0, & \text{elsewhere} \end{cases}$$

where $i = 1, 2, 3, 4$, and $f_c = n_c/T$ for some fixed integer n_c.

(a) What is the dimensionality, N, of the space spanned by this set of signals?

(b) Find a set of orthonormal basis functions to represent this set of signals.

(c) Using the expansion,

$$s_i(t) = \sum_{j=1}^{N} s_{ij}\phi_j(t), \qquad i = 1, 2, 3, 4$$

find the coefficients s_{ij}.

(d) Plot the locations of $s_i(t)$, $i = 1, 2, 3, 4$, in the signal space, using the results of parts (b) and (c).

Problem 8.3 Given two real-valued vectors $\mathbf{x}$ and $\mathbf{y}$, prove Schwarz's inequality, which, using vector notation, states that

$$|\mathbf{x}^T\mathbf{y}| \le \|\mathbf{x}\| \, \|\mathbf{y}\|$$

where $|(\mathbf{x}^T\mathbf{y})|$ is the absolute value of the inner product of $\mathbf{x}$ and $\mathbf{y}$, and $\|\mathbf{x}\|$ and $\|\mathbf{y}\|$ are the lengths of $\mathbf{x}$ and $\mathbf{y}$, respectively. *Hint:* Consider the inequality $\|\mathbf{x} \pm a\mathbf{y}\|^2 \ge 0$, and find an appropriate choice for the scalar a.

Problem 8.4 Consider a random process $X(t)$ expanded in the form

$$X(t) = \sum_{j=1}^{N} X_j\phi_j(t) + W'(t), \qquad 0 \le t \le T$$

where $W'(t)$ is a remainder noise term. The $\{\phi_j(t) \,|\, j = 1, 2, \ldots, N\}$ form an orthonormal set, and the X_j are defined by

$$X_j = \int_0^T X(t)\phi_j(t) \, dt$$

Let $W'(t_k)$ denote a random variable obtained by observing $W'(t)$ at time $t = t_k$. Show that

$$E[X_j W'(t_k)] = 0, \quad \begin{cases} j = 1, 2, \ldots, N \\ 0 \le t_k \le T \end{cases}$$

Problem 8.5 In the *Bayes' test*, applied to a binary hypothesis testing problem where we have to choose one of two possible hypotheses H_0 and H_1, we minimize the *risk R* defined by:

$$R = C_{00}\, p_0 P(\text{say } H_0|H_0 \text{ is true})$$
$$+ C_{10}\, p_0 P(\text{say } H_1|H_0 \text{ is true})$$
$$+ C_{11}\, p_1 P(\text{say } H_1|H_1 \text{ is true})$$
$$+ C_{01}\, p_1 P(\text{say } H_0|H_1 \text{ is true})$$

The C_{00}, C_{10}, C_{11}, and C_{01} denote the costs assigned to the four possible outcomes of the experiment: the first subscript indicates the hypothesis chosen and the second the hypothesis that was true. Assume that $C_{10} > C_{00}$ and $C_{01} > C_{11}$. The p_0 and p_1 denote the *a priori* probabilities of hypotheses H_0 and H_1, respectively.

(a) Given the observation vector $\mathbf{x}$, show that the partitioning of the observation space so as to minimize the risk R leads to the *likelihood ratio test*:

say H_0 if $\Lambda(\mathbf{x}) < \lambda$

say H_1 if $\Lambda(\mathbf{x}) > \lambda$

where $\Lambda(\mathbf{x})$ is the *likelihood ratio*

$$\Lambda(\mathbf{x}) = \frac{f_{\mathbf{x}}(\mathbf{x}|H_1)}{f_{\mathbf{x}}(\mathbf{x}|H_0)}$$

and λ is the *threshold* of the test defined by

$$\lambda = \frac{p_0(C_{10} - C_{00})}{p_1(C_{01} - C_{11})}$$

(b) What are the cost values for which the Bayes' criterion reduces to the minimum probability of error criterion?

Problem 8.6 A PSK signal is applied to a correlator supplied with a phase reference that lies within ϕ radians of the exact carrier phase. Determine the effect of the phase error ϕ on the average probability of error of the system.

Problem 8.7 The signal vectors $\mathbf{s}_1$ and $\mathbf{s}_2$ are used to represent binary symbols 1 and 0, respectively, in a coherent binary FSK system. The receiver decides in favor of symbol 1 when

$$\mathbf{x}^T\mathbf{s}_1 < \mathbf{x}^T\mathbf{s}_2$$

where $\mathbf{x}^T\mathbf{s}_i$ is the inner product of the observation vector $\mathbf{x}$ and the signal vector $\mathbf{s}_i$, where $i = 1, 2$. Show that this decision rule is equivalent to the condition $x_1 > x_2$, where x_1 and x_2 are the elements of the observation vector $\mathbf{x}$. Assume that the signal vectors $\mathbf{s}_1$ and $\mathbf{s}_2$ have equal energy.

Problem 8.8 Consider a phase-locked loop consisting of a multiplier, loop filter, and voltage-controlled oscillator (VCO). Let the signal applied to the multiplier input be a PSK signal defined by

$$s(t) = A_c \cos[2\pi f_c t + k_p m(t)]$$

where k_p is the phase sensitivity, and the data signal $m(t)$ takes on the value $+1$ for binary symbol 1 and -1 for binary symbol 0. The VCO output is

$$r(t) = A_c \sin[2\pi f_c t + \theta(t)]$$

(a) Evaluate the loop filter output, assuming that this filter removes only modulated components with carrier frequency $2f_c$.

(b) Show that this output is proportional to the data signal $m(t)$ when the loop is phase locked, that is, $\theta(t) = 0$.

Problem 8.9 The signal component of a coherent PSK system is defined by

$$s(t) = A_c k \sin(2\pi f_c t) \pm A_c \sqrt{1 - k^2} \cos(2\pi f_c t)$$

where $0 \le t \le T_b$, and the plus sign corresponds to symbol 1 and the minus sign corresponds to symbol 0. The first term represents a carrier component included for the purpose of synchronizing the receiver to the transmitter.

(a) Draw a signal-space diagram for the scheme described here; what observations can you make about this diagram?

(b) Show that, in the presence of additive white Gaussian noise of zero mean and power spectral density $N_0/2$, the average probability of error is

$$P_e = \frac{1}{2} \operatorname{erfc}\left(\sqrt{\frac{E_b}{N_0}(1 - k^2)}\right)$$

where

$$E_b = \tfrac{1}{2} A_c^2 T_b$$

(c) Suppose that 10 percent of the transmitted signal power is allocated to the carrier component. Determine the E_b/N_0 required to realize a probability of error equal to 10^{-4}.

(d) Compare this value of E_b/N_0 with that required for a conventional PSK system with the same probability of error.

Problem 8.10 An FSK system transmits binary data at the rate of 2.5×10^6 bits per second. During the course of transmission, white Gaussian noise of zero mean and power spectral density 10^{-20} watts per hertz is added to the signal. In the absence of noise, the amplitude of the received sinusoidal wave for digit 1 or 0 is 1 microvolt. Determine the average probability of symbol error for the following system configurations:

(a) Coherent binary FSK.

(b) Coherent MSK.

(c) Noncoherent binary FSK.

Problem 8.11

(a) In a coherent FSK system, the signals $s_1(t)$ and $s_2(t)$ representing symbols

1 and 0, respectively, are defined by

$$s_1(t),\ s_2(t)\ =\ A_c \cos\left[2\pi\left(f_c \pm \frac{\Delta f}{2}\right)t\right], \qquad 0 \le t \le T_b$$

Assuming that $f_c > \Delta f$, show that the correlation coefficient of the signals $s_1(t)$ and $s_2(t)$ is approximately given by

$$\rho = \frac{\displaystyle\int_0^{T_b} s_1(t)s_2(t)\ dt}{\displaystyle\int_0^{T_b} s_1^2(t)\ dt} \simeq \mathrm{sinc}(2\ \Delta f T_b)$$

(b) What is the minimum value of frequency shift Δf for which the signals $s_1(t)$ and $s_2(t)$ are orthogonal?

(c) What is the value of Δf that minimizes the average probability of symbol error?

(d) For the value of Δf obtained in part (c), determine the increase in E_b/N_0 required so that this coherent FSK system has the same noise performance as a coherent binary PSK system.

Problem 8.12   A binary FSK signal with *discontinuous phase* is defined by

$$s(t) = \begin{cases} \sqrt{\dfrac{2E_b}{T_b}}\cos\left[2\pi\left(f_c + \dfrac{\Delta f}{2}\right)t + \theta_1\right] & \text{for symbol 1} \\[4mm] \sqrt{\dfrac{2E_b}{T_b}}\cos\left[2\pi\left(f_c - \dfrac{\Delta f}{2}\right)t + \theta_2\right] & \text{for symbol 0} \end{cases}$$

where E_b is the signal energy per bit, T_b is the bit duration, and θ_1 and θ_2 are sample values of uniformly distributed random variables over the interval 0 to 2π. In effect, the two oscillators supplying the transmitted frequencies $f_c \pm \Delta f/2$ operate independently of each other. Assume that $f_c \gg \Delta f$.

(a) Evaluate the power spectral density of the FSK signal.

(b) Show that for frequencies far removed from the carrier frequency f_c, the power spectral density falls off as the inverse square of frequency.

Problem 8.13   Set up a block diagram for the generation of Sunde's FSK signal $s(t)$ with continuous phase by using the representation given in Eq. (8.213), which is reproduced here:

$$s(t) = \sqrt{\frac{2E_b}{T_b}}\cos\left(\frac{\pi t}{T_b}\right)\cos(2\pi f_c t) \mp \sqrt{\frac{2E_b}{T_b}}\sin\left(\frac{\pi t}{T_b}\right)\sin(2\pi f_c t)$$

Problem 8.14   In the on–off keying version of an ASK system, symbol 1 is represented by transmitting a sinusoidal carrier of amplitude $\sqrt{2E_b/T_b}$, where E_b is the signal energy per bit and T_b is the bit duration. Symbol 0 is represented by switching off the carrier. Assume that symbols 1 and 0 occur with equal probability.

For an AWGN channel, determine the average probability of error for this ASK system, assuming:

(a) Coherent reception.

(b) Noncoherent reception, operating with a large value of bit energy-to-noise density ratio E_b/N_0.

Note: When x is large, the modified Bessel function of the first kind of zero order may be approximated as follows (see Appendix 4):

$$I_0(x) \simeq \frac{\exp(x)}{\sqrt{2\pi x}}$$

Problem 8.15 The purpose of a *radar system* is basically to detect the presence of a target and to extract useful information about the target. Suppose that in such a system, hypothesis H_0 is that there is no target present, so that the received signal $x(t) = w(t)$, where $w(t)$ is white Gaussian noise of zero mean and power spectral density $N_0/2$. For hypothesis H_1, a target is present, and $x(t) = w(t) + s(t)$, where $s(t)$ is an echo produced by the target. Assume that $s(t)$ is completely known. Evaluate:

(a) The *probability of false alarm* defined as the probability that the receiver decides that a target is present when it is not.

(b) The *probability of detection* defined as the probability that the receiver decides that a target is present when it is.

Problem 8.16 Binary data are transmitted over a microwave link at the rate of 10^6 bits per second and the power spectral density of the noise at the receiver input is 10^{-10} watts per hertz. Find the average carrier power required to maintain an average probability of error $P_e \le 10^{-4}$ for (a) coherent binary PSK, and (b) DPSK.

Problem 8.17 The values of E_b/N_0 required to realize an average probability of symbol error $P_e = 10^{-4}$ using coherent binary PSK and coherent FSK (conventional) systems are equal to 7.2 and 13.5, respectively. Using the approximation (see Appendix 7)

$$\text{erfc}(u) \simeq \frac{1}{\sqrt{\pi u}} \exp(-u^2)$$

determine the separation in the values of E_b/N_0 for $P_e = 10^{-4}$, using

(a) Coherent binary PSK and DPSK.
(b) Coherent binary PSK and QPSK.
(c) Coherent binary FSK (conventional) and noncoherent binary FSK.
(d) Coherent binary FSK (conventional) and coherent MSK.

Problem 8.18 The binary sequence 1100100010 is applied to the DPSK transmitter of Fig. 8.30a.

(a) Sketch the resulting waveform at the transmitter output.

(b) Applying this waveform to the DPSK receiver of Fig. 8.30b, show that, in the absence of noise, the original binary sequence is reconstructed at the receiver output.

Problem 8.19

(a) Given the input binary sequence 1100100010, sketch the waveforms of the in-phase and quadrature components of a modulated wave obtained by using the QPSK based on the signal set of Fig. 8.18.

(b) Sketch the QPSK waveform itself for the input binary sequence specified in part (a).

Problem 8.20   Let P_{eI} and P_{eQ} denote the probabilities of symbol error for the in-phase and quadrature channels of a narrow-band system. Show that the average probability of symbol error for the overall system is given by

$$P_e = P_{eI} + P_{eQ} - P_{eI}P_{eQ}$$

Problem 8.21   There are two ways of detecting an MSK signal. One way is to use a coherent receiver to take full account of the phase information content of the MSK signal. Another way is to use a noncoherent receiver and disregard the phase information. The second method offers the advantage of simplicity of implementation, at the expense of a degraded noise performance. By how many decibels do we have to increase the bit energy-to-noise density ratio E_b/N_0 in the second case so as to realize an average probability of symbol error equal to 10^{-5} in both cases?

Problem 8.22

(a) Sketch the waveforms of the in-phase and quadrature components of the MSK signal in response to the input binary sequence 1100100010.

(b) Sketch the MSK waveform itself for the binary sequence specified in part (a).

Problem 8.23   In a special form of quadriphase-shift keying known as the *offset QPSK*, the in-phase data stream is delayed relative to the quadrature data stream by half a symbol period $T/2$.

(a) What is the average probability of symbol error for an offset QPSK system?

(b) What is the power spectral density of an offset QPSK signal produced by a random binary sequence in which symbols 1 and 0 (represented by ± 1) are equally likely, and the symbols in different time slots are statistically independent and identically distributed?

(c) What are the similarities between the offset QPSK and MSK, and what features distinguish them?

Problem 8.24   In Section 8.18 we compared the noise performances of coherent binary PSK, coherent binary FSK, QPSK, MSK, DPSK, and noncoherent FSK by using the bit error rate as the basis of comparison. In this problem we take a different viewpoint and use the average probability of symbol error, P_e, to do the comparison. Plot P_e versus E_b/N_0 for each of these schemes, and comment on your results.

Problem 8.25   In Section 8.16 we derived the formula for the bit error rate of noncoherent binary FSK as a special case of noncoherent orthogonal modulation. In this problem we revisit this issue. As before, we assume that the binary symbol represented by signal $s_1(t)$ is transmitted. According to the material presented in Section 8.16, we note the following:

- The random variable L_2 represented by the sample value l_2 of Eq. (8.157) is Rayleigh distributed.
- The random variable L_1 represented by the sample value l_1 of Eq. (8.163) is Rician-distributed.

The Rayleigh and Rician distributions are discussed in Chapter 4. Using the probability distributions defined in that chapter, derive the formula of Eq. (8.176) for the BER of noncoherent binary FSK.

Problem 8.26 Equation (8.186) is an approximate formula for the average probability of symbol error for coherent *M*-ary PSK. This formula was derived using the union bound in light of the signal-space diagram of Fig. 8.33*b*. Given that message point m_1 was transmitted, show that the approximate formula of Eq. (8.186) may be derived directly from Fig. 8.33*b*.

Problem 8.27 Consider an *M*-ary QAM with an in-phase and a quadrature component as described in Section 8.13. The signal constellation of each of these two components has a geometry similar to that of *M*-ary PAM with a corresponding number of amplitude levels. Using this relationship, derive the formula for P'_e in Eq. (8.195) that defines the average probability of symbol error for either component of *M*-ary QAM.

Problem 8.28 Figure P8.2*a* shows a noncoherent receiver using a matched filter for the detection of a sinusoidal signal of known frequency but random phase, in the presence of additive white Gaussian noise. An alternative implementation of this receiver is its mechanization in the frequency domain as a *spectrum analyzer receiver*, as in Fig. P8.2*b*, where the correlator computes the finite time autocorrelation function $R_x(\tau)$ defined by

$$R_x(\tau) = \int_0^{T-\tau} x(t)x(t + \tau)\ dt, \qquad 0 \leq \tau \leq T$$

Show that the square-law envelope detector output sampled at time $t = T$ in Fig. P8.2*a* is twice the spectral output of the Fourier transformer sampled at frequency $f = f_c$ in Fig. P8.2*b*.

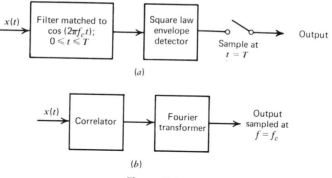

(a)

(b)

Figure P 8-2

Problem 8.29 The *noise equivalent bandwidth* of a bandpass signal is defined as the value of bandwidth that satisfies the relation

$$2BS(f_c) = P/2$$

where $2B$ is the noise equivalent bandwidth centered around the mid-band frequency f_c, $S(f_c)$ is the maximum value of the power spectral density of the signal at $f = f_c$, and P is the average power of the signal. Show that the noise equivalent bandwidths of binary PSK, QPSK, and MSK are as follows:

Type of Modulation	Noise Bandwidth/Bit Rate
Binary PSK	1.0
QPSK	0.5
MSK	0.62

Note: You may use the definite integrals in Table A7.1 of Appendix 7.

CHAPTER 9

Spread-Spectrum Modulation

9.1 INTRODUCTION

A major issue of concern in the study of digital communictions as considered in the previous two chapters is that of providing for the efficient utilization of bandwidth and power. Notwithstanding the importance of these two primary communication resources, there are situations where it is necessary to sacrifice their efficient utilization in order to meet certain other design objectives. For example, the system may be required to provide a form of *secure* communication in a *hostile* environment such that the transmitted signal is not easily detected or recognized by unwanted listeners. This requirement is catered to by a class of signaling techniques known collectively as *spread-spectrum modulation.*

The primary advantage of a spread-spectrum communication system is its ability to reject *interference* whether it be the *unintentional* interference by another user simultaneously attempting to transmit through the channel, or the *intentional* interference by a hostile transmitter attempting to jam the transmission.

The definition of spread-spectrum modulation[1] may be stated in two parts:

1. *Spread spectrum is a means of transmission in which the data sequence occupies a bandwidth in excess of the minimum bandwidth necessary to send it.*

2. *The spectrum spreading is accomplished before transmission through the use of a code that is independent of the data sequence. The same code is used in the receiver (oper-*

ating in synchronism with the transmitter) to despread the received signal so that the original data sequence may be recovered.

Although standard modulation techniques such as frequency modulation and pulse-code modulation do satisfy part 1 of this definition, they are not spread-spectrum techniques because they do not satisfy part 2 of the definition.

Spread-spectrum modulation was originally developed for military applications, where resistance to jamming (interference) is of major concern. However, there are civilian applications that also benefit from the unique characteristics of spread-spectrum modulation. For example, it can be used to provide *multipath rejection* in a ground-based mobile radio environment. Yet another application is in *multiple-access* communications in which a number of independent users are required to share a common channel without an external synchronizing mechanism; here, for example, we may mention a ground-based mobile radio environment involving mobile vehicles that must communicate with a central station.

In this chapter, we discuss principles of spread-spectrum modulation, with emphasis on direct-sequence and frequency-hopping techniques. In a *direct-sequence spread-spectrum* technique, two stages of modulation are used. First, the incoming data sequence is used to modulate a wide-band code. This code transforms the narrow-band data sequence into a noiselike wide-band signal. The resulting wide-band signal undergoes a second modulation using a phase-shift keying technique. In a *frequency-hop spread-spectrum* technique, on the other hand, the spectrum of a data-modulated carrier is widened by changing the carrier frequency in a pseudo-random manner. For their operation, both of these techniques rely on the availability of a noiselike spreading code called a *pseudo-random* or *pseudo-noise sequence*. Since such a sequence is basic to the operation of spread-spectrum modulation, it is logical that we begin our study by describing the generation and properties of pseudo-noise sequences.

9.2 PSEUDO-NOISE SEQUENCES

A *pseudo-noise (PN) sequence* is a periodic binary sequence with a noiselike waveform that is usually generated by means of a *feedback shift register*, a general block diagram of which is shown in Fig. 9.1. A feedback shift register consists of an ordinary *shift register* made up of m flip-flops (two-state memory stages) and a *logic circuit* that are interconnected to form a multiloop *feedback* circuit. The flip-flops in the shift register are regulated by a single timing *clock*. At each pulse (tick) of the clock, the *state* of each flip-flop is shifted to the next one down the line. With each clock pulse the logic circuit computes a Boolean function of the

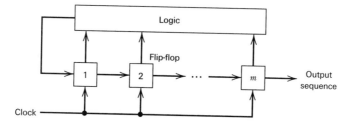

Figure 9.1 Feedback shift register.

states of the flip-flops. The result is then fed back as the input to the first flip-flop, thereby preventing the shift register from emptying. The PN sequence so generated is determined by the length m of the shift register, its initial state, and the feedback logic.

Let $s_j(k)$ denote the state of the jth flip-flop after the kth clock pulse; this state may be represented by symbol 0 or 1. The state of the shift register after the kth clock pulse is then defined by the set $\{s_1(k), s_2(k), \ldots, s_m(k)\}$, where $k \geq 0$. For the initial state, k is zero. From the definition of a shift register, we have

$$s_j(k + 1) = s_{j-1}(k), \qquad \begin{cases} k \geq 0 \\ 1 \leq j \leq m \end{cases} \tag{9.1}$$

where $s_0(k)$ is the input applied to the first flip-flop after the kth clock pulse. According to the configuration described in Fig. 9.1, $s_0(k)$ is a Boolean function of the individual states $s_1(k), s_2(k), \ldots, s_m(k)$. For a specified length m, this Boolean function uniquely determines the subsequent sequence of states and therefore the PN sequence produced at the output of the final flip-flop in the shift register. With a total number of m flip-flops, the number of possible states of the shift register is at most 2^m. It follows therefore that the PN sequence generated by a feedback shift register must eventually become *periodic* with a period of at most 2^m.

A feedback shift register is said to be *linear* when the feedback logic consists entirely of *modulo-2 adders*. In such a case, the *zero state* (e.g., the state for which all the flip-flops are in state 0) is *not* permitted. We say so because for a zero state, the input $s_0(k)$ produced by the feedback logic would be 0, the shift register would then continue to remain in the zero state, and the output would therefore consist entirely of 0s. Consequently, the period of a PN sequence produced by a linear feedback shift register with m flip-flops cannot exceed $2^m - 1$. When the period is exactly $2^m - 1$, the PN sequence is called a *maximum-length-sequence* or simply *m-sequence*.

Example 1

Consider the linear feedback shift register shown in Fig. 9.2, involving three flip-flops. The input s_0 applied to the first flip-flop is equal to the modulo-2 sum of s_1 and s_3. It is assumed that the initial state of the shift register is 100 (reading the contents of the three flip-flops from left to right). Then, the succession of states will be as follows:

$$100, 110, 111, 011, 101, 010, 001, 100, \ldots.$$

The output sequence (the last position of each state of the shift register) is therefore

$$00111010 \ldots$$

which repeats itself with period $2^3 - 1 = 7$.

Note that the choice of 100 as the initial state is an arbitrary one. Any of the other six permissible states could serve equally well as an initial state. The resulting output sequence would then simply experience a cyclic shift.

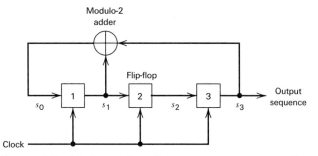

Figure 9.2 Maximum-length sequence generator for $m = 3$.

Properties of Maximum-Length Sequences[2]

Maximum-length sequences have many of the properties possessed by a truly *random binary sequence*. A random binary sequence is a sequence in which the presence of binary symbol 1 or 0 is equally probable. Some properties of maximum-length sequences are as follows:

1. *In each period of a maximum-length sequence, the number of 1s is always one more than the number of 0s.* This property is called the *balance property*.

2. *Among the runs of 1s and of 0s in each period of a maximum-length sequence, one-half the runs of each kind are of length one, one-fourth are of length two, one-eighth are of length three, and so on as long as these fractions represent meaningful numbers of runs.* This property is called the *run property*. By a "run" we mean a subsequence of identical symbols (1s or 0s) within one period of the sequence. The length of this subsequence is the length of the run. For a maximum-length sequence generated by a linear feedback shift register of length m, the total number of runs is $(N + 1)/2$, where $N = 2^m - 1$.

3. *The autocorrelation function of a maximum-length sequence is periodic and binary-valued.* This property is called the *correlation property*.

   The *period* of a maximum-length sequence is defined by

   $$N = 2^m - 1 \tag{9.2}$$

   where m is the length of the shift register. Let binary symbols 0 and 1 of the sequence be denoted by the levels -1 and $+1$, respectively. Let $c(t)$ denote the resulting waveform of the maximum-length sequence, as illustrated in Fig. 9.3a for $N = 7$. The period of the waveform $c(t)$ is (based on terminology used in subsequent sections)

   $$T_b = NT_c \tag{9.3}$$

   where T_c is the duration assigned to symbol 1 or 0 in the maximum-length sequence. By definition, the autocorrelation function of a periodic signal $c(t)$ of period T_b is

   $$R_c(\tau) = \frac{1}{T_b} \int_{-T_b/2}^{T_b/2} c(t)\,c(t - \tau)\,dt \tag{9.4}$$

   where the lag τ lies in the interval $(-T_b/2, T_b/2)$; Eq. (9.4) is a special case of Eq. (4.80). Applying this formula to a maximum-length sequence repre-

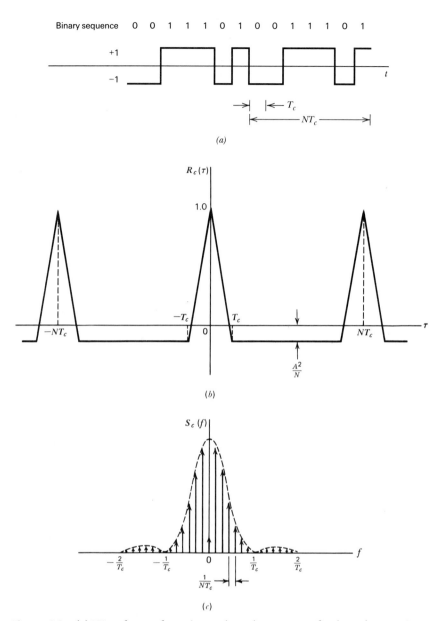

Figure 9.3 (a) Waveform of maximum-length sequence for length $m = 3$ or period $N = 7$. (b) Autocorrelation function. (c) Power spectral density. All three parts refer to the output of the feedback shift register of Fig. 9.2.

sented by $c(t)$, we get

$$
R_c(\tau) = \begin{cases} 1 - \dfrac{N+1}{NT_c}|\tau|, & |\tau| \le T_c \\[2mm] -\dfrac{1}{N}, & \text{for the remainder of the period} \end{cases} \tag{9.5}
$$

This result is shown plotted in Fig. 9.3b for the case of $m = 3$ or $N = 7$.

From Chapter 2 we recall that periodicity in the time domain is transformed into uniform sampling in the frequency domain. This interplay between the time and frequency domains is borne out by the power spectral density of the maximum-length wave $c(t)$. Specifically, taking the Fourier transform of Eq. (9.5), we get the sampled spectrum

$$S_c(f) = \frac{1}{N^2}\,\delta(f) + \frac{1+N}{N^2} \sum_{\substack{n=-\infty \\ n \neq 0}}^{\infty} \mathrm{sinc}^2\left(\frac{n}{N}\right) \delta\left(f - \frac{n}{NT_c}\right) \qquad (9.6)$$

which is shown plotted in Fig. 9.3c for $m = 3$ or $N = 7$.

Comparing the results of Fig. 9.3 for a maximum-length sequence with the corresponding results shown in Fig. 4.17 for a random binary sequence, we may make the following observations:

- For a period of the maximum-length sequence, the autocorrelation function $R_c(\tau)$ is somewhat similar to that of a random binary wave.
- The waveforms of both sequences have the same envelope, $\mathrm{sinc}^2(fT)$, for their power spectral densities. The fundamental difference between them is that whereas the random binary sequence has a continuous spectral density characteristic, the corresponding characteristic of a maximum-length sequence consists of delta functions spaced $1/NT_c$ Hz apart.

As the shift-register length m, or equivalently, the period N of the maximum-length sequence is increased, the maximum-length sequence becomes increasingly similar to the random binary sequence. Indeed, in the limit, the two sequences become identical when N is made infinitely large. However, the price paid for making N large is an increasing storage requirement, which imposes a practical limit on how large N can actually be made.

Choosing a Maximum-Length Sequence

Now that we understand the properties of a maximum-length sequence and the fact that we can generate it using a linear feedback shift register, the key question that we need to address is: How do we find the feedback logic for a desired period N? The answer to this question is to be found in the theory of error-control codes, which is covered in the next chapter. The task of finding the required feedback logic is made particularly easy for us by virtue of the extensive tables of the necessary feedback connections for varying shift-register lengths that have been compiled in the literature. In Table 9.1, we present the sets of maximum (feedback) taps pertaining to shift-register lengths $m = 2, 3, \ldots, 8$.[3] Note that as m increases, the number of alternative schemes (codes) is enlarged. Also, for every set of feedback connections shown in this table, there is an "image" set that generates an identical maximum-length code, reversed in time sequence.

The particular sets identified with an asterisk in Table 9.1 correspond to *Mersenne prime length sequences*, for which the period N is a prime number.

Example 2

Consider a maximum-length sequence requiring the use of a linear feedback-shift register of length $m = 5$. For feedback taps, we select the set [5, 2] from

Table 9.1   Maximum-Length Sequences of Shift-Register Lengths 2 Through 8

Shift-Register Length, m	Feedback Taps
2*	[2, 1]
3*	[3, 1]
4	[4, 1]
5*	[5, 2], [5, 4, 3, 2], [5, 4, 2, 1]
6	[6, 1], [6, 5, 2, 1], [6, 5, 3, 2]
7*	[7, 1], [7, 3], [7, 3, 2, 1], [7, 4, 3, 2], [7, 6, 4, 2], [7, 6, 3, 1], [7, 6, 5, 2], [7, 6, 5, 4, 2, 1], [7, 5, 4, 3, 2, 1]
8	[8, 4, 3, 2], [8, 6, 5, 3], [8, 6, 5, 2], [8, 5, 3, 1], [8, 6, 5, 1], [8, 7, 6, 1], [8, 7, 6, 5, 2, 1], [8, 6, 4, 3, 2, 1]

Table 9.1. The corresponding configuration of the code generator is shown in Fig. 9.4a. Assuming that the initial state is 10000, the evolution of one period of the maximum-length sequence generated by this scheme is shown in Table 9.2a, where we see that the generator returns to the initial 10000 after 31 iterations; that is, the period is 31, which agrees with the value obtained from Eq. (9.2).

Suppose next we select another set of feedback taps from Table 9.1, namely, [5, 4, 2, 1]. The corresponding code generator is thus as shown in Fig. 9.4b. For the initial state 10000, we now find that the evolution of the maximum-length sequence is as shown in Table 9.2b. Here again, the generator returns to the

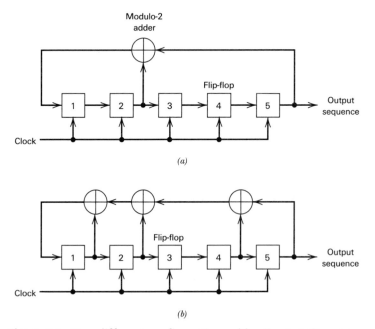

Figure 9.4   Two different configurations of feedback shift register of length $m = 5$. (a) Feedback connections [5, 2]. (b) Feedback connections [5, 4, 2, 1].

Table 9.2a Evolution of the Maximum-Length Sequence Generated by the Feedback-Shift Register of Fig. 9.4a

Feedback Symbol	State of Shift Register					Output Symbol
	1	0	0	0	0	
0	0	1	0	0	0	0
1	1	0	1	0	0	0
0	0	1	0	1	0	0
1	1	0	1	0	1	0
1	1	1	0	1	0	1
1	1	1	1	0	1	0
0	0	1	1	1	0	1
1	1	0	1	1	1	0
1	1	1	0	1	1	1
0	0	1	1	0	1	1
0	0	0	1	1	0	1
0	0	0	0	1	1	0
1	1	0	0	0	1	1
1	1	1	0	0	0	1
1	1	1	1	0	0	0
1	1	1	1	1	0	0
1	1	1	1	1	1	0
0	0	1	1	1	1	1
0	0	0	1	1	1	1
1	1	0	0	1	1	1
1	1	1	0	0	1	1
0	0	1	1	0	0	1
1	1	0	1	1	0	0
0	0	1	0	1	1	0
0	0	0	1	0	1	1
1	1	0	0	1	0	1
0	0	1	0	0	1	0
0	0	0	1	0	0	1
0	0	0	0	1	0	0
0	0	0	0	0	1	0
1	1	0	0	0	0	1

Code: 0000101011101100011111001101001

initial state 10000 after 31 iterations, and so it should. But the maximum-length sequence generated is different from that shown in Table 9.2a.

Clearly, the code generator of Fig. 9.4a has an advantage over that of Fig. 9.4b, as it requires fewer feedback connections.

Table 9.2b Evolution of the Maximum-Length
Sequence Generated by the Feedback-
Shift Register of Fig. 9.4b

Feedback Symbol	State of Shift Register					Output Symbol
	1	0	0	0	0	
1	1	1	0	0	0	0
0	0	1	1	0	0	0
1	1	0	1	1	0	0
0	0	1	0	1	1	0
1	1	0	1	0	1	1
0	0	1	0	1	0	1
0	0	0	1	0	1	0
1	1	0	0	1	0	1
0	0	1	0	0	1	0
0	0	0	1	0	0	1
0	0	0	0	1	0	0
1	1	0	0	0	1	0
0	0	1	0	0	0	1
1	1	0	1	0	0	0
1	1	1	0	1	0	0
1	1	1	1	0	1	0
1	1	1	1	1	0	1
1	1	1	1	1	1	0
0	0	1	1	1	1	1
1	1	0	1	1	1	1
1	1	1	0	1	1	1
0	0	1	1	0	1	1
0	0	0	1	1	0	1
1	1	0	0	1	1	0
1	1	1	0	0	1	1
1	1	1	1	0	0	1
0	0	1	1	1	0	0
0	0	0	1	1	1	0
0	0	0	0	1	1	1
0	0	0	0	0	1	1
1	1	0	0	0	0	1

Code: 000011010100100010111110110011

9.3   A NOTION OF SPREAD SPECTRUM

An important attribute of spread-spectrum modulation is that it can provide protection against externally generated interfering (jamming) signals with finite power. The jamming signal may consist of a fairly powerful broadband noise or multitone waveform that is directed at the receiver for the purpose of disrupting communications. Protection against jamming waveforms is provided by purposely making the information-bearing signal occupy a bandwidth far in excess of the minimum bandwidth necessary to transmit it. This has the effect of making the transmitted signal assume a noiselike appearance so as to blend into the background. The transmitted signal is thus enabled to propagate through the channel undetected by anyone who may be listening. We may therefore think of spread spectrum as a method of "camouflaging" the information-bearing signal.

One method of widening the bandwidth of an information-bearing (data) sequence involves the use of *modulation.* Let $\{b_k\}$ denote a binary data sequence, and $\{c_k\}$ denote a pseudo-noise (PN) sequence. Let the waveforms $b(t)$ and $c(t)$ denote their respective nonreturn-to-zero (polar) representations in terms of two levels equal in amplitude and opposite in polarity, namely, ± 1. We will refer to $b(t)$ as the information-bearing (data) signal, and to $c(t)$ as the PN signal. The desired modulation is achieved by applying the data signal $b(t)$ and the PN signal $c(t)$ to a product modulator or multiplier, as in Fig. 9.5*a.* We know from Fourier transform theory that multiplication of two signals produces a signal whose spectrum equals the convolution of the spectra of the two component signals. Thus, if the message signal $b(t)$ is narrow band and the PN signal $c(t)$ is wide band, *the product (modulated) signal $m(t)$ will have a spectrum that is nearly the same as the wideband PN signal.* In other words, in the context of our present application, the PN sequence performs the role of a *spreading code.*

By multiplying the information-bearing signal $b(t)$ by the PN signal $c(t)$, each information bit is "chopped" up into a number of small time increments, as illustrated in the waveforms of Fig. 9.6. These small time increments are commonly referred to as *chips.*

For *baseband* transmission, the product signal $m(t)$ represents the *transmitted signal.* We may thus express the transmitted signal as

$$m(t) = c(t)b(t) \tag{9.7}$$

The received signal $r(t)$ consists of the transmitted signal $m(t)$ plus an additive *interference* denoted by $i(t)$, as shown in the channel model of Fig. 9.5*b.* Hence,

$$
\begin{aligned}
r(t) &= m(t) + i(t) \\
&= c(t)b(t) + i(t)
\end{aligned}
\tag{9.8}
$$

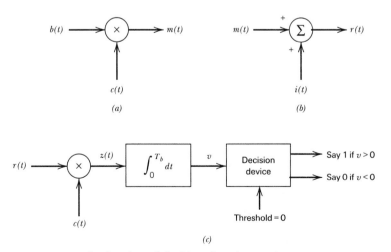

Figure 9.5 Idealized model of baseband spread-spectrum system. (*a*) Transmitter. (*b*) Channel. (*c*) Receiver.

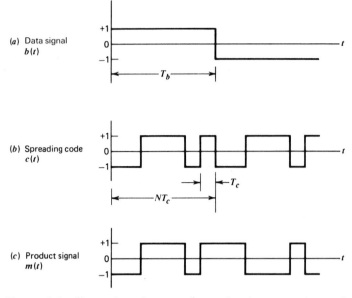

Figure 9.6 Illustrating the waveforms in the transmitter of Fig. 9.5a.

To recover the original message signal $b(t)$, the received signal $r(t)$ is applied to a *demodulator* that consists of a multiplier followed by an integrator, and a decision device, as in Fig. 9.5c. The multiplier is supplied with a locally generated PN sequence that is an exact *replica* of that used in the transmitter. Moreover, we assume that the receiver operates in perfect *synchronism* with the transmitter, which means that the PN sequence in the receiver is lined up exactly with that in the transmitter. The multiplier output in the receiver is therefore given by

$$z(t) = c(t)r(t)$$
$$= c^2(t)b(t) + c(t)i(t) \tag{9.9}$$

Equation (9.9) shows that the data signal $b(t)$ is multiplied *twice* by the PN signal $c(t)$, whereas the unwanted signal $i(t)$ is multiplied only *once*. The PN signal $c(t)$ alternates between the levels -1 and $+1$, and the alternation is destroyed when it is squared; hence,

$$c^2(t) = 1 \qquad \text{for all } t \tag{9.10}$$

Accordingly, we may simplify Eq. (9.9) as

$$z(t) = b(t) + c(t)i(t) \tag{9.11}$$

We thus see from Eq. (9.11) that the data signal $b(t)$ is reproduced at the multiplier output in the receiver, except for the effect of the interference represented by the additive term $c(t)i(t)$. Multiplication of the interference $i(t)$ by the locally generated PN signal $c(t)$ means that the spreading code will affect the interference just as it did the original signal at the transmitter. We now observe that the data component $b(t)$ is narrow band, whereas the spurious component $c(t)i(t)$ is wide band. Hence, by applying the multiplier output to a baseband (low-pass) filter with a bandwidth just large enough to accommodate the recovery of the data signal $b(t)$, most of the power in the spurious component $c(t)i(t)$ is filtered out. The effect of the interference $i(t)$ is thus significantly reduced at the receiver output.

In the receiver shown in Fig. 9.5c the low-pass filtering action is actually performed by the integrator that evaluates the area under the signal produced at the multiplier output. The integration is carried out for the bit interval $0 \leq t \leq T_b$, providing the sample value v. Finally, a decision is made by the receiver: If v is greater than the threshold of zero, the receiver says that binary symbol 1 of the original data sequence was sent in the interval $0 \leq t \leq T_b$, and if v is less than zero, the receiver says that symbol 0 was sent; if v is exactly zero the receiver makes a random guess in favor of 1 or 0.

In summary, the use of a spreading code (with pseudo-random properties) in the transmitter produces a wide-band transmitted signal that appears *noiselike* to a receiver that has *no* knowledge of the spreading code. From the discussion presented in Section 9.2, we note that (for a prescribed data rate) the longer we make the period of the spreading code, the closer will the transmitted signal be to a truly random binary wave, and the harder it is to detect. Naturally, the price we have to pay for the improved protection against interference is increased transmission bandwidth, system complexity, and processing delay. However, when our primary concern is the security of transmission, these are not unreasonable costs to pay.

9.4 DIRECT-SEQUENCE SPREAD SPECTRUM WITH COHERENT BINARY PHASE-SHIFT KEYING

The spread-spectrum technique described in the previous section is referred to as *direct-sequence spread spectrum.* The discussion presented there was in the context of baseband transmission. To provide for the use of this technique in passband transmission over a satellite channel, for example, we may incorporate *coherent binary phase-shift keying* (PSK) into the transmitter and receiver, as shown in Fig. 9.7. The transmitter of Fig. 9.7a first converts the incoming binary data sequence $\{b_k\}$ into an NRZ waveform $b(t)$, which is followed by two stages of modulation. The first stage consists of a product modulator or multiplier with the data signal $b(t)$ (representing a data sequence) and the PN signal $c(t)$ (representing the PN sequence) as inputs. The second stage consists of a binary PSK modulator. The transmitted signal $x(t)$ is thus a *direct-sequence spread binary phase-shift-keyed* (DS/BPSK) *signal.* The phase modulation $\theta(t)$ of $x(t)$ has one of two values, 0 and π, depending on the polarities of the message signal $b(t)$ and PN signal $c(t)$ at time t in accordance with the truth table of Table 9.3.

Figure 9.8 illustrates the waveforms for the second stage of modulation. Part of the modulated waveform shown in Fig. 9.6c is reproduced in Fig. 9.8a; the waveform shown here corresponds to one period of the PN sequence. Figure 9.8b shows the waveform of a sinusoidal carrier, and Fig. 9.8c shows the DS/BPSK waveform that results from the second stage of modulation.

The receiver, shown in Fig. 9.7b, consists of two stages of demodulation. In the first stage, the received signal $y(t)$ and a locally generated carrier are applied

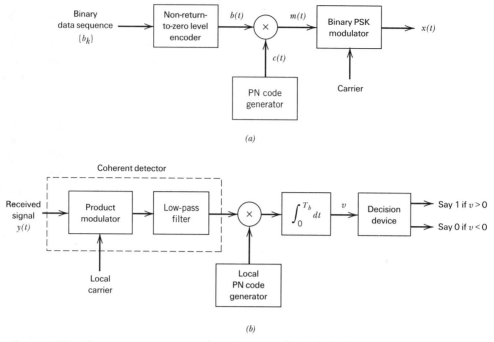

Figure 9.7 Direct-sequence spread coherent phase-shift keying. (a) Transmitter. (b) Receiver.

to a product modulator followed by a low-pass filter whose bandwidth is equal to that of $m(t)$. This stage of the demodulation process reverses the phase-shift keying applied to the transmitted signal. The second stage of demodulation performs spectrum despreading by multiplying the low-pass filter output by a locally generated replica of the PN signal $c(t)$, followed by integration over a bit interval $0 \leq t \leq T_b$, and finally decision making in the manner described in Section 9.3.

Model for Analysis

In the normal form of the transmitter, shown in Fig. 9.7a, the spectrum spreading is performed prior to phase modulation. For the purpose of analysis, however, we find it more convenient to interchange the order of these operations, as shown in the model of Fig. 9.9. We are permitted to do this because the spectrum spreading and the binary phase-shift keying are both linear operations; likewise for the phase demodulation and spectrum despreading. But for the interchange of operations to be feasible, it is important to synchronize the incoming data

Table 9.3 Truth Table for Phase Modulation $\theta(t)$, Radians

		Polarity of Data Sequence $b(t)$ at Time t	
		+	−
Polarity of PN	+	0	π
sequence $c(t)$ at time t	−	π	0

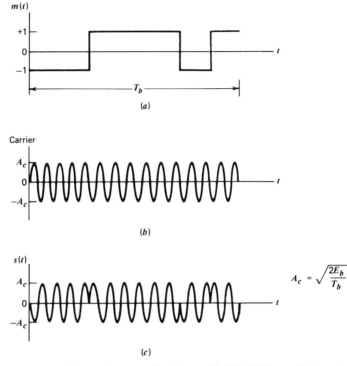

Figure 9.8   (a) Product signal m(t) = c(t)a(t). (b) Sinusoidal carrier.
(c) DS/BPSK signal.

sequence and the PN sequence. The model of Fig. 9.9 also includes representations of the channel and the receiver. In this model, it is assumed that the interference $j(t)$ limits performance, so that the effect of channel noise may be
ignored. Accordingly, the channel output is given by

$$y(t) = x(t) + j(t)$$
$$= c(t)s(t) + j(t)$$

(9.12)

where $s(t)$ is the binary PSK signal, and $c(t)$ is the PN signal. In the channel
model included in Fig. 9.9, the interfering signal is denoted by $j(t)$. This notation

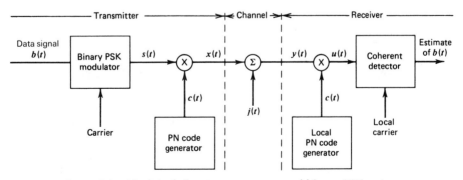

Figure 9.9   Model of direct-sequence spread binary PSK system.

is chosen purposely to be different from that used for the interference in Fig. 9.5b. The channel model in Fig. 9.9 is passband in spectral content, whereas that in Fig. 9.5b is in baseband form.

In the receiver, the received signal $y(t)$ is first multiplied by the PN signal $c(t)$ yielding an output that equals the coherent detector input $u(t)$. Thus,

$$
\begin{aligned}
u(t) &= c(t)y(t) \\
&= c^2(t)s(t) + c(t)j(t) \\
&= s(t) + c(t)j(t)
\end{aligned}
\tag{9.13}
$$

In the last line of Eq. (9.13), we have noted that, by design, the PN signal $c(t)$ satisfies the property described in Eq. (9.10), reproduced here for convenience:

$$
c^2(t) = 1 \qquad \text{for all } t
$$

Equation (9.13) shows that the coherent detector input $u(t)$ consists of a binary PSK signal $s(t)$ embedded in additive code-modulated interference denoted by $c(t)j(t)$. The modulated nature of the latter component forces the interference signal (jammer) to spread its spectrum such that the detection of information bits at the receiver output is afforded increased reliability.

Synchronization

For its proper operation, a spread-spectrum communication system requires that the locally generated PN sequence used in the receiver to despread the received signal be *synchronized* to the PN sequence used to spread the transmitted signal in the transmitter.[4] A solution to the synchronization problem consists of two parts: *acquisition* and *tracking*. In acquisition, or *coarse* synchronization, the two PN codes are aligned to within a fraction of a chip in as short a time as possible. Once the incoming PN code has been acquired, tracking, or *fine* synchronization, takes place. Typically, PN acquisition proceeds in two steps. First, the received signal is multiplied by a locally generated PN code to produce a measure of *correlation* between it and the PN code used in the transmitter. Next, an appropriate *decision-rule and search strategy* is used to process the measure of correlation so obtained to determine whether the two codes are in synchronism and what to do if they are not. As for tracking, it is accomplished using phase-lock techniques very similar to those used for the local generation of coherent carrier references. The principal difference between them lies in the way in which phase discrimination is implemented.

9.5   SIGNAL-SPACE DIMENSIONALITY AND PROCESSING GAIN

Having developed a conceptual understanding of spread-spectrum modulation and a method for its implementation, we are ready to undertake a detailed mathematical analysis of the technique. The approach we have in mind is based on the signal-space theoretic ideas of Chapter 8. In particular, we develop signal-space representations of the transmitted signal and the interfering signal (jammer).

In this context, consider the set of orthonormal basis functions:

$$\phi_k(t) = \begin{cases} \sqrt{\dfrac{2}{T_c}}\cos(2\pi f_c t), & kT_c \le t \le (k+1)T_c \\ 0, & \text{otherwise} \end{cases} \tag{9.14}$$

$$\tilde{\phi}_k(t) = \begin{cases} \sqrt{\dfrac{2}{T_c}}\sin(2\pi f_c t), & kT_c \le t \le (k+1)T_c \\ 0, & \text{otherwise} \end{cases} \tag{9.15}$$

$$k = 0, 1, \ldots, N-1$$

where T_c is the *chip duration*, and N is the number of chips per bit. Accordingly, we may describe the transmitted signal $x(t)$ for the interval of an information bit as follows:

$$\begin{aligned} x(t) &= c(t)s(t) \\ &= \pm\sqrt{\frac{2E_b}{T_b}}\,c(t)\cos(2\pi f_c t) \\ &= \pm\sqrt{\frac{E_b}{N}}\sum_{k=0}^{N-1} c_k\phi_k(t), \qquad 0 \le t \le T_b \end{aligned} \tag{9.16}$$

where E_b is the signal energy per bit; the plus sign corresponds to information bit 1, and the minus sign corresponds to information bit 0. The code sequence $\{c_0, c_1, \ldots, c_{N-1}\}$ denotes the PN sequence, with $c_k = \pm 1$. The transmitted signal $x(t)$ is therefore N-dimensional in that it requires a minimum of N orthonormal functions for its representation.

Consider next the representation of the interfering signal (jammer), $j(t)$. Ideally, the jammer likes to place all of its available energy in exactly the same N-dimensional signal space as the transmitted signal $x(t)$; otherwise, part of its energy goes to waste. However, the best that the jammer can hope to know is the transmitted signal bandwidth. Moreover, there is no way that the jammer can have knowledge of the signal phase. Accordingly, we may represent the jammer by the general form

$$j(t) = \sum_{k=0}^{N-1} j_k\phi_k(t) + \sum_{k=0}^{N-1} \tilde{j}_k\tilde{\phi}_k(t), \qquad 0 \le t \le T_b \tag{9.17}$$

where

$$j_k = \int_0^{T_b} j(t)\phi_k(t)\,dt, \qquad k = 0, 1, \ldots, N-1 \tag{9.18}$$

and

$$\tilde{j}_k = \int_0^{T_b} j(t)\tilde{\phi}_k(t)\,dt, \qquad k = 0, 1, \ldots, N-1 \tag{9.19}$$

Thus, the interference $j(t)$ is $2N$-dimensional; that is, it has twice the number of dimensions required for representing the transmitted DS/BPSK signal $x(t)$. In terms of the representation given in Eq. (9.17), we may express the average power of the interference $j(t)$ as follows:

$$
\begin{aligned}
J &= \frac{1}{T_b} \int_0^{T_b} j^2(t)\ dt \\
&= \frac{1}{T_b} \sum_{k=0}^{N-1} j_k^2 + \frac{1}{T_b} \sum_{k=0}^{N-1} \tilde{j}_k^2
\end{aligned}
\tag{9.20}
$$

Moreover, due to lack of knowledge of signal phase, the best strategy a jammer can apply is to place equal energy in the cosine and sine coordinates defined in Eqs. (9.18) and (9.19); hence, we may safely assume

$$
\sum_{k=0}^{N-1} j_k^2 = \sum_{k=0}^{N-1} \tilde{j}_k^2
\tag{9.21}
$$

Correspondingly, we may simplify Eq. (9.20) as

$$
J = \frac{2}{T_b} \sum_{k=0}^{N-1} j_k^2
\tag{9.22}
$$

Our aim is to tie these results together by finding the signal-to-noise ratios measured at the input and output of the DS/BPSK receiver in Fig. 9.9. To do this, we use Eq. (9.13) to express the coherent detector output as

$$
\begin{aligned}
v &= \sqrt{\frac{2}{T_b}} \int_0^{T_b} u(t)\cos(2\pi f_c t)\ dt \\
&= v_s + v_{cj}
\end{aligned}
\tag{9.23}
$$

where the components v_s and v_{cj} are due to the despread binary PSK signal, $s(t)$, and the spread interference, $c(t)j(t)$, respectively. These two components are defined as follows:

$$
v_s = \sqrt{\frac{2}{T_b}} \int_0^{T_b} s(t)\cos(2\pi f_c t)\ dt
\tag{9.24}
$$

and

$$
v_{cj} = \sqrt{\frac{2}{T_b}} \int_0^{T_b} c(t)j(t)\cos(2\pi f_c t)\ dt
\tag{9.25}
$$

Consider first the component v_s due to the signal. The despread binary PSK signal $s(t)$ equals

$$
s(t) = \pm \sqrt{\frac{2E_b}{T_b}} \cos(2\pi f_c t), \qquad 0 \le t \le T_b
\tag{9.26}
$$

where the plus sign corresponds to information bit 1, and the minus sign corresponds to information bit 0. Hence, assuming that the carrier frequency f_c is an integer multiple of $1/T_b$, we have

$$v_s = \pm\sqrt{E_b} \tag{9.27}$$

Consider next the component v_{cj} due to interference. Expressing the PN signal $c(t)$ in the explicit form of a sequence, $\{c_0, c_1, \ldots, c_{N-1}\}$, we may rewrite Eq. (9.25) in the corresponding form

$$v_{cj} = \sqrt{\frac{2}{T_b}} \sum_{k=0}^{N-1} c_k \int_{kT_c}^{(k+1)T_c} j(t)\cos(2\pi f_c t) \, dt \tag{9.28}$$

Using Eq. (9.14) for $\phi_k(t)$, and then Eq. (9.18) for the coefficient j_k, we may redefine v_{cj} as

$$
\begin{aligned}
v_{cj} &= \sqrt{\frac{T_c}{T_b}} \sum_{k=0}^{N-1} c_k \int_0^{T_b} j(t)\phi_k(t) \, dt \\
&= \sqrt{\frac{T_c}{T_b}} \sum_{k=0}^{N-1} c_k j_k
\end{aligned}
\tag{9.29}
$$

We next approximate the PN sequence as an *independent and identically distributed (i.i.d.) binary sequence*. We emphasize the implication of this approximation by recasting Eq. (9.29) in the form

$$V_{cj} = \sqrt{\frac{T_c}{T_b}} \sum_{k=0}^{N-1} C_k j_k \tag{9.30}$$

where V_{cj} and C_k are random variables with sample values v_{cj} and c_k, respectively. In Eq. (9.30), the jammer is assumed to be fixed. With the C_k treated as i.i.d. random variables, we find that the probability of the event $C_k = \pm 1$ is

$$P(C_k = 1) = P(C_k = -1) = \tfrac{1}{2} \tag{9.31}$$

Accordingly, the mean of the random variable V_{cj} is zero since, for fixed k, we have

$$
\begin{aligned}
E[C_k j_k | j_k] &= j_k P(C_k = 1) - j_k P(C_k = -1) \\
&= \tfrac{1}{2} j_k - \tfrac{1}{2} j_k \\
&= 0
\end{aligned}
\tag{9.32}
$$

For a fixed vector $\mathbf{j}$, representing the set of coefficients $j_0, j_1, \ldots, j_{N-1}$, the variance of V_{cj} is given by

$$\operatorname{var}[V_{cj}|\mathbf{j}] = \frac{1}{N}\sum_{k=0}^{N-1} j_k^2 \tag{9.33}$$

Since the *spread factor* $N = T_b/T_c$, we may use Eq. (9.22) to express this variance in terms of the average interference power J as

$$\text{var}[V_{cj}|\mathbf{j}] = \frac{JT_c}{2} \tag{9.34}$$

Thus, the random variable V_{cj} has zero mean and variance $JT_c/2$.

From Eq. (9.27), we note that the signal component at the coherent detector output (during each bit interval) equals $\pm\sqrt{E_b}$, where E_b is the signal energy per bit. Hence, the peak instantaneous power of the signal component is E_b. Accordingly, we may define the *output signal-to-noise ratio* as the instantaneous peak power E_b divided by the variance of the equivalent noise component in Eq. (9.34). We thus write

$$(\text{SNR})_O = \frac{2E_b}{JT_c} \tag{9.35}$$

The average signal power at the receiver input equals E_b/T_b. We thus define an *input signal-to-noise* ratio as

$$(\text{SNR})_I = \frac{E_b/T_b}{J} \tag{9.36}$$

Hence, eliminating E_b/J between Eqs. (9.35) and (9.36), we may express the output signal-to-noise ratio in terms of the input signal-to-noise ratio as

$$(\text{SNR})_O = \frac{2T_b}{T_c}(\text{SNR})_I \tag{9.37}$$

It is customary practice to express signal-to-noise ratios in decibels. We may thus write Eq. (9.37) in the equivalent form

$$10\log_{10}(\text{SNR})_O = 10\log_{10}(\text{SNR})_I + 3 + 10\log_{10}(\text{PG}), \text{ dB} \tag{9.38}$$

where

$$\text{PG} = \frac{T_b}{T_c} \tag{9.39}$$

The 3-dB term on the right-hand side of Eq. (9.38) accounts for the gain in SNR that is obtained through the use of coherent detection (which presumes exact knowledge of the signal phase by the receiver). This gain in SNR has nothing to do with the use of spread spectrum. Rather, it is the last term, $10\log_{10}(\text{PG})$, that accounts for the *gain in SNR obtained by the use of spread spectrum*. The ratio PG, defined in Eq. (9.39), is therefore referred to as the *processing gain*. Specifically, it represents the gain achieved by processing a spread-spectrum signal over an unspread signal. Note that both the processing gain PG and the spread factor N (i.e., PN sequence length) equal the ratio T_b/T_c. Thus, the longer we make the PN sequence (or, correspondingly, the smaller the chip time T_c is), the larger will the processing gain be.

9.6 PROBABILITY OF ERROR

Let the coherent detector output v in the direct-sequence spread BPSK system of Fig. 9.9 denote the sample value of a random variable V. Let the equivalent noise component v_{cj} produced by external interference denote the sample value of a random variable V_{cj}. Then, from Eqs. (9.23) and (9.27) we deduce that

$$V = \pm\sqrt{E_b} + V_{cj} \tag{9.40}$$

where E_b is the transmitted signal energy per bit. The plus sign refers to sending symbol (information bit) 1, and the minus sign refers to sending symbol 0. The decision rule used by the coherent detector of Fig. 9.9 is to declare that the received bit in an interval $(0, T_b)$ is 1 if the detector output exceeds a threshold of zero, and that it is 0 if the detector output is less than the threshold; if the detector output is exactly zero, the receiver makes a random guess in favor of 1 or 0. With both information bits assumed equally likely, we find that (because of the symmetric nature of the problem) the average probability of error P_e is the same as the conditional probability of (say) the receiver making a decision in favor of symbol 1, given that symbol 0 was sent. That is,

$$\begin{aligned} P_e &= P(V > 0 | \text{symbol 0 was sent}) \\ &= P(V_{cj} > \sqrt{E_b}) \end{aligned} \tag{9.41}$$

Naturally, the probability of error P_e depends on the random variable V_{cj} defined by Eq. (9.30). According to this definition, V_{cj} is the sum of N identically distributed random variables. Hence, from the *central limit theorem*, we deduce that for large N, the random variable V_{cj} assumes a Gaussian distribution. Indeed, the spread factor or PN sequence length N is typically large in the direct-sequence spread-spectrum systems encountered in practice.

Earlier we evaluated the mean and variance of V_{cj}; see Eqs. (9.32) and (9.34). We may therefore state that the equivalent noise component V_{cj} contained in the coherent detector output may be approximated as a Gaussian random variable with zero mean and variance $JT_c/2$, where J is the average interference power and T_c is the chip duration. With this approximation at hand, we may then proceed to calculate the probability of the event $V_{cj} > \sqrt{E_b}$, and thus express the average probability of error in accordance with Eq. (9.41) as

$$P_e \simeq \frac{1}{2}\text{erfc}\left(\sqrt{\frac{E_b}{JT_c}}\right) \tag{9.42}$$

This simple formula, invoking the Gaussian assumption, is appropriate for DS/BPSK binary systems with large spread factor N.

Antijam Characteristics

It is informative to compare Eq. (9.42) with the formula for the average probability of error for a coherent binary PSK system reproduced here for convenience of presentation [see Eq. (8.101)]

$$P_e = \frac{1}{2}\text{erfc}\left(\sqrt{\frac{E_b}{N_0}}\right) \tag{9.43}$$

Based on this comparison, we see that insofar as the calculation of bit error rate in a direct-sequence spread binary PSK system is concerned, the interference may be treated as wide-band noise of power spectral density $N_0/2$, defined by

$$\frac{N_0}{2} = \frac{JT_c}{2} \tag{9.44}$$

This relation is simply a restatement of an earlier result given in Eq. (9.34).

Since the signal energy per bit $E_b = PT_b$, where P is the average signal power and T_b is the bit duration, we may express the signal energy per bit-to-noise spectral density ratio as

$$\frac{E_b}{N_0} = \left(\frac{T_b}{T_c}\right)\left(\frac{P}{J}\right) \tag{9.45}$$

Using the definition of Eq. (9.39) for the processing gain PG we may reformulate this result as

$$\frac{J}{P} = \frac{\text{PG}}{E_b/N_0} \tag{9.46}$$

The ratio J/P is termed the *jamming margin*. Accordingly, the jamming margin and the processing gain, both expressed in decibels, are related by

$$(\text{Jamming margin})_{\text{dB}} = (\text{Processing gain})_{\text{dB}} - 10\log_{10}\left(\frac{E_b}{N_0}\right)_{\text{min}} \tag{9.47}$$

where $(E_b/N_0)_{\text{min}}$ is the minimum value needed to support a prescribed average probability of error.

EXAMPLE 3

A spread-spectrum communication system has the following parameters:

$$\text{Information bit duration, } T_b = 4.095 \text{ ms}$$

$$\text{PN chip duration, } T_c = 1 \text{ }\mu\text{s}$$

Hence, using Eq. (9.39) we find that the processing gain is

$$\text{PG} = 4095$$

Correspondingly, the required PN sequence is $N = 4095$, and the shift-register length is $m = 12$.

For a satisfactory reception, we may assume that the average probability of error is not to exceed 10^{-5}. From the formula for a coherent binary PSK receiver, we find that $E_b/N_0 = 10$ yields an average probability of error equal to 0.387×10^{-5}. Hence, using this value for E_b/N_0, and the value calculated for the processing gain, we find from Eq. (9.47) that the jamming margin is

$$(\text{Jamming margin})_{dB} = 10 \log_{10} 4095 - 10 \log_{10}(10)$$
$$= 36.1 - 10$$
$$= 26.1 \text{ dB}$$

That is, information bits at the receiver output can be detected reliably even when the noise or interference at the receiver input is up to 409.5 times the received signal power. Clearly, this is a powerful advantage against interference (jamming), which is realized through the clever use of spread-spectrum modulation.

9.7 FREQUENCY-HOP SPREAD SPECTRUM

In the type of spread-spectrum systems discussed previously, the use of a PN sequence to modulate a phase-shift-keyed signal achieves *instantaneous* spreading of the transmission bandwidth. The ability of such a system to combat the effects of jammers is determined by the processing gain of the system, which is a function of the PN sequence length. The processing gain can be made larger by employing a PN sequence with narrow chip duration, which, in turn, permits a greater transmission bandwidth and more chips per bit. However, the capabilities of physical devices used to generate the PN spread-spectrum signals impose a practical limit on the attainable processing gain. Indeed, it may turn out that the processing gain so attained is still not large enough to overcome the effects of some jammers of concern, in which case we have to resort to other methods. One such alternative method is to force the jammer to cover a wider spectrum by *randomly hopping* the data-modulated carrier from one frequency to the next. In effect, the spectrum of the transmitted signal is spread *sequentially* rather than instantaneously; the term "sequentially" refers to the pseudo-random-ordered sequence of frequency hops.

The type of spread spectrum in which the carrier hops randomly from one frequency to another is called *frequency-hop (FH) spread spectrum.* A common modulation format for FH systems is that of *M-ary frequency-shift keying* (MFSK). The combination of these two techniques is referred to simply as FH/MFSK. (A description of *M*-ary FSK was presented in Chapter 8).

Since frequency hopping does not cover the entire spread spectrum instantaneously, we are led to consider the rate at which the hops occur. In this context, we may identify two basic (technology-independent) characterizations of frequency hopping:

1. *Slow-frequency hopping,* in which the *symbol rate* R_s of the MFSK signal is an integer multiple of the *hop rate* R_h. That is, several symbols are transmitted on each frequency hop.

2. *Fast-frequency hopping,* in which the hop rate R_h is an integer multiple of the MFSK symbol rate R_s. That is, the carrier frequency will change or hop several times during the transmission of one symbol.

Obviously, slow-frequency hopping and fast-frequency hopping are the converse of one another. In the following, these two characterizations of frequency hopping are considered in turn.

Slow-Frequency Hopping

Figure 9.10*a* shows the block diagram of an FH/MFSK transmitter, which involves *frequency modulation* followed by *mixing*. First, the incoming binary data are applied to an *M*-ary FSK modulator. The resulting modulated wave and the output from a digital *frequency synthesizer* are then applied to a mixer that consists of a multiplier followed by a band-pass filter. The filter is designed to select the sum frequency component resulting from the multiplication process as the transmitted signal. In particular, successive *k*-bit segments of a PN sequence drive the

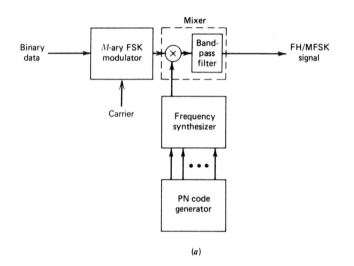

(*a*)

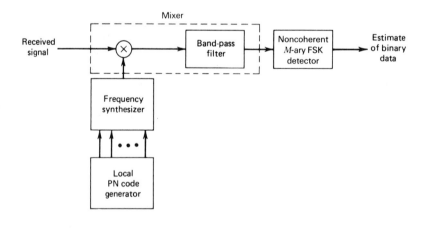

(*b*)

Figure 9.10 Frequency hop spread *M*-ary frequency-shift keying. (*a*) Transmitter. (*b*) Receiver.

frequency synthesizer, which enables the carrier frequency to hop over 2^k distinct values. On a single hop, the bandwidth of the transmitted signal is the same as that resulting from the use of a conventional MFSK with an alphabet of $M = 2^K$ orthogonal signals. However, for a complete range of 2^K frequency hops, the transmitted FH/MFSK signal occupies a much larger bandwidth. Indeed, with present-day technology, FH bandwidths on the order of several GHz are attainable, which is an order of magnitude larger than that achievable with direct-sequence spread spectra. An implication of these large FH bandwidths is that coherent detection is possible only within each hop, because frequency synthesizers are unable to maintain phase coherence over successive hops. Accordingly, most frequency-hop spread-spectrum communication systems use noncoherent *M*-ary modulation schemes.

In the receiver depicted in Fig. 9.10*b*, the frequency hopping is first removed by *mixing* (down-converting) the received signal with the output of a local frequency synthesizer that is synchronously controlled in the same manner as that in the transmitter. The resulting output is then band-pass filtered, and subsequently processed by a *noncoherent M-ary FSK detector*. To implement this *M*-ary detector, we may use a bank of *M* noncoherent matched filters, each of which is matched to one of the MFSK tones. (Noncoherent matched filters were described in Chapter 8). An estimate of the original symbol transmitted is obtained by selecting the largest filter output.

An individual FH/MFSK tone of shortest duration is referred to as a *chip*; this terminology should not be confused with that used in Section 9.4 describing DS/BPSK . The *chip rate*, R_c, for an FH/MFSK system is defined by

$$R_c = \max(R_h, R_s) \qquad (9.48)$$

where R_h is the *hop rate*, and R_s is the *symbol rate*.

A slow FH/MFSK signal is characterized by having multiple symbols transmitted per hop. Hence, each symbol of a slow FH/MFSK signal is a chip. Correspondingly, in a slow FH/MFSK system, the bit rate R_b of the incoming binary data, the symbol rate R_s of the MFSK signal, the chip rate R_c, and the hop rate R_h are related by

$$R_c = R_s = \frac{R_b}{K} \geq R_h \qquad (9.49)$$

where $K = \log_2 M$.

At each hop, the MFSK tones are separated in frequency by an integer multiple of the chip rate $R_c = R_s$, ensuring their orthogonality. The implication of this condition is that any transmitted symbol will not produce any crosstalk in the other $M - 1$ noncoherent matched filters constituting the MFSK detector of the receiver in Fig. 9.10*b*. By "crosstalk" we mean the spillover from one filter output into an adjacent one. The resulting performance of the slow FH/MFSK system is the same as that for the noncoherent detection of conventional (unhopped) MFSK signals in additive white Gaussian noise. Thus the interfering (jamming) signal has an effect on the FH/MFSK receiver, in terms of average probability of symbol error, equivalent to that of additive white Gaussian noise on a conventional noncoherent M-ary FSK receiver experiencing no interference.

Assuming that the jammer decides to spread its average power J over the entire frequency-hopped spectrum, the jammer's effect is equivalent to an AWGN with power spectral density $N_0/2$, where $N_0 = J/W_c$ and W_c is the FH bandwidth. The spread-spectrum system is thus characterized by the *symbol energy-to-noise spectral density ratio*:

$$\frac{E}{N_0} = \frac{P/J}{W_c/R_s} \tag{9.50}$$

where the ratio P/J is the reciprocal of the jamming margin. The other ratio is the processing gain of the slow FH/MFSK system, defined by

$$\begin{aligned} \text{PG} &= \frac{W_c}{R_s} \\ &= 2^k \end{aligned} \tag{9.51}$$

That is, the processing gain (expressed in decibels) is equal to $10 \log_{10} 2^k = 3k$, where k is the length of the PN segment employed to select a frequency hop.

This result assumes that the jammer spreads its power over the entire FH spectrum. However, if the jammer decides to concentrate on just a few of the hopped frequencies, then the processing gain realized by the receiver would be less than $3k$ decibels.

EXAMPLE 4

Figure 9.11*a* illustrates the variation of the frequency of a slow FH/MFSK signal with time for one complete period of the PN sequence. The period of the PN sequence is $2^4 - 1 = 15$. The FH/MFSK signal has the following parameters:

Number of bits per MFSK symbol	$K = 2$
Number of MFSK tones	$M = 2^K = 4$
Length of PN segment per hop	$k = 3$
Total number of frequency hops	$2^k = 8$

In this example, the carrier is hopped to a new frequency after transmitting two symbols or equivalently, four information bits. Figure 9.11*a* also includes the input binary data, and the PN sequence controlling the selection of FH carrier frequency. It is noteworthy that although there are eight distinct frequencies available for hopping, only three of them are utilized by the PN sequence.

Figure 9.11*b* shows the variation of the dehopped frequency with time. This variation is recognized to be the same as that of a conventional MFSK signal produced by the given input data.

Fast-frequency Hopping

A fast FH/MFSK system differs from a slow FH/MFSK system in that there are multiple hops per *M*-ary symbol. Hence, in a fast FH/MFSK system, each hop is a chip. In general, fast-frequency hopping is used to defeat a smart jammer's tactic that involves two functions: measurement of the spectral content of the

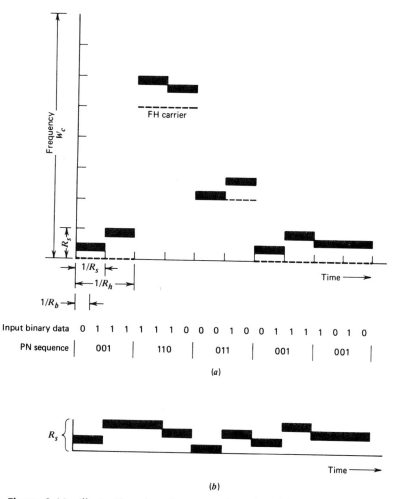

Figure 9.11 Illustrating slow-frequency hopping. (a) Frequency variation for one complete period of the PN sequence. (b) Variation of
the dehopped frequency with time.

transmitted signal, and retuning of the interfering signal to that portion of the
frequency band. Clearly, to overcome the jammer, the transmitted signal must
be hopped to a new carrier frequency *before* the jammer is able to complete the
processing of these two functions.

For data recovery at the receiver, noncoherent detection is used. However,
the detection procedure is quite different from that used in a slow FH/MFSK
receiver. In particular, two procedures may be considered:

1. For each FH/MFSK symbol, separate decisions are made on the K frequency-
   hop chips received, and a simple rule based on *majority vote* is used to make
   an estimate of the dehopped MFSK symbol.

2. For each FH/MFSK symbol, likelihood functions are computed as functions
   of the total signal received over K chips, and the larger one is selected.

A receiver based on the second procedure is optimum in the sense that it minimizes the average probability of symbol error for a given E_b/N_0.

EXAMPLE 5

Figure 9.12*a* illustrates the variation of the transmitted frequency of a fast FH/MFSK signal with time. The signal has the following parameters:

Number of bits per MFSK symbol $K = 2$

Number of MFSK tones $M = 2^K = 4$

Length of PN segment per hop $k = 3$

Total number of frequency hops $2^k = 8$

In this example, each MFSK symbol has the same number of bits and chips; that is, the chip rate R_c is the same as the bit rate R_b. After each chip, the carrier frequency of the transmitted MFSK signal is hopped to a different value, except for few occasions when the *k*-chip segment of the PN sequence repeats itself.

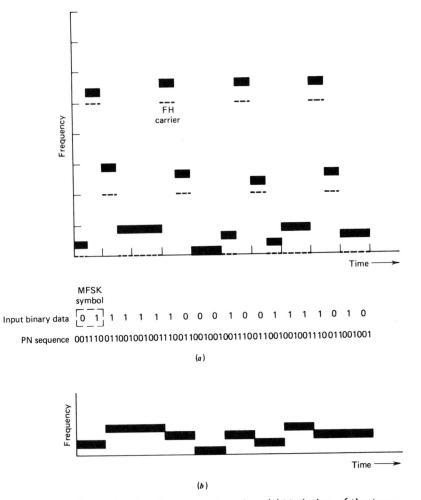

Input binary data 0 1 1 1 1 1 0 0 0 1 0 0 1 1 1 1 0 1 0

PN sequence 0011100110010010011100110010010011100110010010011100110010001

(a)

(b)

Figure 9.12 Illustrating fast-frequency hopping. (*a*) Variation of the transmitter frequency with time. (*b*) Variation of the dehopped frequency with time.

Figure 9.12*b* depicts the time variation of the frequency of the dehopped MFSK signal, which is the same as that in Example 4.

9.8   CODE-DIVISION MULTIPLEXING

The discussion of spread-spectrum modulation has thus far been in the context of a single user. We may support a multitude of users over a common communication channel in a spread-spectrum environment by using a technique known as *code-division multiplexing* (CDM), which relies on the use of PN codes with different generators for the individual users. Code-division multiplexing does not require the bandwidth allocation of frequency-division multiplexing (discussed in Chapter 3), nor the time synchronization needed in time-division multiplexing (discussed in Chapter 6).

However, an important phenomenon that arises in a code-division multiplex system is the partial cross-correlation of PN sequences, which manifests itself as a result of cross-talk between any two users sharing a common radio environment.

Partial Cross-Correlation of PN Sequences

Consider the arrangement described in Fig. 9.13 involving a pair of users. It is assumed that the carrier of user i has the same frequency and phase as that of user j, which corresponds to a worst-case situation. The issue to be addressed is the effect of a *bit-timing error* in the absence of noise. To be specific, we wish to evaluate the *interference* produced at the receiver output of user j due to user i acting alone. Let the message signal $b_i(t)$, assumed to be in polar form, represent the raw data sequence supplied by user i. Let the time difference τ denote the misalignment in the position of a bit in the message signal $b_i(t)$, measured with respect to the timing information built into the design of the system belonging to user j. Then, following through the scheme of Fig. 9.13, stage by stage, we find that the interference evaluated at the input of the decision device in the receiver of user j is given by (except for a scaling factor)

$$v_j(\tau)\Big|_{b_j(t)=0} = \int_0^{T_b} b_i(t-\tau)c_j(t)c_i(t-\tau)\,dt$$
$$= \pm\int_0^{T_b} c_j(t)c_i(t-\tau)\,dt \tag{9.52}$$

In the second line of Eq. (9.52), we have assumed that

$$b_i(t-\tau) = \pm 1, \quad \begin{cases} 0 \le t \le T_b \\ 0 \le \tau \le T_b \end{cases} \tag{9.53}$$

For obvious reasons, we may rewrite Eq. (9.52) in the following form:

$$v_j(\tau)\big|_{b_j(t)=0} = \pm T_b R_{ji}(\tau) \tag{9.54}$$

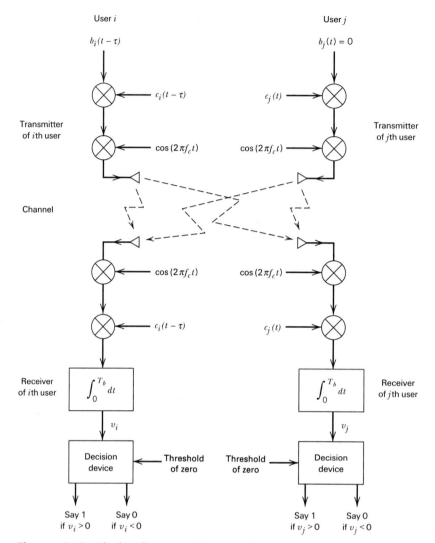

Figure 9.13 Block diagram illustrating the partial cross-correlation phenomenon.

where

$$R_{ji}(\tau) = \frac{1}{T_b} \int_0^{T_b} c_j(t)\, c_i(t - \tau)\; dt \tag{9.55}$$

The time-averaged quantity $R_{ji}(\tau)$ is called the *partial cross-correlation function* of the two PN sequences $c_j(t)$ and $c_i(t)$ that pertain to the two users.

For the interference between the users i and j in a code-division multiple access to be zero, we require that the partial cross-correlation function $R_{ji}(\tau)$ be zero for all τ. In general, however, maximum-length sequences do not have good cross-correlation properties, as illustrated in the following computer experiment.

COMPUTER EXPERIMENT

Consider two PN sequences of period $N = 63$. One sequence has the feedback taps [6, 1], and the other sequence has the feedback taps [6, 5, 2, 1], which are picked in accordance with Table 9.1. Both sequences have the same autocorrelation function $R_c(\tau)$, which is readily determined using Eq. (9.4); the result so obtained is shown plotted in Fig. 9.14a.

The calculation of the partial cross-correlation function $R_{ji}(\tau)$, however, becomes more complicated as we increase the period N. To perform this calculation, we may use computer simulation for varying τ inside the interval $(0, T_b)$. The results of this computation are shown in Fig. 9.14b for the maximum-length sequence [6, 5, 2, 1] and its mirror image [6, 5, 4, 1], and in Fig. 9.14c for the maximum-length sequences [6, 1] and [6, 5, 2, 1]. The latter two figures confirm the poor cross-correlation property of maximum-length sequences, compared to their autocorrelation function.

Gold Sequences[5]

The fact that, in general, maximum-length sequences do not exhibit a good partial cross-correlation makes them unsuitable for code-division multiplexing of multiple users. To overcome this difficulty, we may use a particular class of PN sequences called *Gold sequences* (*codes*). Figure 9.15 shows an example of a Gold code generator, which involves the modulo-2 addition of two maximum-length sequences of period $N = 2^m - 1$ for the case of $m = 6$. Provided that the maximum-length sequences are carefully chosen, the cross-correlation of Gold sequences is only three-valued. Table 9.4 presents a summary of the cross-correlation values of Gold sequences and their relative frequencies. From this

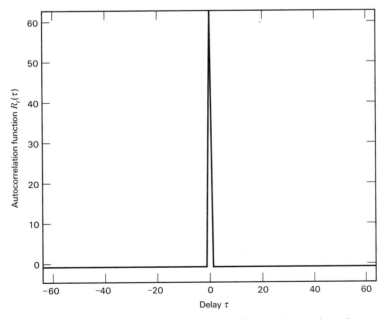

Figure 9.14a Autocorrelation function for maximum-length sequences [6, 1], [6, 5, 2, 1] and the latter's mirror image [6, 5, 4, 1].

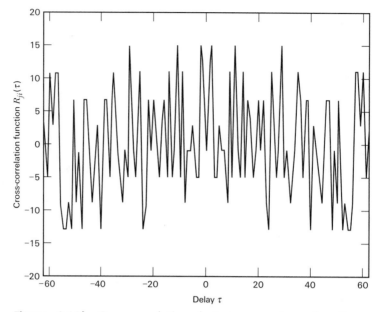

Figure 9.14*b* Cross-correlation between maximum-length sequence [6, 5, 2, 1] and its mirror image [6, 5, 4, 1].

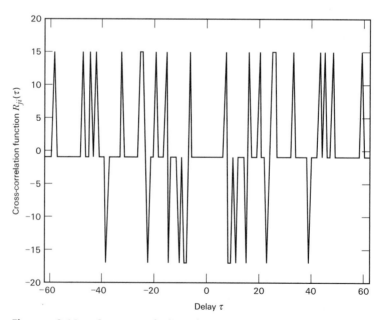

Figure 9.14*c* Cross-correlation between maximum-length sequences [6, 1] and [6, 5, 2, 1].

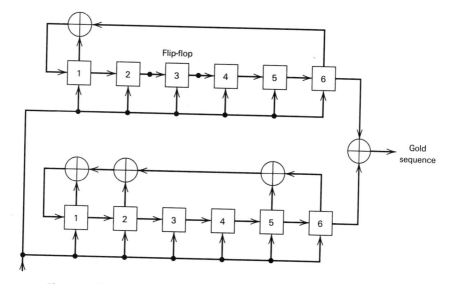

Figure 9.15   Gold code generator; length of sequence is $2^m - 1$.

table we see that when the shift-register length m is even and not divisible by 4, the Gold sequences are more often of very low cross-correlation, namely, $-1/N$, for 75 percent of the code words correlated. The good cross-correlation properties of Gold codes make them suitable candidates for code-division multiplexing systems.

9.9   SUMMARY AND DISCUSSION

Direct-sequence M-ary phase shift keying (DS/MPSK) and *frequency-hop M-ary frequency shift-keying* (FH/MFSK) represent two principal categories of spread-spectrum communications. Both of them rely on the use of a pseudo-noise (PN) sequence, which is applied differently in the two categories.

In a DS/MPSK system, the PN sequence makes the transmitted signal assume a noiselike appearance by spreading its spectrum over a broad range of frequencies simultaneously. For the phase-shift keying, we may use binary PSK (i.e.,

Table 9.4   Three-Level Cross-Correlation Properties of Gold Sequences

Shift-Register Length, m	Period (Code Length) N	Cross-Correlation	Frequency of Occurrence
m odd	$N = 2^m - 1$	$-1/N$	~0.50
		$-(2^{(m+1)/2} + 1)/N$	~0.25
		$(2^{(m+1)/2} - 1)/N$	~0.25
m even and not divisible by 4	$N = 2^m - 1$	$-1/N$	~0.75
		$-(2^{(m+2)/2} + 1)/N$	~0.125
		$(2^{(m+2)/2} - 1)/N$	~0.125

$M = 2$) with a single carrier, as discussed in Section 9.4. Alternatively, we may use QPSK (i.e., $M = 4$), in which case the data are transmitted using a pair of carriers in phase quadrature. The usual motivation for using QPSK is to provide for improved bandwidth efficiency; see Chapter 8. In a spread-spectrum system, bandwidth efficiency is usually not of prime concern. Rather, the use of QPSK is motivated by the fact that it is less sensitive to some types of interference (jamming).

In an FH/MFSK system, the PN sequence makes the carrier hop over a number of frequencies in a pseudo-random manner, with the result that the spectrum of the transmitted signal is spread in a sequential manner.

Naturally, the direct-sequence and frequency-hop spectrum-spreading techniques may be employed in a single system. The resulting system is referred to as a *hybrid* DS/FH *spread-spectrum system*. The reason for seeking a hybrid approach is that advantages of both the direct-sequence and frequency-hop spectrum-spreading techniques are realized in the same system.

A discussion of spread-spectrum communications would be incomplete without some reference to jammer waveforms. The jammers encountered in practice include the following types:

1. *The barrage noise jammer*, which consists of band-limited white Gaussian noise of high average power. The barrage noise jammer is a brute-force jammer that does not exploit any knowledge of the antijam communication system except for its spread bandwidth.

2. *The partial-band noise jammer*, which consists of noise whose total power is evenly spread over some frequency band that is a subset of the total spread bandwidth. Owing to the smaller bandwidth, the partial-band noise jammer is easier to generate than the barrage noise jammer.

3. *The pulsed noise jammer*, which involves transmitting wide-band noise of power

$$J_{peak} = \frac{J}{p}$$

for a fraction p of the time, and nothing for the remaining fraction $1 - p$ of the time. The average noise power equals J.

4. *The single-tone jammer*, which consists of a sinusoidal wave whose frequency lies inside the spread bandwidth; as such, it is the easiest of all jamming signals to generate.

5. *The multitone jammer*, which is the tone equivalent of the partial-band noise jammer.

In addition to these five, many other kinds of jamming waveforms occur in practice. In any event, there is no single jamming waveform that is worst for all spread-spectrum systems, and there is no single spread-spectrum system that is best against all possible jamming waveforms.

NOTES AND REFERENCES

1. The definition of spread-spectrum modulation presented in the Introduction is adapted from Pickholtz, Schilling, and Milstein (1982). This paper presents a tutorial review of the theory of spread-spectrum communications. For introductory papers on

the subject, see Viterbi (1979), and Cook and Marsh (1983). For books on the subject, see Dixon (1984), Holmes (1982), Ziemer and Peterson (1985, pp. 327–649), Cooper and McGillem (1986, pp. 269–411), and Simon, Omura, Scholtz, and Levitt (1985, Volumes I, II, and III). The three-volume book by Simon et al. is the most exhaustive treatment of spread-spectrum communications available in the open literature. The development of spread-spectrum communications dates back to about the mid-1950s. For a historical account of these techniques, see Scholtz (1982). This latter paper traces the origins of spread-spectrum communications back to the 1920s. Much of the historical material presented in this paper is reproduced in Chapter 2, Volume I, of the book by Simon et al. The book edited by Dixon (1976) provides a collection of some important papers on the subject of spread-spectrum modulation and related topics.

2. For further details on maximum-length sequences, see Golomb (1964, pp. 1–32), Simon, Omura, Scholtz, and Levitt (1985, pp. 283–295), and Peterson and Weldon (1972). The last reference includes an extensive list of polynomials for generating maximum-length sequences; see also Dixon (1984). For a tutorial paper on pseudo-noise sequences, see Sarwate and Pursley (1980).

3. Table 9.1 is extracted from the book by Dixon (1984, pp. 81–83), where feedback connections of maximum-length sequences are tabulated for shift-register length m extending up to 89.

4. For detailed discussion of the synchronization problem in spread-spectrum communications, see Ziemer and Peterson (1985, Chapters 9 and 10) and Simon et al. (1985, Volume III).

5. The original papers on Gold sequences are Gold (1967, 1968). A detailed discussion of Gold sequences is presented in Holmes (1982); Table 9.4 is taken from the latter reference.

PROBLEMS

Problem 9.1 A pseudo-noise (PN) sequence is generated using a feedback shift register of length $m = 4$. The chip rate is 10^7 chips per second. Find the following parameters:

(a) PN sequence length.
(b) Chip duration of the PN sequence.
(c) PN sequence period.

Problem 9.2 Figure P9.1 shows a four-stage feedback shift register. The initial state of the register is 1000. Find the output sequence of the shift register.

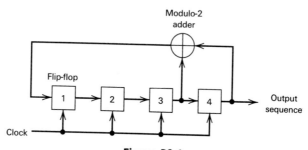

Figure P9.1

Problem 9.3 For the feedback shift register given in Problem 9.2, demonstrate the balance property and run property of a PN sequence. Also, calculate and plot the autocorrelation function of the PN sequence produced by this shift register.

Problem 9.4 Referring to Table 9.1, develop the maximum-length codes for the three feedback taps [6, 1], [6, 5, 2, 1], and [6, 5, 3, 2], whose period is $N = 63$.

Problem 9.5 Figure P9.2 shows the modular multitap version of the linear feedback shift-register shown in Fig. 9.4b. Demonstrate that the PN sequence generated by this scheme is exactly the same as that described in Table 9.2b.

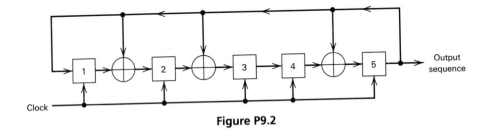

Figure P9.2

Problem 9.6 Show that the truth table given in Table 9.3 can be constructed by combining the following two steps:

(a) The message signal $b(t)$ and PN signal $c(t)$ are added modulo-2.
(b) Symbols 0 and 1 at the modulo-2 adder output are represented by phase shifts of 0 and 180 degrees, respectively.

Problem 9.7 A single-tone jammer

$$j(t) = \sqrt{2J}\cos(2\pi f_c t + \theta)$$

is applied to a DS/BPSK system. The N-dimensional transmitted signal $x(t)$ is described by Eq. (9.16). Find the $2N$ coordinates of the jammer $j(t)$.

Problem 9.8 The processing gain of a spread-spectrum system may be expressed as the ratio of the spread bandwidth of the transmitted signal to the despread bandwidth of the received signal. Justify this statement for the DS/BPSK system.

Problem 9.9 A direct-sequence spread binary phase-shift keying system uses a feedback shift register of length 19 for the generation of the PN sequence. Calculate the processing gain of the system.

Problem 9.10 In a DS/BPSK system, the feedback shift register used to generate the PN sequence has length $m = 19$. The system is required to have an average probability of symbol error due to externally generated interfering signals that does not exceed 10^{-5}. Calculate the following system parameters in decibels:

(a) Processing gain.
(b) Antijam margin.

Problem 9.11 A slow FH/MFSK system has the following parameters:

Number of bits per MFSK symbol = 4

Number of MFSK symbols per hop = 5

Calculate the processing gain of the system.

Problem 9.12 A fast FH/MFSK system has the following parameters:

Number of bits per MFSK symbol = 4

Number of hops per MFSK symbol = 4

Calculate the processing gain of the system.

COMPUTER EXPERIMENT

Continuing with the computer experiment described in Section 9.8, do the following:

(a) Compute the partial cross-correlation function of a PN sequence with feedback taps [5, 2] and its image sequence defined by the feedback taps [5, 3].

(b) Repeat the computation for the PN sequence with feedback taps [5, 2] and the PN sequence with feedback taps [5, 4, 2, 1].

(c) Repeat the computation for the PN sequence with feedback taps [5, 4, 3, 2] and the PN sequence with feedback taps [5, 4, 2, 1].

The feedback taps [5, 2], [5, 4, 3, 2], and [5, 4, 2, 1] are possible taps for a maximum-length sequence of period 31, in accordance with Table 9.1.

CHAPTER 10

Fundamental Limits in Information Theory

10.1 INTRODUCTION

As mentioned in Chapter 1 and reiterated along the way, the purpose of a communication system is to carry information-bearing baseband signals from one place to another over a communication channel. In preceding chapters of the book, we have described a variety of modulation schemes for accomplishing this objective. But what do we mean by the term "information"? To address this issue, we need to invoke *information theory*.[1] This broadly based mathematical discipline has made fundamental contributions, not only to communications, but also computer science, statistical physics, statistical inference, and probability and statistics.

In the context of communications, information theory deals with mathematical modeling and analysis of a communication system rather than with physical sources and physical channels. In particular, it provides answers to two fundamental questions (among others):

- What is the irreducible complexity below which a signal cannot be compressed?
- What is the ultimate transmission rate for reliable communication over a noisy channel?

The answers to these questions lie in the entropy of a source and the capacity of a channel, respectively. "Entropy" is defined in terms of the probabilistic behavior of a source of information; it is so named in deference to the parallel use of this concept in thermodynamics. "Capacity" is defined as the intrinsic ability of a channel to convey information; it is naturally related to the noise characteristics of the channel. A remarkable result that emerges from information theory is that if the entropy of the source is less than the capacity of the channel, then error-free communication over the channel can be achieved.

This chapter is devoted to a discussion of information theory. Mathematical definitions are introduced for the entropy of an information source and the capacity of a noisy channel, and theorems involving the use of these terms are formulated. These theorems define the *fundamental limits* on the performance of a communication system by specifying (1) the minimum number of bits per symbol required to fully represent the source, and (2) the maximum rate at which information transmission can take place over the channel. In light of this theoretical framework, we revisit some of the modulation schemes studied in preceding chapters of the book. Moreover, the theorems provide the mathematical foundations for signal compression (discussed in this chapter) and error-control coding schemes (discussed in the next chapter).

10.2   Uncertainty, Information, and Entropy

Suppose that a *probabilistic experiment* involves the observation of the output emitted by a discrete source during every unit of time (signaling interval). The source output is modeled as a discrete random variable, S, which takes on symbols from a fixed finite *alphabet*

$$\mathcal{S} = \{s_0, s_1, \ldots, s_{K-1}\} \tag{10.1}$$

with probabilities

$$P(S = s_k) = p_k, \quad k = 0, 1, \ldots, K - 1 \tag{10.2}$$

Of course, this set of probabilities must satisfy the condition

$$\sum_{k=0}^{K-1} p_k = 1 \tag{10.3}$$

We assume that the symbols emitted by the source during successive signaling intervals are statistically independent. A source having the properties just described is called a *discrete memoryless source*, memoryless in the sense that the symbol emitted at any time is independent of previous choices.

Can we find a measure of how much "information" is produced by such a source? To answer this question, we note that the idea of information is closely related to that of "uncertainty" or "suprise," as described next.

Consider the event $S = s_k$, describing the emission of symbol s_k by the source with probability p_k, as defined in Eq. (10.2). Clearly, if the probability $p_k = 1$ and $p_i = 0$ for all $i \neq k$, then there is no "surprise" and therefore no "information" when symbol s_k is emitted, since we know what the message from the source must be. If, on the other hand, the source symbols occur with different probabilities, and the probability p_k is low, then there is more "surprise" and therefore "information" when symbol s_k is emitted by the source than when symbol s_i, $i \neq k$, with higher probability is emitted. Thus, the words "uncertainty," "surprise," and "information" are all related. Before the event $S = s_k$ occurs, there is an amount of uncertainty. When the event $S = s_k$ occurs there is an amount of surprise. After the occurrence of the event $S = s_k$, there is gain in the amount of information, the essence of which may be viewed as the *resolution of uncertainty*. Moreover, the amount of information is related to the *inverse* of the probability of occurrence.

We define the amount of information gained after observing the event $S = s_k$, which occurs with probability p_k, as the *logarithmic* function[2]

$$I(s_k) = \log\left(\frac{1}{p_k}\right) \tag{10.4}$$

This definition exhibits the following important properties that are intuitively satisfying:

1.

$$I(s_k) = 0 \quad \text{for} \quad p_k = 1 \tag{10.5}$$

Obviously, if we are absolutely *certain* of the outcome of an event, even before it occurs, there is *no* information gained.

2.

$$I(s_k) \geq 0 \quad \text{for} \quad 0 \leq p_k \leq 1 \tag{10.6}$$

That is to say, the occurrence of an event $S = s_k$ either provides some or no information, but never brings about a *loss* of information.

3.

$$I(s_k) > I(s_i) \quad \text{for} \quad p_k < p_i \tag{10.7}$$

That is, the less probable an event is, the more information we gain when it occurs.

4. $I(s_k s_l) = I(s_k) + I(s_l)$ if s_k and s_l are statistically independent.

The base of the logarithm in Eq. (10.4) is quite arbitrary. Nevertheless, it is the standard practice today to use a logarithm to base 2. The resulting unit of information is called the *bit* (a contraction of *bi*nary digi*t*). We thus write

$$\begin{aligned} I(s_k) &= \log_2\left(\frac{1}{p_k}\right) \\ &= -\log_2 p_k \quad \text{for } k = 0, 1, \ldots, K-1 \end{aligned} \tag{10.8}$$

When $p_k = 1/2$, we have $I(s_k) = 1$ bit. Hence, *one bit is the amount of information that we gain when one of two possible and equally likely (i.e., equiprobable) events occurs.* Note that the information $I(s_k)$ is positive, since the logarithm of a number less than one, such as a probability, is negative.

The amount of information $I(s_k)$ produced by the source during an arbitrary signaling interval depends on the symbol s_k emitted by the source at that time. Indeed, $I(s_k)$ is a discrete random variable that takes on the values $I(s_0)$,

$I(s_1), \ldots, I(s_{K-1})$ with probabilities $p_0, p_1, \ldots, p_{K-1}$ respectively. The mean of $I(s_k)$ over the source alphabet $\mathcal{S}$ is given by

$$H(\mathcal{S}) = E[I(s_k)]$$

$$= \sum_{k=0}^{K-1} p_k I(s_k) \qquad (10.9)$$

$$= \sum_{k=0}^{K-1} p_k \log_2 \left(\frac{1}{p_k} \right)$$

The important quantity $H(\mathcal{S})$ is called the *entropy*[3] of a discrete memoryless source with source alphabet $\mathcal{S}$. It is a measure of the *average information content per source symbol*. Note that the entropy $H(\mathcal{S})$ depends only on the probabilities of the symbols in the alphabet $\mathcal{S}$ of the source. Thus, the symbol $\mathcal{S}$ in $H(\mathcal{S})$ is not an argument of a function but rather a label for a source.

Some Properties of Entropy

Consider a discrete memoryless source whose mathematical model is defined by Eqs. (10.1) and (10.2). The entropy $H(\mathcal{S})$ of such a source is bounded as follows:

$$0 \leq H(\mathcal{S}) \leq \log_2 K \qquad (10.10)$$

where K is the *radix* (number of symbols) of the alphabet $\mathcal{S}$ of the source. Furthermore, we may make two statements:

1. $H(\mathcal{S}) = 0$, if and only if the probability $p_k = 1$ for some k, and the remaining probabilities in the set are all zero; this lower bound on entropy corresponds to *no uncertainty*.
2. $H(\mathcal{S}) = \log_2 K$, if and only if $p_k = 1/K$ for all k (i.e., all the symbols in the alphabet $\mathcal{S}$ are *equiprobable*); this upper bound on entropy corresponds to *maximum uncertainty*.

To prove these properties of $H(\mathcal{S})$, we proceed as follows. First, since each probability p_k is less than or equal to unity, it follows that each term $p_k \log_2(1/p_k)$ in Eq. (10.9) is always nonnegative, and so $H(\mathcal{S}) \geq 0$. Next, we note that the product term $p_k \log_2(1/p_k)$ is zero if, and only if, $p_k = 0$ or 1. We therefore deduce that $H(\mathcal{S}) = 0$ if, and only if, $p_k = 0$ or 1, that is, $p_k = 1$ for some k and all the rest are zero.

This completes the proofs of the lower bound in Eq. (10.10) and statement (1).

To prove the upper bound in Eq. (10.10) and statement (2), we make use of a property of the natural logarithm:

$$\ln x \leq x - 1, \qquad x \geq 0 \qquad (10.11)$$

This inequality can be readily verified by plotting the functions $\ln x$ and $(x - 1)$ versus x, as shown in Fig. 10.1. Here we see that the line $y = x - 1$ always lies above the curve $y = \ln x$. The equality holds *only* at the point $x = 1$, where the line is tangential to the curve.

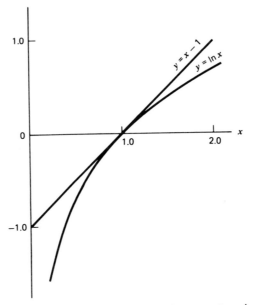

Figure 10.1   Plots of the functions $x - 1$ and $\ln x$.

To proceed with the proof, consider first any two probability distributions $\{p_0, p_1, \ldots, p_{K-1}\}$ and $\{q_0, q_1, \ldots, q_{K-1}\}$ on the alphabet $\mathcal{S} = \{s_0, s_1, \ldots, s_{K-1}\}$ of a discrete memoryless source. We may then write

$$\sum_{k=0}^{K-1} p_k \log_2\left(\frac{q_k}{p_k}\right) = \frac{1}{\ln 2} \sum_{k=0}^{K-1} p_k \ln\left(\frac{q_k}{p_k}\right)$$

Hence, using the inequality of Eq. (10.11), we get

$$\sum_{k=0}^{K-1} p_k \log_2\left(\frac{q_k}{p_k}\right) \leq \frac{1}{\ln 2} \sum_{k=0}^{K-1} p_k\left(\frac{q_k}{p_k} - 1\right)$$

$$\leq \frac{1}{\ln 2} \sum_{k=0}^{K-1} (q_k - p_k)$$

$$\leq \frac{1}{\ln 2}\left(\sum_{k=0}^{K-1} q_k - \sum_{k=0}^{K-1} p_k\right) = 0$$

We thus have the *fundamental inequality*

$$\sum_{k=0}^{K-1} p_k \log_2\left(\frac{q_k}{p_k}\right) \leq 0 \tag{10.12}$$

where the equality holds only if $q_k = p_k$ for all k.

Suppose we next put

$$q_k = \frac{1}{K}, \qquad k = 0, 1, \ldots, K - 1 \tag{10.13}$$

which corresponds to an alphabet $\mathcal{S}$ with *equiprobable* symbols. The entropy of a discrete memoryless source with such a characterization equals

$$\sum_{k=0}^{K-1} q_k \log_2\left(\frac{1}{q_k}\right) = \log_2 K \qquad (10.14)$$

Also, the use of Eq. (10.13) in Eq. (10.12) yields

$$\sum_{k=0}^{K-1} p_k \log_2\left(\frac{1}{p_k}\right) \leq \log_2 K$$

Equivalently, the entropy of a discrete memoryless source with an arbitrary probability distribution for the symbols of its alphabet $\mathcal{S}$ is bounded as

$$H(\mathcal{S}) \leq \log_2 K$$

Thus, $H(\mathcal{S})$ is always less than or equal to $\log_2 K$. The equality holds only if the symbols in the alphabet $\mathcal{S}$ are equiprobable, as in Eq. (10.13). This completes the proof of Eq. (10.10) and statements (1) and (2).

EXAMPLE 1 Entropy of Binary Memoryless Source

To illustrate the properties of $H(\mathcal{S})$, we consider a binary source for which symbol 0 occurs with probability p_0 and symbol 1 with probability $p_1 = 1 - p_0$. We assume that the source is memoryless so that successive symbols emitted by the source are statistically independent.

The entropy of such a source equals

$$\begin{aligned} H(\mathcal{S}) &= -p_0 \log_2 p_0 - p_1 \log_2 p_1 \\ &= -p_0 \log_2 p_0 - (1 - p_0)\log_2(1 - p_0) \text{ bits} \end{aligned} \qquad (10.15)$$

We note that

1. When $p_0 = 0$, the entropy $H(\mathcal{S}) = 0$; this follows from the fact that $x \log x \to 0$ as $x \to 0$.
2. When $p_0 = 1$, the entropy $H(\mathcal{S}) = 0$.
3. The entropy $H(\mathcal{S})$ attains its maximum value, $H_{max} = 1$ bit, when $p_1 = p_0 = 1/2$, that is, symbols 1 and 0 are equally probable.

The function of p_0 given on the right-hand side of Eq. (10.15) is frequently encountered in information-theoretic problems. It is therefore customary to assign a special symbol to this function. Specifically, we define

$$\mathcal{H}(p_0) = -p_0 \log_2 p_0 - (1 - p_0)\log_2(1 - p_0) \qquad (10.16)$$

We refer to $\mathcal{H}(p_0)$ as the *entropy function*. The distinction between Eq. (10.15) and Eq. (10.16) should be carefully noted. The $H(\mathcal{S})$ of Eq. (10.15) gives the entropy of a discrete memoryless source with source alphabet $\mathcal{S}$. The $\mathcal{H}(p_0)$ of Eq. (10.16), on the other hand, is a function of the prior probability p_0 defined on the interval $[0, 1]$. Accordingly, we may plot the entropy function $\mathcal{H}(p_0)$ versus p_0, defined on the interval $[0, 1]$, as in Fig. 10.2. The curve in Fig. 10.2 highlights the observations made under points 1, 2, and 3.

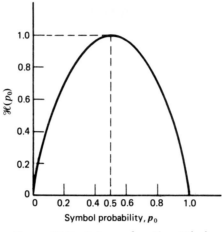

Figure 10.2 Entropy function $\mathcal{H}(p_0)$.

Extension of a Discrete Memoryless Source

In discussing information-theoretic concepts, we often find it useful to consider *blocks* rather than individual symbols, with each block consisting of n successive source symbols. We may view each such block as being produced by an *extended source* with a source alphabet $\mathcal{S}^n$ that has K^n *distinct* blocks, where K is the number of distinct symbols in the source alphabet $\mathcal{S}$ of the original source. In the case of a discrete memoryless source, the source symbols are statistically independent. Hence, the probability of a source symbol in $\mathcal{S}^n$ is equal to the product of the probabilities of the n source symbols in $\mathcal{S}$ constituting the particular source symbol in $\mathcal{S}^n$. We may thus intuitively expect that $H(\mathcal{S}^n)$, the entropy of the extended source, is equal to n times $H(\mathcal{S})$, the entropy of the original source. That is, we may write

$$H(\mathcal{S}^n) = nH(\mathcal{S}) \tag{10.17}$$

EXAMPLE 2

Consider a discrete memoryless source with source alphabet $\mathcal{S} = \{s_0, s_1, s_2\}$ with respective probabilities

$$p_0 = \tfrac{1}{4}$$

$$p_1 = \tfrac{1}{4}$$

$$p_2 = \tfrac{1}{2}$$

Hence, the use of Eq. (10.9) yields the entropy of the source as

$$H(\mathcal{S}) = p_0 \log_2\left(\frac{1}{p_0}\right) + p_1 \log_2\left(\frac{1}{p_1}\right) + p_2 \log_2\left(\frac{1}{p_2}\right)$$

$$= \frac{1}{4}\log_2(4) + \frac{1}{4}\log_2(4) + \frac{1}{2}\log_2(2)$$

$$= \frac{3}{2} \text{ bits}$$

Table 10.1 Alphabet Particulars of Second-Order Extension of a Discrete Memoryless Source

Symbols of $\mathcal{S}^2$	σ_0	σ_1	σ_2	σ_3	σ_4	σ_5	σ_6	σ_7	σ_8
Corresponding sequences of symbols of $\mathcal{S}$	$s_0 s_0$	$s_0 s_1$	$s_0 s_2$	$s_1 s_0$	$s_1 s_1$	$s_1 s_2$	$s_2 s_0$	$s_2 s_1$	$s_2 s_2$
Probability $p(\sigma_i)$, $i = 0, 1, \ldots, 8$	$\frac{1}{16}$	$\frac{1}{16}$	$\frac{1}{8}$	$\frac{1}{16}$	$\frac{1}{16}$	$\frac{1}{8}$	$\frac{1}{8}$	$\frac{1}{8}$	$\frac{1}{4}$

Consider next the second-order extension of the source. With the source alphabet $\mathcal{S}$ consisting of three symbols, it follows that the source alphabet $\mathcal{S}^2$ of the extended source has nine symbols. The first row of Table 10.1 presents the nine symbols of $\mathcal{S}^2$, denoted as $\sigma_0, \sigma_1, \ldots, \sigma_8$. The second row of the table presents the compositions of these nine symbols in terms of the corresponding sequences of source symbols s_0, s_1, and s_2, taken two at a time. The probabilities of the nine source symbols of the extended source are presented in the last row of the table. Accordingly, the use of Eq. (10.9) yields the entropy of the extended source as

$$H(\mathcal{S}^2) = \sum_{i=0}^{8} p(\sigma_i) \log_2 \frac{1}{p(\sigma_i)}$$

$$= \frac{1}{16} \log_2(16) + \frac{1}{16} \log_2(16) + \frac{1}{8} \log_2(8) + \frac{1}{16} \log_2(16)$$

$$+ \frac{1}{16} \log_2(16) + \frac{1}{8} \log_2(8) + \frac{1}{8} \log_2(8) + \frac{1}{8} \log_2(8) + \frac{1}{4} \log_2(4)$$

$$= 3 \text{ bits}$$

We thus see that $H(\mathcal{S}^2) = 2H(\mathcal{S})$ in accordance with Eq. (10.17).

10.3 SOURCE-CODING THEOREM

An important problem in communications is the *efficient* representation of data generated by a discrete source. The process by which this representation is accomplished is called *source encoding*. The device that performs the representation is called a *source encoder*. For the source encoder to be *efficient*, we require knowledge of the statistics of the source. In particular, if some source symbols are known to be more probable than others, then we may exploit this feature in the generation of a *source code* by assigning *short* code words to *frequent* source symbols, and *long* code words to *rare* source symbols. We refer to such a source code as a *variable-length code*. The *Morse code* is an example of a variable-length code. In the Morse code, the letters of the alphabet and numerals are encoded into streams of *marks* and *spaces*, denoted as dots "." and dashes "-", respectively. Since in the English language, the letter E occurs more frequently than the letter Q, for example, the Morse code encodes E into a single dot ".", the shortest code word in the code, and it encodes Q into "--.-", the longest code word in the code.

Our primary interest is in the development of an efficient source encoder that satisfies two functional requirements:

1. The code words produced by the encoder are in *binary* form.
2. The source code is *uniquely decodable*, so that the original source sequence can be reconstructed perfectly from the encoded binary sequence.

Consider then the scheme shown in Fig. 10.3, which depicts a discrete memoryless source whose output s_k is converted by the source encoder into a block of 0s and 1s, denoted by b_k. We assume that the source has an alphabet with K different symbols, and that the kth symbol s_k occurs with probability p_k, $k = 0$, $1, \ldots, K - 1$. Let the binary code word assigned to symbol s_k by the encoder have length l_k, measured in bits. We define the average code-word length, $\overline{L}$, of the source encoder as

$$\overline{L} = \sum_{k=0}^{K-1} p_k l_k \qquad (10.18)$$

In physical terms, the parameter $\overline{L}$ represents the *average number of bits per source symbol* used in the source encoding process. Let $L_{\min}$ denote the *minimum* possible value of $\overline{L}$. We then define the *coding efficiency* of the source encoder as

$$\eta = \frac{L_{\min}}{\overline{L}} \qquad (10.19)$$

With $\overline{L} \geq L_{\min}$, we clearly have $\eta \leq 1$. The source encoder is said to be *efficient* when η approaches unity.

But how is the minimum value $L_{\min}$ determined? The answer to this fundamental question is embodied in Shannon's first theorem: the *source-coding theorem*,[4] which may be stated as follows:

Given a discrete memoryless source of entropy $H(\mathcal{S})$, the average code-word length $\overline{L}$ for any distortionless source encoding is bounded as

$$\overline{L} \geq H(\mathcal{S}) \qquad (10.20)$$

Accordingly, the entropy $H(\mathcal{S})$ represents a *fundamental limit* on the average number of bits per source symbol necessary to represent a discrete memoryless source in that it can be made as small as, but no smaller than, the entropy $H(\mathcal{S})$. Thus with $L_{\min} = H(\mathcal{S})$, we may rewrite the efficiency of a source encoder in terms of the entropy $H(\mathcal{S})$ as

$$\eta = \frac{H(\mathcal{S})}{\overline{L}} \qquad (10.21)$$

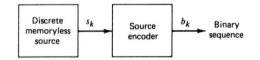

Figure 10.3   Source encoding.

10.4 DATA COMPACTION

A common characteristic of signals generated by physical sources is that, in their natural form, they contain a significant amount of information that is *redundant,* the transmission of which is therefore wasteful of primary communication resources. For *efficient* signal transmission, the *redundant information should be removed from the signal prior to transmission.* This operation, with *no* loss of information, is ordinarily performed on a signal in digital form, in which case we refer to it as *data compaction* or *lossless data compression.* The code resulting from such an operation provides a representation of the source output that is not only efficient in terms of the average number of bits per symbol but also exact in the sense that the original data can be reconstructed with no loss of information. The entropy of the source establishes the fundamental limit on the removal of redundancy from the data. Basically, data compaction is achieved by assigning short descriptions to the most frequent outcomes of the source output and longer descriptions to the less frequent ones.

In this section, we discuss some source-coding schemes for data compaction. We begin the discussion by describing a type of source code known as a prefix code, which is not only decodable but also offers the possibility of realizing an average code-word length that can be made arbitrarily close to the source entropy.

Prefix Coding

Consider a discrete memoryless source of source alphabet $\{s_0, s_1, \ldots, s_{K-1}\}$ and source statistics $\{p_0, p_1, \ldots, p_{K-1}\}$. For a source code representing the output of this source to be of practical use, the code has to be uniquely decodable. This restriction ensures that for each finite sequence of symbols emitted by the source, the corresponding sequence of code words is different from the sequence of code words corresponding to any other source sequence. We are specifically interested in a special class of codes satisfying a restriction known as the *prefix condition.* To define the prefix condition, let the code word assigned to source symbol s_k be denoted by $(m_{k_1}, m_{k_2}, \ldots, m_{k_n})$, where the individual elements $m_{k_1}, \ldots, m_{k_n}$ are 0s and 1s, and n is the code-word length. The initial part of the code word is represented by the elements $m_{k_1}, \ldots, m_{k_i}$ for some $i \leq n$. Any sequence made up of the initial part of the code word is called a *prefix* of the code word. A *prefix code* is defined as a code in which no code word is the prefix of any other code word.

To illustrate the meaning of a prefix code, consider the three source codes described in Table 10.2. Code I is not a prefix code since the bit 0, the code

Table 10.2   Illustrating the Definition of a Prefix Code

Source Symbol	Probability of Occurrence	Code I	Code II	Code III
s_0	0.5	0	0	0
s_1	0.25	1	10	01
s_2	0.125	00	110	011
s_3	0.125	11	111	0111

word for s_0, is a prefix of 00, the code word for s_2. Likewise, the bit 1, the code word for s_1, is a prefix of 11, the code word for s_3. Similarly, we may show that code III is not a prefix code, but code II is.

To decode a sequence of code words generated from a prefix source code, the *source decoder* simply starts at the beginning of the sequence and decodes one code word at a time. Specifically, it sets up what is equivalent to a *decision tree*, which is a graphical portrayal of the code words in the particular source code. For example, Fig. 10.4 depicts the decision tree corresponding to code II in Table 10.2. The tree has an *initial state* and four *terminal states* corresponding to source symbols s_0, s_1, s_2, and s_3. The decoder always starts at the initial state. The first received bit moves the decoder to the terminal state s_0 if it is 0, or else to a second decision point if it is 1. In the latter case, the second bit moves the decoder one step further down the tree, either to terminal state s_1 if it is 0, or else to a third decision point if it is 1, and so on. Once each terminal state emits its symbol, the decoder is reset to its initial state. Note also that each bit in the received encoded sequence is examined only once. For example, the encoded sequence 1011111000 . . . is readily decoded as the source sequence $s_1 s_3 s_2 s_0 s_0$. . . . The reader is invited to carry out this decoding.

A prefix code has the important property that it is *always* uniquely decodable. Indeed, if a prefix code has been constructed for a discrete memoryless source with source alphabet $\{s_0, s_1, \ldots, s_{K-1}\}$ and source statistics $\{p_0, p_1, \ldots, p_{K-1}\}$ and the code word for symbol s_k has length l_k, $k = 0, 1, \ldots, K - 1$, then the code-word lengths of the code satisfy a certain inequality known as the *Kraft–McMillan inequality*.[5] In mathematical terms, we may state that

$$\sum_{k=0}^{K-1} 2^{-l_k} \leq 1 \tag{10.22}$$

where the factor 2 refers to the radix (number of symbols) in the binary alphabet. Conversely, we may state that if the code-word lengths of a code for a discrete

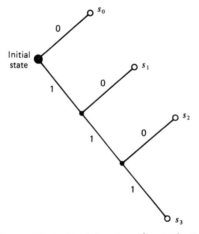

Figure 10.4 Decision tree for code II of Table 10.1.

memoryless source satisfy the Kraft–McMillan inequality, then a prefix code with these code-word lengths can be constructed.

Although all prefix codes are uniquely decodable, the converse is not true. For example, code III in Table 10.2 does not satisfy the prefix condition, and yet it is uniquely decodable since the bit 0 indicates the beginning of each code word in the code.

Prefix codes are distinguished from other uniquely decodable codes by the fact that the end of a code word is always recognizable. Hence, the decoding of a prefix can be accomplished as soon as the binary sequence representing a source symbol is fully received. For this reason, prefix codes are also referred to as *instantaneous codes*.

Given a discrete memoryless source of entropy $H(\mathscr{L})$, the average code-word length $\overline{L}$ of a prefix code is bounded as follows:

$$H(\mathscr{L}) \leq \overline{L} < H(\mathscr{L}) + 1 \qquad (10.23)$$

The left-hand bound of (10.23) is satisfied with equality under the condition that symbol s_k is emitted by the source with probability

$$p_k = 2^{-l_k} \qquad (10.24)$$

where l_k is the length of the code word assigned to source symbol s_k. We then have

$$\sum_{k=0}^{K-1} 2^{-l_k} \leq \sum_{k=0}^{K-1} p_k = 1$$

Under this condition, the Kraft–McMillan inequality of Eq. (10.22) implies that we can construct a prefix code, such that the length of the code word assigned to source symbol s_k is l_k. For such a code, the average code-word length is

$$\overline{L} = \sum_{k=0}^{K-1} \frac{l_k}{2^{l_k}} \qquad (10.25)$$

and the corresponding entropy of the source is

$$\begin{aligned} H(\mathscr{L}) &= \sum_{k=0}^{K-1} \left(\frac{1}{2^{l_k}} \right) \log_2(2^{l_k}) \\ &= \sum_{k=0}^{K-1} \frac{l_k}{2^{l_k}} \end{aligned} \qquad (10.26)$$

Hence, in this special (rather meretricious) case, we find from Eqs. (10.25) and (10.26) that the prefix code is *matched* to the source in that $\overline{L} = H(\mathscr{L})$.

But how do we match the prefix code to an arbitrary discrete memoryless source? The answer to this problem lies in the use of an *extended code*. Let $\overline{L}_n$ denote the average code-word length of the extended prefix code. For a uniquely decodable code, $\overline{L}_n$ is the smallest possible. From Eq. (10.23), we deduce that

$$H(\mathscr{L}^n) \leq \overline{L}_n < H(\mathscr{L}^n) + 1 \qquad (10.27)$$

Substituting Eq. (10.17) for an extended source in Eq. (10.27), we get

$$nH(\mathcal{L}) \le \overline{L}_n < nH(\mathcal{L}) + 1$$

or, equivalently,

$$H(\mathcal{L}) \le \frac{\overline{L}_n}{n} < H(\mathcal{L}) + \frac{1}{n} \tag{10.28}$$

In the limit, as n approaches infinity, the lower and upper bounds in Eq. (10.28) converge, as shown by

$$\lim_{n \to \infty} \frac{1}{n} \overline{L}_n = H(\mathcal{L}) \tag{10.29}$$

We may therefore state that by making the order n of an extended prefix source encoder large enough, we can make the code faithfully represent the discrete memoryless source $\mathcal{L}$ as closely as desired. In other words, the average code-word length of an extended prefix code can be made as small as the entropy of the source provided the extended code has a high enough order, in accordance with the source-coding theorem. However, the price we have to pay for decreasing the average code-word length is increased decoding complexity, which is brought about by the high order of the extended prefix code.

Huffman Coding

We next describe an important class of prefix codes known as Huffman codes. The basic idea behind *Huffman coding*[6] is to assign to each symbol of an alphabet a sequence of bits roughly equal in length to the amount of information conveyed by the symbol in question. The end result is a source code whose average code-word length approaches the fundamental limit set by the entropy of a discrete memoryless source, namely, $H(\mathcal{L})$. The essence of the *algorithm* used to synthesize the Huffman code is to replace the prescribed set of source statistics of a discrete memoryless source with a simpler one. This *reduction* process is continued in a step-by-step manner until we are left with a final set of only two source statistics (symbols), for which (0, 1) is an optimal code. Starting from this trivial code, we then work backward and thereby construct the Huffman code for the given source.

Specifically, the Huffman *encoding algorithm* proceeds as follows:

1. The source symbols are listed in order of decreasing probability. The two source symbols of lowest probability are assigned a 0 and a 1. This part of the step is referred to as a *splitting* stage.

2. These two source symbols are regarded as being *combined* into a new source symbol with probability equal to the sum of the two original probabilities. (The list of source symbols, and therefore source statistics, is thereby *reduced* in size by one.) The probability of the new symbol is placed in the list in accordance with its value.

3. The procedure is repeated until we are left with a final list of source statistics (symbols) of only two for which a 0 and a 1 are assigned.

The code for each (original) source symbol is found by working backward and tracing the sequence of 0s and 1s assigned to that symbol as well as its successors.

EXAMPLE 3

The five symbols of the alphabet of a discrete memoryless source and their probabilities are shown in the left-most two columns of Fig. 10.5a. Following through the Huffman algorithm, we reach the end of the computation in four steps, resulting in the *Huffman tree* shown in Fig. 10.5a. The code words of the Huffman code for the source are tabulated in Fig. 10.5b. The average code-word length is therefore

$$\overline{L} = 0.4(2) + 0.2(2) + 0.2(2) + 0.1(3) + 0.1(3)$$

$$= 2.2$$

The entropy of the specified discrete memoryless source is calculated as follows [see Eq. (10.9)]:

$$H(\mathscr{L}) = 0.4 \log_2\left(\frac{1}{0.4}\right) + 0.2 \log_2\left(\frac{1}{0.2}\right) + 0.2 \log_2\left(\frac{1}{0.2}\right)$$

$$+ 0.1 \log_2\left(\frac{1}{0.1}\right) + 0.1 \log_2\left(\frac{1}{0.1}\right)$$

$$= 0.52877 + 0.46439 + 0.46439 + 0.33219 + 0.33219$$

$$= 2.12193$$

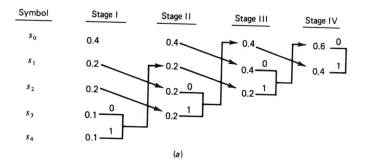

(a)

Symbol	Probability	Code word
s_0	0.4	00
s_1	0.2	10
s_2	0.2	11
s_3	0.1	010
s_4	0.1	011

(b)

Figure 10.5 (a) Example of the Huffman encoding algorithm. (b) Source code.

For the example at hand, we may make two observations:

1. The average code-word length $\bar{L}$ exceeds the entropy $H(\mathcal{L})$ by only 3.67 percent.
2. The average code-word length $\bar{L}$ does indeed satisfy Eq. (10.23).

It is noteworthy that the Huffman encoding process (i.e., the Huffman tree) is not unique. In particular, we may site two variations in the process that are responsible for the nonuniqueness of the Huffman code. First, at each splitting stage in the construction of a Huffman code, there is arbitrariness in the way a 0 and a 1 are assigned to the last two source symbols. Whichever way the assignments are made, however, the resulting differences are trivial. Second, ambiguity arises when the probability of a *combined* symbol (obtained by adding the last two probabilities pertinent to a particular step) is found to equal another probability in the list. We may proceed by placing the probability of the new symbol as *high* as possible, as in Example 3. Alternatively, we may place it as *low* as possible. (It is presumed that whichever way the placement is made, high or low, it is consistently adhered to throughout the encoding process.) But this time, noticeable differences arise in that the code words in the resulting source code can have different lengths. Nevertheless, the average code-word length remains the same.

As a measure of the variability in code-word lengths of a source code, we define the *variance* of the average code-word length $\bar{L}$ over the ensemble of source symbols as

$$\sigma^2 = \sum_{k=0}^{K-1} p_k (l_k - \bar{L})^2 \qquad (10.30)$$

where $p_0, p_1, \ldots, p_{K-1}$ are the source statistics, and l_k is the length of the code word assigned to source symbol s_k. It is usually found that when a combined symbol is moved as high as possible, the resulting Huffman code has a significantly smaller variance σ^2 than when it is moved as low as possible. On this basis, it is reasonable to choose the former Huffman code over the latter.

In Example 3, a combined symbol was moved as high as possible. In Example 4, presented next, a combined symbol is moved as low as possible. Thus, by comparing the results of these two examples, we are able to appreciate the subtle differences and similarities between the two Huffman codes.

EXAMPLE 4

Consider again the same discrete memoryless source described in Example 3. This time, however, we move the probability of a combined symbol as low as possible. The resulting Huffman tree is shown in Fig. 10.6a. Working backward through this tree and tracing through the various steps, we find that the code words of this second Huffman code for the source are as tabulated in Fig. 10.6b. The average code-word length for the second Huffman code is therefore

$$\bar{L} = 0.4(1) + 0.2(2) + 0.2(3) + 0.1(4) + 0.1(4)$$

$$= 2.2$$

which is exactly the same as that for the first Huffman code of Example 3. However, as remarked earlier, the individual code words of the second Huffman code

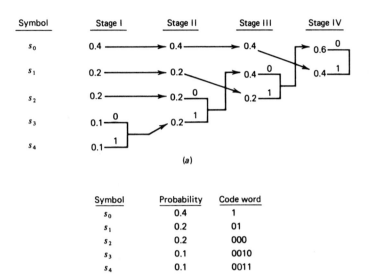

Figure 10.6 (a) Example illustrating nonuniqueness of the Huffman encoding algorithm. (b) Another source code.

have different lengths, compared to the corresponding ones of the first Huffman code.

The use of Eq. (10.30) yields the variance of the first Huffman code obtained in Example 3 as

$$\sigma_1^2 = 0.4(2 - 2.2)^2 + 0.2(2 - 2.2)^2 + 0.2(2 - 2.2)^2$$
$$+ 0.1(3 - 2.2)^2 + 0.1(3 - 2.2)^2$$
$$= 0.16$$

On the other hand, for the second Huffman code obtained in this example, we have from Eq. (10.30):

$$\sigma_2^2 = 0.4(1 - 2.2)^2 + 0.2(2 - 2.2)^2 + 0.2(3 - 2.2)^2$$
$$+ 0.1(4 - 2.2)^2 + 0.1(4 - 2.2)^2$$
$$= 1.36$$

These results confirm that the minimum variance Huffman code is obtained by moving the probability of a combined symbol as high as possible.

Lempel–Ziv Coding

A drawback of the Huffman code is that it requires knowledge of a probabilistic model of the source; unfortunately, in practice, source statistics are not always known *a priori*. Moreover, in modeling text we find that storage requirements prevent the Huffman code from capturing the higher-order relationships between words and phrases, thereby compromising the efficiency of the code. To overcome these practical limitations, we may use the *Lempel–Ziv algorithm*,[7] which is intrinsically *adaptive* and simpler to implement than Huffman coding.

Basically, encoding in the Lempel–Ziv algorithm is accomplished by *parsing the source data stream into segments that are the shortest subsequences not encountered previously.* To illustrate this simple and yet elegant idea, consider the example of an input binary sequence specified as follows:

$$000101110010100101\ldots$$

It is assumed that the binary symbols 0 and 1 are already stored in that order in the code book. We thus write

Subsequences stored: 0, 1

Data to be parsed: 00010111001010 0101 . . .

The encoding process begins at the left. With symbols 0 and 1 already stored, the *shortest subsequence* of the data stream encountered for the first time and not seen before is 00; and so we write

Subsequences stored: 0, 1, 00

Data to be parsed: 0101110010100101 . . .

The second, shortest subsequence not seen before is 01; accordingly, we go on to write

Subsequences stored: 0, 1, 00, 01

Data to be parsed: 01110010100101 . . .

The next, shortest subsequence not encountered previously is 011; hence, we write

Subsequences stored: 0, 1, 00, 01, 011

Data to be parsed: 10010100101 . . .

We continue in the manner described here until the given data stream has been completely parsed. Thus, for the example at hand, we get the *code book* of binary subsequences shown in the second row of Fig. 10.7.

The first row shown in this figure merely indicates the numerical positions of the individual subsequences in the code book. We now recognize that the first subsequence of the data stream, 00, is made up of the concatenation of the *first* code book entry, 0, with itself; it is therefore represented by the number 11. The second subsequence of the data stream, 01, consists of the *first* code book entry,

Numerical positions:	1	2	3	4	5	6	7	8	9
Subsequences:	0	1	00	01	011	10	010	100	101
Numerical representations:			11	12	42	21	41	61	62
Binary encoded blocks:			0010	0011	1001	0100	1000	1100	1101

Figure 10.7   Illustrating the encoding process performed by the Lempel–Ziv algorithm on the binary sequence 000101110010100101 . . . .

0, concatenated with the *second* code book entry, 1; it is therefore represented by the number 12. The remaining subsequences are treated in a similar fashion. The complete set of numerical representations for the various subsequences in the code book is shown in the third row of Fig. 10.7. As a further example illustrating the composition of this row, we note that the subsequence 010 consists of the concatenation of the subsequence 01 in position 4 and symbol 0 in position 1; hence, the numerical representation 41. The last row shown in Fig. 10.7 is the binary encoded representations of the different subsequences of the data stream.

The last symbol of each subsequence in the code book (i.e., the second row of Fig. 10.7) is an *innovation symbol*, which is so called in recognition of the fact that its appendage to a particular subsequence distinguishes it from all previous subsequences stored in the code book. Correspondingly, the last bit of each uniform block of bits in the binary encoded representation of the data stream (i.e., the fourth row in Fig. 10.7) represents the innovation symbol for the particular subsequence under consideration. The remaining bits provide the equivalent binary representation of the "pointer" to the *root subsequence* that matches the one in question except for the innovation symbol.

The decoder is just as simple as the encoder. Specifically, it uses the pointer to identify the root subsequence and then appends the innovation symbol. Consider, for example, the binary encoded block 1101 in position 9. The last bit, 1, is the innovation symbol. The remaining bits, 110, point to the root subsequence 10 in position 6. Hence, the block 1101 is decoded into 101, which is correct.

From the example described here, we note that, in contrast to Huffman coding, the Lempel–Ziv algorithm uses fixed-length codes to represent a variable number of source symbols; this feature makes the Lempel–Ziv code suitable for synchronous transmission. In practice, fixed blocks of 12 bits long are used, which implies a code book of 4096 entries.

For a long time, Huffman coding was unchallenged as the algorithm of choice for data compaction. In recent years, however, the Lempel–Ziv algorithm has taken over almost completely from the Huffman algorithm. The Lempel–Ziv algorithm is now the standard algorithm for file compression. When it is applied to ordinary English text, the Lempel–Ziv algorithm achieves a compaction of approximately 55 percent. This is to be contrasted with a compaction of approximately 43 percent achieved with Huffman coding. The reason for this behavior is that, as mentioned previously, Huffman coding does not take advantage of the intercharacter redundancies of the language. On the other hand, the Lempel–Ziv algorithm is able to do the best possible compaction of text (within certain limits) by working effectively at higher levels.

10.5 DISCRETE MEMORYLESS CHANNELS

Up to this point in the chapter, we have been preoccupied with discrete memoryless sources responsible for information generation. We next consider the issue of information transmission, with particular emphasis on reliability. We start the discussion by considering a discrete memoryless channel, the counterpart of a discrete memoryless source.

A *discrete memoryless channel* is a statistical model with an input X and an output Y that is a *noisy* version of X; both X and Y are random variables. Every unit of time, the channel accepts an input symbol X selected from an alphabet $\mathscr{X}$ and, in response, it emits an output symbol Y from an alphabet $\mathscr{Y}$. The channel is said to be "discrete" when both of the alphabets $\mathscr{X}$ and $\mathscr{Y}$ have *finite* sizes. It is said to be "memoryless" when the current output symbol depends *only* on the current input symbol and *not* any of the previous ones.

Figure 10.8 depicts a view of a discrete memoryless channel. The channel is described in terms of an *input alphabet*

$$\mathscr{X} = \{x_0, x_1, \ldots, x_{J-1}\}, \tag{10.31}$$

an *output alphabet*,

$$\mathscr{Y} = \{y_0, y_1, \ldots, y_{K-1}\}, \tag{10.32}$$

and a set of *transition probabilities*

$$p(y_k|x_j) = P(Y = y_k|X = x_j) \qquad \text{for all } j \text{ and } k \tag{10.33}$$

Naturally, we have

$$0 \le p(y_k|x_j) \le 1 \qquad \text{for all } j \text{ and } k \tag{10.34}$$

Also, the input alphabet $\mathscr{X}$ and output alphabet $\mathscr{Y}$ need not have the same size. For example, in channel coding, the size K of the output alphabet $\mathscr{Y}$ may be larger than the size J of the input alphabet $\mathscr{X}$; thus, $K \ge J$. On the other hand, we may have a situation in which the channel emits the same symbol when either one of two input symbols is sent, in which case we have $K \le J$.

A convenient way of describing a discrete memoryless channel is to arrange the various transition probabilities of the channel in the form of a matrix as follows:

$$\mathbf{P} = \begin{bmatrix} p(y_0|x_0) & p(y_1|x_0) & \cdots & p(y_{K-1}|x_0) \\ p(y_0|x_1) & p(y_1|x_1) & \cdots & p(y_{K-1}|x_1) \\ \vdots & \vdots & & \vdots \\ p(y_0|x_{J-1}) & p(y_1|x_{J-1}) & \cdots & p(y_{K-1}|x_{J-1}) \end{bmatrix} \tag{10.35}$$

The J-by-K matrix $\mathbf{P}$ is called the *channel matrix*. Note that each *row* of the channel matrix $\mathbf{P}$ corresponds to a *fixed channel input*, whereas each column of the matrix

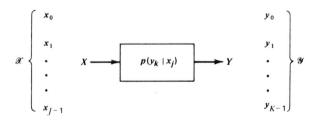

Figure 10.8.   Discrete memoryless channel.

corresponds to a *fixed channel output*. Note also that a fundamental property of the channel matrix **P**, as defined here, is that the sum of the elements along any row of the matrix is always equal to one; that is,

$$\sum_{k=0}^{K-1} p(y_k|x_j) = 1 \qquad \text{for all } j \tag{10.36}$$

Suppose now that the inputs to a discrete memoryless channel are selected according to the *probability distribution* $\{p(x_j), j = 0, 1, \ldots, J - 1\}$. In other words, the event that the channel input $X = x_j$ occurs with probability

$$p(x_j) = P(X = x_j) \qquad \text{for } j = 0, 1, \ldots, J - 1 \tag{10.37}$$

Having specified the random variable X denoting the channel input, we may now specify the second random variable Y denoting the channel output. The *joint probability distribution* of the random variables X and Y is given by

$$\begin{aligned} p(x_j, y_k) &= P(X = x_j, Y = y_k) \\ &= P(Y = y_k|X = x_j)P(X = x_j) \\ &= p(y_k|x_j)p(x_j) \end{aligned} \tag{10.38}$$

The *marginal probability distribution* of the output random variable Y is obtained by averaging out the dependence of $p(x_j, y_k)$ on x_j, as shown by

$$\begin{aligned} p(y_k) &= P(Y = y_k) \\ &= \sum_{j=0}^{J-1} P(Y = y_k|X = x_j)P(X = x_j) \\ &= \sum_{j=0}^{J-1} p(y_k|x_j)p(x_j) \qquad \text{for } k = 0, 1, \ldots, K - 1 \end{aligned} \tag{10.39}$$

The probabilities $p(x_j)$ for $j = 0, 1, \ldots, J - 1$, are known as the *a priori probabilities* of the various input symbols. Equation (10.39) states that if we are given the input *a priori* probabilities $p(x_j)$ and the channel matrix [i.e., the matrix of transition probabilities $p(y_k|x_j)$], then we may calculate the probabilities of the various output symbols, the $p(y_k)$.

EXAMPLE 5 Binary Symmetric Channel

The *binary symmetric channel* is of great theoretical interest and practical importance. It is a special case of the discrete memoryless channel with $J = K = 2$. The channel has two input symbols ($x_0 = 0$, $x_1 = 1$) and two output symbols ($y_0 = 0$, $y_1 = 1$). The channel is symmetric because the probability of receiving a 1 if a 0 is sent is the same as the probability of receiving a 0 if a 1 is sent. This conditional probability of error is denoted by p. The *transition probability diagram* of a binary symmetric channel is as shown in Fig. 10.9.

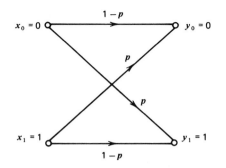

Figure 10.9 Transition probability diagram of binary symmetric channel.

10.6 MUTUAL INFORMATION

Given that we think of the channel output Y (selected from alphabet $\mathcal{Y}$) as a noisy version of the channel input X (selected from alphabet $\mathcal{X}$), and that the entropy $H(\mathcal{X})$ is a measure of the prior uncertainty about X, how can we measure the uncertainty about X after observing Y? To answer this question, we extend the ideas developed in Section 10.2 by defining the *conditional entropy* of X selected from alphabet $\mathcal{X}$, given that $Y = y_k$. Specifically, we write

$$H(\mathcal{X}|Y = y_k) = \sum_{j=0}^{J-1} p(x_j|y_k) \log_2\left[\frac{1}{p(x_j|y_k)}\right] \tag{10.40}$$

This quantity is itself a random variable that takes on the values $H(\mathcal{X}|Y = y_0), \ldots, H(\mathcal{X}|Y = y_{K-1})$ with probabilities $p(y_0), \ldots, p(y_{K-1})$, respectively. The mean of entropy $H(\mathcal{X}|Y = y_k)$ over the output alphabet $\mathcal{Y}$ is therefore given by

$$\begin{aligned}
H(\mathcal{X}|\mathcal{Y}) &= \sum_{k=0}^{K-1} H(\mathcal{X}|Y = y_k)p(y_k) \\
&= \sum_{k=0}^{K-1}\sum_{j=0}^{J-1} p(x_j|y_k)p(y_k) \log_2\left[\frac{1}{p(x_j|y_k)}\right] \\
&= \sum_{k=0}^{K-1}\sum_{j=0}^{J-1} p(x_j,y_k) \log_2\left[\frac{1}{p(x_j|y_k)}\right]
\end{aligned} \tag{10.41}$$

where, in the last line, we have made use of the relation

$$p(x_j,y_k) = p(x_j|y_k)p(y_k) \tag{10.42}$$

The quantity $H(\mathcal{X}|\mathcal{Y})$ is called a *conditional entropy*. It represents *the amount of uncertainty remaining about the channel input after the channel output has been observed.*

Since the entropy $H(\mathcal{X})$ represents our uncertainty about the channel input *before* observing the channel output, and the conditional entropy $H(\mathcal{X}|\mathcal{Y})$ represents our uncertainty about the channel input *after* observing the channel output, it follows that the difference $H(\mathcal{X}) - H(\mathcal{X}|\mathcal{Y})$ must represent our uncertainty about the channel input that is *resolved* by observing the channel output.

This important quantity is called the *mutual information* of the channel. Denoting the mutual information by $I(\mathcal{X};\mathcal{Y})$, we may thus write

$$I(\mathcal{X};\mathcal{Y}) = H(\mathcal{X}) - H(\mathcal{X}|\mathcal{Y}) \tag{10.43}$$

Similarly, we may write

$$I(\mathcal{Y};\mathcal{X}) = H(\mathcal{Y}) - H(\mathcal{Y}|\mathcal{X}) \tag{10.44}$$

where $H(\mathcal{Y})$ is the entropy of the channel output and $H(\mathcal{Y}|\mathcal{X})$ is the conditional entropy of the channel output given the channel input.

Properties of Mutual Information

The mutual information $I(\mathcal{X};\mathcal{Y})$ has the following important properties.

PROPERTY 1

The mutual information of a channel is symmetric; that is

$$I(\mathcal{X};\mathcal{Y}) = I(\mathcal{Y};\mathcal{X}) \tag{10.45}$$

where the mutual information $I(\mathcal{X};\mathcal{Y})$ is a measure of the uncertainty about the channel input that is resolved by *observing* the channel output, and the mutual information $I(\mathcal{Y};\mathcal{X})$ is a measure of the uncertainty about the channel output that is resolved by *sending* the channel input.

To prove this property, we first use the formula for entropy and then use Eqs. (10.36) and (10.38), in that order, to express $H(\mathcal{X})$ as

$$
\begin{aligned}
H(\mathcal{X}) &= \sum_{j=0}^{J-1} p(x_j) \log_2\left[\frac{1}{p(x_j)}\right] \\
&= \sum_{j=0}^{J-1} p(x_j) \log_2\left[\frac{1}{p(x_j)}\right] \sum_{k=0}^{K-1} p(y_k|x_j) \\
&= \sum_{j=0}^{J-1}\sum_{k=0}^{K-1} p(y_k|x_j)p(x_j) \log_2\left[\frac{1}{p(x_j)}\right] \\
&= \sum_{j=0}^{J-1}\sum_{k=0}^{K-1} p(x_j,y_k) \log_2\left[\frac{1}{p(x_j)}\right]
\end{aligned}
\tag{10.46}
$$

Hence, substituting Eqs. (10.41) and (10.46) into Eq. (10.43) and then combining terms we get

$$I(\mathcal{X};\mathcal{Y}) = \sum_{j=0}^{J-1}\sum_{k=0}^{K-1} p(x_j,y_k) \log_2\left[\frac{p(x_j|y_k)}{p(x_j)}\right] \tag{10.47}$$

From *Bayes' rule* for conditional probabilities, we have [see Eqs. (10.38) and (10.42)]

$$\frac{p(x_j|y_k)}{p(x_j)} = \frac{p(y_k|x_j)}{p(y_k)} \tag{10.48}$$

Hence, substituting Eq. (10.48) into Eq. (10.47), and interchanging the order of summation, we may write

$$I(\mathcal{X};\mathcal{Y}) = \sum_{k=0}^{K-1} \sum_{j=0}^{J-1} p(x_j,y_k) \log_2\left[\frac{p(y_k|x_j)}{p(y_k)}\right]$$

$$= I(\mathcal{Y};\mathcal{X}) \tag{10.49}$$

which is the desired result.

PROPERTY 2

The mutual information is always nonnegative; that is

$$I(\mathcal{X};\mathcal{Y}) \geq 0 \tag{10.50}$$

To prove this property, we first note from Eq. (10.42) that

$$p(x_j|y_k) = \frac{p(x_j,y_k)}{p(y_k)} \tag{10.51}$$

Hence, a substituting Eq. (10.51) into Eq. (10.47), we may express the mutual information of the channel as

$$I(\mathcal{X};\mathcal{Y}) = \sum_{j=0}^{J-1} \sum_{k=0}^{K-1} p(x_j,y_k) \log_2\left(\frac{p(x_j,y_k)}{p(x_j)p(y_k)}\right) \tag{10.52}$$

Next, a direct application of the fundamental inequality [defined by Eq. (10.12)] yields the desired result

$$I(\mathcal{X};\mathcal{Y}) \geq 0$$

with equality if, and only if,

$$p(x_j,y_k) = p(x_j)p(y_k) \qquad \text{for all } j \text{ and } k \tag{10.53}$$

Property 2 states that *we cannot lose information, on the average, by observing the output of a channel.* Moreover, the average mutual information is zero if, and only if, the input and output symbols of the channel are statistically independent, as in Eq. (10.53).

PROPERTY 3

The mutual information of a channel is related to the joint entropy of the channel input and channel output by

$$I(\mathcal{X};\mathcal{Y}) = H(\mathcal{X}) + H(\mathcal{Y}) - H(\mathcal{X},\mathcal{Y}) \tag{10.54}$$

where the joint entropy $H(\mathcal{X},\mathcal{Y})$ is defined by

$$H(\mathcal{X};\mathcal{Y}) = \sum_{j=0}^{J-1} \sum_{k=0}^{K-1} p(x_j,y_k) \log_2\left(\frac{1}{p(x_j,y_k)}\right) \tag{10.55}$$

To prove Eq. (10.54), we first rewrite the definition for the joint entropy $H(\mathcal{X},\mathcal{Y})$ as

$$H(\mathcal{X};\mathcal{Y}) = \sum_{j=0}^{J-1} \sum_{k=0}^{K-1} p(x_j,y_k) \log_2\left[\frac{p(x_j)p(y_k)}{p(x_j,y_k)}\right]$$
$$+ \sum_{j=0}^{J-1} \sum_{k=0}^{K-1} p(x_j,y_k) \log_2\left[\frac{1}{p(x_j)p(y_k)}\right] \quad (10.56)$$

The first double summation term on the right-hand side of Eq. (10.56) is recognized as the negative of the mutual information of the channel $I(\mathcal{X};\mathcal{Y})$, previously given in Eq. (10.52). As for the second summation term, we manipulate it as follows:

$$\sum_{j=0}^{J-1} \sum_{k=0}^{K-1} p(x_j,y_k) \log_2\left[\frac{1}{p(x_j)p(y_k)}\right] = \sum_{j=0}^{J-1} \log_2\left[\frac{1}{p(x_j)}\right] \sum_{k=0}^{K-1} p(x_j,y_k)$$
$$+ \sum_{k=0}^{K-1} \log_2\left[\frac{1}{p(y_k)}\right] \sum_{j=0}^{J-1} p(x_j,y_k)$$
$$= \sum_{j=0}^{J-1} p(x_j) \log_2\left[\frac{1}{p(x_j)}\right] \quad (10.57)$$
$$+ \sum_{k=0}^{K-1} p(y_k) \log_2\left[\frac{1}{p(y_k)}\right]$$
$$= H(\mathcal{X}) + H(\mathcal{Y})$$

Accordingly, using Eqs. (10.52) and (10.57) in Eq. (10.56), we get the result

$$H(\mathcal{X},\mathcal{Y}) = -I(\mathcal{X};\mathcal{Y}) + H(\mathcal{X}) + H(\mathcal{Y}) \quad (10.58)$$

Rearranging terms in this equation, we get the result given in Eq. (10.54), thereby confirming Property 3.

We conclude our discussion of the mutual information of a channel by providing a diagramatic interpretation of Eqs. (10.43), (10.44), and (10.54). The interpretation is given in Fig. 10.10. The entropy of channel input X is represented by the circle on the left. The entropy of channel output Y is represented by the circle on the right. The mutual information of the channel is represented by the overlap between these two circles.

10.7 CHANNEL CAPACITY

Consider a discrete memoryless channel with input alphabet $\mathcal{X}$, output alphabet $\mathcal{Y}$, and transition probabilities $p(y_k|x_j)$. The mutual information of the channel is defined by the first line of Eq. (10.49), which is reproduced here for convenience:

$$I(\mathcal{X};\mathcal{Y}) = \sum_{j=0}^{J-1} \sum_{k=0}^{K-1} p(x_j,y_k) \log_2\left[\frac{p(y_k|x_j)}{p(y_k)}\right]$$

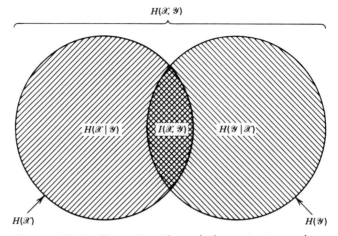

Figure 10.10 Illustrating the relations among various channel parameters.

Here we note that [see Eq. (10.38)]

$$p(x_j, y_k) = p(y_k|x_j)p(x_j)$$

Also, from Eq. (10.39), we have

$$p(y_k) = \sum_{j=0}^{J-1} p(y_k|x_j)p(x_j)$$

From these three equations we see that it is necessary for us to know the input probability distribution $\{p(x_j)|j = 0, 1, \ldots, J - 1\}$ so that we may calculate the mutual information $I(\mathcal{X};\mathcal{Y})$. The mutual information of a channel therefore depends not only on the channel but also on the way in which the channel is used.

The input probability distribution $\{p(x_j)\}$ is obviously independent of the channel. We can then maximize the average mutual information $I(\mathcal{X};\mathcal{Y})$ of the channel with respect to $\{p(x_j)\}$. Hence, *we define the channel capacity of a discrete memoryless channel as the maximum average mutual information $I(\mathcal{X};\mathcal{Y})$ in any single use of the channel (i.e., signaling interval), where the maximization is over all possible input probability distributions $\{p(x_j)\}$ on $\mathcal{X}$.* The channel capacity is commonly denoted by C. We thus write

$$C = \max_{\{p(x_j)\}} I(\mathcal{X};\mathcal{Y}) \tag{10.59}$$

The channel capacity C is measured in *bits per channel use.*

Note that the channel capacity C is a function only of the transition probabilities $p(y_k|x_j)$, which define the channel. The calculation of C involves maximization of the average mutual information $I(\mathcal{X};\mathcal{Y})$ over J variables [i.e., the input probabilities $p(x_0), \ldots, p(x_{J-1})$] subject to two constraints:

$$p(x_j) \geq 0 \text{ for all } j$$

and

$$\sum_{j=0}^{J-1} p(x_j) = 1$$

In general, the variational problem of finding the channel capacity C is a challenging task.

EXAMPLE 6 Binary Symmetric Channel (Revisited)

Consider again the *binary symmetric channel,* which is described by the *transition probability diagram* of Fig. 10.9. This diagram is uniquely defined by the conditional probability of error p.

The entropy $H(X)$ is maximized when the channel input probability $p(x_0) = p(x_1) = 1/2$, where x_0 and x_1 are each 0 or 1. The mutual information $I(\mathscr{X};\mathscr{Y})$ is similarly maximized, so that we may write

$$C = I(\mathscr{X};\mathscr{Y})|_{p(x_0) = p(x_1) = \frac{1}{2}}$$

From Fig. 10.9, we have

$$p(y_0|x_1) = p(y_1|x_0) = p$$

and

$$p(y_0|x_0) = p(y_1|x_1) = 1 - p$$

Therefore, substituting these channel transition probabilities into Eq. (10.49) with $J = K = 2$, and then setting the input probability $p(x_0) = p(x_1)$ in accordance with Eq. (10.59), we find that the capacity of the binary symmetric channel is

$$C = 1 + p \log_2 p + (1 - p)\log_2(1 - p) \tag{10.60}$$

Using the definition of the entropy function given in Eq. (10.16), we may reduce Eq. (10.60) to

$$C = 1 - H(p)$$

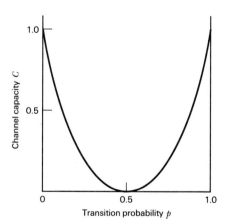

Figure 10.11 Variation of channel capacity of a binary symmetric channel with transition probability p.

The channel capacity C varies with the probability of error (transition probability) p as shown in Fig. 10.11, which is symmetric about $p = 1/2$. Comparing the curve in this figure with that in Fig. 10.2, we may make the following observations:

1. When the channel is *noise free*, permitting us to set $p = 0$, the channel capacity C attains its maximum value of one bit per channel use, which is exactly the information in each channel input. At this value of p, the entropy function $H(p)$ attains its minimum value of zero.
2. When the conditional probability of error $p = 1/2$ due to noise, the channel capacity C attains its minimum value of zero, whereas the entropy function $H(p)$ attains its maximum value of unity; in such a case the channel is said to be *useless*.

10.8 CHANNEL CODING THEOREM

The inevitable presence of *noise* in a channel causes discrepancies (errors) between the output and input data sequences of a digital communication system. For a relatively noisy channel, the probability of error may have a value as high as 10^{-2}, which means that (on the average) 99 out of 100 transmitted bits are received correctly. For many applications, this *level of reliability* is found to be far from adequate. Indeed, a probability of error equal to 10^{-6} or even lower is often a necessary requirement. To achieve such a high level of performance, we may have to resort to the use of channel coding.

The design goal of channel coding is to increase the resistance of a digital communication system to channel noise. Specifically, *channel coding* consists of *mapping* the incoming data sequence into a channel input sequence, and *inverse mapping* the channel output sequence into an output data sequence in such a way that the overall effect of channel noise on the system is minimized. The first mapping operation is performed in the transmitter by a *channel encoder*, whereas the inverse mapping operation is performed in the receiver by a *channel decoder*, as shown in the block diagram of Fig. 10.12; to simplify the exposition, we have not included source encoding (before channel encoding) and source decoding (after channel decoding) in Fig. 10.12.

The channel encoder and channel decoder in Fig. 10.12 are both under the designer's control and should be designed to optimize the overall effectiveness of the communication system. The approach taken is to introduce *redundancy* in the channel encoder so as to reconstruct the original source sequence as accurately as possible. Thus, in a rather loose sense, we may view channel coding as the *dual* of source coding in that the former introduces controlled redundancy to improve reliability, whereas the latter reduces redundancy to improve efficiency.

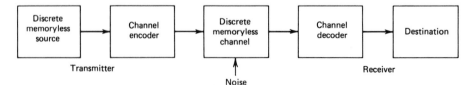

Figure 10.12 Block diagram of digital communication system.

The subject of channel coding is treated in detail in Chapter 11. For the purpose of our present discussion, it suffices to confine our attention to *block codes*. In this class of codes, the message sequence is subdivided into sequential blocks each k bits long, and each k-bit block is *mapped* into an n-bit block, where $n > k$. The number of redundant bits added by the encoder to each transmitted block is $n - k$ bits. The ratio k/n is called the *code rate*. Using r to denote the code rate, we may thus write

$$r = \frac{k}{n}$$

where, of course, r is less than unity.

The accurate reconstruction of the original source sequence at the destination requires that the *average probability of symbol error* be arbitrarily low. This raises the following important question: Does there exist a sophisticated channel coding scheme such that the probability that a message bit will be in error is less than any positive number ε (i.e., as small as we want it), and yet the channel coding scheme is efficient in that the code rate need not be too small? The answer to this fundamental question is an emphatic "yes." Indeed, the answer to the question is provided by Shannon's second theorem in terms of the channel capacity C, as described in what follows. Up until this point, *time* has not played an important role in our discussion of channel capacity. Suppose then the discrete memoryless source in Fig. 10.12 has the source alphabet $\mathcal{S}$ and entropy $H(\mathcal{S})$ bits per source symbol. We assume that the source emits symbols once every T_s seconds. Hence, the *average information rate* of the source is $H(\mathcal{S})/T_s$ bits per second. The decoder delivers decoded symbols to the destination from the source alphabet $\mathcal{S}$ and at the same source rate of one symbol every T_s seconds. The discrete memoryless channel has a channel capacity equal to C bits per use of the channel. We assume that the channel is capable of being used once every T_c seconds. Hence, the *channel capacity per unit time* is C/T_c bits per second, which represents the maximum rate of information transfer over the channel. We are now ready to state Shannon's second theorem, known as the channel coding theorem.

Specifically, the *channel coding theorem*[8] for a discrete memoryless channel is stated in two parts as follows.

(i) Let a discrete memoryless source with an alphabet $\mathcal{S}$ have entropy $H(\mathcal{S})$ and produce symbols once every T_s seconds. Let a discrete memoryless channel have capacity C and be used once every T_c seconds. Then, if

$$\frac{H(\mathcal{S})}{T_s} \leq \frac{C}{T_c} \tag{10.61}$$

there exists a coding scheme for which the source output can be transmitted over the channel and be reconstructed with an arbitrarily small probability of error. The parameter C/T_c is called the critical rate. *When Eq. (10.61) is satisfied with the equality sign, the system is said to be signaling at the critical rate.*
(ii) Conversely, if

$$\frac{H(\mathcal{S})}{T_s} > \frac{C}{T_c}$$

it is not possible to transmit information over the channel and reconstruct it with an arbitrarily small probability of error.

The channel coding theorem is the single most important result of information theory. The theorem specifies the channel capacity C as a *fundamental limit* on the rate at which the transmission of reliable error-free messages can take place over a discrete memoryless channel.

It is important to note that the channel coding theorem does not show us how to construct a good code. Rather, the theorem can be characterized as an *existence proof* in the sense that it tells us that if the condition of Eq. (10.61) is satisfied, then good codes do exist. In Chapter 11, we describe several good codes suitable for use with discrete memoryless channels.

Application of the Channel Coding Theorem to Binary Symmetric Channels

Consider a discrete memoryless source that emits equally likely binary symbols (0s and 1s) once every T_s seconds. With the source entropy equal to one bit per source symbol (see Example 1), the information rate of the source is $(1/T_s)$ bits per second. The source sequence is applied to a channel encoder with *code rate* r. The channel encoder produces a symbol once every T_c seconds. Hence, the *encoded symbol transmission rate* is $(1/T_c)$ symbols per second. The channel encoder engages a binary symmetric channel once every T_c seconds. Hence, the channel capacity per unit time is (C/T_c) bits per second, where C is determined by the prescribed channel transition probability p in accordance with Eq. (10.60). Accordingly, the channel coding theorem [part (i)] implies that if

$$\frac{1}{T_s} \leq \frac{C}{T_c} \tag{10.62}$$

the probability of error can be made arbitrarily low by the use of a suitable channel encoding scheme. But the ratio T_c/T_s equals the code rate of the channel encoder:

$$r = \frac{T_c}{T_s} \tag{10.63}$$

Hence, we may restate the condition of Eq. (10.62) as

$$r \leq C \tag{10.64}$$

That is, for $r \leq C$, there exists a code (with code rate less than or equal to C) capable of achieving an arbitrarily low probability of error.

Example 7

In this example, we present a graphical interpretation of the channel coding theorem. We also bring out a surprising aspect of the theorem by taking a look at a simple coding scheme.

Consider first a binary symmetric channel with transition probability

$p = 10^{-2}$. For this value of p, we find from Eq. (10.60) that the channel capacity $C = 0.9192$. Hence, from the channel coding theorem, we may state that for any $\varepsilon > 0$ and $r \leq 0.9192$, there exists a code of large enough length n and code rate r, and an appropriate decoding algorithm, such that when the coded bit stream is sent over the given channel, the average probability of channel decoding error is less than ε. This result is illustrated in Fig. 10.13, where we have plotted the average probability of error versus the code rate r. In this figure, we have arbitrarily set the limiting value $\varepsilon = 10^{-8}$.

To put the significance of this result in perspective, consider next a simple coding scheme that involves the use of a *repetition code*, in which each bit of the message is repeated several times. Let each bit (0 or 1) be repeated n times, where $n = 2m + 1$ is an odd integer. For example, for $n = 3$, we transmit 0 and 1 as 000 and 111, respectively. Intuitively, it would seem logical to use a *majority*

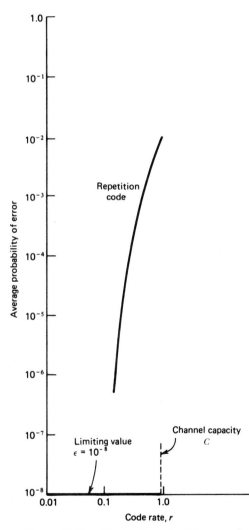

Figure 10.13 Illustrating significance of the channel coding theorem.

Table 10.3 Average Probability
of Error for Repetition Code

Code Rate, $r = 1/n$	Average Probability of Error, P_e
1	10^{-2}
$\frac{1}{3}$	3×10^{-4}
$\frac{1}{5}$	10^{-6}
$\frac{1}{7}$	4×10^{-7}
$\frac{1}{9}$	10^{-8}
$\frac{1}{11}$	5×10^{-10}

rule for decoding, which operates as follows: *If in a block of n received bits (representing one bit of the message), the number of 0s exceeds the number of 1s, the decoder decides in favor of a 0. Otherwise, it decides in favor of a 1.* Hence, an error occurs when $m + 1$ or more bits out of $n = 2m + 1$ bits are received incorrectly. Because of the assumed symmetric nature of the channel, the *average probability of error P_e* is independent of the *a priori* probabilities of 0 and 1. Accordingly, we find that P_e is given by (see Problem 10.24)

$$P_e = \sum_{i=m+1}^{n} \binom{n}{i} p^i (1 - p)^{n-i} \tag{10.65}$$

where p is the transition probability of the channel.

Table 10.3 gives the average probability of error P_e for a repetition code, which is calculated by using Eq. (10.65) for different values of the code rate r. The values given here assume the use of a binary symmetric channel with transition probability $p = 10^{-2}$. The improvement in reliability displayed in Table 10.3 is achieved at the cost of decreasing code rate. The results of this table are also shown plotted as the curve labeled "repetition code" in Fig. 10.13. This curve illustrates the *exchange of code rate for message reliability*, which is a characteristic of repetition codes.

This example highlights the unexpected result presented to us by the channel coding theorem. The result is that it is not necessary to have the code rate r approach zero (as in the case of repetition codes) so as to achieve more and more reliable operation of the communication link. The theorem merely requires that the code rate be less than the channel capacity C.

10.9 DIFFERENTIAL ENTROPY AND MUTUAL INFORMATION FOR CONTINUOUS ENSEMBLES

The sources and channels considered in our discussion of information-theoretic concepts thus far have involved ensembles of random variables that are *discrete* in amplitude. In this section, we extend some of these concepts to *continuous* random variables and random vectors. The motivation for doing so is to pave the way for the description of another fundamental limit in information theory, which we take up in the next section.

Consider a continuous random variable X with the *probability density function* $f_X(x)$. By analogy with the entropy of a discrete random variable, we introduce the following definition:

$$h(X) = \int_{-\infty}^{\infty} f_X(x) \log_2 \left[\frac{1}{f_X(x)} \right] dx \qquad (10.66)$$

We refer to $h(X)$ as the *differential entropy of X* to distinguish it from the ordinary or *absolute entropy*. We do so in recognition of the fact that although $h(X)$ is a useful mathematical quantity to know, it is *not* in any sense a measure of the randomness of X. Nevertheless, we justify the use of Eq. (10.66) in what follows. We begin by viewing the continuous random variable X as the limiting form of a discrete random variable that assumes the value $x_k = k\Delta x$, where $k = 0, \pm 1, \pm 2, \ldots$, and Δx approaches zero. By definition, the continuous random variable X assumes a value in the interval $[x_k, x_k + \Delta x]$ with probability $f_X(x_k)\Delta x$. Hence, permitting Δx to approach zero, the ordinary entropy of the continuous random variable X may be written in the limit as follows:

$$
\begin{aligned}
H(X) &= \lim_{\Delta x \to 0} \sum_{k=-\infty}^{\infty} f_X(x_k) \, \Delta x \log_2 \left(\frac{1}{f_X(x_k) \, \Delta x} \right) \\
&= \lim_{\Delta x \to 0} \left[\sum_{k=-\infty}^{\infty} f_X(x_k) \log_2 \left(\frac{1}{f_X(x_k)} \right) \Delta x - \log_2 \Delta x \sum_{k=-\infty}^{\infty} f_X(x_k) \, \Delta x \right] \\
&= \int_{-\infty}^{\infty} f_X(x) \log_2 \left(\frac{1}{f_X(x)} \right) dx - \lim_{\Delta x \to 0} \log_2 \Delta x \int_{-\infty}^{\infty} f_X(x) \, dx \\
&= h(X) - \lim_{\Delta x \to 0} \log_2 \Delta x
\end{aligned}
$$

$$(10.67)$$

where, in the last line, we have made use of Eq. (10.66) and the fact that the total area under the curve of the probability density function $f_X(x)$ is unity. In the limit as Δx approaches zero, $-\log_2 \Delta x$ approaches infinity. This means that the entropy of a continuous random variable is infinitely large. Intuitively, we would expect this to be true, because a continuous random variable may assume a value anywhere in the interval $(-\infty, \infty)$ and the uncertainty associated with the variable is on the order of infinity. We avoid the problem associated with the term $\log_2 \Delta x$ by adopting $h(X)$ as a differential entropy, with the term $-\log_2 \Delta x$ serving as reference. Moreover, since the information transmitted over a channel is actually the difference between two entropy terms that have a common reference, the information will be the same as the difference between the corresponding differential entropy terms. We are therefore perfectly justified in using the term $h(X)$, defined in Eq. (10.66), as the differential entropy of the continuous random variable X.

When we have a continuous random vector $\mathbf{X}$ consisting of n random variables $X_1, X_2, \ldots, X_n$, we define the differential entropy of $\mathbf{X}$ as the *n-fold integral*

$$h(\mathbf{X}) = \int_{-\infty}^{\infty} f_{\mathbf{X}}(\mathbf{x}) \log_2 \left[\frac{1}{f_{\mathbf{X}}(\mathbf{x})} \right] d\mathbf{x} \qquad (10.68)$$

where $f_{\mathbf{X}}(\mathbf{x})$ is the *joint probability density function* of $\mathbf{X}$.

EXAMPLE 8   Uniform Distribution

Consider a random variable X uniformly distributed over the interval $(0,\ a)$. The probability density function of X is

$$f_X(x) = \begin{cases} \dfrac{1}{a}, & 0 < x < a \\ 0, & \text{otherwise} \end{cases}$$

Applying Eq. (10.66) to this distribution, we get

$$\begin{aligned} h(X) &= \int_0^a \frac{1}{a} \log(a)\ dx \\ &= \log a \end{aligned} \tag{10.69}$$

Note that $\log a < 0$ for $a < 1$. Thus, this example shows that, unlike a discrete random variable, the differential entropy of a continuous random variable can be negative.

EXAMPLE 9   Gaussian Distribution

Consider an arbitrary pair of random variables X and Y, whose probability density functions are respectively denoted by $f_Y(x)$ and $f_X(x)$ where x is merely a dummy variable. Adapting the fundamental inequality of Eq. (10.12) to the situation at hand, we may write[9]

$$\int_{-\infty}^{\infty} f_Y(x)\ \log_2 \left(\frac{f_X(x)}{f_Y(x)} \right)\ dx \le 0 \tag{10.70}$$

or equivalently

$$-\int_{-\infty}^{\infty} f_Y(x)\log_2 f_Y(x)\ dx \le -\int_{-\infty}^{\infty} f_Y(x)\log_2 f_X(x)\ dx \tag{10.71}$$

The quantity on the left-hand side of Eq. (10.71) is the differential entropy of the random variable Y; hence,

$$h(Y) \le -\int_{-\infty}^{\infty} f_Y(x)\log_2 f_X(x)\ dx \tag{10.72}$$

Suppose now the random variables X and Y are described as follows:

• The random variables X and Y have the *same mean* μ and the *same variance* σ^2.
• The random variable X is *Gaussian distributed* as shown by

$$f_X(x) = \frac{1}{\sqrt{2\pi}\sigma}\exp\left(-\frac{(x - \mu)^2}{2\sigma^2} \right) \tag{10.73}$$

Hence, substituting Eq. (10.73) into (10.72), and changing the base of the logarithm from 2 to e, we get

$$h(Y) \le -\log_2 e \int_{-\infty}^{\infty} f_Y(x)\left(-\frac{(x - \mu)^2}{2\sigma^2} - \ln(\sqrt{2\pi}\sigma) \right)\ dx \tag{10.74}$$

We now recognize the following properties of the random variable Y (given that its mean is μ and its variance is σ^2):

$$\int_{-\infty}^{\infty} f_Y(x) \, dx = 1$$

$$\int_{-\infty}^{\infty} (x - \mu)^2 f_Y(x) \, dx = \sigma^2$$

We may therefore simplify Eq. (10.74) as

$$h(Y) \leq \tfrac{1}{2} \log_2(2\pi e \sigma^2) \qquad (10.75)$$

The quantity on the right-hand side of Eq. (10.75) is in fact the differential entropy of the Gaussian random variable X:

$$h(X) = \tfrac{1}{2} \log_2(2\pi e \sigma^2) \qquad (10.76)$$

Finally, combining Eqs. (10.75) and (10.76), we may write

$$h(Y) \leq h(X), \qquad \begin{cases} X\text{: Gaussian random variable} \\ Y\text{: another random variable} \end{cases} \qquad (10.77)$$

where equality holds if, and only if, $Y = X$.

We may now summarize the results of this important example as follows:

1. *For a finite variance σ^2, the Gaussian random variable has the largest differential entropy attainable by any random variable.*

2. *The entropy of a Gaussian random variable X is uniquely determined by the variance of X (i.e., it is independent of the mean of X).*

Indeed, it is because of Property 1 that the Gaussian channel model is so widely used in the study of digital communication systems.

Mutual Information

Consider next a pair of continuous random variables X and Y. By analogy with Eq. (10.49), we define the *mutual information* between the random variables X and Y as follows:

$$I(X;Y) = \int_{-\infty}^{\infty} \int_{-\infty}^{\infty} f_{X,Y}(x,y) \, \log_2 \left[\frac{f_X(x|y)}{f_X(x)} \right] dx \, dy \qquad (10.78)$$

where $f_{X,Y}(x,y)$ is the joint probability density function of X and Y, and $f_X(x|y)$ is the conditional probability density function of X, given that $Y = y$. Also, by analogy with Eqs. (10.45), (10.50), (10.43) and (10.44) we find that the mutual information $I(X;Y)$ has the following properties:

1. $I(X;Y) = I(Y;X)$ (10.79)

2. $I(X;Y) \geq 0$ (10.80)

3. $I(X;Y) = h(X) - h(X|Y)$ (10.81)

      $= h(Y) - h(Y|X)$

The parameter $h(X)$ is the differential entropy of X; likewise for $h(Y)$. The parameter $h(X|Y)$ is the *conditional differential entropy* of X, given Y; it is defined by the double integral

$$h(X|Y) = \int_{-\infty}^{\infty} \int_{-\infty}^{\infty} f_{X,Y}(x,y) \, \log_2 \left[\frac{1}{f_X(x|y)} \right] \, dx \, dy \qquad (10.82)$$

The parameter $h(Y|X)$ is the conditional differential entropy of Y, given X; it is defined in a manner similar to $h(X|Y)$.

10.10 INFORMATION CAPACITY THEOREM

In this section, we use the idea of average mutual information to formulate the information capacity theorem for *band-limited, power-limited Gaussian channels*. To be specific, consider a zero-mean stationary process $X(t)$ that is band-limited to B Hertz. Let X_k, $k = 1, 2, \ldots, K$, denote the continuous random variables obtained by uniform sampling of the process $X(t)$ at the Nyquist rate of $2B$ samples per second. These samples are transmitted in T seconds over a noisy channel, also band-limited to B Hertz. Hence, the number of samples, K, is given by

$$K = 2BT \qquad (10.83)$$

We refer to X_k as a sample of the *transmitted signal*. The channel output is perturbed by *additive white Gaussian noise* of zero mean and power spectral density $N_0/2$. The noise is band-limited to B Hertz. Let the continuous random variables Y_k, $k = 1, 2, \ldots, K$ denote samples of the received signal, as shown by

$$Y_k = X_k + N_k, \qquad k = 1, 2, \ldots, K \qquad (10.84)$$

The noise sample N_k is Gaussian with zero mean and variance given by

$$\sigma^2 = N_0 B \qquad (10.85)$$

We assume that the samples Y_k, $k = 1, 2, \ldots, K$ are statistically independent.

A channel for which the noise and the received signal are as described in Eqs. (10.84) and (10.85) is called a *discrete-time, memoryless Gaussian channel*. It is modeled as in Fig. 10.14. To make meaningful statements about the channel, however, we have to assign a *cost* to each channel input. Typically, the transmitter is *power limited*; it is therefore reasonable to define the cost as

$$E[X_k^2] = P, \qquad k = 1, 2, \ldots, K \qquad (10.86)$$

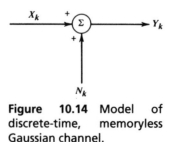

Figure 10.14 Model of discrete-time, memoryless Gaussian channel.

where P is the *average transmitted power*. The *power-limited Gaussian channel* described herein is of not only theoretical but also practical importance in that it models many communication channels, including radio and satellite links.

The *information capacity* of the channel is defined as the maximum of the mutual information between the channel input X_k and the channel output Y_k over all distributions on the input X_k that satisfy the power constraint of Eq. (10.86). Let $I(X_k;Y_k)$ denote the average mutual information between X_k and Y_k. We may then define the information capacity of the channel as

$$C = \max_{f_{X_k}(x)} \{I(X_k;Y_k) : E[X_k^2] = P\} \qquad (10.87)$$

where the maximization is performed with respect to $f_{X_k}(x)$, the probability density function of X_k.

The average mutual information $I(X_k;Y_k)$ can be expressed in one of the two equivalent forms shown in Eq. (10.81). For the purpose at hand, we use the second line of this equation and so write

$$I(X_k;Y_k) = h(Y_k) - h(Y_k|X_k) \qquad (10.88)$$

Since X_k and N_k are independent random variables, and their sum equals Y_k, as in Eq. (10.84), we find that the conditional differential entropy of Y_k, given X_k, is equal to the differential entropy of N_k (see Problem 10.28):

$$h(Y_k|X_k) = h(N_k) \qquad (10.89)$$

Hence, we may rewrite Eq. (10.88) as

$$I(X_k;Y_k) = h(Y_k) - h(N_k) \qquad (10.90)$$

Since $h(N_k)$ is independent of the distribution of X_k, maximizing $I(X_k;Y_k)$ in accordance with Eq. (10.87) requires maximizing $h(Y_k)$, the differential entropy of sample Y_k of the received signal. For $h(Y_k)$ to be maximum, Y_k has to be a Gaussian random variable (see Example 9). That is, the samples of the received signal represent a noiselike process. Next, we observe that since N_k is Gaussian by assumption, the sample X_k of the transmitted signal must be Gaussian too. We may therefore state that the maximization specified in Eq. (10.87) is attained by choosing the samples of the transmitted signal from a noiselike process of average power P. Correspondingly, we may reformulate Eq. (10.87) as

$$C = I(X_k;Y_k) : X_k \text{ Gaussian}, \qquad E[X_k^2] = P \qquad (10.91)$$

where the mutual information $I(X_k;Y_k)$ is defined in accordance with Eq. (10.90).

For the evaluation of the information capacity C, we proceed in three stages:

1. The variance of sample Y_k of the received signal equals $P + \sigma^2$. Hence, the use of Eq. (10.76) yields the differential entropy of Y_k as

$$h(Y_k) = \tfrac{1}{2} \log_2[2\pi e(P + \sigma^2)] \qquad (10.92)$$

2. The variance of the noise sample N_k equals σ^2. Hence, the use of Eq. (10.76) yields the differential entropy of N_k as

$$h(N_k) = \tfrac{1}{2} \log_2(2\pi e\sigma^2) \qquad (10.93)$$

3. Substituting Eqs. (10.92) and (10.93) into Eq. (10.90) and recognizing the definition of information capacity given in Eq. (10.91), we get the desired result:

$$C = \frac{1}{2} \log_2\left(1 + \frac{P}{\sigma^2}\right) \text{ bits per transmission} \qquad (10.94)$$

With the channel used K times for the transmission of K samples of the process $X(t)$ in T seconds, we find that the information *capacity per unit time* is (K/T) times the result given in Eq. (10.94). The number K equals $2BT$, as in Eq. (10.83). Accordingly, we may express the information capacity per transmission as

$$C = B \log_2\left(1 + \frac{P}{N_0 B}\right) \text{ bits per second} \qquad (10.95)$$

where we have used Eq. (10.85) for the noise variance σ^2.

Based on the formula of Eq. (10.95), we may now state Shannon's third (and most famous) theorem, the *information capacity theorem*,[10] as follows:

The information capacity of a continuous channel of bandwidth B Hertz, perturbed by additive white Gaussian noise of power spectral density $N_0/2$ and limited in bandwidth to B, is given by

$$C = B \log_2\left(1 + \frac{P}{N_0 B}\right) \text{ bits per second}$$

where P is the average transmitted power.

The information capacity theorem is one of the most remarkable results of information theory for, in a single formula, it highlights most vividly the interplay among three key system parameters: channel bandwidth, average transmitted power (or, equivalently, average received signal power), and noise power spectral density at the channel output.

The theorem implies that, for given average transmitted power P and channel bandwidth B, we can transmit information at the rate C bits per second, as defined in Eq. (10.95), with arbitrarily small probability of error by employing sufficiently complex encoding systems. It is not possible to transmit at a rate higher than C bits per second by any encoding system without a definite probability of error. Hence, the channel capacity theorem defines the *fundamental limit* on the rate of error-free transmission for a power-limited, band-limited Gaussian channel. To approach this limit, however, the transmitted signal must have statistical properties approximating those of white Gaussian noise.

Sphere Packing[11]

To provide a plausible argument supporting the information capacity theorem, suppose that we use an encoding scheme that yields K code words, one for each sample of the transmitted signal. Let n denote the length (i.e., the number of bits) of each code word. It is presumed that the coding scheme is designed to produce an acceptably low probability of symbol error. Furthermore, the code words satisfy the power constraint; that is, the average power contained in the transmission of each code word with n bits is nP, where P is the average power per bit.

Suppose that any code word in the code is transmitted. The received vector of n bits is Gaussian distributed with mean equal to the transmitted code word and variance equal to $n\sigma^2$, where σ^2 is the noise variance. With high probability, the received vector lies inside a sphere of radius $\sqrt{n\sigma^2}$, centered on the transmitted code word. This sphere is itself contained in a larger sphere of radius $\sqrt{n(P + \sigma^2)}$, where $n(P + \sigma^2)$ is the average power of the received vector.

We may thus visualize the picture portrayed in Fig. 10.15. With everything inside a small sphere of radius $\sqrt{n\sigma^2}$ assigned to the code word on which it is centered, it is reasonable to say that when this particular code word is transmitted, the probability that the received vector will lie inside the correct "decoding" sphere is high. The key question is: How many decoding spheres can be packed inside the larger sphere of received vectors? In other words, how many code words can we in fact choose? To answer this question, we first recognize that the volume of an n-dimensional sphere of radius r may be written as $A_n r^n$, where A_n is a scaling factor. We may therefore make the following statements:

- The volume of the sphere of received vectors is $A_n[n(P + \sigma^2)]^{n/2}$.
- The volume of the decoding sphere is $A_n(n\sigma^2)^{n/2}$.

Accordingly, it follows that the maximum number of *nonintersecting* decoding spheres that can be packed inside the sphere of possible received vectors is

$$\frac{A_n[n(P + \sigma^2)]^{n/2}}{A_n(n\sigma^2)^{n/2}} = \left(1 + \frac{P}{\sigma^2}\right)^{n/2}$$

$$= 2^{(n/2)\log_2(1 + P/\sigma^2)}$$

(10.96)

Taking the logarithm of this result to base 2, we readily see that the maximum number of bits per transmission for a low probability of error is indeed as defined previously in Eq. (10.94).

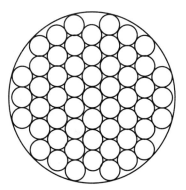

Figure 10.15 Illustrating the sphere-packing problem.

10.11 IMPLICATIONS OF THE INFORMATION CAPACITY THEOREM

Now that we have an intuitive feel for the information capacity theorem, we may go on to discuss its implications in the context of a Gaussian channel that is limited in both power and bandwidth. For the discussion to be useful, however, we need an ideal framework against which the performance of a practical communication system can be assessed. To this end, we introduce the notion of an *ideal system* defined as one that transmits data at a bit rate R_b equal to the information capacity C. We may then express the average transmitted power as

$$P = E_b C \qquad (10.97)$$

where E_b is the transmitted energy per bit. Accordingly, the ideal system is defined by the equation

$$\frac{C}{B} = \log_2\left(1 + \frac{E_b}{N_0}\frac{C}{B}\right) \qquad (10.98)$$

Equivalently, we may define the *signal energy-per-bit to noise power spectral density ratio* E_b/N_0 in terms of the *bandwidth efficiency* C/B for the ideal system as

$$\frac{E_b}{N_0} = \frac{2^{C/B} - 1}{C/B} \qquad (10.99)$$

A plot of bandwidth efficiency R_b/B versus E_b/N_0 is called the *bandwidth-efficiency diagram*. A generic form of this diagram is displayed in Fig. 10.16, where the curve labeled "capacity boundary" corresponds to the ideal system for which $R_b = C$. Based on Fig. 10.16, we can make the following observations:

1. For *infinite bandwidth*, the ratio E_b/N_0 approaches the limiting value

$$\left(\frac{E_b}{N_0}\right)_\infty = \lim_{B\to\infty}\left(\frac{E_b}{N_0}\right)$$
$$= \ln 2 = 0.693 \qquad (10.100)$$

This value is called the *Shannon limit*. Expressed in decibels, it equals -1.6 dB. The corresponding limiting value of the channel capacity is obtained by letting the channel bandwidth B in Eq. (10.95) approach infinity; we thus find that

$$C_\infty = \lim_{B\to\infty} C$$
$$= \frac{P}{N_0}\log_2 e \qquad (10.101)$$

2. The *capacity boundary*, defined by the curve for the critical bit rate $R_b = C$, separates combinations of system parameters that have the potential for supporting error-free transmission ($R_b < C$) from those for which error-free transmission is not possible ($R_b > C$). The latter region is shown shaded in Fig. 10.16.

3. The diagram highlights potential *trade-offs* among E_b/N_0, R_b/B, and probability of symbol error P_e. In particular, we may view movement of the oper-

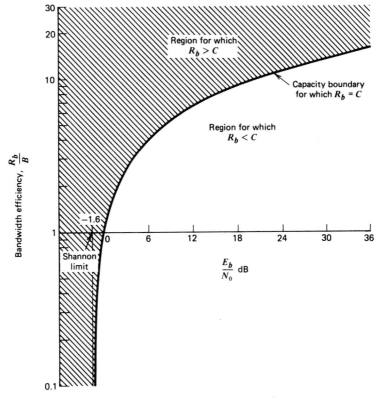

Figure 10.16 Bandwidth-efficiency diagram.

ating point along a horizontal line as trading P_e versus E_b/N_0 for a fixed R_b/B. On the other hand, we may view movement of the operating point along a vertical line as trading P_e versus R_b/B for a fixed E_b/N_0.

The Shannon limit may also be exhibited in terms of the E_b/N_0 required by the ideal system for error-free transmission to be possible. Specifically, we describe the ideal system by writing

$$P_e = \begin{cases} 0, & (E_b/N_0) \geq \ln 2 \\ 1, & (E_b/N_0) < \ln 2 \end{cases} \qquad (10.102)$$

This viewpoint of the ideal system is described in Fig. 10.17. The boundary shown here between error-free transmission and unreliable transmission with possible errors, defined by the Shannon limit, corresponds to the capacity boundary in Fig. 10.16.

EXAMPLE 10 *M*-ary PCM

In this example, we look at an *M*-ary PCM system in light of the channel capacity theorem under the assumption that the system operates above the error threshold. That is, the average probability of error due to channel noise is negligible.

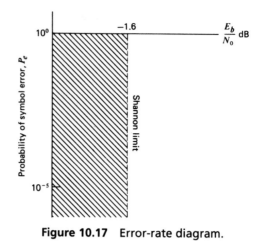

Figure 10.17   Error-rate diagram.

We assume that the M-ary PCM system uses a code word consisting of n code elements, each having one of M possible discrete amplitude levels; hence the name "M-ary." From Chapter 6 we recall that for a PCM system to operate above the error threshold, there must be provision for a noise margin that is sufficiently large to maintain a negligible error rate due to channel noise. This, in turn, means there must be a certain separation between these M discrete amplitude levels. Call this separation $k\sigma$, where k is a constant and $\sigma^2 = N_0 B$ is the noise variance measured in a channel bandwidth B. The number of amplitude levels M is usually an integer power of 2. The average transmitted power will be least if the amplitude range is symmetrical about zero. Then the discrete amplitude levels, normalized with respect to the separation $k\sigma$, will have the values $\pm 1/2$, $\pm 3/2, \ldots, \pm (M-1)/2$. We assume that these M different amplitude levels are equally likely. Accordingly, we find that the average transmitted power is given by

$$P = \frac{2}{M}\left[\left(\frac{1}{2}\right)^2 + \left(\frac{3}{2}\right)^2 + \cdots + \left(\frac{M-1}{2}\right)^2\right](k\sigma)^2$$
$$= k^2\sigma^2\left(\frac{M^2-1}{12}\right) \tag{10.103}$$

Suppose that the M-ary PCM system described herein is used to transmit a message signal with its highest frequency component equal to W Hertz. The signal is sampled at the Nyquist rate of $2W$ samples per second. We assume that the system uses a quantizer of the midrise type, with L equally likely representation levels. Hence, the probability of occurrence of any one of the L representation levels is $1/L$. Correspondingly, the amount of information carried by a single sample of the signal is $\log_2 L$ bits. With a maximum sampling rate of $2W$ samples per second, the maximum rate of information transmission of the PCM system, measured in bits per second, is given by

$$R_b = 2W\log_2 L \text{ bits per second} \tag{10.104}$$

Since the PCM system uses a code word consisting of n code elements, each having one of M possible discrete amplitude values, we have M^n different possible code words. For a unique encoding process, we require

$$L = M^n \qquad (10.105)$$

Clearly, the rate of information transmission in the system is unaffected by the use of an encoding process. We may therefore eliminate L between Eqs. (10.104) and (10.105) to obtain

$$R_b = 2Wn \log_2 M \text{ bits per second} \qquad (10.106)$$

Equation (10.103) defines the average transmitted power required to maintain an M-ary PCM system operating above the error threshold. Hence, solving this equation for the number of discrete amplitude levels, M, we get

$$M = \left(1 + \frac{12P}{k^2 N_0 B} \right)^{1/2} \qquad (10.107)$$

where $\sigma^2 = N_0 B$ is the variance of the channel noise measured in a bandwidth B. Therefore, substituting Eq. (10.107) into Eq. (10.106), we obtain

$$R_b = Wn \log_2 \left(1 + \frac{12P}{k^2 N_0 B} \right) \qquad (10.108)$$

The channel bandwidth B required to transmit a rectangular pulse of duration $1/2nW$ (representing a code element in the code word) is given by (see Chapter 7)

$$B = \kappa nW \qquad (10.109)$$

where κ is a constant with a value between 1 and 2. Using the minimum possible value $\kappa = 1$, we find that the channel bandwidth $B = nW$. We may thus rewrite Eq. (10.108) as

$$R_b = B \log_2 \left(1 + \frac{12P}{k^2 N_0 B} \right) \qquad (10.110)$$

The *ideal system* is described by Shannon's channel capacity theorem, given in Eq. (10.95). Hence, comparing Eq.(10.110) with Eq. (10.95), we see that they are identical if the average transmitted power in the PCM system is increased by the factor $k^2/12$, compared with the ideal system. Perhaps the most interesting point to note about Eq. (10.110) is that the form of the equation is right: *Power and bandwidth in a PCM system are exchanged on a logarithmic basis, and the information capacity C is proportional to the channel bandwidth B.*

EXAMPLE 11   *M*-ary PSK and *M*-ary FSK

In this example, we compare the bandwidth-power exchange capabilities of M-ary PSK and M-ary FSK signals in light of Shannon's information capacity theorem. Consider first a coherent M-ary PSK system that employs a *nonorthogonal* set of M phase-shifted signals for the transmission of binary data. Each signal in the set represents a symbol with $\log_2 M$ bits. Using the definition of null-to-null bandwidth, we may express the bandwidth efficiency of M-ary PSK as follows [see Eq. (8.231)]:

$$\frac{R_b}{B} = \frac{\log_2 M}{2}$$

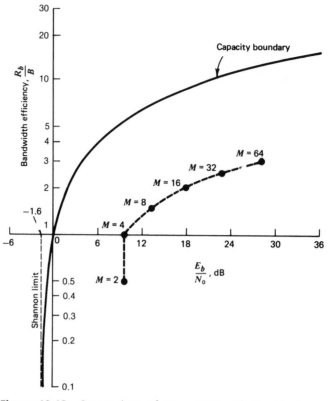

Figure 10.18 Comparison of *M*-ary PSK with the ideal system for $P_e = 10^{-5}$.

In Fig. 10.18, we show the operating points for different numbers of phase levels $M = 2, 4, 8, 16, 32, 64$. Each point corresponds to an average probability of symbol error $P_e = 10^{-5}$. In the figure we have also included the capacity boundary for the ideal system. We observe from Fig. 10.18 that as M is increased, the bandwidth efficiency is improved, but the value of E_b/N_0 required for error-free transmission moves away from the Shannon limit.

Consider next a coherent *M*-ary FSK system that uses an *orthogonal* set of M frequency-shifted signals for the transmission of binary data, with the separation between adjacent signal frequencies set at $1/2T$, where T is the symbol period. As with the *M*-ary PSK, each signal in the set represents a symbol with $\log_2 M$ bits. The bandwidth efficiency of *M*-ary FSK is as follows [see Eq. (8.231)]:

$$\frac{R_b}{B} = \frac{2 \log_2 M}{M}$$

In Fig. 10.19, we show the operating points for different numbers of frequency levels $M = 2, 4, 8, 16, 32, 64$ for an average probability of symbol error $P_e = 10^{-5}$. In the figure, we have also included the capacity boundary for the ideal system. We see that increasing M in (orthogonal) *M*-ary FSK has the opposite effect to that in (nonorthogonal) *M*-ary PSK. In particular, as M is increased, which is equivalent to increased bandwidth requirement, the operating point moves closer to the Shannon limit.

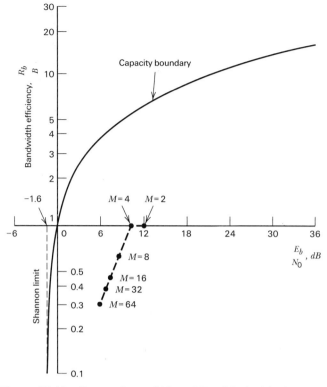

Figure 10.19 Comparison of *M*-ary FSK with the ideal system for $P_e = 10^{-5}$.

10.12 RATE DISTORTION THEORY

In Section 10.3 we introduced the source coding theorem for a discrete memoryless source, according to which the average code-word length must be at least as large as the source entropy for perfect coding (i.e., perfect representation of the source). However, in many practical situations there are constraints that force the coding to be imperfect, thereby resulting in unavoidable *distortion*. For example, constraints imposed by a communication channel may place an upper limit on the permissible code rate and therefore average code-word length assigned to the information source. As another example, the information source may have a continuous amplitude as in the case of speech, and the requirement is to quantize the amplitude of each sample generated by the source to permit its representation by a code word of finite length as in pulse-code modulation. In such cases, the problem is referred to as *source coding with a fidelity criterion*, and the branch of information theory that deals with it is called *rate distortion theory*.[12] Rate distortion theory finds applications in two types of situations:

- Source coding where the permitted coding alphabet cannot exactly represent the information source, in which case we are forced to do *data compression.*
- Information transmission at a rate greater than channel capacity.

Accordingly, rate distortion theory may be viewed as a natural extension of Shannon's coding theorems.

Rate Distortion Function

Consider a discrete memoryless source defined by an M-ary alphabet $X : \{x_i | i = 1, 2, \ldots, M\}$, which consists of a set of statistically independent symbols together with the associated symbol probabilities $\{p_i | i = 1, 2, \ldots, M\}$. Let R be the average code rate in bits per code word. The representation code words are taken from another alphabet $Y : \{y_j | j = 1, 2, \ldots, N\}$. The source coding theorem states that this second alphabet provides a perfect representation of the source provided that $R > H$, where H is the source entropy. But if we are forced to have $R < H$, then there is unavoidable distortion and therefore loss of information.

Let $p(x_i, y_j)$ denote the joint probability of occurrence of source symbol x_i and representation symbol y_j. From Bayes' rule, we have

$$p(x_i, y_j) = p(y_j | x_i) p(x_i) \tag{10.111}$$

where $p(y_j | x_i)$ is a transition probability. Let $d(x_i, y_j)$ denote a measure of the cost incurred in representing the source symbol x_i by the symbol y_j; the quantity $d(x_i, y_j)$ is referred to as a *single-letter distortion measure*. The statistical average of $d(x_i, y_j)$ over all possible source symbols and representation symbols is given by

$$\bar{d} = \sum_{i=1}^{M} \sum_{j=1}^{N} p(x_i) p(y_j | x_i) \, d(x_i, y_j) \tag{10.112}$$

Note that the average distortion $\bar{d}$ is a nonnegative continuous function of the transition probabilities $p(y_j | x_i)$ that are determined by the source encoder–decoder pair.

A conditional probability assignment $p(y_j | x_i)$ is said to be *D-admissible* if and only if the average distortion $\bar{d}$ is less than or equal to some acceptable value D. The set of all D-admissible conditional probability assignments is denoted by

$$P_D = \{p(y_j | x_i) : \bar{d} \leq D\} \tag{10.113}$$

For each set of transition probabilities, we have an average mutual information

$$I(X; Y) = \sum_{i=1}^{M} \sum_{j=1}^{N} p(x_i) p(y_j | x_i) \, \log\left(\frac{p(y_j | x_i)}{p(y_j)}\right) \tag{10.114}$$

A *rate distortion function* $R(D)$ is defined as *the smallest coding rate possible for which the average distortion is guaranteed not to exceed D*. Specifically, for a fixed D we write[13]

$$R(D) = \min_{p(y_j | x_i) \in P_D} I(X; Y) \tag{10.115}$$

subject to the constraint

$$\sum_{j=1}^{N} p(y_j | x_i) = 1 \qquad \text{for } i = 1, 2, \ldots, M \tag{10.116}$$

The rate distortion function $R(D)$ is measured in units of bits if the base-2 logarithm is used in Eq. (10.114). Intuitively, we expect the distortion D to decrease

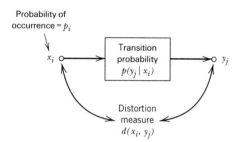

Figure 10.20 Summary of rate distortion theory.

as the rate distortion function $R(D)$ is increased. We may say conversely that tolerating a large distortion D permits the use of a smaller rate for coding and/or transmission of information.

Figure 10.20 summarizes the main parameters of rate distortion theory. In particular, given the source symbols $\{x_i\}$ and their probabilities $\{p_i\}$ and given a definition of the single-letter distortion measure $d(x_i, y_j)$, the calculation of the rate distortion function $R(D)$ involves finding the conditional probability assignment $p(y_j|x_i)$ subject to certain constraints imposed on $p(y_j|x_i)$. This is a variational problem, the solution of which is unfortunately not straightforward in general.

EXAMPLE 12 Gaussian Source

Consider a discrete-time, memoryless Gaussian source with zero mean and variance σ^2. Let x denote the value of a sample generated by such a source. Let y denote a quantized version of x that permits a finite representation of it. The *squared error distortion*

$$d(x,y) = (x - y)^2 \tag{10.117}$$

provides a distortion measure that is widely used for continuous alphabets. The rate distortion function for the Gaussian source with squared error distortion, as described herein, is given by

$$R(D) = \begin{cases} \dfrac{1}{2}\log\left(\dfrac{\sigma^2}{D}\right), & 0 \le D \le \sigma^2 \\ 0, & D > \sigma^2 \end{cases} \tag{10.118}$$

In this case, we see that $R(D) \to \infty$ as $D \to 0$, and $R(D) = 0$ for $D = \sigma^2$.

10.13 COMPRESSION OF INFORMATION

Rate distortion theory naturally leads us to consider the idea of *signal compression* that involves a purposeful or unavoidable reduction in the information content of data from a continuous or discrete source. Specifically, we may think of a *signal compressor* as a device that supplies a code with the least number of symbols for the representation of the source output, subject to a permissible or acceptable *distortion*. The signal compressor thus retains the essential information content of the source output by blurring fine details in a deliberate but controlled manner. Accordingly, signal compression is a *lossy* operation in the sense that the source entropy is reduced (i.e., information is lost), irrespective of the type of source being considered.

In the case of a discrete source, the reason for using data compression is to encode the source output at a rate smaller than the source entropy. By so doing, the source coding theorem is violated, which means that exact reproduction of the original data is *no longer* possible.

In the case of a continuous source, the entropy is infinite, and therefore a signal compression code must always be used to encode the source output at a finite rate. Consequently, it is impossible to digitally encode an analog signal with a finite number of bits without producing some distortion. This statement is in perfect accord with the idea of pulse-code modulation, which was studied in Chapter 6. There it was shown that quantization, which is basic to the analog-to-digital conversion process in pulse-code modulation, always introduces distortion (known as quantization noise) into the transmitted signal. A quantizer may therefore be viewed as a signal compressor.

The uniform and nonuniform quantizers considered in Chapter 6 are said to be *scalar quantizers* in the sense that they deal with samples of the analog signal (i.e., continuous source output) one at a time. Each sample is converted into a quantized value, with the conversion being independent from sample to sample. A scalar quantizer is a rather simple signal compressor, which makes it attractive for practical use. Yet it can provide a surprisingly good performance; this is especially so if nonuniform quantization is used.

There is another class of quantizers known as *vector quantizers* that use blocks of consecutive samples of the source output to form vectors, each of which is treated as a single entity. The essential operation in a vector quantizer is the quantization of a random vector[14] by encoding it as a binary code word. The vector is encoded by comparing it with a *codebook* consisting of a set of stored reference vectors known as *code vectors* or *patterns*. Each pattern in the codebook is used to represent input vectors that are identified by the encoder to be similar to the particular pattern, subject to the maximization of an appropriate fidelity criterion. The encoding process in a vector quantizer may thus be viewed as a *pattern matching operation.*

Let N be the number of reference patterns in the codebook, k be the dimension of the vector (i.e., the number of samples in each block), and r be the coded transmission rate in bits per sample. These three parameters are related as follows:

$$r = \frac{\log_2 N}{k} \tag{10.119}$$

Then, assuming that the size of the codebook is sufficiently large, the signal-to-quantizing noise ratio (SNR) for the vector quantizer is given by

$$10 \log_{10} (\text{SNR}) = 6\left(\frac{\log_2 N}{k}\right) + C_k \text{ dB} \tag{10.120}$$

where C_k is a constant (expressed in dB) that depends on the dimension k. According to Eq. (10.120), the SNR for a vector quantizer increases approximately at the rate of $6/k$ dB for each doubling of the codebook size. Equivalently, we may state that the SNR increases by 6 dB per unit increase in rate (bits per sample) as in the standard PCM using a uniform scalar quantizer. The advantage of the vector quantizer over the scalar quantizer is that its constant term C_k has a higher value, because the vector quantizer optimally exploits the correlations among the samples constituting a vector. Specifically, the constant C_k increases with the dimension k, approaching the ultimate rate-distortion limit for a given source of information. However, the improvement in SNR is attained at the cost of increased encoding complexity, which grows exponentially with the dimension k for a specified rate r; unfortunately, this is the main obstacle to the use of vector quantization in practice.

A final comment on the compression of information is in order. When the output of a source of information is compressed, the resulting data stream usually contains redundant bits. These redundant bits may be removed by using a lossless algorithm such as Huffman coding or Lempel–Ziv coding for the purpose of data compaction. Another method of removing redundant bits is to use some form of prediction as, for example, in differential pulse-code modulation (DPCM) applied to the encoding of speech and video signals.

10.14 SUMMARY AND DISCUSSION

In this chapter we have established four fundamental limits on different aspects of a communication system. The limits are embodied in the source coding theorem, the channel coding theorem, the information capacity theorem, and the rate distortion function.

The *source coding theorem*, Shannon's first theorem, provides the mathematical tool for assessing *data compaction*, that is, *lossless data compression* of data generated by a discrete memoryless source. The theorem tells us that we can make the average number of binary code elements (bits) per source symbol as small as, but no smaller than, the entropy of the source measured in bits. The *entropy* of a source is a function of the probabilities of the source symbols that constitute the alphabet of the source. Since entropy is a measure of uncertainty, the entropy is maximum when the associated probability distribution generates maximum uncertainty.

The *channel coding theorem*, Shannon's second theorem, is the most surprising as well as the single most important result of information theory. For a *binary symmetric channel*, the channel coding theorem tells us that for any *code rate r* less than or equal to the *channel capacity C*, codes do exist such that the average probability of error is as small as we want it. A binary symmetric channel is the simplest form of a discrete memoryless channel. It is symmetric because the probability of receiving a 1 if a 0 is sent is the same as the probability of receiving a 0 if a 1 is sent. This probability, the probability that an error will occur, is termed a *transition probability*. The transition probability p is determined not only by the additive noise at the channel output but also by the kind of receiver used. The value of p uniquely defines the channel capacity C.

Shannon's third remarkable theorem, the *information capacity theorem*, tells us that there is a maximum to the rate at which any communication system can operate reliably (i.e., free of errors) when the system is constrained in power.

This maximum rate is called the *information capacity*, measured in bits per second. When the system operates at a rate greater than the information capacity, it is condemned to a high probability of error, regardless of the choice of signal set used for transmission or the receiver used for processing the received signal.

Finally, the rate distortion function provides the mathematical tool for signal compression (i.e., solving the problem of source coding with a fidelity criterion): The rate distortion function can be applied to a discrete as well as continuous memoryless source.

Data compression and data compaction, in that order, are only two of three elements that constitute the *dissection* of source coding. From the introductory material presented in Chapter 1, we recall that there is also *data encryption* to be considered; see Fig. 1.4. The use of data encryption in the transmitter and its inverse, *decryption*, in the receiver is invoked for the purpose of *secret communication* between a source of information and an end user. Some basic aspects of cryptography, covering both encryption and decryption, follow quite naturally from information theory, as discussed in Appendix 10. Other issues relating to cryptography are also discussed in that appendix.

One last comment is in order. Shannon's information theory, as presented in this chapter, has been entirely in the context of memoryless sources and channels. The theory can be extended to deal with sources and channels with *memory*, in which case a symbol of interest depends on preceding symbols; however, the level of exposition needed to do this is beyond the scope of this book.

NOTES AND REFERENCES

1. According to Lucky (1989), the first mention of the phrase "information theory" by Shannon occurs in a 1945 memorandum entitled "A Mathematical Theory of Cryptography." It is rather curious that the phrase "information theory" was never used in the classic 1948 paper by Shannon, which laid down the foundations of information theory. For an introductory treatment of information theory, see Chapter 2 of Lucky (1989) and the paper by Wyner (1981); see also the books of Adámek (1991), Hamming (1980), and Abramson (1963). For more advanced treatments of the subject, see the books of Cover and Thomas (1991), Blahut (1987), McEliece (1977), and Gallager (1968). For a collection of papers on the development of information theory (including the 1948 classic paper by Shannon), see Slepian (1974).

2. The use of a logarithmic measure of information was first suggested by Hartley (1928); however, Hartley used logarithms to base 10.

3. In statistical physics, the entropy of a physical system is defined by (Reif, 1967, p. 147)

$$\mathcal{L} = k \ln \Omega$$

where k is Boltzmann's constant, Ω is the number of states accessible to the system, and ln denotes the natural logarithm. This entropy has the dimensions of energy because its definition involves the constant k. In particular, it provides a *quantitative measure of the degree of randomness of the system*. Comparing the entropy of statistical physics with that of information theory, we see that they have a similar form. For a detailed discussion of the relation between them, see Pierce (1961, pp. 184–207) and Brillouin (1962).

4. For the original proof of the source coding theorem, see Shannon (1948). A general proof of the source coding theorem is also given in the following books: Viterbi and

Omura (1979, pp. 13–19), McEliece (1977, Chapter 3), and Gallager (1968, pp. 38–55). The source coding theorem is also referred to in the literature as the *noiseless coding theorem,* noiseless in the sense that it establishes the condition for error-free encoding to be possible.

5. For proof of the Kraft–McMillan inequality, see Cover and Thomas (1991, pp. 82–84), Blahut (1990, pp. 298–299), and McEliece (1977, pp. 239–240). For a proof of Eq. (10.23), see Cover and Thomas (1991, pp. 87–88), Blahut (1990, pp. 300–301), and McEliece (1977, pp. 241–242).

6. The Huffman code is named after its inventor: D. A. Huffman (1952). For a readable account of Huffman coding and its use in data compaction, see Adámek (1991).

7. The original papers on the Lempel–Ziv algorithm are Ziv and Lempel (1977, 1978). For readable descriptions of the Lempel–Ziv algorithm, see Lucky (1989, pp. 118–122), Blahut (1990, pp. 314–319), and Gitlin, Hayes, and Weinstein (1992, pp. 120–122). For the application of the Lempel–Ziv algorithm to the compaction of English text, see Lucky (1989, pp. 122–128) and the paper by Welch (1984); see also the review paper by Weiss and Shremp (1993).

8. The channel coding theorem is also known as the *noisy coding theorem.* The original proof of the theorem is given in Shannon (1948). A proof of the theorem is also presented in Hamming (1980, Chapters 9 and 10) in sufficient detail so that a general appreciation of relevant results is developed. The second part of the theorem is referred to in the literature as *the converse to the coding theorem.* A proof of this theorem is presented in the following references: Viterbi and Omura (1979, pp. 28–34) and Gallager (1968, pp. 76–82).

9. The quantity

$$\int_{-\infty}^{\infty} f_Y(x) \log_2 \left(\frac{f_X(x)}{f_Y(x)} \right) dx$$

on the left-hand side of Eq. (10.70) is called relative entropy or the Kullback–Leibler measure; see Kullback (1968).

10. Shannon's information capacity theorem is also referred to in the literature as the *Shannon–Hartley law* in recognition of early work by Hartley on information transmission (Hartley, 1928). In particular, Hartley showed that the amount of information that can be transmitted over a given channel is proportional to the product of the channel bandwidth and the time of operation.

11. A lucid exposition of sphere packing is presented in Cover and Thomas (1991, pp. 242–243); see also Wozencraft and Jacobs (1965, pp. 323–341).

12. For a complete treatment of rate distortion theory, see the book by Berger (1971); this subject is also treated in somewhat less detail in Cover and Thomas (1991), McEliece (1977), and Gallager (1968).

13. For the derivation of Eq. (10.115), see Cover and Thomas (1991, p. 345). An algorithm for computation of the rate distortion function $R(D)$ defined in Eq. (10.115) is described in Blahut (1987, pp. 220–221) and Cover and Thomas (1991, pp. 364–367).

14. For the early papers on vector quantization, see Gersho (1979) and Linde, Buzo, and Gray (1980). For a tutorial review of vector quantization, see Gray (1984). Equation (10.120), defining the SNR for a vector quantizer, is discussed in Gersho and Cuperman (1983).

PROBLEMS

Problem 10.1 Let p denote the probability of some event. Plot the amount of information gained by the occurrence of this event for $0 \leq p \leq 1$.

Problem 10.2 A source emits one of four possible symbols during each signaling interval. The symbols occur with the probabilities:

$$p_0 = 0.4$$
$$p_1 = 0.3$$
$$p_2 = 0.2$$
$$p_3 = 0.1$$

Find the amount of information gained by observing the source emitting each of these symbols.

Problem 10.3 A source emits one of four symbols s_0, s_1, s_2, and s_3 with probabilities 1/3, 1/6, 1/4, and 1/4, respectively. The successive symbols emitted by the source are statistically independent. Calculate the entropy of the source.

Problem 10.4 Let X represent the outcome of a single roll of a fair die. What is the entropy of X?

Problem 10.5 The sample function of a Gaussian process of zero mean and unit variance is uniformly sampled and then applied to a uniform quantizer having the input–output amplitude characteristic shown in Fig. P10.1. Calculate the entropy of the quantizer output.

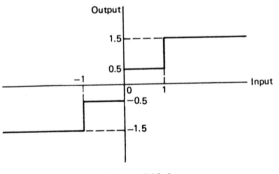

Figure P10.1

Problem 10.6 Consider a discrete memoryless source with source alphabet $\mathcal{L} = \{s_0, s_1, \ldots, s_{K-1}\}$ and source statistics $\{p_0, p_1, \ldots, p_{K-1}\}$. The nth extension of this source is another discrete memoryless source with source alphabet $\mathcal{L}^n = \{\sigma_0, \sigma_1, \ldots, \sigma_{M-1}\}$, where $M = K^n$. Let $P(\sigma_i)$ denote the probability of σ_i.

(a) Show that

$$\sum_{i=0}^{M-1} P(\sigma_i) = 1$$

which is to be expected.

(b) Show that

$$\sum_{i=0}^{M-1} P(\sigma_i) \log_2\left(\frac{1}{p_{i_k}}\right) = H(\mathcal{L}), \qquad k = 1, 2, \ldots, n$$

where p_{i_k} is the probability of symbol s_{i_k}, and $H(\mathcal{L})$ is the entropy of the original source.

(c) Hence, show that

$$H(\mathcal{L}^n) = \sum_{i=0}^{M-1} P(\sigma_i)\log_2\frac{1}{P(\sigma_i)}$$
$$= nH(\mathcal{L})$$

Problem 10.7 Consider a discrete memoryless source with source alphabet $\mathcal{L} = \{s_0, s_1, s_2\}$ and source statistics $\{0.7, 0.15, 0.15\}$.

(a) Calculate the entropy of the source.

(b) Calculate the entropy of the second-order extension of the source.

Problem 10.8 It may come as a surprise, but the number of bits needed to store text is much less than that required to store its spoken equivalent. Can you explain the reason for it?

Problem 10.9 Consider a discrete memoryless source whose alphabet consists of K equiprobable symbols.

(a) Explain why the use of a fixed-length code for the representation of such a source is about as efficient as any code can be.

(b) What conditions have to be satisfied by K and the code length for the coding efficiency to be 100 percent?

Problem 10.10 Consider the four codes listed below:

Symbol	Code I	Code II	Code III	Code IV
s_0	0	0	0	00
s_1	10	01	01	01
s_2	110	001	011	10
s_3	1110	0010	110	110
s_4	1111	0011	111	111

Two of these four codes are prefix codes. Identify them, and construct their individual decision trees.

Problem 10.11 Consider a sequence of letters of the English alphabet with their probabilities of occurrence as given here:

Letter	a	i	l	m	n	o	p	y
Probability	0.1	0.1	0.2	0.1	0.1	0.2	0.1	0.1

Compute two different Huffman codes for this alphabet. Hence, for each of the two codes, find the average code-word length and the variance of the average code-word length over the ensemble of letters.

Problem 10.12   A discrete memoryless source has an alphabet of seven symbols whose probabilities of occurrence are as described here:

Symbol	s_0	s_1	s_2	s_3	s_4	s_5	s_6
Probability	0.25	0.25	0.125	0.125	0.125	0.0625	0.0625

Compute the Huffman code for this source, moving a "combined" symbol as high as possible. Explain why the computed source code has an efficiency of 100 percent.

Problem 10.13   Consider a discrete memoryless source with alphabet $\{s_0, s_1, s_2\}$ and statistics $\{0.7, 0.15, 0.15\}$ for its output.

(a) Apply the Huffman algorithm to this source. Hence, show that the average code-word length of the Huffman code equals 1.3 bits/symbol.

(b) Let the source be extended to order two. Apply the Huffman algorithm to the resulting extended source, and show that the average code-word length of the new code equals 1.1975 bits/symbol.

(c) Compare the average code-word length calculated in part (b) with the entropy of the original source.

Problem 10.14   Figure P10.2 shows a Huffman tree. What is the code word for each of the symbols A, B, C, D, E, F, and G represented by this Huffman tree? What are their individual code-word lengths?

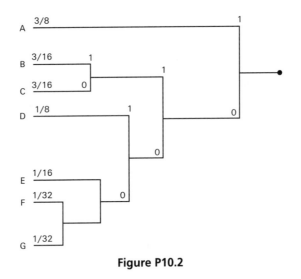

Figure P10.2

Problem 10.15   A computer executes four instructions that are designated by the code words (00, 01, 10, 11). Assuming that the instructions are used independently with probabilities (1/2, 1/8, 1/8, 1/4), calculate the percentage by which the number of bits used for the instructions may be reduced by the use of an optimum source code. Construct a Huffman code to realize the reduction.

Problem 10.16   Consider the following binary sequence

11101001100010110100 . . .

Use the Lempel–Ziv algorithm to encode this sequence. Assume that the binary symbols 0 and 1 are already in the codebook.

Problem 10.17 Consider the transition probability diagram of a binary symmetric channel shown in Fig. 10.9. The input binary symbols 0 and 1 occur with equal probability. Find the probabilities of the binary symbols 0 and 1 appearing at the channel output.

Problem 10.18 Repeat the calculation in Problem 10.17, assuming that the input binary symbols 0 and 1 occur with probabilities 1/4 and 3/4, respectively.

Problem 10.19 Consider a binary symmetric channel characterized by the transition probability p. Plot the mutual information of the channel as a function of p_1, the a priori probability of symbol 1 at the channel input; do your calculations for the transition probability $p = 0, 0.1, 0.2, 0.3, 0.5$.

Problem 10.20 Figure 10.11 depicts the variation of the channel capacity of a binary symmetric channel with the transition probability p. Use the results of Problem 10.19 to explain this variation.

Problem 10.21 Consider the binary symmetric channel described in Fig. 10.9. Let p_0 denote the probability of choosing binary symbol $x_0 = 0$, and let $p_1 = 1 - p_0$ denote the probability of choosing binary symbol $x_1 = 1$. Let p denote the transition probability of the channel.

(a) Show that the average mutual information between the channel input and channel output is given by

$$I(\mathscr{X};\mathscr{Y}) = \mathscr{H}(z) - \mathscr{H}(p)$$

where

$$\mathscr{H}(z) = z \log_2\left(\frac{1}{z}\right) + (1 - z)\log_2\left(\frac{1}{1 - z}\right)$$

$$z = p_0 p + (1 - p_0)(1 - p)$$

and

$$\mathscr{H}(p) = p \log_2\left(\frac{1}{p}\right) + (1 - p)\log_2\left(\frac{1}{1 - p}\right)$$

(b) Show that the value of p_0 that maximizes $I(\mathscr{X};\mathscr{Y})$ is equal to 1/2.

(c) Show that the channel capacity equals

$$C = 1 - \mathscr{H}(p)$$

Problem 10.22 Two binary symmetric channels are connected in cascade, as shown in Fig. P10.3. Find the overall channel capacity of the cascaded connection, assuming that both channels have the same transition probability diagram shown in Fig. 10.9.

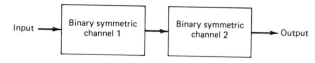

Figure P10.3

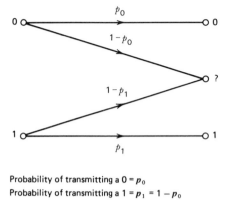

Probability of transmitting a $0 = p_0$
Probability of transmitting a $1 = p_1 = 1 - p_0$

Figure P10.4

Problem 10.23 The *binary erasure channel* is described in Fig. P10.4. The inputs are labeled 0 and 1, and the outputs are labeled 0, 1, and ?. Find the capacity of the channel.

Problem 10.24 Consider a digital communication system that uses a *repetition code* for the channel encoding/decoding. In particular, each transmission is repeated n times, where $n = 2m + 1$ is an odd integer. The decoder operates as follows. If in a block of n received bits, the number of 0s exceeds the number of 1s, the decoder decides in favor of a 0. Otherwise, it decides in favor of a 1. An error occurs when $m + 1$ or more transmissions out of $n = 2m + 1$ are incorrect. Assume a binary symmetric channel.

(a) For $n = 3$, show that the average probability of error is given by

$$P_e = 3p^2(1 - p) + p^3$$

where p is the transition probability of the channel.

(b) For $n = 5$, show that the average probability of error is given by

$$P_e = 10p^3(1 - p)^2 + 5p^4(1 - p) + p^5$$

(c) For the general case, show that the average probability of error is given by

$$P_e = \sum_{i=m+1}^{n} \binom{n}{i} p^i (1 - p)^{n-i}$$

Problem 10.25 Let $X_1, X_2, \ldots, X_n$ denote the elements of a Gaussian vector **X**. The X_i are independent with mean μ_i and variance σ_i^2, $i = 1, 2, \ldots, n$. Show that the differential entropy of the vector **X** equals

$$h(\mathbf{X}) = \frac{n}{2} \log_2[2\pi e(\sigma_1^2 \sigma_2^2 \ldots \sigma_n^2)^{1/n}]$$

What does $h(\mathbf{X})$ reduce to if the variances are equal?

Problem 10.26 A continuous random variable X is constrained to a peak magnitude M; that is, $-M < X < M$.

(a) Show that the differential entropy of X is maximum when it is uniformly distributed, as shown by

$$f_X(x) = \begin{cases} 1/2M, & -M < x \le M \\ 0, & \text{otherwise} \end{cases}$$

(b) Show that the maximum differential entropy of X is $\log_2 2M$.

Problem 10.27 Prove the properties given in Eqs. (10.79) to (10.81) for the mutual information $I(X;Y)$.

Problem 10.28 Consider the continuous random variable Y defined by

$$Y = X + N$$

where X and N are statistically independent. Show that the conditional differential entropy of Y, given X, equals

$$h(Y|X) = h(N)$$

where $h(N)$ is the differential entropy of N.

Problem 10.29 A voice-grade channel of the telephone network has a bandwidth of 3.4 kHz.

(a) Calculate the information capacity of the telephone channel for a signal-to-noise ratio of 30 dB.

(b) Calculate the minimum signal-to-noise ratio required to support information transmission through the telephone channel at the rate of 9600 b/s.

Problem 10.30 Alphanumeric data are entered into a computer from a remote terminal through a voice-grade telephone channel. The channel has a bandwidth of 3.4 kHz and output signal-to-noise ratio of 20 dB. The terminal has a total of 128 symbols. Assume that the symbols are equiprobable and the successive transmissions are statistically independent.

(a) Calculate the information capacity of the channel.

(b) Calculate the maximum symbol rate for which error-free transmission over the channel is possible.

Problem 10.31 A black-and-white television picture may be viewed as consisting of approximately 3×10^5 elements, each of which may occupy one of 10 distinct brightness levels with equal probability. Assume that (1) the rate of transmission is 30 picture frames per second and (2) the signal-to-noise ratio is 30 dB.

  Using the information capacity theorem, calculate the minimum bandwidth required to support the transmission of the resultant video signal.

Note: As a matter of interest, commercial television transmissions actually employ a bandwidth of 4.2 MHz, which fits into an allocated bandwidth of 6 MHz.

Problem 10.32 Equation (10.120) for the signal-to-noise ratio (SNR) of a vector quantizer includes the SNR formula of Eq. (6.43) for standard pulse-code modulation as a special case for which $k = 1$. Justify the validity of this inclusion.

Problem 10.33 All practical data compression and data transmission schemes lie between two limits set by the rate distortion function and the channel capacity theorem. Both of these theorems involve the notion of average mutual information, but in different ways. Elaborate on the issues raised by these two statements.

CHAPTER 11

Error-Control Coding

11.1 INTRODUCTION

The task facing the designer of a digital communication system is that of providing a cost-effective facility for transmitting information from one end of the system at a rate and a level of reliability and quality that are acceptable to a user at the other end. The two key system parameters available to the designer are transmitted signal power and channel bandwidth. These two parameters, together with the power spectral density of receiver noise, determine the signal energy per bit-to-noise power density ratio E_b/N_0. In Chapter 8, we showed that this ratio uniquely determines the bit error rate for a particular modulation scheme. Practical considerations usually place a limit on the value that we can assign to E_b/N_0. Accordingly, in practice, we often arrive at a modulation scheme and find that it is not possible to provide acceptable data quality (i.e., low enough error performance). For a fixed E_b/N_0, the only practical option available for changing data quality from problematic to acceptable is to use *error-control coding*.[1]

Another practical motivation for the use of coding is to reduce the required E_b/N_0 for a fixed bit error rate. This reduction in E_b/N_0 may, in turn, be exploited to reduce the required transmitted power or reduce the hardware costs by requiring a smaller antenna size in the case of radio communications.

Error control for data integrity may be exercised by means of *forward error correction* (FEC). Figure 11.1*a* shows the model of a digital communication system using such an approach. The discrete source generates information in the form of binary symbols. The *channel encoder* in the transmitter accepts message bits and adds *redundancy* according to a prescribed rule, thereby producing encoded data at a higher bit rate. The *channel decoder* in the receiver exploits the redundancy to decide which message bits were actually transmitted. The combined goal of the channel encoder and decoder is to minimize the effect of channel noise. That is, the number of errors between the channel encoder input (derived from the source) and the channel decoder output (delivered to the user) is minimized.

The addition of redundancy in the coded messages implies the need for increased transmission bandwidth. Moreover, the use of error-control coding adds *complexity* to the system, especially for the implementation of decoding operations in the receiver. Thus, the design trade-offs in the use of error-control coding to achieve acceptable error performance include considerations of bandwidth and system complexity.

There are many different error-correcting codes (with roots in diverse mathematical disciplines) that we can use. Historically, these codes have been classified into *block codes* and *convolutional codes*. The distinguishing feature for the classification is the presence or absence of *memory* in the encoders for the two codes.

To generate an (n,k) block code, the channel encoder accepts information in successive k-bit *blocks*; for each block, it adds $n-k$ redundant bits that are algebraically related to the k message bits, thereby producing an overall encoded block of n bits, where $n > k$. The n-bit block is called a *code word*, and n is called the *block length* of the code. The channel encoder produces bits at the rate $R_0 = (n/k)R_s$, where R_s is the bit rate of the information source. The dimensionless ratio $r = k/n$ is called the *code rate*, where $0 < r < 1$. The bit rate R_0, coming out of the encoder, is called the *channel data rate*. Thus, the code rate is a dimensionless ratio, whereas the data rate produced by the source and the channel data rate are both measured in bits per second.

In a convolutional code, the encoding operation may be viewed as the *discrete-time convolution* of the input sequence with the impulse response of the encoder. The duration of the impulse response equals the memory of the encoder. Accordingly, the encoder for a convolutional code operates on the incoming message sequence, using a "sliding window" equal in duration to its own memory. This, in turn, means that in a convolutional code, unlike a block code, the channel encoder accepts message bits as a continuous sequence and thereby generates a continuous sequence of encoded bits at a higher rate.

In the model depicted in Fig. 11.1*a*, the operations of channel coding and modulation are performed separately. When, however, bandwidth efficiency is of major concern, the most effective method of implementing forward error-control correction coding is to combine it with modulation as a single function, as shown in Fig. 11.1*b*. In such an approach, coding is redefined as a process of imposing certain patterns on the transmitted signal.

The bulk of the material presented in this chapter relates to channel coding techniques suitable for forward error correction (FEC). There is, however, another major approach known as *automatic-repeat-request (ARQ)*, which is also widely used for solving the error-control problem. The philosophy of ARQ is quite

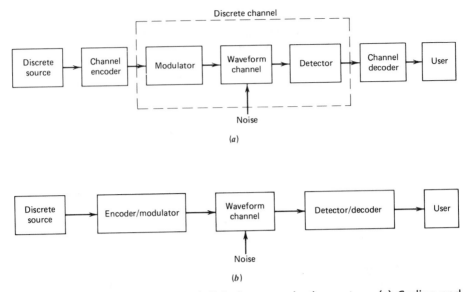

Figure 11.1 Simplified models of digital communication system. (*a*) Coding and modulation performed separately. (*b*) Coding and modulation combined.

different from that of FEC. Specifically, ARQ utilizes redundancy for the sole purpose of error detection. Upon detection, the receiver requests a repeat transmission, which necessitates the use of a return path (feedback channel).

11.2 DISCRETE MEMORYLESS CHANNELS

Returning to the model of Fig. 11.1*a*, the waveform channel is said to be memoryless if the detector output in a given interval depends only on the signal transmitted in that interval, and not on any previous transmission. Under this condition, we may model the combination of the modulator, the waveform channel, and the detector as a *discrete memoryless channel*. Such a channel is completely described by the set of transition probabilities $p(j|i)$, where i denotes a modulator input symbol, j denotes a demodulator output symbol, and $p(j|i)$ denotes the probability of receiving symbol j, given that symbol i was sent. (Discrete memoryless channels were described previously at some length in Section 10.5.)

The simplest discrete memoryless channel results from the use of binary input and binary output symbols. When binary coding is used, the modulator has only the binary symbols 0 and 1 as inputs. Likewise, the decoder has only binary inputs if binary quantization of the demodulator output is used, that is, a *hard decision* is made on the demodulator output as to which symbol was actually transmitted. In this situation, we have a *binary symmetric channel* (BSC) with a *transition probability diagram* as shown in Fig. 11.2. The binary symmetric channel, assuming a channel noise modeled as additive white Gaussian noise (AWGN) channel, is completely described by the *transition probability p*. The majority of coded digital communication systems employ binary coding with hard-decision decoding, due to the simplicity of implementation offered by such an approach.

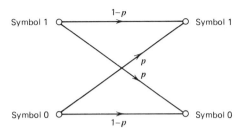

Figure 11.2 Transition probability diagram of binary symmetric channel.

The use of hard decisions prior to decoding causes an irreversible loss of information in the receiver. To reduce this loss, *soft-decision* coding is used. This is achieved by including a multilevel quantizer at the demodulator output, as illustrated in Fig. 11.3a for the case of binary PSK signals. The input–output characteristic of the quantizer is shown in Fig. 11.3b. The modulator has only the binary symbols 0 and 1 as inputs, but the demodulator output now has an alphabet with Q symbols. Assuming the use of the quantizer as described in Fig. 11.3b, we have $Q = 8$. Such a channel is called a *binary input Q-ary output discrete memoryless channel*. The corresponding channel transition probability diagram is shown in Fig. 11.3c. The form of this distribution, and consequently the decoder performance, depends on the location of the representation levels of the quantizer, which in turn depends on the signal level and noise variance. Accordingly, the demodulator must incorporate automatic gain control if an effective multilevel quantizer is to be realized. Moreover, the use of soft decisions complicates the implementation of the decoder. Nevertheless, soft-decision decoding offers significant improvement in performance over hard-decision decoding.

The Channel Coding Theorem Revisited

In Chapter 10, we established the concept of *channel capacity*, which, for a discrete memoryless channel, represents the maximum amount of information transmitted per channel use. The *channel coding theorem* states that if a discrete memoryless channel has capacity C and a source generates information at a rate less than C, then there exists a coding technique such that the output of the source may be transmitted over the channel with an arbitrarily low probability of symbol error. For the special case of a binary symmetric channel, the theorem tells us that if the code rate r is less than the channel capacity C, then it is possible to find a code that achieves error-free transmission over the channel. Conversely, it is not possible to find such a code if the code rate r is greater than the channel capacity C.

The channel coding theorem thus specifies the channel capacity C as a *fundamental limit* on the rate at which the transmission of reliable (error-free) messages can take place over a discrete memoryless channel. The issue that matters is not the signal-to-noise ratio, so long as it is large enough, but how the channel input is encoded.

The most unsatisfactory feature of the channel coding theorem, however, is its nonconstructive nature. The theorem asserts the *existence of good codes* but does not tell us how to find them. We are still faced with the task of finding a good

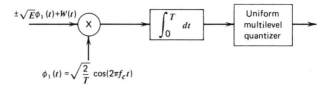

$$\phi_1(t) = \sqrt{\frac{2}{T}} \cos(2\pi f_c t)$$

(a)

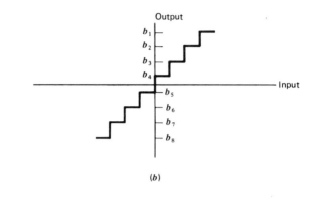

(b)

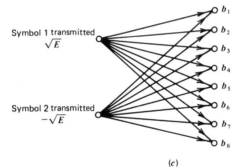

(c)

Figure 11.3 Binary input Q-ary output discrete memory-less channel. (a) Receiver. (b) Transfer characteristic of multilevel quantizer. (c) Channel transition probability diagram. Parts (b) and (c) are illustrated for 8 levels.

code that ensures reliable transmission of information over the channel. The error-control coding techniques presented in this chapter provide different methods of achieving this important system requirement.

The codes described here are *binary codes*, for which the alphabet consists only of symbol 0 and 1. In such a code, the encoding and decoding functions involve the binary arithmetic operations of *modulo-2 addition and multiplication* performed on code words in the code. The rules for binary arithmetic are summarized in Appendix 9.

Throughout this chapter, we use an ordinary plus sign ($+$) to denote modulo-2 addition. The use of this terminology will not lead to confusion, since the whole chapter relies on binary arithmetic. In so doing, we avoid the use of a special symbol $\oplus$, as we did in preceding chapters.

11.3 LINEAR BLOCK CODES

A code is said to be *linear* if any two code words in the code can be added in modulo-2 arithmetic to produce a third code word in the code. Consider then an (n, k) linear block code in which k bits of the n code bits are always identical to the message sequence to be transmitted. The $n - k$ bits in the remaining portion are computed from the message bits in accordance with a prescribed encoding rule that determines the mathematical structure of the code. Accordingly, these $n - k$ bits are referred to as *generalized parity check bits* or simply *parity bits.* Block codes in which the message bits are transmitted in unaltered form are called *systematic codes.* For applications requiring *both* error detection and error correction, the use of systematic block codes simplifies implementation of the decoder.

Let $m_0, m_1, \ldots, m_{k-1}$ constitute a block of k arbitrary message bits. Thus we have 2^k distinct message blocks. Let this sequence of message bits be applied to a linear block encoder, producing an n-bit code word whose elements are denoted by $c_0, c_1, \ldots, c_{n-1}$. Let $b_0, b_1, \ldots, b_{n-k-1}$ denote the ($n - k$) parity bits in the code word. For the code to possess a systematic structure, a code word is divided into two parts, one of which is occupied by the message bits and the other by the parity bits. Clearly, we have the option of sending the message bits of a code word before the parity bits, or vice versa. The former option is illustrated in Fig. 11.4, and its use is assumed in the sequel.

According to the representation of Fig. 11.4, the ($n - k$) left-most bits of a code word are identical to the corresponding parity bits, and the k right-most bits of the code word are identical to the corresponding message bits. We may therefore write

$$c_i = \begin{cases} b_i, & i = 0, 1, \ldots, n - k - 1 \\ m_{i+k-n}, & i = n - k, \ n - k + 1, \ldots, n - 1 \end{cases} \tag{11.1}$$

The ($n - k$) parity bits are *linear sums* of the k message bits, as shown by the generalized relation

$$b_i = p_{0i}\, m_0 + p_{1i}\, m_1 + \cdots + p_{k-1,i}\, m_{k-1} \tag{11.2}$$

where the coefficients are defined as follows:

$$p_{ij} = \begin{cases} 1 & \text{if } b_i \text{ depends on } m_j \\ 0 & \text{otherwise} \end{cases} \tag{11.3}$$

The coefficients p_{ij} are chosen in such a way that the rows of the generator matrix are linearly independent and the parity equations are *unique.*

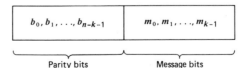

$b_0, b_1, \ldots, b_{n-k-1}$	$m_0, m_1, \ldots, m_{k-1}$
Parity bits	Message bits

Figure 11.4 Structure of code word.

The system of Eqs. (11.1) and (11.2) defines the mathematical structure of the (n, k) linear block code. This system of equations may be rewritten in a compact form using matrix notation. To proceed with this reformulation, we define the 1-by-k *message vector* $\mathbf{m}$, the 1-by-$(n - k)$ parity vector $\mathbf{b}$, and the 1–by-n code vector $\mathbf{c}$ as follows:

$$\mathbf{m} = [m_0, m_1, \ldots, m_{k-1}] \tag{11.4}$$

$$\mathbf{b} = [b_0, b_1, \ldots, b_{n-k-1}] \tag{11.5}$$

$$\mathbf{c} = [c_0, c_1, \ldots, c_{n-1}] \tag{11.6}$$

Note that all three vectors are *row vectors*. The use of row vectors is adopted in this chapter for the sake of being consistent with the notation commonly used in the coding literature. We may thus rewrite the set of simultaneous equations defining the parity bits in the compact matrix form:

$$\mathbf{b} = \mathbf{mP} \tag{11.7}$$

where $\mathbf{P}$ is the k-by-$(n - k)$ *coefficient matrix* defined by

$$\mathbf{P} = \begin{bmatrix} p_{00} & p_{01} & \cdots & p_{0,n-k-1} \\ p_{10} & p_{11} & \cdots & p_{1,n-k-1} \\ \vdots & \vdots & & \vdots \\ p_{k-1,0} & p_{k-1,1} & \cdots & p_{k-1,n-k-1} \end{bmatrix} \tag{11.8}$$

where p_{ij} is 0 or 1.

From the definitions given in Eqs. (11.4)–(11.6), we see that $\mathbf{c}$ may be expressed as a partitioned row vector in terms of the vectors $\mathbf{m}$ and $\mathbf{b}$ as follows:

$$\mathbf{c} = [\mathbf{b} \mid \mathbf{m}] \tag{11.9}$$

Hence, substituting Eq. (11.7) in Eq. (11.9) and factoring out the common message vector $\mathbf{m}$, we get

$$\mathbf{c} = \mathbf{m}[\mathbf{P} \mid \mathbf{I}_k] \tag{11.10}$$

where $\mathbf{I}_k$ is the k-by-k *identity matrix*:

$$\mathbf{I}_k = \begin{bmatrix} 1 & 0 & \cdots & 0 \\ 0 & 1 & \cdots & 0 \\ \vdots & \vdots & & \vdots \\ 0 & 0 & \cdots & 1 \end{bmatrix} \tag{11.11}$$

Define the k-by-n *generator matrix*

$$\mathbf{G} = [\mathbf{P} \mid \mathbf{I}_k] \tag{11.12}$$

The generator matrix $\mathbf{G}$ of Eq. (11.12) is said to be in the *echelon canonical form* in that its k rows are linearly independent; that is, it is not possible to express any row of the matrix $\mathbf{G}$ as a linear combination of the remaining rows. Using the definition of the generator matrix $\mathbf{G}$, we may simplify Eq. (11.10) as

$$\mathbf{c} = \mathbf{mG} \tag{11.13}$$

The full set of code words, referred to simply as *the code*, is generated in accordance with Eq. (11.13) by letting the message vector $\mathbf{m}$ range through the set of all 2^k binary k-tuples (1-by-k vectors). Moreover, the sum of any two code words is another code word. This basic property of linear block codes is called *closure*. To prove its validity, consider a pair of code vectors $\mathbf{c}_i$ and $\mathbf{c}_j$ corresponding to a pair of message vectors $\mathbf{m}_i$ and $\mathbf{m}_j$, respectively. Using Eq. (11.13) we may express the sum of $\mathbf{c}_i$ and $\mathbf{c}_j$ as

$$\mathbf{c}_i + \mathbf{c}_j = \mathbf{m}_i\mathbf{G} + \mathbf{m}_j\mathbf{G}$$
$$= (\mathbf{m}_i + \mathbf{m}_j)\mathbf{G}$$

The modulo-2 sum of $\mathbf{m}_i$ and $\mathbf{m}_j$ represents a new message vector. Correspondingly, the modulo-2 sum of $\mathbf{c}_i$ and $\mathbf{c}_j$ represents a new code vector.

There is another way of expressing the relationship between the message bits and parity-check bits of a linear block code. Let $\mathbf{H}$ denote an $(n - k)$-by-n matrix, defined as

$$\mathbf{H} = [\mathbf{I}_{n-k} \mid \mathbf{P}^T] \tag{11.14}$$

where $\mathbf{P}^T$ is an $(n - k)$-by-k matrix, representing the transpose of the coefficient matrix $\mathbf{P}$, and $\mathbf{I}_{n-k}$ is the $(n - k)$-by-$(n - k)$ identity matrix. Accordingly, we may perform the following multiplication of partitioned matrices:

$$\mathbf{HG}^T = [\mathbf{I}_{n-k} \mid \mathbf{P}^T]\begin{bmatrix}\mathbf{P}^T \\ \hline \mathbf{I}_k\end{bmatrix}$$
$$= \mathbf{P}^T + \mathbf{P}^T$$

where we have used the fact that multiplication of a rectangular matrix by an identity matrix of compatible dimensions leaves the matrix unchanged. In modulo-2 arithmetic, we have $\mathbf{P}^T + \mathbf{P}^T = \mathbf{0}$, where $\mathbf{0}$ denotes an $(n - k)$-by-k null matrix (i.e., a matrix that has zeros for all of its elements). Hence,

$$\mathbf{HG}^T = \mathbf{0} \tag{11.15}$$

Equivalently, we have $\mathbf{GH}^T = \mathbf{0}$. Postmultiplying both sides of Eq. (11.13) by $\mathbf{H}^T$, the transpose of $\mathbf{H}$, and then using Eq. (11.15), we get

$$\mathbf{cH}^T = \mathbf{mGH}^T$$
$$= \mathbf{0} \tag{11.16}$$

The matrix $\mathbf{H}$ is called the *parity-check matrix* of the code, and the set of equations specified by Eq. (11.16) are called *parity-check equations*.

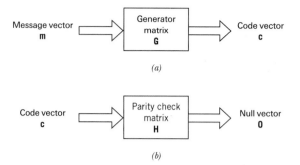

Figure 11.5 Block diagram representations of the generator equation (11.13) and the parity-check equation (11.16).

The generator equation (11.13) and the parity-check detector equation (11.16) are basic to the description and operation of a linear block code. These two equations are depicted in the form of block diagrams in Fig. 11.5a and 11.5b, respectively.

EXAMPLE 1   Repetition Codes

Repetition codes represent the simplest type of linear block codes. In particular, a single message bit is encoded into a block of n identical bits, producing an $(n, 1)$ block code. Such a code allows provision for a variable amount of redundancy. There are only two code words in the code: an all-zero code word and an all-one code word.

Consider, for example, the case of a repetition code with $k = 1$ and $n = 5$. In this case, we have four parity bits that are the same as the message bit. Hence, the identity matrix $\mathbf{I}_k = 1$, and the coefficient matrix $\mathbf{P}$ consists of a 1-by-4 vector that has 1 for all of its elements. Correspondingly, the generator matrix equals a row vector of all 1s, as shown by

$$\mathbf{G} = [1 \quad 1 \quad 1 \quad 1 \;\vdots\; 1]$$

The transpose of the coefficient matrix $\mathbf{P}$, namely, matrix $\mathbf{P}^T$, consists of a 4-by-1 vector that has 1 for all of its elements. The identity matrix $\mathbf{I}_{n-k}$ consists of a 4-by-4 matrix. Hence, the parity-check matrix equals

$$\mathbf{H} = \begin{bmatrix} 1 & 0 & 0 & 0 & \vdots & 1 \\ 0 & 1 & 0 & 0 & \vdots & 1 \\ 0 & 0 & 1 & 0 & \vdots & 1 \\ 0 & 0 & 0 & 1 & \vdots & 1 \end{bmatrix}$$

Since the message vector consists of a single binary symbol, 0 or 1, it follows from Eq. (11.13) that there are only two code words: 00000 and 11111 in the (5, 1) repetition code, as expected. Note also that $\mathbf{HG}^T = \mathbf{0}$, modulo-2, in accordance with Eq. (11.15).

Syndrome Decoding

The generator matrix $\mathbf{G}$ is used in the encoding operation at the transmitter. On the other hand, the parity-check matrix $\mathbf{H}$ is used in the decoding operation at the receiver. In the context of the latter operation, let $\mathbf{r}$ denote the 1-by-n *received vector* that results from sending the code vector $\mathbf{c}$ over a noisy channel. We express the vector $\mathbf{r}$ as the sum of the original code vector $\mathbf{c}$ and a vector $\mathbf{e}$, as shown by

$$\mathbf{r} = \mathbf{c} + \mathbf{e} \tag{11.17}$$

The vector $\mathbf{e}$ is called the *error vector* or *error pattern*. The ith element of $\mathbf{e}$ equals 0 if the corresponding element of $\mathbf{r}$ is the same as that of $\mathbf{c}$. On the other hand, the ith element of $\mathbf{e}$ equals 1 if the corresponding element of $\mathbf{r}$ is different from that of $\mathbf{c}$, in which case an error is said to have occurred in the ith location. That is, for $i = 1, 2, \ldots, n$, we have

$$e_i = \begin{cases} 1 & \text{if an error has occurred in the } i\text{th location} \\ 0 & \text{otherwise} \end{cases} \tag{11.18}$$

The receiver has the task of decoding the code vector $\mathbf{c}$ from the received vector $\mathbf{r}$. The algorithm commonly used to perform this decoding operation starts with the computation of a 1-by-$(n - k)$ vector called the *error-syndrome vector* or simply the *syndrome*.[2] The importance of the syndrome lies in the fact that it depends only upon the error pattern.

Given a 1-by-n received vector $\mathbf{r}$, the corresponding syndrome is formally defined as

$$\mathbf{s} = \mathbf{r}\mathbf{H}^T \tag{11.19}$$

Accordingly, the syndrome has the following important properties.

PROPERTY 1

The syndrome depends only on the error pattern, and not on the transmitted code word.

To prove this property, we first use Eqs. (11.17) and (11.19) and then Eq. (11.16) to obtain

$$\begin{aligned} \mathbf{s} &= (\mathbf{c} + \mathbf{e})\mathbf{H}^T \\ &= \mathbf{c}\mathbf{H}^T + \mathbf{e}\mathbf{H}^T \\ &= \mathbf{e}\mathbf{H}^T \end{aligned} \tag{11.20}$$

Hence, the parity-check matrix $\mathbf{H}$ of a code permits us to compute the syndrome $\mathbf{s}$, which depends only upon the error pattern $\mathbf{e}$.

PROPERTY 2

All error patterns that differ by a code word have the same syndrome.

For k message bits, there are 2^k distinct code vectors denoted as c_i, $i = 0$, $1, \ldots, 2^k - 1$. Correspondingly, for any error pattern e, we define the 2^k distinct vectors e_i as

$$e_i = e + c_i, \qquad i = 0, 1, \ldots, 2^k - 1 \qquad (11.21)$$

The set of vectors $\{e_i, i = 0, 1, \ldots, 2^k - 1\}$ so defined is called a *coset* of the code. In other words, a coset has exactly 2^k elements that differ at most by a code vector. Thus, an (n, k) linear block code has 2^{n-k} possible cosets. In any event, multiplying both sides of Eq. (11.21) by the matrix H^T, we get

$$
\begin{aligned}
e_i H^T &= e H^T + c_i H^T \\
&= e H^T
\end{aligned}
\qquad (11.22)
$$

which is independent of the index i. Accordingly, we may state that each coset of the code is characterized by a unique syndrome.

We may put Properties 1 and 2 in perspective by expanding Eq. (11.20). Specifically, with the matrix H having the systematic form given in Eq. (11.14), where the matrix P is itself defined by Eq. (11.8), we find from Eq. (11.20) that the $(n - k)$ elements of the syndrome s are linear combinations of the n elements of the error pattern e as shown by

$$
\begin{aligned}
s_0 &= e_0 + e_{n-k}p_{00} + e_{n-k+1}p_{10} + \cdots + e_{n-1}p_{k-1,0} \\
s_1 &= e_1 + e_{n-k}p_{01} + e_{n-k+1}p_{11} + \cdots + e_{n-1}p_{k-1,1} \\
&\vdots \\
s_{n-k-1} &= e_{n-k-1} + e_{n-k}p_{0,n-k-1} + \cdots + e_{n-1}p_{k-1,n-k-1}
\end{aligned}
\qquad (11.23)
$$

This set of $(n - k)$ linear equations clearly shows that the syndrome contains information about the error pattern and may therefore be used for error detection. However, it should be noted that the set of equations is *underdetermined* in that we have more unknowns than equations. Accordingly, there is *no* unique solution for the error pattern. Rather there are 2^k error patterns that satisfy Eq. (11.23) and therefore result in the same syndrome, in accordance with Property 2 and Eq. (11.22); the true error pattern is just one of the 2^k possible solutions. In other words, the information contained in the syndrome s about the error pattern s is *not* enough for the decoder to compute the exact value of the transmitted code vector. Nevertheless, knowledge of the syndrome s reduces the search for the true error pattern e from 2^n to 2^{n-k} possibilities. In particular, the decoder has the task of making the best selection from the coset corresponding to s.

Minimum Distance Considerations

Consider a pair of code vectors c_1 and c_2 that have the same number of elements. The *Hamming distance* $d(c_1, c_2)$ between such a pair of code vectors is defined as the number of locations in which their respective elements differ.

The *Hamming weight* $w(c)$ of a code vector c is defined as the number of nonzero elements in the code vector. Equivalently, we may state that the Hamming weight of a code vector is the distance between the code vector and the all-zero code vector.

The *minimum distance* $d_{\min}$ of a linear block code is defined as the smallest Hamming distance between any pair of code vectors in the code. That is, the minimum distance is the same as the smallest Hamming weight of the difference between any pair of code vectors. From the closure property of linear block codes, the sum (or difference) of two code vectors is another code vector. Accordingly, we may state that *the minimum distance of a linear block code is the smallest Hamming weight of the nonzero code vectors in the code.*

The minimum distance $d_{\min}$ is related to the structure of the parity-check matrix **H** of the code in a fundamental way. From Eq. (11.16) we know that a linear block code is defined by the set of all code vectors for which $\mathbf{cH}^T = \mathbf{0}$, where $\mathbf{H}^T$ is the transpose of the parity-check matrix **H**. Let the matrix **H** be expressed in terms of its columns as follows:

$$\mathbf{H} = [\mathbf{h}_1, \mathbf{h}_2, \ldots, \mathbf{h}_n] \qquad (11.24)$$

Then, for a code vector **c** to satisfy the condition $\mathbf{cH}^T = \mathbf{0}$, the vector **c** must have 1s in such positions that the corresponding rows of $\mathbf{H}^T$ sum to the zero vector **0**. However, by definition, the number of 1s in a code vector is the Hamming weight of the code vector. Moreover, the smallest Hamming weight of the nonzero code vectors in a linear block code equals the minimum distance of the code. Hence, *the minimum distance of a linear block code is defined by the minimum number of rows of the matrix $\mathbf{H}^T$ whose sum is equal to the zero vector.*

The minimum distance of a linear block code, $d_{\min}$, is an important parameter of the code. Specifically, it determines the error-correcting capability of the code. Suppose an (n, k) linear block code is required to detect and correct all error patterns (over a binary symmetric channel), and whose Hamming weight is less than or equal to t. That is, if a code vector $\mathbf{c}_i$ in the code is transmitted and the received vector is $\mathbf{r} = \mathbf{c}_i + \mathbf{e}$, we require that the decoder output $\hat{\mathbf{c}} = \mathbf{c}_i$, whenever the error pattern **e** has a Hamming weight $w(\mathbf{e}) \le t$. We assume that the 2^k code vectors in the code are transmitted with equal probability. The best strategy for the decoder then is to pick the code vector closest to the received vector **r**, that is, the one for which the Hamming distance $d(\mathbf{c}_i, \mathbf{r})$ is the smallest. With such a strategy, the decoder will be able to detect and correct all error patterns of Hamming weight $w(\mathbf{e}) \le t$, provided that the minimum distance of the code is equal to or greater than $2t + 1$. We may demonstrate the validity of this requirement by adopting a geometric interpretation of the problem. In particular, the 1-by-n code vectors and the 1-by-n received vector are represented as points in an n-dimensional space. Suppose that we construct two spheres, each of radius t, around the points that represent code vectors $\mathbf{c}_i$ and $\mathbf{c}_j$. Let these two spheres be disjoint, as depicted in Fig. 11.6a. For this condition to be satisfied, we require that $d(\mathbf{c}_i, \mathbf{c}_j) \ge 2t + 1$. If then the code vector $\mathbf{c}_i$ is transmitted and the Hamming distance $d(\mathbf{c}_i, \mathbf{r}) \le t$, it is clear that the decoder will pick $\mathbf{c}_i$ as it is the code vector closest to the received vector **r**. If, on the other hand, the Hamming distance $d(\mathbf{c}_i, \mathbf{c}_j) \le 2t$, the two spheres around $\mathbf{c}_i$ and $\mathbf{c}_j$ intersect, as depicted in Fig. 11.6b. Here we see that if $\mathbf{c}_i$ is transmitted, there exists a received vector **r** such that the Hamming distance $d(\mathbf{c}_i, \mathbf{r}) \le t$, and yet **r** is as close to $\mathbf{c}_j$ as it is to $\mathbf{c}_i$. Clearly, there is now the possibility of the decoder picking the vector $\mathbf{c}_j$, which is wrong. We thus conclude that *an (n,k) linear block code has the power to correct all*

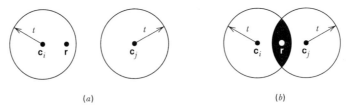

Figure 11.6 (a) Hamming distance $d(\mathbf{c}_i,\mathbf{c}_j) \geq 2t + 1$. (b) Hamming distance $d(\mathbf{c}_i,\mathbf{c}_j) < 2t$.

error patterns of weight t or less if, and only if,

$$d(\mathbf{c}_i,\mathbf{c}_j) \geq 2t + 1 \qquad \text{for all } \mathbf{c}_i \text{ and } \mathbf{c}_j$$

By definition, however, the smallest distance between any pair of code vectors in a code is the minimum distance of the code, $d_{\min}$. We may therefore state that *an (n,k) linear block code of minimum distance $d_{\min}$ can correct up to t errors if, and only if,*

$$t \leq \lfloor \tfrac{1}{2}(d_{\min} - 1)\rfloor \tag{11.25}$$

where $\lfloor\ \rfloor$ *denotes the largest integer* less than or equal to the enclosed quantity. Equation (11.25) gives the error-correcting capability of a linear block code a quantitative meaning.

Syndrome Decoding

We are now ready to describe a syndrome-based decoding scheme for linear block codes. Let $\mathbf{c}_1, \mathbf{c}_2, \ldots, \mathbf{c}_{2^k}$ denote the 2^k code vectors of an (n,k) linear block code. Let $\mathbf{r}$ denote the received vector, which may have one of 2^n possible values. The receiver has the task of partitioning the 2^n possible received vectors into 2^k disjoint subsets $D_1, D_2, \ldots, D_{2^k}$ in such a way that the ith subset D_i corresponds to code vector $\mathbf{c}_i$ for $1 \leq i \leq 2^k$. The received vector $\mathbf{r}$ is decoded into $\mathbf{c}_i$ if it is in the ith subset. For the decoding to be correct, $\mathbf{r}$ must be in the subset that belongs to the code vector $\mathbf{c}_i$ that was actually sent.

The 2^k subsets described herein constitute a *standard array* of the linear block code. To construct it, we may exploit the linear structure of the code by proceeding as follows:

1. The 2^k code vectors are placed in a row with the all-zero code vector $\mathbf{c}_1$ as the left-most element.
2. An error pattern $\mathbf{e}_2$ is picked and placed under $\mathbf{c}_1$, and a second row is formed by adding $\mathbf{e}_2$ to each of the remaining code vectors in the first row; it is important that the error pattern chosen as the first element in a row not have previously appeared in the standard array.
3. Step 2 is repeated until all the possible error patterns have been accounted for.

Figure 11.7 illustrates the structure of the standard array so constructed. The 2^k columns of this array represent the disjoint subsets $D_1, D_2, \ldots, D_{2^k}$. The 2^{n-k} rows of the array represent the cosets of the code, and their first elements $\mathbf{e}_2, \ldots, \mathbf{e}_{2^{n-k}}$ are called *coset leaders*.

$$
\begin{array}{cccccc}
\mathbf{c}_1 = \mathbf{0} & \mathbf{c}_2 & \mathbf{c}_3 & \cdots & \mathbf{c}_i & \cdots & \mathbf{c}_{2^k} \\
\mathbf{e}_2 & \mathbf{c}_2 + \mathbf{e}_2 & \mathbf{c}_3 + \mathbf{e}_2 & \cdots & \mathbf{c}_i + \mathbf{e}_2 & \cdots & \mathbf{c}_{2^k} + \mathbf{e}_2 \\
\mathbf{e}_3 & \mathbf{c}_2 + \mathbf{e}_3 & \mathbf{c}_3 + \mathbf{e}_3 & \cdots & \mathbf{c}_i + \mathbf{e}_3 & \cdots & \mathbf{c}_{2^k} + \mathbf{e}_3 \\
\vdots & \vdots & \vdots & & \vdots & & \vdots \\
\mathbf{e}_j & \mathbf{c}_2 + \mathbf{e}_j & \mathbf{c}_3 + \mathbf{e}_j & \cdots & \mathbf{c}_i + \mathbf{e}_j & \cdots & \mathbf{c}_{2^k} + \mathbf{e}_j \\
\vdots & \vdots & \vdots & & \vdots & & \vdots \\
\mathbf{e}_{2^{n-k}} & \mathbf{c}_2 + \mathbf{e}_{2^{n-k}} & \mathbf{c}_3 + \mathbf{e}_{2^{n-k}} & & \mathbf{c}_i + \mathbf{e}_{2^{n-k}} & \cdots & \mathbf{c}_{2^k} + \mathbf{e}_{2^{n-k}}
\end{array}
$$

Figure 11.7 Standard array for an (n,k) block code.

For a given channel, the probability of decoding error is minimized when the most likely error patterns (i.e., those with the largest probability of occurrence) are chosen as the coset leaders. In the case of a binary symmetric channel, the smaller the Hamming weight of an error pattern the more likely it is to occur. Accordingly, the standard array should be constructed with each coset leader having the minimum Hamming weight in its coset.

We may now describe a decoding procedure for a linear block code:

1. For the received vector $\mathbf{r}$, compute the syndrome $\mathbf{s} = \mathbf{r}\mathbf{H}^T$.
2. Within the coset characterized by the syndrome $\mathbf{s}$, identify the coset leader (i.e., the error pattern with the largest probability of occurrence); call it $\mathbf{e}_0$.
3. Compute the code vector

$$\mathbf{c} = \mathbf{r} + \mathbf{e}_0 \tag{11.26}$$

as the decoded version of the received vector $\mathbf{r}$.

This procedure is called *syndrome decoding*.

EXAMPLE 2 Hamming Codes[3]

Consider a family of (n,k) linear block codes that have the following parameters:

Block length: $n = 2^m - 1$

Number of message bits: $k = 2^m - m - 1$

Number of parity bits: $n - k = m$

where $m \geq 3$. These are the so-called *Hamming codes*.

Consider, for example, the $(7, 4)$ Hamming code with $n = 7$ and $k = 4$, corresponding to $m = 3$. The generator matrix of the code must have a structure that conforms to Eq. (11.12). The following matrix represents an appropriate generator matrix for the $(7, 4)$ Hamming code:

$$
\mathbf{G} = \left[\begin{array}{ccc|cccc}
1 & 1 & 0 & 1 & 0 & 0 & 0 \\
0 & 1 & 1 & 0 & 1 & 0 & 0 \\
1 & 1 & 1 & 0 & 0 & 1 & 0 \\
1 & 0 & 1 & 0 & 0 & 0 & 1
\end{array}\right]
$$

$$\underbrace{}_{\mathbf{P}} \quad \underbrace{}_{\mathbf{I}_k}$$

Table 11.1 Code Words of a (7, 4) Hamming Code

Message Word	Code Word	Weight of Code Word	Message Word	Code Word	Weight of Code Word
0 0 0 0	0 0 0 0 0 0 0	0	1 0 0 0	1 1 0 1 0 0 0	3
0 0 0 1	1 0 1 0 0 0 1	3	1 0 0 1	0 1 1 1 0 0 1	4
0 0 1 0	1 1 1 0 0 1 0	4	1 0 1 0	0 0 1 1 0 1 0	3
0 0 1 1	0 1 0 0 0 1 1	3	1 0 1 1	1 0 0 1 0 1 1	4
0 1 0 0	0 1 1 0 1 0 0	3	1 1 0 0	1 0 1 1 1 0 0	4
0 1 0 1	1 1 0 0 1 0 1	4	1 1 0 1	0 0 0 1 1 0 1	3
0 1 1 0	1 0 0 0 1 1 0	3	1 1 1 0	0 1 0 1 1 1 0	4
0 1 1 1	0 0 1 0 1 1 1	4	1 1 1 1	1 1 1 1 1 1 1	7

The corresponding parity-check matrix is given by

$$
\mathbf{H} = \underbrace{\begin{bmatrix} 1 & 0 & 0 \\ 0 & 1 & 0 \\ 0 & 0 & 1 \end{bmatrix}}_{\mathbf{I}_{n-k}} \Bigg| \underbrace{\begin{bmatrix} 1 & 0 & 1 & 1 \\ 1 & 1 & 1 & 0 \\ 0 & 1 & 1 & 1 \end{bmatrix}}_{\mathbf{P}^T}
$$

With $k = 4$, there are $2^k = 16$ distinct message words, which are listed in Table 11.1. For a given message word, the corresponding code word is obtained by using Eq. (11.13). Thus, the application of this equation results in the 16 code words listed in Table 11.1.

In Table 11.1, we have also listed the Hamming weights of the individual code words in the (7, 4) Hamming code. Since the smallest of the Hamming weights for the nonzero code words is 3, it follows that the minimum distance of the code is 3. Indeed, Hamming codes have the property that the minimum distance $d_{\min} = 3$, independent of the value assigned to the number of parity bits m.

To illustrate the relation between the minimum distance $d_{\min}$ and the structure of the parity-check matrix $\mathbf{H}$, consider the code word 0110100. In the matrix multiplication defined by Eq. (11.16), the nonzero elements of this code word "sift" out the second, third, and fifth columns of the matrix $\mathbf{H}$ yielding

$$
\begin{bmatrix} 0 \\ 1 \\ 0 \end{bmatrix} + \begin{bmatrix} 0 \\ 0 \\ 1 \end{bmatrix} + \begin{bmatrix} 0 \\ 1 \\ 1 \end{bmatrix} = \begin{bmatrix} 0 \\ 0 \\ 0 \end{bmatrix}
$$

We may perform similar calculations for the remaining 14 nonzero code words. We thus find that the smallest number of columns in $\mathbf{H}$ that sums to zero is 3, confirming the earlier statement that $d_{\min} = 3$.

An important property of Hamming codes is that they satisfy the condition of Eq. (11.25) with the equality sign, assuming that $t = 1$. This means that Hamming codes are *single-error correcting binary perfect codes*.

Assuming single-error patterns, we may formulate the seven coset leaders listed in the right-hand column of Table 11.2. The corresponding syndromes,

Table 11.2 Decoding
Table for the (7, 4)
Hamming Code Defined
in Table 11.1

Syndrome	Error Pattern
0 0 0	0 0 0 0 0 0 0
1 0 0	1 0 0 0 0 0 0
0 1 0	0 1 0 0 0 0 0
0 0 1	0 0 1 0 0 0 0
1 1 0	0 0 0 1 0 0 0
0 1 1	0 0 0 0 1 0 0
1 1 1	0 0 0 0 0 1 0
1 0 1	0 0 0 0 0 0 1

listed in the left-hand column, are calculated in accordance with Eq. (11.20). The zero syndrome signifies no transmission errors.

Suppose, for example, the code vector [1110010] is sent, and the received vector is [1100010] with an error in the third bit. Using Eq. (11.19), the syndrome is calculated to be

$$\mathbf{s} = [1100010] \begin{bmatrix} 1 & 0 & 0 \\ 0 & 1 & 0 \\ 0 & 0 & 1 \\ 1 & 1 & 0 \\ 0 & 1 & 1 \\ 1 & 1 & 1 \\ 1 & 0 & 1 \end{bmatrix}$$

$$= [0 \quad 0 \quad 1]$$

From Table 11.2 the corresponding coset leader (i.e., error pattern with the highest probability of occurrence) is found to be [0010000], indicating correctly that the third bit of the received vector is erroneous. Thus, adding this error pattern to the received vector, in accordance with Eq. (11.26), yields the correct code vector actually sent.

Dual Code

Given a linear block code, we may define its *dual* as follows. Taking the transpose of both sides of Eq. (11.15), we have

$$\mathbf{GH}^T = \mathbf{0}$$

where $\mathbf{H}^T$ is the transpose of the parity-check matrix of the code, and $\mathbf{0}$ is a new zero matrix. This equation suggests that every (n,k) linear block code with generator matrix $\mathbf{G}$ and parity-check matrix $\mathbf{H}$ has a *dual code* with parameters $(n, n - k)$, generator matrix $\mathbf{H}$ and parity-check matrix $\mathbf{G}$.

11.4   CYCLIC CODES

Cyclic codes form a subclass of linear block codes. Indeed, many of the important linear block codes discovered to date are either cyclic codes or closely related to cyclic codes. An advantage of cyclic codes over most other types of codes is that they are easy to encode. Furthermore, cyclic codes possess a well-defined mathematical structure, which has led to the development of very efficient decoding schemes for them.

A binary code is said to be a *cyclic code* if it exhibits two fundamental properties:

1. *Linearity property:*   *The sum of any two code words in the code is also a code word.*
2. *Cyclic property:*   *Any cyclic shift of a code word in the code is also a code word.*

Property 1 restates the fact that a cyclic code is a linear block code (i.e., it can be described as a parity-check code). To restate Property 2 in mathematical terms, let the n-tuple $(c_0, c_1, \ldots, c_{n-1})$ denote a code word of an (n,k) linear block code. The code is a cyclic code if the n-tuples

$$(c_{n-1}, c_0, \ldots, c_{n-2}),$$
$$(c_{n-2}, c_{n-1}, \ldots, c_{n-3}),$$
$$\vdots$$
$$(c_1, c_2, \ldots, c_{n-1}, c_0)$$

are all code words in the code.

To develop the algebraic properties of cyclic codes, we use the elements $c_0, c_1, \ldots, c_{n-1}$ of a code word to define the *code polynomial*

$$c(X) = c_0 + c_1 X + c_2 X^2 + \cdots + c_{n-1} X^{n-1} \qquad (11.27)$$

where X is an indeterminate. Naturally, for binary codes, the coefficients are 1s and 0s. Each power of X in the polynomial $c(X)$ represents a one-bit *shift* in time. Hence, multiplication of the polynomial $c(X)$ by X may be viewed as a shift to the right. The key question is: How do we make such a shift *cyclic*? The answer to this question is addressed next.

Let the code polynomial $c(X)$ be multiplied by X^i, yielding

$$\begin{aligned}
X^i c(X) &= X^i(c_0 + c_1 X + \cdots + c_{n-i-1}X^{n-i-1} + c_{n-i}X^{n-i} \\
&\quad + \cdots + c_{n-1}X^{n-1}) \\
&= c_0 X^i + c_1 X^{i+1} + \cdots + c_{n-i-1}X^{n-1} + c_{n-i}X^n \\
&\quad + \cdots + c_{n-1}X^{n+i-1} \qquad (11.28) \\
&= c_{n-i}X^n + \cdots + c_{n-1}X^{n+i-1} + c_0 X^i + c_1 X^{i+1} \\
&\quad + \cdots + c_{n-i-1}X^{n-1}
\end{aligned}$$

where, in the last line, we have merely rearranged terms. Recognizing, for example, that $c_{n-i} + c_{n-i} = 0$ in modulo-2 addition, we may manipulate the first i terms of Eq. (11.28) as follows:

$$X^i c(X) = c_{n-i} + \cdots + c_{n-1}X^{i-1} + c_0 X^i + c_1 X^{i+1} + \cdots + c_{n-i-1}X^{n-1}$$
$$+ c_{n-i}(X^n + 1) + \cdots + c_{n-1}X^{i-1}(X^n + 1) \qquad (11.29)$$

Next, we introduce the following definitions:

$$c^{(i)}(X) = c_{n-i} + \cdots + c_{n-1}X^{i-1} + c_0 X^i + c_1 X^{i+1} \qquad (11.30)$$
$$+ \cdots + c_{n-i-1}X^{n-1}$$

$$q(X) = c_{n-i} + c_{n-i+1}X + \cdots + c_{n-1}X^{i-1} \qquad (11.31)$$

Accordingly, Eq. (11.29) is reformulated in the compact form

$$X^i c(X) = q(X)(X^n + 1) + c^{(i)}(X) \qquad (11.32)$$

The polynomial $c^{(i)}(X)$ is recognized as the code polynomial of the code word $(c_{n-i}, \ldots, c_{n-1}, c_0, c_1, \ldots, c_{n-i-1})$ obtained by applying i cyclic shifts to the code word $(c_0, c_1, \ldots, c_{n-i-1}, c_{n-i}, \ldots, c_{n-1})$. Moreover, from Eq. (11.32) we readily see that $c^{(i)}(X)$ is the remainder that results from dividing $X^i c(X)$ by $(X^n + 1)$. We may thus formally state the cyclic property in polynomial notation as follows: *If $c(X)$ is a code polynomial, then the polynomial*

$$c^{(i)}(X) = X^i c(X) \bmod (X^n + 1) \qquad (11.33)$$

is also a code polynomial for any cyclic shift i; the term "mod" is the abbreviation for "modulo." The special form of polynomial multiplication described in Eq. (11.33) is referred to as *multiplication modulo $X^n + 1$*. In effect, the multiplication is subject to the constraint $X^n = 1$, the application of which restores the polynomial $X^i c(X)$ to order $n - 1$ for all $i < n$. (Note that in modulo-2 arithmetic, $X^n + 1$ has the same value as $X^n - 1$.)

Generator Polynomial

The polynomial $X^n + 1$ and its factors play a major role in the generation of cyclic codes. Let $g(X)$ be a polynomial of degree $n - k$ that is a factor of $X^n + 1$; as such, $g(X)$ is *the polynomial of least degree in the code*. In general, $g(X)$ may be expanded as follows:

$$g(X) = 1 + \sum_{i=1}^{n-k-1} g_i X^i + X^{n-k} \qquad (11.34)$$

where the coefficient g_i is equal to 0 or 1. According to this expansion, the polynomial $g(X)$ has two terms with coefficient 1 separated by $n - k - 1$ terms. The polynomial $g(X)$ is called the *generator polynomial* of a cyclic code. A cyclic code is uniquely determined by the generator polynomial $g(X)$ in that each code polynomial in the code can be expressed in the form of a polynomial product as follows:

$$c(X) = a(X)g(X) \qquad (11.35)$$

where $a(X)$ is a polynomial in X with degree $k - 1$. The $c(X)$ so formed satisfies the condition of Eq. (11.33) since $g(X)$ is a factor of $X^n + 1$.

Suppose we are given the generator polynomial $g(X)$ and the requirement is to encode the message sequence $(m_0, m_1, \ldots, m_{k-1})$ into an (n,k) *systematic cyclic code*. That is, the message bits are transmitted in unaltered form, as shown by the following structure for a code word (see Fig. 11.4):

$$\underbrace{(b_0, b_1, \ldots, b_{n-k-1},}_{n-k \text{ parity bits}} \quad \underbrace{m_0, m_1, \ldots, m_{k-1})}_{k \text{ message bits}}$$

Let the *message polynomial* be defined by

$$m(X) = m_0 + m_1 X + \cdots + m_{k-1} X^{k-1} \tag{11.36}$$

and let

$$b(X) = b_0 + b_1 X + \cdots + b_{n-k-1} X^{n-k-1} \tag{11.37}$$

According to Eq.(11.1), we want the code polynomial to be in the form

$$c(X) = b(X) + X^{n-k} m(X) \tag{11.38}$$

Hence, the use of Eqs. (11.35) and (11.38) yields

$$a(X)g(X) = b(X) + X^{n-k} m(X)$$

Equivalently, in light of modulo-2 addition, we may write

$$\frac{X^{n-k} m(X)}{g(X)} = a(X) + \frac{b(X)}{g(X)} \tag{11.39}$$

Equation (11.39) states that the polynomial $b(X)$ is the *remainder* left over after dividing $X^{n-k} m(X)$ by $g(X)$.

We may now summarize the steps involved in the encoding procedure for an (n,k) cyclic code assured of a systematic structure. Specifically, we proceed as follows:

1. Multiply the message polynomial $m(X)$ by X^{n-k}.
2. Divide $X^{n-k} m(X)$ by the generator polynomial $g(X)$, obtaining the remainder $b(X)$.
3. Add $b(X)$ to $X^{n-k} m(X)$, obtaining the code polynomial $c(X)$.

Parity-Check Polynomial

An (n,k) cyclic code is uniquely specified by its generator polynomial $g(X)$ of order $(n-k)$. Such a code is also uniquely specified by another polynomial of degree k, which is called the *parity-check polynomial*, defined by

$$h(X) = 1 + \sum_{i=1}^{k-1} h_i X^i + X^k \tag{11.40}$$

where the coefficients h_i are 0 or 1. The parity-check polynomial $h(X)$ has a form similar to the generator polynomial in that there are two terms with coefficient 1, but separated by $k - 1$ terms.

The generator polynomial $g(X)$ is equivalent to the generator matrix **G** as a description of the code. Correspondingly, the parity-check polynomial, denoted by $h(X)$, is an equivalent representation of the parity-check matrix **H**. We thus find that the matrix relation $\mathbf{H}\mathbf{G}^T = \mathbf{O}$ presented in Eq. (11.15) for linear block codes corresponds to the relationship

$$g(X)h(X)\bmod(X^n + 1) = 0 \tag{11.41}$$

Accordingly, we may state that *the generator polynomial $g(X)$ and the parity-check polynomial $h(X)$ are factors of the polynomial $X^n + 1$,* as shown by

$$g(X)h(X) = X^n + 1 \tag{11.42}$$

This property provides the basis for selecting the generator or parity-check polynomial of a cyclic code. In particular, we may state that if $g(X)$ is a polynomial of degree $(n - k)$ and it is also a factor of $X^n + 1$, then $g(X)$ is the generator polynomial of an (n,k) cyclic code. Equivalently, we may state that if $h(X)$ is a polynomial of degree k and it is also a factor of $X^n + 1$, then $h(X)$ is the parity-check polynomial of an (n,k) cyclic code.

A final comment is in order. Any factor of $X^n + 1$ with degree $(n - k)$, the number of parity bits, can be used as a generator polynomial. For large values of n, the polynomial $X^n + 1$ may have many factors of degree $n - k$. Some of these polynomial factors generate good cyclic codes, whereas some of them generate bad cyclic codes. The issue of how to select generator polynomials that produce good cyclic codes is very difficult to resolve. Indeed, coding theorists have expended much effort in the search for good cyclic codes.

Generator and Parity-Check Matrices

Given the generator polynomial $g(X)$ of an (n,k) cyclic code, we may construct the generator matrix **G** of the code by noting that the k polynomials $g(X)$, $Xg(X), \ldots, X^{k-1}g(X)$ span the code. Hence, the n-tuples corresponding to these polynomials may be used as rows of the k-by-n generator matrix **G**.

However, the construction of the parity-check matrix **H** of the cyclic code from the parity-check polynomial $h(X)$ requires special attention, as described here. Multiplying Eq. (11.42) by $a(x)$ and then using Eq. (11.35), we obtain

$$c(X)h(X) = a(X) + X^n a(X) \tag{11.43}$$

The polynomials $c(X)$ and $h(X)$ are themselves defined by Eqs. (11.27) and (11.40), respectively, which means that their product on the left-hand side of Eq. (11.43) contains terms with powers extending up to $n + k - 1$. On the other hand, the polynomial $a(X)$ has degree $k - 1$ or less, the implication of which is that the powers of $X^k, X^{k+1}, \ldots, X^{n-1}$ do *not* appear in the polynomial on the right-hand side of Eq. (11.43). Thus, setting the coefficients of $X^k, X^{k-1}, \ldots,$ X^{n-1} in the expansion of the product polynomial $c(X)h(X)$ equal to zero, we obtain the following set of $n - k$ equations:

$$\sum_{i=j}^{j+k} c_i h_{k+j-i} = 0 \qquad \text{for } 0 \le j \le n - k - 1 \qquad (11.44)$$

Comparing Eq. (11.44) with the corresponding relation of Eq. (11.16), we may make the following important observation: The coefficients of the parity-check polynomial $h(X)$ involved in the polynomial multiplication described in Eq. (11.44) are arranged in *reversed* order with respect to the coefficients of the parity-check matrix **H** involved in forming the inner product of vectors described in Eq. (11.16). This observation suggests that we define the *reciprocal of the parity-check polynomial* as follows:

$$X^k h(X^{-1}) = X^k\left(1 + \sum_{i=1}^{k-1} h_i X^{-i} + X^{-k}\right)$$
$$= 1 + \sum_{i=1}^{k-1} h_{k-i} X^i + X^k \qquad (11.45)$$

which is also a factor of $X^n + 1$. The n-tuples pertaining to the $(n - k)$ polynomials $X^k h(X^{-1})$, $X^{k+1} h(X^{-1})$, ..., $X^{n-1} h(X^{-1})$ may now be used as rows of the $(n - k)$-by-n parity-check matrix **H**.

In general, the generator matrix **G** and the parity-check matrix **H** constructed in the manner described here are not in their systematic forms. They can be put into their systematic forms by performing simple operations on their respective rows, as illustrated in Example 3 presented later in the section.

Encoder for Cyclic Codes

Earlier we showed that the encoding procedure for an (n,k) cyclic code in systematic form involves three steps: (1) multiplication of the message polynomial $m(X)$ by X^{n-k}, (2) division of $X^{n-k}m(X)$ by the generator polynomial $g(X)$ to obtain the remainder $b(X)$, and (3) addition of $b(X)$ to $X^{n-k}m(X)$ to form the desired code polynomial. These three steps can be implemented by means of the encoder shown in Fig. 11.8, consisting of a *linear feedback shift register* with $(n - k)$ stages.

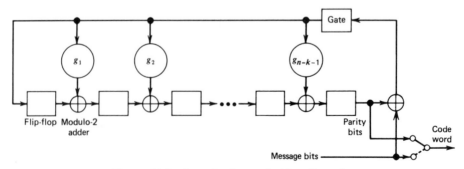

Figure 11.8 Encoder for an (n,k) cyclic code.

The boxes in Fig. 11.8 represent *flip-flops* or *unit-delay elements*. The flip-flop is a device that resides in one of two possible states denoted by 0 and 1. An *external clock* (not shown in Fig. 11.8) controls the operation of all the flip-flops. Every time the clock ticks, the contents of the flip-flops (initially set to the state 0) are shifted out in the directionof the arrows. In addition to the flip-flops, the encoder of Fig. 11.8 includes a second set of logic elements, namely, *adders*, which compute the modulo-2 sums of their respective inputs. Finally, the *multipliers* multiply their respective inputs by the associated coefficients. In particular, if the coefficient $g_i = 1$, the multiplier is just a direct "connection." If, on the other hand, the coefficient $g_i = 0$, the multiplier is "no connection."

The operation of the encoder shown in Fig. 11.8 proceeds as follows:

1. The gate is switched on. Hence, the k message bits are shifted into the channel. As soon as the k message bits have entered the shift register, the resulting $(n - k)$ bits in the register form the parity bits [recall that the parity bits are the same as the coefficients of the remainder $b(X)$].

2. The gate is switched off, thereby breaking the feedback connections.

3. The contents of the shift register are shifted out into the channel.

Calculation of the Syndrome

Suppose the code word $(c_0, c_1, \ldots, c_{n-1})$ is transmitted over a noisy channel, resulting in the received word $(r_0, r_1, \ldots, r_{n-1})$. From Section 11.3, we recall that the first step in the coding of a linear block code is to calculate the syndrome for the received word. If the syndrome is zero, there are no transmission errors in the received word. If, on the other hand, the syndrome is nonzero, the received word contains transmission errors that require correction.

In the case of a cyclic code in systematic form, the syndrome can be calculated easily. Let the received word be represented by a polynomial of degree $n - 1$ or less, as shown by

$$r(X) = r_0 + r_1 X + \cdots + r_{n-1} X^{n-1} \tag{11.46}$$

Let $q(X)$ denote the quotient and $s(X)$ denote the remainder, which are the results of dividing $r(X)$ by the generator polynomial $g(X)$. We may therefore express $r(X)$ as follows:

$$r(X) = q(X)g(X) + s(X) \tag{11.47}$$

The remainder $s(X)$ is a polynomial of degree $n - k - 1$ or less, which is the result of interest. It is called the *syndrome polynomial* because its coefficients make up the $(n - k)$-by-1 syndrome $\mathbf{s}$.

Figure 11.9 shows a *syndrome calculator* that is identical to the encoder of Fig. 11.8 except for the fact that the received bits are fed into the $(n - k)$ stages of the feedback shift register from the left. As soon as all the received bits have been shifted into the shift register, its contents define the syndrome $\mathbf{s}$.

The syndrome polynomial $s(X)$ has the following useful properties that follow from the definition given in Eq. (11.47).

1. *The syndrome of a received word polynomial is also the syndrome of the corresponding error polynomial.*

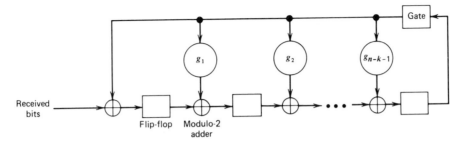

Figure 11.9 Syndrome calculator.

Given that a cyclic code with polynomial $c(X)$ is sent over a noisy channel, the received word polynomial is defined by

$$r(X) = c(X) + e(X) \qquad (11.48)$$

where $e(X)$ is the *error polynomial*. Equivalently, we may write

$$e(X) = r(X) + c(X) \qquad (11.49)$$

Hence, substituting Eqs. (11.35) and (11.47) in (11.49), we get

$$e(X) = u(X)g(X) + s(X) \qquad (11.50)$$

where the quotient is $u(X) = a(X) + q(X)$. Equation (11.50) shows that $s(X)$ is also the syndrome of the error polynomial $e(X)$. The implication of this property is that when the syndrome polynomial $s(X)$ is nonzero, the presence of transmission errors in the received word is detected.

2. *Let $s(X)$ be the syndrome of a received word polynomial $r(X)$. Then, the syndrome of $Xr(X)$, a cyclic shift of $r(X)$, is $Xs(X)$.*

Applying a cyclic shift to both sides of Eq. (11.47), we get

$$Xr(X) = Xq(X)g(X) + Xs(X) \qquad (11.51)$$

from which we readily see that $Xs(X)$ is the remainder of the division of $Xr(X)$ by $g(X)$. Hence, the syndrome of $Xr(X)$ is $Xs(X)$ as stated. We may generalize this result by stating that if $s(X)$ is the syndrome of $r(X)$, then $X^i s(X)$ is the syndrome of $X^i r(X)$.

3. *The syndrome polynomial $s(X)$ is identical to the error polynomial $e(X)$, assuming that the errors are confined to the $(n - k)$ parity-check bits of the received word polynomial $r(X)$.*

The assumption made here is another way of saying that the degree of the error polynomial $e(X)$ is less than or equal to $(n - k - 1)$. Since the generator polynomial $g(X)$ is of degree $(n - k)$, by definition, it follows that Eq. (11.50) can only be satisfied if the quotient $u(X)$ is zero. In other words, the error polynomial $e(X)$ and the syndrome polynomial $s(X)$ are one and the same. The implication of Property 3 is that, under the aforementioned conditions, error correction can be accomplished simply by adding the syndrome polynomial $s(X)$ to the received word polynomial $r(X)$.

EXAMPLE 3   Hamming Codes Revisited

To illustrate the issues relating to the polynomial representation of cyclic codes, we consider the generation of a (7, 4) cyclic code. With the block length $n = 7$, we start by factorizing $X^7 + 1$ into three *irreducible polynomials*:

$$X^7 + 1 = (1 + X)(1 + X^2 + X^3)(1 + X + X^3)$$

By an "irreducible polynomial" we mean a polynomial that cannot be factored using only polynomials with coefficients from the binary field. An irreducible polynomial of degree m is said to be *primitive* if the smallest positive integer n for which the polynomial divides $X^n + 1$ is $n = 2^m - 1$. For the example at hand, the two polynomials $(1 + X^2 + X^3)$ and $(1 + X + X^3)$ are primitive. Let us take

$$g(X) = 1 + X + X^3$$

as the generator polynomial, whose degree equals the number of parity bits. This means that the parity-check polynomial is given by

$$h(X) = (1 + X)(1 + X^2 + X^3)$$
$$= 1 + X + X^2 + X^4$$

whose degree equals the number of message bits $k = 4$.

Next, we illustrate the procedure for the construction of a code word by using this generator polynomial to encode the message sequence 1001. The corresponding message polynomial is given by

$$m(X) = 1 + X^3$$

Hence, multiplying $m(X)$ by $X^{n-k} = X^3$, we get

$$X^{n-k}m(X) = X^3 + X^6$$

The second step is to divide $X^{n-k}m(X)$ by $g(X)$, the details of which (for the example at hand) are given below:

$$
\begin{array}{r}
X^3 + X \\
X^3 + X + 1 \overline{\smash{\big)}\, X^6 \qquad\qquad + X^3} \\
\underline{X^6 \qquad + X^4 + X^3} \\
X^4 \\
\underline{X^4 \qquad\quad + X^2 + X} \\
X^2 + X
\end{array}
$$

Note that in this long division we have treated subtraction the same as addition, since we are operating in modulo-2 arithmetic. We may thus write

$$\frac{X^3 + X^6}{1 + X + X^3} = X + X^3 + \frac{X + X^2}{1 + X + X^3}$$

That is, the quotient $a(X)$ and remainder $b(X)$ are as follows, respectively:

$$a(X) = X + X^3$$
$$b(X) = X + X^2$$

Hence, from Eq. (11.38) we find that the desired code polynomial is

$$c(X) = b(X) + X^{n-k}m(X)$$
$$= X + X^2 + X^3 + X^6$$

The code word is therefore 0111001. The four right-most bits, 1001, are the specified message bits. The three left-most bits, 011, are the parity-check bits. The code word thus generated is exactly the same as the corresponding one shown in Table 11.1 for a (7, 4) Hamming code.

We may generalize this result by stating that *any cyclic code generated by a primitive polynomial is a Hamming code of minimum distance 3.*

We next show that the generator polynomial $g(X)$ and the parity-check polynomial $h(X)$ uniquely specify the generator matrix $\mathbf{G}$ and the parity-check matrix $\mathbf{H}$, respectively.

To construct the 4-by-7 generator matrix $\mathbf{G}$, we start with four polynomials represented by $g(X)$ and three cyclic-shifted versions of it, as shown by

$$g(X) = 1 + X + X^3$$
$$Xg(X) = X + X^2 + X^4$$
$$X^2g(X) = X^2 + X^3 + X^5$$
$$X^3g(X) = X^3 + X^4 + X^6$$

The polynomials $g(X)$, $Xg(X)$, $X^2g(X)$, and $X^3g(X)$ represent code polynomials in the (7, 4) Hamming code. If the coefficients of these polynomials are used as the elements of the rows of a 4-by-7 matrix, we get the following generator matrix:

$$\mathbf{G} = \begin{bmatrix} 1 & 1 & 0 & 1 & 0 & 0 & 0 \\ 0 & 1 & 1 & 0 & 1 & 0 & 0 \\ 0 & 0 & 1 & 1 & 0 & 1 & 0 \\ 0 & 0 & 0 & 1 & 1 & 0 & 1 \end{bmatrix}$$

Clearly, the generator matrix $\mathbf{G}$ so constructed is not in systematic form. We can put it into a systematic form by adding the first row to the third row, and adding the sum of the first two rows to the fourth row. These manipulations result in the new generator matrix

$$\mathbf{G} = \begin{bmatrix} 1 & 1 & 0 & 1 & 0 & 0 & 0 \\ 0 & 1 & 1 & 0 & 1 & 0 & 0 \\ 1 & 1 & 1 & 0 & 0 & 1 & 0 \\ 1 & 0 & 1 & 0 & 0 & 0 & 1 \end{bmatrix}$$

This generator matrix is exactly the same as that in Example 2.

We next show how to construct the 3-by-7 parity-check matrix $\mathbf{H}$ from the parity-check polynomial $h(X)$. To do this, we first take the *reciprocal* of $h(X)$, namely, $X^4h(X^{-1})$. For the problem at hand, we form three polynomials represented by $X^4h(X^{-1})$ and two shifted versions of it, as shown by

$$X^4h(X^{-1}) = 1 + X^2 + X^3 + X^4$$
$$X^5h(X^{-1}) = X + X^3 + X^4 + X^5$$
$$X^6h(X^{-1}) = X^2 + X^4 + X^5 + X^6$$

Using the coefficients of these three polynomials as the elements of the rows of

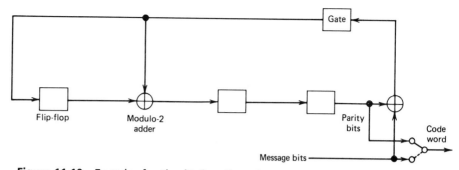

Figure 11.10 Encoder for the (7,4) cyclic code generated by $g(X) = 1 + X + X^3$.

the 3-by-7 parity-check matrix, we get

$$\mathbf{H} = \begin{bmatrix} 1 & 0 & 1 & 1 & 1 & 0 & 0 \\ 0 & 1 & 0 & 1 & 1 & 1 & 0 \\ 0 & 0 & 1 & 0 & 1 & 1 & 1 \end{bmatrix}$$

Here again we see that the matrix **H** is not in systematic form. To put it into a systematic form, we add the third row to the first row to obtain

$$\mathbf{H} = \begin{bmatrix} 1 & 0 & 0 & 1 & 0 & 1 & 1 \\ 0 & 1 & 0 & 1 & 1 & 1 & 0 \\ 0 & 0 & 1 & 0 & 1 & 1 & 1 \end{bmatrix}$$

This parity-check matrix is exactly the same as that of Example 2.

Figure 11.10 shows the encoder for the (7, 4) cyclic Hamming code generated by the polynomial $g(X) = 1 + X + X^3$. To illustrate the operation of this encoder, consider the message sequence (1001). The contents of the shift register are modified by the incoming message bits as in Table 11.3. After four shifts, the contents of the shift register, and therefore the parity bits, are (011). Accordingly, appending these parity bits to the message bits (1001), we get the code word (0111001); this result is exactly the same as that determined earlier in the example.

Figure 11.11 shows the corresponding syndrome calculator for the (7, 4) Hamming code. Let the transmitted code word be (0111001) and the received

Table 11.3 Contents of the Shift Register in the Encoder of Fig. 11.10 for Message Sequence (1001)

Shift	Input	Register Contents
		0 0 0 (initial state)
1	1	1 1 0
2	0	0 1 1
3	0	1 1 1
4	1	0 1 1

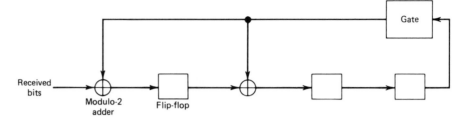

Figure 11.11 Syndrome calculator for the (7,4) cyclic code generated by the polynomial $g(X) = 1 + X + X^3$.

word be (0110001); that is, the middle bit is in error. As the received bits are fed into the shift register, initially set to zero, its contents are modified as in Table 11.4. At the end of the seventh shift, the syndrome is identified from the contents of the shift register as 110. Since the syndrome is nonzero, the received word is in error. Moreover, from Table 11.2, we see that the error pattern corresponding to this syndrome is 0001000. This indicates that the error is in the middle bit of the received word, which is indeed the case.

EXAMPLE 4   Maximum-Length Codes

For any positive integer $m \geq 3$, there exists a *maximum-length code* with the following parameters:

$$\text{Block length:} \qquad n = 2^m - 1$$

$$\text{Number of message bits:} \quad k = m$$

$$\text{Minimum distance:} \qquad d_{\min} = 2^{m-1}$$

Maximum-length codes are generated by polynomials of the form

$$g(X) = \frac{1 + X^n}{h(X)} \tag{11.52}$$

Table 11.4   Contents of the Syndrome Calculator in Fig. 11.11 for the Received Word 0110001

Shift	Input Bit	Contents of Shift Register
		000 (initial state)
1	1	100
2	0	010
3	0	001
4	0	110
5	1	111
6	1	001
7	0	110

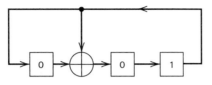

Figure 11.12 Encoder for the (7,3) maximum length code; the initial state of the encoder is shown in the figure.

where $h(X)$ is any primitive polynomial of degree m. Earlier we stated that any cyclic code generated by a primitive polynomial is a Hamming code of minimum distance 3 (see Example 3). It follows therefore that maximum-length codes are the *dual* of Hamming codes.

The polynomial $h(X)$ defines the feedback connections of the encoder. The generator polynomial $g(X)$ defines one period of the maximum-length code, assuming that the encoder is in the initial state 00 ... 01. To illustrate these points, consider the example of a $(7, 3)$ maximum-length code, which is the dual of the $(7, 4)$ Hamming code described in Example 3. Thus, choosing

$$h(X) = 1 + X + X^3$$

we find that the generator polynomial of the $(7, 3)$ maximum-length code is

$$g(X) = 1 + X + X^2 + X^4$$

Figure 11.12 shows the encoder for the $(7, 3)$ maximum-length code, the feedback connections of which are exactly the same as those shown in Fig. 9.2. The period of the code is $n = 7$. Thus, assuming that the encoder is in the initial state 001, as indicated in Fig. 11.12, we find the output sequence is described by

$$\underbrace{1 \quad 0 \quad 0}_{\text{initial state}} \quad \underbrace{1 \quad 1 \quad 1 \quad 0 \quad 1 \quad 0 \quad 0}_{g(X) = 1 + X + X^2 + X^4}$$

This result may be readily validated by cycling through the encoder of Fig. 11.12. Note that if we were to choose the other primitive polynomial

$$h(X) = 1 + X^2 + X^3$$

for the $(7, 3)$ maximum-length code, we would simply get the "image" of the code described above, and the output sequence would be "reversed" in time.

Other Cyclic Codes

We conclude the discussion of cyclic codes by presenting the characteristics of three other important classes of cyclic codes.

Cyclic Redundancy Check Codes Cyclic codes are extremely well-suited for *error detection*. We make this statement for two reasons. First, they can be designed to detect many combinations of likely errors. Second, the implementation of both encoding and error-detecting circuits is practical. It is for these reasons that virtually all error-detecting codes used in practice are of the cyclic-code type.

Table 11.5   CRC Codes

Code	Generator Polynomial, $g(X)$	$n - k$
CRC-12 code	$1 + X + X^2 + X^3 + X^{11} + X^{12}$	12
CRC-16 code	$1 + X^2 + X^{15} + X^{16}$	16
CRC-CCITT code	$1 + X^5 + X^{12} + X^{16}$	16

A cyclic code used for error-detection is referred to as *cyclic redundancy check (CRC) code*.

We define an *error burst* of length B in an n-bit received word as a contiguous sequence of B bits in which the first and last bits of any number of intermediate bits are received in error. Binary (n,k) CRC codes are capable of detecting the following error patterns:

1. All error bursts of length $n - k$ or less.
2. A fraction of error bursts of length equal to $n - k + 1$; the fraction equals $1 - 2^{-(n-k-1)}$.
3. A fraction of error bursts of length greater than $n - k + 1$; the fraction equals $1 - 2^{-(n-k-1)}$.
4. All combinations of $d_{\min} - 1$ (or fewer) errors.
5. All error patterns with an odd number of errors if the generator polynomial $g(X)$ for the code has an even number of nonzero coefficients.

Table 11.5 presents the generator polynomials of three CRC codes that have become international standards. All three contain $1 + X$ as a prime factor. The CRC-12 code is used for 6-bit characters, and the other two codes are used for 8-bit characters.

Bose–Chaudhuri–Hocquenghem (BCH) Codes[4]   One of the most important and powerful classes of linear block codes are *BCH codes*, which are cyclic codes with a wide variety of parameters. The most common binary BCH codes, known as *primitive BCH codes*, are characterized for any positive integers m (equal to or greater than 3) and t [less than $(2^m - 1)/2$] by the following parameters:

Block length: $n = 2^m - 1$

Number of message bits: $k \geq n - mt$

Minium distance: $d_{\min} \geq 2t + 1$

Each BCH code is a *t-error correcting code* in that it can detect and correct up to t random errors per code word. The Hamming single-error correcting codes can be described as BCH codes. The BCH codes offer flexibility in the choice of code parameters, namely, block length and code rate. Furthermore, for block lengths of a few hundred bits or less, the BCH codes are among the best known codes of the same block length and code rate.

A detailed treatment of the construction of BCH codes is beyond the scope of our present discussion. To provide a feel for their capability, we present in Table 11.6, the code parameters and generator polynomials for binary block

Table 11-6 Binary BCH Codes of Length up to $2^5 - 1$

n	k	t	Generator Polynomial								
7	4	1								1	011
15	11	1								10	011
15	7	2							111	010	001
15	5	3						10	100	110	111
31	26	1								100	101
31	21	2						11	101	101	001
31	16	3				1	000	111	110	101	111
31	11	5		101	100	010	011	011	010	101
31	6	7	11	001	011	011	110	101	000	100	111

Notations: n = block length

k = number of message bits

t = maximum number of detectable errors

The high-order coefficients of the generator polynomial $g(X)$ are at the left.

BCH codes of length up to $2^5 - 1$. For example, suppose we wish to construct the generator polynomial for (15, 7) BCH code. From table 11.6 we have (111 010 001) for the coefficients of the generator polynomial; hence, we may write

$$g(X) = X^8 + X^7 + X^6 + X^4 + 1$$

Reed–Solomon Codes[5] The *Reed–Solomon codes* are an important subclass of *nonbinary* BCH codes; they are often abbreviated as RS codes. The encoder for an RS code differs from a binary encoder in that it operates on multiple bits rather than individual bits. Specifically, an RS (n,k) code is used to encode m-bit symbols into blocks consisting of $n = 2^m - 1$ symbols, that is, $m(2^m - 1)$ bits, where $m \geq 1$. Thus, the encoding algorithm expands a block of k symbols to n symbols by adding $n - k$ redundant symbols. When m is an integer power of two, the m-bit symbols are called *bytes*. A popular value of m is 8; indeed, 8-bit RS codes are extremely powerful.

A t-error-correcting RS code has the following parameters:

Block length: $n = 2^m - 1$ symbols

Message size: k symbols

Parity-check size: $n - k = 2t$ symbols

Minimum distance: $d_{\min} = 2t + 1$ symbols

The block length of the RS code is one less than the size of a code symbol, and the minimum distance is one greater than the number of parity-check symbols. The RS codes make highly efficient use of redundancy, and block lengths and symbol sizes can be adjusted readily to accommodate a wide range of message sizes. Moreover, the RS codes provide a wide range of code rates that can be chosen to optimize performance. Finally, efficient decoding techniques are available for use with RS codes, which is one more reason for their wide application.

11.5   CONVOLUTIONAL CODES[6]

In block coding, the encoder accepts a k-bit message block and generates an n-bit code word. Thus, code words are produced on a block-by-block basis. Clearly, provision must be made in the encoder to buffer an entire message block before generating the associated code word. There are applications, however, where the message bits come in *serially* rather than in large blocks, in which case the use of a buffer may be undesirable. In such situations, the use of *convolutional coding* may be the preferred method. A convolutional encoder operates on the incoming message sequence continuously in a serial manner.

The encoder of a binary convolutional code with rate $1/n$, measured in bits per symbol, may be viewed as a *finite-state machine* that consists of an M-stage shift register with prescribed connections to n modulo-2 adders, and a multiplexer that serializes the outputs of the adders. An L-bit message sequence produces a coded output sequence of length $n(L + M)$ bits. The *code rate* is therefore given by

$$r = \frac{L}{n(L + M)} \quad \text{bits/symbol} \tag{11.53}$$

Typically, we have $L \gg M$. Hence, the code rate simplifies to

$$r \simeq \frac{1}{n} \quad \text{bits/symbol} \tag{11.54}$$

The *constraint length* of a convolutional code, expressed in terms of message bits, is defined as the number of shifts over which a single message bit can influence the encoder output. In an encoder with an M-stage shift register, the *memory* of the encoder equals M message bits, and $K = M + 1$ shifts are required for a message bit to enter the shift register and finally come out. Hence, the constraint length of the encoder is K.

Figure 11.13a shows a convolutional encoder with $n = 2$ and $K = 3$. Hence, the code rate of this encoder is $1/2$. The encoder of Fig. 11.13a operates on the incoming message sequence, one bit at a time.

We may generate a binary convolutional code with rate k/n by using k separate shift registers with prescribed connections to n modulo-2 adders, an input multiplexer and an output multiplexer. An example of such an encoder is shown in Fig. 11.13b, where $k = 2$, $n = 3$, and the two shift registers have $K = 2$ each. The code rate is $2/3$. In this second example, the encoder processes the incoming message sequence two bits at a time.

The convolutional codes generated by the encoders of Fig. 11.13 are *nonsystematic* codes. Unlike block coding, the use of nonsystematic codes is ordinarily preferred over systematic codes in convolutional coding.

Each path connecting the output to the input of a convolutional encoder may be characterized in terms of its *impulse response*, defined as the response of that path to a symbol 1 applied to its input, with each flip-flop in the encoder set initially in the zero state. Equivalently, we may characterize each path in terms of a *generator polynomial*, defined as the *unit-delay transform* of the impulse response. To be specific, let the *generator sequence* $(g_0^{(i)}, g_1^{(i)}, g_2^{(i)}, \ldots, g_M^{(i)})$ denote the impulse response of the ith path, where the coefficients $g_0^{(i)}, g_1^{(i)}, g_2^{(i)}, \ldots,$

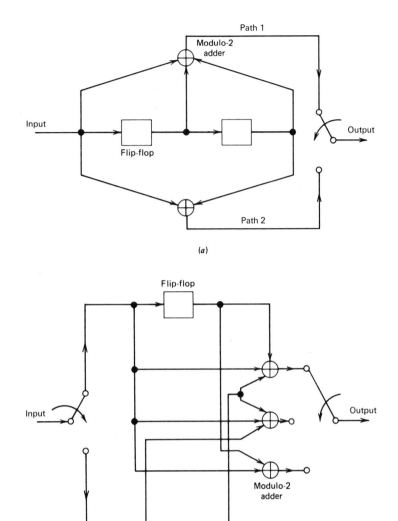

Figure 11.13 (a) Constraint length-3, rate-$\frac{1}{2}$ convolutional encoder. (b) Constraint length-2, rate-$\frac{2}{3}$ convolutional encoder.

$g_M^{(i)}$ equal 0 or 1. Correspondingly, the *generator polynomial* of the ith path is defined by

$$g^{(i)}(D) = g_0^{(i)} + g_1^{(i)}D + g_2^{(i)}D^2 + \cdots + g_M^{(i)}D^M \qquad (11.55)$$

where D denotes the unit-delay variable. The complete convolutional encoder is described by the set of generator polynomials $\{g^{(1)}(D), g^{(2)}(D), \ldots, g^{(n)}(D)\}$. Traditionally, different variables are used for the description of convolutional and cyclic codes, with D being commonly used for convolutional codes and X for cyclic codes.

EXAMPLE 5

Consider the convolutional encoder of Fig. 11.13a, which has two paths numbered 1 and 2 for convenience of reference. The impulse response of path 1 is (1, 1, 1). Hence, the corresponding generator polynomial is given by

$$g^{(1)}(D) = 1 + D + D^2$$

The impulse response of path 2 is (101). Hence, the corresponding generator polynomial is given by

$$g^{(2)}(D) = 1 + D^2$$

For the message sequence (10011), say, we have the polynomial representation

$$m(D) = 1 + D^3 + D^4$$

As with Fourier transformation, convolution in the time domain is transformed into multiplication in the D domain. Hence, the output polynomial of path 1 is given by

$$\begin{aligned} c^{(1)}(D) &= g^{(1)}(D)\,m(D) \\ &= (1 + D + D^2)(1 + D^3 + D^4) \\ &= 1 + D + D^2 + D^3 + D^6 \end{aligned}$$

From this we immediately deduce that the output sequence of path 1 is (1111001). Similarly, the output polynomial of path 2 in Fig. 11.13a is

$$\begin{aligned} c^{(2)}(D) &= g^{(2)}(D)\,m(D) \\ &= (1 + D^2)(1 + D^3 + D^4) \\ &= 1 + D^2 + D^3 + D^4 + D^5 + D^6 \end{aligned}$$

The output sequence of path 2 is therefore (1011111). Finally, multiplexing the two output sequences of paths 1 and 2, we get the encoded sequence

$$\mathbf{c} = (11, 10, 11, 11, 01, 01, 11)$$

Note that the message sequence of length $L = 5$ bits produces an encoded sequence of length $n(L + K - 1) = 14$ bits. Note also that for the shift register to be restored to its zero initial state, a terminating sequence of $K - 1 = 2$ zeros is appended to the last input bit of the message sequence. The terminating sequence of $K - 1$ zeros is called the *tail of the message.*

Code Tree, Trellis, and State Diagram

Traditionally, the structural properties of a convolutional encoder are portrayed in graphical form by using any one of three equivalent diagrams: code tree, trellis, and state diagram. We will use the convolutional encoder of Fig. 11.13a as a running example to illustrate the insights that each one of these three diagrams can provide.

We begin the discussion with the *code tree* of Fig. 11.14. Each branch of the tree represents an input symbol, with the corresponding pair of output binary symbols indicated on the branch. The convention used to distinguish the input binary symbols 0 and 1 is as follows. An input 0 specifies the upper branch of a bifurcation, whereas input 1 specifies the lower branch. A specific *path* in the

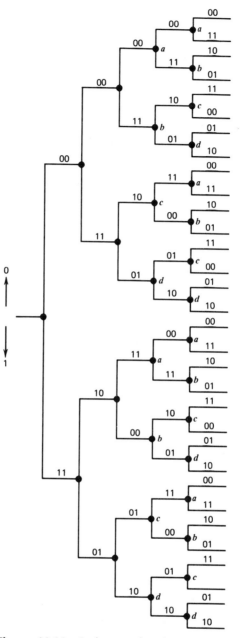

Figure 11.14 Code tree for the convolutional encoder of Fig. 11.13a.

tree is traced from left to right in accordance with the input (message) sequence. The corresponding coded symbols on the branches of that path constitute the input (message) sequence. Consider, for example, the message sequence (10011) applied to the input of the encoder of Fig. 11.13a. Following the procedure just described, we find that the corresponding encoded sequence is (11, 10, 11, 11, 01), which agrees with the first 5 pairs of bits in the encoded sequence $\{c_i\}$ derived in Example 5.

From the diagram of Fig. 11.14, we observe that the tree becomes *repetitive* after the first three branches. Indeed, beyond the third branch, the two nodes labeled *a* are identical, and so are all the other node pairs that are identically labeled. We may establish this repetitive property of the tree by examining the associated encoder of Fig. 11.13*a*. The encoder has memory $M = K - 1 = 2$ message bits. Hence, when the third message bit enters the encoder, the first message bit is shifted out of the register. Consequently, after the third branch, the message sequences ($100\ m_3 m_4 \ldots$) and ($000\ m_3 m_4 \ldots$) generate the same code symbols, and the pair of nodes labeled *a* may be joined together. The same reasoning applies to other nodes. Accordingly, we may collapse the code tree of Fig. 11.14 into the new form shown in Fig. 11.15, called a *trellis*. It is so called since a trellis is a treelike structure with remerging branches. The convention used in Fig. 11.15 to distinguish between input symbols 0 and 1 is as follows. A code branch produced by an input 0 is drawn as a solid line, whereas a code branch produced by an input 1 is drawn as a dashed line. As before, each input (message) sequence corresponds to a specific path through the trellis. For example, we readily see from Fig. 11.15 that the message sequence (10011) produces the encoded output sequence (11, 10, 11, 11, 01), which agrees with our previous result.

A trellis is more instructive than a tree in that it brings out explicitly the fact that the associated convolutional encoder is a finite-state machine. We define the *state* of a convolutional encoder of rate $1/n$ as the $(K - 1)$ message bits stored in the encoder's shift register. At time j, the portion of the message sequence containing the most recent K bits is written as ($m_{j-K+1}, \ldots, m_{j-1}, m_j$), where m_j is the *current* bit. The $(K - 1)$-bit state of the encoder at time j is therefore written simply as ($m_{j-1}, \ldots, m_{j-K+2}, m_{j-K+1}$). In the case of the simple convolutional encoder of Fig. 11.13*a* we have $(K - 1) = 2$. Hence, the state of this encoder can assume any one of four possible values, as described in Table 11.7. The trellis contains $(L + K)$ *levels*, where L is the length of the incoming message sequence, and K is the constraint length of the code. The levels of the trellis are labeled as $j = 0, 1, \ldots, L + K - 1$ in Fig. 11.15 for $K = 3$. Level j is also referred to as *depth* j; both terms are used interchangeably. The first $(K - 1)$ levels correspond to the encoder's departure from the initial state *a*, and the last $(K - 1)$ levels correspond to the encoder's return to the state *a*. Clearly, not all the states can be reached in these two portions of the trellis.

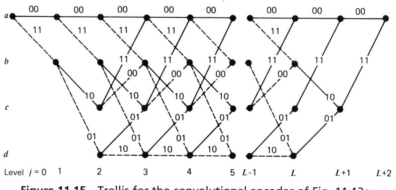

Figure 11.15 Trellis for the convolutional encoder of Fig. 11.13*a*.

Table 11.7 State Table for the Convolutional Encoder of Fig. 11.13*a*

State	Binary Description
a	00
b	10
c	01
d	11

However, in the central portion of the trellis, for which the level j lies in the range $K - 1 \leq j \leq L$, all the states of the encoder are reachable. Note also that the central portion of the trellis exhibits a fixed periodic structure.

Consider next a portion of the trellis corresponding to times j and $j + 1$. We assume that $j \geq 2$ for the example at hand, so that it is possible for the current state of the encoder to be *a, b, c,* or *d.* For convenience of presentation, we have reproduced this portion of the trellis in Fig. 11.16*a.* The left nodes represent the four possible current states of the encoder, whereas the right nodes

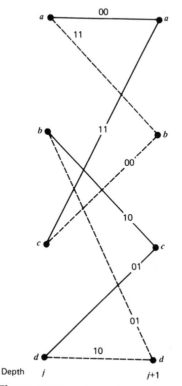

Depth   *j*   *j*+1

Figure 11.16*a* A portion of the central part of the trellis for the encoder of Fig. 11.13*a*.

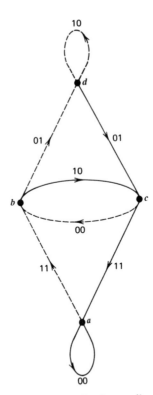

Figure 11.16b State diagram of the convolutional encoder of Fig. 11.13a.

represent the next states. Clearly, we may coalesce the left and right nodes. By so doing, we obtain the *state diagram* of the encoder, shown in Fig. 11.16b. The nodes of the figure represent the four possible states of the encoder, with each node having two incoming branches and two outgoing branches. A transition from one state to another in response to input 0 is represented by a solid branch, whereas a transition in response to input 1 is represented by a dashed branch. The binary label on each branch represents the encoder's output as it moves from one state to another. Suppose, for example, the current state of the encoder is (01). The application of input 1 to the encoder of Fig. 11.13a results in the state (10) and the encoded output (00). Accordingly, with the help of this state diagram, we may readily determine the output of the encoder of Fig. 11.13a for any incoming message sequence. We simply start at state a, the all-zero initial state, and walk through the state diagram in accordance with the message sequence. We follow a solid branch if the input is a 0 and a dashed branch if it is a 1. As each branch is traversed, we output the corresponding binary label on the branch. Consider, for example, the message sequence (10011). For this input we follow the path *abcabd*, and therefore output the sequence (11, 10, 11, 11, 01), which agrees exactly with our earlier result. Thus, the input–output relation of a convolutional encoder is also completely described by its state diagram.

11.6 MAXIMUM LIKELIHOOD DECODING OF CONVOLUTIONAL CODES

Now that we understand the operation of a convolutional encoder, the next issue to be considered is the decoding of a convolutional code. In this section we first describe the underlying theory of maximum likelihood decoding, and then present an efficient algorithm for its practical implementation.

Let **m** denote a *message vector*, and **c** denote the corresponding *code vector* applied by the encoder to the input of a discrete memoryless channel. Let **r** denote the *received vector*, which may differ from the transmitted code vector due to channel noise. Given the received vector **r**, the decoder is required to make an *estimate* $\hat{\mathbf{m}}$ of the message vector. Since there is a one-to-one correspondence between the message vector **m** and the code vector **c**, the decoder may equivalently produce an estimate $\hat{\mathbf{c}}$ of the code vector. We may then put $\hat{\mathbf{m}} = \mathbf{m}$ if and only if $\hat{\mathbf{c}} = \mathbf{c}$. Otherwise, a *decoding error* is committed in the receiver. The *decoding rule* for choosing the estimate $\hat{\mathbf{c}}$, given the received vector **r**, is said to be optimum when the *probability of decoding error* is minimized. From the material presented in Chapter 8, we may state that for equiprobable messages, the probability of decoding error is minimized if the estimate $\hat{\mathbf{c}}$ is chosen to maximize the *log-likelihood function*. Let $p(\mathbf{r}|\mathbf{c})$ denote the conditional probability of receiving **r**, given that **c** was sent. The log-likelihood function equals $\ln p(\mathbf{r}|\mathbf{c})$. The *maximum likelihood decoder* or decision rule is described as follows:

$$\begin{array}{c} \textit{Choose the estimate } \hat{\mathbf{c}} \textit{ for which the} \\ \textit{log-likelihood function } \ln p(\mathbf{r}|\mathbf{c}) \textit{ is maximum.} \end{array} \qquad (11.56)$$

Consider now the special case of a binary symmetric channel. In this case, both the transmitted code vector **c** and the received vector **r** represent binary sequences of length N, say. Naturally, these two sequences may differ from each other in some locations because of errors due to channel noise. Let c_i and r_i denote the ith elements of **c** and **r**, respectively. We then have

$$p(\mathbf{r}|\mathbf{c}) = \prod_{i=1}^{N} p(r_i|c_i) \qquad (11.57)$$

Correspondingly, the log-likelihood function equals

$$\ln p(\mathbf{r}|\mathbf{c}) = \sum_{i=1}^{N} \ln p(r_i|c_i) \qquad (11.58)$$

Let

$$p(r_i|c_i) = \begin{cases} p, & \text{if } r_i \neq c_i \\ 1 - p, & \text{if } r_i = c_i \end{cases} \qquad (11.59)$$

Suppose also that the received vector **r** differs from the transmitted code vector **c** in exactly d positions. The number d is the *Hamming distance* between vectors **r** and **c**. Then, we may rewrite the log-likelihood function in Eq. (11.58) as

$$\ln p(\mathbf{r}|\mathbf{c}) = d \ln p + (N - d)\ln(1 - p)$$
$$= d \ln\left(\frac{p}{1 - p}\right) + N \ln(1 - p) \qquad (11.60)$$

In general, the probability of an error occurring is low enough for us to assume $p < 1/2$. We also recognize that $N \ln(1 - p)$ is a constant for all $\mathbf{c}$. Accordingly, we may restate the maximum-likelihood decoding rule for the binary symmetric channel as follows:

> *Choose the estimate $\hat{\mathbf{c}}$ that minimizes the Hamming distance*   (11.61)
> *between the received vector $\mathbf{r}$ and the transmitted vector $\mathbf{c}$.*

That is, for the binary symmetric channel, the maximum-likelihood decoder reduces to a *minimum distance decoder*. In such a decoder, the received vector $\mathbf{r}$ is compared with each possible transmitted code vector $\mathbf{c}$, and the particular one closest to $\mathbf{r}$ is chosen as the correct transmitted code vector. The term "closest" is used in the sense of minimum number of differing binary symbols (i.e., Hamming distance) between the code vectors under investigation.

The Viterbi Algorithm[7]

The equivalence between maximum likelihood decoding and minimum distance decoding for a binary symmetric channel implies that we may decode a convolutional code by choosing a path in the code tree whose coded sequence differs from the received sequence in the fewest number of places. Since a code tree is equivalent to a trellis, we may equally limit our choice to the possible paths in the trellis representation of the code. The reason for preferring the trellis over the tree is that the number of nodes at any level of the trellis does not continue to grow as the number of incoming message bits increases; rather, it remains constant at 2^{K-1}, where K is the constraint length of the code.

Consider, for example, the trellis diagram of Fig 11.15 for a convolutional code with rate $r = 1/2$ and constraint length $K = 3$. We observe that at level $j = 3$, there are two paths entering any of the four nodes in the trellis. Moreover, these two paths will be identical onward from that point. Clearly, a minimum distance decoder may make a decision at that point as to which of those two paths to retain, without any loss of performance. A similar decision may be made at level $j = 4$, and so on. This sequence of decisions is exactly what the *Viterbi algorithm* does as it walks through the trellis. The algorithm operates by computing a *metric* or discrepancy for every possible path in the trellis. The metric for a particular path is defined as the Hamming distance between the coded sequence represented by that path and the received sequence. Thus, for each node (state) in the trellis of Fig. 11.15 the algorithm compares the two paths entering the node. The path with the lower metric is retained, and the other path is discarded. This computation is repeated for every level j of the trellis in the range $M \le j \le L$, where $M = K - 1$ is the encoder's memory and L is the length of the incoming message sequence. The paths that are retained by the algorithm are called *survivor* or *active paths*. For a convolutional code of constraint length $K = 3$, for example, no more than $2^{K-1} = 4$ survivor paths and their metrics will ever be stored. This list of 2^{K-1} paths is always guaranteed to contain the maximum-likelihood choice.

A difficulty that may arise in the application of the Viterbi algorithm is the possibility that when the paths entering a state are compared, their metrics are found to be identical. In such a situation, we make the choice by flipping a fair coin (i.e., simply make a guess).

In summary, the Viterbi algorithm is a maximum-likelihood decoder, which is optimum for an AWGN channel. It proceeds in a step-by-step fashion as follows:

Initialization. Label the left-most state of the trellis (i.e., the all-zero state at level 0) as 0, since there is no discrepancy at this point in the computation. *Computation step $j + 1$.* Let $j = 0, 1, 2, \ldots$, and suppose that at the previous step j we have done two things:

- Identified all survivor paths.
- Stored the survivor path and its metric for each state of the trellis.

Then, at level (clock time) $j + 1$ compute the metric for all the paths entering each state of the trellis by adding the metric of the incoming branches to the metric of the connecting survivor path from level j. Hence, for each state, identify the path with the lowest metric as the survivor of step $j + 1$, thereby updating the computation.

Final Step. Continue the computation until the algorithm completes its forward search through the trellis and therefore reaches the termination node (i.e., all-zero state), at which time it makes a decision on the maximum likelihood path. Then, like a block decoder, the sequence of symbols associated with that path is released to the destination as the decoded version of the received sequence.

However, when the received sequence is very long (near infinite), the storage requirement of the Viterbi algorithm becomes too high, and some compromises must be made. The approach usually taken is to "truncate" the path memory of the decoder as described here. A *decoding window* of length ℓ is specified, and the algorithm operates on a corresponding frame of the received sequence, always stopping after ℓ steps. A decision is then made on the "best" path and the symbol associated with the first branch on that path is released to the user. The symbol associated with the last branch of the path is dropped. Next, the decoding window is moved forward one time interval, and a decision on the next code frame is made, and so on. The decoding decisions made in this way are no longer truly maximum likelihood, but they can be made almost as good provided that the decoding window is long enough. Experience and analysis have shown that satisfactory results are obtained if the decoding window length ℓ is on the order of 5 times the constraint length K of the convolutional code or more.

EXAMPLE 6

Suppose that the encoder of Fig. 11.13a generates an all-zero sequence that is sent over a binary symmetric channel, and that the received sequence is (0100010000 ...). There are two errors in the received sequence due to noise in the channel: one in the second bit and the other in the sixth bit. We wish to show that this double-error pattern is correctable through the application of the Viterbi decoding algorithm.

In Fig. 11.17, we show the results of applying the algorithm for level $j = 1$, 2, 3, 4, 5. We see that for $j = 2$ there are (for the first time) four paths, one for each of the four states of the encoder. The figure also includes the metric of each path for each level in the computation.

In the left side of Fig. 11.17, for $j = 3$ we show the paths entering each of the states, together with their individual metrics. In the right side of the figure,

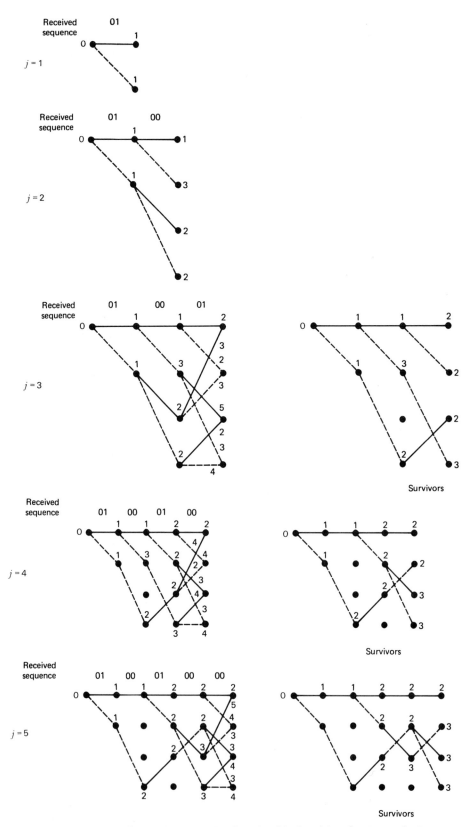

Figure 11.17 Illustrating steps in the Viterbi algorithm for Example 6.

we show the four survivors that result from application of the algorithm for level $j = 3, 4, 5$.

Examining the four survivors in Fig. 11.17 for $j = 5$, we see that the all-zero path has the smallest metric and will remain the path of smallest metric from this point forward. This clearly shows that the all-zero sequence is the maximum likelihood choice of the Viterbi decoding algorithm, which agrees exactly with the transmitted sequence.

EXAMPLE 7

Suppose next that the received sequence is $(1100010000 \ldots)$, which contains three errors compared to the transmitted all-zero sequence.

In Fig. 11.18, we show the results of applying the Viterbi decoding algorithm for $j = 1, 2, 3, 4$. We see that in this example the correct path has been eliminated by level $j = 3$. Clearly, a triple-error pattern is uncorrectable by the Viterbi algorithm when applied to a convolutional code of rate 1/2 and constraint length $K = 3$. The exception to this rule is a triple-error pattern spread over a time span longer than one constraint length, in which case it is very likely to be correctable.

Free Distance of a Convolutional Code

The performance of a convolutional code depends not only on the decoding algorithm used but also on the distance properties of the code. In this context, the most important single measure of a convolutional code's ability to combat channel noise is the free distance, denoted by d_{free}. The *free distance* of a convolutional code is defined as *the minimum Hamming distance between any two code words in the code.* A convolutional code with free distance d_{free} can correct t errors if and only if d_{free} is greater than $2t$.

The free distance can be obtained quite simply from the state diagram of the convolutional encoder. Consider, for example, Fig. 11.16b, which shows the state diagram of the encoder of Fig. 11.13a. Any nonzero code sequence corresponds to a complete path beginning and ending at the 00 state (i.e., node a). We thus find it useful to split this node in the manner shown in the modified state diagram of Fig. 11.19, which may be viewed as a *signal-flow graph* with a single input and a single output. A signal-flow graph consists of *nodes* and directed *branches*; it operates by the following rules:

1. A branch multiplies the signal at its input node by the *transmittance* characterizing that branch.
2. A node with incoming branches *sums* the signals produced by all of those branches.
3. The signal at a node is applied equally to all the branches outgoing from that node.
4. The *transfer function* of the graph is the ratio of the output signal to the input signal.

Returning to the signal-flow graph of Fig. 11.19, we note that the exponent of D on a branch in this graph describes the Hamming weight of the encoder output corresponding to that branch. The exponent of L is always equal to one, since the length of each branch is one. Let $T(D,L)$ denote the transfer function of the

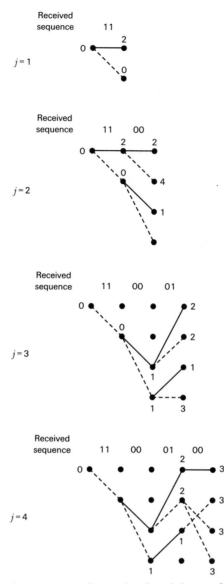

Figure 11.18 Illustrating breakdown of the Viterbi algorithm in Example 7.

signal-flow graph, with D and L playing the role of dummy variables. For the example of Fig. 11.19, we may readily use rules 1, 2, and 3 to obtain the following input-output relations:

$$\left. \begin{array}{r} b = D^2 L a_0 + L c \\ c = DLb + DLd \\ d = DLb + DLd \\ a_1 = D^2 L c \end{array} \right\} \qquad (11.62)$$

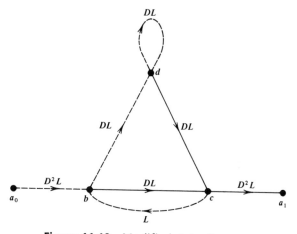

Figure 11.19   Modified state diagram.

where a_0, b, c, d, and a_1 denote the node signals of the graph. Solving the set of Eqs. (11.62) for the ratio a_1/a_0, we thus find that the transfer function of the graph in Fig. 11.19 is given by

$$T(D,L) = \frac{D^5 L^3}{1 - DL(1 + L)} \tag{11.63}$$

Using the binomial expansion, we may equivalently write

$$T(D,L) = D^5 L^3 \sum_{i=0}^{\infty} \left(DL(1 + L)\right)^i \tag{11.64}$$

Setting $L = 1$ in Eq. (11.64), we thus get the *distance transfer function* expressed in the form of a power series:

$$T(D) = D^5 + 2D^6 + 4D^7 + \cdots \tag{11.65}$$

Since the free distance is the minimum Hamming distance between any two code words in the code and the distance transfer function $T(D)$ enumerates the number of code words that are a given distance apart, it follows that the exponent of the first term in the expansion of $T(D)$ defines the free distance. Thus the convolutional code of Fig. 11.13a has a free distance $d_{\text{free}} = 5$.

This result indicates that up to two errors in the received sequence are correctable, for two or fewer transmission errors will cause the received sequence to be at most at a Hamming distance of 2 from the transmitted sequence but at least at a Hamming distance of 3 from any other code sequence in the code. In other words, in spite of the presence of any pair of transmission errors, the received sequence remains closer to the transmitted sequence than any other possible code sequence. However, this statement is no longer true if there are three or more *closely spaced* transmission errors in the received sequence. These observations confirm the results reported earlier in Examples 6 and 7.

In using the distance transfer function $T(D)$ to calculate the free distance of a convolutional code, it is assumed that the power series in the unit-delay variable

D representing $T(D)$ is *convergent* (i.e., its sum has a "finite" value). This assumption is required to justify the expansion given in Eq. (11.65) for the convolutional code of Fig. 11.13a. However, there is no guarantee that $T(D)$ is always convergent. When $T(D)$ is nonconvergent, an infinite number of decoding errors are caused by a finite number of transmission errors; the convolutional code is then subject to catastrophic error propagation, and the code is called a *catastrophic code*.[8] In this context it is noteworthy that a *systematic* convolutional code cannot be catastrophic. Unfortunately, for a prescribed constraint length K, the free distances that can be attained with systematic convolutional codes using the schemes of Fig. 11.13 are usually smaller than for the case of nonsystematic convolutional codes, as indicated in Table 11.8.

Asymptotic Coding Gain[9]

The transfer function of the modified encoder state diagram may be used to evaluate a *bound on the bit error rate* for a given decoding scheme; details of this evaluation are, however, beyond the scope of our present discussion. Here we simply summarize the results for two special channels, the binary symmetric channel and the binary-input additive white Gaussian noise (AWGN) channel, assuming the use of binary phase-shift keying (PSK) with coherent detection.

1. *Binary symmetric channel.* The binary symmetric channel may be modeled as an additive white Gaussian noise channel with binary phase-shift keying (PSK) as the modulation and with hard-decision demodulation. The transition probability p of the binary symmetric channel is then equal to the bit error rate (BER) for the uncoded binary PSK system. For large values of E_b/N_0, the ratio of signal energy per bit-to-noise spectral density, the bit error rate for binary PSK without coding is dominated by the exponential factor $\exp(-E_b/N_0)$. On the other hand, the bit error rate for the same modulation scheme with convolutional coding is dominated by the exponential factor $\exp(-d_{\text{free}}rE_b/2N_0)$, where r is the code rate and d_{free} is the free distance of the convolutional code. Therefore, as a figure of merit for measuring the improvement in error performance made by the use of coding with hard-decision decoding, we may use the exponents to define the

Table 11.8 Maximum Free Distances
Attainable with Systematic and
Nonsystematic Convolutional
Codes of Rate 1/2

Constraint Length K	Systematic	Nonsystematic
2	3	3
3	4	5
4	4	6
5	5	7
6	6	8
7	6	10
8	7	10

asymptotic coding gain (in decibels) as follows:

$$G_a = 10 \log_{10}\left(\frac{d_{\text{free}}r}{2}\right) \text{ dB} \tag{11.66}$$

2. *Binary-input AWGN channel.* Consider next the case of a memoryless binary-input AWGN channel with no output quantization [i.e., the output amplitude lies in the interval $(-\infty, \infty)$]. For this channel, theory shows that for large values of E_b/N_0 the bit error rate for binary PSK with convolutional coding is dominated by the exponential factor $\exp(-d_{\text{free}}rE_b/N_0)$, where the parameters are as previously defined. Accordingly, in this case, we find that the asymptotic coding gain is defined by

$$G_a = 10 \log_{10}(d_{\text{free}}r) \text{ dB} \tag{11.67}$$

From Eqs. (11.66) and (11.67) we see that the asymptotic coding gain for the binary-input AWGN channel is greater than that for the binary symmetric channel by 3 dB. In other words, for large E_b/N_0, the transmitter for a binary symmetric channel must generate an additional 3 dB of signal energy (or power) over that for a binary-input AWGN channel if we are to achieve the same error performance. Clearly, there is an advantage to be gained by permitting an unquantized demodulator output instead of making hard decisions. This improvement in performance, however, is attained at the cost of increased decoder complexity due to the requirement for accepting analog inputs.

The asymptotic coding gain for a binary-input AWGN channel is approximated to within about 0.25 dB by a binary input Q-ary output discrete memoryless channel with the number of representation levels $Q = 8$. This means that we may avoid the need for an analog decoder by using a soft-decision decoder that performs finite output quantization (typically, $Q = 8$), and yet realize a performance close to the optimum.

11.7 TRELLIS-CODED MODULATION[10]

In the traditional approach to channel coding described in the preceding sections of the chapter, encoding is performed separately from modulation in the transmitter; likewise for decoding and detection in the receiver. Moreover, error control is provided by transmitting additional redundant bits in the code, which has the effect of lowering the information bit rate per channel bandwidth. That is, bandwidth efficiency is traded for increased power efficiency.

To attain a more effective utilization of the available bandwidth and power, coding and modulation have to be treated as a single entity. We may deal with this new situation by redefining coding as *the process of imposing certain patterns on the transmitted signal.* Indeed, this definition includes the traditional idea of parity coding.

Trellis codes for band-limited channels result from the treatment of modulation and coding as a *combined* entity rather than as two separate operations. The combination is itself referred to as *trellis-coded modulation* (TCM). This form of signaling has three basic features:

1. The number of signal points in the constellation used is larger than what is required for the modulation format of interest with the same data rate; the

additional points allow redundancy for forward error-control coding without sacrificing bandwidth.

2. Convolutional coding is used to introduce a certain dependency between successive signal points, such that only certain *patterns* or *sequences of signal points* are permitted.

3. Soft-decision decoding is performed in the receiver, in which the permissible sequence of signals is modeled as a trellis structure; hence, the name "trellis codes."

This latter requirement is the result of using a larger signal constellation. By increasing the size of the constellation the probability of symbol error increases for a fixed signal-to-noise ratio. Hence, with hard-decision demodulation we would face a loss before we begin. Performing soft-decision decoding on the combined code and modulation trellis ameleorates this problem.

In the presence of AWGN, maximum likelihood decoding of trellis codes consists of finding that particular path through the trellis with *minimum squared Euclidean distance* to the received sequence. Thus, in the design of trellis codes, the emphasis is on maximizing the Euclidean distance between code vectors (or, equivalently, code words) rather than maximizing the Hamming distance of an error-correcting code. The reason for this approach is that, except for conventional coding with binary PSK and QPSK, maximizing the Hamming distance is not the same as maximizing the squared Euclidean distance. Accordingly, in what follows, the Euclidean distance is adopted as the distance measure of interest. Moreover, while a more general treatment is possible, the discussion is (by choice) confined to the case of *two-dimensional constellations of signal points*. The implication of such a choice is to restrict the development of trellis codes to multilevel amplitude and/or phase modulation schemes such as *M*-ary PSK and *M*-ary QAM.

The approach used to design this type of trellis codes involves partitioning an *M*-ary constellation of interest successively into 2, 4, 8, . . . subsets with size $M/2$, $M/4$, $M/8$, . . . , and having progressively larger increasing minimum Euclidean distance between their respective signal points. Such a design approach by *set partitioning* represents the "key idea" in the construction of efficient coded modulation techniques for band-limited channels.

In Fig. 11.20, we illustrate the partioning procedure by considering a circular constellation that corresponds to 8-PSK. The figure depicts the constellation itself and the 2 and 4 subsets resulting from two levels of partitioning. These subsets share the common property that the minimum Euclidean distances between their individual points follow an increasing pattern: $d_0 < d_1 < d_2$.

Figure 11.21 illustrates the partitioning of a rectangular constellation corresponding to 16-QAM. Here again we see that the subsets have increasing within-subset Euclidean distances: $d_0 < d_1 < d_2 < d_3$.

Based on the subsets resulting from successive partitioning of a two-dimensional constellation, we may devise relatively simple and yet highly effective coding schemes. Specifically, to send *n* bits/symbol with *quadrature modulation* (i.e., one that has in-phase and quadrature components), we start with a two-dimensional constellation of 2^{n+1} signal points appropriate for the modulation format of interest; a circular grid is used for *M*-ary PSK, and a rectangular one for *M*-ary QAM. In any event, the constellation is partitioned into 4 or 8 subsets. One or two incoming bits per symbol enter a rate-1/2 or rate-2/3 binary con-

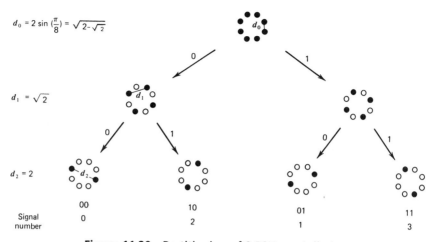

$$d_0 = 2 \sin \left(\frac{\pi}{8}\right) = \sqrt{2 - \sqrt{2}}$$

$$d_1 = \sqrt{2}$$

$$d_2 = 2$$

Signal
number

Figure 11.20   Partitioning of 8-PSK constellation.

volutional encoder, respectively; the resulting two or three coded bits per symbol
determine the selection of a particular subset. The remaining uncoded data bits
determine which particular point from the selected subset is to be signaled. This
class of trellis codes is known as *Ungerboeck codes*.

Since the modulator has memory, we may use the Viterbi algorithm to perform maximum likelihood sequence detection at the receiver. Each branch in
the trellis of the Ungerboeck code corresponds to a subset rather than an individual signal point. The first step in the detection is to determine the signal point
within each subset that is closest to the received signal point in the Euclidean
sense. The signal point so determined and its metric (i.e., the squared Euclidean
distance between it and the received point) may be used thereafter for the
branch in question, and the Viterbi algorithm may then proceed in the usual
manner.

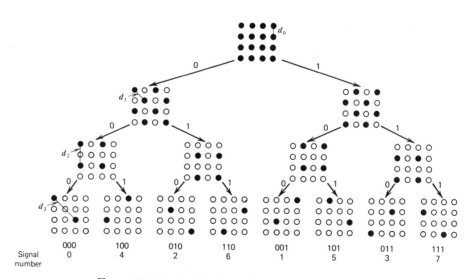

Signal
number

Figure 11.21   Partitioning of 16-QAM constellation.

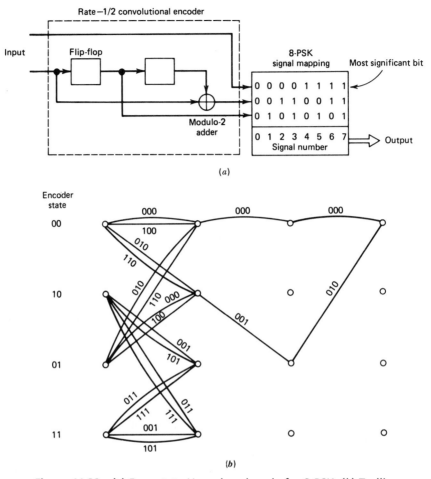

Figure 11.22 (a) Four-state Ungerboeck code for 8-PSK. (b) Trellis.

Ungerboeck Codes for 8-PSK

The scheme of Fig. 11.22a depicts the simplest Ungerboeck 8-PSK code for the transmission of 2 bits/symbol. The scheme uses a rate-1/2 convolutional encoder; the corresponding trellis of the code is shown in Fig. 11.22b, which has four states. Note that the most significant bit of the incoming binary word is left uncoded. Therefore, each branch of the trellis may correspond to two different output values of the 8-PSK modulator or, equivalently, to one of the four 2-point subsets shown in Fig. 11.20. The trellis of Fig. 11.22b also includes the minimum distance path.

The scheme of Fig. 11.23a depicts another Ungerboeck 8-PSK code for transmitting 2 bits/sample; it is next in the level of complexity. This second scheme uses a rate-2/3 convolutional encoder. Therefore, the corresponding trellis of the code has eight states, as shown in Fig. 11.23b. In this case, both bits of the incoming binary word are encoded. Hence, each branch of the trellis corresponds to a specific output value of the 8-PSK modulator. The trellis of Fig. 11.23b also includes the minimum distance path.

Figures 11.22b and 11.23b also include the encoder states. In Fig. 11.22, the

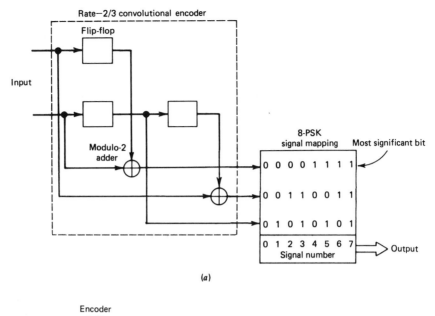

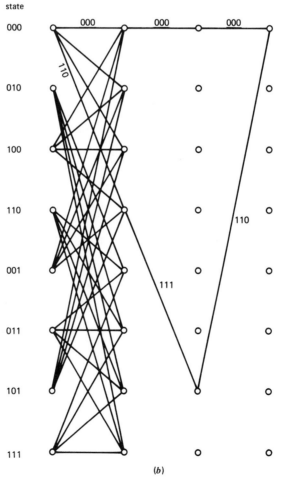

Figure 11.23 (*a*) Eight-state Ungerboeck code for 8-PSK. (*b*) Trellis.

state of the encoder is defined by the contents of the two-stage shift register. On the other hand, in Fig. 11.23 it is defined by the content of the single-stage (top) shift register followed by that of the two-stage (bottom) shift register.

Asymptotic Coding Gain

Following the discussion in Section 11.6, we define the *asymptotic coding gain* of Ungerboeck codes as

$$G_a = 10 \log_{10}\left(\frac{d_{\text{free}}^2}{d_{\text{ref}}^2}\right) \tag{11.68}$$

where d_{free} is the *free Euclidean distance* of the code and d_{ref} is the minimum Euclidean distance of an uncoded modulation scheme operating with the same signal energy per bit. For example, by using the Ungerboeck 8-PSK code of Fig. 11.22a, the signal constellation has 8 message points, and we send 2 message bits per point. Hence, uncoded transmission requires a signal constellation with 4 message points. We may therefore regard uncoded 4-PSK as the reference for the Ungerboeck 8-PSK code of Fig. 11.22a.

The Ungerboeck 8-PSK code of Fig. 11.22a achieves an asymptotic coding gain of 3 dB, calculated as follows:

1. Each branch of the trellis in Fig. 11.22b corresponds to a subset of two antipodal signal points. Hence, the free Euclidean distance d_{free} of the code can be no larger than the Euclidean distance d_2 between the antipodal signal points of such a subset. We may therefore write

$$d_{\text{free}} = d_2 = 2$$

   where the distance d_2 is shown defined in Fig. 11.24a; see also Fig. 11.20.

2. The minimum Euclidean distance of an uncoded QPSK, viewed as a reference operating with the same signal energy per bit, equals (see Fig. 11.24b)

$$d_{\text{ref}} = \sqrt{2}$$

Hence, as previously stated, the use of Eq. (11.68) yields an asymptotic coding gain of $10 \log_{10} 2 = 3$ dB.

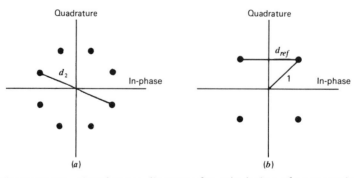

Figure 11.24 Signal-space diagrams for calculation of asymptotic coding gain.

Table 11.9 Asymptotic Coding Gain of Ungerboeck 8-PSK Codes, with Respect to Uncoded 4-PSK

Number of states	4	8	16	32	64	128	256	512
Coding gain (dB)	3	3.6	4.1	4.6	4.8	5	5.4	5.7

The asymptotic coding gain achievable with Ungerboeck codes increases with the number of states in the convolutional encoder. Table 11.9 presents the asymptotic coding gain (in dB) for Ungerboeck 8-PSK codes for increasing number of states, expressed with respect to uncoded 4-PSK. Note that improvements on the order of 6 dB require codes with a very large number of states.

11.8 CODING FOR COMPOUND-ERROR CHANNELS

Many real communication channels exhibit a mixture of *independent* and *burst error statistics*. We refer to such channels as *compound-error channels*. In *telephone channels*, for example, bursts of errors result from *impulse noise* on circuits due to lightning and transients in central office switching equipment. In *radio channels*, bursts of errors are produced by atmospherics, multipath fading, and interferences from other users of the frequency band.

An effective method for error protection over compound-error channels is based on *automatic-repeat-request*[11] (ARQ), which is by far the oldest and most widely used scheme for error control in data communication systems. In Fig. 11.25, we show the block diagram of a basic ARQ system. The system requires the use of a *feedback channel* for requesting message retransmission. As usual, encoded bits are applied to a modulator for transmission over the channel, and the received bits are demodulated and decoded. The system includes *transmit and receive controllers* that exchange information via the feedback channel. The simplest ARQ system employs a *start and stop strategy*, which proceeds as follows:

1. A block of data is encoded into a code word for transmission over the communication channel.

2. The transmitter stops and waits until it receives (via the feedback channel) acknowledgment of correct reception of the code word or a request for retransmission.

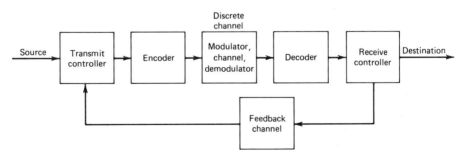

Figure 11.25 Block diagram of an ARQ system.

Clearly, this strategy requires only error detection (rather than both error detection and correction) in the receiver. Virtually error-free data transmission can be attained by the proper choice of a code for error detection; cyclic redundancy check (CRC) codes are well-suited for this application. Moreover, the feedback channel has to be practically noiseless. This is not a severe restriction. Since the need for information transfer over the feedback channel is at a low rate, a substantial amount of redundancy can be used to ensure reliable feedback transmission.

Another straightforward and effective method to apply coding on a burst-error channel is to use *interleaving*. With this method, the channel is effectively transformed into an independent-error channel for which many forward-error correction coding techniques are applicable. The block diagram of such a system is shown in Fig. 11.26. The transmitter includes an encoder followed by an *interleaver* that scrambles the encoded data stream in a deterministic manner. Specifically, successive bits (or symbols) transmitted over the channel are separated as widely as possible. In the receiver, a *deinterleaver* is used to perform the inverse operation. That is, the received data are unscrambled, so that the decoding operation may proceed properly. Whereas the original data sequence goes through interleaving in the transmitter and then deinterleaving in the receiver, error bursts are processed by the deinterleaver only. Accordingly, after deinterleaving, error bursts that occur on the channel are spread out in the data sequence to be decoded, thereby spanning many code words. The combination of interleaving and forward-error correction thus provides an effective means of combating the effect of error bursts.

A variation of the interleaving method described in Fig. 11.26 is commercially exploited in the *compact disc digital audio system*.[12] In this system, optical recording is used to produce a master disc, from which compact discs are manufactured by galvanic processing. Production of the master disc and its replication into compact discs may be viewed as the "channel" of the digital audio system. An important characteristic of the channel is that it exhibits a burstlike error behavior, for which several factors are responsible. First, small unwanted particles or air bubbles trapped in the plastic material cause errors to occur when the recorded information is optically read out. Second, fingerprints or scratches on the disc and surface roughness cause additional error bursts. For protection against channel errors, the recorded digital information includes parity bytes derived in two Reed–Solomon encoders. Moreover, interleaving is used to spread the errors out. In particular, the data streams entering the first encoder, those between the two encoders, and those leaving the second encoder are scrambled by means of sets of delay lines. Accordingly, burst errors will, after deinterleaving on playback, be spread over a longer time interval, so that they can be corrected more easily. The two Reed–Solomon (n,k) codes have the values (32, 28) and (28, 24), so that each uses four parity bytes. Hence, each code can correct up to four error/erasure bytes.

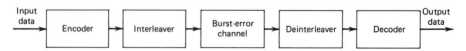

Figure 11.26 Block diagram of interleaving method for burst-error channel.

11.9 SUMMARY AND DISCUSSION

In this chapter, we studied error-control coding techniques that have established themselves as an essential ingredient for reliable communication over noisy channels. The effect of errors occurring during transmission is reduced by adding redundancy to the data prior to transmission. The redundancy is used to enable a decoder in the receiver to detect and correct errors.

In FEC codes, decoding is usually more complex than encoding. These codes may be divided into block codes and convolutional codes. In *block coding*, the encoder splits the incoming data stream into blocks and processes each block individually by adding redundancy in accordance with a prescribed algorithm. Likewise, the decoder processes each block individually; it corrects errors by exploiting the redundancy.

Many of the important block codes are *cyclic codes*. The encoders for these codes can be implemented using linear shift registers with feedback. Moreover, because of the inherent algebraic structure of cyclic codes, the decoders for them are conceptually quite simple. Important examples of cyclic codes are *Hamming codes, maximum-length codes, BCH codes,* and *Reed–Solomon codes*. In particular, Reed–Solomon codes are very powerful nonbinary codes, capable of combatting combinations of both random and burst errors. Also, virtually all codes used for error detection are cyclic codes; such codes are called *cyclic redundancy check (CRC) codes*.

In *convolutional coding*, the encoder processes the incoming data stream continuously. For decoding, the *Viterbi algorithm* is most frequently employed. The Viterbi algorithm exploits the trellis representation of a convolutional code, providing a recursive procedure for maximum likelihood sequence estimation. The complexity of the Viterbi algorithm, however, confines its application to convolutional codes with a constraint length K not exceeding 11. At conventional bit rates, the value $K = 7$ is usually considered appropriate. Historically, the Viterbi algorithm has been implemented in software form on a computer or microprocessor. However, with continuing improvements in very-large-scale integrated (VLSI) circuit technology, it is becoming increasingly economical to implement Viterbi decoders in hardware form.

In the traditional approach to forward-error correction, coding is performed separately from modulation, and bandwidth efficiency is reduced by the addition of redundancy. In *trellis-coded modulation*, on the other hand, the operations of coding and modulation are combined so as to permit significant coding gains over conventional uncoded multilevel modulation without sacrificing bandwidth efficiency. In an important class of trellis codes known as *Ungerboeck codes*, multilevel (amplitude and/or phase) modulation is combined with convolutional coding. At the receiver, the Viterbi algorithm is used to perform a maximum-likelihood sequence detection. Thus, coding gains of 3 to 6 dB are attained at bandwidth efficiencies equal to or larger than 2 bits per second per Hertz. These are the values for which operation on many band-limited channels is desired.

NOTES AND REFERENCES

1. For an introductory discussion of error correction by coding, see Chapter 2 of Lucky (1989); see also the book by Adámek (1991), and the paper by Bhargava (1983).

The classic book on error-control coding is Peterson and Weldon (1972). Error-control coding is also discussed in the classic book of Gallager (1968). The books of Lin and Costello (1983), Micheleson and Levesque (1985), and MacWilliams and Sloane (1977) are also devoted to error-control coding. For a collection of key papers in the development of coding theory, see the book edited by Berlekamp (1974).

2. In medicine, the term "syndrome" is used to describe a pattern of symptoms that aids in the diagnosis of a disease. In coding, the error pattern plays the role of the disease and parity-check failure that of a symptom. This use of "syndrome" was coined by Hagelbarger (1959).

3. The first error-correcting codes (known as Hamming codes) were invented by Hamming at about the same time as the conception of information theory by Shannon; for details see the classic paper by Hamming (1950).

4. For a description of BCH codes and their decoding algorithms, see Lin and Costello (1983, pp. 141–183) and MacWilliams and Sloane (1977, pp. 257–293). Table 11.6 is adapted from Lin and Costello (1983).

5. The Reed–Solomon codes are named in honor of their inventors: see their classic 1960 paper. For details of Reed–Solomon codes, see MacWilliams and Sloane (1977, pp. 294–306).

6. Convolutional codes were first introduced, as an alternative to block codes, by P. Elias (1955).

7. In a classic paper, Viterbi (1967) proposed a decoding algorithm for convolutional codes that has become known as the *Viterbi algorithm.* The algorithm was recognized by Forney (1972, 1973) to be a maximum likelihood decoder. Readable accounts of the Viterbi algorithm are presented in Lin and Costello (1983), Blahut (1990), and Adámek (1991).

8. Catastrophic convolutional codes are discussed in Benedetto, Biglieri, and Castellani (1987). Table 11.8 is adapted from their book.

9. For details of the evaluation of asymptotic coding gain for binary symmetric and binary-input AWGN channels, see Viterbi and Omura (1979, pp. 242–252) and Lin and Costello (1983, pp. 322–329).

10. Trellis-coded modulation was invented by G. Ungerboeck; its historical evolution is described in Ungerboeck (1982). An illuminating discussion of trellis-coded modulation is presented in Blahut (1990). For a detailed treatment of the subject, see the books by Biglieri, Divsalar, McLane, and Simon (1991) and Anderson, Aulin, and Sundberg (1986). See also the special issue of the *IEEE Communications Magazine* (December 1991), devoted to coded modulation. Table 11.9 is adapted from Ungerboeck (1982).

Forney and Eyuboğlu (1991) describe techniques that combine equalization and trellis-coded modulation, which approach the information capacity of high-SNR bandlimited channels (e.g., telephone channels); Shannon's information capacity was described in Chapter 10. This is a significant achievement.

11. For a survey of various ARQ schemes, see Lin, Costello, and Miller (1984).

12. For a description of the various signal-processing operations used in compact disc digital audio systems, see the paper by Peek (1985).

PROBLEMS

Problem 11.1 Consider a binary input Q-ary output discrete memoryless channel. The channel is said to be symmetric if the channel transition probability $p(j|i)$ satisfies the condition:

$$p(j|0) = p(Q - 1 - j|1), \quad j = 0, 1, \ldots, Q - 1$$

Suppose that the channel input symbols 0 and 1 are equally likely. Show that the channel output symbols are also equally likely; that is,

$$p(j) = \frac{1}{Q}, \qquad j = 0, 1, \ldots, Q - 1$$

Problem 11.2   Consider the quantized demodulator for binary PSK signals shown in Fig. 11.3*a*. The quantizer is a four-level quantizer, normalized as in Fig. P11.1. Evaluate the transition probabilities of the binary input-quaternary output discrete memoryless channel so characterized. Hence, show that it is a symmetric channel. Assume that the transmitted signal energy per bit is E_b, and the additive white Gaussian noise has zero mean and power spectral density $N_0/2$.

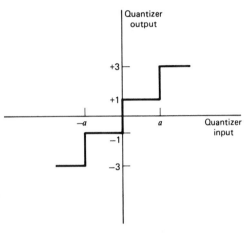

Figure P11.1

Problem 11.3   Consider a binary input AWGN channel, in which the binary symbols 1 and 0 are equally likely. The binary symbols are transmitted over the channel by means of phase-shift keying. The code symbol energy is E, and the AWGN has zero mean and power spectral density $N_0/2$. Show that the channel transition probability is given by

$$p(y|0) = \frac{1}{\sqrt{2\pi}} \exp\left[-\frac{1}{2} \left(y + \sqrt{\frac{2E}{N_0}} \right)^2 \right], \qquad -\infty < y < \infty$$

Problem 11.4   In a *single-parity-check code*, a single parity bit is appended to a block of k message bits $(m_1, m_2, \ldots, m_k)$. The single parity bit b_1 is chosen so that the code word satisfies the *even parity rule*:

$$m_1 + m_2 + \cdots + m_k + b_1 = 0, \qquad \text{mod } 2$$

For $k = 3$, set up the 2^k possible code words in the code defined by this rule.

Problem 11.5   Compare the parity-check matrix of the (7, 4) Hamming code considered in Example 2 with that of a (4, 1) repetition code.

Problem 11.6   Consider the (7, 4) Hamming code of Example 2. The generator matrix **G** and the parity-check matrix **H** of the code are described in that example. Show that these two matrices satisfy the condition

$$\mathbf{H}\mathbf{G}^T = \mathbf{0}$$

Problem 11.7

(a) For the (7, 4) Hamming code described in Example 2, construct the eight code words in the dual code.

(b) Find the minimum distance of the dual code determined in part (a).

Problem 11.8   Consider the (5, 1) repetition code of Example 1. Evaluate the syndrome **s** for the following error patterns:

(a) All five possible single-error patterns

(b) All 10 possible double-error patterns

Problem 11.9   For an application that requires error detection *only*, we may use a *nonsystematic* code. In this problem, we explore the generation of such a cyclic code. Let $g(X)$ denote the generator polynomial, and $m(X)$ denote the message polynomial. We define the code polynomial $c(X)$ simply as

$$c(X) = m(X)g(X)$$

Hence, for a given generator polynomial, we may readily determine the code words in the code. To illustrate this procedure, consider the generator polynomial for a (7, 4) Hamming code:

$$g(X) = 1 + X + X^3$$

Determine the 16 code words in the code, and confirm the nonsystematic nature of the code.

Problem 11.10   The polynomial $1 + X^7$ has two primitive factors, namely, $1 + X + X^3$ and $1 + X^2 + X^3$. In Example 3, we used $1 + X + X^3$ as the generator polynomial for a (7, 4) Hamming code. In this problem, we consider the adoption of $1 + X^2 + X^3$ as the generator polynomial. This should lead to a (7, 4) Hamming code that is different from the code analyzed in Example 3. Develop the encoder and syndrome calculator for the generator polynomial:

$$g(X) = 1 + X^2 + X^3$$

Compare your results with those in Example 3.

Problem 11.11   Consider the (7, 4) Hamming code defined by the generator polynomial

$$g(X) = 1 + X + X^3$$

The code word 0111001 is sent over a noisy channel, producing the received word 0101001. Determine the syndrome polynomial $s(X)$ for this received word, and show that it is identical to the error polynomial $e(X)$.

Problem 11.12   The generator polynomial of a (15, 11) Hamming code is defined by

$$g(X) = 1 + X + X^4$$

Develop the encoder and syndrome calculator for this code, using a systematic form for the code.

Problem 11.13   Consider the (15, 4) maximum-length code that is the dual of the (15, 11) Hamming code of Problem 11.12. Do the following:

(a) Find the feedback connections of the encoder, and compare your result with those of Table 10.1 presented in the previous chapter.

(b) Find the generator polynomial $g(X)$; hence, determine the output sequence assuming the initial state 0001. Confirm the validity of your result by cycling the initial state through the encoder.

Problem 11.14 Consider the (31, 15) Reed–Solomon code.

(a) How many bits are there in a symbol of the code?

(b) What is the block length in bits?

(c) What is the minimum distance of the code?

(d) How many symbols in error can the code correct?

Problem 11.15 A convolutional encoder has a single-shift register with two stages, (i.e., constraint length $K = 3$), three modulo-2 adders, and an output multiplexer. The generator sequences of the encoder are as follows:

$$g^{(1)} = (1, 0, 1)$$

$$g^{(2)} = (1, 1, 0)$$

$$g^{(3)} = (1, 1, 1)$$

Draw the block diagram of the encoder.

Note: For problems 11.16 to 11.23 that are to follow, the same message sequence 10111 . . . is used so that we may compare the outputs of different encoders for the same input.

Problem 11.16 Consider the rate $r = 1/2$, constraint length $K = 2$ convolutional encoder of Fig. P11.2. The code is systematic. Find the encoder output produced by the message sequence 10111. . . .

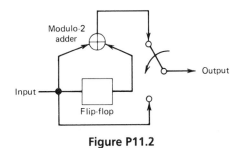

Figure P11.2

Problem 11.17 Figure P11.3 shows the encoder for a rate $r = 1/2$, constraint length $K = 4$ convolutional code. Determine the encoder output produced by the message sequence 10111. . . .

Problem 11.18 Consider the encoder of Fig. 11.13*b* for a rate $r = 2/3$, constraint length $K = 2$ convolutional code. Determine the code sequence produced by the message sequence 10111. . . .

Problem 11.19 Construct the code tree for the convolutional encoder of Fig. P11.2. Trace the path through the tree that corresponds to the message sequence 10111 . . . , and compare the encoder output with that determined in Problem 11.16.

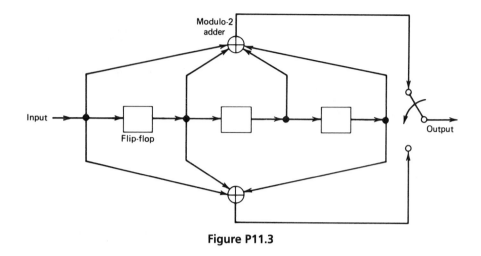

Figure P11.3

Problem 11.20   Construct the code tree for the encoder of Fig. P11.3. Trace the path through the tree that corresponds to the message sequence 10111. . . . Compare the resulting encoder output with that found in Problem 11.17.

Problem 11.21   Construct the trellis diagram for the encoder of Fig. P11.3, assuming a message sequence of length 5. Trace the path through the trellis corresponding to the message sequence 10111. . . . Compare the resulting encoder output with that found in Problem 11.17.

Problem 11.22   Construct the state diagram for the encoder of Fig. P11.3. Starting with the all-zero state, trace the path that corresponds to the message sequence 10111 . . . , and compare the resulting code sequence with that determined in Problem 11.17.

Problem 11.23   Consider the encoder of Fig. 11.13*b*.

(a)   Construct the state diagram for this encoder.
(b)   Starting from the all-zero state, trace the path that corresponds to the message sequence 10111. . . . Compare the resulting code sequence with that determined in Problem 11.18.

Problem 11.24   By viewing the minimum shift keying (MSK) scheme as a finite-state machine, construct the trellis diagram for MSK. (A description of MSK is presented in Chapter 8.)

Problem 11.25   The trellis diagram of a rate-1/2, constraint length-3 convolutional code is shown in Fig. P11.4. The all-zero sequence is transmitted, and the received sequence is 100010000. . . . Using the Viterbi algorithm, compute the decoded sequence.

Problem 11.26   Consider a rate-1/2, constraint length-7 convolutional code with free distance $d_{\text{free}} = 10$. Calculate the asymptotic coding gain for the following two channels:

(a)   Binary symmetric channel
(b)   Binary input AWGN channel

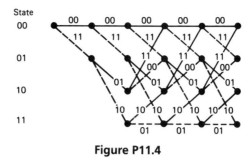

State

Figure P11.4

Problem 11.27 In Section 11.6 we described the Viterbi algorithm for maximum-likelihood decoding of a convolutional code. Another application of the Viterbi algorithm is for maximum-likelihood demodulation of a received sequence corrupted by intersymbol interference due to a dispersive channel. Figure P11.5 shows the trellis diagram for intersymbol interference, assuming a binary data sequence. The channel is discrete, described by the finite impulse response $(1, 0.1)$. The received sequence is $(1.0, -0.3, -0.7, 0, \ldots)$. Use the Viterbi algorithm to determine the maximum likelihood decoded version of this sequence.

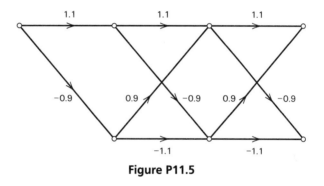

Figure P11.5

Problem 11.28 Figure P11.6 depicts the signal constellation of a hybrid amplitude/phase modulation scheme. Partition this constellation into eight subsets. At each stage of the partitioning, indicate the within-subset (shortest) Euclidean distance.

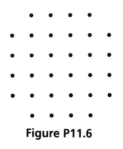

Figure P11.6

CHAPTER 12

Advanced Communication Systems

12.1 INTRODUCTION

Much of the material on communication theory presented in previous chapters of the book has been based on a particular idealization of the communication channel, namely, a *channel model limited in bandwidth and corrupted by additive white Gaussian noise.* The *classical communication theory* so developed is mathematically elegant, providing a sound introduction to the ever-expanding field of communication systems. An example of a physical channel that is well represented by such a model is the satellite communications channel. It is therefore befitting that we devote the first topic in this final chapter of the book to a discussion of satellite communications.

The next topic covered in this chapter on advanced communication systems is mobile radio. The mobile radio channel deviates from the idealized channel model mentioned above due to the presence of multipath, which is a non-Gaussian form of signal-dependent interference that arises because of reflections of the transmitted signal from fixed and moving objects. The presence of multipath raises practical difficulties in the use of a mobile radio channel and complicates its mathematical analysis. The mobile radio channel is basic to the operation of indoor and outdoor forms of wireless communications.

The third topic we have picked for discussion in this chapter is optical communications. Optical fiber transmission has a major advantage over traditional transmission media in that a single-mode fiber is essentially transparent. Specifically, the potentially usable bandwidth of a single-mode fiber is almost limitless, and there is no apparent noise to be concerned about. These unusual channel characteristics make it difficult to fit optical communications into Shannon's classical model of information theory.

The spectacular advances in single-mode fibers and related optical devices are being exploited to build a *photonic network* (i.e., an all-optical fiber network) that will eventually bring fiber to the home. This is the so-called *broadband integrated services digital network* (B-ISDN), which offers users a flexible, all-purpose, two-way transmission network that is much more advanced than the telephone network in use today. Thus, the section on broadband ISDN follows naturally from the one on optical communications.

We conclude the chapter and the book by presenting an emerging view of the telecommunications environment. Needless to say, the telecommunications environment is undergoing some profound changes brought on by spectacular advances made in lightwave technology, computers, and very-large-scale integrated (VLSI) circuit technology.

12.2 SATELLITE COMMUNICATIONS[1]

In a geostationary satellite communications system, a message signal is transmitted from an Earth station via an *uplink* to a satellite, amplified in a *transponder* (i.e., electronic circuitry) on board the satellite, and then retransmitted from the satellite via a *downlink* to another Earth station, as illustrated in Fig. 12.1. The most popular frequency band for satellite communications is 6 GHz (C-band) for the uplink and 4 GHz for the downlink. The use of this frequency band offers the following advantages:

- Relatively inexpensive microwave equipment.
- Low attenuation due to rainfall; rainfall is the primary atmospheric cause of signal degradation.
- Insignificant sky background noise; the sky background noise (due to random noise emissions from galactic, solar, and terrestrial sources) reaches its lowest level between 1 and 10 GHz.

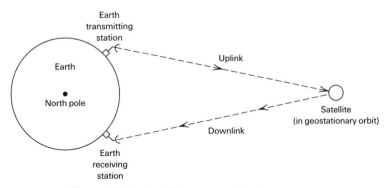

Figure 12.1 Satellite communications system.

However, radio interference limits the applications of communication satellites operating in the 6/4 GHz band, because the transmission frequencies of this band coincide with those used for terrestrial microwave systems. This problem is eliminated in the more powerful "second-generation" communication satellites that operate in the 14/12 GHz band (i.e., Ku-band); moreover, the use of these higher frequencies makes it possible to build smaller and therefore less expensive antennas.

The block diagram of Fig. 12.2 shows the basic components of a single transponder channel of a typical communications satellite. Specifically, the receiving antenna output of the uplink is applied to the cascade connection of the following components:

• *Bandpass filter*, designed to separate the received signal from among the different radio channels.
• *Low-noise amplifier.*
• *Frequency down-converter*, the purpose of which is to convert the received radio frequency (RF) signal to the desired down-link frequency.
• *Traveling-wave tube amplifier*, which provides high gain over a wide band of frequencies. In a traveling-wave tube, an electromagnetic signal travels along a helix (i.e., a spring-shaped coil of wire), while electrons in a high-voltage beam travel through the helix at a speed close to that of the signal wave; the net result is the transfer of power from the electrons to the wave, which grows rapidly as the signal wave travels down the helix.

The channel configuration shown in Fig. 12.2 uses a single frequency translation. Other channel configurations do the frequency conversion from the uplink to the downlink frequency in two stages: down-conversion to an intermediate frequency, followed by amplification, and then up-conversion to the desired transmit frequency.

Propagation time delay becomes particularly pronounced in a satellite channel because of the large distances involved. Specifically, speech signals sent by satellite incur a round-trip transmission delay of approximately 270 ms. Hence, for speech signals, any impedance mismatch at the receiving end of a satellite link results in an *echo* of the speaker's voice, which is heard back at the transmitting end after a round-trip delay of approximately 540 ms. We may overcome this problem by using an *echo canceller*, which is a device that subtracts an estimate of the echo from the return path; estimation of the echo is performed by means of a special filter that adapts itself to the changing channel characteristics.

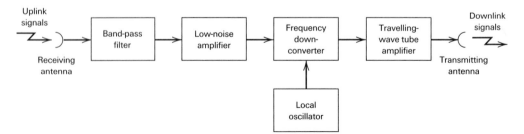

Figure 12.2 Block diagram of transponder.

The satellite channel is closely represented by an *additive white Gaussian noise (AWGN) model*, which applies to both the uplink and downlink portions of the satellite communications system. Accordingly, the material presented in Chapter 9 on passband systems for the transmission of data, with emphasis on digital modulation techniques, is directly applicable to digital satellite communications.

Multiple Access and Broadcasting

A satellite transponder differs from a conventional microwave line-of-sight repeater in that many earth stations can access the satellite from widely different locations on earth at the same time or nearly so. Such a capability is called *multiple access*. The purpose of multiple access is to permit the communication resources of the satellite to be shared by a large number of users seeking to communicate with each other. For obvious reasons, it is desirable that the sharing of satellite resources be accomplished without causing serious interference to each other. In this context, we may identify four basic types of multiple access techniques:

1. *Frequency-division multiple access* (FDMA). In this technique, disjoint sub-bands of frequencies are allocated to the different users on a continuous-time basis. In order to reduce interference between users allocated adjacent channel bands, *guard bands* are used to act as buffer zones, as illustrated in Fig. 12.3a. These guard bands are necessary because of the impossibility of achieving ideal filtering. Moreover, recognizing that nonlinearity of the sat-

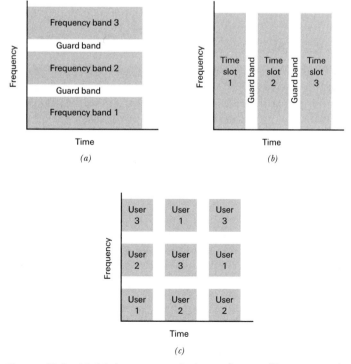

Figure 12.3 Multiple-access techniques for satellite communications.

ellite transponder is the primary cause of interference between users, the traveling-wave tube amplifier in the transponder is purposely operated below capacity. Consequently, the power efficiency of the FDMA technique is reduced, because of the necessary *power backoff* of the traveling-wave tube amplifier.

2. *Time-division multiple access* (TDMA). In this second technique, each user is allocated the full spectral occupancy of the satellite transponder, but only for a short duration of time called a *time slot*. As shown in Fig. 12.3*b*, buffer zones in the form of *guard times* are inserted between the assigned time slots. This is done to reduce interference between users by allowing for time uncertainty that arises due to system imperfections, especially in synchronization schemes. An advantage of TDMA over FDMA is that the satellite transponder is able to operate close to full power efficiency by permitting the traveling-wave tube amplifier to run into saturation.

3. *Code-division multiple access* (CDMA). In FDMA, the satellite resource is shared by dividing it along the frequency coordinate into disjoint frequency bands, as illustrated in Fig. 12.3*a*. In TDMA, the satellite resource is shared by dividing it along the time coordinate into disjoint time slots, as illustrated in Fig. 12.3*b*. These two techniques constitute the most popular choices for multiple access applications. In Fig. 12.3*c*, we illustrate another technique for sharing the satellite resource by using a hybrid combination of FDMA and TDMA, which represents a specific form of code-division multiple access (CDMA). Specifically, *frequency hopping* may be employed to ensure that during each successive time slot, the frequency bands assigned to the users are reordered in an essentially random manner. For example, during time slot 1, user 1 occupies frequency band 1, user 2 occupies frequency band 2, user 3 occupies frequency band 3, and so on. During time slot 2, user 1 hops to frequency band 3, user 2 hops to frequency band 1, user 3 hops to frequency band 2, and so on. Such an arrangement has the appearance of the users playing a game of musical chairs. An important advantage of CDMA over both FDMA and TDMA is that it can provide for *secure* communications. In the type of CDMA illustrated in Fig. 12.3*c*, the frequency hopping mechanism can be implemented through the use of a *pseudo-noise (PN) sequence*, which is a cyclic code with noiselike characteristics; PN sequences were discussed in Chapters 9 and 11.

4. *Space-division multiple access* (SDMA). In this multiple access technique, resource allocation is achieved by exploiting the spatial separation of earth stations. In particular, *multibeam antennas* are used to separate radio signals by pointing them along different directions. This is made possible by means of *onboard switching* designed to select the proper antenna beam for radio transmission. Thus, different Earth stations are enabled to access the satellite transponder simultaneously on the same frequency or in the same time slot.

All these multiple access techniques share a common feature: allocating the communication resources of the satellite through the use of disjointedness (or orthogonality in a loose sense) in time, frequency, or space.

Another channel characteristic unique to satellite transmission is the capability of *broadcasting*. A distinguishing feature of broadcasting satellites is their high power transmission to inexpensive receivers. This characteristic is exploited in the use of *direct broadcast satellites* (DBS), designed for convenient home reception of television services.

The emphasis in satellite communications is on broad-area coverage. In the next section, we consider wireless communications where the emphasis is on mobility.

12.3 MOBILE RADIO

The term "mobile radio" as used in this book is meant to encompass indoor or outdoor forms of *wireless communications* where a radio transmitter or receiver is capable of being moved, regardless of whether it actually moves or not. Due to the complex and variable nature of the mobile radio channel, it is not feasible to use a deterministic approach for its characterization. Rather, it is necessary to resort to the use of practical measurements and statistical analysis.[2] The aim of such a study is to quantify two factors of primary concern:

1. *Median signal strength,* which enables us to predict the minimum power needed to radiate from the transmitter so as to provide an acceptable quality of coverage over a predetermined service area.
2. *Signal variability,* which characterizes the fading nature of the channel.

Our specific interest in wireless communications is in the context of *cellular radio* that has the inherent capability of building mobility into the telephone network. With such a capability, a user can move freely within a service area and simultaneously communicate with any telephone subscriber in the world. An idealized model of the cellular radio system, illustrated in Fig. 12.4, consists of an array of hexagonal *cells,* with a *base station* located at the center of each cell; a typical cell has a radius of 1 to 12 miles. The function of the base stations is to act as an interface between *mobile subscribers* and the cellular radio system. The base stations are themselves connected to a *switching center* by dedicated wirelines.

The mobile switching center has two important roles. First, it acts as the interface between the cellular radio system and the public switched telephone network. Second, it performs overall supervision and control of the mobile com-

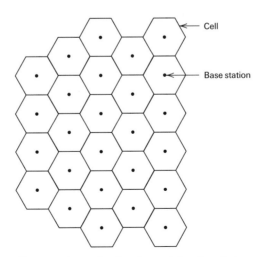

Figure 12.4 Idealized model of cellular radio.

munications. It performs the latter function by monitoring the signal-to-noise ratio of a call in progress, as measured at the base station in communication with the mobile subscriber involved in the call. When the SNR falls below a prescribed threshold, which happens when the mobile subscriber leaves its cell or when the radio channel fades, it is switched to another base station. This switching process, called a *handover* or *handoff*, is designed to move a mobile subscriber from one base station to another during a call in a transparent fashion, that is, without interruption of service.

The cellular concept relies on two essential features, as described here:

1. *Frequency reuse.* The term "frequency reuse" refers to the use of radio channels on the same carrier frequency to cover different areas, which are physically separated from each other sufficiently to ensure that co-channel interference is not objectionable. Thus, instead of covering an entire local area from a single transmitter with high power at a high elevation, frequency reuse makes it possible to achieve two commonsense objectives: keep the transmitted power from each base station to a minimum, and position the antennas of the base stations just high enough to provide for the area coverage of the respective cells.

2. *Cell splitting.* When the demand for service exceeds the number of channels allocated to a particular cell, cell splitting is used to handle the additional growth in traffic within that particular cell. Specifically, cell splitting involves a revision of cell boundaries, so that the local area formerly regarded as a single cell can now contain a number of smaller cells and utilize the channel complements of these new cells. The transmitter power and the antenna height of the new base stations are correspondingly reduced, and the same set of frequencies are reused in accordance with a new plan.

For a hexagonal model of the cellular radio system, we may exploit the basic properties of hexagonal cellular geometry to lay out a radio channel assignment plan that determines which channel set should be assigned to which cell. We begin with two integers i and j ($i \geq j$), called *shift parameters*, which are predetermined in some manner. We note that with a hexagonal cellular geometry there are six "chains" of hexagons that emanate from each hexagon and that extend in different directions. Thus, starting with any cell as a reference, we find the nearest *co-channel cells* by proceeding as follows:

• Move i cells along any chain of hexagons, turn counterclockwise 60 degrees, and move j cells along the chain that lies on this new direction. The jth cells so located and the reference cell constitute the set of co-channel cells.

This procedure is repeated for a different reference cell, until all the cells in the system are covered. Figure 12.5 illustrates the application of this procedure for a single reference cell and the example of $i = 2$ and $j = 2$.

In North America, the band of radio frequencies assigned to the cellular system is 800–900 MHz. The subband 824–849 MHz is used to receive signals from the mobile units, and the subband 869–894 MHz is used to transmit signals to the mobile units. The use of these relatively high frequencies has the beneficial feature of providing a good portable coverage by penetrating buildings. In Europe and elsewhere, the base–mobile and mobile–base subbands are reversed.

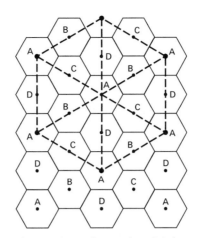

Figure 12.5 Illustrating the determination of co-channel cells.

Propagation Effects

The major propagation problems encountered in the use of cellular radio in built-up areas are due to the fact that the antenna of a mobile unit may lie well below the surrounding buildings. Simply put, there is no "line-of-sight" path to the base station. Instead, radio propagation takes place mainly by way of scattering from the surfaces of the surrounding buildings and by diffraction over and/or around them, as illustrated in Fig. 12.6. The important point to note from Fig. 12.6 is that energy reaches the receiving antenna via more than one path. Accordingly, we speak of a *multipath phenomenon* in that the various incoming radio waves reach their destination from different directions and with different time delays.

To understand the nature of the multipath phenomenon, consider first a "static" multipath environment involving a stationary receiver and a transmitted

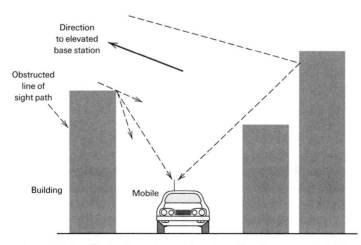

Figure 12.6 Illustrating the mechanism of radio propagation in urban areas. (From Parsons, 1992.)

signal that consists of a narrow-band signal (e.g., unmodulated sinusoidal carrier). Let it be assumed that two attenuated versions of the transmitted signal arrive sequentially at the receiver. The effect of the differential time delay is to introduce a relative phase shift between the two components of the received signal. We may then identify one of two extreme cases that can arise:

- The relative phase shift is zero, in which case the two components add constructively, as illustrated in Fig. 12.7a.
- The relative phase shift is 180 degrees, in which case the two components add destructively, as illustrated in Fig. 12.7b.

We may also use *phasors* to demonstrate the constructive and destructive effects of multipath, as shown in Figs. 12.8a and 12.8b, respectively. Note that in the static multipath environment described herein, the amplitude of the received signal does not vary with time.

Consider next a "dynamic" multipath environment in which the receiver is in motion and two versions of the transmitted narrow-band signal reach the receiver via paths of different lengths. Due to motion of the receiver, there is a continuous change in the length of each propagation path. Hence, the relative phase shift between the two components of the received signal is a function of spatial location of the receiver. As the receiver moves, we now find that the received amplitude (envelope) is no longer constant as was the case in a static environment; rather, it varies with distance, as illustrated in Fig. 12.9. At the top of this figure, we have also included the phasor relationships for the two components of the received signal at various locations of the receiver. Figure 12.9 shows that there is constructive addition at some locations, and almost complete cancellation at some other locations. This phenomenon is referred to as *signal fading*.

In a mobile radio environment encountered in practice, there may of course be a multitude of propagation paths with different lengths, and their contributions to the received signal could combine in a variety of ways. The net result is that the envelope of the received signal varies with location in a complicated fashion, as shown by the experimental record of received signal envelope in an urban area that is presented in Fig. 12.10. This figure clearly displays the fading nature of the received signal. The received signal envelope in Fig. 12.10 is measured in dBm. The unit dBm is defined as $10 \log_{10}(P/P_0)$, with P denoting

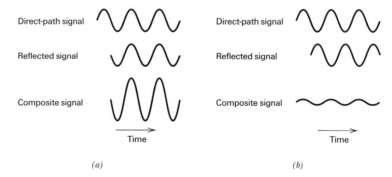

Figure 12.7 (a) Constructive and (b) destructive forms of the multipath phenomenon for sinusoidal signals.

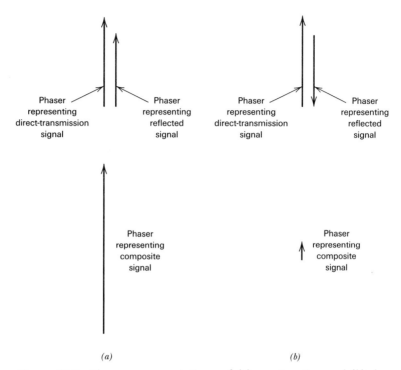

Figure 12.8 Phasor representations of (*a*) constructive and (*b*) de-structive forms of multipath.

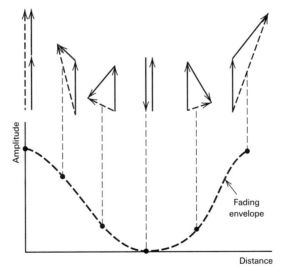

Figure 12.9 Illustrating how the envelope fades as two incoming signals combine with different phases. (From Parsons, 1992.)

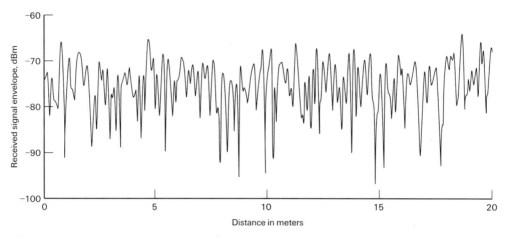

Figure 12.10 Experimental record of received signal envelope in an urban area. (From Parsons, 1992.)

the power being measured and $P_0 = 1$ milliwatt. In the case of Fig. 12.10, P is the instantaneous power in the received signal envelope.

Signal fading is essentially a *spatial phenomenon* that manifests itself in the time domain as the receiver moves. These variations can be related to the motion of the receiver as follows. To be specific, consider the situation illustrated in Fig. 12.11, where the receiver is assumed to be moving along the line AA' with a constant velocity v. It is also assumed that the received signal is due to a radio wave from a scatterer labeled S. Let Δt denote the time taken for the receiver to move from point A to A'. Using the notation described in Fig. 12.11, the incremental change in the path length of the radio wave is deduced to be

$$\Delta \ell = d \cos \alpha$$
$$= v \, \Delta t \cos \alpha \tag{12.1}$$

where α is the spatial angle between the incoming radio wave and the direction of motion of the receiver. Correspondingly, the change in the phase angle of

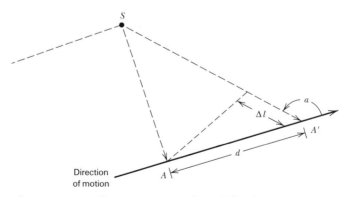

Figure 12.11 Illustrating Doppler shift. (From Parsons, 1992.)

the received signal at point A' with respect to that at point A is given by

$$\Delta\phi = -\frac{2\pi}{\lambda}\Delta\ell$$

$$= -\frac{2\pi v \, \Delta t}{\lambda}\cos\alpha \tag{12.2}$$

where λ is the radio wavelength. The apparent change in frequency, or the *Doppler-shift*, is therefore

$$\nu = -\frac{1}{2\pi}\frac{\Delta\phi}{\Delta t}$$

$$= \frac{v}{\lambda}\cos\alpha \tag{12.3}$$

The Doppler-shift ν is positive (resulting in an increase in frequency) when the radio waves arrive from ahead of the mobile unit, and it is negative when the radio waves arrive from behind the mobile unit.

The narrow-band characterization of the multipath environment considered above is appropriate for radio mobile transmissions where the signal bandwidth is very small compared to the reciprocal of the spread in propagation path delays. Multipath in such an environment results in two effects: rapid fading of the received signal envelope and a spread in Doppler shifts in the received spectrum. Real-life signals radiated in a mobile radio environment may, however, occupy a bandwidth wide enough to require more detailed considerations of the effects of multipath propagation on the received signal. In what follows, we present a statistical characterization of a mobile radio channel.

Fading Multipath Channels

Consider a mobile radio channel with multiple propagation paths. In accordance with the complex notation described in Chapter 2, we may express the transmitted band-pass signal as

$$s(t) = \mathrm{Re}[\tilde{s}(t)\exp(j2\pi f_c t)] \tag{12.4}$$

where $\tilde{s}(t)$ is the complex (low-pass) envelope of $s(t)$, and f_c is a nominal carrier frequency. Since the channel is time varying due to multipath effects, the impulse response of the channel is delay dependent and therefore a time-varying function. Let the impulse response of the channel be expressed as

$$h(\tau;t) = \mathrm{Re}[\tilde{h}(\tau;t)\exp(j2\pi f_c t)] \tag{12.5}$$

where $\tilde{h}(\tau;t)$ is the (low-pass) complex impulse response of the channel, and τ is a delay variable. The complex impulse response $\tilde{h}(\tau;t)$ is called the *input delay-spread function* of the channel. The (low-pass) complex envelope of the channel output is defined by the convolution integral

$$\tilde{s}_o(t) = \int_{-\infty}^{\infty} \tilde{s}(t-\tau)\tilde{h}(\tau;t) \, d\tau \tag{12.6}$$

In general, the behavior of a mobile radio channel can be described only in statistical terms. For analytic purposes, the delay-spread function $\tilde{h}(\tau;t)$ may thus be modeled as a *zero-mean complex-valued Gaussian process*. Then, at any time t the envelope $|\tilde{h}(\tau;t)|$ is Rayleigh distributed, and the channel is referred to as a *Rayleigh fading channel*. When, however, the mobile radio environment includes *fixed* scatterers, we are no longer justified in using a zero-mean model to describe the input delay-spread function $\tilde{h}(\tau;t)$. In such a case, it is more appropriate to use a Rician distribution to describe the envelope $|\tilde{h}(\tau;t)|$, and the channel is referred to as a *Rician fading channel*. The Rayleigh and Rician distributions for a real-valued random process were considered in Chapter 4. In the discussion presented in this chapter, we consider only a Rayleigh fading channel.

The *time-varying transfer function* of the channel is defined as the Fourier transform of the input delay-spread function $\tilde{h}(\tau;t)$ with respect to the delay variable τ, as shown by

$$\tilde{H}(f;t) = \int_{-\infty}^{\infty} \tilde{h}(\tau;t)\exp(-j2\pi f\tau)\ d\tau \tag{12.7}$$

where f denotes the frequency variable. The time-varying transfer function $\tilde{H}(f;t)$ may be viewed as a frequency transmission characteristic of the channel.

For a statistical characterization of the channel, we make the following assumptions:

- The input delay-spread function $\tilde{h}(\tau;t)$ is a *zero-mean, complex-valued Gaussian process*. Our interest is confined to short-term fading; it is therefore reasonable to assume that $\tilde{h}(\tau;t)$ is also *wide-sense stationary*. Since Fourier transformation is linear, the time-varying transfer function $\tilde{H}(f;t)$ has similar statistics.

- The channel is an *uncorrelated scattering channel*, which means that contributions from scatterers with different propagation delays are uncorrelated.

Consider then the autocorrelation function of the input delay-spread function $\tilde{h}(\tau;t)$. Since $\tilde{h}(\tau;t)$ is complex valued, we use the following definition for the autocorrelation function (see Appendix 8):

$$R_{\tilde{h}}(\tau_1,t_1;\tau_2,t_2) = E[\tilde{h}^*(\tau_1;t_1)\tilde{h}(\tau_2;t_2)] \tag{12.8}$$

where E is the statistical expectation operator, the asterisk denotes complex conjugation, τ_1 and τ_2 are the propagation delays of the two paths involved in the calculation, and t_1 and t_2 are the times at which the outputs of the two paths are observed. Invoking wide-sense stationarity in the time variable t and uncorrelated scattering in the time-delay variable τ, we may reformulate the autocorrelation function of $\tilde{h}(\tau;t)$ as

$$\begin{aligned} R_{\tilde{h}}(\tau_1,\tau_2;\Delta t) &= E[\tilde{h}^*(\tau_1;t)\tilde{h}(\tau_2;t+\Delta t)] \\ &= r_{\tilde{h}}(\tau_1;\Delta t)\ \delta(\tau_1-\tau_2) \end{aligned} \tag{12.9}$$

where Δt is the difference between the observation times, and $\delta(\tau_1-\tau_2)$ is a Dirac delta function. Using τ in place of τ_1, the remaining function in Eq. (12.9) is redefined by

$$r_{\tilde{h}}(\tau;\Delta t) = E[\tilde{h}(\tau;t)\tilde{h}^*(\tau;t+\Delta t)] \tag{12.10}$$

The function $r_{\tilde{h}}(\tau,\Delta t)$ is called the *multipath autocorrelation profile* of the channel.

Consider next a statistical characterization of the channel in terms of the complex-valued, time-varying transfer function $\tilde{H}(f;t)$. The autocorrelation function of $\tilde{H}(f;t)$ is defined by

$$R_{\tilde{H}}(f_1,t_1;f_2,t_2) = E[\tilde{H}*(f_1;t_1)\tilde{H}(f_2;t_2)] \qquad (12.11)$$

where f_1 and f_2 represent two frequencies in the spectrum of a transmitted signal. The autocorrelation function $R_{\tilde{H}}(f_1,t_1;f_2,t_2)$ provides a statistical measure of the extent to which the signal is distorted by transmission through the channel. From Eqs. (12.7), (12.8), and (12.11) we find that the autocorrelation functions $R_{\tilde{H}}(f_1,t_1;f_2,t_2)$ and $R_{\tilde{h}}(\tau_1,t_1;\tau_2,t_2)$ are related by a form of two-dimensional Fourier transformation as follows:

$$R_{\tilde{H}}(f_1,t_1;f_2,t_2) = \int_{-\infty}^{\infty}\int_{-\infty}^{\infty} R_{\tilde{h}}(\tau_1,t_1;\tau_2,t_2)\exp[j2\pi(f_1\tau_1 - f_2\tau_2)]\,d\tau_1\,d\tau_2 \quad (12.12)$$

Invoking wide-sense stationarity in the time domain, we may reformulate Eq. (12.11) as

$$R_{\tilde{H}}(f_1,f_2;\Delta t) = E[\tilde{H}*(f_1;t)\tilde{H}(f_2;t + \Delta t)] \qquad (12.13)$$

This definition suggests that the autocorrelation function $R_{\tilde{H}}(f_1,f_2;\Delta t)$ may be measured by pairs of spaced tones to carry out cross-correlation measurements on the resulting channel outputs. Such a measurement presumes wide-sense stationarity in the time domain. If we also assume wide-sense stationarity in the frequency domain, we may go one step further and write

$$\begin{aligned} R_{\tilde{H}}(f,f + \Delta f;\Delta t) &= r_{\tilde{H}}(\Delta f;\Delta t) \\ &= E[\tilde{H}*(f;t)\tilde{H}(f + \Delta f;t + \Delta t)] \end{aligned} \qquad (12.14)$$

This specialized form of the autocorrelation function of $\tilde{H}(f;t)$ is in fact the Fourier transform of the multipath autocorrelation profile $r_{\tilde{h}}(\tau;\Delta t)$ with respect to the delay-time variable τ, as shown by

$$r_{\tilde{H}}(\Delta f;\Delta t) = \int_{-\infty}^{\infty} r_{\tilde{h}}(\tau;\Delta t)\exp(-j2\pi\tau\Delta f)\,d\tau \qquad (12.15)$$

The function $r_{\tilde{H}}(\Delta f;\Delta t)$ is called the *spaced-frequency spaced-time correlation function* of the channel.

Finally, we introduce a function $S(\tau;\nu)$ that forms a Fourier-transform pair with the multipath autocorrelation profile $r_{\tilde{h}}(\tau;\Delta t)$ with respect to the variable Δt, as shown by

$$S(\tau;\nu) = \int_{-\infty}^{\infty} r_{\tilde{h}}(\tau;\Delta t)\exp(-j2\pi\nu\,\Delta t)\,d(\Delta t) \qquad (12.16)$$

and

$$r_{\tilde{h}}(\tau;\Delta t) = \int_{-\infty}^{\infty} S(\tau;\nu)\exp(j2\pi\nu\,\Delta t)\,d\nu \qquad (12.17)$$

The function $S(\tau;\nu)$ may also be defined in terms of $r_{\tilde{H}}(\Delta f;\Delta t)$ by applying a form of double Fourier transformation: a Fourier transform with respect to the time variable Δt and an inverse Fourier transform with respect to the frequency variable Δf. That is to say,

$$S(\tau;\nu) = \int_{-\infty}^{\infty} \int_{-\infty}^{\infty} r_{\tilde{H}}(\Delta f;\Delta t)\exp(-j2\pi\nu\,\Delta t)\exp(j2\pi\tau\,\Delta f)\,d(\Delta t)\,d(\Delta f)$$

(12.18)

Figure 12.12 displays the functional relationships between $r_{\tilde{h}}(\tau;\Delta t)$, $r_{\tilde{H}}(\Delta f;\Delta t)$, and $S(\tau;\nu)$ in terms of the Fourier transform and its inverse.

The function $S(\tau;\nu)$ is called the *scattering function* of the channel. For a physical interpretation of it, consider the transmission of a single tone of frequency f' (relative to the carrier). The complex envelope of the resulting filter output is

$$\tilde{s}_o(t) = \exp(j2\pi f't)\tilde{H}(f';t)$$

(12.19)

The autocorrelation function of $\tilde{s}_o(t)$ is

$$E\left[\tilde{s}_o^*(t)\tilde{s}_o(t+\Delta t))\right] = \exp(j2\pi f'\Delta t)E[\tilde{H}^*(f';t)\tilde{H}(f';t+\Delta t]$$
$$= \exp(j2\pi f'\Delta t)r_{\tilde{H}}(0;\Delta t)$$

(12.20)

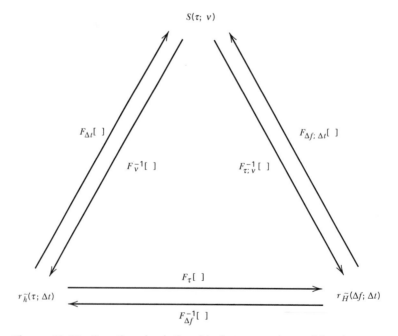

Figure 12.12 Functional relationship between the multipath autocorrelation profile $r_{\tilde{h}}(\tau;\Delta t)$; the spaced-frequency spaced-time correlation function $r_{\tilde{H}}(\Delta f;\Delta t)$, and the scattering function $S(\tau;\nu)$.

where, in the last line, we have made use of Eq. (12.14). Putting $\Delta f = 0$ in Eq. (12.15), and then using Eq. (12.17), we may write

$$r_{\tilde{H}}(0;\Delta t) = \int_{-\infty}^{\infty} r_{\tilde{h}}(\tau;\Delta t) \ d\tau$$

$$= \int_{-\infty}^{\infty} \left[\int_{-\infty}^{\infty} S(\tau;\nu) \ d\tau \right] \exp(j2\pi\nu \ \Delta t) \ d\nu$$

(12.21)

Hence, we may view the integral

$$\int_{-\infty}^{\infty} S(\tau;\nu) \ d\tau$$

as the power spectral density of the channel output relative to the frequency f' of the transmitted tone, and with the Doppler shift ν acting as the frequency variable. Generalizing this result, we may state that the scattering function $S(\tau;\nu)$ provides a statistical measure of the output power of the channel, expressed as a function of the time delay τ and the Doppler shift ν. When both τ and ν are significant, the channel is said to be *doubly spread.*

Delay Spread and Doppler Spread

Putting $\Delta t = 0$ in Eq. (12.10), we may write

$$P_{\tilde{h}}(\tau) = r_{\tilde{h}}(\tau;0)$$

$$= E[|\tilde{h}(\tau;t)|^2]$$

(12.22)

The function $P_{\tilde{h}}(\tau)$ describes the intensity (averaged over the fading fluctuations) of the scattering process at propagation delay τ. Accordingly, $P_{\tilde{h}}(\tau)$ is called the *delay power spectrum* or the *multipath intensity profile* of the channel. The delay power spectrum may also be defined in terms of the scattering function $S(\tau;\nu)$ by averaging it over all Doppler shifts. Specifically, putting $\Delta t = 0$ in Eq. (12.17) and then using the first line of Eq. (12.22), we may write

$$P_{\tilde{h}}(\tau) = \int_{-\infty}^{\infty} S(\tau;\nu) \ d\nu$$

(12.23)

Figure 12.13 shows an example of a delay power spectrum that depicts a typical plot of the power spectral density versus excess delay; the excess delay is measured with respect to the time delay for the shortest echo path. Note, as in Fig. 12.10, the power is measured in dBm. The "threshold level" included in Fig. 12.13 defines the power level below which the receiver fails to operate satisfactorily.

Two statistical moments of $P_{\tilde{h}}(\tau)$ of interest are the *average delay*, τ_{av}, and the *delay spread*, σ_{τ}. The average delay is defined as the first central moment (i.e.,

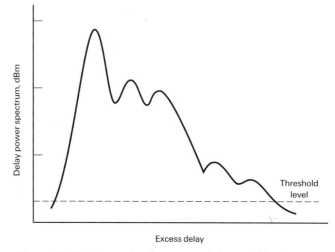

Figure 12.13 Example of a power-delay profile for a mobile radio channel. (From Parsons, 1992.)

the mean) of $P_{\bar{h}}(\tau)$, as shown by

$$\tau_{\mathrm{av}} = \frac{\int_0^\infty \tau P_{\bar{h}}(\tau)\ d\tau}{\int_0^\infty P_{\bar{h}}(\tau)\ d\tau} \tag{12.24}$$

The delay spread is defined as the square root of the second central moment of $P_{\bar{h}}(\tau)$, as shown by

$$\sigma_\tau = \left(\frac{\int_0^\infty (\tau - \tau_{\mathrm{av}})^2 P_{\bar{h}}(\tau)\ d\tau}{\int_0^\infty P_{\bar{h}}(\tau)\ d\tau}\right)^{1/2} \tag{12.25}$$

The reciprocal of the delay spread σ_τ is a measure of the *coherence bandwidth* of the channel. Two sinusoidal components with a frequency separation greater than the coherence bandwidth are treated differently by the channel. In particular, when a message signal is transmitted through a mobile radio channel and the coherence bandwidth of the channel is small compared to the message bandwidth, the received signal is distorted severely by the channel; in such a case, the channel is said to be *frequency selective*. If, on the other hand, the coherence bandwidth of the channel is large compared to the message bandwidth, the channel is said to be *frequency nonselective*.

Consider next the issue of relating the Doppler effects to time variations of the channel. For this purpose, we first set $\Delta f = 0$, which corresponds to the transmission of a single tone (of some appropriate frequency) over the channel. The spaced-frequency spaced-time correlation function of the channel then reduces to $r_{\tilde{H}}(0;\Delta t)$. Hence, evaluating the Fourier transform of this function with respect to the time variable Δt, we may write

$$S_{\tilde{H}}(\nu) = \int_{-\infty}^{\infty} r_{\tilde{H}}(0;\Delta t)\exp(-j2\pi\nu\,\Delta t)\,d(\Delta t) \qquad (12.26)$$

The function $S_{\tilde{H}}(\nu)$ defines the power spectrum of the channel expressed as a function of the Doppler shift ν; it is therefore called the *Doppler spectrum* of the channel. The Doppler spectrum may also be defined in terms of the scattering function by averaging it over all possible propagation delays, as shown by

$$S_{\tilde{H}}(\nu) = \int_{-\infty}^{\infty} S(\tau;\nu)\,d\tau \qquad (12.27)$$

The Doppler shift ν may assume positive and negative values with equal likelihood. The mean Doppler shift is therefore zero. The square root of the second moment of the Doppler spectrum is thus defined by

$$\sigma_\nu = \left(\frac{\int_{-\infty}^{\infty} \nu^2 S_{\tilde{H}}(\nu)\,d\nu}{\int_{-\infty}^{\infty} S_{\tilde{H}}(\nu)\,d\nu} \right)^{1/2} \qquad (12.28)$$

The parameter σ_ν provides a measure of the width of the Doppler spectrum; it is therefore called the *Doppler spread* of the channel. The reciprocal of the Doppler spread is called the *coherence time* of the channel. A slowly varying channel has a small Doppler spread or, equivalently, a large coherence time.

Another useful parameter that is often used in measurements is the *fade rate* of the channel. For a Rayleigh fading channel, the average fade rate is given by

$$f_e = 1.475\sigma_\nu \text{ crossings per second} \qquad (12.29)$$

As the name implies, the fade rate provides a measure of the rapidity of fading of the channel.

Some typical values encountered in a mobile radio environment are as follows:

- The delay spread σ_τ amounts to about 20 μs.
- The Doppler spread σ_ν due to motion of a vehicle may extend up to 40–80 Hz.

Binary Signaling over a Rayleigh Fading Channel

In Chapter 8, we determined the average probability of symbol error for the transmission of binary data over a channel corrupted by additive white Gaussian noise. In a mobile radio environment, we have an additional effect to consider, namely, the fluctuations in the amplitude and phase of the received signal due to multipath effects. To be specific, consider the transmission of binary data over a Rayleigh fading channel, for which the (low-pass) complex envelope of the received signal may be modeled as follows:

$$\tilde{x}(t) = \alpha \exp(-j\phi)\tilde{s}(t) + \tilde{w}(t) \tag{12.30}$$

where $\tilde{s}(t)$ is the complex envelope of the transmitted (band-pass) signal, α is a Rayleigh-distributed random variable describing attenuation in transmission, ϕ is a uniformly distributed random variable describing phase-shift in transmission, and $\tilde{w}(t)$ is a complex-valued white Gaussian noise process. It is assumed that the channel is nonselective and slowly fading, so that we can estimate the phase-shift ϕ from the received signal without error. Suppose then that coherent binary phase-shift keying is used to do the data transmission. Under the condition that α is fixed or constant over a bit interval, we may adapt Eq. (8.102) for the situation at hand by expressing the average probability of symbol error (i.e., bit error rate) due to the additive white Gaussian noise acting alone as follows:

$$P_e(\gamma) = \tfrac{1}{2}\mathrm{erfc}(\sqrt{\gamma}) \tag{12.31}$$

where γ is an attenuated version of the transmitted signal energy per bit-to-noise spectral density ratio E_b/N_0, as shown by

$$\gamma = \frac{\alpha^2 E_b}{N_0} \tag{12.32}$$

Now, insofar as a mobile radio channel is concerned, we may view $P_e(\gamma)$ as a conditional probability given that α is fixed. Thus, to evaluate the average probability of symbol error in the combined presence of fading and noise, we must average $P_e(\gamma)$ over all possible values of γ, as shown by

$$P_e = \int_0^\infty P_e(\gamma) f(\gamma) \, d\gamma \tag{12.33}$$

where $f(\gamma)$ is the probability density function of γ. From Eq. (12.32) we note that γ depends on the squared value of α. Since α is Rayleigh distributed, we find that γ has a *chi-square distribution* with two degrees of freedom (see Problem 4.3). In particular, we may express the probability density function of γ as

$$f(\gamma) = \frac{1}{\gamma_0}\exp\left(-\frac{\gamma}{\gamma_0}\right), \qquad \gamma \geq 0 \tag{12.34}$$

The term γ_0 is the *mean value of the received signal energy per bit-to-noise spectral density ratio*, defined by

$$\begin{aligned} \gamma_0 &= E[\gamma] \\[2mm] &= \frac{E_b}{N_0} E[\alpha^2] \end{aligned} \tag{12.35}$$

where $E[\alpha^2]$ is the mean-square value of the Rayleigh-distributed random variable α.

Substituting Eqs. (12.31) and (12.34) into (12.33), and carrying out the integration, we get the final result

$$P_e = \frac{1}{2}\left(1 - \sqrt{\frac{\gamma_0}{1 + \gamma_0}}\right) \tag{12.36}$$

Equation (12.36) defines the bit error rate for coherent binary phase-shift keying (PSK) over a slow, nonselective Rayleigh fading channel. Following a similar approach, we may derive the corresponding bit error rates for coherent binary frequency-shift keying (FSK), binary differential phase-shift keying (DPSK), and noncoherent binary FSK. The results of these evaluations are summarized in Table 12.1. In Fig. 12.14, we have used the exact formulas of Table 12.1 to plot the bit error rate versus γ_0 expressed in decibels. For the sake of comparison, we have also included in Fig. 12.14 plots for the bit error rates of coherent binary PSK and noncoherent binary FSK for a nonfading channel. We see that Rayleigh fading results in a severe degradation in the noise performance of a digital passband transmission system, the degradation being measured in tens of decibels of additional mean signal-to-noise ratio compared to a nonfading channel for the same bit error rate. In particular, for large γ_0 we may derive the approximate formulas given in the last column of Table 12.1, according to which the asymptotic decrease in the bit error rate with the average signal energy per bit-to-noise spectral density ratio γ_0 follows an *inverse* law. This behavior is dramatically different from the case of a nonfading channel, for which the asymptotic decrease in the bit error rate with γ_0 follows an *exponential* law.

The practical implication of this difference is that in a mobile radio environment, we have to provide a large increase in mean signal-to-noise ratio (relative to a nonfading environment), so as to ensure a bit error rate that is low enough for practical use. To meet such a requirement, we have to increase the transmitted power, antenna size, and so on, which can be costly in terms of implementation. Alternatively, we may utilize special modulation and reception techniques that are less vulnerable to fading effects. Among these techniques,

Table 12.1 Bit Error Rates for Binary Signaling Over a Slow, Nonselective Rayleigh Fading Channel

Type of Signaling	Exact Formula for the Bit Error Rate P_e	Approximate Formula for the Bit Error Rate, Assuming Large γ_0
Coherent binary PSK	$\frac{1}{2}\left(1 - \sqrt{\frac{\gamma_0}{1 + \gamma_0}}\right)$	$\frac{1}{4\gamma_0}$
Coherent binary FSK	$\frac{1}{2}\left(1 - \sqrt{\frac{\gamma_0}{2 + \gamma_0}}\right)$	$\frac{1}{2\gamma_0}$
Binary DPSK	$\frac{1}{2(1 + \gamma_0)}$	$\frac{1}{2\gamma_0}$
Noncoherent binary FSK	$\frac{1}{2 + \gamma_0}$	$\frac{1}{\gamma_0}$

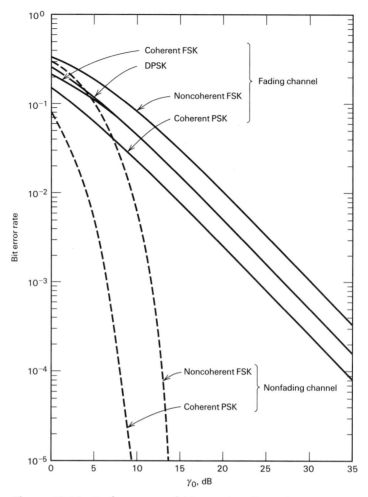

Figure 12.14 Performance of binary signaling schemes on a Rayleigh fading channel, shown as continuous curves.

the best known and most widely used are the multiple-receiver combining techniques referred to collectively as diversity, a brief discussion of which is presented next.

Diversity Techniques

The basic notion of *diversity* may be described as follows. If several replicas of the message signal can be transmitted simultaneously over independently fading channels, then there is a good likelihood that at least one of the received signals will not be degraded by fading. There are several methods for making such a provision. In the context of our present discussion, the following diversity techniques are of particular interest:

- Frequency diversity
- Time (signal-repetition) diversity
- Space diversity

In *frequency diversity*, the message signal is transmitted using several carriers that are spaced sufficiently apart from each other to provide independently fading versions of the signal. This may be accomplished by choosing a frequency spacing equal to or larger than the coherence bandwidth of the channel.

In *time diversity*, the same message signal is transmitted in different time slots, with the spacing between successive time slots being equal to or greater than the coherence time of the channel. Time diversity may be likened to the use of a repetition code for error-control coding.

In *space diversity*, multiple transmitting or receiving antennas are used, with the spacing between adjacent antennas being chosen so as to assure the independence of fading events; this may be satisfied by spacing the adjacent antennas by at least 10 times the radio wavelength.

Given that by one of these means we create L independently fading channels, we may then use a *linear diversity combining structure* involving L separate receivers, as depicted in Fig. 12.15. The system is designed to compensate only for *short-term effects* of a fading channel. Moreover, it is assumed that *noise-free estimates* of the channel attenuation factors $\{\alpha_\ell\}$ and the channel phase-shifts $\{\phi_\ell\}$ are available. Then, the linear combiner achieves optimum performance for binary data transmission (discussed here for the purpose of illustration) by proceeding as follows: The output of the kth matched filter in the ℓth receiver, $\tilde{v}_{\ell k}(t)$, is multiplied by $\alpha_\ell \exp(j\phi_\ell)$ that represents the complex conjugate of the ℓth channel gain, where $\ell = 1, 2, \ldots, L$, and $k = 0, 1$. Thus, the linear combiner results in two output complex envelopes defined by

$$\tilde{v}_k(t) = \sum_{\ell=1}^{L} \alpha_\ell \exp(j\phi_\ell)\tilde{v}_{\ell k}(t), \qquad k = 0, 1 \qquad (12.37)$$

according to which $\alpha_\ell \exp(j\phi_\ell)$ plays the role of a *weighting factor*. One output complex envelope $\tilde{v}_0(t)$ corresponds to the transmission of symbol 0, and the other $\tilde{v}_1(t)$ corresponds to the transmission of symbol 1. The real parts of $\tilde{v}_0(t)$ and $\tilde{v}_1(t)$ are then used in the decision-making process. The situation described here applies to binary FSK. In the case of binary PSK, only a single matched filter is needed, in which case the linear combiner produces a single output complex

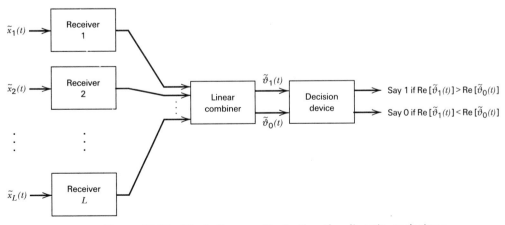

Figure 12.15 Block diagram illustrating the diversity technique.

envelope. Here again, however, the real part of the combiner output is used in the decision-making process.

In the linear combiner described herein, the "instantaneous" output signal-to-noise ratio (SNR) is the sum of the instantaneous SNRs on the individual diversity branches (channels). This optimum form of a linear combiner is therefore referred to as a *maximal-ratio combiner*.

Figure 12.16 shows the noise performance of coherent binary PSK, binary DPSK, and noncoherent binary FSK for $L = 2, 4$ independently fading channels. For the sake of comparison, we have also included in this figure the corresponding graphs for a fading channel with no diversity (i.e., $L = 1$). Figure 12.16 clearly illustrates the effectiveness of diversity as a means of mitigating the short-term effects of fading.

The diversity techniques described above operate on the basis of creating a number of distinguishable and independently fading channels. A more sophisticated approach for achieving diversity, known as *multipath diversity*, involves the

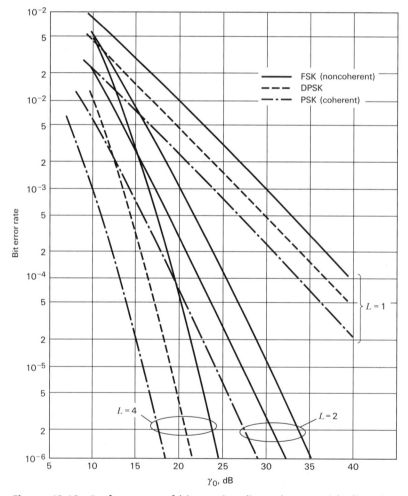

Figure 12.16 Performance of binary signaling schemes with diversity. (From Proakis, 1989, with permission of McGraw-Hill).

transmission of a wideband signal whose bandwidth is much larger than the coherence bandwidth of the channel. This is precisely the form of signaling used in spread spectrum modulation, where a maximum-length sequence with noise-like properties is exploited in the generation of a transmitted signal with a large bandwidth. Note, however, that it is the impulselike correlation property of the maximum-length sequence that is important. In particular, by cross-correlating the received signal with a replica of the maximum-length sequence used in trans-mission, as in the direct sequence spread-spectrum binary PSK system (discussed in Chapter 9), the multipath component made up of delayed versions of the desired signal is treated by the receiver in exactly the same way as any other uncorrelated interfering signal. This statement assumes that the multipath com-ponent is delayed with respect to the desired signal component by more than one chip (code bit) duration. To be specific, let W denote the transmitted signal bandwidth, enabling the correlation receiver to achieve a time resolution of $1/W$. This means that in a multipath environment the number of resolvable signal components is $T_c W$, where T_c is the coherence time of the channel. Since the coherence time T_c is the reciprocal of the Doppler spread σ_ν, we may view mul-tipath diversity involving the use of a wideband signal of bandwidth W as the equivalent of frequency diversity with the number of independently fading chan-nels equal to W/σ_ν.

12.4 OPTICAL COMMUNICATIONS[3]

Optical fibers have unique characteristics that make them highly attractive as a transmission medium, offering the following advantages (see Chapter 1):

- *Enormous potential bandwidth*
- *Low transmission losses*
- *Immunity to electromagnetic interference*
- *Small size and weight*
- *Ruggedness and flexibility*

These unique properties of optical fibers have fueled phenomenal advances in lightwave systems technology during the past two decades, which have, in turn, revolutionized long-distance communications.

Types of Optical Fiber

In Fig. 12.17 we show the cross sections of three important types of optical fibers. Above each cross section a typical profile of the *index of refraction* for the fiber is also presented; "profile" refers to the value of refractive index considered as a function of distance from the axis of the fiber. Figure 12.17 includes typical dimensions so as to provide some idea of the physical size of the fibers considered here.

A *step index fiber* is an optical fiber whose core has a constant index of re-fraction n_1 and whose cladding has a slightly lower index of refraction n_2. Figure 12.17a shows the refractive index profile of a *single-mode step index fiber*. A fiber *mode* represents a pattern of standing waves formed by the propagating field across the core diameter of the fiber. The fiber is said to operate in a single-

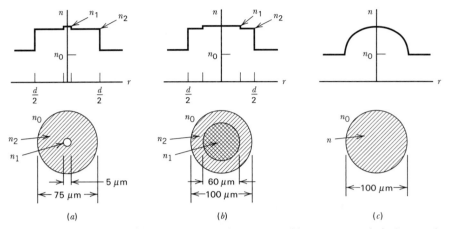

Figure 12.17 Cross-sectional drawings for various fiber types and their associated index of refraction profiles. (*a*) Single-mode step index fiber. (*b*) Multimode step index fiber. (*c*) Graded index fiber. (Adapted from Miller et al., 1973, with permission of the IEEE).

mode fasion if there is only one half-cycle of standing waves across the fiber's core diameter; if there are many half-cycles of such standing waves, the fiber is said to be operating as a *multimode* fiber.

With a step index fiber of the single-mode type, only the central ray can propagate in the core, as illustrated in Fig. 12.18*a*. Accordingly, this type of optical fiber provides the ultimate in transmission bandwidth, which makes it the ideal candidate for long-distance communications. The primary mechanism for bandwidth limitation in a single-mode fiber is *material dispersion,* which stems from the inherent dependence of the refractive index of the core material on the operating frequency. Material dispersion in a single-mode fiber manifests itself in a manner similar to dispersion in an ordinary telephone channel, albeit on a different time scale.

From a mechanical perspective, a multimode fiber is preferred over a single-mode fiber as it has a much larger core diameter and it is therefore easier to splice and to couple segments together with lower loss. Another attractive feature of a multimode fiber is that because of its large core diameter, it is fairly easy to couple it to an optical source. Figure 12.17*b* shows the refractive index profile of a *multimode step index fiber.* In this case, rays of light launched into the core by an optical source are confined by a mechanism of *total internal reflection,* which guides the rays along the axis of the fiber, as illustrated in Fig. 12.18*b*. For total

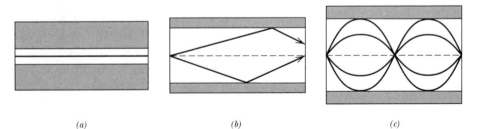

Figure 12.18 Ray paths for different fiber types.

internal reflection to occur, the angle of incidence ϕ measured with respect to the normal to the interface between the core and the cladding has to be greater than the *critical angle* ϕ_c, which is defined in accordance with Snell's law:

$$\sin\phi_c = \frac{n_2}{n_1} \qquad (12.38)$$

When $\phi > \phi_c$, the light is reflected back into the originating dielectric medium at the same angle to the normal, as illustrated in Fig. 12.19. In practice, there is some field extending into the cladding. However, this extraneous field decays almost exponentially with distance from the core boundary; it may therefore be made negligibly small at the outer surface of the fiber.

The practical limitation of a multimode step index fiber is produced by *modal dispersion* or *multipath dispersion*. This dispersive phenomenon is absent altogether in a single-mode fiber; it arises in a multimode step index fiber because the various rays inside the core have different lengths, as illustrated in Fig. 12.18b, and therefore experience different propagation times. Specifically, the rays with small angles of incidence reach the receiver before those other rays with large angles of incidence. The propagation time of a ray may be viewed as a random variable with statistical averages of its own. The *mean time delay* is defined as the average of the various propagation times of the different rays. The variance of the propagation time is defined as the square of the difference between the propagation time and the mean time delay, averaged over the ensemble of propagation times. The square root of the variance (i.e., the standard deviation) of the propagation time is called the *delay spread* of the channel. The effect of delay spread becomes apparent when a stream of digital data is transmitted through the fiber at a higher bit rate. When a narrow pulse is launched into the fiber by the transmitter, modal dispersion causes it to spread into a broader pulse. Consequently, a sequence of short pulses (resulting from the use of a high bit rate) would tend to overlap in time, which makes it difficult or even impossible to distinguish them individually.

One way to reduce the pulse spreading effect is to use a *graded index fiber* with a refractive index profile, as shown in Fig. 12.17c. The graded index fiber differs from the step index fiber in that the refractive index of the core is not constant but rather variable, with its highest value occurring at the axis of the fiber and tapering off roughly parabolically toward the core–cladding interface. The core of a graded index fiber acts as a long, continuous converging lens, which has the effect of making the rays launched into the fiber propagate along

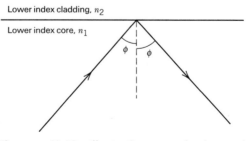

Figure   12.19   Illustrating   total   internal reflection.

helical paths, as illustrated in Fig. 12.18c. Indeed, through a careful selection of the refractive index profile of the fiber, all the guided waves can be arranged to propagate at nearly the same axial speed. By so doing, modal dispersion in the fiber is reduced. Consequently, graded index fibers have a useful transmission bandwidth that may be an order of magnitude greater than that achievable using multimode step index fibers. Moreover, although they are not capable of supporting transmission bandwidths available with single-mode fibers, multimode graded index fibers have the practical advantage of a larger core diameter coupled with bandwidths that make them suitable for long-distance communications.

Another major limitation of optical fibers of all types is the *attenuation* or *transmission loss*, which refers to the loss of optical energy as a light pulse propagates along the fiber. There are two basic physical mechanisms responsible for transmission loss in an optical fiber:

- *Absorption*, which arises due to interaction of the propagating light with impurities in the silica glass, thereby resulting in the dissipation of some of the transmitted optical power as heat in the waveguide.
- *Scattering*, which is produced by geometrical imperfections in the waveguide, and which causes light to be redirected out of the waveguide.

The overall attenuation produced by these mechanisms is called *fiber loss*, which increases exponentially with distance. It also varies with wavelength in the manner described in Fig. 12.20. The first transmission window at 850 nanometres (nm) produces an attenuation of approximately 2.5 dB/km. The second transmission window at 1260 nm produces an attenuation of approximately 0.5 dB/km. The third transmission window at 1570 nm produces an attenuation of approximately 0.20 dB/km. Accordingly, 1300 nm and 1550 nm wavelengths have been chosen for long-distance lightwave transmission systems.

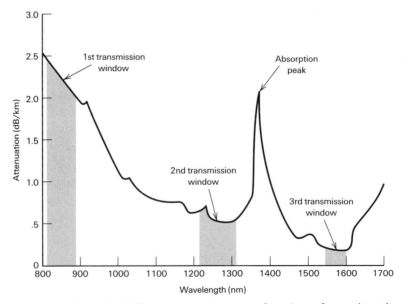

Figure 12.20 Optical fiber attenuation as a function of wavelength. (From Nellist, 1992, with permission of the IEEE.)

Multiplicative Noise in Optical Fibers

An important characteristic of optical fiber links is that their noise problems are many orders of magnitude less severe than those encountered in traditional communication channels. Unlike moving electrons that tend to interact with each other, *moving photons do not interact.* Accordingly, there is *no* noise generated inside an optical fiber. However, a light-wave system has several sources of *multiplicative noise*, which is a signal-dependent form of noise that arises due to random fluctuations of the end-to-end transmission loss across the fiber. Multiplicative noise is different from *additive noise* (e.g., thermal noise at the front end of a receiver) in that it disappears when the transmitted signal is switched off.

There are two types of multiplicative noise in an optical fiber link, namely, *modal noise* and *mode partition noise.* Modal noise occurs when the fiber supports many modes of propagation, as in the case of multimode fibers. On the other hand, mode partition noise arises when the optical source emits several frequencies in a simultaneous fashion or rapid succession. Mode partition noise is particularly noticeable in single-mode fiber systems that use lasers emitting light at many different frequencies.

Sources and Detectors

The transmission of information in the form of light propagating within an optical fiber requires the construction of an *optical communication system.* The principal components of the digital version of such a system are shown in Fig. 12.21. The source encoder in the transmitter is used to convert the message signal from an analog source of information into a stream of bits. The source decoder in the receiver is used to reconstruct the message signal for delivery to the user of information. The source encoder and source decoder are of electrical design. The optical components of the system are represented by the optical source in the transmitter, the optical fiber as the transmission medium, and the optical

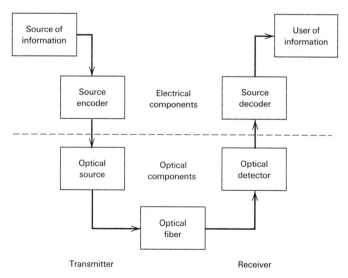

Figure 12.21 Block diagram of optical communication system.

detector in the receiver part of the system. The transmitter emits pulses of optical *power*, with each pulse being "on" or "off" in accordance with the source output. Note that power is a baseband quantity that varies at the bit rate and *not* the optical frequency.

For the *optical source*, we may use an injection laser diode (ILD) or a light emitting diode (LED). The ILD and LED are both solid-state semiconductor devices that can be modulated by varying the electrical current used to power the devices. In a system design, the choice of the light source determines the optical signal power available for transmission. The light source has a driver that typically consists of a high-current–low-voltage device; the light source is thus turned on and off by switching the drive current on and off in response to the coded signal.

The on–off light pulses produced by the transmitter are launched into the optical fiber waveguide. The *collector efficiency* of the fiber depends on its core diameter and acceptance solid angle. The acceptance solid angle refers to the range of angles captured in the core of the fiber via total internal reflection; the acceptance angle expressed in radians defines the *numerical aperture* of the optical fiber. We thus have to account for a *source-to-fiber coupling loss* that varies over a wide range, depending on the particular combination of light source and optical fiber selected. During the course of propagation along the fiber, a light pulse also suffers fiber loss, the causes of which were pointed out earlier. In addition, as mentioned previously, dispersion (material or modal in origin) causes a short pulse to spread into a broader pulse as it propagates along the optical fiber.

At the receiver, the *optical detector* converts the pulses of optical power emerging from the fiber into electrical pulses. One approach for the implementation of the detector is to have the incident optical power illuminate a semiconductor diode, resulting in the generation of mobile carriers (holes and electrons) through the absorption of photons. These mobile carriers, in turn, flow in the presence of an applied electric field toward their respective majority sides (*p*- and *n*-type material), thereby producing an observable current. This is the principle of operation of *p-i-n* and avalanche photodiodes. The choice of optical detector and its associated circuitry determines the *receiver sensitivity*. It is also important to recognize that since the optical power is modulated at the transmitter, and since the optical fiber waveguide operates linearly on the propagation power, the optical detector behaves as a *linear device that converts power to current*.

From this discussion, it is apparent that a lightwave transmission link differs from its metallic wire or coaxial cable counterpart in that power, rather than current (both baseband quantities), propagates through the optical fiber waveguide.

In the design of a lightwave transmission link, two separate factors have to be considered: *transmission bandwidth* and *signal losses*. The transmission bandwidth of an optical fiber is determined by a dispersive mechanism in the fiber. This, in turn, limits the feasible data rate or the rate at which light pulses can propagate through the optical fiber. The signal losses are contributed by source-to-fiber coupling loss, fiber loss, fiber-to-fiber loss (due to joining the fiber by a permanent splice), and fiber-to-detector loss. Knowing these losses, we may formulate a *link budget* that accounts for every loss of power as the transmitted signal makes its way across the fiber link to its destination. Hence, given the optical power available from the light source, we may readily determine the power available at the detector.

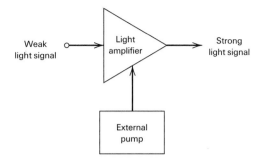

Figure 12.22 Wide-band lightwave amplifier.

Lightwave Amplifiers

The allowable distance of transmission over an optical fiber can be limited by pulse spreading as well as by power loss. To accommodate communication over long distances, the conventional approach is to use *repeaters* placed at regular intervals along the communication link, with each repeater consisting of an optical detector followed by electronic amplification, which is then followed by retransmission in optical form. Such an approach is adequate as long as the fiber link is required to carry a single bit stream and the data rate is less than a few gigabits per second. However, with data rates increasing and the use of a single optical fiber for many wavelength-multiplexed bit streams proving attractive, the need for an all-optical network has taken on a sense of urgency. To accommodate such a provision, we require wide-band, purely *lightwave amplifiers*.

At present, it appears that a practical proposition is to use an *erbium-doped fiber amplifier*. Figure 12.22 shows the block diagram of such a wide-band amplifier, with power from an external pump being used to drive a process that leads to the transformation of a weak input light signal into a strong output light signal.

12.5 BROADBAND INTEGRATED SERVICES DIGITAL NETWORK[4]

The discussion of optical communications naturally leads us to consider its impact on network aspects of the telecommunications environment. In this context, a sensible starting point is ISDN.

The *integrated services digital network* (ISDN) refers to an all-digital network that can provide telephone, data, and other services for users in homes and offices, both locally and around the world. ISDN has two key features: universal access and integrated user services. This network is in the process of being implemented. Notwithstanding the implementation of ISDN, in the late 1980s attention in the telecommunications standards community shifted to a second-generation ISDN, known as *broadband ISDN* (B-ISDN). This latter network is an all-purpose digital network of the future that will provide many capabilities to users that are not available in the current "narrow-band" ISDN.

The goal of B-ISDN is to define a user interface and network that will meet a wide range of needs as described here:

- *Interactive and distributive services.* Interactive services involve a two-way exchange of information between two subscribers or between a subscriber and a service provider; these include conversational services, messaging services, and retrieval services. On the other hand, distributive services involve information transfer primarily in one direction, from a service provider to a subscriber; these include broadcast services.

- *Continuous and bursty traffic.* Continuous traffic (e.g., voice and interactive video) may require guaranteed bandwidth so as to meet prescribed performance requirements (i.e., low information loss and delay). On the other hand, some bursty data traffic characterized by short and sporadic transmissions may be handled more cost effectively with a low grade of service, which is done by sharing network resources on a statistical basis.

- *Point-to-point and point-to-multipoint connections.* Some services such as voice require the use of point-to-point connections, whereas other services such as multimedia communications benefit from parallel connections between end points.

The transmission structure of broadband ISDN involves data rates in excess of 45 Mb/s, possibly extending up to 2.5 Gb/s (and beyond). The optical fiber is the only transmission medium that can sustain such high data rates. Accordingly, the introduction of B-ISDN depends on the pace at which optical fiber links are built.

Another distinctive feature of B-ISDN is the way in which the switching mechanism is implemented. For this function, a new user–network interface protocol known as the asynchronous transfer mode (ATM) has been proposed by the Consultative Committee for International Telegraph and Telephone (CCITT). The *asynchronous transfer mode* (ATM) is a high-bandwidth, low-delay, packetlike technique used for switching and multiplexing; it is independent of the physical means of transport. The "low-delay" feature of the technique is needed to support real-time services such as voice. The "high-bandwidth" feature is required to handle video.

ATM allows for the transport of digital information in the form of small, fixed-size packets called *cells*. This form of *fast packet switching* is made possible by exploiting advances in microelectronics, advances in protocols and software, and switch architectures with more parallelism. The use of cells makes it possible to provide "bandwidth on demand," in the sense that they are available on demand to users needing them. Each cell has a header with a label called the *connection identifier*, which explicitly associates the cell with a specified virtual channel on a physical link; a *virtual cell*, referring to certain logical connections in ATM, is the basic unit of switching in B-ISDN. In today's packet-switching network, packets are combined with other traffic using byte-to-byte multiplexing at the physical layer of the open systems interconnection (OSI) reference model (see Section 1.11 on communication networks). We similarly find that in ATM, flexible handling of dynamically varying loads is provided by means of cell-by-cell multiplexing with each cell having a labeled header used to identify which cells belong to which components of the aggregate traffic stream.

The asynchronous nature of ATM arises from the fact that the cells are not produced with reference to a fixed synchronous cycle; rather, they are generated as required by the application at hand. ATM was adopted over a *synchronous transfer mode* (STM) to overcome two limitations of the synchronous approach:

- STM does not provide a flexible interface for meeting a wide variety of needs.
- STM complicates the switching system when multiple high data rates are used.

In contrast, ATM switches provide the users a great deal of flexibility and opportunity for the efficient sharing of *network resources* (e.g., bandwidth, buffers, and processing horsepower). The first issue to be considered in the study of ATM is therefore the efficient way in which network resources are used. Second, there is the issue of *quality of service* (QOS), which is measured in terms of two parameters:

- *Cell loss ratio*, defined as the ratio of the number of cells lost in transport across the network to the total number of cells pumped into the network.
- *Propagation delay*, defined as the time taken for the cells of a particular service to propagate across the network.

For example, voice traffic is delay sensitive but tolerant to cell loss, high-speed data traffic used for file transfers is loss sensitive but delay insensitive, whereas interactive video traffic is both loss sensitive and delay sensitive. The challenge in the use of ATM is the provision of an adequate quality of service for a wide range of traffic, while at the same time making efficient use of network resources. These requirements can be satisfied only by having a traffic management strategy. In particular, an effective mechanism for *congestion control* in an ATM-based transport network is needed; congestion refers to a condition of sustained cell loss. Needless to say, congestion control is a very active research topic and likely to remain so for a long while.

SONET

After their generation, the ATM cells are structured for transport across the network. The proposed transmission option for B-ISDN is to place the cells in a synchronous time-division multiplex envelope. According to this option, the bit stream at the interface has an external frame structure based on a synchronous optical network (SONET), which is the next generation transmission hierarchy. SONET[5] was originally proposed by BellCore and standardized by the American National Standards Institute (ANSI). SONET defines the rates and format for the optical transmission of digital information. In particular, the SONET specification defines a hierarchy of standardized data rates extending from 51.84 Mb/s to 9.953 Gb/s.

The SONET system hierarchy consists of four layers, as described here:

1. *Photonic layer.* This is the physical layer, which includes specifications on types of optical fiber that may be used, required minimum laser powers, dispersion characteristics of the transmitting lasers, and required sensitivity of the receivers.
2. *Section layer.* This second layer creates the basic SONET frames and converts electronic signals into photonic ones; it also has monitoring capabilities.
3. *Line layer.* This third layer is responsible for synchronization and multiplexing of data into SONET frames; it also performs protection and maintenance functions and switching.
4. *Path layer.* This final layer of SONET is responsible for end-to-end transport of data at the appropriate rate.

The basic SONET building block is the STS-1 frame. It consists of 810 octets, which are transmitted once every 125 μs for an overall data rate of 51.84 Mb/s.

These basic units are synchronously byte-interleaved to generate any desired multiple of the basic data rate. For example, the STS-3 signal is obtained by interleaving three STS-1 frames for a data rate of 155.52 MB/s.

12.6　AN EMERGING VIEW OF TELECOMMUNICATIONS[6]

The full implementation of B-ISDN would lead to the evolution of an "information society" dominated by *multimedia communication systems* that are expected to support all conceivable types of communication media. Moreover, it would be possible to combine and integrate the communication media for individual users and their applications as necessary.

Consider, for example, users with access to large computers in remote locations who would like to bring the high-quality graphics and video capabilities of those computers to their desktop workstation. Today's desktop workstations equal yesterday's mainframe computers in computing power and operating speed, thanks to the spectacular advances that have been made in microelectronics. With computing power of that magnitude at their disposal, it is therefore only natural for users of desktop workstations to demand more flexible interfaces and easy accesses to local telephone networks.

Desktop workstations are also playing new roles in interpersonal communications, enabling users to interact with each other from their homes or offices by exchanging information in different media. These *computer-based desktop conference systems* support real-time conversations among the participants through a coordinated exchange of voice, video, and computer data (e.g., text, graphics, and spreadsheets). The conference participants can be close to one another, exchanging information over local area networks, or they can be geographically separated, in which case the exchange of information takes place over wide area networks. In any event, the common goal of computer-based, desktop conference systems is to emulate important attributes of face-to-face conversations, sitting in the comfort of the users' own homes or offices.

In a similar vein, there is a general trend among business investment corporations toward the development of high-capacity private communication networks interconnecting corporate centers in different geographic locations. These networks meet a distinct business need to have access to and transport large volumes of data from one location to another, and to do it all in a highly secure fashion. The construction of this kind of communication network is the outcome of a natural desire on the part of business customers, regardless of their size, to exercise control over their own telecommunication services.

The availability of B-ISDN, coupled with rapid advances in storage technologies, will make it feasible to provide *multimedia-on-demand information services*. The services offered by a *multimedia server* may be viewed as somewhat analogous to those of a neighborhood videotape rental store. The multimedia server digitally stores multimedia information such as entertainment movies, educational documentaries, and advertisements on a large assortment of extremely high-capacity storage devices (e.g., optical and magnetic disks). These storage devices are accessible with short seek time and are permanently on line. Subscribers can select multimedia information, and the multimedia server has the ability to satisfy their requests in an interactive manner. The architectural vision of a multimedia server as described herein is feasible within the next several years rather than decades.

Multimedia communications, illustrated by the different examples described above, offer new opportunities, but also create important requirements and challenges of their own for the networks used to do the information transmission.

As our modern society evolves, so does the insatiable appetite for personal communications, the provision of which may be viewed as the ideal telecommunication service. Broadly speaking, a *personal communication network* (PCN) is based on a concept similar to that of a cellular mobile radio network in the sense that both systems rely on an array of cells served by individual base stations. However, they differ from each other in a fundamental respect: A cellular mobile radio network merely provides *terminal mobility*, whereas the development of a personal communication network would permit *personal mobility*. The goal of PCN is to define a user interface and network that will make it possible for a single telephone unit and telephone number to follow the user wherever he or she goes. On the road, the portable unit (terminal) acts like a cellular phone. At the home or office, the portable terminal handshakes with the network in place, and normal services are thereby restored.

In conclusion, it is sometimes said that, from the beginning, the evolution of the telecommunications environment took the wrong path, with telephonic communications taking place over wired lines and television receivers getting their signals over the air. However, it appears that this evolutionary course may be reversed, at least in part, with ever-increasing provisions being made by the telecommunications industry for wired television and portable telephony. Such a trend will get a further boost with the introduction of B-ISDN and PCNs.

NOTES AND REFERENCES

1. For detailed treatment of satellite communications and related issues, see the following books: Sklar (1988), Pratt and Bostian (1986), Wu (1984), Bhargava et al. (1981), and Spilker, Jr. (1977). The first, third, fourth, and fifth books emphasize the use of satellites for digital communications. The book by Pratt and Bostian presents a broad treatment of satellite communications, emphasizing such diverse topics as radio-wave propagation, antennas, orbital mechanics, signal processing, and radio electronics.

2. The classic paper on the characterization of randomly time-varying channels is Bello (1963). For a comprehensive treatment of the mobile radio propagation channel, see the book by Parsons (1992). This book presents the fundamentals of VHF and UHF propagation, propagation over irregular terrain and in built-up areas, and a statistical characterization of the mobile radio channel. The statistical characterization of a mobile radio channel is also discussed in Proakis (1989); this book provides a readable account of the effect of fading on the error performance of Rayleigh fading channels, and a good discussion of diversity techniques. For a full treatment of the subject, see Chapters 9 through 11 by Stein in the book edited by Schwartz, Bennett, and Stein (1966). For an original treatment of cellular radio, see the paper by MacDonald (1979). For overview papers on the subject, see Oetting (1983) and Steele (1989).

3. For a readable account of optical fibers and their applications to communications, see the following books: Senior (1992), Personick (1985), and Green, Jr. (1993). In particular, the latter book presents an up-to-date and comprehensive treatment of the subject, including a chapter devoted to lightwave amplifiers. For a thorough and precise analysis of the propagation of light waves in an optical fiber, we need to treat it

as a dielectric waveguide and use Maxwell's equations to carry out the analysis; such an analysis is highly mathematical in nature. For a readable account of the analysis, see Chapter 3 of Green, Jr. (1992). The book edited by Henry and Personick (1990) presents a collection of papers on fiber components, optical amplifiers, receivers, and optical communication systems.

4. For detailed treatments of broadband ISDN, see the book by Stallings (1992). As we move into the gigabit world, with the introduction of B-ISDN, attention may have to be given to the "latency versus bandwidth" trade-off; for a discussion of this issue, see the paper by Kleinrock (1992).

5. The CCITT has defined an alternative international standard to SONET for the transport of digital information-bearing signals, which is known as SDH (synchronous digital hierarchy). In SDH the basic bit rate is 156 Mb/s, and multiples of this bit rate are accepted. For a discussion of optical communication network trends, including SDH, see the Special Issue of *Proceedings of the IEEE*, vol. 81, November 1993.

6. For readable accounts of multimedia communications, see the Special Issue of *IEEE Communications Magazine* (vol. 30, May 1992). For overview papers on the different aspects of PCNs, see the Special Issue of *IEEE Communications Magazine* (June 1992). For discussion of the issues involved in realizing a global information network, see the Special Issue of *IEEE Communications Magazine* (October 1992).

Speech and Television as Sources of Information

A1.1 SPEECH

The *speech production process*[1] may be viewed as a form of filtering, in which a *sound source excites a vocal tract filter.* The vocal tract consists of a tube of nonuniform cross-sectional area, beginning at the *glottis* (i.e., the opening between the vocal cords) and ending at the *lips,* as outlined by the dashed lines in Fig. A1.1, which shows a sagittal plane X-ray photograph of a human vocal system. Depending on the mode of excitation provided by the source, the sounds constituting a speech signal may be classified into two distinct types:

1. *Voiced sounds,* for which the source of excitation is pulselike and periodic; in this case, the speech signal is produced by forcing air (from the lungs) through the glottis with the vocal cords vibrating in a relaxation oscillation.
2. *Unvoiced sounds,* for which the source of excitation is noiselike (i.e., random); in this second case, the speech signal is produced by forming a constriction in the vocal tract toward the mouth end and forcing a continuous stream of air through the constriction at a high velocity.

As sound, be it voiced or unvoiced, propagates along the vocal tract, the spectrum is shaped by the frequency selectivity of the vocal tract. This effect is similar to the resonance phenomenon observed in organ pipes. In the context

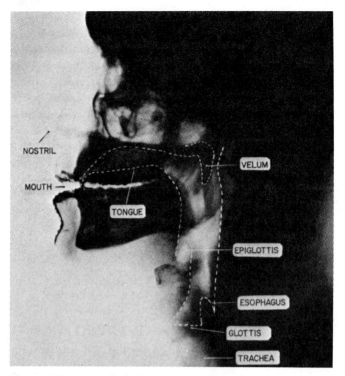

Figure A1.1 Saggital plane X-ray of the human vocal apparatus. (From Rabiner and Schafer, 1978, with permission of the IEEE.)

of speech production, the resonant frequencies of the vocal tract are called *formant frequencies* or simply *formants*.

A speech signal is *nonstationary* in the sense that its characteristics vary with time. This behavior is illustrated in Fig. A1.2. Figure A1.2*a* shows the waveform of a speech signal, with amplitude of the signal plotted as a function of time. Several important characteristics of speech signals may be observed from this waveform:

- Voiced sounds during vowels are characterized by quasi-periodicity, low-frequency content, and large amplitude.
- Unvoiced sounds or fricatives are characterized by randomness, high-frequency content, and relatively low amplitude.
- The transition between voiced and unvoiced sounds is gradual.

Another insightful way of displaying the nonstationary character of speech signals is to use a *spectrogram*[2]. The spectrogram is a two-dimensional display, with time and frequency representing the horizontal and vertical axes, respectively. (A formal definition of the spectogram in the context of time-frequency analysis is presented in Appendix 3.) The darkness of the display is proportional to the signal energy. Figure A1.2*b* shows the spectrogram for the utterance described in Fig. A1.2*a*. The following characteristics of speech signals may be noted from the spectrogram of Fig. A1.2*b*:

- The formants (i.e., peaks of the spectrum) show up as dark horizontal bands.

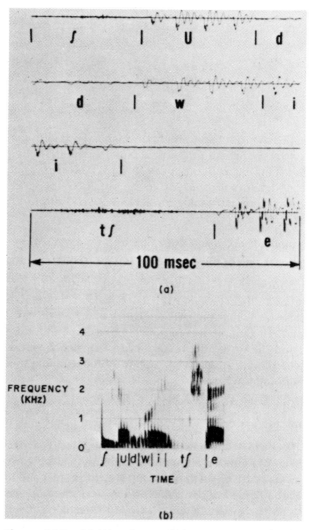

Figure A1.2   (a) Waveform of the utterance "Should we cha(se)"; (b) corresponding spectrogram. (From Rabiner and Schafer, 1978, with permission of Prentice-Hall.)

- Voiced sounds appear as vertical striations.
- Unvoiced sounds appear as rectangular dark patterns randomly punctuated with light spots.

Terminal-Analog Model

From the foregoing discussion, we note that the characterization of a speech signal depends on whether the speech is voiced or unvoiced. Moreover, the vocal tract imposes its resonances on the sound excitation so as to generate different sounds of speech. These observations suggest that we may use a *terminal-analog model* for speech production. The basic idea of the model is that the vocal tract and radiation effects may be accounted for by a linear time-varying system, as depicted in Fig. A1.3. It is assumed that the parameters of the system are related

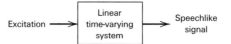

Figure A1.3 Basic idea of terminal-analog model.

to the speech production process in such a way that the output of the system has the desired speechlike properties. Thus, the model is "equivalent" to the physical model insofar as its output behavior is concerned; but the internal structure of the model does *not* conform to the physics of the speech production process.

For the terminal-analog model of Fig. A1.3 to produce a speechlike signal, it must satisfy two requirements:

- The mode of excitation must vary with time.
- The linear system must exhibit resonant frequencies that also vary with time.

The nature of these time variations is illustrated in Fig. A1.2. However, an important point that emerges from a detailed examination of the waveforms shown in this figure is that the characteristics of a speech signal vary relatively slowly with time. In general, it is reasonable to assume that the properties of the excitation and the vocal tract remain essentially unchanged for durations of 10–20 ms. Accordingly, we may construct a terminal-analog model of speech production by doing two things:

- Use an excitation that depends on whether the speech is voiced or unvoiced.
- Permit the parameters of the linear system to vary slowly with time.

The linear system must account for the filtering action of the vocal tract and the effects of radiation at the lips. Assuming that these two effects are separable, we may take care of them by having the linear system composed of the cascade connection of a *vocal tract model* and a *radiation model*, with only the vocal tract model having its parameters controlled externally, as shown in Fig. A1.4. Note also that this cascade connection is common to both voiced and unvoiced speech.

To complete the terminal-analog model, we need to specify the mode of excitation used to drive the linear system. In the case of voiced speech, we may represent the excitation by using an *impulse-train generator* that produces a sequence of impulses (i.e., pulses of very short duration), whose spacing is chosen to equal the desired *pitch period* (i.e., fundamental period of the voiced speech

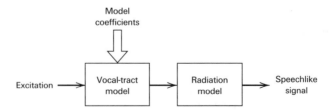

Figure A1.4 Formulation of linear time–varying system for speech production as the cascade connection of two models.

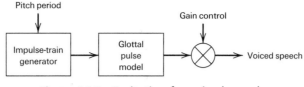

Figure A1.5 Excitation for voiced speech.

signal). This signal, in turn, excites a *glottal pulse model* so as to produce a waveform similar to that of a glottal signal; remember that the vocal tract begins at the glottis. We may thus represent the excitation for voiced speech as shown in Fig. A1.5, where a *gain control* is included in order to control the intensity of the voiced excitation. The excitation for unvoiced speech is much simpler; it merely consists of a random *noise generator* characterized by a flat power spectrum.

Putting all the above-mentioned ideas together, we get the complete model shown in Fig. A1.6. This model includes a controlled *switch* that selects a voiced or unvoiced source of excitation, depending on the type of speech signal being modeled. Virtually all practical speech synthesizers use a terminal-analog model having the general structure shown in Fig. A1.6.

A1.2 TELEVISION

The generaton of a television signal involves the use of a form of spatial sampling called *raster scanning*, the essence of which is described in Chapter 1. The reproduction quality of a TV picture is limited by two basic factors:

1. The number of lines available in a raster scan, which limits resolution of the picture in the vertical direction.
2. The channel bandwidth available for transmitting the video signal, which limits resolution of the picture in the horizontal direction.

For each direction, *resolution* is expressed in terms of the maximum number of lines alternating between black and white that can be resolved in the TV image along the pertinent direction by a human observer.

Consider first the image resolution in the vertical direction, denoted by R_v. It is tempting to equate the vertical resolution R_v to the total number of scan

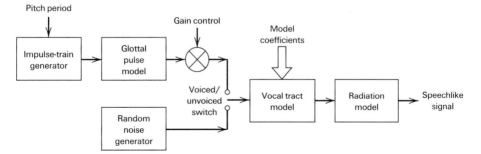

Figure A1.6 Complete model of speech production process.

lines per frame minus those lines in the vertical interval that are not used for display. In practice, however, this is not so, because the scanning process that changes the image into a video signal in the camera (at the transmitter) and then reconstructs the image on the display (at the receiver) is in reality a *sampling process*. (The sampling process is discussed in detail in Chapter 6.) Consequently, the vertical resolution in a TV picture is reduced not only by the vertical retrace, but also by *aliasing* that can result from the improper use of sampling; aliasing refers to a high-frequency component of a signal taking on the identity of a low-frequency component in the sampled version of the signal. In the context of television, we have

$$R_v = k(N - 2N_{vr}) \qquad \text{(A1.1)}$$

where N is the *total* number of raster scan lines in a frame and N_{vr} is the number of lines per field that are lost during the vertical retrace. The fact that the vertical resolution R_v in Eq. (A1.1) is a fraction of $(N - 2N_{vr})$ is called the *Kell effect*; correspondingly, k is called the *Kell factor*. Normally, the Kell factor ranges between 0.6 and 0.7. Let a denote the raster height. Then, we may express the vertical resolution in a TV picture in terms of *horizontal lines per unit distance* as

$$\frac{R_v}{a} = \frac{k}{a}(N - 2N_{vr}) \quad \text{lines/unit distance} \qquad \text{(A1.2)}$$

Consider next the horizontal resolution, denoted as R_h; this resolution is expressed in terms of the maximum number of lines that can be resolved in a TV picture along the horizontal direction. To determine R_h, we assume that the picture elements or *pixels* are arranged as alternate black and white squares along the scanning line. The corresponding video signal is a square wave with a fundamental frequency equal to the video bandwidth. Since there are two pixels per cycle of the square wave, we may express the horizontal resolution of a TV picture as

$$R_h = 2B(T - T_{hr}) \qquad \text{(A1.3)}$$

where B is the *video bandwidth*, T is the total duration of one scanning line, and T_{hr} is the duration of a horizontal retrace.

Let b denote the raster width. We may then express the horizontal resolution of a TV picture in terms of *vertical lines per unit distance* as

$$\frac{R_h}{b} = \frac{2B}{b}(T - T_{hr}) \quad \text{lines/unit distance} \qquad \text{(A1.4)}$$

A natural choice for the video bandwidth B is to make the vertical resolution equal to the horizontal resolution, as shown by

$$\frac{R_v}{a} = \frac{R_h}{b} \qquad \text{(A1.5)}$$

Hence, using Eqs. (A1.2), (A1.4), and (A1.5) to solve for the bandwidth B, we get the desired result, namely,

$$B = \frac{k}{2}\left(\frac{b}{a}\right)\left(\frac{N - 2N_{vr}}{T - T_{hr}}\right) \tag{A1.6}$$

The ratio of raster width b to raster height a is called the *aspect ratio*.

In the NTSC (National Television System Committee) system, which is the North American Standard, we have the following parameter values:

Aspect ratio $= \dfrac{b}{a} = \dfrac{4}{3}$

Total lines per frame $= N = 525$

Vertical retrace $= N_{vr} = 21$ lines/field

Kell factor $= k = 0.7$

Total line time $= T = 63.5\ \mu s$

Horizontal retrace time $= T_{hr} = 10\ \mu s$

Substituting these values in Eq. (A1.6), we get the video bandwidth:

$$B = 4.21\ \text{MHz}$$

This result is very close to the actual maximum frequency in the standard video signal, which is 4.2 MHz.

Color Television

The transmission of *color* in commercial TV broadcasting is based on the premise that all colors found in nature can be approximated by mixing three additive primary colors: *red, green,* and *blue*. We refer to them as "primary" colors, because none of them is a mixture of the other two. These three primary colors are represented by the video signals $m_R(t)$, $m_G(t)$, and $m_B(t)$, respectively. To conserve bandwidth and also produce a picture that can be viewed on a conventional black-and-white (monochrome) television receiver, the transmission of these three primary colors is accomplished by observing that they can be uniquely represented by any three signals that are independent linear combinations of $m_R(t)$, $m_G(t)$, and $m_B(t)$. In the standard color-television system, the three signals that are transmitted have the form

$$m_L(t) = 0.30m_R(t) + 0.59m_G(t) + 0.11m_B(t)$$
$$m_I(t) = 0.60m_R(t) - 0.28m_G(t) - 0.32m_B(t) \tag{A1.7}$$
$$m_Q(t) = 0.21m_R(t) - 0.52m_G(t) + 0.31m_B(t)$$

The signal $m_L(t)$ is called the *luminance signal*; when received on a conventional monochrome television receiver, it produces a black-and-white version of the color picture. The signals $m_I(t)$ and $m_Q(t)$ are called the *chrominance signals*; they indicate the way the color of the picture departs from shades of gray. With $m_L(t)$, $m_I(t)$, and $m_Q(t)$ defined as before, we have by simultaneous solution:

$$m_R(t) = m_L(t) - 0.96m_I(t) + 0.62m_Q(t)$$
$$m_G(t) = m_L(t) - 0.28m_I(t) - 0.64m_Q(t) \tag{A1.8}$$
$$m_B(t) = m_L(t) - 1.10m_I(t) + 1.70m_Q(t)$$

The luminance signal $m_L(t)$ is assigned the entire 4.2 MHz bandwidth. Owing to certain properties of human vision, tests show that if the nominal bandwidths of the chrominance signals $m_I(t)$ and $m_Q(t)$ are 1.6 MHz and 0.6 MHz, respectively, then satisfactory color reproduction is possible.

High-Definition Television

High-definition television (HDTV)[3] is in reality high-fidelity color television. In HDTV, the image quality is an improvement of the NTSC system by a quantum leap. In particular, the HDTV system has approximately twice as much luminance definition both vertically and horizontally as the conventional NTSC system, which makes the total number of pixels in the image four times as great. Moreover, the use of a larger screen in HDTV increases the number of pixels by another quarter. The bandwidth of the HDTV source is about 22 MHz. HDTV provides the viewer with a feeling of realism and involvement that is unattainable otherwise. This result is achieved by using a variety of advanced signal processing techniques.

However, for HDTV to be widely acceptable, two requirements are critical. First, there should be *receiver compatibility*, which means that the signal must be able to feed an HDTV and NTSC TV simultaneously and be received on the NTSC receiver with substantially the same picture quality as that achievable by conventional means. Meanwhile, the HDTV realizes the full benefits afforded to it, including increased resolution. Second, a bandwidth, of no more than twice the 6 MHz per channel for NTSC TV broadcast should be required. Thus, compression and compaction of the much wider bandwidth HDTV signal must be done.

Improved vertical resolution is catered to by using 1050 lines, that is, twice as many scan lines as in NTSC. Both interlaced and progressive scanning have been proposed. The Kell factor approaches unity (it is usually taken as 0.9) for progressive scanning.

NOTES AND REFERENCES

1. For detailed treatment of the speech production process and the characterization of speech signals see O'Shaughnessy (1987), Chapter 3 of Rabiner and Schafer (1978), and Flanagan (1972).

2. The classic reference for the spectrogram, displaying the nonstationary character of speech signals, is the book by Potter, Kopp, and Green (1947).

3. From a historical perspective, research into high-definition wide-screen television started in Japan in 1968; the outstanding contributor there is Takashi Fugio. The material presented on HDTV in Section A1.2 is based on Benson and Fink (1991). For additional references on the subject, see the papers by Schreiber (1987) and Rzeszewski (1990). The January 1993 Special Issue of the IEEE Journal *Selected Areas in Communications* is devoted to HDTV and related matters.

Fourier Series

In this appendix we review the formulation of the Fourier series and develop the Fourier transform as a generalization of the Fourier series.

Let $g_{T_0}(t)$ denote a *periodic signal* with period T_0. By using a *Fourier series expansion* of this signal, we are able to resolve it into an infinite sum of sine and cosine terms. The expansion may be expressed in the trigonometric form:

$$g_{T_0}(t) = a_0 + 2 \sum_{n=1}^{\infty} [a_n \cos(2\pi n f_0 t) + b_n \sin(2\pi n f_0 t)] \qquad \text{(A2.1)}$$

where f_0 is the *fundamental frequency*:

$$f_0 = \frac{1}{T_0} \qquad \text{(A2.2)}$$

The coefficients a_n and b_n represent the amplitudes of the cosine and sine terms, respectively. The quantity $n f_0$ represents the nth harmonic of the fundamental frequency f_0. Each of the terms $\cos(2\pi n f_0 t)$ and $\sin(2\pi n f_0 t)$ is called a *basis function*. These basis functions form an *orthogonal* set over the interval T_0 in that they satisfy the following set of relations:

$$\int_{-T_0/2}^{T_0/2} \cos(2\pi m f_0 t)\cos(2\pi n f_0 t)\ dt = \begin{cases} T_0/2, & m = n \\ 0, & m \neq n \end{cases} \qquad \text{(A2.3)}$$

$$\int_{-T_0/2}^{T_0/2} \cos(2\pi m f_0 t)\sin(2\pi n f_0 t)\ dt = 0, \qquad \text{for all } m \text{ and } n \quad \text{(A2.4)}$$

$$\int_{-T_0/2}^{T_0/2} \sin(2\pi m f_0 t)\sin(2\pi n f_0 t)\ dt = \begin{cases} T_0/2, & m = n \\ 0, & m \neq n \end{cases} \qquad \text{(A2.5)}$$

To determine the coefficient a_0, we integrate both sides of Eq. (A2.1) over a complete period. We thus find that a_0 is the *mean value* of the periodic signal $g_{T_0}(t)$ over one period, as shown by the *time average*

$$a_0 = \frac{1}{T_0} \int_{-T_0/2}^{T_0/2} g_{T_0}(t)\ dt \qquad \text{(A2.6)}$$

To determine the coefficient a_n, we multiply both sides of Eq. (A2.1) by $\cos(2\pi n f_0 t)$ and integrate over the interval $-T_0/2$ to $T_0/2$. Then, using Eqs. (A2.3) and (A2.4) we find that

$$a_n = \frac{1}{T_0} \int_{-T_0/2}^{T_0/2} g_{T_0}(t)\cos(2\pi n f_0 t)\ dt, \qquad n = 1, 2, \ldots \quad \text{(A2.7)}$$

Similarly, we find that

$$b_n = \frac{1}{T_0} \int_{-T_0/2}^{T_0/2} g_{T_0}(t)\sin(2\pi n f_0 t)\ dt, \qquad n = 1, 2, \ldots \quad \text{(A2.8)}$$

A basic question that arises at this point is the following: Given a periodic signal $g_{T_0}(t)$ of period T_0, how do we know that the Fourier series expansion of Eq. (A2.1) is *convergent* in that the infinite sum of terms in this expansion is exactly equal to $g_{T_0}(t)$? To resolve this issue, we have to show that for the coefficients a_0, a_n, and b_n calculated in accordance with Eqs. (A2.6) to (A2.8), this series will indeed converge to $g_{T_0}(t)$. In general, for a periodic signal $g_{T_0}(t)$ of arbitrary waveform, there is no guarantee that the series of Eq. (A2.1) will converge to $g_{T_0}(t)$ or that the coefficients a_0, a_n, and b_n will even exist. A rigorous proof of convergence of the Fourier series is beyond the scope of this book. Here we simply state that a periodic signal $g_{T_0}(t)$ can be expanded in a Fourier series if the signal $g_{T_0}(t)$ satisfies the *Dirichlet conditions*:

1. The function $g_{T_0}(t)$ is single-valued within the interval T_0.
2. The function $g_{T_0}(t)$ has at most a finite number of discontinuities in the interval T_0.
3. The function $g_{T_0}(t)$ has a finite number of maxima and minima in the interval T_0.
4. The function $g_{T_0}(t)$ is absolutely integrable, that is,

$$\int_{-T_0/2}^{T_0/2} |g_{T_0}(t)|\ dt < \infty$$

The Dirichlet conditions are satisfied by the periodic signals usually encountered in communication systems. At a point of one-dimensional discontinuity, the Fourier series converges to the average value just to the left of the point and the value just to the right of the point.

Complex Exponential Fourier Series

The Fourier series of Eq. (A2.1) can be put into a much simpler and more elegant form with the use of complex exponentials. We do this by substituting in Eq. (A2.1) the exponential forms for the cosine and sine, namely:

$$\cos(2\pi n f_0 t) = \frac{1}{2}[\exp(j2\pi n f_0 t) + \exp(-j2\pi n f_0 t)]$$

$$\sin(2\pi n f_0 t) = \frac{1}{2j}[\exp(j2\pi n f_0 t) - \exp(-j2\pi n f_0 t)]$$

We thus obtain

$$g_{T_0}(t) = a_0 + \sum_{n=1}^{\infty} [(a_n - jb_n)\exp(j2\pi n f_0 t)$$
$$+ (a_n + jb_n)\exp(-j2\pi n f_0 t)] \quad (A2.9)$$

Let c_n denote a complex coefficient related to a_n and b_n by

$$c_n = \begin{cases} a_n - jb_n, & n > 0 \\ a_0, & n = 0 \\ a_n + jb_n, & n < 0 \end{cases} \quad (A2.10)$$

Then, we may simplify Eq. (A2.9) as follows:

$$g_{T_0}(t) = \sum_{n=-\infty}^{\infty} c_n \exp(j2\pi n f_0 t) \quad (A2.11)$$

where

$$c_n = \frac{1}{T_0} \int_{-T_0/2}^{T_0/2} g_{T_0}(t)\exp(-j2\pi n f_0 t)\, dt \quad n = 0, \pm 1, \pm 2, \dots \quad (A2.12)$$

The series expansion of Eq. (A2.11) is referred to as the *complex exponential Fourier series*. The c_n are called the *complex Fourier coefficients*. Equation (A2.12) states that, given a periodic signal $g_{T_0}(t)$, we may determine the complete set of complex Fourier coefficients. On the other hand, Eq. (A2.11) states that, given this set of coefficients, we may reconstruct the original periodic signal $g_{T_0}(t)$ exactly. From the mathematics of real and complex analysis, Eq. (A2.12) is an *inner product* of the signal with the *basis functions* $\exp(j2\pi n f_0 t)$, by whose linear combination all square integrable functions can be expressed using Eq. (A2.11).

According to this representation, a periodic signal contains all frequencies (both positive and negative) that are harmonically related to the fundamental.

The presence of negative frequencies is simply a result of the fact that the mathematical model of the signal as described by Eq. (A2.11) requires the use of negative frequencies. Indeed, this representation also requires the use of complex-valued basis functions, namely, $\exp(j2\pi n f_0 t)$, which have no physical meaning either. The reason for using complex-valued basis functions and negative frequency components is merely to provide a compact mathematical description of a periodic signal, which is well-suited for both theoretical and practical work.

Discrete Spectrum

The representation of a periodic signal by a Fourier series is equivalent to resolution of the signal into its various harmonic components. Thus, using the complex exponential Fourier series, we find that a periodic signal $g_{T_0}(t)$ with period T_0 has components at frequencies 0, $\pm f_0$, $\pm 2f_0$, $\pm 3f_0$, ..., and so forth, where $f_0 = 1/T_0$ is the fundamental frequency. That is, while the signal $g_{T_0}(t)$ exists in the time domain, we may say that its frequency-domain description consists of components at frequencies 0, $\pm f_0$, $\pm 2f_0$, ..., called the *spectrum*. If we specify the periodic signal $g_{T_0}(t)$, we can determine its spectrum; conversely, if we specify the spectrum, we can determine the corresponding signal. This means that a periodic signal $g_{T_0}(t)$ can be specified in two equivalent ways:

1. A *time-domain representation*, where $g_{T_0}(t)$ is defined as a function of time.
2. A *frequency-domain representation*, where the signal is defined in terms of its spectrum.

Although these two descriptions are separate aspects of a given phenomenon, they are not independent of each other, but are related, as Fourier theory shows.

In general, the Fourier coefficient c_n is a complex number, and so we may express it in the form:

$$c_n = |c_n|\exp[j\arg(c_n)] \tag{A2.13}$$

The $|c_n|$ defines the amplitude of the nth harmonic component of the periodic signal $g_{T_0}(t)$, so that a plot of $|c_n|$ versus frequency yields the *discrete amplitude spectrum* of the signal. A plot of $\arg(c_n)$ versus frequency yields the *discrete phase spectrum* of the signal. We refer to the spectrum as a *discrete spectrum* because both the amplitude and phase of c_n have nonzero values only for discrete frequencies that are integer (both positive and negative) multiples of the fundamental frequency.

For a real-valued periodic function $g_{T_0}(t)$, we find, from the definition of the Fourier coefficient c_n given by Eq. (A2.12), that

$$c_{-n} = c_n^* \tag{A2.14}$$

where c_n^* is the complex conjugate of c_n. We therefore have

$$|c_{-n}| = |c_n| \tag{A2.15}$$

and

$$\arg(c_{-n}) = -\arg(c_n) \tag{A2.16}$$

That is, the amplitude spectrum of a real-valued periodic signal is symmetric (an even function of n), and the phase spectrum is antisymmetric (an odd function of n) about the vertical axis passing through the origin.

EXAMPLE   Periodic Pulse Train

Consider a periodic train of rectangular pulses of duration T and period T_0, as shown in Fig. A2.1. For convenience of analysis, the origin has been chosen to coincide with the center of the pulse. This signal may be described analytically over one period, as follows:

$$g_{T_0}(t) = \begin{cases} A, & -\dfrac{T}{2} \le t \le \dfrac{T}{2} \\ 0, & \text{for the remainder of the period} \end{cases} \tag{A2.17}$$

Using Eq. (A2.12) to evaluate the complex Fourier coefficient c_n, we get

$$\begin{aligned} c_n &= \frac{1}{T_0} \int_{-T/2}^{T/2} A \exp(-j2\pi n f_0 t)\, dt \\ &= \frac{A}{n\pi} \sin\left(\frac{n\pi T}{T_0}\right), \qquad n = 0, \pm 1, \pm 2, \ldots \end{aligned} \tag{A2.18}$$

where T/T_0 is termed the *duty cycle*.

We may simplify notation by using the *sinc function*:

$$\text{sinc}(\lambda) = \frac{\sin(\pi\lambda)}{\pi\lambda} \tag{A2.19}$$

Thus, we may rewrite Eq. (A2.18) as follows:

$$c_n = \frac{TA}{T_0}\text{sinc}\left(\frac{nT}{T_0}\right) = \frac{TA}{T_0}\text{sinc}(f_n T) \tag{A2.20}$$

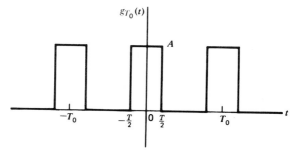

Figure A2.1   Periodic train of rectangular pulses of amplitude A, duration T, and period T_0.

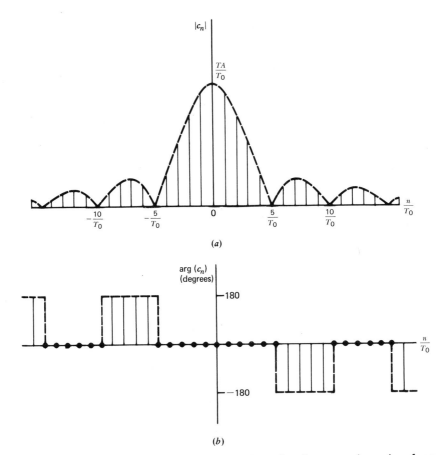

Figure A2.2 Discrete spectrum of a periodic train of rectangular pulses for a duty cycle $T/T_0 = 0.2$. (a) Amplitude spectrum. (b) Phase spectrum.

In Fig. A2.2, we have plotted the amplitude spectrum $|c_n|$ and phase spectrum $\arg(c_n)$ versus the discrete frequency $f_n = n/T_0$ for a duty cycle T/T_0 equal to 0.2. Based on this figure, we may note the following:

1. The line spacing in the amplitude spectrum in Fig. A2.2a is determined by the period T_0.

2. The envelope of the amplitude spectrum is determined by the pulse amplitude A, pulse duration T, and duty cycle T/T_0.

3. Zero-crossings occur in the envelope of the amplitude spectrum at frequencies that are integer multiples of $1/T$.

4. The phase spectrum takes on the values 0 degrees and ± 180 degrees, depending on the polarity of $\text{sinc}(nT/T_0)$; in Fig. A2.2b we have used both 180 degrees and -180 degrees to preserve antisymmetry.

A2.1 FOURIER TRANSFORM

In the previous section, we used the Fourier series to represent a periodic signal. We now wish to develop a similar representation for a signal $g(t)$ that is nonperiodic in terms of complex exponential signals. In order to do this, we first construct a periodic function $g_{T_0}(t)$ of period T_0 in such a way that $g(t)$ defines one cycle of this periodic function, as illustrated in Fig. A2.3. In the limit, we let the period T_0 become infinitely large, so that we may write

$$g(t) = \lim_{T_0 \to \infty} g_{T_0}(t) \tag{A2.21}$$

Representing the periodic function $g_{T_0}(t)$ in terms of the complex exponential form of the Fourier series, we have

$$g_{T_0}(t) = \sum_{n=-\infty}^{\infty} c_n \exp\left(\frac{j2\pi nt}{T_0}\right) \tag{A2.22}$$

where

$$c_n = \frac{1}{T_0} \int_{-T_0/2}^{T_0/2} g_{T_0}(t) \exp\left(-\frac{j2\pi nt}{T_0}\right) dt \tag{A2.23}$$

We have purposely written the exponents as shown in Eqs. (A2.22) and (A2.23) because we wish to let T_0 approach infinity in accordance with Eq. (A2.21). Define

$$\Delta f = \frac{1}{T_0} \tag{A2.24}$$

$$f_n = \frac{n}{T_0} \tag{A2.25}$$

and

$$G(f_n) = c_n T_0 \tag{A2.26}$$

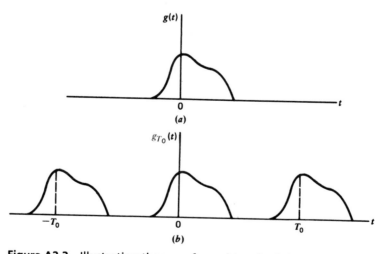

Figure A2.3 Illustrating the use of an arbitrarily defined function of time to construct a periodic waveform. (a) Arbitrarily defined function of time $g(t)$. (b) Periodic waveform $g_{T_0}(t)$ based on $g(t)$.

Thus, making this change of notation in the Fourier series representation of $g_{T_0}(t)$ given in Eqs. (A2.22) and (A2.23), we get the following relations for the interval $-T_0/2 \le t \le T_0/2$:

$$g_{T_0}(t) = \sum_{n=-\infty}^{\infty} G(f_n)\exp(j2\pi f_n t)\ \Delta f \qquad\qquad (A2.27)$$

where

$$G(f_n) = \int_{-T_0/2}^{T_0/2} g_{T_0}(t)\exp(-j2\pi f_n t)\ dt \qquad\qquad (A2.28)$$

Now let the period T_0 approach infinity or, equivalently, its reciprocal Δf approach zero. Then we find that, in the limit, the discrete frequency f_n approaches the continuous frequency variable f, and the discrete sum in Eq. (A2.27) becomes an integral defining the area under a continuous function of frequency f, namely, $G(f)\exp(j2\pi f t)$. Also, as T_0 approaches infinity, the function $g_{T_0}(t)$ approaches $g(t)$. Therefore, in the limit, Eqs. (A2.27) and (A2.28) become, respectively,

$$g(t) = \int_{-\infty}^{\infty} G(f)\exp(j2\pi f t)\ df \qquad\qquad (A2.29)$$

where

$$G(f) = \int_{-\infty}^{\infty} g(t)\exp(-j2\pi f t)\ dt \qquad\qquad (A2.30)$$

We have thus achieved our aim of representing an arbitrarily defined signal $g(t)$ in terms of exponential functions over the entire interval $(-\infty < t < \infty)$. Given the function $g(t)$, Eq. (A2.30) defines the Fourier transform $G(f)$. Conversely, Eq. (A2.29) defines the inverse Fourier transform of $G(f)$.

Time–Frequency Representations of Signals

A3.1   INTRODUCTION

In Chapter 2, we used two alternative methods of signal representation. One is the description of the signal as a function of time; the other is Fourier analysis. These two methods work perfectly well if the signal of interest is known to be stationary. In this sense, a *stationary signal* may be described as one whose spectral characteristics do *not* change with time. If, however, the signal is *nonstationary*, as is frequently the case in practice, then it is advantageous to have a description of the signal that involves *both* time and frequency. We need only to look at ourselves to realize that such a description is naturally built into our daily experiences—especially the auditory and visual perceptions. It is therefore important that we have some understanding of the extremely useful subject of time–frequency analysis.

As the name implies, *time–frequency analysis*[1] involves mapping a signal (i.e., a one-dimensional function of time) into an image (i.e., a two-dimensional function of time and frequency) that displays the temporal localization of the signal's spectral components. In conceptual terms, we may think of the mapping as a *time-varying spectral representation* of the signal. This representation is analogous to a musical score, with time and frequency representing the two principal axes. Such a method of signal representation is extremely useful, because the two-dimensional image produced by the analysis is potentially easier to "visualize" and therefore to understand than the original time function or its Fourier transform. In particular, the values of the time–frequency representation of the signal provide an indication of the specific times at which certain spectral components of the signal are observed.

Basically, there are two classes of time–frequency representations of signals: *linear* and *quadratic* (nonlinear). In this appendix, we concern ourselves with linear representations only. Specifically, we present brief expositions of the short-time Fourier transform and the wavelet transform, in that order. These two methods are linear because they both satisfy the principle of superposition. Linearity is a highly desirable property in certain applications that involve the processing of signals with a multitude of components as, for example, in speech recognition or video compression.

A3.2   SHORT-TIME FOURIER TRANSFORM[2]

The short-time Fourier transform (STFT) is a natural extension of the Fourier transform. To be specific, let $g(t)$ denote a signal that is assumed to be "stationary" when viewed through a *temporal window* $w(t)$ that is of limited extent; in general, $w(t)$ is comlex valued. The *short-time Fourier transform* of the signal $g(t)$ is then defined as the Fourier transform of the *windowed signal* $g(t) w^*(t - \tau)$, where τ is the center position of the window and the asterisk denotes complex conjugation. That is, we have

$$\text{STFT}(\tau,f) = \int_{-\infty}^{\infty} g(t) w^*(t - \tau) \exp(-j2\pi ft) \, dt \qquad (A3.1)$$

which is linear in the signal $g(t)$. Accordingly, the short-time Fourier transform maps the one-dimensional function $g(t)$ into the two-dimensional function $\text{STFT}(\tau,f)$. The parameter f plays a role similar to that of frequency in the ordinary Fourier transform. Moreover, for a given $g(t)$, the result obtained by computing $\text{STFT}(\tau,f)$ is dependent on the choice of the window $w(t)$. Many different window shapes are used in practice. Typically, they are symmetric, unimodal, and smooth; examples include a Gaussian window or a raised cosine (single period).

In mathematical terms, the integral of Eq. (A3.1) can be considered as the *inner (scalar) product* of the signal $g(t)$ with *a two-parameter family of basis functions*, $w(t - \tau) \exp(j2\pi ft)$, for varying τ and f. It is important to note that, in general, these basis functions do *not* constitute an orthonormal set.

Many of the properties of the Fourier transform are carried over to the short-time Fourier transform. In particular, we may mention the following two signal-preserving properties:

1.   *The short-time Fourier transform preserves time shifts, except for a linear modulation.* That is, if $\text{STFT}(\tau,f)$ is the short-time Fourier transform of the signal $g(t)$, then the short-time Fourier transform of the time-shifted signal $g(t - t_0)$ is given by $\exp(-j2\pi ft_0) \, \text{STFT}(\tau - t_0, f)$.

2.   *The short-time Fourier transform preserves frequency shifts.* That is, if $\text{STFT}(\tau,f)$ is the short-time Fourier transform of the signal $g(t)$, then the short-time

Fourier transform of the modulated signal $g(t)\exp(-j2\pi f_0 t)$ is given by STFT $(\tau, f - f_0)$.

Spectrogram

The squared modulus of the short-term Fourier transform of a signal $g(t)$ is called a *spectrogram*, as shown by

$$SPEC(\tau, f) = |STFT(\tau, f)|^2$$

$$= \left| \int_{-\infty}^{\infty} g(t)w^*(t - \tau)\exp(-j2\pi ft) \, dt \right|^2 \qquad (A3.2)$$

The spectrogram represents a simple and yet powerful extension of the classical Fourier theory. In physical terms, it provides a measure of the signal energy in the time–frequency plane. Indeed, the spectrogram is used extensively in the analysis of speech signals and other nonstationary signals for the physical insight it provides.

Implementations of the Short-Time Fourier Transform

Figure A3.1a shows a computational implementation of the short-time Fourier transform that follows directly from the defining equation (A3.1). The box labeled $F[\ ]$ represents a Fourier transformer that operates on its input with respect to the variable t. By shifting the window $w(t)$ along the time axis (i.e., assigning different values to the parameter τ), localizations of the signal's spectral components can be computed at different times. Figure A3.1b shows the corresponding *low-pass filter implementation* of the short-time Fourier transform; according to this latter representation, Eq. (A3.1) is expressed as the convolution

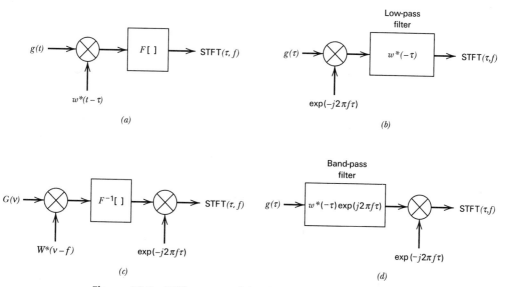

Figure A3.1 Different models of the short-time Fourier transform.

of the linearly modulated signal $g(\tau)\exp(-j2\pi f\tau)$ and the filter's impulse response $w^*(-\tau)$; in this convolution operation, time t plays the role of a dummy variable.

For another interpretation of the short-time Fourier transform, we may rewrite Eq. (A3.1) in the equivalent form

$$\text{STFT}(\tau,f) = \exp(-j2\pi f\tau)[g(\tau) \bigstar w^*(-\tau)\exp(j2\pi f\tau)] \qquad (A3.3)$$

where $\bigstar$ denotes convolution, and $w^*(-\tau)\exp(j2\pi f\tau)$ represents a *linearly modulated impulse response*. Recognizing that convolution in the time domain is transformed into multiplication in the frequency domain, we may go one step further and redefine the short-time Fourier transform of the signal $g(t)$ as

$$\text{STFT}(\tau,f) = \exp(-j2\pi f\tau) \int_{-\infty}^{\infty} G(\nu) W^*(\nu-f)\exp(j2\pi\nu\tau) \, d\nu \qquad (A3.4)$$

where the frequency ν is a dummy variable, $G(\nu)$ is the Fourier transform of $g(t)$, and the *spectral window* $W(\nu)$ is the Fourier transform of the temporal window $w(t)$. Equation (A3.4) shows that, except for the linear phase factor $\exp(-j2\pi f\tau)$, the short-time Fourier transform of the signal $g(t)$ can also be defined as the inverse Fourier transform of the *windowed spectrum* $G(\nu) W^*(f-\nu)$. We thus have a second computational implementation of the short-time Fourier transform as shown in Fig. A3.1c, where the box labeled $F^{-1}[\ \ ]$ is an inverse Fourier transformer that operates on its input with respect to the variable ν. By sliding the spectral window $W(\nu)$ along the frequency axis (i.e., assigning different values to the parameter f), we may compute the localizations of the signal's characteristics in the τ domain for different frequencies. Figure A3.1d shows the corresponding *band-pass filter implementation* of the short-time Fourier transform. The impulse response of the filter is the modulated window $w^*(-\tau)\exp(j2\pi f\tau)$, with the parameter f viewed as the "carrier frequency." Equivalently, we may think of f as the "midband frequency" of the band-pass filter. Typically, f is assigned a discrete set of values $\{nf_0|n = 1,2,\ldots\}$, in which case the model of A3.1d leads to a *filter bank* with each member acting as an "analysis filter" that computes the short-time Fourier transform for a specific value of the frequency f.

Time–Frequency Resolution of the Short-Time Fourier Transform

The "time-domain" computational interpretation of the short-time Fourier transform depicted in Fig. A3.1a may be viewed as the *dual* of the "frequency-domain" computational interpretation shown in Fig. A.31c. The nature of this duality is illustrated graphically in Fig. A3.2. According to the picture portrayed here, the short-time Fourier transform partitions the *time–frequency plane* into a succession of semiinfinite vertical strips representing the Fourier transform of windowed segments of the signal or, equivalently, a succession of semiinfinite horizontal strips representing the band-pass filter bank analysis of the signal.

The time–frequency duality of the short-time Fourier transform, which is inherited from the Fourier transform, points to a basic shortcoming of the transformation. To be specific, consider a pair of purely sinusoidal signals whose frequencies are spaced Δf Hertz apart. We may then ask: What is the minimum

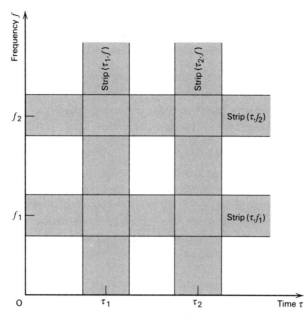

Figure A3.2 Time–frequency plane of the short-time Fourier transform; times τ_1 and τ_2 specify two different positions of the temporal window $w(t)$, whereas f_1 and f_2 are two different midband frequencies of the spectral window $W(\nu)$. For a Gaussian function, $w(t)$ and $W(\nu)$ have the same general envelope.

value of Δf that the short-time Fourier transform can resolve? To answer this question, we may use the root-mean-square bandwidth of Eq. (2.58) to formulate Δf as

$$\Delta f = \left[\frac{\displaystyle\int_{-\infty}^{\infty} \nu^2 |W(\nu)|^2 \, d\nu}{\displaystyle\int_{-\infty}^{\infty} |W(\nu)|^2 \, d\nu} \right]^{1/2} \tag{A3.5}$$

where $W(\nu)$ is the Fourier transform of the window $w(t)$. Consider next the dual problem involving a pair of very short pulses spaced $\Delta\tau$ seconds apart. Here also we may ask: What is the minimum value of the $\Delta\tau$ that the short-time Fourier transform can resolve? To answer this second question, we may use the root-mean-square duration of Eq. (2.59) to express $\Delta\tau$ as

$$\Delta\tau = \left[\frac{\displaystyle\int_{-\infty}^{\infty} t^2 |w(t)|^2 \, dt}{\displaystyle\int_{-\infty}^{\infty} |w(t)|^2 \, dt} \right]^{1/2} \tag{A3.6}$$

From the discussion presented in Section 2.5, we now recognize that the time resolution $\Delta\tau$ and the frequency resolution Δf cannot be chosen arbitrarily.

Rather, the time–bandwidth product imposes the following lower bound (see Problem 2.15):

$$\Delta \tau \, \Delta f \geq \frac{1}{4\pi} \qquad\qquad (A3.7)$$

This relationship is referred to as the *uncertainty principle* from analogy with statistical quantum mechanics. Clearly, insofar as the time–frequency resolution of the short-time Fourier transform is concerned, the best that we can do is to satisfy Eq. (A3.7) with the equality sign. This requirement can indeed be realized by choosing a *Gaussian window*, see Problem 2.15. Note, however, that once the Gaussian window has been specified, the time–frequency resolution capability of the short-time Fourier transform is *fixed* over the entire time–frequency plane. This point is well illustrated in Fig. A3.3a, where the time–frequency plane is partitioned into *tiles* of the same shape and size. The implication of this uniform tiling is that the filter bank based on the band-pass model of Fig. A3.1d is made up of *constant-bandwidth* filters.

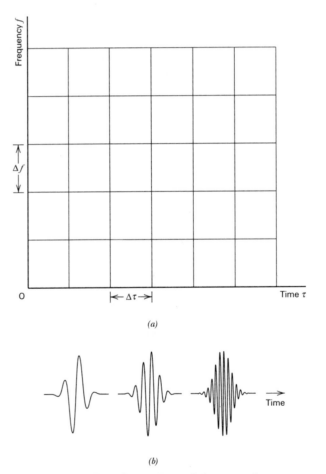

(a)

(b)

Figure A3.3   (a) Uniform tiling of the time–frequency plane by the short-time Fourier transform. (b) Real parts of associated basis functions.

Figure A3.3*b* displays the real parts of associated basis functions of the short-time Fourier transform. All the basis functions have exactly the same duration, since the same window is used for all frequencies.

A3.3   WAVELET TRANSFORM[3]

To overcome the time–frequency resolution limitation of the short-time Fourier transform, we need a form of mapping that has the ability to trade off time resolution for frequency resolution, and vice versa. One such method is known as the wavelet transform. Given a nonstationary signal $g(t)$, the *wavelet transform* is defined as the inner product of $g(t)$ with the two-parameter family of basis functions denoted by

$$\psi_{\tau,a}(t) = |a|^{-1/2}\psi\left(\frac{t-\tau}{a}\right) \tag{A3.8}$$

where a is a *scale factor* (also referred to as a *dilation parameter*) and τ is a *time delay*. Thus, in mathematical terms, the wavelet transform of the signal $g(t)$ is defined by

$$WT(\tau,a) = |a|^{-1/2}\int_{-\infty}^{\infty} g(t)\psi^*\left(\frac{t-\tau}{a}\right)dt \tag{A3.9}$$

In wavelet analysis the basis function $\psi_{\tau,a}(t)$ is an oscillating function; there is therefore no need to use sines and cosines (waves) as in Fourier analysis. Note also that wavelet transformation is *invertible* in that if we are given the wavelet transform $WT(\tau,a)$, the original signal $g(t)$ can be *reconstructed* without loss of information.

The basis functions $\psi_{\tau,a}(t)$, are called *wavelets*; they constitute the building blocks of a wavelet analysis. According to the defining equation (A3.8), the wavelets are scaled and translated versions of a prototype $\psi(t)$, called the *basic wavelet* or *mother wavelet*. The delay parameter τ gives the position of the wavelet $\psi_{\tau,a}(t)$, while the scale factor a governs its frequency content. For $|a| \ll 1$, the wavelet $\psi_{\tau,a}(t)$ is a very highly concentrated and shrunken version of the basic wavelet $\psi(t)$, with frequency content mostly in the high-frequency range. On the other hand, for $|a| \gg 1$ the wavelet $\psi_{\tau,a}(t)$ is very much spread out and has mostly low frequencies.

In the windowed Fourier analysis, the goal is to measure the *local frequency* content of a signal. On the other hand, in wavelet analysis, we measure the *similarity* between the signal and the wavelet $\psi_{\tau,a}(t)$ for varying τ and a. The dilations by $1/a$ result in several *magnifications* of the signal, with distinct *resolutions*.

Just as the short-time Fourier transform has signal-preserving properties of its own, so does the wavelet transform:

1.   *The wavelet transform preserves time shifts.* That is, if $WT(\tau,a)$ is the wavelet transform of the signal $g(t)$, then $WT(\tau - t_0,a)$ is the wavelet transform of the time-shifted signal $g(t - t_0)$.

2.   *The wavelet transform preserves time scaling.* That is, if $WT(\tau,a)$ is the wavelet

transform of the signal $g(t)$, then the wavelet transform of the time-scaled signal $|a_0|^{1/2} g(a_0 t)$ is $\mathrm{WT}(a_0\tau, a/a_0)$.

However, unlike the short-time Fourier transform, the wavelet transform does *not* preserve frequency shifts.

The wavelet transform performs a time-scale analysis. Thus, its squared modulus is called a *scalogram*, as shown by

$$\mathrm{SCAL}(\tau,a) = |\mathrm{WT}(\tau,a)|^2 \tag{A3.10}$$

The scalogram represents a distribution of the energy of the signal in the time-scale plane. In both the scalogram and spectrogram, phase information is lost; neither one of them can therefore be inverted in general. Also, both are bilinear functions of the signal under analysis, with the result that "cross-terms" appear as interference patterns, which are undesirable.

Time–Frequency Resolution of the Wavelet Transform

In general, the basic wavelet $\psi(t)$ can be any *band-pass* function. To establish a connection with the modulated window in the short-time Fourier transform, we choose

$$\psi(t) = w(t)\exp(-j2\pi f_0 t) \tag{A3.11}$$

The window $w(t)$ is typically a low-pass function. Thus, Eq. (A3.11) describes the basic wavelet $\psi(t)$ as a complex, linearly modulated signal whose frequency content is concentrated essentially around the "carrier" frequency f_0. Equivalently, we may think of the basic wavelet $\psi(t)$ as the impulse response of a band-pass filter with "midband" frequency f_0.

The scale factor a of the wavelet $\psi_{\tau,a}(t)$ is *inversely* related to its carrier (midband) frequency f, as shown by

$$a = \frac{f_0}{f} \tag{A3.12}$$

Since, by definition, a wavelet is a scaled version of the same prototype, it follows that we may also write

$$\frac{\Delta f}{f} = Q \tag{A3.13}$$

where Δf is the frequency resolution of the wavelet $\psi_{\tau,a}(t)$ and Q is a constant. Choosing the window $w(t)$ to be a Gaussian function as discussed in Section A3.1, we may use Eq. (A3.7) with the equality sign to express the time resolution of the wavelet $\psi_{\tau,a}(t)$ as

$$\Delta\tau = \frac{1}{4\pi\,\Delta f} \tag{A3.14}$$
$$= \frac{1}{4\pi Q f}$$

In light of Eqs. (A3.13) and (A3.14), we may now formally state the time–frequency resolution properties of the wavelet transform as follows:

- The time resolution $\Delta\tau$ varies *inversely* with the midband frequency f of the analyzing wavelet; hence it can be made arbitrarily small at high frequencies.
- The frequency resolution Δf varies *linearly* with the midband frequency f of the analyzing wavelet; hence, it can be made arbitrarily small at low frequencies.

Thus, the wavelet transform is particularly well suited for the analysis of signals with high-frequency transients superimposed on longer-lived low-frequency components.

In Section A3.2, we stated that the short-time Fourier transform has a *fixed* resolution, as shown in Fig. A3.3*a*. In contrast, the wavelet transform has a *multiresolution* capability, as shown in Fig. A3.4*a*. Here we see that the wavelet transform partitions the time–frequency plane into tiles of the same area, but with varying widths and heights that depend on f. Thus, unlike the short-time Fourier transform, the wavelet transform provides a trade-off between time and fre-

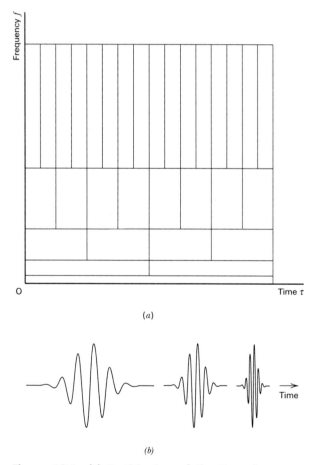

(*a*)

(*b*)

Figure A3.4   (*a*) Partitioning of the time–frequency plane by the wavelet transform. (*b*) Real parts of associated basis functions.

quency resolutions, which are represented by the widths and heights of the tiles, respectively (i.e., narrower widths/heights correspond to better resolution).

Figure A3.4*b* displays the real parts of basis functions of the wavelet transform. We see that every time the basis function is compressed by a factor of two, its midband (carrier) frequency is increased by the same factor.

Constant-Q Filter Bank Interpretation of the Wavelet Transform

Viewing Δf as the bandwidth of an analyzing wavelet, we may in turn use Eq. (A3.13) to view the wavelet transform as a bank of band-pass (analysis) filters with a *constant-Q factor*. Since the Q-factor is defined as the ratio of a filter's bandwidth to its midband frequency, it follows that each time the midband frequency of a particular analysis filter is increased, the bandwidth of the filter is increased by a corresponding amount. Thus, whereas the frequency responses of the analysis filters in the short-time Fourier transform are regularly spaced along the frequency axis as in Fig. A3.5*a*, in the case of the wavelet transform they are regularly spaced on a *logarithmic* frequency scale as shown in Fig. 3.5*b*. Such a nonuniform spacing of filters closely matches the human ear's decreasing frequency resolution with increasing frequency.

Orthonormal Wavelets

There is another form of wavelet transforms that uses *orthonormal basis of wavelets*[4] with good localization in both time and frequency; this latter class of wavelet transforms has *no* analog in the windowed Fourier case. It turns out that the orthonormal basis of wavelets is related to special filters for *subband coding*. These filters lead to exact waveform reconstruction without aliasing and without am-

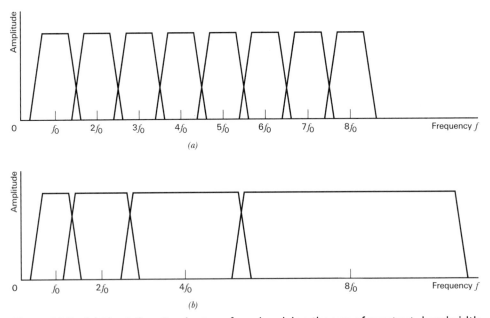

Figure A3.5 (*a*) Short-time Fourier transform involving the use of constant–bandwidth filters. (*b*) Wavelet transform involving the use of constant-Q factor filters.

plitude and phase distortion; the idea of subband coding was discussed briefly in Section 6.13.

A3.4 CONCLUDING REMARKS

The short-time Fourier transform and the wavelet transform constitute two important tools for the time–frequency analysis[5] of nonstationary signals (i.e., signals whose spectral characteristics vary with time). In their own ways, they map a nonstationary signal into a two-dimensional image that helps us think in physical terms. The short-time Fourier transform and wavelet transform have analogous formulations, and they both lend themselves to "fast" implementation in discrete form. The main difference between them is that the *time–frequency tiles* produced by the short-time Fourier transform are all of the same shape and size, whereas those produced by the wavelet transform are of different shapes but constant area. Thus, the wavelet transform handles frequency logarithmically rather than linearly, resulting in a time–frequency analysis with constant $\Delta f / f$ (i.e., constant Q).

An elegant generalization of wavelets is the idea of *wave packets*[6] (also referred to as *wavelet packets*), which are made up of orthonormal basis functions. The main feature of wave packet tiling is a frequency resolution that may be "adapted" to the signal at hand, as illustrated in Fig. A3.6. In the example shown here, we see that best frequency resolution is achieved at midband frequencies, and the frequency resolution at high frequencies is better than that at low frequencies; the associated time resolutions are proportional in a corresponding way.

In conclusion, wavelets and wave packets are powerful signal-processing tools that are of recent origin. Perhaps their biggest potential for application is in speech and image compression.

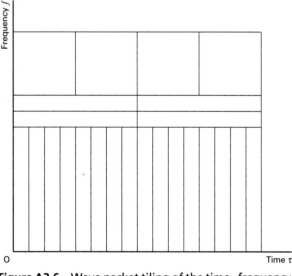

Figure A3.6 Wave packet tiling of the time–frequency plane.

Notes and References

1. Gabor (1946) was the first to emphasize the limitation of the Fourier transform in dealing with nonstationary signals; in his 1946 classic paper, Gabor proposed a transform that bears his name for the time–frequency analysis of nonstationary signals.

2. The short-time Fourier transform follows naturally from the Gabor transform.

3. The foundations of the wavelet transform were laid down by a team of French researchers led by J. Morlet (an engineer in oil prospecting), A. Grossmann (a physicist in quantum mechanics), and Y. Meyer (a mathematician) in the mid 1980s; they named their joint work "ondelettes" (wavelets).

4. Daubechies (1992) describes the construction of a special class of filters for the generation of orthonormal basis of wavelets; these filters are known as *Daubechies filters*.

5. For a review paper on the short-time Fourier transform, see Hlawatsch and Boudreaux-Bartels (1992); this paper also includes a review of quadratic methods of time–frequency analysis. A detailed treatment of short-time Fourier transform theory and its application to speech signals is presented in the book by Riley (1988).

    For a review paper on wavelet analysis, see Rioul and Vetterli (1991). For a commentary of the relevance of wavelet analysis in science and technology, see Meyer (1993). For excellent insights into computational aspects of wavelet transforms, see Strang (1992, 1993).

6. For a detailed treatment of wave packets and their application to high-resolution still picture compression, see Wickerhauser and Coifman (1992).

Bessel Functions

A Bessel function of the first kind of order n and argument x, commonly denoted by $J_n(x)$, is defined by

$$J_n(x) = \frac{1}{2\pi} \int_{-\pi}^{\pi} \exp(jx \sin\theta - jn\theta)\, d\theta \qquad \text{(A4.1)}$$

or, equivalently,

$$J_n(x) = \frac{1}{\pi} \int_0^{\pi} \cos(x \sin\theta - n\theta)\, d\theta \qquad \text{(A4.2)}$$

Just as the trigonometric functions can be expanded in power series, so can the Bessel function $J_n(x)$ be expanded in a power series:

$$J_n(x) = \sum_{m=0}^{\infty} \frac{(-1)^m (\tfrac{1}{2}x)^{n+2m}}{m!\,(n+m)!} \qquad \text{(A4.3)}$$

In particular, for $n = 0$, we have

$$J_0(x) = 1 - \frac{x^2}{2^2} + \frac{x^4}{2^2 \cdot 4^2} - \frac{x^6}{2^2 \cdot 4^2 \cdot 6^2} + \cdots \qquad \text{(A4.4)}$$

for $n = 1$,

$$J_1(x) = \frac{x}{2} - \frac{x^3}{2^2 \cdot 4} + \frac{x^5}{2^2 \cdot 4^2 \cdot 6} - \cdots \tag{A4.5}$$

and for $n = 2$,

$$J_2(x) = \frac{x^2}{2 \cdot 4} - \frac{x^4}{2^2 \cdot 4 \cdot 6} + \frac{x^6}{2^2 \cdot 4^2 \cdot 6 \cdot 8} - \cdots \tag{A4.6}$$

and so on for higher values of n.

The Bessel function $J_n(x)$ has the following properties:

1. $$J_n(x) = (-1)^n J_{-n}(x) \tag{A4.7}$$

To prove this relation, we replace θ by $(\pi - \theta)$ in Eq. (A4.2). Then, noting that $\sin(\pi - \theta) = \sin\theta$, we get

$$J_n(x) = \frac{1}{\pi} \int_0^\pi \cos(x \sin\theta + n\theta - n\pi) \, d\theta$$

$$= \frac{1}{\pi} \int_0^\pi [\cos(n\pi)\cos(x \sin\theta + n\theta) + \sin(n\pi)\sin(x \sin\theta + n\theta)] \, d\theta$$

For integer values of n, we have

$$\cos(n\pi) = (-1)^n$$

$$\sin(n\pi) = 0$$

Therefore,

$$J_n(x) = \frac{(-1)^n}{\pi} \int_0^\pi \cos(x \sin\theta + n\theta) \, d\theta \tag{A4.8}$$

From Eq. (A4.2), we also find that by replacing n with $-n$:

$$J_{-n}(x) = \frac{1}{\pi} \int_0^\pi \cos(x \sin\theta + n\theta) \, d\theta \tag{A4.9}$$

The desired result follows immediately from Eqs. (A4.8) and (A4.9).

2. $$J_n(x) = (-1)^n J_n(-x) \tag{A4.10}$$

This relation is obtained by replacing x with $-x$ in Eq. (A4.2), and then using Eq. (A4.8).

3. $$J_{n-1}(x) + J_{n+1}(x) = \frac{2n}{x} J_n(x) \tag{A4.11}$$

This *recurrence formula* is useful in constructing tables of Bessel coefficients.

4. For small values of x, we have

$$J_n(x) \simeq \frac{x^n}{2^n n!} \tag{A4.12}$$

This relation is obtained simply by retaining the first term in the power series of Eq. (A4.3) and ignoring the higher-order terms. Thus, when x is small, we have

$$J_0(x) \simeq 1$$

$$J_1(x) \simeq \frac{x}{2} \qquad\qquad (A4.13)$$

$$J_n(x) \simeq 0 \qquad \text{for } n > 1$$

5. For large values of x, we have

$$J_n(x) \simeq \sqrt{\frac{2}{\pi x}} \cos\left(x - \frac{\pi}{4} - \frac{n\pi}{2} \right) \qquad\qquad (A4.14)$$

This shows that for large values of x, the Bessel function $J_n(x)$ behaves like a sine wave with progressively decreasing amplitude.

6. With x real and fixed, $J_n(x)$ approaches zero as the order n goes to infinity.

7.
$$\sum_{n=-\infty}^{\infty} J_n(x)\exp(jn\phi) = \exp(jx\sin\phi) \qquad\qquad (A4.15)$$

To prove this property, consider the sum $\sum_{n=-\infty}^{\infty} J_n(x)\exp(jn\phi)$ and use the formula of Eq. (A4.1) for $J_n(x)$ to obtain

$$\sum_{n=-\infty}^{\infty} J_n(x)\exp(jn\phi) = \frac{1}{2\pi} \sum_{n=-\infty}^{\infty} \exp(jn\phi) \int_{-\pi}^{\pi} \exp(jx\sin\theta - jn\theta)\, d\theta$$

Interchanging the order of integration and summation:

$$\sum_{n=-\infty}^{\infty} J_n(x)\exp(jn\phi) = \frac{1}{2\pi} \int_{-\pi}^{\pi} d\theta\, \exp(jx\sin\theta) \sum_{n=-\infty}^{\infty} \exp[jn(\phi-\theta)]$$

$$(A4.16)$$

From Example 11 of Chapter 2, we note that

$$\delta(\phi-\theta) = \frac{1}{2\pi} \sum_{n=-\infty}^{\infty} \exp[jn(\phi-\theta)], \qquad -\pi \le \phi - \theta \le \pi \quad (A4.17)$$

Therefore, substituting Eq. (A4.17) in (A4.16) and using the sifting property of a delta function, we get

$$\sum_{n=-\infty}^{\infty} J_n(x)\exp(jn\phi) = \int_{-\pi}^{\pi} \exp(jx\sin\theta)\delta(\phi-\theta)\, d\theta$$

$$= \exp(jx\sin\phi)$$

which is the desired result.

8.
$$\sum_{n=-\infty}^{\infty} J_n^2(x) = 1 \qquad \text{for all } x \qquad\qquad (A4.18)$$

To prove this property, we may proceed as follows. We observe that $J_n(x)$ is real. Hence, multiplying Eq. (A4.1) by its complex conjugate and summing over all possible values of n, we get

$$\sum_{n=-\infty}^{\infty} J_n^2(x)$$

$$= \frac{1}{(2\pi)^2} \sum_{n=-\infty}^{\infty} \int_{-\pi}^{\pi} \int_{-\pi}^{\pi} \exp(jx\sin\theta - jn\theta - jx\sin\phi + jn\phi)\, d\theta\, d\phi$$

Interchanging the order of double integration and summation:

$$\sum_{n=-\infty}^{\infty} J_n^2(x)$$

$$= \frac{1}{(2\pi)^2} \int_{-\pi}^{\pi} \int_{-\pi}^{\pi} d\theta\, d\phi \, \exp[jx(\sin\theta - \sin\phi)] \sum_{n=-\infty}^{\infty} \exp[jn(\phi - \theta)] \quad \text{(A4.19)}$$

Substituting Eq. (A4.17) in (A4.19) and using the sifting property of a delta function, we finally get

$$\sum_{n=-\infty}^{\infty} J_n^2(x) = \frac{1}{2\pi} \int_{-\pi}^{\pi} d\theta = 1$$

which is the desired result.

Many of these properties of the Bessel function $J_n(x)$ may also be illustrated in numerical terms by referring to Table A4.1.

A4.1 MODIFIED BESSEL FUNCTION OF THE FIRST KIND

Suppose in Eq. (A4.15) that we replace jx with x and the angle ϕ with $\theta - \pi/2$, and in the course of so doing we define a new function $I_n(x)$ that is related to $J_n(x)$ as follows:

$$I_n(x) = j^{-n} J_n(jx) \quad \text{(A4.20)}$$

We thus obtain

$$\sum_{n=-\infty}^{\infty} I_n(x)\exp(jn\theta) = \exp(x\cos\theta) \quad \text{(A4.21)}$$

From this relation it follows that

$$I_n(x) = \frac{1}{2\pi} \int_{-\pi}^{\pi} \exp(x\cos\theta)\cos(n\theta)\, d\theta \quad \text{(A4.22)}$$

The function $I_n(x)$ defined in Eq. (A4.22) is a real function of x, and it is known as the *modified Bessel function of the first kind of order n*.

In contrast to the Bessel function $J_n(x)$, the function $I_n(x)$ is not of the oscillatory type, but rather its behavior is similar to that of an exponential function.

Table A4.1 Table of Bessel Functions[a]

$J_n(x)$

$n \backslash x$	0.5	1	2	3	4	6	8	10	12
0	0.9385	0.7652	0.2239	−0.2601	−0.3971	0.1506	0.1717	−0.2459	0.0477
1	0.2423	0.4401	0.5767	0.3391	−0.0660	−0.2767	0.2346	0.0435	−0.2234
2	0.0306	0.1149	0.3528	0.4861	0.3641	−0.2429	−0.1130	0.2546	−0.0849
3	0.0026	0.0196	0.1289	0.3091	0.4302	0.1148	−0.2911	0.0584	0.1951
4	0.0002	0.0025	0.0340	0.1320	0.2811	0.3576	−0.1054	−0.2196	0.1825
5	—	0.0002	0.0070	0.0430	0.1321	0.3621	0.1858	−0.2341	−0.0735
6		—	0.0012	0.0114	0.0491	0.2458	0.3376	−0.0145	−0.2437
7			0.0002	0.0025	0.0152	0.1296	0.3206	0.2167	−0.1703
8			—	0.0005	0.0040	0.0565	0.2235	0.3179	0.0451
9				0.0001	0.0009	0.0212	0.1263	0.2919	0.2304
10				—	0.0002	0.0070	0.0608	0.2075	0.3005
11					—	0.0020	0.0256	0.1231	0.2704
12						0.0005	0.0096	0.0634	0.1953
13						0.0001	0.0033	0.0290	0.1201
14						—	0.0010	0.0120	0.0650

[a]For more extensive tables of Bessel functions, see Watson (1966, pp. 666–697), and Abramowitz and Stegun (1965, pp. 358–406).

The function $I_n(x)$ can be expressed in the form of a power series as:

$$I_n(x) = \sum_{m=0}^{\infty} \frac{(\frac{1}{2}x)^{n+2m}}{m!(m+n)!} \qquad \text{(A4.23)}$$

For the special case of $n = 0$, Eq. (A4.22) reduces to

$$I_0(x) = \frac{1}{2\pi} \int_{-\pi}^{\pi} \exp(x\cos\theta)\, d\theta \qquad \text{(A4.24)}$$

For small values of x, we have

$$I_0(x) \simeq 1, \qquad \text{(A4.25)}$$

while for large values of x,

$$I_0(x) \simeq \frac{\exp(x)}{\sqrt{2\pi x}} \qquad \text{(A4.26)}$$

Schwarz's Inequality

Let $g_1(t)$ and $g_2(t)$ be functions of the real variable t in the interval $a \le t \le b$. We assume that $g_1(t)$ and $g_2(t)$ satisfy the conditions

$$\int_a^b |g_1(t)|^2 \, dt < \infty \tag{A5.1}$$

$$\int_a^b |g_2(t)|^2 \, dt < \infty \tag{A5.2}$$

Then, according to *Schwarz's inequality*, we have

$$\left| \int_a^b g_1(t) g_2(t) \, dt \right|^2 \le \int_a^b |g_1(t)|^2 \, dt \int_a^b |g_2(t)|^2 \, dt \tag{A5.3}$$

To prove this inequality, we first form the integral[1]

$$\int_a^b [\lambda g_1^*(t) + g_2^*(t)][\lambda g_1(t) + g_2(t)] \, dt = \lambda^2 A + \lambda (B + B^*) + C \tag{A5.4}$$

---

[1] Our proof follows Thomas (1969, pp. 619–620).

where λ is a real variable, the asterisk signifies complex conjugation, and

$$A = \int_a^b |g_1(t)|^2 \, dt \geq 0 \tag{A5.5}$$

$$B = \int_a^b g_1^*(t) g_2(t) \, dt \tag{A5.6}$$

$$C = \int_a^b |g_2(t)|^2 \, dt \geq 0 \tag{A5.7}$$

The integral of Eq. (A5.4) exists, is real, and is a nonnegative function of λ, say $f(\lambda)$. Since $f(\lambda)$ is nonnegative, it must have no real roots except possibly a double root. From the quadratic formula, we must then have

$$(B + B^*)^2 \leq 4AC \tag{A5.8}$$

Note that $(B + B^*)/2$ is equal to the real part of B. On substituting Eqs. (A5.5) to (A5.7) in (A5.8), we get

$$\left\{ \int_a^b [g_1^*(t)g_2(t) + g_1(t)g_2^*(t)] \, dt \right\}^2 \leq 4 \int_a^b |g_1(t)|^2 \, dt \int_a^b |g_2(t)|^2 \, dt \tag{A5.9}$$

This is the most general form of Schwarz's inequality that is appropriate for complex functions $g_1(t)$ and $g_2(t)$. For the case when both $g_1(t)$ and $g_2(t)$ are real, we have

$$g_1^*(t)g_2(t) + g_1(t)g_2^*(t) = 2g_1(t)g_2(t) \tag{A5.10}$$

and Eq. (A5.3) follows immediately.

Note that equality is obtained [aside from the trivial case where both $g_1(t)$ and $g_2(t)$ are zero] when the double root exists in Eq. (A5.4); that is, when

$$\lambda g_1(t) + g_2(t) = \lambda g_1^*(t) + g_2^*(t) = 0 \tag{A5.11}$$

Since λ is real, $g_1(t)$ and $g_2(t)$ are linearly related. Looking at the problem from a slightly different viewpoint, we see that there is a real value of λ for which Eq. (A5.4) is zero and for which its first derivative with respect to λ vanishes; that is.

$$2\lambda A + (B + B^*) = 0 \tag{A5.12}$$

or

$$\lambda = -\frac{B + B^*}{2A} = -\frac{\int_a^b [g_1^*(t)g_2(t) + g_1(t)g_2^*(t)] \, dt}{2 \int_a^b |g_1(t)|^2 \, dt} \tag{A5.13}$$

This relation holds if, and only if,

$$g_2(t) = -\lambda g_1(t) \tag{A5.14}$$

This last relationship is equivalent to Eq. (A5.11).

Noise Figure

A convenient measure of the noise performance of a linear two-port device is furnished by the so-called *noise figure*. Consider a linear two-port device connected to a signal source of internal impedance $Z(f) = R(f) + jX(f)$ at the input, as in Fig. A6.1. The noise voltage $v(t)$ represents the thermal noise associated with the internal resistance $R(f)$ of the source. The output noise of the device is made up of two contributions, one due to the source and the other due to the device itself. We define *the available output noise power in a band of width Δf centered at frequency f as the maximum average noise power in this band, obtainable at the output of the device.* The maximum noise power that the two-port device can deliver to an external load is obtained when the load impedance is the complex conjugate of the output impedance of the device, that is, when the resistance is matched and the reactance is tuned out. We define *the noise figure of the two-port device as the ratio of the total available output noise power (due to the device and the source) per unit bandwidth to the portion thereof due solely to the source.*

Let the spectral density of the total available noise power of the device output be $S_{NO}(f)$, and the spectral density of the available noise power due to the source at the device input be $S_{NS}(f)$. Also let $G(f)$ denote the *available power gain* of the two-port device, defined as *the ratio of the available signal power at the output of the device to the available signal power of the source when the signal is a sinusoidal wave*

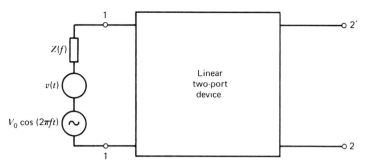

Figure A6.1 Linear two-port device.

of frequency f. Then we may express the noise figure *F* of the device as

$$F = \frac{S_{NO}(f)}{G(f)S_{NS}(f)} \tag{A6.1}$$

If the device were noise free, $S_{NO}(f) = G(f)S_{NS}(f)$, and the noise figure would then be unity. In a physical device, however, $S_{NO}(f)$ is larger than $G(f)S_{NS}(f)$, so that the noise figure is always larger than unity. The noise figure is commonly expressed in decibels, that is, as $10 \log_{10} F$.

The noise figure may also be expressed in an alternative form. Let $P_S(f)$ denote the available signal power from the source, which is the maximum average signal power that can be obtained. For the case of a source providing a single-frequency signal component with open-circuit voltage $V_0 \cos(2\pi f t)$, the available signal power is obtained when the load connected to the source is

$$Z^*(f) = R(f) - jX(f)$$

Under this condition, we find that

$$P_S(f) = \left[\frac{V_0}{2R(f)}\right]^2 R(f)$$
$$= \frac{V_0^2}{4R(f)} \tag{A6.2}$$

The available signal power at the output of the device is therefore

$$P_O(f) = G(f)P_S(f) \tag{A6.3}$$

Then, multiplying both the numerator and denominator of the right-hand side of Eq. (A6.1) by $P_S(f)\,\Delta(f)$, we obtain

$$F = \frac{P_S(f)S_{NO}(f)\,\Delta f}{G(f)P_S(f)S_{NS}(f)\,\Delta f}$$
$$= \frac{P_S(f)S_{NO}(f)\,\Delta f}{P_O(f)S_{NS}(f)\,\Delta f} \tag{A6.4}$$
$$= \frac{\rho_S(f)}{\rho_O(f)}$$

where

$$\rho_S(f) \;=\; \frac{P_S(f)}{S_{NS}(f)\,\Delta f} \qquad\qquad (A6.5)$$

$$\rho_O(f) \;=\; \frac{P_O(f)}{S_{NO}(f)\,\Delta f} \qquad\qquad (A6.6)$$

We refer to $\rho_S(f)$ as the *available signal-to-noise ratio of the source* and to $\rho_O(f)$ as the *available signal-to-noise ratio at the device output*, both measured in a narrow band of width Δf centered at f. Since the noise figure is always greater than unity, it follows from Eq. (A6.4) that the signal-to-noise ratio always decreases with amplification, which is a significant result.

The noise figure F is a function of the operating frequency f; it is therefore referred to as the *spot noise figure*. In contrast, we may define an *average noise figure* F_0 of a two-port device as the ratio of the total noise power at the device output to the output noise power due solely to the source. That is,

$$F_0 \;=\; \frac{\displaystyle\int_{-\infty}^{\infty} S_{NO}(f)\;df}{\displaystyle\int_{-\infty}^{\infty} G(f)\,S_{NS}(f)\;df} \qquad\qquad (A6.7)$$

It is apparent that in the case of thermal noise in the input circuit with $R(f)$ constant, and constant gain throughout a fixed band with zero gain at other frequencies, the spot noise figure F and the average noise figure F_0 are identical.

A6.1   EQUIVALENT NOISE TEMPERATURE

A disadvantage of the noise figure F is that when it is used to compare low-noise devices, the values obtained are all close to unity, which makes the comparison rather difficult. In such cases, it is preferable to use the *equivalent noise temperature*. Consider a linear two-port device whose input resistance is matched to the internal resistance of the source as shown in Fig. A6.2. In this diagram, we have also included the noise voltage generator associated with the internal resistance R_s of the source. The mean-square value of this noise voltage is $4kTR_s\,\Delta f$, where k is Boltzmann's constant. Hence, the available noise power at the device input is

$$N_1 \;=\; kT\,\Delta f \qquad\qquad (A6.8)$$

Let N_d denote the noise power contributed by the two-port device to the total available output noise power N_2. We define N_d as

$$N_d \;=\; GkT_e\,\Delta f \qquad\qquad (A6.9)$$

where G is the available power gain of the device and T_e is its equivalent noise temperature. Then it follows that the total output noise power is

$$\begin{aligned} N_2 &= GN_1 + N_d \qquad\qquad (A6.10)\\ &= Gk(T + T_e)\,\Delta f \end{aligned}$$

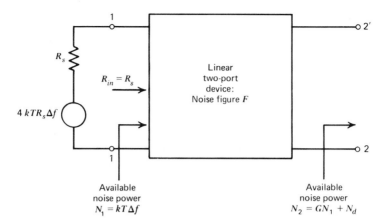

Available
noise power
$N_1 = kT\Delta f$

Available
noise power
$N_2 = GN_1 + N_d$

Figure A6.2 Linear two-port device matched to the internal resistance of a source connected to the input.

The noise figure of the device is therefore

$$F = \frac{N_d}{N_2} = \frac{T + T_e}{T} \qquad (A6.11)$$

Solving for the equivalent noise temperature:

$$T_e = T(F - 1) \qquad (A6.12)$$

The noise figure F is measured under matched input conditions, and with the noise source at temperature T.

A6.2 CASCADE CONNECTION OF TWO-PORT NETWORKS

It is often necessary to evaluate the noise figure of a cascade connection of two-port networks whose individual noise figures are known. Consider Fig. A6.3, consisting of a pair of two-port networks of noise figures F_1 and F_2 and power gains G_1 and G_2, connected in cascade. It is assumed that the devices are matched, and that the noise figure F_2 of the second network is defined assuming an input noise power N_1.

At the input of the first network, we have a noise power N_1 contributed by the source, plus an equivalent noise power $(F_1 - 1)N_1$ contributed by the network itself. The output noise power from the first network is therefore $F_1 N_1 G_1$.

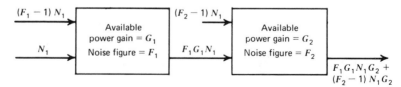

Figure A6.3 A cascade of two noisy two-port networks.

Added to this noise power at the input of the second network, we have the equivalent extra power $(F_2 - 1)N_1$ contributed by the second network itself. The output noise power from this second network is therefore equal to $F_1 G_1 N_1 G_2 + (F_2 - 1)N_1 G_2$. We may consider the noise figure F as the ratio of the actual output noise power to the output noise power assuming the networks to be noiseless. We may therefore express the overall noise figure of the cascade connection of Fig. A6.3 as

$$
\begin{aligned}
F &= \frac{F_1 G_1 N_1 G_2 + (F_2 - 1)N_1 G_2}{N_1 G_1 G_2} \\
&= F_1 + \frac{F_2 - 1}{G_1}
\end{aligned}
\tag{A6.13}
$$

The result may be readily extended to the cascade connection of any number of two-port networks, as shown by

$$
F = F_1 + \frac{F_2 - 1}{G_1} + \frac{F_3 - 1}{G_1 G_2} + \frac{F_4 - 1}{G_1 G_2 G_3} + \cdots
\tag{A6.14}
$$

where $F_1, F_2, F_3, \ldots$ are the individual noise figures, and $G_1, G_2, G_3, \ldots$ are the available power gains, respectively. Equation (A6.14) shows that if the first stage of the cascade connection in Fig. A6.3 has a high gain, the overall noise figure F is dominated by the noise figure of the first stage.

Correspondingly, we may express the overall equivalent noise temperature of the cascade connection of any number of noisy two-port networks as follows:

$$
T_e = T_1 + \frac{T_2}{G_1} + \frac{T_3}{G_1 G_2} + \frac{T_4}{G_1 G_2 G_3} + \cdots
\tag{A6.15}
$$

where $T_1, T_2, T_3, \ldots$ are the equivalent noise temperatures of the individual networks, and $G_1, G_2, G_3, \ldots$ are the available power gains, respectively. Equation (A6.15) is known as the *Friis formula*. Here again we note that if the gain G_1 of the first stage is high, the equivalent noise temperature T_e is dominated by that of the first stage.

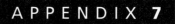

Error Function

The *error function*, denoted by $\mathrm{erf}(u)$, is defined in a number of different ways in the literature. We shall use the following definition:

$$\mathrm{erf}(u) \;=\; \frac{2}{\sqrt{\pi}} \int_0^u \exp(-z^2)\, dz \tag{A7.1}$$

The error function has two useful properties:

1. $\mathrm{erf}(-u) \;=\; -\mathrm{erf}(u)$ (A7.2)
   This is known as the *symmetry relation.*

2. As u approaches infinity, $\mathrm{erf}(u)$ approaches unity; that is

$$\frac{2}{\sqrt{\pi}} \int_0^\infty \exp(-z^2)\, dz \;=\; 1 \tag{A7.3}$$

The *complementary error function* is defined by

$$\mathrm{erfc}(u) \;=\; \frac{2}{\sqrt{\pi}} \int_u^\infty \exp(-z^2)\, dz \tag{A7.4}$$

Table A7.1   The Error Function[a]

u	$\mathrm{erf}(u)$	u	$\mathrm{erf}(u)$
0.00	0.00000	1.10	0.88021
0.05	0.05637	1.15	0.89612
0.10	0.11246	1.20	0.91031
0.15	0.16800	1.25	0.92290
0.20	0.22270	1.30	0.93401
0.25	0.27633	1.35	0.94376
0.30	0.32863	1.40	0.95229
0.35	0.37938	1.45	0.95970
0.40	0.42839	1.50	0.96611
0.45	0.47548	1.55	0.97162
0.50	0.52050	1.60	0.97635
0.55	0.56332	1.65	0.98038
0.60	0.60386	1.70	0.98379
0.65	0.64203	1.75	0.98667
0.70	0.67780	1.80	0.98909
0.75	0.71116	1.85	0.99111
0.80	0.74210	1.90	0.99279
0.85	0.77067	1.95	0.99418
0.90	0.79691	2.00	0.99532
0.95	0.82089	2.50	0.99959
1.00	0.84270	3.00	0.99998
1.05	0.86244	3.30	0.999998

[a] The error function is tabulated extensively in several references; see for example, Abramowitz and Stegun (1965, pp. 297–316).

It is related to the error function as follows:

$$\mathrm{erfc}(u) = 1 - \mathrm{erf}(u) \qquad (A7.5)$$

Table A7.1 gives values of the error function $\mathrm{erf}(u)$ for u in the range 0 to 3.3.

A7.1   BOUNDS ON THE COMPLEMENTARY ERROR FUNCTION

Substituting $u - x$ for z in Eq. (A7.4), we get

$$\mathrm{erfc}(u) = \frac{2}{\sqrt{\pi}} \exp(-u^2) \int_{-\infty}^{0} \exp(2ux)\exp(-x^2) \; dx$$

For any real x, the value of $\exp(-x^2)$ lies between the successive partial sums of the power series

$$1 - \frac{x^2}{1!} + \frac{(x^2)^2}{2!} - \frac{(x^2)^3}{3!} + \cdots$$

Therefore, for $u > 0$, we find, on using $(n + 1)$ terms of this series, that $\mathrm{erfc}(u)$ lies between the values taken by

$$\frac{2}{\sqrt{\pi}} \exp(-u^2) \int_{-\infty}^{0} \left(1 - x^2 + \frac{x^4}{2} - \cdots \pm \frac{x^{2n}}{n!} \right) \exp(2ux)\, dx$$

for even n and for odd n. Putting $2ux = -v$ and using the integral

$$\int_{0}^{\infty} v^n \exp(-v)\, dv = n!$$

we obtain the following *asymptotic expansion* for $\mathrm{erfc}(u)$, assuming $u > 0$:

$$\mathrm{erfc}(u) \simeq \frac{\exp(-u^2)}{\sqrt{\pi} u} \left[1 - \frac{1}{2u^2} + \frac{1 \cdot 3}{2^2 u^4} - \cdots \pm \frac{1 \cdot 3 \cdot 5 \cdots (2n-1)}{2^n u^{2n}} \right]$$

$$(A7.6)$$

For large positive values of u, the successive terms of the series on the right-hand side of Eq. (A7.6) decrease very rapidly. We thus deduce two simple bounds on $\mathrm{erfc}(u)$, one lower and the other upper, as shown by[1]

$$\frac{\exp(-u^2)}{\sqrt{\pi} u} \left(1 - \frac{1}{2u^2} \right) < \mathrm{erfc}(u) < \frac{\exp(-u^2)}{\sqrt{\pi} u} \qquad (A7.7)$$

For large positive u, a second bound on the complementary error function $\mathrm{erfc}(u)$ is obtained by omitting the multiplying factor $1/u$ in the upper bound of Eq. (A7.7):

$$\mathrm{erfc}(u) < \frac{\exp(-u^2)}{\sqrt{\pi}} \qquad (A7.8)$$

In Fig. A7.1, we have plotted $\mathrm{erfc}(u)$, the two bounds defined by Eq. (A7.7), and the upper bound of Eq. (A7.8). We see that for $u \geq 1.5$ the bounds on $\mathrm{erfc}(u)$, defined by Eq. (A7.7), become increasingly tight.

A7.2 THE *Q*-FUNCTION

Consider a *standardized* Gaussian random variable X of zero mean and unit variance. The probability that an observed value of the random variable X will be greater than v is given by the *Q-function*:

$$Q(v) = \frac{1}{\sqrt{2\pi}} \int_{v}^{\infty} \exp\left(-\frac{x^2}{2} \right) \qquad (A7.9)$$

---

[1]The derivation of Eq. (A7.7) follows Blachman (1966).

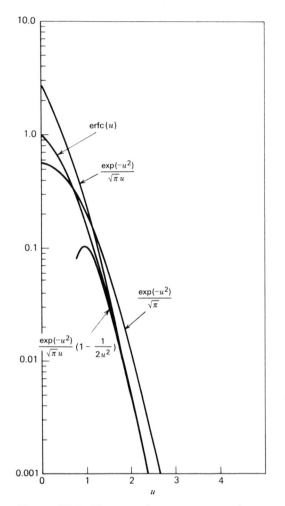

Figure A7.1 The complementary error function and its bounds.

The Q-function defines the area under the standardized Gaussian tail. Inspection of Eqs. (A7.4) and (A7.9) reveals that the Q-function is related to the complementary error function as follows:

$$Q(v) = \frac{1}{2} \, \text{erfc}\left(\frac{v}{\sqrt{2}}\right) \tag{A7.10}$$

Conversely, putting $u = v/\sqrt{2}$, we have

$$\text{erfc}(u) = 2Q(\sqrt{2}u) \tag{A7.11}$$

APPENDIX 8

Statistical Characterization of Complex Random Processes

In this appendix, we extend some of the important statistical properties of Chapter 4 to deal with a complex-valued random process. Such a process arises in the equivalent low-pass representation of a narrow-band random process.

Let $X(t)$ denote a narrow-band random process with a nominal carrier frequency f_c. In light of the material presented in Chapter 2, we may express $X(t)$ as follows:

$$X(t) = \text{Re}[\tilde{X}(t)\exp(j2\pi f_c t)] \tag{A8.1}$$

where $\tilde{X}(t)$ is the (low-pass) complex envelope of $X(t)$. Recognizing that the real part of a complex number is equal to one half times the sum of the number and its complex conjugate, we may rewrite Eq. (A8.1) as follows:

$$X(t) = \tfrac{1}{2}[\tilde{X}(t)\exp(j2\pi f_c t) + \tilde{X}^*(t)\exp(-j2\pi f_c t)] \tag{A8.2}$$

where the asterisk denotes complex conjugation. From the random process theory presented in Chapter 4, the autocorrelation function of the real-valued process $X(t)$ is defined by

$$R_X(t_1, t_2) = E[X(t_1)X(t_2)] \tag{A8.3}$$

where t_1 and t_2 are two observation times. Hence, substituting Eq. (A8.2) in (A8.3) and expanding terms, we get

$$
\begin{aligned}
R_X(t_1,t_2) &= \tfrac{1}{4}E[\tilde{X}(t_1)\tilde{X}^*(t_2)]\exp(-j2\pi f_c(t_2 - t_1)) \\
&\quad + \tfrac{1}{4}E[\tilde{X}^*(t_1)\tilde{X}(t_2)]\exp(j2\pi f_c(t_2 - t_1)) \\
&\quad + \tfrac{1}{4}E[\tilde{X}(t_1)\tilde{X}(t_2)]\exp(j2\pi f_c(t_1 + t_2)) \\
&\quad + \tfrac{1}{4}E[\tilde{X}^*(t_1)\tilde{X}^*(t_2)]\exp(-j2\pi f_c(t_1 + t_2)) \\
&= \tfrac{1}{2}\mathrm{Re}\{E[\tilde{X}^*(t_1)\tilde{X}(t_2)]\exp(j2\pi f_c(t_2 - t_1))\} \\
&\quad + \tfrac{1}{2}\mathrm{Re}\{E[\tilde{X}(t_1)\tilde{X}(t_2)]\exp(j2\pi f_c(t_1 + t_2))\}
\end{aligned}
\tag{A8.4}
$$

Thus, in order to specify the autocorrelation function of a real-valued random process $X(t)$ in terms of its complex envelope $\tilde{X}(t)$, we need two expectations:

$$
E[\tilde{X}^*(t_1)\tilde{X}(t_2)]
$$

and

$$
E[\tilde{X}(t_1)\tilde{X}(t_2)]
$$

Fortunately, however, a narrow-band process is so constituted in practice that we have[1]

$$
E[\tilde{X}(t_1)\tilde{X}(t_2)] = 0
\tag{A8.5}
$$

Indeed, using Eq. (A8.4) we may readily show that the condition of Eq. (A8.5) is necessary for the process $X(t)$ to be wide-sense stationary. Accordingly, we may simplify Eq. (A8.4) as follows:

$$
R_X(t_1,t_2) = \tfrac{1}{2}\mathrm{Re}[R_{\tilde{X}}(t_1,t_2)\exp(j2\pi f_c(t_2 - t_1))]
\tag{A8.6}
$$

where $R_{\tilde{X}}(t_1,t_2)$ is the autocorrelation function of the complex envelope $\tilde{X}(t)$, defined by

$$
R_{\tilde{X}}(t_1,t_2) = E[\tilde{X}^*(t_1)\tilde{X}(t_2)]
\tag{A8.7}
$$

In the case of a wide-sense stationary, narrow-band process $X(t)$, we may rewrite Eq. (A8.7) in the form

$$
R_{\tilde{X}}(\tau) = E[\tilde{X}^*(t)\tilde{X}(t + \tau)]
\tag{A8.8}
$$

The autocorrelation function $R_{\tilde{X}}(\tau)$ has similar properties to those of $R_X(\tau)$, except for one major difference: $R_{\tilde{X}}(\tau)$ is itself complex-valued whereas $R_X(\tau)$ is

---

[1]This point is made in Bello (1963). Sakrison (1968) also discusses the statistical characterization of complex-valued random processes, where it is remarked that the quantity $E[\tilde{X}(t_1)\tilde{X}(t_2)]$ is so seldom used that no notation is adopted for it.

real valued. In particular, we have

$$R_{\tilde{X}}(-\tau) = R_{\tilde{X}}^{*}(\tau) \tag{A8.9}$$

That is, $R_{\tilde{X}}(\tau)$ exhibits *conjugate symmetry*.

The power spectral density of the complex envelope $\tilde{X}(t)$ is defined as the Fourier transform of the autocorrelation function $R_{\tilde{X}}(\tau)$, as shown by

$$S_{\tilde{X}}(f) = \int_{-\infty}^{\infty} R_{\tilde{X}}(\tau) \exp(-j2\pi f\tau) \, d\tau \tag{A8.10}$$

The important point to note here is that the power spectral density $S_{\tilde{X}}(f)$ is a *real-valued quantity* by virtue of the conjugate symmetry property described in Eq. (A8.9). However, unlike the power spectral density of the real-valued process $X(t)$, the power spectral density $S_{\tilde{X}}(f)$ defined in Eq. (A8.10) is *not* in general an even function of the frequency f. Except for this difference, the two power spectral densities $S_{\tilde{X}}(f)$ and $S_{X}(f)$ have similar properties.

Binary
Arithmetic

In binary input–binary output coded systems, the alphabet consists only of symbols 0 and 1. A *field* is described simply as a set of elements in which we can perform addition, subtraction, multiplication, and division without leaving the set. The field in which we are interested is a *binary field* that contains only two elements: 0 and 1. In this appendix, we discuss arithmetic over the binary field.

In binary arithmetic we use *modulo-2 addition* and *multiplication*. The rules for modulo-2 addition are as follows:

$$0 + 0 = 0$$
$$1 + 0 = 1$$
$$0 + 1 = 1$$
$$1 + 1 = 0$$

where the plus sign is intended to mean modulo-2 addition. There are no "carries" in modulo-2 addition. Hence, the sum of an odd number of 1s equals 1, and the sum of an even number of 1s equals 0. Note that since $1 + 1 = 0$, we may write $1 = -1$. Accordingly, in binary arithmetic, *subtraction is the same as addition*.

For modulo-2 multiplication, we have the following rules:

$$0 \times 0 = 0$$
$$1 \times 0 = 0$$
$$0 \times 1 = 0$$
$$1 \times 1 = 1$$

Division is trivial in that $1 \div 1 = 1, 0 \div 1 = 0$, and division by 0 is not permitted.

It is of interest to note that modulo-2 addition is simply the EXCLUSIVE-OR operation in logic. Correspondingly, modulo-2 multiplication is the AND operation in logic.

Cryptography

Secrecy is certainly important to the security or integrity of information transmission. Indeed, the need for secure communications is more profound than ever, recognizing that the conduct of much of our commerce, business, and personal matters is being carried out today through the medium of computers, which has replaced the traditional medium of papers.

Cryptology is the umbrella term used to describe the science of secret communications; it is derived from the Greek *kryptos* and *logos* which mean "hidden" and "word," respectively.[1] The subject matter of cryptology may be partitioned neatly into *cryptography* and *cryptanalysis*. Cryptography deals with the transformations of a message into coded form by *encryption* and the recovery of the original message by *decryption*. The original message to be encrypted (enciphered) is called the *plaintext*, and the result produced by encryption is called a *cryptogram* or *ciphertext*; the latter two terms are used interchangeably. The set of data transformations used to do the encryption is called a *cipher*; normally, the transformations are parameterized by one or more *keys*. *Cryptanalysis*, on the other hand, deals with how to undo cryptographic communications by breaking a cipher or forging coded signals that may be accepted as genuine.

Cryptographic systems offer three important services:

1. *Secrecy*, which refers to the denial of access to information by unauthorized users.
2. *Authenticity*, which refers to the validation of the source of a message.
3. *Integrity*, which refers to the assurance that a message was not modified by accidental or deliberate means in transit.

A conventional cryptographic system relies on the use of a single piece of private and necessarily secret information known as the *key*; hence, conventional cryptography is referred to as *single-key cryptography* or *secret-key cryptography*.[2] This form of cryptography operates on the premise that the key is known to the encrypter (sender) and by the decrypter (receiver) but to no others; the assumption is that once the message is encrypted, it is (probably) impossible to do the decryption without knowledge of the key.

Public-key cryptography,[3] also called *two-key cryptography*, differs from conventional cryptography in that there is no longer a single secret key shared by two users. Rather, each user is provided with key material of one's own, and the key material is divided into two portions: a public component and a private component. The public component generates a public transformation, and the private component generates a private transformation. But, of course, the private transformation must be kept secret for secure communication between the two users.

A10.1 SECRET-KEY CRYPTOGRAPHY

Basically, the flow of information in a secret-key cryptographic system is as shown in Fig. A10.1. The message source generates a plaintext message, which is encrypted into a cryptogram at the transmitting end of the system. The cryptogram is sent to an *authorized user* at the receiving end over an "insecure" channel; a channel is considered insecure if its security is inadequate for the needs of its users. It is assumed that in the course of transmission the cryptogram may be intercepted by an *enemy cryptanalyst*[4] (i.e., would-be intruder into a cryptographic system). The requirement is to do the encryption in such a way that the enemy is prevented from learning the contents of the plaintext message.

In abstract terms, a *cryptographic system or cipher (for short) is defined as a set of invertible transformations of the plaintext space (i.e., the set of possible plaintext messages) into the cryptogram space (i.e., the set of all possible cryptograms)*. Each particular transformation corresponds to encryption (enciphering) of a plaintext with a partic-

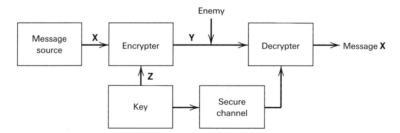

Figure A10.1 Block diagram of secret-key cryptographic system.

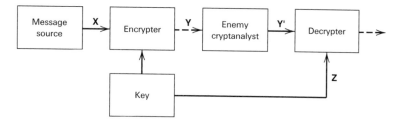

Figure A10.2 Illustrating the intrusion of an enemy cryptanalyst.

ular key. The invertibility of the transformation means that unique decryption (deciphering) of the cryptogram is possible when the key is known. Let **X** denote the plaintext message, **Y** denote the cryptogram, and **Z** denote the key. Let **F** denote the invertible transformation producing the cryptogram **Y**, as follows:

$$\mathbf{Y} = \mathbf{F}(\mathbf{X},\mathbf{Z}) = \mathbf{F}_z(\mathbf{X}) \tag{A10.1}$$

The transformation is intended to make the cryptogram **Y** useless to the enemy. At the receiving end of the system, the cryptogram **Y** is decrypted with the inverse transformation $\mathbf{F}^{-1}$ to recover the original plaintext message **X**, as shown by

$$\mathbf{F}^{-1}(\mathbf{Y},\mathbf{Z}) = \mathbf{F}_z^{-1}(\mathbf{Y}) = \mathbf{F}_z^{-1}(\mathbf{F}_z(\mathbf{X})) = \mathbf{X} \tag{A10.2}$$

In physical terms, the cryptographic system consists of a set of instructions, a piece of physical hardware, or a computer program. In any event, the system is designed to have the capability of encrypting the plaintext (and, of course, decrypting the resulting cryptogram) in a variety of ways; the particular way chosen to do the actual encryption is determined by the specific key.

The security of the system resides in the secret nature of the key, which requires that the key must be delivered to the receiver over a *secure channel* (e.g., registered mail, courier service) as implied in Fig. A10.1. The cryptographic system depicted in this figure provides a solution to the *secrecy problem*, preventing an enemy from extracting information from messages transmitted over an insecure communication channel. Cryptography also provides a solution to the *authentication problem*, preventing an enemy cryptanalyst from impersonating the message sender. In this second situation, the enemy cryptanalyst is the one who originates a "fraudulent" cryptogram **Y'** that is delivered to the receiver (decrypter), as shown in Fig. A10.2. The authentic cryptogram **Y** is shown as a dashed input to the enemy cryptanalyst, indicating that the enemy produces the fraudulent cryptogram **Y'** without ever seeing the authentic one. The receiver may be able to recognize **Y'** as fraudulent by decrypting it with the correct key **Z**; hence, the line from the receiver output to the destination is shown dashed to suggest rejection of the fraudulent cryptogram **Y'** by the receiver.

A10.2 BLOCK AND STREAM CIPHERS

Much as error-correcting codes are classified into block codes and convolutional codes, cryptographic systems (ciphers) may be classified into two broad classes: *block ciphers* and *stream ciphers*. Block ciphers operate in a purely combinatorial

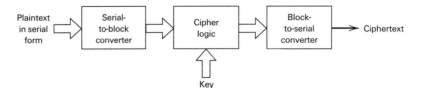

Figure A10.3   Block diagram of a block cipher.

fashion on large blocks of plaintext, whereas stream ciphers process the plaintext in small pieces (i.e., characters or bits).

Figure A10.3 shows the generic form of a block cipher. The plaintext (consisting of serial data) is divided into large blocks, each of which is usually made up of a fixed number of bits. Successive blocks of the plaintext are enciphered (encrypted) using the same secret key, otherwise independently; the resulting enciphered blocks are finally converted into serial form. Thus, a particular plaintext block identical to a previous such block gives rise to an identical ciphertext block. Specifically, each bit of a particular ciphered block is chosen to be a function of all the bits of the associated plaintext block and the key; the goal of a block cipher is to have no specific bit of the plaintext ever appear in the ciphertext directly.

Block ciphers operate with a fixed transformation applied to large blocks of plaintext data, on a block-by-block basis. In contrast, a stream cipher operates on the basis of a time-varying transformation applied to individual bits of the plaintext. The most popular stream ciphers are the so-called *binary additive stream ciphers*, the generic form of which is shown in Fig. A10.4. In such a cipher, the secret key is used to control a *keystream generator* that emits a binary sequence called the *keystream*, whose length is much larger than that of the key. Let x_n, y_n, and z_n denote the plaintext bit, ciphertext bit, and keystream bit at time n, respectively. The ciphertext bits are then determined by simple modulo-2 addition of the plaintext bits and the keystream bits, as shown by

$$y_n = x_n \oplus z_n, \qquad n = 1, 2, \ldots, N \qquad (A10.3)$$

where N is the length of the keystream. Since addition and subtraction in modulo-2 arithmetic are exactly the same, Eq. (A10.3) also implies the following

$$x_n = y_n \oplus z_n, \qquad n = 1, 2, \ldots, N \qquad (A10.4)$$

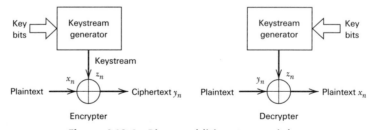

Figure A10.4   Binary additive stream cipher.

We thus see that in binary additive stream ciphers, identical devices can be used to perform encryption and decryption, as shown in Fig. A10.4. The secret key is chosen according to some probability distribution. To provide secure encryption, the keystream should resemble a coin-tossing (i.e., completely random) sequence as closely as possible.

Block ciphers are normally designed in such a way that a small change in an input block of plaintext produces a major change in the resulting output. This *error propagation* property of block ciphers is valuable in authentication in that it makes it improbable for an enemy cryptanalyst to modify encrypted data, unless knowledge of the key is available. On the other hand, a binary additive stream cipher has *no* error propagation; the decryption of a distorted bit in the ciphertext affects only the corresponding bit of the resulting output.

Stream ciphers are generally better suited for the secure transmission of data over error-prone communication channels; they are used in applications where high data rates are a requirement (as in secure video, for example) or when a minimal transmission delay is essential.[5]

Requirement for Secrecy

In cryptography, a fundamental assumption is that an enemy cryptanalyst has knowledge of the entire mechanism used to perform encryption, except for the secret key. We may identify the following forms of attack that may be attempted by the enemy cryptanalyst, depending on the availability of additional knowledge:

1. *Ciphertext-only attack* is a cryptanalytic attack in which the enemy cryptanalyst has access to part or all of the ciphertext.
2. *Known-plaintext attack* is a cryptanalytic attack in which the enemy cryptanalyst has knowledge of some ciphertext–plaintext pairs formed with the actual secret key.
3. *Chosen-plaintext attack* is a cryptanalytic attack in which the enemy cryptanalyst is able to submit any chosen plaintext message and receive in return the correct ciphertext for the actual secret key.
4. *Chosen-ciphertext attack* is a cryptanalytic attack in which the enemy cryptanalyst is able to choose an arbitrary ciphertext and find the correct result for its decryption.

A ciphertext-only attack occurs frequently in practice. In this form of attack, an enemy cryptanalyst uses only knowledge of the statistical structure of the language in use (e.g., in English the letter *e* occurs with a probability of 13 percent, and the letter *q* is always followed by *u*) and knowledge of some probable words (e.g., a letter probably begins with ''Dear Sir/Madam:''). A known-plaintext attack may take place by virtue of the standard computer formats used in programming languages and data generation. In any case, the ciphertext-only attack is viewed as the weakest threat to which a cryptographic system can be subjected, and any system that succumbs to it is therefore considered totally insecure. Thus, for a cryptographic system to provide secrecy, at the minimum it should be immune to ciphertext-only attacks; ideally, it should also be immune to known-plaintext attacks.

A10.3 INFORMATION-THEORETIC APPROACH

In the *Shannon model of cryptography,* named in recognition of Shannon's 1949 landmark paper on the information-theoretic approach to secrecy systems, the enemy cryptanalyst is assumed to have unlimited time and computing power. But the enemy is presumably restricted to a ciphertext-only attack. Cryptanalysis in the Shannon model is defined as the process of finding the secret key, given the cryptogram (ciphertext) and the a priori probabilities of the various plaintexts and keys. The secrecy of the system is considered *broken* when the enemy cryptanalyst performs decryption successfully, obtaining a *unique* solution to the cryptogram.[6]

Let $\mathbf{X} = (X_1, X_2, \ldots, X_N)$ denote an N-bit plaintext message, and $\mathbf{Y} = (Y_1, Y_2, \ldots, Y_N)$ denote the corresponding N-bit cryptogram; that is, both the plaintext and the cryptogram have the same number of bits. It is assumed that the secret key $\mathbf{Z}$ used to construct the cryptogram is drawn according to some probability distribution. The uncertainty about $\mathbf{X}$ is expressed by the entropy $H(\mathbf{X})$, and the uncertainty about $\mathbf{X}$ given knowledge of $\mathbf{Y}$ is expressed by the conditional entropy $H(\mathbf{X}|\mathbf{Y})$. The *mutual information* between $\mathbf{X}$ and $\mathbf{Y}$ is defined by

$$I(\mathbf{X};\mathbf{Y}) = H(\mathbf{X}) - H(\mathbf{X}|\mathbf{Y}) \tag{A10.5}$$

The mutual information $I(\mathbf{X};\mathbf{Y})$ represents a basic measure of security (secrecy) in the Shannon model.

Perfect Security

Assuming that an enemy cryptanalyst can observe only the cryptogram $\mathbf{Y}$, it seems appropriate that we define the *perfect security* of a cryptographic system to mean that the plaintext $\mathbf{X}$ and the cryptogram $\mathbf{Y}$ are statistically independent. In other words, we have

$$I(\mathbf{X};\mathbf{Y}) = 0 \qquad \text{for all } N \tag{A10.6}$$

Then, using Eq. (A10.5), we find that the condition for perfect security may be rewritten as

$$H(\mathbf{X}|\mathbf{Y}) = H(\mathbf{X}) \tag{A10.7}$$

Equation (A10.7) states that the best an enemy cryptanalyst can do, given the cryptogram $\mathbf{Y}$, is to guess the plaintext message $\mathbf{X}$ according to the probability distribution of all possible messages.

Given the secret key $\mathbf{Z}$, we note that

$$\begin{aligned} H(\mathbf{X}|\mathbf{Y}) &\leq H(\mathbf{X},\mathbf{Z}|\mathbf{Y}) \\ &= H(\mathbf{Z}|\mathbf{Y}) + H(\mathbf{X}|\mathbf{Y},\mathbf{Z}) \end{aligned} \tag{A10.8}$$

The conditional entropy $H(\mathbf{X}|\mathbf{Y},\mathbf{Z})$ is zero if and only if $\mathbf{Y}$ and $\mathbf{Z}$ together uniquely determine $\mathbf{X}$; this is indeed a valid assumption when the decryption process is performed with knowledge of the secret key $\mathbf{Z}$. Hence, we may simplify Eq. (A10.8) as follows:

$$H(\mathbf{X}|\mathbf{Y}) \leq H(\mathbf{Z}|\mathbf{Y})$$
$$\leq H(\mathbf{Z}) \tag{A10.9}$$

Thus, substituting Eq. (A10.9) in (A10.7), we find that for a cryptographic system to provide perfect security, the following condition must be satisfied

$$H(\mathbf{Z}) \geq H(\mathbf{X}) \tag{A10.10}$$

The inequality of Eq. (A10.10) is *Shannon's fundamental bound for perfect security*; it states that for perfect security, the uncertainty of a secret key $\mathbf{Z}$ must be at least as large as the uncertainty of the plaintext $\mathbf{X}$ that is concealed by the key.

For the case when the plaintext and key alphabets are of the same size, the use of Shannon's bound for perfect security yields the following result: The key must be at least as long as the plaintext. The conclusion to be drawn from this result is that the length of the *secret key* needed to build a perfectly secure cryptographic system may be impractically large for most applications. Nevertheless, perfect security has a place in the practical picture: It may be used when the number of possible messages is small or in cases where the greatest importance is attached to perfect security.

A well-known, perfectly secure cipher is the *one-time pad*[7] (sometimes called the *Vernam cipher*), which is used for unconventional applications such as two users communicating on a hotline with high confidentiality requirements. The one-time pad is a stream cipher for which the key is the same as the keystream, as shown in Fig. A10.5. For encryption the input consists of two components: a message represented by a sequence of message bits $\{x_n | n = 1, 2, \ldots\}$, and a key represented by a sequence of statistically independent and uniformly distributed bits $\{z_n | n = 1, 2, \ldots\}$. The resultant cipher $\{y_n | n = 1, 2, \ldots\}$ is obtained by the modulo-2 addition of the two input sequences, as shown by

$$y_n = x_n \oplus z_n, \qquad n = 1, 2, \ldots$$

Consider, for example, the binary message sequence 00011010 and the binary key sequence 01101001. The modulo-2 addition of these two sequences is written as follows

Message:	00011010
Key:	01101001
Cipher:	01110011

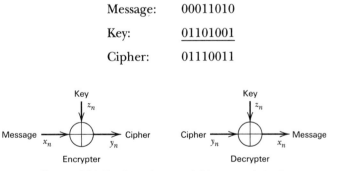

Figure A10.5   One-time pad (Vernam cipher).

In the encryption rule described here, key bit 1 interchanges 0s and 1s in the message sequence, and key bit 0 leaves the message bits unchanged. The message sequence is recovered simply by modulo-2 addition of the binary cipher and key sequences, as shown by

Cipher:	01110011
Key:	01101001
Message:	00011010

The one-time pad is perfectly secure, because the mutual information between the message and the cipher is zero; it is therefore completely undecipherable.

Unicity Distance

Consider now the practical case of an imperfect cipher and ask the question: When can an enemy cryptanalyst break the cipher? As the amount of intercepted text increases, intuitively we expect that a point may be reached at which it becomes possible for an enemy cryptanalyst with unlimited time and computing power to find the key and thus break the cipher. This critical point in the Shannon model is called the *unicity distance*, which is formally defined as the smallest N such that the conditional entropy $H(\mathbf{Z}|Y_1, Y_2, \ldots, Y_N)$ is approximately zero. For a particular kind of "random cipher," the unicity distance is approximately given by[8]

$$N_0 \simeq \frac{H(\mathbf{Z})}{r \log L_y} \tag{A10.11}$$

where $H(\mathbf{Z})$ is the entropy of the key $\mathbf{Z}$, and L_y is the size of the ciphertext alphabet. The parameter r is the *percentage redundancy* of the message information contained in the N-bit ciphertext; it is itself defined by

$$r = 1 - \frac{H(\mathbf{X})}{N \log L_y} \tag{A10.12}$$

where $H(\mathbf{X})$ is the entropy of the plaintext $\mathbf{X}$. In most cryptographic systems, the size L_y of the ciphertext alphabet is the same as the size L_x of the plaintext alphabet; in such a case, r is just the percentage redundancy of the plaintext itself. Although the derivation of Eq. (A10.11) assumes a certain well-defined "random cipher," it can be used to estimate the unicity distance for ordinary types of ciphers, which is the routine practice today.

Let K be the number of digits in the key $\mathbf{Z}$ that are chosen from an alphabet of size L_z; then, we may express the entropy of the key $\mathbf{Z}$ as follows:

$$H(\mathbf{Z}) \leq \log(L_z^K) = K \log L_z \tag{A10.13}$$

with equality if and only if the key is completely random. Let the size L_z of the key alphabet be the same as the size L_y of the ciphertext alphabet, and let the key be chosen completely at random to maximize the unicity distance. Then, substituting Eq. (A10.13) with equality in Eq. (A10.11), we get the simple result

$$N_0 \simeq \frac{K}{r} \qquad\qquad (A10.14)$$

To illustrate the application of Eq. (A10.14), consider a cryptographic system with $L_x = L_y = L_z$, which is used for the encryption of English text. The percentage redundancy r for typical English text is about 75 percent. Hence, according to Eq. (A10.14), an enemy cryptanalyst can break the cipher after intercepting only about 1.333K bits of ciphertext data, where K is the key size.

However, it is important to note that an imperfect cipher that is potentially breakable can still be of practical value. When the intercepted ciphertext contains sufficient information to satisfy Eq. (A10.11), there is no guarantee that an enemy cryptanalyst with limited computational resources can actually break the cipher. Specifically, it is possible for the cipher to be designed in such a way that the task of the cryptanalysis, though known to be attainable with a finite amount of computation, is so overwhelming that it will literally exhaust the physical computing resources of the universe. In such a case, the imperfect cipher is said to be *computationally secure.*

Role of Data Compression in Cryptography

Lossless data compression or *data compaction is a useful tool in cryptography.* We say this because data compaction removes redundancy, thereby increasing the unicity distance N_0 in accordance with Eq. (A10.11). To exploit this idea, data compaction is used prior to encryption in the transmitter, and the redundant information is reinserted after decryption in the receiver; the net result is that the authorized user at the receiver output sees no difference, and yet the information transmission has been made more secure. It would be tempting to consider the use of perfect data compaction to remove all redundancy, thereby transforming a message source into a completely random source and resulting in $N_0 = \infty$ with any key size. Unfortunately, we do not have a device capable of performing perfect data compaction on realistic message sources, nor is it likely that there will ever be such a device. It is therefore futile to rely on data compaction alone for data security. Nevertheless, limited data compaction tends to increase security, which is the reason why cryptographers view data compression as a useful trick.

Diffusion and Confusion

In the Shannon model of cryptography, two methods suggest themselves as general principles to guide the design of practical ciphers. The methods are called *diffusion* and *confusion,* the aims of which (by themselves or together) are to frustrate a statistical analysis of ciphertext by the enemy and therefore make it extremely difficult to break the cipher.

In the method of diffusion, the statistical structure of the plaintext is hidden by spreading out the influence of a single bit in the plaintext over a large number of bits in the ciphertext. This spreading has the effect of forcing the enemy to intercept a tremendous amount of material for the determination of the statistical structure of the plaintext, since the structure is evident only in many blocks, each one of which has a very small probability of occurrence. In the method of confusion, the data transformations are designed to complicate the determination of the way in which the statistics of the ciphertext depend on the statistics of the plaintext. Thus, a good cipher uses a combination of diffusion and confusion.

For a cipher to be of practical value, however, it must not only be difficult to break the cipher by an enemy cryptanalyst, but also it should be easy to encrypt and decrypt data given knowledge of the secret key. We may satisfy these two design objectives using a *product cipher*, based on the notion of "divide and conquer." Specifically, the implementation of a strong cipher is accomplished as a succession of simple component ciphers, each of which contributes a modest amount of diffusion and confusion to the overall makeup of the cipher. Product ciphers are often built using substitution ciphers and transposition ciphers as basic components; these simple ciphers are described here:

1. *Substitution cipher.* In a substitution cipher each letter of the plaintext is replaced by a fixed substitute, usually also a letter from the same alphabet, with the particular substitution rule being determined by the secret key. Thus the plaintext

$$\mathbf{X} = (x_1, x_2, x_3, x_4, \ldots)$$

where $x_1, x_2, x_3, \ldots$ are the successive letters, is transformed into the ciphertext

$$\begin{aligned} \mathbf{Y} &= (y_1, y_2, y_3, y_4, \ldots) \\ &= (f(x_1), f(x_2), f(x_3), f(x_4), \ldots) \end{aligned} \tag{A10.15}$$

where $f(\cdot)$ is a function with an inverse. When the substitutes are letters, the key is a permutation of the alphabet. Consider, for example, the ciphertext alphabet of Fig. A10.6, where we see that the first letter Y is the substitute for A, the second letter D is the substitute for B, and so on. The use of a substitution cipher results in confusion.

2. *Transposition cipher.* In a transposition cipher, the plaintext is divided into groups of fixed period d and the same permutation is applied to each group, with the particular permutation rule being determined by the secret key. For example, consider the permutation rule described in Fig. A10.7, for which the period is $d = 4$. According to this cipher, letter x_1 is moved from position 1 in the plaintext to position 4 in the ciphertext. Thus, the plaintext

$$\mathbf{X} = (x_1, x_2, x_3, x_4, x_5, x_6, x_7, x_8, \ldots)$$

is transformed into the ciphertext

$$\mathbf{Y} = (x_3, x_4, x_2, x_1, x_7, x_8, x_6, x_5, \ldots)$$

Although the single-letter statistics of the ciphertext $\mathbf{Y}$ are the same as those of the plaintext $\mathbf{X}$, the higher-order statistics are changed. The use of a transposition cipher results in diffusion.

By interleaving the simple substitutions and transpositions and repeating the interleaving process many times, it is possible to build a strong cipher equipped with good diffusion and confusion.

Plaintext
letters
ABCDEFGHIJKLMNOPQRSTUVWXYZ

Ciphertext
letters
YDUBHNACSVXELPFMKQJRWGOZIT

Figure A10.6 Substitution cipher

Plaintext
letters
x_1 x_2 x_3 x_4

Ciphertext
letters
x_3 x_4 x_2 x_1

Figure A10.7 Transposition cipher

EXAMPLE

Consider the plaintext message

THE KING IS DEAD LONG LIVE THE KING

Using the permuted alphabet described in Fig. A10.6 for the substitution cipher, this plaintext is transformed into the ciphertext

RCHXSPASJBHYBEFPAESGHRCHXSPA

Suppose next we apply the permutation rule described in Fig. A10.7 for the transposition cipher; accordingly, the ciphertext resulting from the substitution cipher is further transformed into

HXCRASPSHYBJFBEBSGEACHRHPASX

which has no resemblance to the original plaintext.

A10.4 THE DATA ENCRYPTION STANDARD

The *data encryption standard* (DES)[9] is certainly the best known, and arguably the most widely used, secret-key cryptoalgorithm; the term "algorithm" is used to describe a sequence of computations. The basic DES algorithm can be used for both data encryption and data authentication. It is the standard cryptoalgorithm for data storage and mail systems, electronic funds transfers (retail and whole-sale), and electronic business data interchange.

The DES algorithm is a strong block cipher that operates on 64-bit blocks of plaintext data and uses a 56-bit key; it is designed in accordance with Shannon's methods of diffusion and confusion. Essentially the same algorithm is used for encryption and decryption. The overall transformations employed in the DES algorithm may be written as $P^{-1}\{F[P(\mathbf{X})]\}$, where $\mathbf{X}$ is the plaintext, P is a certain permutation, and the function F combines substitutions and transpositions. The function F is itself obtained by cascading a certain function f, with each stage of the cascade referred to as a *round*.

The flow-chart of Fig. A10.8 shows the details of the DES algorithm for encryption. After a certain initial permutation, a plaintext of 64 bits is divided into a left-half L_0 and a right-half R_0, each of which is 32 bits long. The algorithm then performs 16 rounds of a key-dependent computation, with the ith round of the computation described as follows:

$$L_i = R_{i-1} \qquad i = 1, 2, \ldots, 16 \qquad \text{(A10.16)}$$

$$R_i = L_{i-1} \oplus f(R_{i-1}, Z_i) \qquad i = 1, 2, \ldots, 16 \qquad \text{(A10.17)}$$

On the right-hand side of Eq. (A10.17), the addition is modulo 2 and each Z_i is a different 48-bit block of the key used in round i. The function $f(.\,,.)$ is a function with a 32-bit output. The result of the 16th round is reversed, obtaining the sequence $R_{16}L_{16}$. This 32-bit sequence is input into a final permutation P^{-1} to produce the 64-bit ciphertext. The aim is that after 16 rounds of key-dependent computations, the patterns in the original plaintext are undetectable in the ciphertext. From Eqs. (A10.16) and (A10.17), we note that for decryption the function $f(.\,,.)$ need not be invertible, because (L_{i-1}, R_{i-1}) can be recovered from (L_i, R_i) simply as follows:

$$R_{i-1} = L_i \qquad i = 1, 2, \ldots, 16 \qquad \text{(A10.18)}$$

$$L_{i-1} = R_i \oplus f(L_i, Z_i) \qquad i = 1, 2, \ldots, 16 \qquad \text{(A10.19)}$$

Equation (A10.19) holds even if the function $f(.\,,.)$ is a many-to-one function (i.e., it does not have a unique inverse).

Figure A10.9 shows the flowchart for computing the function $f(.\,,.)$. The 32-bit block R is first expanded into a new 48-bit block R' by repeating the edge bits of each successive 4-bit byte (i.e., the bits numbered 1, 4, 5, 9, 12, 15, ..., 28, 32). Thus, given the 32-bit block R written as

$$R = \underbrace{r_1 r_2 r_3 r_4}_{\substack{\text{first} \\ \text{4-bit byte}}} \quad \underbrace{r_5 r_6 r_7 r_8}_{\substack{\text{second} \\ \text{4-bit byte}}} \quad \cdots \quad \underbrace{r_{29} r_{30} r_{31} r_{32}}_{\substack{\text{eighth} \\ \text{4-bit byte}}}$$

we construct the expanded 48-bit block R' as follows:

$$R = \underbrace{r_{32} r_1 r_2 r_3 r_4 r_5}_{\substack{\text{first} \\ \text{6-bit byte}}} \quad \underbrace{r_4 r_5 r_6 r_7 r_8 r_9}_{\substack{\text{second} \\ \text{6-bit byte}}} \quad \cdots \quad \underbrace{r_{28} r_{29} r_{30} r_{31} r_{32} r_1}_{\substack{\text{eighth} \\ \text{6-bit byte}}}$$

The 48-bit blocks R' and Z_i are added modulo 2, and the resultant is divided into eight 6-bit bytes. Let these bytes be denoted by $B_1, B_2, \ldots, B_8$. We thus write

$$B_1 B_2 \cdots B_8 = R' \oplus Z_i \qquad \text{(A10.20)}$$

Each 6-bit byte B_i is input to a substitution box S_i in the form of a look-up table, producing a 4-bit output $S_i(B_i)$. Each output bit of the substitution box $S_i(B_i)$ is a Boolean function of the 6-bit byte B_i. The eight outputs $S_1(B_1)$, $S_2(B_2)$, ...,

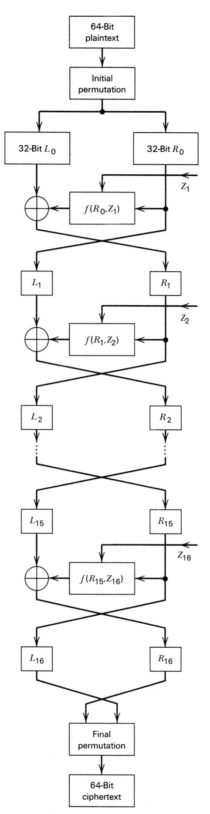

Figure A10.8 Data encryption standard. (From Diffie and Hellman, 1979, with permission of the IEEE.)

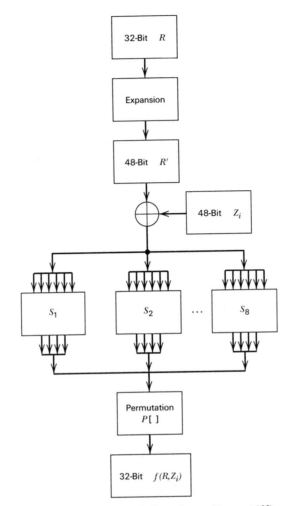

Figure A10.9 $f(R, K)$ flowchart. (From Diffie and Hellman, 1979, with permission of the IEEE.)

$S_8(B_8)$ are arranged into a single 32-bit block that is input to the permutation box denoted by $P[\cdot]$. The permuted output so produced is the desired 32-bit function $f(R, Z_i)$, as shown by

$$f(R, Z_i) = P[S_1(B_1)S_2(B_2) \cdots S_8(B_8)] \qquad (A10.21)$$

The 48-bit block Z_i for the ith iteration uses a different subset of the 64-bit key Z_0. The procedure used to determine each Z_i is called the *key-schedule calculation*, the flowchart of which is shown in Fig. A10.10. The key Z_0 has eight parity bits in positions 8, 16, . . . , 64, which are used for error detection in their respective 8-bit bytes; the errors of concern may arise in the generation, distribution, and storage of the key Z_0. The permuted choice 1 disregards the parity bits of Z_0 and then permutes the remaining 56 bits that are loaded into two 28-bit shift registers, each with 24 taps. The 48 taps of the two shift registers are subjected to 16 iterations of computation, with each iteration involving one or two cyclic left shifts followed by a permutation, referred to as permuted choice 2.

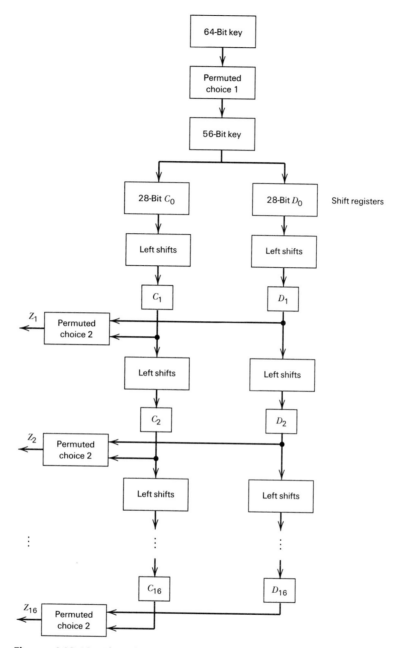

Figure A10.10 Flowchart for the key-schedule calculation. (From Diffie and Hellman, 1979, with permission of the IEEE.)

The outputs resulting from these 16 iterations provide the different 48-bit blocks $Z_1, Z_2, \ldots, Z_{16}$ of the key used in iteration $1, 2, \ldots, 16$, respectively.

Despite all the claims to the contrary, it appears that no one has yet demonstrated a fundamental weakness of the DES algorithm. Notwithstanding all the controversy surrounding its use, perhaps the most significant contribution of the DES algorithm is the fact that it has been instrumental in raising the level of interest in using cryptography as a mechanism for secure computer networks.

A10.5 PUBLIC-KEY CRYPTOGRAPHY[10]

For a pair of users to engage in cryptographic communication over an insecure channel, it is necessary for the users to exchange key information prior to communication. The requirement for a secure distribution of keys among authorized users applies to all cryptographic systems, regardless of their type. In conventional cryptography, the users employ a physically secure channel (e.g., courier service or registered mail) for key distribution. However, the use of such a supplementary channel points to a major limitation of conventional cryptography. Needless to say, the use of courier service or registered mail for key distribution is costly, inconvenient, low-bandwidth, and slow; also, it is not always secure.

The problem of key distribution is particularly accentuated in large communication networks, where the number of possible connections grows as $(n^2 - n)/2$ for n users. For large n, the cost of key distribution becomes prohibitive. Thus, in the development of large, secure communication networks, we are compelled to rely on the use of insecure channels for both exchange of key information and subsequent secure communication. This constraint raises a fundamental question: How can key information be exchanged securely over an insecure channel? In *public-key cryptography*, this seemingly difficult issue is resolved by making some key material "public" and thereby considerably simplifying the task of key management. This is in direct contrast to conventional cryptography, where the key is kept completely secret from an enemy cryptanalyst.

A public key cryptographic system is described by two sets of algorithms that compute invertible functions (transformations). Let these two sets of algorithms be denoted by $\{E_z\}$ and $\{D_z\}$ that are indexed by z. The invertible transformations computed by these algorithms may be written as follows

$$E_z: \quad f_z(x) = y \tag{A10.22}$$

$$D_z: \quad f_z^{-1}(y) = x \tag{A10.23}$$

where x is a certain input message in the domain of some function f_z indexed by z, and y is the corresponding cryptogram in the range of f_z. A fundamental requirement of the system is that the function f_z must be a *trapdoor one-way function*. The term "one-way" refers to the fact that for x in the domain of f_z, it must be easy to compute $f_z(x)$ from knowledge of the algorithm E_z, but for a certain cryptogram y in the range of f_z, an enemy cryptanalyst must find it extremely difficult to compute the inverse $f_z^{-1}(y)$. On the other hand, an authorized user in possession of the associated algorithm D_z would find it easy to compute the inverse $f_z^{-1}(y)$. Thus the *private key* (algorithm) D_z provides a "trapdoor" that makes the problem of inverting the function f_z appear extremely difficult from the viewpoint of the cryptanalyst, but easy for the (sole authorized) possessor of D_z. Since knowledge of the key (algorithm) E_z does not by itself make it possible to compute the inverse of f_z, it may be made *public*; hence, the name "public-key cryptography."

The notion emerging from the description of a public-key cryptographic system presented herein is that the keys come in inverse pairs (i.e., public key and private key), and that each pair of keys has two basic properties:

1. *Whatever message is encrypted with one of the keys can be decrypted with the other key.*
2. *Given knowledge of the public key, it is computationally infeasible to find the secret key.*

The use of public-key cryptography as described herein makes it possible to solve the secrecy problem as follows. Subscribers to a secure communication system list their public keys in a "telephone directory" along with their names and addresses. A subscriber can then send a private message to another subscriber simply by looking up the public key of the addressee and using the key to encrypt the message. The encrypted message (i.e., ciphertext) can only be read by the holder of that particular public key. In fact, should the original message (i.e., plaintext) be lost, even its sender would find it extremely difficult to recover the message from the ciphertext.

The key management of public-key cryptography makes it well suited for the development of large, secure communication networks. Indeed, it has evolved from a simple concept to a mainstay of cryptographic technology. It is being considered for implementation in hundreds of thousands of secure telephones and secure data communication systems.

Diffie-Hellman Public Key Distribution

In a simple and yet elegant system known as the *Diffie-Hellman public key-distribution system*, use is made of the fact that it is easy to calculate a discrete exponential but difficult to calculate a discrete logarithm. To be more specific, consider the *discrete exponential function*

$$Y = \alpha^X \bmod p \qquad \text{for } 1 \leq X \leq p - 1 \tag{A10.24}$$

where the arithmetic is performed modulo p. The α is an integer that should be *primitive* (i.e., all powers of α generate all the elements mod p relatively prime to $p - 1$). Correspondingly, X is referred to as the *discrete logarithm* of Y to the base α, mod p, as shown by

$$X = \log_\alpha Y \bmod p \qquad \text{for } 1 \leq Y \leq p - 1 \tag{A10.25}$$

The calculation of Y from X is easy, using the trick of square-and-multiply. For example, for $X = 16$ we have

$$Y = \alpha^{16} = (((\alpha^2)^2)^2)^2$$

On the other hand, the problem of calculating X from Y is much more difficult.

In the Diffie-Hellman public key-distribution system, all users are presumed to know both α and p. A user i, say, selects an independent random number X_i uniformly from the set of integers $\{1, 2, \ldots, p\}$ that is kept as a *private secret*. But the discrete exponential

$$Y_i = \alpha^{X_i} \bmod p \tag{A10.26}$$

is deposited in a *public directory* with the user's name and address. Every other user of the system does the same thing. Now, suppose that users i and j wish to communicate privately. To proceed, user i fetches Y_j from the public directory and uses the private secret X_i to compute

$$
\begin{aligned}
K_{ji} &= (Y_j)^{X_i} \bmod p \\
&= (\alpha^{X_j})^{X_i} \bmod p \\
&= \alpha^{X_j X_i} \bmod p
\end{aligned}
\tag{A10.27}
$$

In a similar way, user j computes K_{ij}. But we have

$$
K_{ji} = K_{ij}
\tag{A10.28}
$$

Accordingly, users i and j arrive at K_{ji} as the *secret key* in a conventional crypto-system. Another user must compute K_{ji} using the information Y_i and Y_j obtained from the public directory, applying the alternative formula

$$
K_{ji} = (Y_j)^{\log_e Y_i} \bmod p
\tag{A10.29}
$$

Apparently, there is no other method for an enemy to find the secret key K_{ji}; however, there is no proof for it. In light of what we said earlier, Eq. (A10.29) is difficult to calculate as it involves a discrete logarithm, whereas Eq. (A10.27) is easy to calculate as it involves a discrete exponential. Thus, security of the system depends on the difficulty encountered in computing a discrete logarithm.

The Diffie-Hellman public key-distribution system is the oldest system in its class; nevertheless, it is still generally considered to be one of the most secure and practical public key-distribution systems.

A10.6 THE RIVEST–SHAMIR–ADLEMAN SYSTEM

To develop a public-key cryptographic system is no easy task. Indeed, numerous such systems have been proposed in the literature, but unfortunately most of them have proven to be insecure. To date, the most successful implementation of public-key cryptography is the *Rivest–Shamir–Adleman (RSA) system*,[11] which uses ideas from classical number theory. It is considered to be one of the most secure cryptographic systems in that it has withstood many attempts by experts in the field to break it.

The RSA algorithm is a block cipher based on the fact that finding a random prime number of large size (e.g., 100 digit) is computationally easy, but factoring the product of two such numbers is currently considered computationally in-feasible. Specifically, the computation of parameters specific to the RSA algo-rithm proceeds as follows:

1. Choose two *very large prime numbers*, p and q, at random; the prime numbers have to be fairly carefully chosen as some prime numbers lead to a very weak system.
2. Multiply the numbers p and q, obtaining the product

$$
pq = n
\tag{A10.30}
$$

Find the *Euler totient function* of n, using the formula

$$
\phi(n) = (p - 1)(q - 1)
\tag{A10.31}
$$

Equation (A10.31) follows from the definition of the Euler totient function $\phi(n)$ as the number of positive integers i less than n, such that the *greatest common divisor* of i and n is equal to one.

3. Let e be a positive integer less than $\phi(n)$, such that the greatest common divisor of e and $\phi(n)$ is equal to one. Hence, find a positive integer d less than $\phi(n)$, such that

$$de = 1 \bmod \phi(n) \tag{A10.32}$$

The RSA trapdoor one-way function is then defined simply by computing the *discrete exponentiation*

$$f_z(x) = x^e = y \bmod n \tag{A10.33}$$

The values of n and e constitute the public key; hence, publishing the easy-to-find algorithm E_z to compute the function f_z amounts just to publishing the numbers n and e.

The prime numbers p and q constitute the private key. Since d is related to p and q, possession of the easy-to-find (when one knows the trapdoor z) algorithm D_z to compute the inverse function f_z^{-1} amounts just to knowing p and q. In particular, the inverse function is defined by

$$f_z^{-1}(y) = y^d \bmod n \tag{A10.34}$$

The decrypting exponent d is found using Eq. (A10.32), which is equivalent to the statement (in ordinary integer arithmetic) that

$$de = \phi(n)Q + 1 \tag{A10.35}$$

for some integer Q. Note that $\phi(n)$ is itself related to p and q by Eq. (A10.31). Since $y = x^e$, we may use Eqs. (A10.32) and (A10.33) to write

$$
\begin{aligned}
y^d &= x^{de} \\
&= x^{\phi(n)Q+1} \\
&= \left(\left(x^{\phi(n)} \right)^Q \right) x
\end{aligned}
\tag{A10.36}
$$

We now make use of a celebrated theorem of Euler, which states that for any positive integers x and n with $x < n$, we have

$$x^{\phi(n)} = 1 \bmod n \tag{A10.37}$$

Hence, the use of Eq. (A10.37) in (A10.36) yields the desired decryption:

$$y^d = x \tag{A10.38}$$

We thus see that finding the inverse function f_z^{-1} is easy, given knowledge of the prime numbers p and q.

The *security* of the RSA cryptoalgorithm rests on the premise that any method of inverting the function f_z is *equivalent to factoring* $n = pq$. This equivalence raises the question: Is an attack by factoring n computationally feasible? It appears that the answer is no, provided that the prime numbers p and q are on the order of 100 decimal digits each and that there is no revolutionary breakthrough in factoring algorithms. For example, today it takes about one day on a supercomputer to factor a number of about 80 decimal digits. To factor a number of 200 decimal digits, it would take 10^8 times that long, which is roughly half a million years!

Digital Signatures[12]

For an electronic mail system to replace the use of ordinary paper mail for business transactions, it must be possible for a user of the system to "sign" an electronic message. The use of a *digital signature* provides proof that the message originated from the sender. To satisfy this requirement, the digital signature must have the following properties:

- The receiver of an electronic message is able to verify the sender's signature.
- The signature is not forgeable.
- The sender of a signed electronic message is unable to disclaim it.

To implement digital signatures using the RSA algorithm, we may proceed as follows. A user in possession of the private key d may sign a given message block m by forming the signature

$$s = m^d \bmod n \qquad (A10.39)$$

It is difficult to compute s unless the private key d is known. Hence, a digital signature defined in accordance with Eq. (A10.39) is difficult to forge. Moreover, the sender of message m cannot deny having sent it, since no one else could have created the signature s. The receiver proceeds by using the public key e to compute

$$s^e = (m^d)^e \bmod n$$
$$= m^{de} \bmod n \qquad (A10.40)$$
$$= m \bmod n$$

where, in the last line, use is made of Eq. (A10.32). Hence, the receiver is able to validate the sender's signature by establishing that the computation of $s^e \bmod n$ produces the same result as the deciphered message m. Thus, the RSA algorithm satisfies all three properties of a digital signature that were described earlier.

A10.7 SUMMARY AND DISCUSSION

Cryptography is a "hot" research area. This statement should not come as a surprise, considering the fact that as we move toward an *information society*, the importance of cryptography as a security mechanism will continue to grow. In this appendix, we have presented an introductory treatment of this highly important subject.

We may classify cryptography into secret-key cryptography and public-key cryptography, depending on whether the key used for the encryption of a message and its decryption is completely secret or partly public. Alternatively, we may classify a cryptographic system into a block cipher or stream cipher, depending on the method of implementation. A block cipher exhibits error propagation, which can prove highly valuable in authentication.

Among the many cryptographic systems developed to date, the data encryption standard (DES) and the Rivest–Shamir–Adleman (RSA) algorithms stand out as the most successful ones. Both of these cryptoalgorithms are block ciphers. They differ from each other in that the DES algorithm involves the use of a secret key whereas the RSA algorithm involves the use of a public key. In a secret-key system, the same key is shared both by the sender and the receiver. On the other hand, in a public-key system, the key is split into two parts: a public key located in the transmitter and a secret key located in the receiver; in the latter system, it is computationally infeasible to recover the plaintext message from its encrypted version without knowledge of the private key.

Although public-key cryptosystems such as RSA provide an effective method for key management, they are inefficient for the bulk encryption of data due to low bandwidths. In contrast, conventional cryptosystems such as DES provide better throughput, but they require key management. This suggests the use of a hybrid approach exploiting the best elements of both cryptosystems as the basis for the practical design of a secure communication system. For example, the RSA algorithm may be used for authentication, and the DES algorithm for encryption.

NOTES AND REFERENCES

1. For an introductory treatment of cryptography, see Chapter 15 of the book by Adámek (1991). For a comprehensive treatment of the many facets of cryptology, see the book edited by Simmons (1992); this book is an expanded edition of a Special Issue of the *Proceedings of the IEEE* (1988) on cryptology. The chapter contributions of the book by Simmons are written by leading authorities on the subject of cryptology. A nice treatment of cryptology is also presented in the book by van Tilborg (1988).

2. The era of scientific secret-key cryptography was ushered in with the publication of a landmark paper by Shannon (1949), which established the connection between cryptography and information theory.

3. The era of public-key cryptography was established with the publication of another landmark paper by Diffie and Hellman (1976), which showed for the first time that it is possible to have secret communications without any transfer of a key between sender and receiver. It was the paper by Diffie and Hellman that sparked the explosion of research interest in cryptology, which has continued ever since. The idea of a public key is patented by Stanford University.

4. The term "enemy cryptanalyst" is commonly used in cryptology to refer to a cryptogram interceptor (eavesdropper); its usage originates from military applications.

5. For a comprehensive treatment of stream ciphers, see Chapter 2 written by R. A. Rueppel in the book *Contemporary Cryptology*, edited by Simmons (1992).

6. For a highly readable account of the Shannon model of cryptography, see the opening chapter by J. L. Massey in the book edited by Simmons (1992).

7. The one-time pad derives its name from its use (shortly before, during, and after World War II) by spies of several governments, who were given a pad of paper with a randomly chosen key and told to use it only for a single encryption. The one-time pad is also known as Vernam's cipher, so named in recognition of its originator, G. S. Vernam.

8. For a complete derivation of Eq. (A10.11), see the original paper by Shannon (1949).

9. The history of the DES algorithm is recounted by M. E. Smid and D. K. Branstad in Chapter 1 of the book edited by Simmons (1992). For a description of the DES algorithm, see Diffie and Hellman (1979). See also the books by Meyer and Matyas (1982) and Torrieri (1992, Chapter 6).

10. For a comprehensive treatment of public-key cryptography, see Chapter 4 by J. Nechvatal in the book edited by Simmons (1992). This book also includes a chapter contribution by W. Diffie that describes the several attempts to devise secure public-key cryptoalgorithms and the gradual evolution of a variety of protocols based on them.

11. The RSA system is patented; it is named in recognition of its originators R. L. Rivest, A. Shamir, and L. Adleman. The original reference for this cryptosystem is Rivest, Shamir, and Adleman (1978).

12. The idea of a digital signature was first discussed by Diffie and Hellman (1976). Its implementation using the RSA algorithm is described by Rivest, Shamir, and Adleman (1978). For a detailed treatment of digital signatures, see Chapter 6 by C. J. Michell, F. Piper, and R. Wild in the book edited by Simmons (1992).

Mathematical Tables

This appendix presents (1) a summary of properties of the Fourier transform, (2) a short table of Fourier-transform pairs, (3) a short table of Hilbert-transform pairs, (4) a list of trigonometric identities, (5) a selected list of series expansions, (6) a selected list of integrals, (7) a list of useful constants, (8) recommended unit prefixes, and (9) the ASCII code.

Table A11.1 Summary of Properties of the Fourier Transform

Property	Mathematical Description
1. Linearity	$ag_1(t) + bg_2(t) \rightleftharpoons aG_1(f) + bG_2(f)$ where a and b are constants
2. Time scaling	$g(at) \rightleftharpoons \dfrac{1}{\|a\|}G\left(\dfrac{f}{a}\right)$ where a is a constant
3. Duality	If $\quad g(t) \rightleftharpoons G(f),$ then $\quad G(t) \rightleftharpoons g(-f)$
4. Time shifting	$g(t - t_0) \rightleftharpoons G(f)\exp(-j2\pi f t_0)$
5. Frequency shifting	$\exp(j2\pi f_c t)g(t) \rightleftharpoons G(f - f_c)$
6. Area under $g(t)$	$\displaystyle\int_{-\infty}^{\infty} g(t)\ dt = G(0)$
7. Area under $G(f)$	$g(0) = \displaystyle\int_{-\infty}^{\infty} G(f)\ df$
8. Differentiation in the time domain	$\dfrac{d}{dt}g(t) \rightleftharpoons j2\pi f G(f)$
9. Integration in the time domain	$\displaystyle\int_{-\infty}^{t} g(\tau)\ d\tau \rightleftharpoons \dfrac{1}{j2\pi f}G(f) + \dfrac{G(0)}{2}\delta(f)$
10. Conjugate functions	If $\quad g(t) \rightleftharpoons G(f),$ then $\quad g^*(t) \rightleftharpoons G^*(-f)$
11. Multiplication in the time domain	$g_1(t)g_2(t) \rightleftharpoons \displaystyle\int_{-\infty}^{\infty} G_1(\lambda)G_2(f - \lambda)\ d\lambda$
12. Convolution in the time domain	$\displaystyle\int_{-\infty}^{\infty} g_1(\tau)g_2(t - \tau)\ d\tau \rightleftharpoons G_1(f)G_2(f)$

Table A11.2 Fourier-Transform Pairs

Time Function	Fourier Transform
$\mathrm{rect}\left(\dfrac{t}{T}\right)$	$T\,\mathrm{sinc}(fT)$
$\mathrm{sinc}(2Wt)$	$\dfrac{1}{2W}\mathrm{rect}\left(\dfrac{f}{2W}\right)$
$\exp(-at)u(t), \quad a>0$	$\dfrac{1}{a+j2\pi f}$
$\exp(-a\lvert t\rvert), \qquad a>0$	$\dfrac{2a}{a^2+(2\pi f)^2}$
$\exp(-\pi t^2)$	$\exp(-\pi f^2)$
$\begin{cases} 1-\dfrac{\lvert t\rvert}{T}, & \lvert t\rvert < T \\ 0, & \lvert t\rvert \ge T \end{cases}$	$T\,\mathrm{sinc}^2(fT)$
$\delta(t)$	1
1	$\delta(f)$
$\delta(t-t_0)$	$\exp(-j2\pi f t_0)$
$\exp(j2\pi f_c t)$	$\delta(f-f_c)$
$\cos(2\pi f_c t)$	$\frac{1}{2}[\delta(f-f_c)+\delta(f+f_c)]$
$\sin(2\pi f_c t)$	$\dfrac{1}{2j}[\delta(f-f_c)-\delta(f+f_c)]$
$\mathrm{sgn}(t)$	$\dfrac{1}{j\pi f}$
$\dfrac{1}{\pi t}$	$-j\,\mathrm{sgn}(f)$
$u(t)$	$\dfrac{1}{2}\delta(f)+\dfrac{1}{j2\pi f}$
$\displaystyle\sum_{i=-\infty}^{\infty}\delta(t-iT_0)$	$\dfrac{1}{T_0}\displaystyle\sum_{n=-\infty}^{\infty}\delta\left(f-\dfrac{n}{T_0}\right)$

Notes: $u(t)$ = unit step function
$\delta(t)$ = Dirac delta function
$\mathrm{rect}(t)$ = rectangular function
$\mathrm{sgn}(t)$ = signum function
$\mathrm{sinc}(t)$ = sinc function

Table A11.3 Hilbert-Transform Pairs[a]

Time Function	Hilbert Transform
$m(t)\cos(2\pi f_c t)$	$m(t)\sin(2\pi f_c t)$
$m(t)\sin(2\pi f_c t)$	$-m(t)\cos(2\pi f_c t)$
$\cos(2\pi f_c t)$	$\sin(2\pi f_c t)$
$\sin(2\pi f_c t)$	$-\cos(2\pi f_c t)$
$\dfrac{\sin t}{t}$	$\dfrac{1-\cos t}{t}$
$\operatorname{rect}(t)$	$-\dfrac{1}{\pi}\ln\left|\dfrac{t-\tfrac{1}{2}}{t+\tfrac{1}{2}}\right|$
$\delta(t)$	$\dfrac{1}{\pi t}$
$\dfrac{1}{1+t^2}$	$\dfrac{t}{1+t^2}$
$\dfrac{1}{t}$	$-\pi\,\delta(t)$

[a]In the first two pairs, it is assumed that $m(t)$ is band limited to the interval $-W \le f \le W$, where $W < f_c$.

Notes: $\delta(t)$: Dirac delta function
   $\operatorname{rect}(t)$: rectangular function
   ln: natural logarithm

Table A11.4 Trigonometric Identities

$$\exp(\pm j\theta) = \cos\theta \pm j\sin\theta$$

$$\cos\theta = \tfrac{1}{2}[\exp(j\theta) + \exp(-j\theta)]$$

$$\sin\theta = \frac{1}{2j}[\exp(j\theta) - \exp(-j\theta)]$$

$$\sin^2\theta + \cos^2\theta = 1$$

$$\cos^2\theta - \sin^2\theta = \cos(2\theta)$$

$$\cos^2\theta = \tfrac{1}{2}[1 + \cos(2\theta)]$$

$$\sin^2\theta = \tfrac{1}{2}[1 - \cos(2\theta)]$$

$$2\sin\theta\cos\theta = \sin(2\theta)$$

$$\sin(\alpha \pm \beta) = \sin\alpha\cos\beta \pm \cos\alpha\sin\beta$$

$$\cos(\alpha \pm \beta) = \cos\alpha\cos\beta \mp \sin\alpha\sin\beta$$

$$\tan(\alpha \pm \beta) = \frac{\tan\alpha \pm \tan\beta}{1 \mp \tan\alpha\tan\beta}$$

$$\sin\alpha\sin\beta = \tfrac{1}{2}[\cos(\alpha - \beta) - \cos(\alpha + \beta)]$$

$$\cos\alpha\cos\beta = \tfrac{1}{2}[\cos(\alpha - \beta) + \cos(\alpha + \beta)]$$

$$\sin\alpha\cos\beta = \tfrac{1}{2}[\sin(\alpha - \beta) + \sin(\alpha + \beta)]$$

Table A11.5 Series Expansions

Taylor series

$$f(x) = f(a) + \frac{f'(a)}{1!}(x - a) + \frac{f''(a)}{2!}(x - a)^2 + \cdots + \frac{f^{(n)}(a)}{n!}(x - a)^n + \cdots$$

where

$$f^{(n)}(a) = \frac{d^n f(x)}{dx^n}\bigg|_{x=a}$$

MacLaurin series

$$f(x) = f(0) + \frac{f'(0)}{1!}x + \frac{f''(0)}{2!}x^2 + \cdots + \frac{f^{(n)}(0)}{n!}x^n + \cdots$$

where

$$f^{(n)}(0) = \frac{d^n f(x)}{dx^n}\bigg|_{x=0}$$

Binomial series

$$(1 + x)^n = 1 + nx + \frac{n(n - 1)}{2!}x^2 + \cdots, \qquad |nx| < 1$$

Exponential series

$$\exp x = 1 + x + \frac{1}{2!}x^2 + \cdots$$

Logarithmic series

$$\ln(1 + x) = x - \tfrac{1}{2}x^2 + \tfrac{1}{3}x^3 - \cdots$$

Trigonometric series

$$\sin x = x - \frac{1}{3!}x^3 + \frac{1}{5!}x^5 - \cdots$$

$$\cos x = 1 - \frac{1}{2!}x^2 + \frac{1}{4!}x^4 - \cdots$$

$$\tan x = x + \frac{1}{3}x^3 + \frac{2}{15}x^5 + \cdots$$

$$\sin^{-1}x = x + \frac{1}{6}x^3 + \frac{3}{40}x^5 + \cdots$$

$$\tan^{-1}x = x - \frac{1}{3}x^3 + \frac{1}{5}x^5 - \cdots, \qquad |x| < 1$$

$$\mathrm{sinc}\,x = 1 - \frac{1}{3!}(\pi x)^2 + \frac{1}{5!}(\pi x)^4 - \cdots$$

Table A11.6   Integrals

Indefinite integrals

$$\int x \sin(ax) \ dx = \frac{1}{a^2}[\sin(ax) - ax\cos(ax)]$$

$$\int x \cos(ax) \ dx = \frac{1}{a^2}[\cos(ax) + ax\sin(ax)]$$

$$\int x \exp(ax) \ dx = \frac{1}{a^2}\exp(ax)(ax - 1)$$

$$\int x \exp(ax^2) \ dx = \frac{1}{2a}\exp(ax^2)$$

$$\int \exp(ax)\sin(bx) \ dx = \frac{1}{a^2 + b^2}\exp(ax)[a\sin(bx) - b\cos(bx)]$$

$$\int \exp(ax)\cos(bx) \ dx = \frac{1}{a^2 + b^2}\exp(ax)[a\cos(bx) + b\sin(bx)]$$

$$\int \frac{dx}{a^2 + b^2 x^2} = \frac{1}{ab}\tan^{-1}\left(\frac{bx}{a}\right)$$

$$\int \frac{x^2 \ dx}{a^2 + b^2 x^2} = \frac{x}{b^2} - \frac{a}{b^3}\tan^{-1}\left(\frac{bx}{a}\right)$$

Definite integrals

$$\int_0^\infty \frac{x \sin(ax)}{b^2 + x^2} \ dx = \frac{\pi}{2}\exp(-ab), \qquad a > 0, \ b > 0$$

$$\int_0^\infty \frac{\cos(ax)}{b^2 + x^2} \ dx = \frac{\pi}{2b}\exp(-ab), \qquad a > 0, \ b > 0$$

$$\int_0^\infty \frac{\cos(ax)}{(b^2 - x^2)^2} \ dx = \frac{\pi}{4b^3}[\sin(ab) - ab\cos(ab)], \qquad a > 0, b > 0$$

$$\int_0^\infty \mathrm{sinc}\ x \ dx = \int_0^\infty \mathrm{sinc}^2\ x \ dx = \frac{1}{2}$$

$$\int_0^\infty \exp(-ax^2) \ dx = \frac{1}{2}\sqrt{\frac{\pi}{a}}, \qquad a > 0$$

$$\int_0^\infty x^2 \exp(-ax^2) \ dx = \frac{1}{4a}\sqrt{\frac{\pi}{a}}, \qquad a > 0$$

Table A11.7   Useful Constants

Physical Constants

Boltzmann's constant	$k = 1.38 \times 10^{-23}$ joule/degree Kelvin
Planck's constant	$h = 6.626 \times 10^{-34}$ joule-second
Electron (fundamental) charge	$q = 1.602 \times 10^{-19}$ coulomb
Speed of light in vacuum	$c = 2.998 \times 10^8$ meters/second
Standard (absolute) temperature	$T_0 = 273$ degree Kelvin
Thermal voltage	$V_T = 0.026$ volt at room temperature
Thermal energy kT at standard temperature	$kT_0 = 3.77 \times 10^{-21}$ joule

One Hertz (Hz) = 1 cycle/second; 1 cycle = 2π radians

One watt (W) = 1 joule/second

Mathematical Constants

Base of natural logarithm	$e = 2.7182818$
Logarithm of e to base 2	$\log_2 e = 1.442695$
Logarithm of 2 to base e	$\ln 2 = 0.693147$
Logarithm of 2 to base 10	$\log_{10} 2 = 0.30103$
Pi	$\pi = 3.1415927$

Table A11.8   Recommended Unit Prefixes

Multiples and Submultiples	Prefixes	Symbols
10^{12}	tera	T
10^9	giga	G
10^6	mega	M
10^3	kilo	K (k)
10^{-3}	milli	m
10^{-6}	micro	μ
10^{-9}	nano	n
10^{-12}	pico	p

Table A11.9 ASCII Code

				7 0	0	0	0	1	1	1	1
				6 0	0	1	1	0	0	1	1
4	3	2	1	5 0	1	0	1	0	1	0	1
0	0	0	0	NUL	DLE	SP	0	@	P	\	p
0	0	0	1	SOH	DC1	!	1	A	Q	a	q
0	0	1	0	STX	DC2	"	2	B	R	b	r
0	0	1	1	ETX	DC3	#	3	C	S	c	s
0	1	0	0	EOT	DC4	$	4	D	T	d	t
0	1	0	1	ENQ	NAK	%	5	E	U	e	u
0	1	1	0	ACK	SYN	&	6	F	V	f	v
0	1	1	1	BEL	ETB	'	7	G	W	g	w
1	0	0	0	BS	CAN	(	8	H	X	h	x
1	0	0	1	HT	EM	)	9	I	Y	i	y
1	0	1	0	LF	SUB	*	:	J	Z	j	z
1	0	1	1	VT	ESC	+	;	K	[	k	{
1	1	0	0	FF	FS	'	<	L	\	l	:
1	1	0	1	CR	GS	−	=	M	]	m	}
1	1	1	0	SO	RS	.	>	N	∧	n	~
1	1	1	1	SI	US	/	?	O	—	o	DEL

ACK	Acknowledge	ENQ	Enquiry	NUL	Null or all zeros	
BEL	Bell or alarm	EOT	End of transmission	RS	Record separator	
BS	Backspace	ESC	Escape	SI	Shift in	
CAN	Cancel	ETB	End of transmission block	SO	Shift out	
CR	Carriage return	ETX	End of text	SOH	Start of heading	
DC1	Device control 1	FF	Form feed	SP	Space	
DC2	Device control 2	FS	File separator	STX	Start of text	
DC3	Device control 3	GS	Group separator	SUB	Substitute	
DC4	Device control 4	HT	Horizontal tab	SYN	Synchronous idle	
DEL	Delete	LF	Line feed	US	Unit separator	
DLE	Data link escape	NAK	Negative acknowledge	VT	Vertical tab	
EM	End of medium					

(From Couch, 1990, with permission of Macmillan.)

Glossary

CONVENTIONS AND NOTATIONS

1. The symbol $|\ |$ means the magnitude of the complex quantity contained within.

2. The symbol $\arg(\ )$ means the phase angle of the complex quantity contained within.

3. The symbol $\operatorname{Re}[\ ]$ means the "real part of" and $\operatorname{Im}[\ ]$ means the "imaginary part of."

4. The symbol $\ln(\ )$ denotes the natural logarithm of the quantity contained within, whereas the logarithm to base a is denoted by $\log_a(\ )$.

5. The use of an asterisk as superscript denotes complex conjugate, e.g., x^* is the complex conjugate of x.

6. The symbol $\rightleftharpoons$ indicates a Fourier-transform pair, e.g., $g(t) \rightleftharpoons G(f)$, where a lowercase letter denotes the time function and a corresponding uppercase letter denotes the frequency function.

7. The symbol $F[\ ]$ indicates the Fourier-transform operation, e.g., $F[g(t)] = G(f)$, and the symbol $F^{-1}[\ ]$ indicates the inverse Fourier-transform operation, e.g., $F^{-1}[G(f)] = g(t)$.

8. The symbol ★ denotes convolution, e.g.,

$$x(t) \; \bigstar \; h(t) \; = \; \int_{-\infty}^{\infty} x(\tau) h(t \, - \, \tau) \, d\tau$$

9. The symbol $\oplus$ denotes modulo-2 addition; except in Chapter 11 where modulo-2 addition is denoted by an ordinary plus sign.

10. The use of subscript T_0 indicates that the pertinent function $g_{T_0}(t)$, say, is a periodic function of time t with period T_0.

11. The use of a hat over a function indicates one of two things:

    (a) the Hilbert transform of a function, e.g., the function $\hat{g}(t)$ is the Hilbert transform of $g(t)$, or

    (b) the estimate of an unknown parameter, e.g., the quantity $\hat{\alpha}(\mathbf{x})$ is an estimate of the unknown parameter α, based on the observation vector $\mathbf{x}$.

12. The use of a tilde over a function indicates the complex envelope of a narrow-band signal, e.g., the function $\tilde{g}(t)$ is the complex envelope of the narrow-band signal $g(t)$.

13. The use of subscript $+$ indicates the pre-envelope of a signal, e.g., the function $g_+(t)$ is the pre-envelope of the signal $g(t)$. We may thus write $g_+(t) = g(t) + j\hat{g}(t)$, where $\hat{g}(t)$ is the Hilbert transform of $g(t)$. The use of subscript $-$ indicates that $g_-(t) = g(t) - j\hat{g}(t) = g_+^*(t)$.

14. The use of subscripts I and Q indicates the in-phase and quadrature components of a narrow-band signal, a narrow-band random process, or the impulse response of a narrow-band filter, with respect to the carrier $\cos(2\pi f_c t)$.

15. For a low-pass message signal, the highest frequency component or message bandwidth is denoted by W. The spectrum of this signal occupies the frequency interval $-W \leq f \leq W$ and is zero elsewhere. For a band-pass signal with carrier frequency f_c, the spectrum occupies the frequency intervals, $f_c - W \leq f \leq f_c + W$ and $-f_c - W \leq f \leq -f_c + W$, and so $2W$ denotes the bandwidth of the signal. The (low-pass) complex envelope of this band-pass signal has a spectrum that occupies the frequency interval $-W \leq f \leq W$.

    For a low-pass filter, the bandwidth is denoted by B. A common definition of filter bandwidth is the frequency at which the amplitude response of the filter drops by 3 dB below the zero-frequency value. For a band-pass filter of mid-band frequency f_c the bandwidth is denoted by $2B$, centered on f_c. The complex low-pass equivalent of this band-pass filter has a bandwidth equal to B.

    The transmission bandwidth of a communication channel, required to transmit a modulated wave, is denoted by B_T.

16. Random variables or random vectors are uppercase (e.g., X or $\mathbf{X}$), and their sample values are lowercase (e.g., x or $\mathbf{x}$).

17. A vertical bar in an expression means "given that," e.g., $f_X(x|H_0)$ is the probability density function of the random variable X, given that hypothesis H_0 is true.

18. The symbol $E[\ ]$ means the expected value of the random variable enclosed within.

19. The symbol var[ ] means the variance of the random variable enclosed within.

20. The symbol cov[ ] means the covariance of the two random variables enclosed within.

21. The average probability of symbol error is denoted by P_e.

    In the case of binary signaling techniques, P_{e0} denotes the conditional probability of error given that symbol 0 was transmitted, and P_{e1} denotes the conditional probability of error given that symbol 1 was transmitted. The *a priori* probabilities of symbols 0 and 1 are denoted by p_0 and p_1, respectively.

22. The symbol $\langle \;\; \rangle$ denotes the time average of the sample function enclosed within.

23. Boldface letter denotes a vector or matrix. The inverse of a square matrix $\mathbf{R}$ is denoted by $\mathbf{R}^{-1}$. The transpose of a vector $\mathbf{w}$ is denoted by $\mathbf{w}^T$.

24. The length of a vector $\mathbf{x}$ is denoted by $\|\mathbf{x}\|$. The Euclidean distance between the vectors $\mathbf{x}_i$ and $\mathbf{x}_j$ is denoted by $d_{ij} = \|\mathbf{x}_i - \mathbf{x}_j\|$.

25. The inner product of two vectors $\mathbf{x}$ and $\mathbf{y}$ is denoted by $\mathbf{x}^T\mathbf{y}$; their outer product is denoted by $\mathbf{xy}^T$

FUNCTIONS

1. Rectangular function:
$$\text{rect}(t) = \begin{cases} 1, & -\tfrac{1}{2} < t < \tfrac{1}{2} \\ 0, & |t| > \tfrac{1}{2} \end{cases}$$

2. Unit step function:
$$u(t) = \begin{cases} 1, & t > 0 \\ 0, & t < 0 \end{cases}$$

3. Signum function:
$$\text{sgn}(t) = \begin{cases} 1, & t > 0 \\ -1, & t < 0 \end{cases}$$

4. Dirac delta function:
$$\delta(t) = 0, \qquad t \neq 0$$
$$\int_{-\infty}^{\infty} \delta(t)\ dt = 1$$

or equivalently
$$\int_{-\infty}^{\infty} g(t)\ \delta(t - t_0)\ dt = g(t_0)$$

5. Sinc function:
$$\text{sinc}(x) = \frac{\sin(\pi x)}{\pi x}$$

6. Sine integral:
$$\text{Si}(u) = \int_{0}^{u} \frac{\sin x}{x}\ dx$$

7. Error function:
$$\text{erf}(u) = \frac{2}{\sqrt{\pi}}\int_0^u \exp(-z^2)\ dz$$

Complementary error function:
$$\text{erfc}(u) = 1 - \text{erf}(u).$$

8. Bessel function of the
first kind of order n:
$$J_n(x) = \frac{1}{2\pi}\int_{-\pi}^{\pi}\exp(jx\sin\theta - jn\theta)\ d\theta$$

9. Modified Bessel function of
the first kind of zero order:
$$I_0(x) = \frac{1}{2\pi}\int_{-\pi}^{\pi}\exp(x\cos\theta)\ d\theta$$

10. Binomial coefficient
$$\binom{n}{k} = \frac{n!}{(n-k)!k!}$$

ABBREVIATIONS

ac:	alternating current
ADPCM:	adaptive differential pulse-code modulation
ANSI:	American National Standards Institute
AM:	amplitude modulation
ARQ:	automatic-repeat-request
ASCII:	American National Standard Code for Information Interchange
ASK:	amplitude-shift keying
ATM:	asynchronous transfer mode
B-ISDN:	broadband ISDN
BER:	bit error rate
BPF:	band-pass filter
BSC:	binary symmetric channel
CCD:	charge-coupled device
CCITT:	Consultative Committee for International Telephone and Telegraph
CDM:	code-division multiplexing
CDMA:	code-division multiple access
CPFSK:	continuous-phase frequency-shift keying
CW:	continuous wave
dB:	decibel
dc:	direct current
DEM:	demodulator
DES:	data encryption standard
DFT:	discrete Fourier transform
DM:	delta modulation
DPCM:	differential pulse-code modulation
DPSK:	differential phase-shift keying

DSB-SC:	double sideband–suppressed carrier
DS/BPSK:	direct sequence/binary phase-shift keying
exp:	exponential
FDM:	frequency-division multiplexing
FDMA:	frequency-division multiple access
FFT:	fast Fourier transform
FH:	frequency hop
FH/MFSK:	frequency hop/M-ary frequency-shift keying
FMFB:	frequency modulator with feedback
FSK:	frequency-shift keying
HDTV:	high definition television
Hz:	Hertz
IDFT:	inverse discrete Fourier transform
IF:	intermediate frequency
ILD:	injection laser diode
I/O:	input/output
ISDN:	integrated services digital network
ISI:	intersymbol interference
ISO:	International Organization for Standardization
LAN:	local-area network
LDM:	linear delta modulation
LED:	light emitting diode
LMS:	least-mean-square
ln:	natural logarithm
log:	logarithm
LPF:	low-pass filter
MAP:	maximum *a posteriori* probability
ms:	millisecond
μs:	microsecond
ML:	maximum likelihood
modem:	modulator–demodulator
MSK:	minimum shift keying
nm:	nanometer
NRZ:	nonreturn-to-zero
NTSC:	National Television Systems Committee
OOK:	on–off keying
OSI:	open systems interconnection
PAM:	pulse-amplitude modulation
PCM:	pulse-code modulation
PCN:	personal communication network
PG:	processing gain
PLL:	phase-locked loop
PN:	pseudo-noise

PSK: phase-shift keying
QAM: quadrature amplitude modulation
QOS: quality of service
QPSK: quadriphase-shift keying
RF: radio frequency
rms: root-mean-square
RS: Reed–Solomon
RS-232   Recommended standard-232 (port)
RSA: Rivest-Shamir-Adelman
RZ: return-to-zero
s: second
SBC: subband coding
SDH: synchronous digital hierarchy
SDMA: space-division multiple access
SDR: signal-to-distortion ratio
SONET: synchronous optical network
SNR: signal-to-noise ratio
STFT: short-time Fourier transform
STM: synchronous transfer mode
TCM: trellis-coded modulation
TDM: time-division multiplexing
TDMA: time-division multiple access
TV: television
UHF: ultra high frequency
VCO: voltage-controlled oscillator
VHF: very high frequency
VLSI: very-large-scale integration
WT: wavelet transform

Bibliography

BOOKS

M. Abramowitz and I. A. Stegun, *Handbook of Mathematical Functions, with Formulas, Graphs, and Mathematical Tables* (New York: Dover Publications, 1965).

N. Abramson, *Information Theory and Coding* (New York: McGraw-Hill, 1963).

J. Adámek, *Foundations of Coding* (New York: Wiley, 1991).

J. B. Anderson, T. Aulin, and C. E. Sundberg, *Digital Phase Modulation* (New York: Plenum, 1986).

J. B. Anderson and S. Mohan, *Source and Channel Coding: An Algorithmic Approach* (Boston, Mass.: Kluwer Academic, 1991).

R. B. Ash, *Information Theory* (New York: Wiley, 1965).

T. C. Bartee (editor), *Data Communications, Networks, and Systems* (Howard W. Sams, 1985).

Bell Telephone Laboratories, *Transmission Systems for Communications* (1971).

Bell Laboratories, *A History of Engineering Science in the Bell System: The Early Years (1875–1925)* (1975).

J. C. Bellamy, *Digital Telephony* (New York: Wiley, 1982).

S. Benedetto, E. Biglieri, and V. Castellani, *Digital Transmission Theory* (Englewood Cliffs, N.J.: Prentice-Hall, 1987).

W. R. Bennett, *Introduction to Signal Transmission* (New York: McGraw-Hill, 1970).

K. B. Benson and D. G. Fink, *HDTV—Advanced Television for the 1990s* (New York: McGraw-Hill, 1991).

T. Berger, *Rate Distortion Theory: A Mathematical Basis for Data Compression* (Englewood Cliffs, N.J.: Prentice-Hall, 1971).

E. R. Berlekamp, *Algebraic Coding Theory* (New York: McGraw-Hill, 1968).

E. R. Berlekamp, editor, *Key Papers in the Development of Coding Theory* (New York: IEEE Press, 1974).

D. Bertsekas and R. Gallagher, *Data Networks* (Englewood Cliffs, N.J.: Prentice-Hall, 1987).

V. K. Bhargava, D. Haccoun, R. Matyas, and P. Nuspl, *Digital Communications by Satellite: Modulation, Multiple Access, and Coding* (New York: Wiley, 1981).

E. Biglieri, D. Divsalar, P. J. Mclane, and M. K. Simon, *Introduction to Trellis-Coded Modulation with Applications* (New York: Macmillan, 1991).

J. A. Bingham, *The Theory and Practice of Modem Design* (New York: Wiley, 1988).

H. S. Black, *Modulation Theory* (Princeton, N.J.: Van Nostrand, 1953).

N. M. Blachman, *Noise and Its Effect on Communication* (New York: McGraw-Hill, 1966).

R. B. Blachman and J. W. Tukey, *The Measurement of Power Spectra, from the Point of View of Communication Engineering* (New York: Dover, 1958).

R. E. Blahut, *Principles and Practice of Information Theory* (Reading, Mass.: Addison-Wesley, 1987).

R. E. Blahut, *Digital Transmission of Information* (Reading, Mass.: Addison-Wesley, 1990).

I. F. Blake, *An Introduction to Applied Probability* (New York: Wiley, 1979).

G. E. P. Box and G. M. Jenkins, *Time Series Analysis: Forecasting and Control* (San Francisco: Holden-Day, 1976).

R. N. Bracewell, *The Fourier Transform and Its Applications*, 2nd ed., rev. (New York: McGraw-Hill, 1986).

E. O. Brigham, *The Fast Fourier Transform* (Englewood Cliffs, N.J.: Prentice-Hall, 1974).

L. Brillouin, *Science and Information Theory*, 2nd ed. (New York: Academic Press, 1962).

G. A. Campbell and R. M. Foster, *Fourier Integrals for Practical Applications*, (Princeton, N.J.: Van Nostrand, 1948).

A. B. Carlson, *Communication Systems: An Introduction to Signals and Noise in Electrical Communication*, 3rd ed. (New York: McGraw-Hill, 1986).

K. W. Cattermole, *Principles of Pulse-code Modulation* (New York: American Elsevier, 1969).

D. C. Champeney, *Fourier Transforms and Their Physical Applications* (London: Academic Press, 1973).

D. N. Chorafas, *Telephony: Today and Tomorrow* (Englewood Cliffs, N.J.: Prentice-Hall, 1984).

G. C. Clark, Jr., and J. B. Cain, *Error-correction Coding for Digital Communications* (New York: Plenum, 1981).

G. R. Cooper and C. D. McGillem, *Probabilistic Methods of Signals and Systems* (New York: Holt, Rinehart & Winston, 1971).

G. R. Cooper and C. D. McGillem, *Modern Communications and Spread Spectrum* (New York: McGraw-Hill, 1986).

L. W. Couch, II, *Digital and Analog Communication Systems*, 3rd ed. (New York: Macmillan, 1990).

T. M. Cover and J. A. Thomas, *Elements of Information Theory* (New York: Wiley, 1991).

H. Cramér and M. R. Leadbetter, *Stationary and Related Stochastic Processes: Sample Function Properties and Their Applications* (New York: Wiley, 1967).

I. Daubechies, *Ten Lectures on Wavelets* (SIAM, 1992).

W. B. Davenport, Jr., and W. L. Root, *An Introduction to the Theory of Random Signals and Noise* (New York: McGraw-Hill, 1958).

W. B. Davenport, Jr., *Probability and Random Processes: An Introduction for Applied Scientists and Engineers* (New York: McGraw-Hill, 1970).

M. dePrycker, *Asynchronous Transfer Mode* (Englewood Cliffs, N.J.: Prentice-Hall, 1991).

R. C. Dixon, *Spread Spectrum Systems*, 2nd ed. (New York: Wiley, 1984).

R. C. Dixon (editor), *Spread Spectrum Techniques* (New York: IEEE Press, 1976).

L. J. Doob, *Stochastic Processes* (New York, Wiley: 1953).

J. J. Downing, *Modulation Systems and Noise* (Englewood Cliffs, N.J.: Prentice-Hall, 1964).

D. F. Elliott and K. R. Rao, *Fast Transforms: Algorithms, Analyses, Applications* (New York: Academic Press, 1982).

K. Feher, *Digital Communications: Microwave Applications* (Englewood Cliffs, N.J.: Prentice-Hall, 1981).

K. Feher (editor), *Advanced Digital Communications: Systems and Signal Processing Techniques* (Englewood Cliffs, N.J.: Prentice-Hall, 1987).

W. Feller, *An Introduction to Probability Theory and its Application* vol. 1, 3rd ed. (New York: Wiley, 1968).

M. J. Feuerstein and T. S. Rappoport, *Wireless Personal Communications* (Boston, Mass.: Kluwer Academic, 1993).

T. L. Fine, *Theories of Probability: An Examination of Foundations* (New York: Academic Press, 1973).

J. L. Flanagan, *Speech Analysis; Synthesis and Perception* (New York/Berlin: Springer-Verlag, 1972).

L. E. Franks (editor), *Data Communication: Fundamentals of Baseband Transmission* (Dowden, Hutchison, and Ross, 1974).

L. E. Franks, *Signal Theory* (Englewood Cliffs, N.J.: Prentice-Hall, 1969).

R. M. Gagliardi, *Introduction to Communications Engineering*, 2nd ed. (New York: Wiley, 1988).

R. G. Gallagher, *Information Theory and Reliable Communication* (New York: Wiley, 1968).

F. M. Gardner, *Phaselock Techniques*, 2nd ed. (New York: Wiley, 1979).

A. Gersho and R. M. Gray, *Vector Quantization and Signal Compression* (Boston, Mass.: Kluwer Academic, 1992).

J. D. Gibson, *Principles of Digital and Analog Communications* (New York: Macmillan, 1989).

A. Gill, *Linear Sequential Circuits; Analysis, Synthesis and Applications* (New York: McGraw-Hill, 1966).

R. D. Gitlin, J. F. Hayes, and S. B. Weinstein, *Data Communications Principles* (New York: Plenum, 1992).

B. Goldberg and H. S. Bennett (editors), *Communications Channels: Characterization and Behavior* (New York: IEEE Press, 1976).

S. Goldman, *Frequency Analysis, Modulation, and Noise* (New York: McGraw-Hill, 1948).

S. W. Golomb (editor), *Digital Communications with Space Applications* (Englewood Cliffs, N.J.: Prentice-Hall, 1964).

S. W. Golomb, *Shift Register Sequences* (San Francisco: Holden–Day, 1967).

D. J. Goodman and S. Nanda (editors), *Third Generation Wireless Information Networks* (Boston, Mass.: Kluwer Academic, 1992).

R. M. Gray and L. D. Davisson, *Random Processes: A Mathematical Approach for Engineers* (Englewood Cliffs, N.J.: Prentice-Hall, 1986).

P. E. Green, Jr., *Computer Network Architectures and Protocols* (New York: Plenum, 1982).

P. E. Green, Jr., *Fiber Optic Networks* (Englewood Cliffs, N.J.: Prentice-Hall, 1993).

W. D. Gregg, *Analog and Digital Communication* (New York: Wiley, 1977).

M. S. Gupta (editor), *Electrical Noise: Fundamentals and Sources* (New York: IEEE Press, 1977).

E. A. Guillemin, *The Mathematics of Circuit Analysis* (New York: Wiley, 1958).

R. W. Hamming, *The Art of Probability for Scientists and Engineers* (Reading, Mass.: Addison-Wesley, 1991).

R. W. Hamming, *Coding and Information Theory* (Englewood Cliffs, N.J.: Prentice-Hall, 1980).

J. F. Hayes, *Modeling and Analysis of Computer Communications Networks* (New York: Plenum, 1984).

S. Haykin, *Communication Systems*, 2nd ed. (New York: Wiley, 1983).

S. Haykin, *Adaptive Filter Theory*, 2nd ed. (Englewood Cliffs, N.J.: Prentice-Hall, 1991).

C. W. Helstrom, *Statistical Theory of Signal Detection* (Elmsford, N.Y.: Pergamon Press, 1968).

C. W. Helstrom, *Probability and Stochastic Processes for Engineers*, 2nd ed. (New York: Macmillan, 1990).

K. Henney (editor), *Radio Engineering Handbook* (New York: McGraw-Hill, 1959).

P. S. Henry and S. D. Personick (editors), *Coherent Lightwave Communications* (New York: IEEE Press).

J. K. Holmes, *Coherent Spread Spectrum Systems* (New York: Wiley, 1982).

W. C. Jakes, Jr. (editor), *Microwave Mobile Communications* (New York: Wiley, 1974).

N. S. Jayant and P. Noll, *Digital Coding of Waveforms: Principles and Applications to Speech and Video* (Englewood Cliffs, N.J.: Prentice-Hall, 1984).

N. S. Jayant (editor), *Waveform Quantization and Coding* (New York: IEEE Press, 1976).

H. Jeffreyes, Sir, *Theory of Probability* (Oxford: Clarendon Press, 1957).

M. C. Jeruchim, B. Balaban, and J. S. Shanmugan, *Simulation of Communication Systems* (New York: Plenum, 1992).

M. Kanefsky, *Communication Techniques for Digital and Analog Signals* (New York: Harper & Row, 1985).

C. K. Kao, *Optical Fiber Systems: Technology, Design and Application* (New York: McGraw-Hill, 1982).

S. M. Kay, *Modern Spectral Estimation: Theory and Applications* (Englewood Cliffs, N.J.: Prentice-Hall, 1988).

R. S. Kennedy, *Fading Dispersive Communication Channels* (New York: Wiley, 1969).

A. Khintchin, *Mathematical Foundations of Information Theory* (Dover, 1957).

L. Kleinrock, *Queueing Systems*, Vols. 1 and 2 (Wiley, 1976).

A. N. Kolmogorov, *Foundations of the Theory of Probability* (New York: Chelsea Publishing, 1956).

I. Korn, *Digital Communications* (Princeton, N.J.: Van Nostrand–Reinhold, 1985).

V. A. Kotel'nikov, *The Theory of Optimum Noise Immenity* (New York: McGraw-Hill, 1960).

S. Kullback, *Information Theory and Statistics* (New York: Dover, 1968).

B. P. Lathi, *Modern Digital and Analog Communication Systems* (New York: Holt, Rinehart & Winston, 1983).

V. B. Lawrence, J. L. LoCicero, and L. B. Milstein (editors), *IEEE Communications Society's Tutorials in Modern Communications* (Computer Science Press, 1983).

P. Lafrance, *Fundamental Concepts in Communication* (Englewood Cliffs, N.J.: Prentice-Hall, 1990).

E. A. Lee and D. G. Messerschmitt, *Digital Communications* (Boston, Mass.: Kluwer Academic, 1988).

Y. W. Lee, *Statistical Theory of Communication* (New York: Wiley, 1960).

W. C. (Y.) Lee, *Mobile Communications Engineering* (New York: McGraw-Hill, 1982).

A. Leon-Garcia, *Probability and Random Processes for Electrical Engineering* (Reading, Mass.: Addison-Wesley, 1989).

M. J. Lighthill, *Introduction to Fourier Analysis and Generalized Functions* (New York: Cambridge University Press, 1958).

S. Lin and D. J. Costello, Jr., *Error Control Coding: Fundamentals and Applications* (Englewood Cliffs, N.J.: Prentice-Hall, 1983).

W. C. Lindsey, *Synchronization Systems in Communication and Control* (Englewood Cliffs, N.J.: Prentice-Hall, 1972).

W. C. Lindsey and M. K. Simon (editors), *Phase-locked Loops and Their Applications* (New York: IEEE Press, 1978).

W. C. Lindsey and M. K. Simon, *Telecommunication Systems Engineering* (Englewood Cliffs, N.J.: Prentice-Hall, 1973).

R. W. Lucky, *Silicon Dreams: Information, Man, and Machine* (New York: St. Martin's Press, 1989).

R. W. Lucky, J. Salz, and E. J. Weldon, Jr., *Principles of Data Communication* (New York: McGraw-Hill, 1968).

F. J. MacWilliams and N. J. A. Sloane, *The Theory of Error-correcting Codes* (Amsterdam: North-Holland, 1977).

J. D. Markel and A. H. Gray, Jr., *Linear Prediction of Speech* (New York/Berlin: Springer-Verlag, 1976).

R. J. Marks, II, *Introduction to Shannon Sampling and Interpolation Theory* (New York/Berlin: Springer-Verlag, 1991).

S. L. Marple, *Digital Spectral Analysis with Applications* (Englewood Cliffs, N.J.: Prentice-Hall, 1987).

R. J. McEliece, *The Theory of Information and Coding: A Mathematical Framework for Communication* (Reading, Mass.: Addison Wesley, 1977).

C. H. Meyer and S. M. Matyas, *Cryptography: A New Dimension in Computer Data Security* (New York: Wiley, 1982).

R. A. Meyers (editor), *Encyclopedia of Telecommunications* (New York: Academic Press, 1989).

H. Meyr and G. Ascheid, *Synchronization in Digital Communications*, vol. I (New York: Wiley, 1990).

A. M. Michelson and A. H. Levesque, *Error-control Techniques for Digital Communication* (New York: Wiley, 1985).

D. Middleton, *An Introduction to Statistical Communication Theory*, (New York: McGraw-Hill, 1960).

J. G. Nellist, *Understanding Telecommunications and Lightwave Systems: An Entry Level Guide* (New York: IEEE Press, 1992).

A. N. Netravali and B. G. Haskell, *Digital Pictures: Representation and Compression* (New York: Plenum, 1988).

A. V. Oppenheim and R. W. Schafer, *Digital Signal Processing* (Englewood Cliffs, N.J.: Prentice-Hall, 1975).

A. V. Oppenheim and R. W. Schafer, *Discrete-Time Signal Processing* (Englewood Cliffs, N.J.: Prentice-Hall, 1989).

D. O'Shaughnessy, *Speech Communication: Human and Machine* (Reading, Mass.: Addison Wesley, 1987).

C. F. J. Overhage (editor), *The Age of Electronics* (New York: McGraw-Hill, 1962).

P. F. Panter, *Modulation, Noise and Spectral Analysis, Applied to Information Transmission* (New York: McGraw-Hill, 1965).

A. Papoulis, *The Fourier Integral and Its Applications* (New York: McGraw-Hill, 1962).

A. Papoulis, *Probability, Random Variables, and Stochastic Processes*, 2nd ed. (New York: McGraw-Hill, 1984).

J. D. Parsons, *The Mobile Radio Propagation Channel* (New York: Wiley, 1992).

E. Parzen, *Stochastic Processes* (San Francisco: Holden–Day, 1962).

P. Z. Peebels, Jr., *Digital Communication Systems* (Englewood Cliffs, N.J.: Prentice-Hall, 1987).

S. D. Personick, *Fiber Optics: Technology and Applications* (New York: Plenum, 1986).

W. W. Peterson and E. J. Weldon, Jr., *Error Correcting Codes*, 2nd ed. (MIT Press, 1972).

J. R. Pierce and A. M. Noll, *Signals: The Science of Telecommunications* (Scientific American Library, 1990).

J. R. Pierce and E. C. Posner, *Introduction to Communication Science and Systems* (New York: Plenum, 1980).

J. R. Pierce, *Symbols, Signals and Noise: The Nature and Process of Communication* (New York: Harper 1961).

H. V. Poor, *An Introduction to Signal Detection and Estimation* (New York/Berlin: Springer-Verlag, 1988).

R. K. Potter, G. A. Kopp, and H. C. Green, *Visible Speech* (New York: Van Nostrand, 1947).

T. Pratt and C. W. Bostian, *Satellite Communications* (New York: Wiley, 1986).

W. K. Pratt, *Digital Image Processing* (New York: Wiley, 1978).

W. H. Press, B. P. Flannery, S. A. Teukolsky, and W. T. VeHerling, (editors), *Numerical Recipes in C: The Art of Scientific Computing* (New York: Cambridge University Press, 1988).

M. B. Priestley, *Spectral Analysis and Time Series* (New York: Academic Press, 1981).

J. G. Proakis, *Digital Communications*, 2nd ed. (New York: McGraw-Hill, 1989).

L. R. Rabiner and B. Gold, *Theory and Application of Digital Signal Processing* (Englewood Cliffs, N.J.: Prentice-Hall, 1975).

L. R. Rabiner and R. W. Schafer, *Digital Processing of Speech Signals* (Englewood Cliffs, N.J.: Prentice-Hall, 1978).

K. R. Rao and P. Yip, *Discrete Cosine Transform: Algorithms, Advantages, and Applications* (New York: Academic Press, 1990).

F. Reif, *Statistical Physics* (New York: McGraw-Hill, 1967).

M. D. Riley, *Speech Time-Frequency Representation* (Reading, M. A.: Kluwer, 1988).

J. H. Roberts, *Angle Modulation: The Theory of System Assessment*, IEE Communication Series 5 (London: Institution of Electrical Engineers, 1977).

H. E. Rowe, *Signals and Noise in Communication Systems* (Princeton, N.J.: Van Nostrand, 1965).

T. S. Rzeszewski (editor), *Television Technology Today* (New York: IEEE Press, 1985).

D. J. Sakrison, *Communication Theory: Transmission of Waveforms and Digital Information* (New York: Wiley, 1968).

W. F. Schreiber, *Fundamentals of Electronic Imaging Systems: Some Aspects of Image Processing*, 3rd ed. (New York/Berlin: Springer-Verlag, 1993).

M. Schwartz, W. R. Bennett, and S. Stein, *Communication Systems and Techniques* (New York: McGraw-Hill, 1966).

M. Schwartz, *Information Transmission, Modulation and Noise: A Unified Approach*, 3rd ed. (New York: McGraw-Hill, 1980).

M. Schwartz, *Telecommunication Networks: Protocols, Modeling, and Analysis* (Reading, Mass: Addison Wesley, 1987).

J. M. Senior, *Optical Fiber Communications: Principles and Practice*, 2nd ed. (Englewood Cliffs, N.J.: Prentice Hall, 1992).

K. S. Shanmugam, *Digital and Analog Communication Systems* (New York: Wiley, 1979).

C. E. Shannon and W. Weaver, *The Mathematical Theory of Communication* (Urbana: University of Illinois Press, 1949).

G. J. Simmons, *Contemporary Cryptology: The Science of Information Integrity* (New York: IEEE Press, 1992).

M. K. Simon, J. K. Omura, R. A. Scholtz, and B. K. Levitt, *Spread Spectrum Communications*, Vols. I, II, and III (Computer Science Press, 1985).

B. Sklar, *Digital Communications: Fundamentals and Applications* (Englewood Cliffs, N.J.: Prentice-Hall, 1988).

D. Slepian (editor), *Key Papers in the Development of Information Theory* (New York: IEEE Press, 1974).

D. R. Smith, *Digital Transmission Systems* (Princeton, N.J.: Van Nostrand–Reinhold, 1985).

J. J. Spilker, Jr., *Digital Communications by Satellite* (Englewood Cliffs, N.J.: Prentice-Hall, 1977).

W. Stallings, *Data and Computer Communications* (New York: Macmillan, 1985).

W. Stallings, *Local Networks*, 2nd ed. (New York: Macmillan, 1987).

W. Stallings, *ISDN and Broadband ISDN*, 2nd ed. (New York: Macmillan, 1992).

H. Stark, F. B. Tuteur, and J. B. Anderson, *Modern Electrical Communications: Analog, Digital and Optical Systems*, 2nd ed. (Englewood Cliffs, N.J.: Prentice-Hall, 1988).

R. Steele, *Delta Modulation Systems* (New York: Wiley, 1975).

S. Stein and J. Jones, *Modern Communication Principles* (New York: McGraw-Hill, 1967).

J. J. Stiffler, *Theory of Synchronous Communications* (Englewood Cliffs, N.J.: Prentice-Hall, 1971).

F. G. Stremler, *Introduction to Communication Systems*, 3rd ed. (Reading, M. A.: Addison-Wesley, 1990).

E. D. Sunde, *Communication Systems Engineering Theory* (New York: Wiley, 1969).

A. S. Tanenbaum, *Computer Networks* (Englewood Cliffs, N.J.: Prentice-Hall, 1981).

H. Taub and D. S. Schilling, *Principles of Communication Systems*, 2nd ed. (New York: McGraw-Hill, 1986).

J. B. Thomas, *An Introduction to Statistical Communication Theory* (New York: Wiley, 1969).

J. B. Thomas, *Introduction to Probability* (New York/Berlin: Springer-Verlag, 1986).

T. M. Thompson, *From Error-correcting Codes Through Sphere Packing to Simple Groups* (The Mathematical Association of America, 1983).

E. C. Titchmarsh, *Introduction to the Theory of Fourier Integrals*, 2nd ed. (London/New York: Oxford University Press, 1950).

D. J. Torrieri, *Principles of Military Communication Systems*, 2nd ed. (Artech House, 1992).

D. G. Tucker, *Modulators and Frequency-Changers* (London: Macdonald, 1953).

G. L. Turin, *Notes on Digital Communications* (Princeton, N.J.: Van Nostrand–Reinhold, 1969).

A. Van der Ziel, *Noise: Source, Characterization, Measurement* (Englewood Cliffs, N.J.: Prentice-Hall, 1970).

H. F. Vanlandingham, *Introduction to Digital Control Systems* (New York: Macmillan, 1985).

H. C. A. van Tilborg, *An Introduction to Cryptology* (Boston, Mass.: Kluwer, 1988).

H. L. Van Trees, *Detection, Estimation, and Modulation Theory*, Part I (New York: Wiley, 1968).

A. J. Viterbi, *Principles of Coherent Communication* (New York: McGraw-Hill, 1966).

A. J. Viterbi and J. K. Omura, *Principles of Digital Communication and Coding* (New York: McGraw-Hill, 1979).

G. N. Watson, *A Treatise on the Theory of Bessel Functions*, 2nd ed. (New York: Cambridge University Press, 1966).

C. L. Weber, *Elements of Detection and Signal Design* (New York: McGraw-Hill, 1968).

A. D. Whalen, *Detection of Signals in Noise* (New York: Academic Press, 1971).

B. Widrow and S. D. Stearns, *Adaptive Signal Processing* (Englewood Cliffs, N.J.: Prentice-Hall, 1985).

N. Wiener, *The Extrapolation, Interpolation, and Smoothing of Stationary Time Series, with Engineering Applications* (New York: Wiley, 1949).

N. Wiener, *The Fourier Integral and Certain of its Applications* (New York: Dover Publications, 1958).

R. A. Williams, *Communication Systems Analysis and Design: A Systems Approach* (Englewood Cliffs, N.J.: Prentice-Hall, 1987).

E. Wong, *Stochastic Processes in Information and Dynamical Systems* (New York: McGraw-Hill, 1971).

P. M. Woodward, *Probability and Information Theory: with Applications to Radar*, 2nd ed. (Elmsford, N.Y.: Pergamon Press, 1964).

J. M. Wozencraft and I. M. Jacobs, *Principles of Communication Engineering* (New York: Wiley, 1965).

W. W. Wu, *Elements of Digital Satellite Communication*, vol. I (New York: Computer Science Press, 1984).

C. R. Wylie and L. C. Barrett, *Advanced Engineering Mathematics*, 5th ed., (New York: McGraw-Hill, 1982).

J. H. Yuen (editor), *Deep Space Telecommunications Systems Engineering* (New York: Plenum, 1983).

R. E. Ziemer and R. L. Peterson, *Digital Communications and Spread Spectrum Systems* (New York: Macmillan, 1985).

R. E. Ziemer and W. H. Tranter, *Principles of Communications*, 3rd ed. (Boston: Houghton Mifflin, 1990).

PAPERS*/REPORTS/PATENTS

M. R. Aaron and D. W. Tufts, "Intersymbol interference and error probability," *IEEE Trans. on Information Theory*, vol. IT-12, pp. 26–34, 1966.

J. E. Abate, "Linear and adaptive delta modulation," *Proceedings of the IEEE*, vol. 55, pp. 298–308, 1967.

S. R. Ahuja and J. R. Ensor, "Coordination and control of multimedia conferencing," *IEEE Communications Magazine*, vol. 30, no. 5, pp. 38–43, 1992.

F. Amoroso, "The bandwidth of digital data signals," *IEEE Communications Magazine*, vol. 18, no. 6, pp. 13–24, 1980.

J. B. Anderson and D. P. Taylor, "A bandwidth-efficient class of signal space codes," *IEEE Transactions on Information Theory*, vol. IT-24, pp. 703–712, 1978.

R. R. Anderson and J. Salz, "Spectra of digital FM," *Bell System Tech. J.*, vol. 44, pp. 1165–1189, 1965.

Note. The following abbreviations are used for some of the journal papers:

IEEE: Institute of Electrical and Electronic Engineers

IEE: Institution of Electrical Engineers (London)

IRE: Institute of Radio Engineers

SIAM: The Society for Industrial and Applied Mathematics

R. Arens, "Complex processes for envelopes of normal noise," *IRE Trans. on Information Theory*, vol. IT-3, pp. 204–207, 1957.

E. H. Armstrong, "A method of reducing disturbances in radio signaling by a system of frequency modulation," *Proceedings of the IRE*, vol. 24, pp. 689–740, 1936.

E. Arthurs and H. Dym, "On the optimum detection of digital signals in the presence of white Gaussian noise—A geometric interpretation and a study of three basic data transmission systems," *IRE Trans. on Communication Systems*, vol. CS-10, pp. 336–372, 1962.

V. Aschoff, "The early history of the binary code," *IEEE Communications Magazine*, vol. 21, no. 1, pp. 4–10, 1983.

M. Austin, "Decision-feedback equalization for digital communication over dispersive channels," *MIT Research Laboratory Electronics Technical Report* 461, 1967.

E. Bedrosian, "The analytic signal representation of modulated waveforms," *Proceedings of the IRE*, vol. 50, pp. 2071–2076, 1962.

P. A. Bello, "Characterization of randomly time-variant linear channels," *IEEE Transactions on Communication Systems*, vol. CS-11, pp. 360–393, 1963.

W. R. Bennett, "Spectra of quantized signals," *Bell System Tech. J.*, vol. 27, pp. 446–472, 1948.

N. Benvenuto, et al., "The 32 kb/s ADPCM coding standard," *AT&T Technical Journal*, vol. 65, pp. 12–22, Sept./Oct. 1986.

V. K. Bhargava, "Forward error correction schemes for digital communications," *IEEE Communications Magazine*, vol. 21, no. 1, pp. 11–19, 1983.

R. C. Bose and D. K. Ray-Chaudhuri, "On a class of error correcting binary group codes," *Information and Control*, vol. 3, pp. 68–79, 1960.

D. G. Brennan, "Linear diversity combining techniques," *Proceedings of the IRE*, vol. 47, pp. 1075–1102, 1959.

D. R. Brillinger, "An introduction to polyspectra," *Annals of Mathematical Statistics*, pp. 1351–1374, 1965.

A. Buzo, A. H. Gray, Jr., R. M. Gray, and J. D. Markel, "Speech coding based upon vector quantization," *IEEE Transactions on Acoustics, Speech, and Signal Processing*, vol. ASSP-28, pp. 562–574, 1980.

C. R. Cahn, "Combined digital phase and amplitude modulation communication systems," *IRE Transactions on Communication Systems*, vol. CS-8, pp. 150–155, 1960.

J. R. Carson, "Notes on the theory of modulation," *Proceedings of the IRE*, vol. 10, pp. 57–64, 1922.

J. R. Carson and T. C. Fry, "Variable frequency electric circuit theory with application to the theory of frequency modulation," *Bell System Tech. J.*, vol. 16, pp. 513–540, 1937.

J. G. Chaffee, "The application of negative feedback to frequency-modulation systems," *Bell System Tech. J.*, vol. 18, pp. 404–437, 1939.

C. E. Cook and H. S. Marsh, "An introduction to spread spectrum," *IEEE Communications Magazine*, vol. 21, no. 2, pp. 8–16, 1983.

J. W. Cooley and J. W. Tukey, "An algorithm for the machine calculation of complex Fourier series," *Math. Comput.*, vol. 19, pp. 297–301, 1965.

J. P. Costas, "Synchronous communications," *Proceedings of the IRE*, vol. 44, pp. 1713–1718, 1956.

J. P. Costas, "Poisson, Shannon, and the radio amateur," *Proceedings of the IRE*, vol. 47, pp. 2058–2068, 1959.

M. G. Crosby, "Frequency modulation noise characteristics," *Proceedings of the IRE*, vol. 25, pp. 472–514, April 1937.

C. C. Cutler, "Differential Quantization of Communication Signals," United States Patent, no. 2-505-361, 1952.

C. L. Dammann, L. D. McDaniel and C. L. Maddox, "D2 channel bank—Multiplexing and coding," *Bell System Tech. J.* vol. 51, pp. 1675–1699, 1972.

R. deBuda, "Coherent demodulation of frequency-shift keying with low deviation ratio," *IEEE Trans. on Communications,* vol. COM-20, pp. 429–535, 1972.

F. E. DeJager, "Deltamodulation, a method of PCM transmission using the 1-unit code," *Phillips Research Reports,* vol. 7, pp. 442–466, 1952.

F. E. DeJager and C. B. Dekker, "Tamed frequency modulation: A novel method to achieve spectrum economy in digital transmission," *IEEE Transactions on Communications,* vol. COM-26, pp. 534–542, 1978.

J. A. Develet, "A threshold criterion for phase-lock demodulation," *Proceedings of the IEEE,* vol. 51, pp. 349–356, 1963.

W. Diffie and M. E. Hellman, "New directions in cryptography," *IEEE Transactions on Information Theory,* vol. IT-22, pp. 644–654, 1976.

W. Diffie and M. E. Hellman, "Privacy and authentication: An introduction to cryptography," *Proceedings of the IEEE,* vol. 67, pp. 397–427, 1979.

M. I. Doelz and E. H. Heald, "Minimum Shift Data Communication System," U. S. Patent no. 2977417, March 1961.

R. M. Dolby, "An audio noise reduction system," *Journal of the Audio Engineering Society,* vol. 15, p. 383, 1967.

J. Dungundji, "Envelopes and pre-envelopes of real wave-forms," *IRE Transactions on Information Theory,* vol. IT-4, pp. 53–57, 1958.

P. Elias, "Coding for noisy channels," *IRE Convention Record,* Part 4, pp. 37–46, March 1955.

L. H. Enloe, "Decreasing the threshold in FM by frequency feedback," *Proceedings of the IRE,* vol. 50, pp. 18–30, 1962.

V. M. Eyuboglu, "Detection of coded modulation signals on linear, severely distorted channels using decision-feedback noise prediction with interleaving," *IEEE Transactions on Communications,* vol. COM-36, pp. 401–409, 1988.

J. L. Flanagan, et al., "Synthetic voices for computers," *IEEE Spectrum,* vol. 7, no. 9, pp. 22–45, 1970.

J. L. Flanagan, M. R. Schroeder, B. S. Atal, R. E. Crochiere, N. S. Jayant, and J. M. Tribolet, "Speech coding," *IEEE Transactions on Communications,* vol. COM-27, pp. 710–737, 1979.

G. D. Forney, Jr., "Maximum likelihood sequence estimation of digital sequences in the presence of intersymbol interference," *IEEE Transactions on Information Theory,* vol. IT-18, pp. 363–378, 1972.

G. D. Forney, Jr., "The Viterbi algorithm," *Proceedings of the IEEE,* vol. 61, pp. 268–278, 1973.

G. D. Forney, Jr., and M. V. Eyuboglu, "Combined equalization and coding using precoding," *IEEE Communications Magazine,* vol. 29, no. 12, pp. 25–34, 1991.

L. E. Franks, "Carrier and bit synchronization in data communications—A tutorial review," *IEEE Transactions on Communications,* vol. COM-28, pp. 1107–1121, 1980.

H. T. Friis, "Noise figures in radio receivers," *Proceedings of the IRE,* vol. 32, pp. 419–422, 1944.

K. E. Fultz and D. B. Penick, "T1 carrier system," *Bell System Tech. J.,* vol. 44, pp. 1405–1451, 1965.

D. Gabor, "Theory of communications," *Journal of IEE* (London), vol. 93, Part III, pp. 429–457, 1946.

W. A. Gardner and L. E. Franks, "Characterization of cyclostationary random signal processes," *IEEE Transactions on Information Theory,* vol. IT-21, pp. 4–14, 1975.

D. A. George, "Matched filters for interfering signals," *IEEE Transactions on Information Theory,* vol. IT-11, pp. 153–154, 1965.

A. Gersho, "Adaptive equalization of highly dispersive channels for data transmission," *Bell System Tech. J.*, vol. 48, pp. 55–70, 1969.

A. Gersho, "Asymptotically optimal block quantization," *IEEE Transactions on Information Theory*, vol. IT-25, pp. 373–380, 1979.

A. Gersho and V. Cuperman, "Vector quantization: A pattern-matching technique for speech coding," *IEEE Communications Magazine*, vol. 21, no. 9, pp. 15–21, 1983.

R. A. Gibby and J. W. Smith, "Some extensions of Nyquist's telegraph transmission theory," *Bell Systems Tech. J.*, vol. 44, pp. 1487–1510, 1965.

R. Gold, "Optimal binary sequences for spread spectrum multiplexing," *IEEE Transactions on Information Theory*, vol. IT-13, pp. 619–621, 1967.

R. Gold, "Maximal recursive sequences with 3-valued recursive cross correlation functions," *IEEE Transactions on Information Theory*, vol. IT-14, pp. 154–156, 1968.

R. M. Gray, "Vector quantization," *IEEE ASSP Magazine*, vol. 1, no. 2, pp. 4–29, 1984.

W. J. Gruen, "Theory of AFC synchronization," *Proceedings of the IRE*, vol. 41, pp. 1043–1048, 1953.

G. T. Hardy, "Personal communication services," *IEEE Communications Magazine*, vol. 30, no. 6, pp. 53–54, 1992.

D. W. Hagelbarger, "Recurrent codes: Easily mechanized, burst-correcting binary codes," *Bell System Tech. J.*, vol. 38, pp. 969–984, 1959.

R. W. Hamming, "Error detecting and error correcting codes," *Bell System Tech. J.*, vol. 29, pp. 147–160, 1950.

J. C. Hancock and R. W. Lucky, "Performance of combined amplitude and phase-modulated communciation systems," *IRE Transactions on Communication Systems*, vol. CS-8, pp. 232–237, 1960.

F. J. Harris, "The discrete Fourier transform applied to time domain signal processing," *IEEE Communications Magazine*, vol. 20, no. 3, pp. 13–22, 1982.

T. V. L. Hartley, "Transmission of information," *Bell System Tech. J.*, vol. 7, pp. 535–563, 1928.

H. H. Henning and J. W. Pan, "D2 channel bank system aspects," *Bell System Tech. J.*, vol. 51, pp. 1641–1657, 1972.

P. S. Henry, "Introduction to lightwave transmission," *IEEE Communications Magazine*, vol. 23, no. 5, pp. 12–16, 1985.

F. S. Hill, Jr., "On time-domain representations for vestigial sideband signals," *Proceedings of the IEEE*, vol. 62, pp. 1032–1033, 1974.

F. Hlawatsch and G. F. Boudreaux-Bartels, "Linear and quadratic time-frequency signal representations," *IEEE Signal Processing Magazine*, vol. 9, no. 4, pp. 21–67, 1992.

D. A. Huffman, "A method for the construction of minimum redundancy codes," *Proceedings of the IRE*, vol. 40, pp. 1098–1101, 1952.

H. Insoe, Y. Yasuda, and J. Murakami, "A telemetering system by code modulation: $\Delta - \Sigma$ modulation," *IRE Transactions on Space Electronics and Telemetry*, vol. SET-8, pp. 204–209, 1962.

J. B. Johnson, "Thermal agitation of electricity in conductors," *Physical Review*, second series, vol. 32, pp. 97–109, 1928.

I. M. Jacobs, "Practical applications of coding," *IEEE Transactions on Information Theory*, vol. IT-20, pp. 305–310, 1974.

N. S. Jayant, "Adaptive delta modulation with a one-bit memory," *Bell System Tech. J.*, vol. 49, pp. 321–342, 1970.

N. S. Jayant, "Digital coding of speech waveforms, PCM, DPCM and DM quantizers," *Proceedings of the IEEE*, vol. 62, pp. 611–632, 1974.

N. S. Jayant, "Coding speech at low bit rates," *IEEE Spectrum*, vol. 23, no. 8, pp. 58–63, 1986.

A. J. Jerri, "The Shannon sampling theorem—Its various extensions and applications: A tutorial review," *Proceedings of the IEEE*, vol. 65, no. 11, pp. 1565–1596, 1977.

P. Kabal and S. Pasupathy, "Partial-response signaling," *IEEE Trans. on Communications*, vol. COM-23, pp. 921–934, 1975.

H. Kaneko, "A unified formulation of segment companding laws and synthesis of codes and digital companders," *Bell System Tech. J.*, vol. 49, pp. 1555–1588, 1970.

A. I. Khintchine, "Korrelationstheorie der stationören stochastischen prozesse," *Mathematiche Annalen*, vol. 109, pp. 415–458, 1934.

L. Kleinrock, "The latency/bandwidth tradeoff in gigabit networks," *IEEE Communications Magazine*, vol. 30, no. 4, pp. 36–40, 1992.

H. Kobayashi, "Correlative level coding and maximum-likelihood decoding," *IEEE Transactions on Information Theory*, vol. IT-17, pp. 586–594, 1971.

E. R. Kretzmer, "Generalization of a technique for binary data communication," *IEEE Transactions on Communication Technology*, vol. COM-14, pp. 67–68, Feb. 1966.

C. F. Kurth, "Generation of single-sideband signals in multiplex communication systems," *IEEE Transactions on Circuits and Systems*, vol. CAS-23, pp. 1–17, Jan. 1976.

W. C. Y. Lee, "Elements of cellular mobile radio systems," *IEEE Transactions on Vehicular Technology*, vol. VT-35, pp. 48–56, 1986.

A. Lender, "The duobinary technique for high-speed data transmission," *IEEE Transactions on Communications and Electronics*, vol. 82, pp. 214–218, May 1963.

A. Lender, "Correlative digital communication techniques," *IEEE Transactions on Communication Technology*, vol. COM-12, pp. 128–135, 1964.

A. Lender, "Correlative level coding for binary-data transmission," *IEEE Spectrum*, vol. 3, no. 2, pp. 104–115, 1966.

S. Lin, et al., "Automatic-repeat-request error control schemes," *IEEE Communications Magazine*, vol. 22, no. 12, pp. 5-16, 1984.

Y. Linde, A. Buzo, and R. M. Gray, "An algorithm for vector quantizer design," *IEEE Trans. on Communications*, vol. COM-28, pp. 84–95, 1980.

D. Linden, "A discussion of sampling theorems," *Proceedings of the IRE*, vol. 47, pp. 1219–1226, 1959.

S. P. Lloyd, "Least squares quantization in PCM," *IEEE Transactions on Information Theory*, vol. IT-28, pp. 129–137, 1982.

R. W. Lucky, "Automatic equalization for digital communication," *Bell System Tech. J.*, vol. 44, pp. 547–588, 1965.

R. W. Lucky, "Techniques for adaptive equalization of digital communication systems," *Bell System Tech. J.*, vol. 45, pp. 255–286, 1966.

R. Lugannani, "Intersymbol interference and probability of error in digital systems," *IEEE Trans. on Information Theory*, vol. IT-15, pp. 682–688, 1969.

V. H. MacDonald "Advanced Mobile Phone Service: The cellular concept," *Bell System Tech. J.*, vol. 58, pp. 15–41, 1979.

J. Max, "Quantizing for minimum distortion," *IRE Trans. on Information Theory*, vol. IT-6, pp. 7–12, 1960.

D. Mennie, "AM Stereo: Five competing options," *IEEE Spectrum*, vol. 15, no. 6, pp. 24–31, 1978.

Y. Meyer, "Book reviews," *Bulletin (New Series) of the American Mathematical Society*, vol. 28, pp. 350–360, 1993.

S. E. Miller, et al., "Research toward optical fiber-transmission systems," *Proceedings of the IEEE*, vol. 61, pp. 1703–1751, 1973.

S. E. Minzer, "Broadband ISDN and asynchronous transfer mode (ATM)," *IEEE Communications Magazine*, vol. 27, no. 9, pp. 17–24, 57, 1989.

P. Monsen, "Feedback equalization for fading dispersive channels," *IEEE Transactions on Information Theory*, vol. IT-17, pp. 56–64, 1971.

E. Murphy, "Whatever happened to AM stereo?," *IEEE Spectrum*, vol. 25, p. 17, 1988.

C. L. Nikas and M. R. Raghuveer, "Bispectrum estimation: A digital signal processing framework," *Proceedings of the IEEE*, vol. 75, pp. 869–891, 1987.

D. O. North "An analysis of the factors which determine signal/noise discrimination in pulsed carrier systems," *Proceedings of the IEEE*, vol. 51, pp. 1016–1027, 1963.

H. Nyquist, "Certain factors affecting telegraph speed," *Bell System Tech. J.*, vol. 3, pp. 324–346, 1924.

H. Nyquist, "Thermal agitation of electric charge in conductors," *Physical Review*, second series, vol. 32, pp. 110–113, 1928.

H. Nyquist, "Certain topics in telegraph transmission theory," *Transactions of the AIEE*, vol. 47, pp. 617–644, Feb. 1928.

J. Oetting, "Cellular mobile radio—An emerging technology," *IEEE Communications Magazine*, vol. 21, no. 8, pp. 10–15, 1983.

B. M. Oliver, J. R. Pierce, and C. E. Shannon, "The philosophy of PCM," *Proceedings of the IRE*, vol. 36, pp. 1324–1331, 1948.

J. B. O'Neal, "Delta modulation quantizing noise analytical and computer simulation results for Gaussian and television input signals," *Bell System Tech. J.*, vol. 45, pp. 117–141, 1966.

J. B. O'Neal, "Predictive quantization (Differential pulse-code modulation) for the transmission of television signals," *Bell System Tech. J.*, vol. 45, pp. 689–721, 1966.

R. E. A. C Paley and N. Wiener, "Fourier transforms in the complex domain," *American Mathematical Society Colloquium Publication*, vol. 19, pp. 16–17, 1934.

S. Pasupathy, "Nyquist's third criterion," *Proceedings of the IEEE*, vol. 62, pp. 860–861, 1974.

S. Pasupathy, "Correlative coding—A bandwidth-efficient signaling scheme," *IEEE Communications Magazine*, vol. 15, no. 4, pp. 4–11, 1977.

S. Pasupathy, "Minimum shift keying—A spectrally efficient modulation," *IEEE Communications Magazine*, vol. 17, no. 4, pp. 14–22, 1979.

J. B. H. Peek, "Communications aspects of the compact disc digital audio system," *IEEE Communications Magazine*, vol. 23, no. 2, pp. 7–15, 1985.

R. L. Pickholtz, D. L. Schilling, and L. B. Milstein, "Theory of Spread-Spectrum Communications—A tutorial," *IEEE Trans. on Communications*, vol. COM-30, pp. 855–884, 1982.

J. G. Proakis, "Advances in equalization for intersymbol interference," *Advances in Communication Systems*, edited by A. J. Viterbi, vol. 4, pp. 123–198, Academic Press, 1975.

S. Qureshi, "Adaptive equalization," *IEEE Communications Magazine*, vol. 20, no. 2, pp. 9–16, March 1982.

S. Qureshi, "Adaptive equalization," *Proceedings of the IEEE*, vol. 73, pp. 1349–1387, 1985.

I. S. Reed and G. Solomon, "Polynomial codes over certain finite fields," *Journal of SIAM*, vol. 8, pp. 300–304, 1960.

A. H. Reeves, "The past, present and future of PCM," *IEEE Spectrum*, vol. 12, no. 5, pp. 58–63, 1975.

S. O. Rice, "Mathematical analysis of random noise," *Bell System Tech J.*, vol. 23, pp. 282–332, 1944; vol. 24, pp. 46–156, 1945.

S. O. Rice, "Statistical properties of a sine-wave plus random noise," *Bell System Tech. J.*, vol. 27, pp. 109–157, 1948.

S. O. Rice, "Noise in FM receivers," in M. Rosenblatt, (editor), *Proceedings of the Symposium on Time Series Analysis* (New York: Wiley, 1963), pp. 395–411.

S. O. Rice, "Envelopes of narrow-band signals," *Proceedings of the IEEE*, vol. 70, pp. 692–699, 1982.

O. Rioul and M. Vetterli, "Wavelets and signal processing," *IEEE Signal Processing Magazine,* vol. 8, no. 10, 1991.

R. L. Rivest, A. Shamir, and L. Adleman, "A method for obtaining digital signatures and public key cryptosystems," *Communications of the Association for Computing Machinery,* vol. 21, pp. 120–126, 1978.

W. L. Root, "Remarks, mostly historical, on signal detection and signal parameter estimation," *Proceedings of the IEEE,* vol. 75, pp. 1446–1457, 1987.

J. Rosenberg, R. E. Kraut, L. Gomez, and C. A. Buzzard, "Multimedia communications for users," *IEEE Communications Magazine,* vol. 30, no. 5, pp. 20–36, 1992.

W. D. Rummler, "A new selective fading model—Application to propagation data," *Bell System Tech. J.,* vol. 58, pp. 1037–1071, 1979.

T. S. Rzeszewski, "A technical assessment of advanced television," *Proceedings of the IEEE,* vol. 78, pp. 789–804, 1990.

D. V. Sarwate and M. B. Pursley, "Crosscorrelation properties of pseudorandom and related sequences," *Proceedings of the IEEE,* vol. 68, pp. 593–619, 1980.

B. Sayar and S. Pasupathy, "Nyquist 3 pulse shaping in continuous phase modulation," *IEEE Transactions on Communications,* vol. COM-35, pp. 57–67, 1987.

H. R. Schindler, "Delta modulation," *IEEE Spectrum,* vol. 7, no. 10, pp. 69–78, 1970.

R. A. Scholtz, "The Origins of spread-spectrum communications," *IEEE Transactions on Communications,* vol. COM-30, p. 822–854, May 1982.

R. A. Scholtz, "Notes on spread-spectrum history," *IEEE Transactions on Communication,* vol. COM-31, pp. 82–84, 1984.

J. S. Schouten, F. DeJager, and J. A. Greefkes, "Delta modulation, a new modulation system for telecommunication," *Phillips Technical Review,* vol. 13, pp. 237–245, 1952.

C. E. Shannon, "A mathematical theory of communication," *Bell System Tech. J.,* vol. 27, pp. 379–423, 623–656, 1948.

C. E. Shannon, "Communication theory of secrecy systems," *Bell System Tech J.,* vol. 28, pp. 656–715, 1949.

C. E. Shannon, "Communication in the presence of noise," *Proceedings of the IRE,* vol. 37, pp. 10–21, 1949.

M. H. Sherif and D. K. Sparrell, "Standards and innovation in telecommunications," *IEEE Communications Magazine,* vol. 30, no. 7, pp. 22–28, 1992.

M. K. Simon and D. Divsalar, "On the implementation and performance of single and double differential detection schemes," *IEEE Trans. on Communications,* vol. 40, pp. 278–291, 1992.

B. Sklar, "A structural overview of digital communications—A tutorial review," Part I, *IEEE Communications Magazine,* vol. 21, no. 5, pp. 4–17, 1983; Part II, *IEEE Communications Magazine,* vol. 21, no. 7, pp. 6–21, 1983.

D. Slepian, "On bandwidth," *Proc. IEEE,* vol. 64, pp. 292–300, 1976.

B. Smith, "Instantaneous compounding of quantized signals," *Bell System Tech. J.,* vol. 36, pp. 653–709, 1957.

E. S. Sousa and S. Pasupathy, "Pulse shape design for teletext data transmission," *IEEE Trans. on Communications,* vol. COM-31, pp. 871–878, 1983.

R. Steele, "The cellular environment of lightweight hand held portables," *IEEE Communications Magazine,* vol. 27, no. 7, pp. 20–29, 1989.

S. Stein, "Unified analysis of certain coherent and noncoherent binary communication systems," *IEEE Transactions on Information Theory,* vol. IT-10, pp. 43–51, 1964.

G. Strang, "The optimal coefficients in Daubechies wavelets," *Physica D,* vol. 60, pp. 239–244, 1992.

G. Strang, "Wavelet transforms versus Fourier transforms," *Bulletin (New Series) of the American Mathematical Society,* vol. 28, pp. 288–305, 1993.

C. E. Sundberg, "Continuous phase modulation," *IEEE Communications Magazine*, vol. 24, no. 4, pp. 25–38, 1986.

D. W. Tufts, "Nyquist's problem—The joint optimiztion of transmitter and receiver in pulse amplitude modulation," *Proceedings of the IEEE*, vol. 53, pp. 248–259, 1965.

G. L. Turin, "An introduction to matched filters," *IRE Transactions on Information Theory*, vol. IT-6, pp. 311–329, 1960.

G. L. Turin, "An introduction to digital matched filters," *Proceedings of the IEEE*, vol. 64, pp. 1092–1112, 1976.

G. Ungerboeck, "Channel coding with multilevel/phase signals," *IEEE Transactions on Information Theory*, IT-28, pp. 55–67, 1982.

G. Ungerboeck, "Trellis-coded modulation with redundant signal sets," Part 1 and 2, *IEEE Communications Magazine*, vol. 25, no. 2, pp. 5–21, 1987.

B. van der Pol, "The fundamental principles of frequency modulation," *Journal of IEE* (London), vol. 93, part III, p. 153–158, 1946.

J. H. Van Vleck and D. Middleton, "A theoretical comparison of visual, aural, and meter reception of pulsed signals in the presence of noise," *Journal of Applied Physics*, vol. 17, pp. 940–971, 1946.

A. J. Viterbi, "Error bounds for convolutional codes and an asymptotically optimum decoding algorithm," *IEEE Trans. on Information Theory*, vol. IT-13, pp. 260–269, 1967.

A. J. Viterbi, "Spread-spectrum communications—Myths and realities," *IEEE Communications Magazine*, vol. 17, no. 3, pp. 11–18, May 1979.

A. J. Viterbi, "When not to spread spectrum—A sequel," *IEEE Communications Magazine*, vol. 23, no. 4, pp. 12–17, 1985.

A. J. Viterbi, "Wireless digital communication: A view based on three lessons learned," *IEEE Communications Magazine*, vol. 29, no. 9, pp. 33–36, 1991.

D. K. Weaver, Jr., "A third method of generation and detection of single-sideband signals," *Proceedings of the IRE*, vol. 44, pp. 1703–1705, 1956.

S. B. Weinstein, "Echo cancellation in the telephone network," *IEEE Communications Magazine*, vol. 15, no. 1, pp. 8–15, 1977.

J. Weiss and D. Schremp, "Putting data on a diet," *IEEE Spectrum*, vol. 30, pp. 36–39, August 1993.

T. A. Welch, "A technique for high performance data compression," *Computer*, vol. 17, no. 6, pp. 8–19, 1984.

M. V. Wickerhauser and R. R. Coifman, "Entropy-based methods for best basis selection," *IEEE Transactions on Information Theory*, vol. 38, pp. 713–718, 1992.

B. Widrow and M. E. Hoff, Jr., "Adaptive switching circuits," *WESCON Convention Record*, Pt. 4, pp. 96–104, 1960.

N. Wiener, "Generalized harmonic analysis," *Acta Mathematica*, vol. 55, p. 117, 1930.

K. A. Wimmer and J. B. Jones, "Global development of PCS," *IEEE Communications Magazine*, vol. 30, no. 6, pp. 22–27, June 1992.

A. D. Wyner, "Fundamental limits in information theory," *Proceedings of the IEEE*, vol. 69, pp. 239–251, 1981.

O. C. Yue, R. Luganani, and S. O. Rice, "Series approximations for the amplitude distribution and density of shot processes," *IEEE Transactions on Communications*, vol. COM-26, pp. 45–54, 1978.

J. Ziv and A. Lempel, "A universal algorithm for sequential data compression," *IEEE Transactions on Information Theory*, vol. IT-23, pp. 337–343, 1977.

J. Ziv and A. Lempel, "Compression of individual sequences via variable-rate coding," *IEEE Transactions on Information Theory*, vol. IT-24, pp. 530–536, 1978.

Index

Adaptive differential pulse-code modulation, *400*

Adaptive subband coding, *402*

Additive white Gaussian noise channel, *477*

A-law, *380*

Amplitude modulation, *122*
  noise in, *322*
  percentage modulation, *124*
  single-tone, *125, 324*
  threshold effect, *325*
  virtues, limitations, and modifications, *131*

Amplitude response, *67*

Amplitude-shift keying, *476, 507*

Analytic signals, *see* Pre-envelope

Angle modulation, *154*

ASCII, *5, 844*

Asynchronous transfer mode, *760*

Autocorrelation function, *243*

Autocovariance function, *243*

Automatic-repeat-request method for error-control coding, *721*

Average probability of symbol error, *477, 495*
  union bound, *496*

Bandwidth, *48, 68*

Bandwidth efficiency, *561*

Bandwidth-noise trade-off, *373*

Baseband pulse transmission,
  binary, *412*
  M-ary, *446*

Baseband signals, *9, 121*

BCH codes, *698*

Bessel functions, *162, 793*
  modified, *796*

Binary arithmetic, *813*

Binary frequency-shift keying, coherent, *511*
  bit error rate, *515*
  power spectrum, *557*

Binary frequency-shift keying, noncoherent, *539*
  bit error rate, *540*

Binary phase-shift keying, *508*
  bit error rate, *510*
  power spectrum, *556*

Binary symmetric channel, *224, 633*

Bipolar return-to-zero signaling, *383*

Bit error rate, *499*

Block codes, *675*
   dual code, *685*
   generator matrix, *676*
   minimum distance considerations, *680*
   parity-check matrix, *677*
   syndrome decoding, *679*
Broadband integrated services digital network,
   759

Campbell's theorem, *270*
Carrier delay, *see* Phase delay
Carson's rule, *165*
Causality, *65*
Central limit theorem, *266*
Channel capacity, *637*
Channel coding theorem, *640, 673*
Characteristic function, *233*
Chi distribution with two degrees of freedom,
   302
Ciphers, block and stream, *817*
Circuit switching, *20*
Code-division multiplexing, *605*
Color television, *771*
Comb filter, *259*
Communication channels, *6*
Communication networks, *19*
Communication process, *1*
Communication resources, primary, *14*
Compact disc, *722*
Complementary error function, *806*
Complex envelope, *85*
Complex exponential function, *55*
Complex impulse response, *91*
Conjugate functions, *43*
Continuous-wave modulation, *121*
   noise in, *313*
Convolutional codes, *700*
   asymptotic coding gain, *714*
   code tree, *702*
   free distance, *711*
   maximum likelihood decoding, *707*
   state diagram, *706*
   trellis, *704*
Convolution integral, *63*
Convolution in the time domain, *46*
Correlation (between random variables), *234*
Correlation-level coding, *434*
   duobinary signaling, *435*
   generalized, *444*
   modified duobinary signaling, *441*
Correlation receiver, *501*
Correlation theorem, *280*
Costas loop, *138*
Cross-correlation function, *247*

Cross-spectral densities, *262*
Cryptography, *815*
   public-key, *830*
   secret-key, *816*
   Shannon model, *820*
Cumulative distribution function, *227*
Cyclic codes, *686*
   generator polynomial, *687*
   parity-check polynomial, *688*
   syndrome calculator, *691*
Cyclic redundancy check codes, *697*

Data compaction, *623*
Data compression, *659*
Data encryption standard, *825*
dc signal, *54*
Decibel, *68*
Delta modulation, *391*
Delta-sigma modulation, *394*
Detection of signals in additive noise,
   with known phase, *491*
   with unknown phase, *503*
Differential coding, *384*
Differential entropy, *644*
   Gaussian distribution, *646*
   uniform distribution, *646*
Differential phase-shift keying, *540*
   bit error rate, *541*
   optimum receiver, *543*
Differential pulse-code modulation, *396*
   processing gain, *398*
Differentiation in the time domain, *40*
Diffie-Hellman public key distribution, *831*
Digital modulation techniques, *507*
Digital passband transmission, *473*
   model of, *474*
Digital signature, *834*
Dirac delta function, *51*
Direct-sequence/binary phase-shift keying, *589*
   jamming margin, *598*
   processing gain, *596*
Dirichlet conditions, *28, 774*
Discrete Fourier transform, *100*
Discrete memoryless channels, *631, 672*
Discrete memoryless source, *615*
   extension of, *620*
Diversity techniques, *750*
Double sideband-suppressed carrier modula-
   tion, *132*
   coherent detection, *136*
   filtering of sidebands, *140*
   noise in, *317*
Duality, *37*

Einstein–Wiener–Khintchine relations, *254*
Energy signals, *10*
Energy spectral density, *47*
Entropy, *617*
Envelope, *90, 122*
Envelope delay, *100*
Envelope detector, *129*
Equalizer, *448*
    adaptive, *452*
    decision-feedback, *459*
    tapped-delay-line, *448*
Equivalent noise temperature, *272, 803*
Ergodic processes, *249*
    ergodicity in the autocorrelation function, *250, 276*
    ergodicity in the mean, *249*
Error control coding, *670*
Error function, *806*
Expected value, *see* Mean, ensemble averaged
Exponential pulse, *32*
Eye pattern, *461*

Fast Fourier transform algorithm, *102*
    decimation-in-frequency, *107*
    decimation-in-time, *107*
Figure of merit, *316*
    noise in, *326*
Filters, *69*
    band-pass, *69, 91, 96*
    low-pass, *69, 71*
Fourier series, *773*
Fourier transform, *27, 778*
    numerical computation of, *100*
    properties of, *33*
Frequency discriminator, *175*
Frequency-division multiplexing, *152*
    transmission bandwidth, *165*
Frequency hop/*M*-ary frequency-shift keying, *599*
    processing gain, *602*
Frequency modulation, *156, 158*
    narrow-band, *159*
    pre-emphasis and de-emphasis, *340*
    threshold effect, *334*
    threshold extension, *338*
    wide-band, *162*
Frequency multiplier, *170*
Frequency shifting, *38*
Frequency-shift keying, *476, 507*
Frequency translation, *151*
Friis formula, *805*
Fundamental inequality, *618*

Gain, *68*
Gaussian distribution, *234, 646*

Gaussian process, *265*
Gaussian pulse, *41*
Geometric interpretation of signals, *483*
Gibb's phenomenon, *72*
Gold sequences, *607*
Gram-Schmidt orthogonalization procedure, *478*
Gray code, *500*
Group delay, *100*

Hamming codes, *683, 693*
Heterodyning, *see* Frequency translation
High definition television, *772*
Hilbert transform, *79*
    properties of, *81*
Huffman coding, *626*

Ideal sampling function, *61*
Impulse response, *62*
Information, *14*
    sources of, *3*
    theory, *614*
Information capacity theorem, *648*
    implications of, *652*
    sphere packing, *650*
In-phase component, *86*
Instantaneous frequency, *155*
Integrated services digital network, *759*
Integrate-and-dump circuit, *418*
Integration in the time domain, *42, 58*
Intersymbol interference, *424*

Kraft-McMillan inequality, *624*

Least-mean-square algorithm, *453*
Lempel–Ziv algorithm, *629*
Lightwave amplifier, *759*
Linearity, *34, 62*
Linear modulation, *88, 199*
Linear Systems, *62*
Line codes, *382*
Low-pass equivalent model, *95*

Manchester code, *383*
M-ary frequency-shift keying, *507, 552*
    bandwidth efficiency, *563*
    information capacity, *656*
M-ary phase-shift keying, *507, 646*
    bandwidth efficiency, *562*
    information capacity, *655*
    power spectrum, *560*
M-ary quadrature amplitude modulation, *549*
Matched filter, *413, 501*
Maximum-length sequences (codes), *581, 696*

Maximum likelihood decoding, *492, 707*
Mean, ensemble-averaged, *230, 242*
Mean-square value, *232*
Minimum shift keying, *523*
  bit error rate, *531*
  phase trellis, *524*
  power spectrum, *558*
  signal-space diagram, *526*
Mixing, *see* Frequency translation
Mobile radio, *7, 735*
  cell splitting, *736*
  frequency reuse, *736*
  propagation effects, *737*
Modulation, *12, 121*
Moments, *232*
  central, *232*
  joint, *234*
M-th power loop, *564*
μ-law, *380*
Multipath channel, *737*
  binary signaling over, *747*
  delay spread, *745*
  Doppler spread, *747*
  fade rate, *747*
  fading, *741*
Multiple access, *733*
  code-division, *734*
  frequency-division, *733*
  space-division, *734*
  time-division, *734*
Multiplication in the time domain, *45*
Mutual information, *634, 647*

Narrow-band noise, *282*
  in-phase and quadrature components, *283*
Noise, *269*
Noise equivalent bandwidth, *281*
Noise figure, *801*
Noncoherent orthogonal modulation, *532*
  average probability of error, *534*
Nonreturn-to-zero signaling, *382*
Nyquist criterion for distortionless baseband transmission, *427*
Nyquist rate, *356*

Observation vector, *490*
On–off signaling, *382*
Open systems interference (OSI) reference model, *21*
Optical communications, *753*
Optical fibers, *7, 753*
  graded index, *755*
  multimode, *754*
  multiplicative noise, *757*
  single-mode, *753*

Packet switching, *20*
Paley–Wiener criterion, *69*
Partial-response signaling, *see* Correlation-level coding
Passband modulation, *see* Linear modulation
Passband signals, *9*
Periodic signals, *10*
  Fourier transform, *59*
Periodogram, *261*
Personal communication network, *763*
Personal computers, *5*
Phase delay, *98*
Phase-locked loop, *181*
  cycle slipping, *191*
  second-order, *187*
Phase modulation, *155*
Phase response, *67*
Phase-shift keying, *476, 507*
Poisson's sum formula, *60*
Poles, *78*
Power signals, *10*
Power spectral density, *252*
Power spectrum, *see* Power spectral density
Pre-envelope, *83*
Prefix coding, *623*
Principle of orthogonality, *456*
Principle of superposition, *62*
Probability density function, *227*
  conditional, *229*
  joint, *229*
Probability theory, *219*
  axioms of probability, *220*
  relative frequency approach, *219*
Pseudo-noise sequences, *579*
Pulse-amplitude modulation, *357*
  flat-top samples, *359*
Pulse-code modulation, *378*
  error rate, *388, 418*
  error threshold, *388*
  information capacity, *653*
  noise considerations, *387*
Pulse-duration modulation, *364, 404, 408*
Pulse modulation, *351*
Pulse-position modulation, *364*
  detection, *368*
  effects of noise, *369*
  generation, *365*

Q-factor, *178*
Q-function, *808*
Quadrature-carrier multiplexing, *139*
Quadrature component, *86*
Quadriphase-shift keying, *507, 517*
  average probability of symbol error, *518*

bit error rate, *521*
power spectrum, *558*
signal-space diagram, *517*
Quantization noise, *375*
Quantization process, *374*
  mid-rise type, *375*
  midtread type, *375*
  nonuniform, *379*
  uniform, *374*

Radio frequency pulse, *38, 90*
Raised cosine spectrum, *431*
Random binary wave, *245, 256*
Random processes, *218, 239*
  complex, *810*
  transmission through linear filters, *250*
Random variables, *226*
  transformations of, *235*
Rayleigh distribution, *294*
Rayleigh's energy theorem, *46*
Receiver model, *314*
Rectangular function, *31*
Reed–Solomon codes, *699, 722*
Return-to-zero signaling, *382*
Regeneration, *384*
Repetition codes, *643, 678*
Rician distribution, *297*
Ring modulator, *133*
Rivest–Shamir–Adleman system, *832*
Rolloff factor, *432*

Sampling process, *352*
Sampling theorem, *356*
Satellite communications, *8, 731*
Schwarz's inequality, *415, 799*
Shannon limit, *652*
Short-time Fourier transform, *782*
  implementations of, *783*
  time-frequency resolution, *784*
Shot noise, *269*
Signal-flow graph, *105, 711*
Signal-to-noise ratio, *316*
Signum function, *35, 56*
Sinc function, *31*
Sinc pulse, *47*
Sine integral, *71*
Single sideband modulation, *147*
  noise in, *319*
  phase distortion, *151*
Sinusoidal functions, *55*
SONET, *761*
Source-coding theorem, *621*
Spectrogram, *766*

Spectrum,
  continuous, *29*
  discrete, *776*
Speech, *3*
  production process, *765*
Spread spectrum modulation, *578*
Squaring loop, *204, 564*
Stability, *65*
Stationarity, *241*
Statistical averages, *230*
Stereo multiplexing:
  FM, *180*
  AM, *205*
Stochastic processes, *see* Random processes
Sufficient statistics, *490*
Sunde's FSK, *512, 557*
Superheterodyne receiver, *195*
Switching modulator, *127*
Synchronization, *385, 564*
  carrier, *564*
  symbol, *566*

Tapped-delay-line filter, *64*
Telephone channel, *6, 448*
Television, *3, 145, 769*
  raster scanning, *3*
Terminal analog model, *767*
Thermal noise, *270*
Time-bandwidth product, *50*
Time-division multiplexing, *362, 385*
Time-frequency analysis, *781*
Time scaling, *36*
Time shifting, *38*
T-1 System, *386*
Transfer function, *66*
Trellis-coded modulation, *715*
  asymptotic coding gain, *720*
Triangular pulse, *41*

Uncertainty principle, *786*
Ungerboeck codes, *718*
Unicity distance, *822*
Uniform distribution, *228, 294, 646*
Unit step function, *32, 57*
Unvoiced sounds, *765*

Variance, *232*
Vector quantization, *660*
Vernam cipher, *821*
Vestigial sideband modulation, *144*
Video bandwidth, *770*
Viterbi algorithm, *708*
Voiced sounds, *765*
Voltage-controlled oscillator, *173*

Wavelet transform, *787*
   time-frequency resolution, *788*
Wave packets, *791*
White noise, *271*
   band-pass filtered, *289*
   low-pass filtered, *273*

RC low-pass filtered, *274*
RLC filtered, *291*
Wiener–Khintchine relations, *see* Einstein–
   Wiener–Khintchine relations

Zeros, *78*